The Handbook of Sex Differences Volume IV

The Handbook of Sex Differences is a four-volume reference work written to assess sex differences, with a primary focus on the human species. Based on the authors' highly influential 2008 book *Sex Differences*, these volumes highlight important new research findings from the last decade and a half alongside earlier findings.

In this, the work's fourth and last volume, two related questions are addressed: Are there universal sex differences (i.e., sex differences found in all societies)? And if the answer is yes, what are they and how can each one be theoretically explained? To answer the first of these two questions, this volume condenses much of the research findings amassed in the book's first three volumes into summary tables. Then, to help identify likely universal sex differences, three versions of social role theory and two versions of evolutionary theory are examined relative to each possible universal sex difference. Consideration is even given to religious scriptures as a sixth type of explanation. In the concluding analyses, 308 likely universal sex differences are identified. No single theory was able to explain all these differences. Nevertheless, the two evolutionary theories were better in this regard than any of the three social role theories, including the recently proposed biosocial version of social role theory.

The Handbook of Sex Differences is of importance for any researcher, student, or professional who requires a comprehensive resource on sex differences.

Lee Ellis is a semi-retired American researcher whose last position was that of a visiting professor in the Department of Anthropology and Sociology, at the University of Malaya, Kuala Lumpur, Malaysia.

Craig T. Palmer is a semi-retired associate professor in the Department of Anthropology at the University of Missouri, Columbia, Missouri, USA.

Rosemary Hopcroft is a professor in the Department of Sociology at the University of North Carolina, Charlotte, North Carolina, USA.

Anthony W. Hoskin is a professor in the Department of Sociology, Social Work, and Criminology, at Idaho State University, Pocatello, Idaho, USA.

Handbook of Sex Differences

Books in this series:

The Handbook of Sex Differences Volume I
Basic Biology
By Lee Ellis, Craig T. Palmer, Rosemary Hopcroft, and Anthony W. Hoskin

The Handbook of Sex Differences Volume II
Cognitive Variables
By Lee Ellis, Craig T. Palmer, Rosemary Hopcroft, and Anthony W. Hoskin

The Handbook of Sex Differences Volume III
Behavioral Variables
By Lee Ellis, Craig T. Palmer, Rosemary Hopcroft, and Anthony W. Hoskin

The Handbook of Sex Differences Volume IV
Identifying Universal Sex Differences
By Lee Ellis, Craig T. Palmer, Rosemary Hopcroft, and Anthony W. Hoskin

The Handbook of Sex Differences Volume IV

Identifying Universal Sex Differences

Lee Ellis
Craig T. Palmer
Rosemary Hopcroft
Anthony W. Hoskin

Routledge
Taylor & Francis Group
NEW YORK AND LONDON

Designed cover image: © Getty Images

First published 2024
by Routledge
605 Third Avenue, New York, NY 10158

and by Routledge
4 Park Square, Milton Park, Abingdon, Oxon, OX14 4RN

Routledge is an imprint of the Taylor & Francis Group, an informa business

ISBN: 978-0-367-43470-0 (hbk)
ISBN: 978-1-032-52114-5 (pbk)
ISBN: 978-1-003-40529-0 (ebk)

DOI: 10.4324/9781003405290

Typeset in Sabon
by MPS Limited, Dehradun

This book stands as a tribute to the thousands of scientists who have conducted research on sex differences. Because of their work and the willingness of many millions of people to provide personal information about themselves, the scientific method has slowly unearthed a fascinating tapestry of findings about how males and females differ (and are sometimes the same).

Contents

Acknowledgments and Appreciation

Throughout the course of compiling research findings for this book, we benefited from the dedicated help of various people. Especially helpful was Stephany Larkin. Under the guidance of the book's first author, over the course of several years, she tirelessly entered thousands of the citations into the appropriate tables, and also entered their corresponding references into an Endnote file.

We also wish to acknowledge that Dr. Scott Wersinger agreed in 2017 to be a contributor to this book. Unfortunately, however, he fell ill and succumbed to cancer before he was able to contribute. Finally, Dr. Petri Kajonius kindly assisted by providing several citations and references to this book.

Preface

Nearly a century ago, a renowned philosopher of science named Bertrand Russell (1927:1–2) wrote: "What passes for knowledge in ordinary life suffers from three defects: It is cocksure, vague, and self-contradictory." While this statement surely has exceptions, it draws attention to the need for scientists, as the most measured arbiters of empirical truth, to be extremely careful when drawing conclusions. Even when doing so, they should periodically recheck these conclusions in light of additional evidence. In this spirit of cautious and tentative reasoning about sex differences, this four-volume work has been assembled and written.

The work's first three volumes set the stage for what this fourth volume will seek to do. We will draw careful conclusion about general patterns in sex differences as indicated by findings from the roughly 40,000 scientific reports cited in these initial volumes. By scientific reports, we mean studies in which samples were drawn and methods were used to measure variables in objective ways that other researchers can replicate. Because sex differences can vary by age and culture, the findings that were summarized throughout the first three volumes categorize them according to basic age groupings and countries in which the samples were obtained.

Throughout this final volume, specific numbers are mentioned frequently. For this reason, the numbers themselves are used rather than spelling them out in the narrative (unless the number happens to appear at the start of a sentence).

Even though the information contained in this work's first three volumes was itself highly condensed, this final volume will be distilled even further. Volume IV consists of five chapters, with two major goals in mind.

The first goal is to objectively identify apparent *universal sex differences* (*USDs*). Toward this end, an entire chapter will present tables listing all of the sex difference variables for with ten or more findings were located along with an objective measure of how consistent the findings were and the number of countries sampled. As explained more below, another entire chapter just examines the variables with nearly perfect agreement regarding there being a sex difference. After these two lengthy chapters (i.e., Chapters 25 and 26), a chapter is presented that contains listings of the variables that have the greatest promise in terms of being actual USDs.

The second purpose of this final volume is to assess how well *theories of sex differences* are able to explain why each apparent USD might exist. In order to accomplish this goal, this volume's initial chapter will present a summary of each one of five theories that have been offered by scientists regarding the possibility of some sex differences being found in all human societies. For more detail regarding the chapter structure of this fourth volume, see below.

Chapter 24 – Theories Pertaining to Why Universal Sex Differences Might Exist. Why would any sex differences be found in all human societies? In other words, what factors would cause males and females, on average, to differ from one another in the same way around the world? From what we have been able to determine, just five scientific theories offer explanations in this regard. Three of these theories emphasize the importance of social roles and are known as social role theories. The remaining two theories are evolutionary in nature.

Besides these five scientific theories of sex differences, it is noteworthy that certain religious scriptures seem to address the possibility of sex differences that might or might not be universal. In an effort to be as inclusive as possible, this chapter makes an attempt to even consider these scriptural passages in our efforts to understand certain sex differences.

Chapter 25 – Sex Difference Variables with Ten or More Findings. This chapter will provide tables that identify and briefly describe all variables listed throughout Volumes I through III with citations to ten or more empirical findings. For each variable contained in these tables, a numerical score (called a *consistency score*, ranging from 0.0 to 100.0) will be provided. (The exact method used to calculate these scores will be described at the beginning of this chapter.) The purpose of this score is to provide an estimate of the extent to which each listed variable with ten or more findings is consist regarding whether males or females exhibit each trait to a greater degree. Also included in this chapter's tables is information about the number of countries represented, the age ranges sampled, and the time periods during which the studies cited were published. Furthermore, for each variable listed, the nature of conclusions reached by any literature reviews or meta-analyses that were located are reported, along with the nature of findings from any studies of the same variable for non-humans.

Chapter 26 – Identifying and Theoretically Explaining Each Potential Universal Sex Difference. In this chapter, every variable with a "nearly perfect" consistency score (i.e., a score of 95.0 or higher, as shown in Chapter 25) is given individual attention regarding the reasonableness of its being deemed a USD. Each assessment is made based not only on the consistency of the evidence and the number of countries sampled, but also in terms of how well at least one of the five theories can provide an explanation for why the variable would be more prevalent in one sex than in the other.

Chapter 27 – Tabular Listings of the Universal Sex Differences in Order of the Evidentiary Strength. As the last step toward identifying USDs, this chapter contains tables listing all of what are termed *likely USDs*. In addition, a special table listing what are called *nearly certain USDs* is provided at the end of the chapter.

Chapter 28 – Epilogue. A brief closing epilogue chapter is offered. It highlights the way forward in continuing the scientific search for USDs.

Throughout this final volume, the tentative nature of all scientific knowledge will be emphasized. Nevertheless, we will show that in several cases, evidence has mounted heavily in favor of just one conclusion about the nature of certain sex differences, no matter what country or time period was sampled. In other words, while we avoid making emphatic statements about the universal status of any sex difference variables, or about the ability of any given theory to explain these sex differences, readers will see that many reasonable conclusions along these lines can now be made.

24 Theories Pertaining to Why Universal Sex Differences Might Exist

Are some traits more common (or more pronounced on average) among males than females (or vice versa) in all human societies? And, if so, why? One of the ways we sought to address these questions was to identify and report the country (or countries) sampled in each and every citation that was made in the thousands of tables presented throughout this book's first three volumes (i.e., in Chapters 1 through 23).

Attention throughout this entire volume turns to carefully identifying universal sex differences (which will be abbreviated as USDs) with a high degree of confidence. As noted in the preface to this final volume, the process of working toward this goal will consist of several steps, roughly corresponding to the volume's first four chapters. In this, the first chapter, a description is provided of five scientific theories that offer explanations for why sex differences exist, especially with respect to the differences being universal. Attention will even be given to the possibility of religious scriptural explanations in this regard.

24.1 Testable Explanations for Sex Differences

In an article on sex differences, sub-titled "Nature, nurture, or God," Neff and Terry-Schmitt (2002:1195) noted that explanations for sex differences can be subsumed under three categories: social, biological, and religious. We will refer to these categories as *sociocultural*, *evolutionary/biological*, and *religious/theological*. These three categories of explanations correspond well with the explanations that will be considered throughout the remainder of this Volume IV. Specifically, six explanations for sex differences that have been offered will now be presented for understanding why sex differences exist, particularly in terms of cognitive or behavioral traits. Five of these explanations are scientific in nature, and one is religious. As we will show, there are three distinguishable scientific explanations of a sociocultural nature and two of an evolutionary/biological nature. Religious explanations will draw on sacred scriptures having to do with how males and females *should* behave.

DOI: 10.4324/9781003405290-1

Given that the evidence considered throughout this four-volume book is empirical in nature, readers may question the wisdom of even considering religious explanations for sex differences in traits. We offer two reasons for doing so. First, religious explanations are not necessarily inconsistent with scientific explanations. Second, a sizable proportion, if not a majority, of people believe that *God-made-us-that-way* is a more reasonable and complete explanation than any scientific theory. Especially in Chapter 26, we will show that religious scriptures may in fact have some bearing on at least some sex difference traits.

One should bear in mind that these three categories of explanations may not be mutually exclusive. For example, sociocultural factors may often interact with evolutionary/biological factors to affect sex differences, either by augmenting or by curtailing their effects. Similarly, both sociocultural and evolutionary/biological factors could be influenced be supernatural influences. Also, some theological teachings could be responsible for sociocultural factors that affect social roles.

This chapter provides descriptions of the six explanations for sex differences. In Chapters 26 and 27, these explanations will be assessed in terms of how well they seem to account for why sex differences in a variety of traits appear to be universal, rather than being found in some cultures but not in others.

24.1.1 The Three Sociocultural Theories

Theories that focus on learning in a sociocultural context as being responsible for sex differences, especially differences in cognitive and behavioral traits, are known as *social role theories*. In essence, these theories assert that, were it not for variations in how cultures treat and train males and females differently, there would be no sex differences in behavior. sex stereotypes and differential (sexist) child rearing practices, males and females would think and behave identically. According to these theories, factors such as genetics, hormones, brain functioning, and evolution have little if any relevance to male-female differences in thought and behavior. The three sociocultural theories are (a) original social role theory, (b) founder effect social role theory, and (c) biosocial role theory.

24.1.1.1 Original Social Role Theory

The most widely recognized theory of sex differences in cognition and behavior is known as *original social role theory*, also sometimes called "core social role theory" (Eagly & Wood 2012:460). This theory assumes that sex differences in behavior are entirely learned within a social context. In other words, if it were not for different learning experiences, boys and girls as well as men and women, would behave indistinguishably. As explained more below, what helps to maintain these sex differences in behavior generation after generations are sex stereotypes.

Dating back at least to the 1970s, the original version of social role theory has been widely espoused by social scientists (Perry & Bussey 1979; Deaux 1985; Eagly 1987; Eagly, Wood & Diekman 2000; Mischel 2015). At the heart of this theory is the assumption that sex differences in how people behave are the result of how cultures socialize males and females differently. In this regard, parents are obviously aware that their baby is either a boy or a girl from birth onward. Whether consciously or not, parents and others in the child's circle of social relationships treat them in accordance with what is considered appropriate for boys and girls. Based on the differential treatment they receive from parents and others who are in their orbit of influence (e.g., teachers), most males and females gradually come to adopt sex-typical sex roles.

Cultural expectations about how males and females think and behave, or at least *should* think and behave, take the form of what are known as *sex role stereotypes*. These stereotypes become established in a particular culture with a high degree of resilience generation after generation and form the bedrock upon which parents and others treat children. One research team described the stereotyping process as follows: "Women and men confirm gender stereotypes in large part because the different roles that they perform place different social demands upon them" (Vogel et al. 2003:520). Others have described the process of sustaining sex roles (or gender roles) over multiple generations as involving "normative expectations" which nearly all children learn to accept as they mature (Eagly & Wood 1991). In other words, most people gradually learn to conform with what others in their culture expect them to become, and these expectations are often distinct for males and for females.

The concept of *social roles* can be thought of as being similar to actors learning parts in a play. Regarding sex, there are essentially two roles or "scripts." Depending on their social surroundings, individuals will come to recognize their own sex and thereby come to adopt the scripts that are appropriate for their particular sex. This generalization, of course, is true not in either-or terms, but in general terms, particularly in societies that are very ridgid in how they conceive of male roles and female roles. Thus, the original social role theory predicts that sex differences in thinking, feeling, and behaving should be greater in more gender-rigid societies than in societies that are more liberal in what they expect of males and females.

The original social role theory has been used to help explain many sex differences in cognition and behavior. Examples include differences in aggressive behavior (Eagly & Steffen 1986), interests in child rearing (Weitz 1977), tendencies to provide care and comfort to others (Rogers 1969), religiosity (Nelsen & Potvin 1981), sexual attitudes and behavior

(Petersen & Hyde 2011), sex stereotyping (Eagly & Steffen 1984), and a host of other social behaviors (Eagly & Wood 1991; Eagly 2013).

Despite the large number of behavioral traits that the original social role theory has been invoked to explain, the theory has shortcomings. One is that the theory offers no explanation for the origins of the sex role stereotypes that are responsible for sex differences in behavior. This theory also has difficulty explaining why *any* sex differences would be culturally universal. In other words, given the tremendous diversity in human cultures (e.g., in languages, religions, and governmental systems), one would not expect to find very many if any sex differences in cognition and behavior in every human culture.

Overall, the original social role theory asserts that, as children become aware of their sex and notice how others of their sex think and behave, nearly all males and females adjust their own ways of thinking and behaving to conform depending on their sex and the culture in which they live. Regarding USDs, the original social role theory implies that few if any sex differences should be universal, especially when distinctly different cultures are compared.

24.1.1.2 The Founder Effect Social Role Theory

In the latter part of the 20th century, Fausto-Sterling (1992) recognized that the original social role theory predicts that few if any universal sex differences in behavior should exist. Without abandoning the view that all sex differences in behavior are learned, she made the following proposal to account for the possibility of some behavioral sex differences existing, "the entire population of the world all evolved from a small progenitor stock" (Fausto-Sterling 1992:199). While she did not specify precisely what was meant by *small progenitor stock*, one can presume that it refers to the first foraging human populations long before migrating out of Africa roughly 50,000 years ago (MC Campbell & Tishkoff 2010; Tierney, deMenocal & Zander 2017).

Fausto-Sterling (1992:199) went on to argue that some behavioral sex differences will have been "faithfully passed down a thousand times over." Unfortunately, she was not specific about what types of sex differences would be faithfully passed down and what ones would not. In any case, Fausto-Sterling's version of social role theory provides a way of preserving the theory's most basic tenet (i.e., that sex differences in behavior are culturally learned, not biologically determined) while still providing an explanation for why some sex differences appear to be found in all human societies. She called this version of social role theory the *founder effect theory*.

The founder effect theory has been criticized on the grounds that it downplays the tremendous diversity in human cultures. In other words, with all the differences in language, technology, religions, and forms of government, it seems hard to accept that sex role stereotypes and sex differences in thoughts and behavior would have been preserved unchanged throughout human history without any biological underpinnings. Along these lines, Browne (2006:152) questioned the founder effect theory by arguing that "the idea that any, let alone many, cognitive and behavioral sex differences would have been preserved in all human cultures simply because those differences happen to have been arbitrarily established in some primordial culture hundreds of thousands of years ago defies credibility."

Probably, the greatest problem with the founder effect theory is that it offers no guidance for identifying *which* sex differences will be preserved in all cultures over thousands of generations and which ones will be free to vary from culture to culture without biology being involved. This is to say that Fausto-Sterling's theory does not identify which sex differences will be "faithfully passed down a thousand times over" and which ones will be culturally unstable. About the only guideposts one can use would involve assuming that the sex differences that will be faithfully passed down would be those exhibited in foraging societies prior to the development of agriculture (about 10,000 years ago).

24.1.1.3 The Biosocial Role Theory

Until the early 1990s, Eagly and Wood were prominent proponents of the original social role theory (Eagly 1987; Eagly & Wood 1991). As the 20th century drew to a close, they switched their advocacy toward a new version of social role theory, one that could account for why some sex differences might be universal. Instead of endorsing Fausto-Sterling's founder effect social role theory, which also predicts universal sex differences, Eagly and Wood proposed what they called the *biosocial role theory*. As described more below, this theory stipulates that certain underlying *biological* differences between the sexes result in very similar sex differences of a behavioral nature in all cultures.

Wood and Eagly (2002) were quick to add that their new theory – also called the *biosocial constructionist theory* (Eagly & Wood 2013:11) – argues that behavioral sex differences are all socially learned. In other words, sex differences in behavior are not due to biological influences except in indirect ways. These indirect ways have to do primarily with the fact that (a) only women can bear children and (b) men tend to be larger and physically stronger than females (Wood & Eagly 2012:58). Due to these basic sex differences, the biosocial role theory asserts that there will be certain inherent divisions of labor, such as males gravitating toward jobs that are more physically strenuous and females being drawn toward occupations involving child care (Eagly & Wood 2012:465). By focusing on how males and females learn different behavioral patterns, these scientists have strongly opposed the idea that evolutionary forces have produced sex differences in behavior (Wood & Eagly 2002:713).

Here is how Wood and Eagly (2002:702) have explained their new theory: "The most important distal determinants of sex-typed roles within a society are (a) the essential sex differences represented by each sex's physical attributes and related behaviors, especially women's childbearing and nursing of infants and men's greater size, speed, and upper-body strength and (b) the contextual factors represented by the social, economic, technological, and ecological forces present in a society." They go on to argue that, "Physical sex differences, in interaction with social and ecological conditions, influence the roles held by men and women because certain activities are more efficiently accomplished by one sex. It can thus be easier for one sex than the other to perform certain activities of daily life under given conditions. The benefits of this greater efficiency emerge because women and men are allied in complementary relationships in societies and engage in a division of labor."

Eagly and Wood (2005:282) state that biosocial role theory leads to the expectation of consistent sex differences "across societies in the activities most closely enabled or constrained by sex-typed physical attributes and reproductive activities." Because of men's size and upper-body strength, for example, men will gravitate toward "tasks requiring speed of locomotion and bursts of strength."

In one publication, Wood and Eagly (2012: 56) state: "Although responsive to local conditions, the division of labor is constrained by women's childbearing and nursing of infants and men's size and strength. Because these biological characteristics influence the efficient performance of many activities in society, they underlie central tendencies in the division of labor as well as its variability across situations, cultures, and history." This theoretical proposal does not attempt to explain *why* males are larger and stronger or *why* only females give birth to offspring. As will be noted more later on, this avoidance is likely to reflect the desire by Wood and Eagly to avoid explaining sex differences in evolutionary terms.

Biosocial role theory has at least two noteworthy limitations. One is that the theory focuses almost entirely on sex differences that occur after childhood. Few if any behavioral sex differences among infants and toddlers should be present, since especially infants would not even yet be aware of their sex (or gender), and few toddlers would have yet learned what sex roles were expected of boys and girls.

Another limitation of the biosocial role theory is that it gives no credence to the possibility of prenatal testosterone exposure influencing sex differences in behavior. As will be documented below, there is considerable evidence that exposing the developing brain to high (male-typical) testosterone has considerable influence on a wide variety of behavioral traits.

Overall, Eagly and Wood's biosocial role theory recognizes that biological factors are relevant to sex differences in behavior, but not in any direct causal sense. Instead, they argue that certain biological sex differences following puberty (such as those pertaining to strength and child bearing) affect behavior patterns of males and females, especially regarding occupations and work. However, these authors argue that sex differences in occupational choices are ultimately due to cultural influences, not evolutionary influences (Wood & Eagly 2002:702). Because certain biological sex differences are present in all cultures, Wood and Eagly's biosocial role theory implies that some sex differences in behavior may end up being universal even though all behavioral sex differences are learned.

The crux of biosocial role theory is that, (a) all sex differences in behavior are learned, and (b) sex differences of a biological nature bias this learning. Specifically, because males tend to be larger and stronger, they are more likely to be engaged in heavy labor, aggression, and warfare (Eagly & Wood 2012:465). And, because only females can give birth, they are more likely to be caregivers, particularly to young children (Eagly & Wood 2012). Thus, from the perspective of this theory, three biological factors – male size, male strength, and female pregnancy and birthing capabilities – help to promote at least a few universal sex differences in behavior. However, all other behavioral sex differences are simply learned through social indoctrination into each culture's unique set of stereotypes.

24.1.2 The Two Evolutionary Theories

In the middle of the 19th century, Charles Darwin (1859) proposed a theory of how life has changed over eons of time from very simple one-celled organisms (such as bacteria) to the wide array of multicellular life forms that exist today (including ourselves). What made this theory especially controversial is that it contradicted the idea that all forms of life were created by God in a three-day timeframe (Larson 2003).

Darwin put forth his theory of evolution primarily to explain what is known as *the fossil record*. This record refers to the dazzling array of life forms embedded in ancient rock deposits. Even though Darwin focused on explaining the fossil record, he came to recognize that his theory had far-reaching implications. For this reason, late in life, Darwin (1871) proposed the concept of *sexual selection*, as a special type of natural selection (that Darwin saw driving the succession of species that occurred over time). *Sexual selection*, Darwin said, occurs *within* species and between and within the sexes, in contrast to most forms of natural selection, that operate more broadly (Clutton-Brock 2017:7 Hönekopp et al. 2007; Puts, Jones & De Bruine 2012).

To give a simple example of *sexual selection*, in nearly all species of mammals, males appear to be larger and stronger than females. Darwin proposed that greater male size and strength helps males compete with each other for mates and to defend territories (which attract mates). In other words, winning fights with other males usually results in the victorious males having more mating opportunities and, therefore, more offspring than the losing males produce (Ellis 1995). Because greater size and strength among females has little effect on the number of offspring they produce, males have been sexually selected for these physical traits. However, by choosing to mate with larger and stronger males, females usually produce male offspring who grow up large and strong themselves, thereby increasing their odds of reproducing in the next generation (Archer 2009).

Along similar lines, tendencies for males of many species (including humans) to be more aggression-prone than females (see Table 25.17.16) are likely due to the fact that dominance is often established through aggression or threats of aggression (Fonberg 1988; Pellegrini & Bartini 2001). Not only does greater size and strength enhance an animal's chances of being dominant, so too does interests and abilities to be aggressive. Again, in most species, the long-term outcome of male aggression often includes increased access to resources and mating opportunities, which in turn, increases the representation of their genes in future generations.

Before examining the two evolutionary theories, it is worth noting that they do not deny that sex differences in behavior have substantial learning elements. Instead, both theories assume that males and females have evolved abilities and interests in learning some types of behavior more readily than other types. In the words, "Although evolutionary theory provides only a part of the explanation for sex differences, that part is fundamental, offers heuristic power, and helps to reorganize factors that otherwise appear disconnected" (Luoto & Varella 2021:8–9).

As a final general comment about Darwin's theory of evolution, it is worth mentioning that the concept of *genes* had not yet been proposed. Only in the early 20th century did scientists recognize the existence of genes and realize that they were an important part of key roles in Darwin's theory regarding both natural selection (Blekhman et al. 2008; Bigham et al. 2010; Rowe & Rundle 2021) and sexual selection (Gangestad & Simpson 2000). Even the interaction of these two forms of selection, along with environmental variables, are now conceptualized as being central to the evolutionary process (Long, Agrawal & Rowe 2012; Kardos et al. 2015; Archer 2019).

24.1.2.1 Sexual Selection Theory

Evolutionary theory with an emphasis on sexual selection has become prominent in guiding research on sex differences in recent decades, especially sex differences that appear to be prevalent all over the world (Geary 2010; Okami & Shackelford 2001; Puts, Jones & De Bruine 2012). For example, as will be documented in the following chapter, studies in many countries have found that males express greater desire for promiscuous sex than females (see 17.7.1.24). This sex difference in preferences has been theoretically explained in terms of sexual selection (Schmitt 2003; Geher & Kaufman 2013). Specifically, males who want to have numerous sex partners (especially before the advent of modern contraception) will have left more offspring in subsequent generations than males who only have a single sex partner.

The logic of sexual selection suggests that the main reason the average female is less interested in mating with numerous sex partners has to do with her lower so-called *reproductive ceiling* (i.e., the number of offspring she can possibly have by mating with numerous partners). Rather than having numerous partners, a female should carefully select mates on the basis of loyalty and ability to secure resources, especially while she gestates and, in the case of mammals, breastfeeds her offspring (Buss & Schmitt 1993). In other words, females have less to gain in reproductive terms from having numerous sex partners; thus, they have been sexually selected for thinking and behaving in different ways about sex.

Overall, sexual selection theory has provided a springboard for many arguments about sex differences not just in physical traits, but also in cognitive and behavioral traits. For example, some evolutionary theorists have argued that, among humans, males have been sexually selected for spatial reasoning so as to better navigate while hunting and throwing projectiles when prey are encountered (Kolakowski & Malina 1974; Geary 2010). More specifics in this regard will be presented in the next chapter.

Unlike scientists who use social role theory to explain sex differences, those who rely on sexual selection theory envision genes playing an important role in producing average differences between males and females. As one proponent argued, "If a sex difference occurs consistently, despite all the variations in learning and socialization practices that occur across cultures, then ... an innate predisposition is probably showing through all the cultural 'noise'" (Lippa 2002:116). Sexual selection theory leads one to expect that if sex differences exist for a particular trait in essentially all societies, then evolutionary forces are likely to be responsible. As noted earlier, social role theorists such as Fausto-Sterling (1992) and Wood and Eagly (2012) do not consider Darwinian evolution as having significant effects on sex differences in behavior, even if those differences are found in all societies.

One criticism of sexual selection theory is that it is little more than a series of after-the-fact just-so stories (Schuett, Tregenza & Dall 2010:232). In other words, researchers discover that there appears to be sex differences in a particular trait, and then concoct a way to argue that the difference has some favorable reproductive consequence for one or the other sex.

Along these lines, one will see in the next chapter that 130 studies worldwide have found that males suffer accidental (including fatal) injuries more than females (although four studies found no significant sex differences in this regard). While it would make no sense to argue that males were being sexually selected for having more accidental injuries than females, evolutionists have argued that males have been favored for taking more risks (Pawlowski, Atwal & Dunbar 2008; Charness & Gneezy 2012; Stern & Madison 2022). As will be explained more later, the gist of the argument would be as follows: Risk-taking can often be associated with obtaining more resources, and more resources often attract mates. If so, high rates of accidental injuries are not being sexually selected per se. Instead, they reflect elevated tendencies to take risks, and are more common in males than in females because risk-taking among males has been favored more than risk-taking among females.

The most noteworthy shortcoming of sexual selection theory is that it only offers explanations for *why* there might be sex differences in various traits. Rarely does it provide account for *how* sex differences would be produced, other than to imply that some type of genetic factors must be involved (Pinker 2002:50). The identity of these genetic factors and how they might be translated into physiological and, ultimately, behavioral traits are not specified by sexual selection theory. To provide more specifics in this regard, one other evolutionary theory has been offered. It is described below.

24.1.2.2 Evolutionary Neuroandrogenic (ENA) Theory

Many evolutionary theorists have recognized that the primary way genes affect cognition and behavior is by making alterations in the brain (e.g., LF Jacobs 1996; MJ Ryan 2021). Brain functioning, in turn, is highly influenced by the presence of hormones, including sex hormones, which are also under substantial genetic control (Ellis 2011a). Such reasoning has led to the development of a theory known as *evolutionary neuroandrogenic (ENA) theory*. The primary purpose of this theory has been to explain sex differences in a wide array of cognitive and behavioral traits (Ellis 2005, 2006, 2011a, 2011b; Genovese 2008; Palermo 2010; DP Schmitt 2015; YC Chan 2016; Winegard & Winegard 2018; Chu et al. 2019; Mazlan et al. 2020).

Two fundamental premises underly ENA theory: First, most sex differences in traits have evolved by sexual selection. Second, the main way this occurs involves genes controlling sex hormones that, in turn, alter brain functioning. The hormones that are the focus of ENA theory are known as *androgens*, sometimes also simply called "male hormones," even though androgens are found in both sexes, albeit in different amounts, depending on age. The basic nature of sex differences in testosterone over the life course are shown in Figure 24.1.

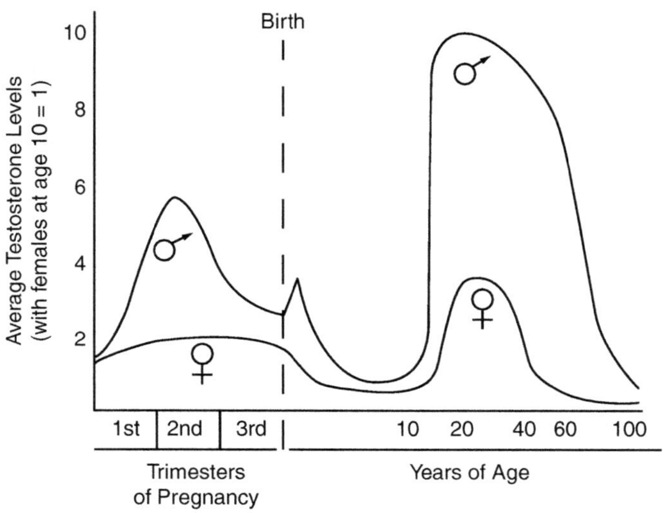

Figure 24.1 Testosterone levels of human males and females from conception through old age.

While androgens are found in both sexes, males have them in greater quantities than do females, especially during fetal development and following the onset of puberty (Davison & Davis 2003; see Figure 24.1). The androgen that is most prevalent in males is *testosterone*, so named because it is primarily produced by the male testes, although female ovaries also produce small amounts as do the adrenal glands in both sexes. Also, one should note that male sex hormones (such as testosterone) and female sex hormones (such as estradiol) can interact with one another in complex ways. To give an example, testosterone can actually be transformed into estradiol (the most common estrogen, which is considered a "female hormone"). This transformation occurs in the presence of a simple enzyme, known as *aromatase* (Czajka-Oraniec & Simpson 2010). This latter point serves to underscore that there is often delicate genetically-influenced biochemistry separating the sexes (Bancroft 2002).

In essence, ENA theory supplements sexual selection theory with the following proposition: Androgens alter brain functioning in ways that produce a variety of average sex differences in traits. These traits not only include physical traits, but also ones of a cognitive and behavioral nature. Individual differences in androgen levels are, in turn, highly influenced by genes, particularly genes on the Y chromosome (which only males have). These genes play a central role in converting would-be female ovaries into testes instead (Carruth, Reisert & Arnold 2002; Savic et al. 2017). Once this conversion occurs early in fetal development, the process of physically masculinizing the body, including the brain, usually occurs over the next several months.

Genes located on chromosomes other than the Y chromosome are also involved in sexual differentiation. For example, a gene on the X chromosome also influences exposure to androgens (Ellis, Hasan, & Hoskin 2021), and complex environmental factors occurring before and after birth that can often alter bodily production of androgens as well (Zaidi 2010; Martel 2013). In general, the effects of androgens on brain and bodily functioning before birth appear to have more impact on lifelong sex differences in cognition and behavior than do androgenic effects following puberty, even though the latter are also significant (Udry, Morris & Kovenock 1995; Udry 2000; Hines, Golombok et al. 2002).

Like sexual selection theory, ENA theory assumes that genetically influenced traits have been sexually selected to promote reproduction in both males and females. In this regard, ENA theory puts a special emphasis on the following evolutionary reasoning when it comes to mammals (including humans): Because females must gestate offspring for long periods of time, they can enhance reproduction by obtaining the assistance of mates who will provide resources to her and her offspring. Male mammals, on the other hand, can reproduce more prolifically by mating with numerous partners. This sex difference has created a "sexual tension" that has woven a complex tapestry of sex differences in numerous brain functioning. These brain functioning patterns manifest themselves in many ways having to do with cognition and behavior.

Both sexual selection theory and ENA theory focus on explaining sex differences that will help males and females pass their genes onto future generations. Where the two theories differ is that ENA theory identifies specific neurohormonal processes that may be involved and even hints at some of the genes that may under these neurohormonal processes. Sexual selection theory does not do this. In other words, ENA theory is essentially an extension of sexual selection theory in which it makes predictions that are much more susceptible to being proven wrong.

Several researchers have begun to discover some of the details about how testosterone affects brain functioning to impact sex differences in cognitive and behavioral traits such as those involving competitiveness, aggression, and risk-taking (Mehta, Jones & Josephs 2008; Volman et al. 2011; Welker, Gruber & Mehta 2015). These studies indicate that, when neurological levels of testosterone are high, key reward-sensitive areas of the brain produce dopamine, a "feel good" neurotransmitter.

High dopamine, in turn, promotes active competitive types of behavior and reduces feelings of self-doubt (BC Campbell et al. 2010; Eisenegger et al. 2017). There is also substantial evidence that, in the brain, testosterone often interacts with cortisol, a major stress hormone, in ways that have complex effects on behavior (Mehta, Jones & Josephs 2008; Dekkers et al. 2019; Knight et al. 2022).

Overall, from a scientific perspective, a "strong theory" is one that generates many hypotheses, all (or nearly all) of which end up being confirmed when subjected to empirical tests. Weak theories, on the other hand, either lead to few testable hypotheses or, when tested, they fail to be supported by most of the empirical evidence. Especially in Chapter 26, attempts will be made to apply each of the five theories to each sex difference variable so as to determine which are the better theories. One will see that, while no single theory explains all of the likely universal sex differences, some theories do much better in this regard than do other theories.

24.1.3 Religious/Theological Explanations

Scientific theories are not advocated as an article of faith, but strictly for the purpose of being empirically tested. To test scientific theories, scientists typically collect data (such as on sex differences) and then see if what a theory predicts is either consistent or inconsistent with the data. (Parenthetically, scientists usually avoid using the word "true" to describe theories. This is because one can never rule out the possibility that some future theory might end up being better at explaining a set of empirical observations than any theory currently available.)

So, despite clear differences between how science and religion seek to understand nature, there are sometimes ways of assessing the relative strength of their explanations. Below are some examples of religious scriptures that have to do with sex differences in traits that can be empirically investigated.

One example of religious teachings having to do with sex differences involves the concept of *patriarchy*. This term, also known as *headship*, refers to the tendency for males to be the primary authority figure or decision-maker, especially in family functioning.

As noted by many theologians, several biblical passages clearly favor patriarchy (e.g., Baloyi 2008; Pillay 2013; Casimir, Chukwuelobe & Ugwu 2014; E Petersen 2016). For example, in Genesis 2:18, God was reported to have said, "It is not good that the man should be alone; I will make him a helper fit for him." Similarly, in Ephesians (5:21–33), Paul commands wives to be subservient to their husbands, because the husband is the head of the wife. To quote the passage, "Wives, submit to your husbands as to the Lord. For the husband is the head of the wife as Christ is the head of the church."

Likewise, 1 Peter (3:1) tells wives to be "submissive to your husbands." In 1 Corinthians 11:3, the following statement is made: "But I want you to realize that the head of every man is Christ, and the head of the woman is man, and the head of Christ is God." This statement is preceded by 1 Corinthians 11:7, where the man is described as being "the image and glory of God; but the woman is the glory of man." While various nuanced interpretations have been offered for these types of passages (see Baloyi 2008), it is hard to escape concluding that several sacred scriptures envision "dominance-submissive relationships and patriarchy as part of God's plan" (Heggen 2019:17).

The Islamic Koran IV:34 also states that men should be in charge of women. The passage stated that "Men have authority over women because Allah has made the one superior to the other." In another passage, one finds the statement that "men have status above women" (Koran II:228).

Regarding sex differences in responsibilities to one another, the Koran (IV:4) states that the husband has primary responsibility for "providing economic support for his wife(s) and family." A similar passage is found in Mormon scriptures. According to the Doctrine and Covenants Section 83: 2, "Women have claim on their husbands for their maintenance." Such edicts could obviously help to account for why more males than females are employed.

Another Koranic verse (IV:34) grants permission to husbands to beat their wives as a way of maintaining their obedience: It states that, "As for the women who show rebellion, you shall first enlighten them, then desert them in beds, and you may beat them as a last resort." This type of Allah-granted permission could help to explain sex differences in spouse abuse, at least in Islamic countries (Pillay 2013; E Petersen 2016). One author also suggested that certain religious teachings could account for why females are more involved in housework than are males (Voicu 2009:158; Voicu 2019).

Returning to the Bible, it addresses the issue of the relative pay for men and women by specifying that women should receive 60% of what men receive (Rhodes 1988:1207). The exact wording appearing in Leviticus (27:1–4) is as follows: "The Lord spoke to Moses and said 'Speak to the Israelites in these words. When a man makes a special vow to the Lord which requires your valuation of living persons, a male between twenty and fifty years old shall be valued at fifty silver shekels, that is shekels by the sacred standard. If it is a female, she shall be valued at thirty shekels."

Regarding the Koran, no statement regarding sex differences in earnings was located. However, one passage stated that, "A male shall inherit twice as much as a female" (Koran IV:11). In the latter portion of Chapter 26, we will discuss the above statements in light of research findings on sex differences in earnings and wealth.

Overall, a substantial number of pronouncements in religious scriptures appear to be relevant to sex differences in various types of behavior. Our purpose in citing them is to assess their potential for helping to explain these differences, especially if they happen to be universal. The value of these and other religious passages (alongside the five scientific theories) will be noted in the following chapters, particularly Chapters 26 and 27.

24.2 Contrasting Science and Theology

Both scientific theories and religious teaching can be used to help understand the universe, and both rely heavily on writing. However, as described below, scientific theories and religious teachings are quite different from one another in at least three respects.

First, religious teachings are usually not intended to be questioned or subjected to careful scrutiny. Scientific theories are proposed for the purpose of being questioned. Thus, it is sometimes said that the best scientific theories ever proposed are those that have not *yet* been proven wrong. While religious scholars might inquire into the authenticity and intended meaning of sacred scriptures, the fundamental truth of these writings is largely a basis on faith, not empirical evidence.

Second, most religions assume the existence of a supernatural being (or beings) that controls nature, with *God* and *Allah* being common names used to refer to this being. Scientists, on the other hand, usually consider the existence of anything outside of nature to be beyond what they can investigate. Consequently, when scientists engage in scientific inquiries, they nearly always *assume* that nature is a "self-contained" system, with their goal being to identify *how* components within this natural system are interrelated.

Third, most religions are organized around a set of sacred writings through which the supernatural has revealed much of its divine plan for the universe, nature, and humanity. These sacred writings also provide guidance to humans regarding how they should conduct their lives. Some of these writings, of course, include information about how males and females are supposed behave in general and in relationship to one another. From a scientific perspective, there are no sacred writings. Instead, everything that has ever been written about a topic can be questioned, with the final arbitrator being one's understanding of the weight of all relevant empirical evidence. As more and more evidence is collected and reported in scientific literature, each theory is rechecked regarding its ability to account for what is found.

24.3 Conclusions

As noted at the beginning of this chapter, explanations for sex differences in traits can be organized into three categories: sociocultural, evolutionary/biological, and religious/theological. In this regard, a sample of U.S.

adolescents and young adults were asked which one of these three types of explanations seemed to best explain why males and females differ in behavior, especially behavior surrounding dominance and submissiveness. This survey indicated that most respondents thought sociocultural factors were most relevant and that religious factors were least relevant (Neff & Terry-Schmitt 2002:1194). However, the respondents may not have been aware of the religious passages cited near the end of Section 24.1, indicating that sacred scriptures have a lot to say about dominance/submissive relationships between the sexes.

Five scientific theories of sex differences bearing on the possibility of these differences being universal were located. Three of these theories explain sex differences in cognitive and behavioral traits predominantly in terms of sociocultural learning influences. They are (a) the original social role theory, (b) the founder-effect social role theory, and (c) the biosocial social role theory. All three of these theories reject the idea that sex differences in cognition and behavior are explainable in evolutionary terms. Instead, they focus on the learning of sociocultural traditions and stereotypes about how males and females are expected to think and behavior.

The remaining two theories are evolutionary in nature. Without denying that learning contributes to sex differences, both of these theories focus on evolutionary forces contributing to sex differences. One of these theories is known as *sexual selection theory*. It relies heavily on Darwin's (1874, 1876) concept of *sexual selection*. According to this concept, males and females have evolved combinations of traits that help each of them to maximize the number of offspring they leave in future generations. These traits may include differences in physical features (such as strength and size) as well as differences in cognitive and behavioral features (such as ones those having to do with competitiveness, risk-taking, and desires to care for children). As a scientific generalization (rather than a cultural stereotype), a major evolved sex difference in behavioral terms appears to be that "women are caregivers and men are providers" (Zhu & Chang 2019:2).

The other evolutionary theory is known as *evolutionary neuroandrogenic (ENA) theory*. This theory accepts Darwin's concept of sexual selection as being central to understanding sex differences. However, it adds an additional proposal to explain *how* these sex differences are produced, especially regarding sex differences of a cognitive and behavioral nature. ENA theory proposes that sex differences result from genes that have programmed the brains of males and females to function differently. These genes have most of their effects by causing males to produce more androgens than do females. In other words, at a physiological level, there would be no sex differences in how humans think and behavior were it not for sex differences in brain exposure to androgens (particularly testosterone).

Saying that sex differences in brain functioning are responsible for sex differences in cognition and behavior does not mean that learning is not important in the process. Instead, average levels of androgens appear to program the brains of both sexes so as to make them more interested in and more capable of learning behavior that helps both males and females to gravitate toward their respective sex roles. These respective sex roles, in turn, help both sexes to contribute genes to the next generation.

After the five scientific theories of sex differences were described, this chapter reviewed various passages from religious scriptures that seem pertinent to sex differences, especially regarding behavior. Some of these passages appear to be sufficiently precise that it will be possible to assess them with empirical data in essentially the same way that scientific theories are assessed.

Compared to this chapter, the upcoming three chapters are lengthy. This is especially true for Chapter 26. The main reason for their lengths is that each chapter will deal with hundreds of variables that may be appropriately considered to be universally found in all (or virtually all) human societies.

25 Sex Difference Variables with Ten or More Findings

The purpose of this chapter is to provide a condensed summary of findings from all tables contained in Chapters 1 through 23 (provided in Volumes I through III) for which at least ten citations were located. Then the basic nature of the findings from each of these tables is described along with what we will call a *consistency score*. As explained in more detail below, this consistency score will provide a quantitative estimate of the degree to which studies on sex differences have reached the same basic conclusion. Furthermore, information will be provided regarding the number of countries sampled for each trait, along with an indication of the time period within which studies were published.

Presentation of evidence in this chapter will follow the order in which the variables were presented in this book's first three volumes. The original table numbers are also reported so that interested readers can return to any of the original tables for more details.

Table Structure

Of the roughly 4,000 variables presented in Chapters 1 through 23 (comprising the book's first three volumes), only about 10% contained citations to ten or more findings. These will be the only variables to be considered in this chapter. To provide concise descriptions of the results for these variables, this chapter will utilize tables with the following headers:

Varia-ble	Orig. table	Number of Findings (and Consistency Scores)						Over view	Coun tries	Time range	Non-human
		I/T	Child	Adol	Adult	WAR	All ages				

The first header simply names the variable involved. Then the second header denotes the number assigned to the original table, all of which are contained in the book's first three volumes. In the next set of headers, abbreviations for the age ranges are provided – i.e., *infant/toddler* (*I/T*),

DOI: 10.4324/9781003405290-2

child, *adolescent* (*adol*), *adult*, and *wide age range* (*WAR*). Also, note that a shaded header is provided for indicating the total number of findings when findings for all age groupings are combined. Parenthetically, a small number of variables have findings for fetal sex differences: these findings will be reported by adding a header in front of the I/T header.

The next two headers have to do with conclusions reached by literature reviews and meta-analyses (*Overview*) and the number of countries sampled. In the case of studies based on multiple countries, these counts appear in parentheses. In the next header, the *Time range* column, reports the first and the most recent year published studies were reported. Finally, the column named *Non-human* provides a snapshot summary of any information regarding findings based on studies of species other than humans.

Abbreviations

Certain abbreviations are used throughout this chapter within each table. Most notably, *M* refers to males and *F* refers to females. Numbers preceding either of these two letters refer to the number of studies reporting that males or females were higher. All numbers appearing in front of the small letter *x* signifies how many studies reported no significant sex difference. Beneath the *Overview* header, the abbreviation *Rev* identifies literature reviews and *Met* refers to meta-analyses.

Consistency Scores

Within each of this chapter's tables, for every variable identified, a *consistency score* is provided. Each consistency score (CS) can range from 0.0 to 100.0. The formula used to calculate each scores is as follows:

$$CS = \left(\frac{\text{Number of Findings for the sex with the greatest number}}{\text{total Number of Findings + Number of Findings for the sex with the fewest number}} \right) \times 100$$

Note that this formula is not a simple proportion. Instead, it effectively doubles the weight of any studies that happen to be opposite of the findings from the majority of studies. Here is a hypothetical illustration: Say that 20 studies of sex differences in a particular trait were located, and that 18 of these studies indicated that females exhibited the trait more than males. If one were to simply report what percentage of studies indicated that females were more, the percentage would be 90%. However, let us assume that one of the two exceptional studies found no significant sex difference, while the other exceptional study reported that males were actually *higher* than females. In this hypothetical example, the consistency score (CS) would be calculated as follows: 18 divided by 18 + 1 + 2 multiplied by 100, or (18/21) x 100, yielding a CS of *85.7*.

In designating consistency scores to be indicative of a likely universal sex difference (USD), we use a CS = 95.0 as the cut-off point. In other words, any consistency score of 95.0 or higher is deemed to be a variable that is likely to be a *USD*. Thus, as consistency scores of 95.0 or higher approach values closer to 100.0, they are increasingly likely to be USDs, especially if the number of relevant studies is well above ten, and the countries sampled is substantial.

25.1 Reproduction, Development, and Morphology: Condensed Findings

This initial section summarizes tables appearing in Chapter 1 for which ten or more findings of sex differences for traits were located. These traits have to do with reproduction (e.g., sex ratios at birth), physical development (e.g., age at puberty), and basic morphologic features (e.g., height and weight).

25.1.1 Sex Ratios

Table 25.1.1 summarizes research findings having to do with what are sometimes known as *primary sex ratio* and *secondary sex ratio*. Primary sex ratio refers to the proportion of males and females found in utero. This is usually assessed with the use of either amniocentesis or ultrasound. The first row of this table shows that most studies have concluded that more males are found in utero, although one study reported changes over the course of gestation.

Regarding the secondary sex ratio (i.e., the proportion of male and female live births), the table's second row summarizes findings from 303 studies, based on samples drawn from 39 countries plus 24 additional studies involving multiple countries. The majority of these studies have indicated that more males than females were born alive, a pattern that is consistent with three literature reviews on the topic. It is worth mentioning that most of the studies that reported more females than males were born alive came from research in which expectant mothers had been exposed to hazardous chemicals during pregnancy. This has been interpreted as suggesting that adverse uterine environments are more likely to induce miscarriages among male fetuses.

The last entry in this table has to do with trends in the secondary sex ratios. Of the 16 pertinent studies that were located, 11 found increases in the proportion of females over time, while 4 reported greater proportions of males being born, and 1 study failed to identify any significant change. Of course, some of the inconsistencies could be due to different time frames that each country used to assess their trends as well as the countries that were sampled.

25.1.2 Sex Ratios According to Parental Social Status

As discussed in Chapter 1, substantial research has been conducted to test the so-called *Trivers-Willard hypothesis*. This hypothesis has been interpreted as

Table 25.1.1 Sex ratios

Variable	*Orig. table*	*Number of sex difference findings (and consistency scores)*							*Overview*	*Countries*	*Time range*	*Non-human*
		Fetal	*Infant*	*Child*	*Adol*	*Adult*	*WAR*	*All ages*				
1. Sex ratio in utero (primary)	1.1.1.1	11M 1 × (91.7)	–	–	–	–	–	11M 1 × (91.7)	–	3	1940–2000	Mixed for various mammals
2. Sex ratio at birth (secondary)	1.1.1.4		286M 7 × 17F (87.4)	–	–	–	–	286M 7 × 17F (87.4)	M: 3Rev	39 (24)	1926–2022	96M 24 × 74F all animal species
3. Trends in sex ratio at birth	1.1.1.5		11F 1 × 4M (55.0)	–	–	–	–	11F 1 × 4M (55.0)	–	3 (4)	1994–2005	–

indicating that greater proportions of males should be born to high-status parents than to low-status parents. Numerous studies have tested this hypothesis with humans using four different indicators of parental social status (i.e., years of education, occupational level, income or wealth, leadership or eminence) along with various composite measures. One can see in Table 25.1.2 that the evidence has been mixed, with just 23 out of the total of 49 studies being consistent with the hypothesis. For the types of social status measures involving ten or more studies, the strongest support has come from income or wealth measures. Corresponding studies of sex ratios and parental dominance are also mixed.

25.1.3 Birth and Reproductive Factors

Table 25.1.3 pertains to birth and reproduction. It shows that most studies of miscarriages have found greater proportions of males being miscarried than females. In the case of stillbirths (and early infant deaths), all 31 studies agree that males are more vulnerable than females, making this a very likely universal sex difference. It is also noteworthy that all studies that were located for non-human mammals have also reported greater proportions of males being miscarried and being stillborn or dying soon after birth than is the case for females.

Numerous studies of birth weight have been published, one dating back to the late 1800s. One can see that the vast majority of these studies indicate that baby boys weigh more than baby girls, and that very similar patterns have been found for several species of non-humans. This is a rather remarkable finding in light of the evidence of higher rates of miscarriage and stillbirths for boys.

The last entry in this table has to do with what is known as *reproductive success*. This variable is usually assessed in terms of the number of offspring an individual has. All but two of the 13 studies of reproductive success indicate that males who have children have greater numbers of children than do females.

25.1.4 Maturation Factors

Table 25.1.4 is concerned with sex differences in rates of maturation. It shows that findings have been extremely consistent in indicating that females mature more rapidly than males in terms of physical maturation and age of pubertal onset. One might also note that studies of various non-human mammals have reached the same conclusion. Therefore, the tendency for females to physically mature more rapidly than males appear to be a universal sex difference for humans and possibly for mammals in general.

Table 25.1.2 Sex ratios according to parental social status

Variable	*Orig. table*	*Measure of Social Status (and Consistency Scores)*						*Overview*	*Countries*	*Time range*	*Non-human*
		Years of Education	*Occupational Level*	*Income or Wealth*	*Leadership or Eminence*	*Multiple Measures*	*All Social Status Measures*				
Sex ratio at birth according to parental status	1.1. 1.7	3M 5 × 3F (21.4)	3M 7 × 2F (21.4)	11M 6 × 2F (52.4)	4M	2M 1x	23M 19 × 7F (41.1)	–	10 (1)	1931–2016	5M 2 × 3F Multiple species

Table 25.1.3 Birth and reproductive factors

Variable	*Orig. table*	*Number of sex difference findings (and consistency scores)*							*Overview*	*Countries*	*Time range*	*Non-human*
		Fetal	*Infant*	*Child*	*Adol*	*Adult*	*WAR*	*All ages*				
1. Miscarriage	1.1. 2.3	43M 2x 3F (84.3)	–	–	–	–	–	43M 2x 3F (84.3)	M: 1Rev	13 (1)	1875- 2018	M: 7 Mammals
2. Stillbirth and early infant death	1.1. 2.4	–	31M (100.0)	–	–	–	–	31M (100.0)	M: 1Rev	10 (3)	1945–2018	M: 2 Mammals
3. Birth weight	1.1. 3.2	–	46M 3x 1F (90.2)	–	–	–	–	46M 3x 1F (90.2)	–	20 (1)	1881–2018	M: 19 x:1
4. Reproductive success	1.1. 4.2	–	–	–	–	–	11M 1x 1F (78.6)	11M 1x 1F (78.6)	M: 1Gen	8 (2)	1967–2009	–

Table 25.1.4 Maturation factors

Variable	*Orig. table*	*Number of sex difference findings (and consistency scores)*							*Overview*	*Countries*	*Time range*	*Non-human*
		Fetal	*I/T*	*Child*	*Adol*	*Adult*	*WAR*	*All ages*				
1. Physical maturation rate	1.1. 6.1	4F	12F (100.0)	6F	8F	2F	9F	41F (100.0)	–	8	1913–2009	4F among mammals
2. Onset of puberty, age at	1.1. 6.2	–	–	–	–	–	39F (100.0)	39F (100.0)	–	11	1935–2006	5F among mammals

25.1.5 Height and Basic Weight Measures

Beginning with height, Table 25.1.5 shows that average sex differences in height vary by age. Only among older adolescents (i.e., over age 15) and adults are all studies in agreement that male height surpasses that of females. It is worth noting that, because humans are unique among living animals regarding upright walking, no directly comparable data were found for other species of living non-humans. However, seven studies of Australopithecus, an extinct upright-walking hominid that eventually led to modern humans, were located. In those studies, sex differences in height were estimated based on skeletal remains. All of these studies concluded that, as adults, males were on average taller than females.

Regarding sex differences in weight (or body size in the case of some studies of non-humans), the last entry in this table shows that findings have been mixed except in the case of adults. For adults, all 72 studies concluded that males exhibit greater overall body weight than do females. There are many comparable studies of other animal species. The only studies summarized in the final column were for primate species most closely rated to humans genetically. One can see that, for adult apes and for extinct Australopithecus, findings have been consistent in indicating that males are larger than females.

25.1.6 Additional Weight and Body Composition Measures

Table 25.1.6 summarizes findings regarding sex differences in various morphological measures having to do with weight, body fat, and muscularity. The first variable is that of the body mass index (BMI), a widely used measure that assesses how much individuals weigh relative to their height. One can see that most studies have found no significant sex differences in the BMI. When there are differences, a majority of studies indicate that, especially among adults, males have a higher BMI than females.

Research having to do with sex differences in both being overweight and being obese is substantial. As shown in the second and third rows, findings for both are quite mixed, with just slight tendencies for males to be overweight and especially to be obese than is the case for females. It is noteworthy that these patterns are inconsistent with conclusions reached by reviews and meta-analyses, all of which indicate that slightly greater proportions of females are overweight and obese than males.

In the fourth row, findings from studies are summarized for people's tendencies to have high proportions of body fat. One can see that, except for three studies that reported no significant differences, 103 studies agreed that the bodies of females contain higher percentages of body fat

Table 25.1.5 Height and basic weight measures

Variable	*Orig. table*	*Number of sex difference findings (and consistency scores)*							*Overview*	*Countries*	*Time range*	*Non-human*
		I/T	*Child*	*Young Adol.*	*Older Adol.*	*Adult*	WAR	*All ages*				
1. Height	1.2. 1.1	6M 2 × 1F	17M 21 × 4F (36.9)	21F 17 × (55.3)	54M (100.0)	167M (100.0)	–	244M 40 × 26F (72.6)	M: 1Rev for adults	40 (18)	1872–2020	M: 7 Australo-pithecus
2. Sitting height	1.2. 1.3	1M 2x	2x	2F 1x	3M	3M	–	7M 5 × 2F (43.7)	–	5 (1)	1885–2005	–
3. Weight (or body size)	1.2. 1.4	10M 2 × (83.3)	18M 8 × 5F (50.0)	8M 5 × 7F (29.6)	13M 1 × (92.9)	72M (100.0)	5M 1x	116M 16 × 12F (73.9)	–	26 (4)	1885–2021	M: 37 Apes, ten Australo-pithecus

Table 25.1.6 Additional weight and body composition measures

Variable	*Orig. table*	*Number of sex difference findings (and consistency scores)*						*Overviews*	*Countries*	*Time range*	*Non-human*
		I/T	*Child*	*Adol*	*Adult*	*WAR*	*All ages*				
1. Body mass index (BMI)	1.2. 2.1	2M 1x	7x 2F	4M 5x 1F (36.4)	39M 40x 19F (33.3)	3M 3x 6F (16.7)	48M 57x 28F (29.8)	–	25 (2)	1993–2019	–
2. Overweight	1.2. 2.7	1x	10F 3x 5M (43.5)	2F 2x 2M	20M 7x 18F (31.7)	6F 3x 3M (40.0)	36F 16x 30M (43.9)	F: 2Rev	21	1960–2012	–
3. Obese	1.2. 2.8	–	6F 4x 4M (33.3)	1F 4x 2M	42F 10x 10M (58.3)	9F 5x 2M (50.0)	58F 33x 18M (45.7)	F: 5Rev 1Met x: 2Rev	27 (3)	1975–2018	–
4. Percent body fat	1.2. 2.13	8F	18F 2x (90.0)	12F **(100.0)**	47F 1x **(97.9)**	18F **(100.0)**	103F 3x **(97.2)**	–	14 (3)	1963–2015	–
5. Proportion of body fat in subcutaneous tissue	1.2. 2.21	2F 1x	1F	2F	5F	1F	10F 1x (90.9)	–	5	1951–2006	–
6. Endomorphic body type	1.2. 3.2	–	–	2F 1x	6F	3F	11F 1x (91.7)	–	8 (1)	1982–2006	–
7. Mesomorphic body type	1.2. 3.3	–	6M 1x	10M 1x (90.9)	5M 2x 2F	3M	24M 4x 2F (75.0)	–	9 (1)	1942–2018	–

than do the bodies of males, making this a likely universal sex difference. The fifth row similarly indicates that the vast majority of studies have found greater proportions of body fat specifically in subcutaneous tissue among females than among males.

The last two entries in this table pertain to measures of the endomorphic body type (i.e., the degree to which one is overweight especially in the mid-section) and the mesomorphic body type (i.e., the degree to which one is highly muscular). One can see that nearly all studies report greater proportions of females to be endomorphic, while most studies have concluded that males are more likely to be mesomorphic.

25.2 Anatomical and Physiological Factors: Condensed Findings

This chapter has to do with a wide range of anatomical and physiological factors not covered in Chapter 1. Otherwise, the main exclusions from this chapter have to do with bodily organs that produce various biochemicals (covered in Chapter 3) and those involving the structure and functioning of the brain (covered in Chapter 4). The summary of findings to be presented here will show that there are a substantial number of anatomical and physiological traits with ten or more findings that appear to exhibit culturally universal sex differences.

25.2.1 Basic Anatomical Factors

Four variables are listed in Table 25.2.1. They have to do with some fairly commonly measured anatomical traits. One can see that most studies have concluded that the waist-to-hip ratio is significantly greater for males than for females. In the case of facial width-to-height ratios, the evidence is quite mixed. However, all research agrees that males have longer fingers or larger hands than do females. This was even found to be true for two studies of non-human primates.

The last variable in this table is known as *2D:4D* and refers to the relative length of the second (pointing) finger when compared to the fourth (ring) finger. This variable has been widely investigated in recent decades, mainly because evidence indicates that it provides a *rough* indication of prenatal exposure to testosterone (Warrington et al. 2018; Sadr et al. 2020). Of course, the average male is exposed to more prenatal testosterone than is the average female. But, even within sex, studies have indicated that individuals with relatively long ring fingers (when compared to their pointing finger) are likely to have been exposed to elevated levels of testosterone during fetal development (Manning et al. 1998). As one can see, most studies have concluded that females have longer pointing fingers relative to their ring fingers than do males, with males typically

Table 25.2.1 Basic anatomical factors

Variable	*Orig. table*	*Number of sex difference findings (and consistency scores)*							*Overview*	*Countries*	*Time range*	*Non-human*
		Fetal	*I/T*	*Child*	*Adol*	*Adult*	*WAR*	*All ages*				
Waist-to-hip ratio	2.1. 1.4	–	–	1M 1x	1M	23M 1F (92.0)	3M	28M 1x 1F (90.3)	–	11 (1)	1990–2013	–
1. Facial width-to-height ratio	2.1. 2.1	–	–	–	–	3M 8x (27.3)	1M 1x 1F	4M 9x 1F (26.7)	M: 1Met x: 1Met	6 (2)	2007–2017	–
2. Finger length or hand size	2.1. 3.3	1M	–	1M	1M	11M (100.0)	–	14M (100.0)	–	8 (1)	1943–2018	2M: Primates
3. 2D:4D	2.1. 3.5	5F	5F 3x	13F 2x (86.7)	6F 1x	180F 18x 2M (90.0)	15F (100.0)	224F 24x 2M (88.9)	F: 1Rev F: 3Met	30 (15)	1875–2021	Mixed

having ring fingers that surpass the length of their pointer fingers. Nevertheless, there are a substantial number of findings that have failed to detect any significant sex differences and a couple of studies that actually reported the contrary. Most, but not all, of the failures to find significant sex differences involved measurement of the left hand or were based on fairly small samples (e.g., less than 100 of each sex). Regarding studies of non-humans, many studies have been reported, and their findings have been extremely inconsistent, even among non-human primates.

25.2.2 Bones and Joints

In Table 25.2.2, four more anatomical factors are listed regarding possible universality in sex differences. One can see that, especially among adults, studies have consistently indicated that males have greater bone density than do females. All of the exceptions simply failed to find significant differences. For studies of sex differences in the loss of bone density (i.e., thinning of the bones) with age, all ten studies were in agreement that female adults experience greater loss on average than do males.

Joint laxity refers to the degree to which joints are highly flexible. The third entry in this table shows that, at least among adults, females have greater laxity than do males. All of the exceptional studies simply failed to identify significant sex differences among children.

The last entry in this table has to do with the distance between the anus and the genitals, a region of the body sometimes also known as the *perineum*. As one can see, all studies agree that AGD is greater among males than females. This conclusion has even been reached without exception in 26 studies of various species of non-human mammals.

25.2.3 Breathing, Strength, and Skin

Sex differences in several basic physiological factors are summarized in Table 25.2.3. Differences in lung capacity generally favor males, especially in adulthood, but not quite to the degree of being a probable universal difference. In the case of muscularity, one can see that all of the evidence supports the conclusion that males are, on average, more muscular, making this an all-but-certain universal sex difference. Similarly, nearly all studies have concluded that males are physically stronger than females. The only exceptions involve the elderly; for them, some studies have found no significant sex differences.

Findings from research pertaining to sex differences in the ability to resist muscle fatigue generally indicate that females are more resistant than males. Nevertheless, the numbers of studies that have reported no

Table 25.2.2 Bones and joints

Variable	*Orig. table*	*Number of sex difference findings (and consistency scores)*							*Overview*	*Countries*	*Time range*	*Non-human*
		Fetal	*I/T*	*Child*	*Adol*	*Adult*	*WAR*	*All ages*				
1. Bone density (or mass)	2.1. 4.1	–	1x	8M 5x (61.5)	15M 3x (83.3)	41M 1x (97.6)	11M (100.0)	75M 10x (88.2)	–	10 (2)	1966–2014	5M: among primates; otherwise mixed
2. Loss of bone density with age	2.1. 4.10	–	–	–	–	10F (100.0)	–	10F (100.0)	–	3	1981–2004	–
Joint laxity	2.1. 4.12	–	–	4F 3x	8F	11F (100.0)	6F	29F 3x (90.6)	F: 1Rev	16 (1)	1978–2012	–
3. Anogenital distance (AGD)	2.1. 5.2	4M	19M (100.0)	–	–	2M	–	25M (100.0)	–	12 (1)	2009–2021	26M: among multiple species

Table 25.2.3 Breathing, strength, and skin

Variable	Orig. table	Number of sex difference findings (and consistency scores)						Overview	Countries	Time range	Non-human
		I/T	Child	Adol	Adult	WAR	All ages				
1. Lung capacity (oxygen intake)	2.1. 8.2	2M	8M 2x (80.0)	7M 1x	18M 1x (94.7)	–	35M 4x (89.7)	–	3	1894–2014	–
2. Muscularity (muscle mass)	2.1. 8.1	3M	–	4M	54M **(100.0)**	7M	68M **(100.0)**	–	9 (1)	1974–2021	18M: among several species
3. Physical strength	2.1. 9.2	2M	3M	14M **(100.0)**	105M 3x **(97.2)**	3M	127M 3x **(97.7)**	M: 1Rev 1Met	22 (3)	1902–2021	2M: among two mammal species
4. Resistance to muscle fatigue	2.1. 9.3	–	–	–	12F 4M (75.0)	–	12F 4M (75.0)	–	5	1975–2006	5F 1M: among rodents
5. Skin color (skin darkness)	2.1. 10.1	–	1M	–	42M 2x **(95.5)**	11M **(100.0)**	54M 2x **(96.4)**	–	22 (8)	1958–2007	–
6. Galvanic skin response	2.1. 10.9	1F	8F 1x 1M (72.7)	–	7F 3M (53.8)	1F	17F 1x 4M (65.4)	–	5	1962–2019	–
7. Larynx (vocal cords) size	2.1. 13.4	3x	6x	11M **(100.0)**	14M **(100.0)**	1x	25M 10x (71.4)	–	3	1925–2001	–

significant differences are great enough to remove this trait from being a candidate for universality.

Regarding skin color, considerable research has reported on sex differences. The table's fifth entry indicates that this trait is a likely universal sex difference, with males having darker skin on average than females.

The galvanic skin response is a measure of skin electrical conductivity. One can see in the sixth entry that most studies indicate that female galvanic skin responses are higher than those of males, but the differences are not sufficient to be considered universal.

In the final entry, sex differences in the size of the larynx (often referred to as the vocal cords) are considered. One can see that, at least for adolescents and adults, all of the evidence supports the conclusion that males have a larger larynx than do females, on average.

25.2.4 Basic Anatomical Factors

Table 25.2.4 has to do with physiological factors surrounding the functioning of the heart. The first three entries involve blood pressure. While there may be small tendencies for males to exhibit higher blood pressure than females under most conditions, the evidence in this regard is insufficient to consider basic blood pressure measures to be universally different between the sexes. A similar conclusion seems warranted in the case of both resting heart rate and heart rate following strenuous exercise, as shown in the fourth and fifth entries.

The table's final entry has to do with what is known as the *QT interval*, which refers to the time laps between the Q wave and the end of the T wave on readings from the electrocardiogram. Generally, longer QT intervals are associated with slower heart rates. Nearly all of the evidence indicates that females have longer QT intervals than do males. This generalization appears to be especially well founded for adults. Most studies of various mammalian species have reached the same conclusion.

25.2.5 Basic Physiological Processes

The final summary table for Chapter 2 is Table 25.2.5. It has to do with two physiological processes: gastric emptying and alcohol metabolism. Gastric emptying refers to the frequency with which one eliminates bodily waste, either in the form of feces or urine. Although there are tendencies for males to do so more often than females, especially regarding feces, there seems to be no difference when urination is considered.

Sex differences in the rate at which alcohol is metabolized following consumption has been assessed in numerous studies, all of which have been limited to adults. One can see that evidence of sex differences among

Table 25.2.4 Basic physiological factors involving the heart

Variable	*Orig. table*	*Number of sex difference findings (and consistency scores)*							*Overview*	*Countries*	*Time range*	*Non-human*
		Fetal	*I/T*	*Child*	*Adol*	*Adult*	*WAR*	*All ages*				
1. Blood pressure in general	2.3. 1.3	–	–	–	1M 1F	20M 5x 6F (54.1)	2M 1x 1F	23M 6x 8F (51.1)	M: 1Rev 1Met	11 (1)	1970–2014	11M 1x: among rodents
2. Diastolic blood pressure	2.3. 1.4	–	–	1F	1F	14M 10x 1F (53.8)	1M 1F	15M 10x 4F (45.5)	–	14	1959–2017	–
3. Systolic blood pressure	2.3. 1.5	–	–	1F	1M	14M 4x 7F (43.7)	1M	16M 4x 8F (44.4)	M: 1Met	13	1972–2019	–
4. Resting heart rate	2.3. 1.11	5M	3M 1x	6F 5x (54.5)	5F 1x 2M	26F 7x 2M (70.3)	1F	38F 13x 10M (62.3)	F: 1Met adults only	10	1920–2021	–
5. Heart rate increase after exercise	2.3. 1.12	1F (by mom)	–	–	–	8F 1x	–	9F 1x (90.0)	–	2	1977–2014	–
6. QT interval	2.3. 1.17	–	–	5F 2x	1F	20F (100.0)	–	26F 2x (92.9)	F: 1Rev for adults	5	1961–2003	6F 1x 1M: among mammal species

Table 25.2.5 Basic physiological processes

Variable	*Orig. table*	*Number of sex difference findings (and consistency scores)*						*Overview*	*Countries*	*Time range*	*Non-human*
		I/T	*Child*	*Adol*	*Adult*	*WAR*	*All ages*				
1. Gastric emptying	2.3. 8.4	–	–	–	8M 3x (72.7)	2M	10M 3x (76.9)	–	5	1987–2000	–
2. Alcohol metabolism	2.3. 3.5	–	–	–	5M 13x 5F (21.7)	–	5M 13x 5F (21.7)	–	3	1971–2003	9F 1x: among rodent species

humans are quite varied, although most studies point toward no significant differences. Nonetheless, studies among various species of rodents largely indicate that females metabolize alcohol more quickly than do males.

25.3 Bodily Fluids, Biochemicals, and Biochemical Receptors: Condensed Findings

Sex differences in all types of bodily fluids, the organs that produce them, as well as with the biochemical receptors for these fluids have been widely studied. Findings for which at least ten observations have been made in this regard are summarized below.

25.3.1 Sex Hormones

Hormones are biochemicals produced in one part of the body (usually a specialized organ) that are then carried by the blood system to other parts of the body, where they have their main effects. The two main categories of hormones that have been studied with respect to possible sex differences are sex hormones and stress hormones. Table 25.3.1 summarizes findings regarding sex differences in these two types of hormones.

Androstenedione is usually classified as a weak androgen (or male-typical hormone). It forms as an intermediate step toward production primarily of testosterone. One can see that six studies found androstenedione to be present in greater amounts among males, although an additional five studies failed to detect significant sex differences. Regarding studies of sex differences among other animals, it is interesting to note that six studies involved spotted hyenas, a species in which females have a vulva resembling a penis. One can see that three of these six studies found higher levels of androstenedione among females rather than males, with the remaining three studies reporting no significant sex difference.

DHEA (*dehydroepiandrosterone*) is a precursor to steroidal hormones such as androstenedione, testosterone, and even estradiol. The second row in the table shows that when there are significant sex differences in this hormone, they favor females; however, most studies have actually reported no significant sex differences in DHEA levels.

A biochemical version of DHEA that is common among mammals is known as *DHEA-S* (with the *S* standing for *sulfate*). This version of DHEA is produced in the outer layer of the adrenal glands. Like DHEA, DHEA-S often provides chemical building blocks for making testosterone and estradiol. In the third row of the table, one can see that nearly all studies have concluded that males produce more DHEA-S than do females, especially among adults. This pattern is surprising given that

Table 25.3.1 Sex hormones

Variable	*Orig. table*	*Number of sex difference findings (and consistency scores)*							*Overview*	*Countries*	*Time range*	*Non-human*
		Fetal	*I/T*	*Child*	*Adol*	*Adult*	*WAR*	*All ages*				
1. Androsten-edione	3.1.1.4	1M	2x	1M 1x	1M	3M 2x	–	6M 5x (54.5)	–	4	1970–2018	3F 3x among spotted hyenas
2. DHEA	3.1.1.6	–	1x	2F 1x	1x	1F 3x	1F	4F 6x (40.0)	–	3	1975–2010	–
3. DHEA-S	3.1.1.7	–	–	–	1x	12M (100.0)	–	12M 1x (92.3)	–	5	1066–2002	–
4. Estrogen	3.1.1.9	3F 4x	1F	1F	4F	7F 3M (53.8)	2F 1x	18F 4x 3M (64.3)	–	8	1941–2016	7F 1x 2M various species
5. Testosterone	3.1.1.14	36M (100.0)	17M 2F (80.9)	4M 3x 2F (44.4)	16M (100.0)	53M (100.0)	3M 1x (75.0)	129M 1x 4F (93.5)	M: 3Gen 3Rev	16 (1)	1966–2021	31M 5x

nearly the opposite sex differences has been found for DHEA without the sulfate linkage (documented in the preceding row).

Estrogen refers to a set of steroidal hormones that are generally considered female-typical hormones. In the table's fourth row, one can see that most studies have indicated that females have higher levels of estrogen (usually sampled in the blood or saliva) than do males. However, there are some exceptions, especially among elderly adults, where higher levels have been found among males.

A widely studied sex hormone is testosterone. It is produced primarily by the testes, although modest amounts are also produced by the ovaries and the adrenal glands. The last row in Table 25.3.1 indicates that males produce more testosterone than do females, especially during fetal development and during adolescence and adulthood. This conclusion is consistent with prior generalizations and literature reviews. Also, studies of other animals have reached the same conclusion.

25.3.2 Stress Hormones

Two recognized stress hormones have been studied extensively regarding possible sex differences. Findings are summarized in Table 25.3.2. Over 50 studies of sex differences in baseline (or resting) levels of cortisol have been published, most involving adults, based on assays drawn from blood or saliva. One can see that most studies report higher resting levels of cortisol among males, at least among adults. This conclusion is not entirely consistent with a meta-analysis. It reported male levels were higher before puberty while levels were higher among females following puberty.

In the case of epinephrine (also known as *adrenaline*), 19 studies point toward higher levels among males. The two exceptions simply failed to identify any significant sex difference.

25.3.3 Biochemicals Other than Hormones

Table 25.3.3 is the last summary table for Chapter 3. It shows that seven biochemicals were found for which at least ten studies of sex differences were identified. The first biochemical is known as *leptin*, a protein (also classified as a hormone) that helps to regulate the production of fat cells in the body. As one can see, except in the first two or three years in life, all studies have concluded that leptin levels in females significantly surpass levels in males.

Turning to cholesterol, many studies have reported on possible sex differences. Some of these studies pertain to cholesterol in general, while most studies provide information about two forms of cholesterol: high-density

Table 25.3.2 Stress hormones

Variable	*Orig. table*	*Number of sex difference findings (and consistency scores)*						*Overview*	*Countries*	*Time range*	*Non-human*
		I/T	*Child*	*Adol*	*Adult*	*WAR*	*All ages*				
1. Cortisol, baseline levels	3.1. 2.4	4x	1M	1F 3x	25M 13x 6F (50.0)	2M 1x	28M 21x 6F (43.9)	M before puberty; F after pub: 1Met	9	1974–2021	1M among domestic pigs
2. Epinephrine (adrenaline)	3.1. 2.9	1M	3M	–	15M 2x (88.2)	–	19M 2x (90.5)	–	5	1972–2014	2F among mammals

Table 25.3.3 Biochemicals other than hormones

Variable	*Orig. table*	*Number of sex difference findings (and consistency scores)*						*Overview*	*Countries*	*Time range*	*Non-human*
		I/T	*Child*	*Adol*	*Adult*	*WAR*	*All ages*				
1. Leptin levels	3.4.1.11	4F 4x	5F	7F	34F (100.0)	7F	57F 4x (93.4)	–	10	1995–2007	–
2. Cholesterol levels in general	3.4.2.3	–	4F 3M	1F 1x	13F 9x 9M (32.5)	2F 1x 2M	20F 11x 5M (39.2)	–	11	1974–2018	1F 1x among mammals
3. High-density lipoprotein (HDL) cholesterol	3.4.2.2	–	2x	1F	10F 1x 1M (76.9)	12F 3M (66.7)	23F 2x 3M (74.2)	–	10	1984–2019	1M 1x among mammals
4. Low-density lipoprotein (LDL) cholesterol	3.4.2.4	–	1M 1F	–	5F 9x 1M (31.2)	5M	7M 9x 6F (25.0)	–	8	1986–2018	1x among mice
5. Glucose levels	3.4.2.7	–	–	–	12M 2x 3F (60.0)	1x	12M 3x 3F (57.1)	–	7	1973–2018	1x among primates
6. Triglycerides	3.4.2.20	–	1F	–	9M 8x 2F (42.8)	2M	11M 8x 3F (44.0)	–	10	1985–2012	–
7. Norepinephrine (noradrenaline)	3.5.1.9	–	–	–	8M 5x (61.5)	–	8M 5x (61.5)	–	2	1988–2014	Complex sex differences

cholesterol (sometimes known as "good cholesterol") and low-density cholesterol (often called "bad cholesterol"). The table shows that findings have been mixed regarding sex differences, with some of the variation possibly depending on age. In the case of high-density cholesterol, the evidence is also mixed.

In the case of glucose, triglycerides, and norepinephrine, findings were all mixed regarding any sex differences. However, they were all leading toward these biochemicals usually being higher among males than females.

25.4 The Brain: Structure and Functioning: Condensed Findings

The human brain is an exceedingly complex structure, and its functioning is likely to be even more intricate than its structure due to the fact that brain functioning reflects not only differences in structure but also all of the changes brought about by experiences and learning. Research findings having to do with sex differences in both the structure of the brain and in its functioning were summarized in Chapter 4. In the section below, a condensation of these summaries is presented.

25.4.1 Major Aspects of Brain Size

Sex differences in six major aspects of brain size are summarized in Table 25.4.1. The first and most obvious of these measures has to do with overall brain volume. Studies of sex differences in brain volume can be traced back to the early 1900s. At that time, the most common way to measure brain volume involved post-mortem removal of the brain from the skull of a cadaver and then submerging it in a vat of water that had been calibrated so that the water level could be measured both before and after the brain was fully submerged. In the latter decades of the 20th century, this type of post-mortem measure has been almost completely replaced by using magnetic resonance imaging (MRI) scans on living individuals.

Based on studies of both water displacement and MRI scans, the table's first row shows that the vast majority of studies have concluded that males have larger brains than do females for all age groups, but especially for adults. The exceptions are worth further explanation. When measuring brain size, some researchers have noted that it is generally the case that larger animals have larger brains (Harvey and Krebs 1990; Sol et al. 2005). Therefore, it is reasonable to adjust for body size when making male-female comparisons in brain size. Some researchers, especially in recent years, have followed this reasoning. Since human adult males tend to be taller and heavier than their female counterparts

Table 25.4.1 Major aspects of brain size

Variable	*Orig. table*	*Number of sex difference findings (and consistency scores)*						*Overview*	*Countries*	*Time range*	*Non-human*
		I/T	*Child*	*Adol*	*Adult*	*WAR*	*All ages*				
1. Brain volume (brain size)	4.1. 1.1	11M 4x (73.3)	12M 1x (92.3)	8M 1x	120M 2x (98.4)	35M 2x (94.6)	186M 10x (94.9)	M: 2Rev 3Met; x: 4Gen	23 (1)	1908–2021	11M 5x among mammals
2. Brain size reduction with age	4.1. 1.5	–	–	–	21M 10x 1F (65.6)	–	21M 10x 1F (65.6)	–	5	1980–2007	–
3. Brain ventricles, size of	4.1. 1.8	–	2M	1M	8M	1M	12M (100.0)	–	6	1995–2011	Mixed findings among mice
4. Cerebral spinal fluid, volume of	4.1. 1.9	–	1F 1x	–	22M 11x (66.7)	1M 1x	23M 13x 1F (60.5)	M: 1Met	11 (1)	1991–2012	–
5. Neocortex (cortex), size of	4.1. 2.1	1M	4M	3M	16M 1x (94.1)	12M (100.0)	36M 1x (97.3)	M: 1Met	8 (1)	1979–2011	3M 1x among rodents
6. Cortical thickness	4.1. 2.5	1x	–	1M 1x	7F 8x 4M (30.4)	1F 5x	8F 14x 5M (25.0)	F: 1Rev	11	1989–2020	3M 1x among rodents

(see Chapter 1), studies that have adjusted for sex differences in body size have sometimes found no significant sex difference in brain size. These are the exceptions shown in the table. Nevertheless, most studies still report slightly greater brain sizes among males. In other words, without adjusting for body size, all studies agree that male brains, on average, are larger; but when sex differences in body size are taken into account, the brain size differences diminish and sometime disappear in statistical terms. One can see in the last column, that most studies of various species of non-human mammals have also concluded that male brains are larger than those of females.

As adults age, their brains tend to lose neurons and thereby decrease slightly in overall size, a phenomenon sometimes known as *neural pruning* (SF Witelson, Beresh and Kigar 2006:393; Jancke, Merrillat et al. 2015). This gradual reduction in neurons mainly involves cells comprising gray matter rather than white matter (Passe et al. 1997), to be discussed later in this chapter on the brain. In the second row of the table on brain size, one can see that nearly all studies have concluded that there are no significant sex differences in the reduction of brain size with age, although, when sex differences are found in brain size reductions with age, nearly all studies indicate that the reductions are greater for males than for females.

The third entry in this table pertains to the size of the brain ventricles, of which humans have four. These ventricles can be thought of as small cavities inside the brain that help to transport cerebral spinal fluid throughout the brain, thereby bathing it with oxygen and nutrients. Of the 12 studies found that reported on sex differences in the total size of the brain ventricles, all of them concluded that they were larger for males than for females. Thus, the size of brain ventricles qualifies as a universal sex difference.

If the brain ventricles are larger in males than in females, one might suspect that the amount of cerebral spinal fluid (CSF) would also be greater. The fourth entry in this table, however, shows that findings have not been entirely consistent in this regard. Specifically, of the 37 studies of sex differences in CSF volume, just 23 studies concluded that males have more than females.

One of the most distinctive aspects of the human brain is the relatively large size of the neocortex (also simply called the *cortex*). While all mammals have a neocortex (whereas most other animals do not), in most mammalian species, the neocortex is small compared to the rest of the brain. In humans, the neocortex comprises well over half of the entire brain. Regarding sex differences, the table's fifth entry shows that 36 out of 37 studies concluded that the neocortex of males is larger than that of females. Thus, nearly all studies have concluded that the neocortex is larger among males than among females.

In the final entry for this table, sex differences in cortical thickness are assessed. The table indicates that most studies failed to find any significant differences in this regard.

25.4.2 Hemispheric Structures

The brain's neocortex has two sides; these are known as the *right hemisphere* and the *left hemisphere*. Substantial research has been devoted to determining if there are sex differences in one or both of these hemispheres. As shown in Table 25.4.2, there are likely to be hemispheric sex differences, although findings are not sufficiently consistent to consider the differences universal.

In the table's first entry, findings are summarized having to do with the relative size of the two hemispheres, i.e., is one hemisphere larger than the other? Most studies indicate that males are more likely to have detectable hemispheric size differences than are females. Then, the second entry addressed the question of which hemisphere is larger. In this case, eight studies have reported that males have larger right hemispheres when compared to the left (while the hemispheres of females are more often of equal size), while two other studies reported no significant sex differences. Perhaps noteworthy is evidence (cited in the row's last column) that similarly inconsistent findings have been reported for rodents.

Eleven studies have examined sex differences in the physical symmetry of the planum temporale. This portion of the cortex includes a region that is very important for comprehending language (called *Wernicke's area*). For nearly all humans, the planum temporale is larger in the left hemisphere than in the right hemisphere, reflecting the fact that the left hemisphere usually dominates over the right in terms of language functioning. However, because females tend to surpass males in terms of language skills (see Chapter 26.2.11.10) and are less likely to stutter (see Chapter 26.2.14.18), considerable research has sought to determine if there might be sex differences in the relative size of the two planum temporals. As shown in the table's third entry, research findings have reached inconsistent conclusions in this regard. The one meta-analysis of sex differences in asymmetry of the planum temporale similarly concluded that no significant differences exist.

The *corpus callosum* refers to the neocortical structure that directly connects the two hemispheres. For more than a century, researchers have sought to determine if the size of this brain structure (or parts thereof) exhibit significant average sex differences. Findings are summarized in the

Table 25.4.2 Hemispheric structures

Variable	*Orig. table*	*Number of sex difference findings (and consistency scores)*							*Overview*	*Countries*	*Time range*	*Non-human*
		Fetal	*I/T*	*Child*	*Adol*	*Adult*	*WAR*	*All ages*				
1. Relative size of the hemispheres	4.1. 4.1	2M 1x	1M	1x	–	16M 6x (72.7)	7M 2x	26M 10x (72.2)	M: 1Rev	9	1977–2009	7M 2x rodents
2. Right hemisphere larger than left	4.1. 4.3	1M	2M	–	–	4M 1F	1M 1F	8M 2F (80.0)	–	4	1984–2007	3M 1F rodents
3. Planum temporale asymmetry	4.1. 4.5	–	–	–	–	3F 4x 3M	1F 1x	4F 4x 3M (28.6)	x: 1Met	5	1994–2015	–
4. Corpus callosum, size of	4.1. 5.1	2F	1M	2x	–	41M 15x 10F (53.9)	1M 2x 4F	42M 19x 6F (56.5)	M: 3Met; x: 1Rev; F: 2Met	8	1906–2020	9M mammals
5. Corpus callosum, size of various components	4.1. 5.2	–	–	–	–	20F 5x 18M (32.8)	6F 2x 1M	26F 7x 19M (36.6)	F: 1Met	6	1906–2007	2M 5x 1F rodents

table's last two rows. Regarding the corpus callosum as a whole, one can see that most studies report that it is larger for males than for females. However, studies that have controlled for the fact that males usually have larger brains overall have concluded that there are either no sex differences or that females may actually have slightly larger corpus callosum than males.

Turning to the size of various components of the corpus callosum, the final row shows that the results are mixed. This is especially true for studies statistically controlling for sex differences in overall brain size. Basically, there is a slight majority of studies indicating that at least some components of the corpus callosum are larger for females than for males.

25.4.3 Subcortical Brain Structures

Findings for parts of the mammalian brain comprising what is known as the *subcortex* are summarized in Table 25.4.3. The first entry in this regard has to do with the amygdala, a region of the brain that helps to regulate emotions. When all age groups are combined, roughly half of the studies indicate that the amygdala of males are larger than those of females, with most of the remainder suggesting that there is no significant sex difference. As was noted earlier, some of the studies statistically controlled for sex differences in overall brain size before making comparisons, while most studies did not do so.

The second subcortical region with ten or more pertinent studies involved the size of the caudate nucleus, a brain region that helps to regulate coordinated movement. One can see that 17 of the 34 findings concluded that the size of the caudate nucleus is larger for males than for females, with 12 of the remaining findings indicating there are no significant sex differences.

In the third row, sex differences in the size of the hippocampus are considered. This brain structure plays key roles in memory retention. When studies of all ages are combined, 14 of 43 findings indicate that males have larger hippocampi than do females, with 19 studies concluding that there are no significant sex differences. Again, some studies controlled for overall brain size before making comparisons, while most did not do so.

Regarding the putamen and the thalamus, the last two rows show that, when sex differences have been found, most indicated that the sizes of both of these brain regions to be larger among males. However, findings have not been sufficiently consistent to elevate the consistency scores to 95.0 or higher.

Table 25.4.3 Subcortical brain structures

Variable	*Orig. table*	*Number of sex difference findings (and consistency scores)*						*Overview*	*Countries*	*Time range*	*Non-human*
		I/T	*Child*	*Adol*	*Adult*	*WAR*	*All ages*				
1. Amygdala, size of	4.1. 6.3	1M	2M 1x	–	11M 11x 3F (39.3)	2M 2x	16M 14x 3F (44.4)	M: 1Gen 1Met; x: 2Met	13	1994–2021	4M rodents; 1x primates
2. Caudate nucleus, size of	4.1. 6.7	1x	2x	–	14M 8x 5F (43.7)	3M 1x	17M 12x 5F (43.6)	M: 1Met	11	1993–2015	1M rodents
3. Hippocampus, size of	4.1. 6.11	1M	2F 2x	1F	9M 14x 7F (23.1)	4F 3x	14F 19x 10M (26.4)	M: 2Met F: 1Gen	12	1994–2019	33M 8x mammals
4. Putamen, size of	4.1. 6.18	1M	1x	–	10M 8x 1F (50.0)	1M 1x 1F	12M 10x 2F (47.2)	M: 1Met	13	1996–2019	–
5. Thalamus, size of	4.1. 6.23	1M	–	–	6M 10x 3F (23.1)	1M	8M 10x 3F (33.3)	M: 1Met	10	2000–2019	–

25.4.4 Gray and White Matter

Throughout the brain, two distinct types of cells (neurons) are found. One type is called *gray matter* while the other is called *white matter*. Besides differing in color, these two types of brain cells can be distinguished in functional terms. Generally, neurons comprising grey matter receive and process incoming sensory stimuli *within* local brain regions. White matter neurons, on the other hand, primarily communicate between different regions of the brain, thereby better integrating the brain's overall functioning. Table 25.4.4 provides four lines of evidence bearing on gray matter and white matter, none of which meet the threshold set for being universal sex differences. Nevertheless, most studies indicate that males have more gray matter overall as well as white matter overall. Females, on the other hand appear to have a higher gray matter to white matter ratio.

25.4.5 Brain Waves

Brain waves are small electrical impulses that occur throughout the brain as it functions. These waves can be detected with electrodes placed on the surface of the skull. Two variables regarding possible sex differences in brain waves were located containing at least ten relevant findings. One can see in Table 25.4.5 that female brain waves appear to be more frequent and/or intense than is the case for brain waves of males. Nevertheless, the findings in this regard do not attain sufficiency to be considered universal sex differences.

25.4.6 Brain Activity and Activation

Table 25.4.6 summarizes findings having to do with sex differences in the amount of brain activity in the brain, including brain activation to various stimuli. The first entry involves activity in the amygdala, a region of the subcortex having to do with emotions. All of the results that were located involved studies of adults. One can see that findings were inconsistent. However, as noted under Overviews, two meta-analyses may offer help in explaining many of these inconsistencies. Specifically, when sexually arousing material were presented to research participants, the amygdalae of males seem to activate more. However, when the fear-promoting stimuli were presented, the amygdalae of females usually activate more.

The table's second entry has to do with whether both hemispheres of the brain activate more or less in unison, as opposed to activating independently. Findings were very inconsistent regarding possible sex differences. As with amygdala functioning discussed above, many of the inconsistencies may be due to tasks being performed when hemispheric activation is being measured. As shown under the overview column, two

Table 25.4.4 Gray and white matter

Variable	*Orig. table*	*Number of sex difference findings (and consistency scores)*						*Overview*	*Countries*	*Time range*	*Non-human*
		I/T	*Child*	*Adol*	*Adult*	*WAR*	*All ages*				
1. Gray matter, volume of	4.2. 3.1	2M	2M	4M 1x	38M 5x 1F (86.4)	13M 1x (92.9)	59M 7x 1F (86.8)	M: 1Met	9 (1)	1994–2019	–
2. Gray matter in specific regions	4.2. 3.2	1M	2M 1x	1M	16M 18F (36.0)	1M 1x 1F	21M 2x 19F (34.4)	M: 1Met	6	1994–2019	–
3. White matter, volume of	4.2. 3.4	2M	1M 3x	9M 2x (81.8)	41M 5x 1F (85.4)	12M 1F (92.3)	65M 10x 1F (85.5)	M: 1Met	17 (2)	1994–2021	–
4. Gray matter to white matter ratio	4.2. 3.10	–	1x	–	24F 4x (85.7)	1F	25F 5x (83.3)	F: 1Rev	8	1987–2018	–

Table 25.4.5 Brain waves

Variable	*Orig. table*	*Number of sex difference findings (and consistency scores)*						*Overview*	*Countries*	*Time range*	*Non-human*
		I/T	*Child*	*Adol*	*Adult*	*WAR*	*All ages*				
1. EEG power (brain wave activity)	4.3. 1.6	–	–	–	15F 3x (83.3)	4F	19F 3x (86.4)	–	4	1953–2013	1F rodent
2. Delta brain wave activity	4.3. 1.6	–	–	–	14F 1x 1M (82.4)	–	14F 1x 1M (82.4)	F: 1Met among the elderly	2	1982–2005	–

Table 25.4.6 Brain activity and activation

Variable	*Orig. table*	*Number of sex difference findings (and consistency scores)*						*Overview*	*Countries*	*Time range*	*Non-human*
		I/T	*Child*	*Adol*	*Adult*	*WAR*	*All ages*				
1. Amygdala activity or activation	4.4. 2.2	–	–	–	9F 7M (66.2)	–	9F 7M (66.2)	M 1Met (sexusl); F: 1Met (fear)	3	2000–2009	–
2. Bilateral brain activation	4.4. 5.1	–	2M 1F	–	8M 5F (61.5)	–	10M 6F (45.5)	x: 2Met (task dependent)	6 (1)	1980- 2017	–
3. Left hemispheric activity/activation	4.4. 5.2	1F	1F	–	14M 4x 1F (70.0)	1F	14M 4x 4F (53.8)	–	9	1981–2010	–
4. Right hemispheric activity/activation in general	4.4. 5.3	2M	1M	–	18M 3x 13F (38.3)	2M 1F	23M 3x 13F (44.2)	M: 1Met (spatial reasoning)	9	1974–2010	–
5. Right hemispheric activity when assessing the emotional states	4.4. 5.7	–	2F 5x 1M	3F 1x	27F 4x 3M (73.0)	–	32F 10x 3M (66.7)	–	2	1924–1989	–
6. Hemispheric specialization in general	4.4. 6.1	1M	2M	–	76M 7x 7F (78.4)	10M 1x (90.9)	89M 8x 7F (80.1)	M: 3 Rev 3Met; x: 2 Rev 1Met	13	1973–2009	–
7. Hemispheric specialization for language	4.4. 6.2	–	2M	3M	15M 1F	–	20M 1F (90.9)	M: 1Met	4	1981–2006	–

meta-analyses noted that the type of tasks being perform can affect bilateral hemispheric activation differently for males and females.

Turning to row 3 in this table, considerable research has investigated sex differences in left hemispheric activation or activity. The left hemisphere is usually more involved in language production and interpretation than is the right hemisphere. One can see that findings are conflicting regarding sex differences in left hemisphere activation. Close examination of the qualifying comments pertaining to the relevant studies suggest that females are more likely than males to activate the left hemisphere when solving spatial reasoning problems, while males are more likely to do so during language tasks.

Row 4 has to do with sex differences in activating the right hemisphere. Findings are mixed. At least some of the inconsistencies seem to be the result of the specific tasks being performed. As noted in the overview column, one meta-analysis concluded that males appear to use the right hemisphere more than females when performing spatial reasoning. However, in row 5, one can see that most studies indicate that females activate the right hemisphere more than males do when dealing with emotional states.

The concept of hemispheric specialization refers to the tendency to shift hemispheres when involved in specific types of tasks, regardless of the nature of those tasks. Row 6 shows that nearly all studies have concluded that males are more hemispheric specialists than are females, and row 7 suggests that this also tends to occur specifically pertinent when undertaking linguistic tasks.

25.5 Physical Health and Illness: Condensed Findings

An enormous amount of research findings on sex differences in variables having to do with physical health and illness has been published. Many of these findings cited ten or more studies. A condensed summary of findings relevant to this line of research is provided in this section.

25.5.1 Morbidity

Morbidity refers to being in a state of ill health. A summary of findings regarding sex differences in measures of morbidity (without identifying the specific nature of the illness involved) is provided in Table 25.5.1. The overall pattern is that females exhibit somewhat more morbidity than do males, except for hospitalizations, where most studies indicate that males are hospitalized more than females. Only in one case, that of seeking or utilizing healthcare services *including ob/gyn* are the vast majority of studies in agreement that morbidity rates are higher among females than among males.

Table 25.5.1 Morbidity

Variable	*Orig. table*	*Number of sex difference findings (and consistency scores)*						*Overview*	*Countries*	*Time range*	*Non-human*
		I/T	*Child*	*Adol*	*Adult*	*WAR*	*All ages*				
1. Overall morbidity (ill health)	5.1. 1.3	3M	2M	–	12F 1x 3M (63.2)	4F	16F 1x 8M (48.5)	F: 1Rev	7	1929–2017	–
2. Sickness/disability days	5.1. 1.4	–	2F 2x 4M	5F 1x	35F 3x (92.1)	35F 1x (**97.2**)	77F 7x 4M (83.7)	–	9	1927–2006	–
3. Seeking or utilizing healthcare services in general	5.1. 2.1	–	–	4F	30F 6x (83.3)	6F	40F 6x (87.0)	F: 1Rev	9	1967–2007	–
4. Seeking or utilizing healthcare services including ob/gyn	5.1. 2.2	–	–	8F 3x (72.7)	40F (**100.0**)	24F (**100.0**)	72F 3x (**96.0**)	–	9 (1)	1957–2005	–
5. Seeking or utilizing healthcare services excluding ob/gyn	5.1. 2.3	2x	2F 4x 1M	3F 1x	35F 3x 2M (83.3)	9F	49F 10x 3M (75.4)	F: 2Rev (post puberty)	10	1966–2004	–
6. Hospital utilization (excluding pregnancy-related)	5.1. 2.6	–	3M	1x	7M 3x 3F (43.7)	2M 1F	12M 4x 4F (50.0)	–	10	1964–2007	–
7. Chronic illness	5.1. 2.8	–	3M 1x	2M	7F 2M	2F 1x	9F 2x 5M (42.8)	–	6	1971–2004	–

25.5.2 Self-Rated Health

Many studies have assessed sex differences in health based on people's self-reports, typically using a five-point Likert scale. The two entries in Table 25.5.2 pertain to traits that correlate with self-rated health. Row 1 builds on considerable evidence that self-rated health is inversely correlated with mortality (Mossey and Shapiro 1982; Svedberg, Lichtenstein and Pedersen 2001). One can see that most studies have found this relationship to be stronger in the case of males than for females although there are many exceptions.

In the table's second row, findings from studies of the relationship between being married and health (usually based on self-ratings) are summarized. Of the 16 findings, 15 concluded that the relationship (which is positive) tends to be stronger for males than for females.

25.5.3 Death

Many studies have examined sex differences in death probabilities at various ages. Findings are summarized in Table 25.5.3. The table shows that most studies indicate that overall infant mortality is greater for males than for females. In the case of what is known as the *sudden infant death syndrome*, however, row 2 shows that all 13 studies concluded that this is more common among baby boys than among baby girls.

Mortality during childhood is usually assessed in terms of deaths between the ages of 2 and 12. One can see in the table's third row that findings are very inconsistent regarding any sex differences.

Row 4 pertains to death rates from puberty onward. For all 18 of these studies, findings agree that males are more likely to die than females throughout this entire age range.

In the last column, research findings having to do with sex differences in the average age at death (sometimes known as life expectancy) are reported. One can see that, of the 297 studies, 273 indicated that females outlive males. It is also worth noting that four published generalizations and five literature reviews all concluded that females usually outlive males.

25.5.4 Accidental Physical Injuries

Table 25.5.4 summarizes numerous studies in which sex differences in persons sustaining accidental physical injuries has been reported. Unless otherwise specified, these studies include both fatal and non-fatal injuries. One can see in the first row that the vast majority of the 134 studies of all types of accidental injuries have concluded that males suffer more than do females. These sex differences are especially pronounced during childhood and adolescence.

Table 25.5.2 Self-rated health

Variable	Orig. table	Number of sex difference findings (and consistency scores)						Overview	Countries	Time range	Non-human
		I/T	Child	Adol	Adult	WAR	All ages				
1. Correspondence between self-rated health and mortality	5.1. 3.2	–	–	–	19M 3x 7F (52.8)	–	19M 3x 7F (52.8)	–	8 (2)	1986–2001	–
2. Relationship between health and being married	5.1. 4.5	–	–	–	15M 1x (93.7)	–	15M 1x (93.7)	–	2	1966–2013	–

Table 25.5.3 Death

Variable	Orig. table	Number of sex difference findings (and consistency scores)						Overview	Countries	Time range	Non-human
		I/T	Child	Adol	Adult	WAR	All ages				
1. Infant mortality in general	5.1. 5.1	62M 26F (54.4)	–	–	–	–	62M 26F (54.4)	–	16 (8)	1923–2012	1M primates; 1x primates; 5F birds
2. Sudden infant death	5.1. 5.2	13M (100.0)	–	–	–	–	13M (100.0)	–	3 (1)	1968- 2004	–
3. Childhood mortality	5.1. 5.3	–	14M 1x 14F (32.6)	–	–	–	14M 1x 14F (32.6)	–	13 (1)	1923–2004	3M birds
4. Mortality following puberty	5.1. 5.4	–	–	5M	5M	8M	18M (100.0)	M: 1Rev	5 (1)	1957–2004	1M primates
5. Age at death (life expectancy)	5.1. 5.5	–	–	–	–	273F 7x 17M (86.9)	273F 7x 17M (86.9)	F: 4Gen 5Rev	34 (78)	1923–2020	Complex but mainly females

Table 25.5.4 Accidental physical injuries

Variable	*Orig. table*	*Number of sex difference findings (and consistency scores)*						*Overview*	*Countries*	*Time range*	*Non-human*
		I/T	*Child*	*Adol*	*Adult*	*WAR*	*All ages*				
1. Accidental injuries and fatalities in general	5.2. 1.1	2M 1x	32M **(100.0)**	13M **(100.0)**	17M 2x 1F (80.9)	66M 2x **(97.1)**	130M 4x **(97.0)**	M: 1Rev 1Met	17 (11)	1964–2011	4M primates
2. Motor vehicle injuries	5.2. 1.2	–	–	10M **(100.0)**	4M 4x 1F	36M 1x 1F (92.3)	50M 5x 1F (87.7)	–	14 (1)	1961–2013	–
3. Motor cycle injuries	5.2. 1.3	–	–	–	2M	9M 1x (90.0)	11M 1x (91.7)	–	7	1990–2012	–
4. Bicycle injuries	5.2. 1.4	–	1M	–	1F 1x	16M **(100.0)**	17M 1x 1F (85.0)	M: 1Rev	7	1985–2015	–
5. Burn injuries	5.2. 1.6	–	5M 1x 1F	–	5M	15M 10F (42.9)	25M 1x 11F (54.3)	M: 4Rev; F: 5Rev	15 (1)	1968–2017	–
6. Injuries caused by other animals	5.2. 1.9	–	–	–	1M	21M 3x (87.5)	22M 3x (88.0)	M: 1Rev	9 (1)	1954–2012	–
7. Accidental drownings	5.2. 1.10	4M	5M	–	–	37M 1F (94.9)	46M 1F **(95.8)**	M: 7Rev	9 (7)	1974–2014	–
8. Accidental injuries from falls (non-fatal)	5.2. 1.11	–	–	–	34F 1x 5M (75.6)	2F 5M	36F 1x 10M (63.2)	M: 1Rev (children)	17	1977–2020	–
9. Fatal accidental injuries from falls	5.2. 1.12	–	1M	–	4M 3F	9M 1x (90.0)	14M 1x 3F (66.6)	–	9	1990–2019	–
10. Accidental lightning strike injuries	5.2. 1.14	–	–	–	–	25M **(100.0)**	25M **(100.0)**	–	14 (3)	1990–2018	–
11. Sports injuries	5.2. 1.15	–	–	2F 1x	10F 4x (71.4)	26M 2x 11F (52.0)	26M 7x 23F (32.9)	M: 2Rev; x: 2Rev; F: 1Rev	8 (1)	1980–2017	–
12. Bone fractures	5.2. 2.1	1M	–	1F	68F 1x 2M (93.2)	6F 9M (25.0)	74F 1x 12M (85.1)	F: 8Rev	19 (1)	1978–2012	1M primates

Three categories of transportation-related injuries are considered in rows 2 through 4. Regarding motor vehicles (i.e., predominantly automobiles), all ten studies of adolescents concluded that males suffer more injuries. However, for studies of adults and wide age ranges (WAR), the results are somewhat less definitive.

Sex differences in burn injuries are summarized in row 5. Findings have been inconsistent, although about twice as many studies indicate that males suffer more than do females. One of the complicating factors in this regard is the location of the burn incidences. Burns to females tend to be more common in the home (especially the kitchen), while burns in other locations are usually more common for males.

Most studies of injuries caused by other animals involve snake bites, particularly among residents of rural areas. Row 6 shows that 22 of the 25 studies all concluded that males suffer more than females, with the remaining three finding no significant difference.

Accidental drownings are typically fatal. As shown in row 7, 46 of the 47 studies concluded that males are more likely to suffer from drownings than are females.

Rows 8 and 9 both involve accidental falls, but the former has to do with non-fatal injuries while the latter involves fatalities. One can see that, in both cases, average sex differences vary. Nonetheless, for non-fatal fall injuries, most studies report that females are more likely to be the victim, while, in the case of fatal falls, males are more often the victim.

In row 10 of this table, accidents caused by lightning strikes are reviewed regarding sex differences. One can see that all 25 located studies concluded that males were more likely to be injured as a result of being struck by lightning than were females, making this a well-established cultural universal sex difference.

In the eleventh row of this table, findings from sports-related injuries are reported. Results are highly inconsistent regarding any sex differences in sports injuries. One can see in the overview column that literature reviews have also come to conflicting conclusions. Many of the inconsistencies can be explained in terms of the types of injuries (e.g., sprains, bone fractures, concussions) and even their seriousness.

Finally, this table provides a summary of the large number of studies that have assessed bone fractures, regardless of their cause. While the findings regarding sex differences have been mixed, row 12 shows that most studies have concluded that females suffer more bone fractures than do males.

25.5.5 Congenital Diseases

Congenital diseases are ones that are normally present at birth. While many of these diseases have been studied regarding sex differences, only three were found to have ten or more pertinent studies. Findings from the first of these three diseases is shown in row 1 of Table 25.5.5. It indicates that 13 studies found more males than females being diagnosed with *Down syndrome*, although two other studies reported no significant sex differences.

Spina bifida refers to the failure of the spinal cord to properly form. Row 2 shows that 11 studies reported the proportion of each sex who exhibit this along with various closely related neuro-tube defects. One can see that nine of these studies concluded that greater proportions of females exhibit these defects, with one of the remaining studies reporting no sex difference and the other finding greater proportions of males.

Tetralogy of fallot has to do with various defects in the formation of the heart. According to row 3, most studies indicate that there are no significant sex differences in this defect, although when differences are found, males are affected more than females.

25.5.6 Cirrhosis of the Liver

Cirrhosis refers to the formation of scar tissue on the liver. It is usually associated with long-term alcohol abuse and sometimes with hepatitis. As shown in Table 25.5.6, research findings having to do with sex differences in this disease are highly inconsistent. Nearly all of the inconsistencies can actually be explained by noting that alcoholism is considerably more prevalent among males than among females. As a result, nearly all of the studies of population-wide incidences of cirrhosis of the liver have concluded that males suffer from this condition more than do females. However, the studies that have restricted their samples to alcoholics have reached the conclusion that cirrhosis of the liver is more prevalent among females, suggesting that female livers may be able to tolerate less long-term heavy alcohol exposure than male livers.

25.5.7 Cancer

Cancer encompasses a set of dreaded diseases that are responsible for millions of deaths every year. Table 25.5.7 provides a summary of studies of sex differences in various forms of cancer for which at least ten studies were located, including studies of all forms of cancer combined. Some of these studies pertain to being diagnosed with cancer, while others involve actually dying from cancer. In the first row, one can see that 36 out of 43 studies of all forms of cancer have concluded that males are more likely to suffer than females.

Table 25.5.5 Congenital diseases

Variable	*Orig. table*	*Number of sex difference findings (and consistency scores)*						*Overview*	*Countries*	*Time range*	*Non-human*
		I/T	*Child*	*Adol*	*Adult*	*WAR*	*All ages*				
1. Down syndrome	5.3. 1.21	–	–	–	–	13M 2x (86.7)	13M 2x (86.7)	–	7 (2)	1965–2014	–
2. Spina bifida and other neurotube defects	5.3. 1.35	–	–	–	–	9F 1x 1M (75.0)	9F 1x 1M (75.0)	–	2 (3)	1973–2014	–
3. Tetralogy of fallot	5.3. 1.54	–	–	–	–	4M 6x (40.0)	4M 6x (40.0)	–	2 (2)	1953–2014	–

Table 25.5.6 Cirrhosis of the liver

Variable	*Orig. table*	*Number of sex difference findings (and consistency scores)*						*Overview*	*Countries*	*Time range*	*Non-human*
		I/T	*Child*	*Adol*	*Adult*	*WAR*	*All ages*				
1. Cirrhosis of the liver	5.3. 2.2	–	–	–	6M 12F (50.0)	12M 2F (75.0)	18M 14F (39.1)	–	8 (1)	1945–2007	–

Table 25.5.7 Cancer

Variable	*Orig. table*	*Number of sex difference findings (and consistency scores)*						*Overview*	*Countries*	*Time range*	*Non-human*
		I/T	*Child*	*Adol*	*Adult*	*WAR*	*All ages*				
1. Cancer in general	5.3. 3.1	–	1M	1M	5M 1x 2F	29M 4F (76.3)	36M 1x 6F (73.5)	M: 1Rev	12 (4)	1912–2012	3M rodents
2. Breast cancer	5.3. 3.5	–	–	–	1F	26F (100.0)	27F (100.0)	F 1Rev	11 (2)	1943–2006	–
3. Colorectal cancer	5.3. 3.7	–	–	–	3M 2x	15M 6x 1F (65.2)	18M 8x 1F (64.3)	–	12 (2)	1961- 2006	–
4. Colorectal cancer survival	5.3. 3.8	–	–	–	–	7F 3x (70.0)	7F 3x (70.0)	–	11 (1)	1961–2005	–
5. Leukemia and non-Hodgkin's lymphoma	5.3. 3.13	–	–	–	–	18M (100.0)	18M (100.0)	–	6 (1)	1961–2003	–
6. Lung cancer	5.3. 3.14	–	–	–	16M 4x 13F (34.8)	23M (100.0)	39M 4x 13F (56.5)	–	12 (4)	1961–2007	–
7. Stomach cancer	5.3. 3.25	–	–	–	1M 1F	16M (100.0)	17M 1F (94.4)	–	9 (3)	1961–2005	–

Regarding breast cancer, while a small proportion of cases involve males, the vast majority only affect females. Thus, the table's second row shows that all 27 studies reached this conclusion.

Colon cancer and cancer of the rectum are often considered together, as we have done here. The third row shows that, while the majority of studies have concluded that males have higher rates than do females, several studies have failed to report significant sex differences, and one actually reported higher rates for females. Ten studies also assessed survival rates from colorectal cancer (which are usually defined as having lived at least two years beyond being diagnosed). One can see that seven of these studies indicated that survival rates for females were higher, while the remaining three studies found no significant sex difference.

Sex differences in being diagnosed with leukemia or non-Hodgkin's lymphoma or dying from either of these closely related forms of cancer were examined by 18 studies. One can see in the fifth row that all of these studies concluded that the rates are higher for males than for females.

In the case of lung cancer, 56 studies of sex differences were located. As shown in row 6, 39 of these studies concluded that males are more likely to suffer from, or die from, this disease than females. It is worth adding that nearly all of the 13 studies that reported higher rates for females were ones that limited their samples to smokers or former smokers.

The final entry in this table was for stomach cancer. One can see that all but one of the 18 studies concluded that stomach cancer was more prevalent among males than among females.

25.5.8 Cardiovascular Diseases

Cardiovascular diseases have to do with diseases of the heart, arteries and veins, sometimes collectively known as the *circulatory system*. A summary of findings of sex differences in nine cardiovascular diseases (or disease-related variables) for which at least ten studies were located is presented in Table 25.5.8. Regarding cardiovascular disease in general, 203 studies were located, the majority of which concluded that males were more likely to be afflicted than females. However, this is in conflict with one meta-analysis that concluded the opposite.

In the case of coronary artery disease, ten studies were located. The second row of the table shows that eight of these studies concluded that females suffer more than do males, albeit the remaining two studies reported no significant sex difference. This is a pattern consistent with one meta-analysis, although the meta-analysis was limited to persons with diabetes.

Table 25.5.8 Cardiovascular diseases

Variable	Orig. table	Number of sex difference findings (and consistency scores)						Overview	Countries	Time range	Non-human
		I/T	Child	Adol	Adult	WAR	All ages				
1. Cardiovascular disease in general	5.3.4.1	–	–	–	135M 8x 36F (62.8)	18M 1x 5F (62.1)	153M 9x 41F (62.7)	F: 1Met (death)	15 (10)	1933–2014	–
2. Coronary artery disease	5.3.4.2	–	–	–	8F 2x (80.0)	–	8F 2x (80.0)	F: 1Met (among diabetics)	2	1980–1996	–
3. Hypertension	5.3.4.3	–	–	–	18F 5x 13M (36.7)	2F 3x 4M	20F 8x 17M (32.3)	–	7	1983–2007	11M rodents
4. Angina pectoris (chest pain)	5.3.4.9	–	–	–	11F 3M (64.7)	–	11F 3M (64.7)	–	3	1972–2003	–
5. Myocardial infarction (heart attack)	5.3.4.22	–	–	–	27F 4x 14M (45.8)	5F 1x 2M	32F 5x 16M (46.4)	–	9 (1)	1965–2003	–
6. Age of first myocardial infarction	5.3.4.23	–	–	–	21M (100.0)	–	21M (100.0)	–	4	1984–1999	–
7. Survival following a myocardial infarction	5.3.4.24	–	–	–	42M 6x 9F (63.6)	4F	42M 6x 13F (56.8)	–	10	1973–2003	2M mouse
8. Stroke	5.3.4.30	–	–	–	13M 6x 13F (28.9)	8M 1x 1F (72.7)	21M 7x 14F (37.5)	M: 3Rev (except among elderly)	9 (3)	1970–2012	–
9. Survival following a stroke	5.3.4.32	–	–	–	7F 2x 3M (46.7)	–	7F 2x 3M (46.7)	–	7	1986–2009	–

Hypertension refers to varying degrees of high blood pressure. One can see in row three that findings are roughly equal in terms of evidence of any sex differences. It is worth adding that 11 studies of rodents all concluded that males have higher rates of hypertension than do females.

People who experience chest pain are said to have *angina pectoris*. As shown in row 4, findings have been mixed, although they largely indicate that females experience angina pectoris more than do males.

Row 5 pertains to sex differences in the frequency of myocardial infarctions (heart attacks). Of the 53 pertinent studies, 32 concluded that females are more frequent sufferers than are males. In the case of the age that persons have their first heart attack, row 6 shows that all 21 studies have indicated that males have their first experience with this life-threatening disease at a younger average age than do females. Sex differences in the likelihood of surviving a myocardial infarction was report in 61 located studies. As shown in row 7, 42 of these studies concluded that males are more likely to survive than females. To explain this confusing set of findings, several factors appear to be involved, one of which seems to involved the fact that females tend to be older when they first experience heart attacks; in fact, women are more likely than men to be elderly and in a relatively poorer health when their first heart attack occurs.

The last two rows in this table have to do with strokes, which are blockages in the arteries carrying blood to the brain. One can see that neither of these studies found consistent sex differences in the incidence of strokes, although most research points toward males having at least slightly more strokes, while females have a somewhat higher probability of surviving after a stroke.

25.5.9 Digestive Diseases and Disorders

Sex differences in three different digestive disorders were located that contained citations to at least ten relevant studies. As shown in Table 25.5.9, the first disease is acid reflex disease. One can see that 17 out of 21 studies concluded that this condition was more common among males than among females.

Sex differences in the frequency of constipation has been reported in 16 studies. Row 2 shows that all but one of these studies concluded that females appear to have constipation more than males.

The final row in this table has to do with an ailment known as irritable bowel syndrome. Of the 40 pertinent studies on sex differences, 26 concluded that this condition is more common in females than in males. (Table 25.5.9).

Table 25.5.9 Digestive diseases and disorders

Variable	*Orig. table*	*Number of sex difference findings (and consistency scores)*						*Overview*	*Countries*	*Time range*	*Non-human*
		I/T	*Child*	*Adol*	*Adult*	*WAR*	*All ages*				
1. Acid reflux disease	5.3. 6.3	–	–	–	16M 3x 1F (76.2)	1M	17M 3x 1F (77.3)	M: 1Rev	8 (1)	1998–2020	–
2. Constipation, frequency of	5.3. 6.7	–	–	–	1x	15F (100.0)	15F 1x (93.7)	–	6	1974- 2012	–
3. Irritable bowel syndrome	5.3. 6.10	–	–	–	26F 6x 8M (54.2)	–	26F 6x 8M (54.2)	F: 1Rev	18	1966–2006	–

25.5.10 Autoimmune Diseases

Sex differences in one general measure and five specific autoimmune diseases were located that had at least ten pertinent studies. The results are presented in Table 25.5.10.

One can see in row 1 that all studies have concluded that, for autoimmune diseases as a whole, females are more often afflicted than are males. This conclusion has also been reached by three literature reviews.

Rows 2, 3, and 4 all pertain to AIDS or HIV infections. One can see that most studies have concluded that AIDS is more common among males, while studies of sex differences in HIV infections have varied considerably, as have studies on HIV-1 RNA viral load.

Studies of sex differences in the prevalence of allergies are summarized in row 5. It indicates that the evidence is fairly mixed, although most studies point toward females being affected more, especially after puberty.

In the case of lupus, row 6 shows that most studies have found female sufferers to be substantially greater than male sufferers. This even appears to be the case for lupus among mice. The exceptions have to do with the most serious forms of this disease, where studies indicate that males predominate.

Findings regarding sex differences in myasthenia gravis are summarized in the last row of this table. One can see that, except for three studies reporting no significant differences, 13 studies concluded that females are affected more by this condition.

25.5.11 Infectious Diseases

Three types of infectious diseases, along with infectious diseases in general, have been studied with regard to sex differences to a degree that at least ten findings were identified. Results are summarized in Table 25.5.10. In the case of infectious diseases overall, row 1 shows that findings of any sex differences have been mixed.

Sex differences in influenza as well as sepsis (rows 2 and 3) largely indicate that males are more likely to contract these diseases than females. Shingles, on the other hand, appears to be more common among females, to the degree sex differences exist (row 4) (Table 25.5.11).

25.5.12 Muscular-Skeletal Diseases

Table 25.5.12 provides a summary of findings regarding sex differences in muscular-skeletal diseases for which at least ten findings were located. Beginning with studies of these diseases in general, row 1 shows that the evidence is quite mixed regarding possible sex differences.

Table 25.5.10 Autoimmune diseases

Variable	*Orig. table*	*Number of sex difference findings (and consistency scores)*						*Overview*	*Countries*	*Time range*	*Non-human*
		I/T	*Child*	*Adol*	*Adult*	WAR	*All ages*				
1. Autoimmune diseases in general	5.3. 7.1	–	–	–	4F	6F	10F (100.0)	F: 3Rev	1 (2)	1986- 2020	–
2. AIDS	5.3. 7.3	1M	1x	–	3M	6.M 1x 1F	9M 2x 1F (69.2)	–	5 (5)	1997–2009	–
3. HIV infections	5.3. 7.4	6F 2x	–	–	1F 3M	3M	7F 2x 6M (33.3)	–	7 (5)	1999–2007	–
4. HIV-1 RNA viral load	5.3. 7.7.	–	–	2F	4F 1x	3M 2x	6F 3x 3F (40.0)	–	7	1990–2001	–
5. Allergies	5.3. 7.9	1F 3M	3F 2x 1M	3F	3F 1x	5F 3x	15F 6x 4M (51.7)	–	10	1969–2002	–
6. Lupus	5.3. 7.19	–	6F	1F	11F 4M (57.9)	10F (100.0)	28F 4M (77.8)	F: 3Rev	6 (1)	1964–2020	2F mouse
7. Myasthenia gravis	5.3. 7.20	–	3x	4F	–	9F	13F 3x (81.2)	–	3	1053–2020	–

Table 25.5.11 Infectious diseases

Variable	*Orig. table*	*Number of sex difference findings (and consistency scores)*						*Overview*	*Countries*	*Time range*	*Non-human*
		I/T	*Child*	*Adol*	*Adult*	*WAR*	*All ages*				
1. Infectious diseases in general	5.3. 8.1	2M	5M 1F	2F	1F	3M 1x 3F	10M 1x 7F (40.0)	M: 1Gen	3 (1)	1928–2003	–
2. Influenza	5.3. 8.8	1M	4M	–	1M 1F	6M	12M 1F (85.7)	–	2 (1)	1933–2004	–
3. Sepsis	5.3. 8.12	1M	–	–	7M 1x	2M	10M 1x (90.9)	–	5	1998–2021	–
4. Shingles	5.3. 8.14	1F	1F	1x	2F 1x	3F 3x	7F 5x (58.3)	–	4	1982–2003	–

Table 25.5.12 Muscular-skeletal diseases

Variable	*Orig. table*	*Number of sex difference findings (and consistency scores)*						*Overview*	*Countries*	*Time range*	*Non-human*
		I/T	*Child*	*Adol*	*Adult*	*WAR*	*All ages*				
1. Muscular-skeletal diseases in general	5.3. 9.1	–	2M	2M	9F 4x 1M (60.0)	3F 3x	12F 7x 5M (41.4)	–	4	1966–2000	–
2. Arthritis in general	5.3. 9.2	–	–	–	13F (100.0)	7F	20F (100.0)	F: 3Rev	7	1976–2020	–
3. Multiple sclerosis	5.3. 9.6	–	1F	1x	10F (100.0)	9F	20F 1x (95.2)	F: 4Rev	10	1981–2020	–
4. Osteoarthritis	5.3 9.7	–	–	–	16F 6x 10M (44.4)	3F	19F 6x 10M (42.2)	–	13	1967–2004	–
5. Osteoporosis	5.3. 9.8	–	–	–	33F (100.0)	2F	35F (100.0)	F: 6Rev	6	1981–2015	3F primates
6. Rheumatoid arthritis	5.3. 9.12	–	–	3F	11F 1x (91.7)	6F	20F 1x (95.2)	F: 1Rev	6	1968–2020	–

In the case of arthritis in general, however, the findings summarized in row 2 shows that the evidence is very consistent, indicating that females are afflicted more than males, particularly for adults. Rows 3, 5, and 6 all suggest that similar conclusions can be reached for multiple sclerosis, osteoporosis, and rheumatoid arthritis (i.e., they are all more prevalent among females than among males). Only in the case of osteoarthritis is the evidence mixed regarding possible sex differences. One can add that in the case of osteoarthritis, studies of three species of non-human primates were also found, and they too indicate that this disease is more common among females than among males.

25.5.13 Neurological Diseases

Four neurological diseases were located for which at least ten findings of sex differences were reported. As shown in Table 25.5.13, row 1 has to do with Alzheimer's disease. While a substantial majority of studies reported higher rates of this disease among females, several other studies failed to document any significant sex differences. In the case of dementia, findings of sex differences are even more inconsistent, although still a majority point toward higher rates among females (row 2). Of course, sex differences in both of these conditions could be at least partially due to the fact that, in most countries, females die at an older age than males, thus allowing them more time to develop these age-related disorders.

In the case of epilepsy, 37 studies of sex differences were found, 29 of which concluded that males have this condition more than do females, albeit four studies reported the opposite (row 3). Especially given that one literature review and one meta-analysis both concluded that males are more often affected by epilepsy (and one rodent study reached the same conclusion), greater male susceptibility to this disorder appears likely.

The final row shows that, with just a few exceptions, most studies have concluded that more males than females are afflicted with Parkinson's disease. This conclusion comports with conclusions reached by one literature review and two meta-analyses.

25.5.14 Respiratory Diseases

Findings from at least ten studies regarding sex differences in respiratory diseases were located for four specific diseases and for such diseases considered as a whole. Results are shown in Table 25.5.14.

Regarding respiratory diseases in general, one can see in row 1 that findings have been mixed regarding any possible sex differences. For asthma and COPD, rows 2 and 3 also indicate that there are virtually no overall sex differences in the prevalence of these two diseases.

Table 25.5.13 Neurological diseases

Variable	Orig. table	Number of sex difference findings (and consistency scores)						Overview	Countries	Time range	Non-human
		I/T	Child	Adol	Adult	WAR	All ages				
1. Alzheimer's disease	5.3. 10.1	–	–	–	46F 13x (80.0)	2F	48F 13x (78.7)	F: 3Rev 2Met	12 (3)	1987–2019	–
2. Dementia	5.3. 10.4	–	–	–	16F 5x 4M (55.2)	–	16F 5x 4M (55.2)	–	7 (2)	1986- 2012	–
3. Epilepsy	5.3. 10.6	1M	5M 1x	–	1x	23M 2x 4F (69.7)	29M 4x 4F (70.7)	M: 1Rev 1Met	13	1934–2015	1M rodent
4. Parkinson's disease	5.3. 10.9	–	–	–	22M 2x 2F (78.6)	1M	23M 2x 2F (79.3)	M: 1Rev 2Met	6 (2)	1976–2015	–

Table 25.5.14 Respiratory diseases

Variable	Orig. table	Number of sex difference findings (and consistency scores)						Overview	Countries	Time range	Non-human
		I/T	Child	Adol	Adult	WAR	All ages				
1. Respiratory diseases in general	5.3. 12.1	–	1M	–	–	8F 1x 6M (38.1)	8F 1x 7M (34.8)	–	4	1961–2012	–
2. Asthma	5.3. 12.2	4M	28M 1x 3F (82.3)	3F 17x 3M (11.5)	25F 10x 1M (67.6)	5F 4x 4M (29.4)	40M 31x 36F (22.3)	–	11 (5)	1978–2020	–
3. Chronic obstructive pulmonary disease (COPD)	5.3. 12.6	–	–	–	8F 5x 5M (30.8)	4M	9M 5x 8F (30.0)	–	6 (2)	1983–2012	–
4. Pneumonia	5.3. 12.10	1M	1M	4M	4M 1F	6M	16M 1F (88.9)	–	3	1941–2007	M: 6 Mouse
5. Tuberculosis	5.3. 12.11	–	2M 1F	4F 10x (28.6)	16M 1x 1F (84.2)	23M 1F (92.0)	45M 11x 2F (75.0)	M: 7Rev	25 (1)	1933–2014	–

In the case of pneumonia and tuberculosis, on the other hand, one can see in rows 4 and 5 that most studies have concluded that males are affected more than females. This conclusion coincides with findings from six studies of mice in the case of pneumonia and with seven literature reviews of human studies regarding tuberculosis.

25.5.15 Other Diseases and Ailments

Table 25.5.15 has to do with a variety of difficult to classify diseases and ailments for which at least ten studies were located regarding possible sex differences. Row 1 has to do with blindness (including severely impaired vision). Of the 35 located studies, 19 concluded that males were more often affected, 13 found no significant sex differences and three reported higher rates in females. Oddly, all three literature reviews on blindness also reported higher female rates.

Findings of sex differences in the prevalence of diabetes, with Type 1 and Type 2 considered together, are summarized in row 2. It indicates that 32 of the 67 studies indicated that males are affected more than females. Nevertheless, 22 of the 67 studies reported the opposite. Some of the inconsistencies is likely due to the severity of the disease and the age at which it is diagnosed. Of seven literature reviews, one can see that six concluded that males are affected more, while the remaining review concluded there were no sex differences.

Rows 3, 4, 5, and 6 all have to do with sex differences in headaches. As one can see, nearly all studies have concluded that females are more likely to suffer from headaches and are especially likely to seek (or obtain) medical treatment from headaches.

In rows 7 through 12, sex differences in experiencing various forms of pain are examined. Regarding pain in general (except in the case of headaches), nearly all studies have concluded that females report greater amounts. Similarly, for abdominal pain, facial and oral pain, muscular-skeletal pain, and back pain, female sufferers outnumber male sufferers. The only exceptions involve results reporting no significant sex differences.

In the last row, results of 11 studies reporting on sex differences in persons seeking (or obtaining) medical treatment for back pain are summarized. Findings in this case largely indicated that there are no significant sex differences, but, when there are differences, they favor both sexes to an equal degree.

25.5.16 Sleep Problems and Disorders

Findings from the four source tables appearing in Chapter 5 that contained ten or more studies involved sleep-related problems are listed in Table 25.5.16. Row 1 has to do with sex differences in the prevalence of

Table 25.5.15 Other diseases and ailments

Variable	Orig. table	Number of sex difference findings (and consistency scores)						Overview	Countries	Time range	Non-human
		I/T	Child	Adol	Adult	WAR	All ages				
1. Blindness or impaired vision	5.3. 13.3	–	–	–	15F 10x 1M (55.5)	4F 3x 2M	19F 13x 3M (50.0)	F: 3Rev	18 (2)	1977–2017	–
2. Diabetes	5.3. 15.2	–	1M	2M 1F	20M 5x 16F (35.1)	9M 8x 5F (33.3)	32M 13x 22F (36.0)	M: 6Rev; x: 1Rev	18 (1)	1974–2020	M: 1 Mouse
3. Headaches in general	5.4. 1.1	–	1M	1F	27F 1M (93.1)	11F 1M (84.6)	39F 3M (86.7)	–	12 (1)	1979–2009	–
4. Headaches except migraines and cluster headaches	5.4. 1.2	–	2x	11F (100.0)	21F 1x 1M (86.5)	31F 3x 2M (81.6)	63F 6x 2M (86.3)	–	21	1952–2004	–
5. Migraine headaches	5.4. 1.3	–	2M 4x	7F 1x	4F 1x	24F 4x (85.7)	35F 10x 2M (71.4)	F: 4Rev	15	1956–2013	–
6. Seeking (or obtaining) medical treatment for headaches	5.4. 1.6	–	–	–	4F	8F	12F (100.0)	–	4	1977–2004	–
7. Pain in general (except headaches)	5.4. 2.1	–	1F 1x	1F	36F 2x (94.7)	4F	42F 2x (95.4)	F: 2Rev	6 (2)	1986–2005	–
8. Abdominal pain	5.4. 2.2	–	1x	3F 1x	1F	5F 1x	9F 3x (75.0)	–	6	1957–1994	–

9. Facial and oral pain	5.4. 2.5	–	–	3F 2x	17F 2x (89.5)	6F 5x (54.5)	26F 9x (74.3)	F: 1Rev	4	1972–2003	–
10. Muscular-skeletal pain	5.4. 2.6	–	1F	1F 4x	21F 4x (84.0)	17F 1x (94.4)	40F 9x (81.6)	F: 2Rev	14	1950–2010	–
11. Back pain	5.4. 2.8	–	–	3F 2x	13F 4x 2M (61.9)	11F 20x 1M (33.3)	29F 26x (52.7)	F: 1Rev	13	1975–2006	–
12. Seeking (or obtaining) medical treatment for back pain	5.4. 2.9	–	–	1F 2x	1M 2x	1M 3x 1F	2M 7x 2F (15.4)	–	6	1983–1993	–

Table 25.5.16 Sleep problems and disorders

Variable	*Orig. table*	*Number of sex difference findings (and consistency scores)*						*Overview*	*Countries*	*Time range*	*Non-human*
		I/T	*Child*	*Adol*	*Adult*	*WAR*	*All ages*				
1. Sleep problems and disorders in general	5.4. 3.1	–	–	3F	13F 4M (61.9)	1M	16F 5M (61.5)	–	7	1993–2018	–
2. Insomnia	5.4. 3.2	–	1M	–	46F 2x (95.8)	6F	48F 2x 1M (92.8)	F: 1Rev 1Met	16 (1)	1976–2018	–
3. Sleep apnea	5.4. 3.5	2M	1M	–	24M (100.0)	7M	34M (100.0)	M: 2Rev	8 (1)	1955–2008	–
4. Nightmares	5.4. 3.7	–	1F	3F	7F 2x	1F	12F (100.0)	F:1Rev	8	1984–2016	–

sleep problems and disorders in general. Out of the 21 pertinent studies, 16 concluded that females have more sleep problems than do males, with the remaining five studies having reported no significant sex differences.

Studies of sex differences in the prevalence of insomnia totaled 51. As shown in row 2, 48 of these studies concluded that females have more trouble sleeping than do males. This pattern is reflected in the conclusions reached by one literature review and by a meta-analysis as well.

In the case of sleep apnea, 34 studies of sex differences were found. Row 3 shows that each and every one of these studies concluded that this disorder is more common among males than among females. Furthermore, two literature reviews reached the same conclusion.

Row 4 has to do with the occurrence of nightmares. One can see that all 12 studies bearing on sex differences in these events concluded that they are more frequent among females than among males. This coincides with findings from a literature review as well.

25.5.17 Residual Ailments

The final summary table having to do with health issues is Table 25.5.17. It contains seven rows, each one pertaining to fairly minor types of ailments. Row 1 has to do with minor ailments and somatoform disorders in general, the latter referring to nonspecific feelings of bodily discomfort. It shows that 22 out of 23 studies have concluded that these types of ill health are more common among females than among males.

Fainting and dizziness have been studied by at least ten studies regarding possible sex differences. Row 2 shows that six of these studies indicated that these conditions are more common among females, while the remaining four reported no significant differences.

Eleven studies of sex differences in incontinence were located. One can see in row 3 that, as adults, seven of these studies concluded that females have this condition more than males.

In Row 4, attention is given to sex differences in motion sickness. As one can see, 32 out of a total of 34 studies concluded that this condition is more common among females than among males, a conclusion that was also reached by one prior literature review.

Fifteen studies of nausea and vomiting were located regarding possible sex differences. Row 5 shows that all of these studies concluded that these experiences are more common among females than among males.

In health surveys, research participants are often asked if they currently have any physical disabilities or limitations. In Row 6, one can see that all 23 studies concluded that females answer affirmatively more than do males.

Table 25.5.17 Residual ailments

Variable	*Orig. table*	*Number of sex difference findings (and consistency scores)*						*Overview*	*Countries*	*Time range*	*Non-human*
		I/T	*Child*	*Adol*	*Adult*	*WAR*	*All ages*				
1. Minor ailments and somatoform disorders in general	5.4.4.1	–	–	–	19F 1x (95.0)	3F	22F 1x (95.6)	–	5	1980–2016	–
2. Fainting and dizziness	5.4.4.5	–	–	–	6F 4x (60.0)	–	6F 4x (60.0)	–	2	1942–1998	–
3. Incontinence	5.4.4.9	–	1M	–	7F 2x 1M	–	7F 2x 2M (53.8)	–	6	1980–2014	–
4. Motion sickness	5.4.4.11	–	3F	1F	25F 2x (92.6)	3F	32F 2x (94.1)	F: 1Rev	10 (2)	1977–2019	Mixed findings
5. Nausea and vomiting in general	5.4.4.12	–	–	1F	14F (100.0)	–	15F (100.0)	–	5	1982–2007	–
6. Physical disabilities and limitations (usually temporary)	5.4.4.15	–	–	2F	21F (100.0)	–	23F (100.0)	F: 1Met	5 (2)	1989–2005	–
7. Tooth decay and dental caries	5.4.4.21	–	7F 2M	2F	8F 4x 4M (40.0)	10F 1x (90.9)	27F 5x 6M (61.4)	F: 3Rev 2Met	16 (2)	1950–2012	–

The final row in this table pertains to tooth decay, usually expressed in the form of dental caries. Out a total of 38 studies, 27 indicated that tooth decay is more common among females than males.

25.6 Responses to Physical and Chemical Factors: Condensed Findings

Chapter 6, as well as Chapters 7 and 8 (all located near the end of Volume I) all cover considerable research relevant to sex differences. However, these three chapters all pertain to sex differences where the number of tables containing ten or more citations involving humans are relatively few. Instead, much more of the research has been based on studies of non-human animals. This is due in part to the fact that much of the pertinent sex differences can only be precisely assessed by way of controlled experiments. In any case, Section 6 of the present chapter (along with Sections 7 and 8) will be among the shortest of all the sections.

25.6.1 Responses to Drug Administration

One type of drug, clozapine (which is most commonly used for treating symptoms of schizophrenia), was investigated by 11 studies regarding possible sex differences in the levels present in the blood after administration. Table 25.6.1 shows that seven of these studies concluded that females exhibited higher circulating levels than males, although the remaining four studies found no significant differences.

25.6.2 Responses to Treatment for Drug Abuse/Addiction

Sex differences in how successfully people addicted to various drugs (except alcohol) are at stopping their abusive and addictive tendencies after receiving professional treatment are shown in row 1 of Table 25.6.2. Most studies report finding no significant sex differences in this regard.

Row 2 of this table involves success in stopping alcohol consumption (at least at excessive levels) among alcohol abusers and alcoholics. It shows that most research has indicated that, after receiving professional treatment, females are more successful than males.

25.6.3 Responses to Surgery

In Table 25.6.3, findings from 19 studies were reported regarding sex differences in death within 30 days after receiving coronary bypass surgery. One can see that the majority of studies have indicated that females have higher probabilities of dying than males.

Table 25.6.1 Responses to drug administration

Variable	*Orig. table*	*Number of sex difference findings (and consistency scores)*						*Overview*	*Countries*	*Time range*	*Non-human*
		I/T	*Child*	*Adol*	*Adult*	*WAR*	*All ages*				
Circulating levels of clozapine after administration	6.1.4.2	–	–	–	7F 4x (63.6)	–	7F 4x (63.6)	–	6	1976–2007	–

Table 25.6.2 Responses to treatment for drug abuse/addiction

Variable	*Orig. table*	*Number of sex difference findings (and consistency scores)*						*Overview*	*Countries*	*Time range*	*Non-human*
		I/T	*Child*	*Adol*	*Adult*	*WAR*	*All ages*				
1. Successful treatment for drug abuse/ addiction	6.12.2.1	–	–	–	5F 9x 1M (31.2)	4x	5F 13x 1M (25.0)	–	3	1980–2004	–
2. Successful treatment for alcohol abuse/ alcoholism	6.12.2.2	–	–	–	8F 2x (80.0)	1x	8F 3x (72.7)	x: 1Rev	1	1980–2005	–

Table 25.6.3 Responses to surgery

Variable	*Orig. table*	*Number of sex difference findings (and consistency scores)*						*Overview*	*Countries*	*Time range*	*Non-human*
		I/T	*Child*	*Adol*	*Adult*	*WAR*	*All ages*				
Death following coronary bypass surgery	6.13.2.2	–	–	–	18F 1x (94.7)	–	18F 1x (94.7)	–	4	1982–2007	–

25.7 Responses to Stress and to Pain: Condensed Findings

As with Section 6, the present Section 7 is brief (as is also the case in the upcoming Section 8). This is because, while there is substantial research on sex differences in responses to stress and to pain, most of it is experimental and based on studies of non-humans. Of course, the focus here, as with the rest of this chapter, is only on variables for which at least ten studies of humans were located.

25.7.1 Effects of Stress

Four variables having to do with the effects of stress were identified in Chapter 7 for which ten or more pertinent studies were located. The first of these four variables that is identified in row 1 of Table 25.7.1 has to do with the effects of stress on adrenaline levels. It indicates that 16 out of 20 studies concluded that males experience more of a rise in adrenaline when subjected to stress than do females. Adrenaline, of course, is a well-known *stress hormone*. The precise nature of the stressors may often explain the inconsistencies in male-female differences in the effects of stress. Also, it is worth noting that the tendency for adrenaline levels to rise more in males than in females in response to most forms of stress is the opposite of what five studies of rodents concluded.

Another so-called stress hormone is *cortisol*. Over 50 studies of sex differences in the effects of stress on this hormone were located. One can see in row 2 that a little less than half of these studies concluded that males experience higher levels of cortisol after being stressed than do females, with most of the exceptions reporting no significant differences. Again, the precise nature of the stressors is likely to be relevant to many of the inconsistencies.

In row 3, sex differences in blood pressure responses to stress are assessed. It shows that, when differences are found, nearly all studies report a greater increase in blood pressure among males, although about a third of the studies reported no significant differences.

The last row in this table has to do with the effects of stress on increases in one's heart rate. In this regard, female heart rates tend to increase more than male rates according to about half of the studies.

25.7.2 Responses to Stress

Considerable research has sought to assess the types of responses people make to stress, i.e., directly confronting the stressor, avoiding it, or talking to friends and relatives about the matter. Table 25.7.2 shows that only two types of responses were located as having ten or more relevant studies. Row 1 has to do with trying to avoid the stress (or the source of the stress) as much as possible. One can see that findings of any sex differences in this type of response are highly inconsistent if they exist at all.

Table 25.7.1 Effects of stress

Variable	*Orig. table*	*Number of Sex Difference Studies (and Consistency Scores)*						*Overview*	*Countries*	*Time range*	*Non-human*
		I/T	*Child*	*Adol*	*Adult*	*WAR*	*All ages*				
1. Effects of stress on adrenaline levels	7.1. 1.2	–	1M	–	14M 3x 1F (73.7)	1M	16M 3x 1F (76.2)	–	5	1978–2001	5F rodents
2. Effect of stress on cortisol levels	7.1. 1.7	1M 1x	2M 10x (16.7)	–	20M 14x 3F (50.0)	2F 1x	23M 26x 3F (41.8)	–	4	1978–2006	–
3. Effects of stress on blood pressure	7.1. 2.1	–	1M 2x	–	27M 9x 1F (71.1)	6M	34M 11x 1F (72.3)	–	1	1976–2004	1M rodents
4. Effects of stress on heart rate	7.1. 2.2	–	–	2F 1M	21F 12x 6M (46.7)	1F 2x 1M	24F 14x 8M (44.4)	F: 1Rev 1Met	7	1978–2019	–

Table 25.7.2 Responses to stress

Variable	*Orig. table*	*Number of Sex Difference Studies (and Consistency Scores)*						*Overview*	*Countries*	*Time range*	*Non-human*
		I/T	*Child*	*Adol*	*Adult*	*WAR*	*All ages*				
1. Adopting an avoidance response to stress	7.1. 7.4	–	1F 2x	1F 3x 2M	5F 3x 1M	–	7F 8x 3M (33.3)	–	2	1978–2004	–
2. Confiding behavior as a coping response to stress	7.1. 7.10	–	–	6F	9F 1x (90.0)	4F	19F 1x (95.0)	F: 1Rev 1Met	4	1974–2000	–

As shown in row 2, most studies have found consistent sex differences in responding to stress by confiding with others about the situation. Specifically, females appear to adopt this approach more than do males. This conclusion comports with conclusions reached by one literature review and one meta-analysis.

25.8 Prenatal Factors: Condensed Findings

As explained in Chapter 8, prenatal factors have to do with events preceding an individual's birth (or hatching). The nature of prenatal factors sometimes directly involves an individual who has not yet been born or hatched, and, other times, it involves the individual's mother. Examples of the latter have to do with mothers consuming various drugs or experiencing stress during pregnancy. While considerable research has been directed toward assessing sex differences in a variety of prenatal variables, the majority of the relevant studies have involved controlled experiments with non-humans. Therefore, this section is very brief.

The only prenatal variable for which at least ten studies of humans were located is presented in Table 25.8.1. There one can see that 14 studies sought to determine if mothers who were carrying male offspring had higher levels of testosterone than mothers who were carrying female offspring. Among the reasons for suspecting that there might be a difference in this regard is that, at least following the first month or so of pregnancy, male fetuses begin to form testes (instead of ovaries) which begin to produce relatively high concentrations of testosterone. Adding complexity to this issue is the fact that the mother produces testosterone of her own (albeit far less than an adult male), and the mother and fetus are sharing all types of chemicals across the placenta throughout pregnancy. One can see that 10 of the 14 relevant studies reported no significant sex differences in maternal levels of testosterone related to the sex of her fetus. The remaining four studies, however, concluded that mothers with male offspring had higher levels of testosterone in their blood system than did mothers carrying female offspring.

25.9 Perceptual Abilities and Motor Functioning: Condensed Findings

The ability to move about in a coordinated fashion requires both the ability to perceive one's environment and to execute actions in response to those perceptions. Are there sex differences regarding either of these abilities? This question is the focus of Chapter 9, which is the first chapter in the second volume of this book. The current section of this chapter

Table 25.8.1 Sex of offspring relative to maternal hormone levels

Variable	*Orig. table*	*Number of sex difference findings (and consistency scores)*							*Overview*	*Countries*	*Time range*	*Non-human*
		Fetal	*I/T*	*Child*	*Adol*	*Adult*	*WAR*	*All ages*				
Sex of offspring and level of testosterone in the mother	8.5.4.1	4M 10x (28.6)	–	–	–	–	–	4M 10x (28.6)	–	3	1971–2020	1M

provides a condensed summary of findings regarding sex differences for traits that have at least ten relevant studies.

25.9.1 Auditory Perceptions

Two variables having to do with sex differences in perceptions of an auditory nature are listed in Table 25.9.1. The first had to do with auditory acuity, or the ability to hear relatively faint sounds. One can see in row 1 that all 13 studies of adults have concluded that females have a keener sense of hearing than do males. However, most studies of infants and toddlers have indicated that there are no significant sex differences in this regard.

Row 2 deals with a phenomenon known as the *right ear advantage*. The concept reflects the fact that, because the left hemisphere tends to be dominant for speech and language processing and that this hemisphere primarily monitors auditory stimuli entering the right ear, most people appear to have more of a right ear bias than a left ear bias when trying to interpret linguistic stimuli. As shown in this row, 15 studies of sex differences concluded that males have more of a right ear advantage than females, although nine other studies concluded that there were no significant sex differences (along with two studies reporting more of an advantage for females).

25.9.2 Tactile Perceptions

Findings from studies of sex differences in tactile perceptions are summarized in Table 25.9.2. Row 1 has to do with tactile discrimination, i.e., the ability to accurately identify objects simply by touching them. One can see that most studies have found no significant sex differences in this ability.

Tactile sensitivity has to do with the ability to detect contact with objects such as a feather or a puff of air on one's skin. As summarized in row 2, findings from 9 out of 13 studies suggest that females are more tactilely sensitive than are males.

25.9.3 Taste and Olfactory Perceptions

Sex differences in the ability to taste and smell for which at least ten studies were found are summarized in Table 25.9.3. Row 1 has to do with taste acuity, i.e., the ability to taste the presence of various substances at low levels of concentration. One can see that 25 out of 29 studies concluded that females appear to be more taste sensitive than males.

Table 25.9.1 Auditory perceptions

Variable	Orig. table	Number of Sex Difference Studies (and Consistency Scores)						Overview	Countries	Time range	Non-human
		I/T	Child	Adol	Adult	WAR	All ages				
1. Auditory acuity	9.1. 1.1	4F 11x 1M (23.5)	5F	–	13F (100.0)	2F	24F 11x 1M (64.9)	–	2	1931–2001	–
2. Right ear advantage	9.1. 1.8	–	5M 2x 2F	–	10M 7x (58.8)	–	15M 9x 2F (53.6)	x: 1Met	7	1963–2005	–

Table 25.9.2 Tactile perceptions

Variable	Orig. table	Number of Sex Difference Studies (and Consistency Scores)						Overview	Countries	Time range	Non-human
		I/T	Child	Adol	Adult	WAR	All ages				
1. Tactile discrimination	9.1. 2.1	–	–	–	3F 4x 1M	2x 1M	3F 6x 2M (23.1)	F: 1Rev x: 1Rev	3	1978–1987	–
2. Tactile sensitivity	9.1. 2.2	7F 2x 1M (63.6)	–	–	2F 1x	–	9F 3x 1M (64.3)	F: 1Rev x: 1Rev	2	1893–1981	–

Table 25.9.3 Taste and olfactory perceptions

Variable	Orig. table	Number of Sex Difference Studies (and Consistency Scores)						Overview	Countries	Time range	Non-human
		I/T	Child	Adol	Adult	WAR	All ages				
1. Taste acuity or sensitivity	9.1. 3.1	1F	2F	–	19F 3x 1M (79.2)	3F	25F 3x 1M (83.3)	F: 1Rev	9 (1)	1888–2018	–
2. Olfactory acuity or sensitivity	9.1. 3.2	1F	1F	1F	43F 7x(86.0)	10F 3x(76.9)	56F 10x (84.8)	F: 1Met	7 (6)	1899–2003	Mixed
3. Odor identification	9.1. 3.3	–	4F 4x	2x	26F 10x (72.2)	2F 3x	32F 19x (62.7)	F: 2Rev 2Met x:1Met	16 (3)	1974–2019	–

In row 2, studies of sex differences in olfactory acuity are summarized. Sixty-six findings were located, 56 of which concluded that females are more sensitive than males, with the remaining ten studies reporting no significant differences.

The third row in this summary table has to do with the ability to accurately identify various odors. Out of 51 findings, 32 indicated that females were more accurate than males, while the remaining 19 reported no significant differences.

25.9.4 Visual Perceptions

In Table 25.9.4, research findings regarding sex differences in three aspects of visual perceptions are summarized. Row 1 has to do with visual acuity. It shows that all but one of the ten identified studies concluded that males have greater visual acuity than do females.

The ability to discriminate or recognize colors were investigated in 35 studies. As shown in row 2, every one of these studies concluded that females were better able to discriminate and recognize colors than were males.

Thirty-one studies of perceptual speed in visual discrimination were found. Twenty-six of these studies concluded that females were faster than males in performing visually based perceptual tasks.

25.9.5 Other Perceptions (Except Pain)

Table 25.9.5 has to do with two types of perceptions not yet covered. Row 1 deals with estimating or perceiving environmental hazards, usually in the form of physical injury risks. Out of 35 studies, 34 indicated that females perceive these risks to be higher than is the case for males.

Twenty-four studies were located regarding possible sex differences in the tendency to confuse right and left. In row 2, one can see that all but one of these studies concluded that females reported having such confusion more often than males.

25.9.6 Pain Perceptions and Tolerance

As shown in Table 25.9.6, three rows pertain to pain perceptions and tolerances. Row 1 summarizes findings of sex differences in pain perceptions that are predominantly experimental in nature. Out of a total of 112 findings, 85 indicated that females perceived pain earlier or at lower intensity levels than was the case for males. As indicated in row 2, this was also true of pain perceptions of various specific types. While the consistency scores for pain perceptions are substantially below the 95.0 cutoff, it is worth mentioning that most studies have found circulating testosterone

Table 25.9.4 Visual perceptions

Variable	*Orig. table*	*Number of Sex Difference Studies (and Consistency Scores)*						*Overview*	*Countries*	*Time range*	*Non-human*
		I/T	*Child*	*Adol*	*Adult*	*WAR*	*All ages*				
1. Visual acuity	9.1. 4.1	–	1M	–	6M 1x	2M	9M 1x(90.0)	M: 1Rev	3	1885–1999	–
2. Color discrimination/ recognition (naming colors)	9.1. 4.5	1F	4F	6F	21F (100.0)	3F	35F(100.0)	–	7 (1)	1903–2004	–
3. Perceptual speed in visual discrimination	9.1. 4.7	–	4F 1x 1M	10F 1M(83.3)	12F 2M(75.0)	–	26F 1x 4M (74.3)	x: 1Met	4	1911–2002	–

Table 25.9.5 Other perceptions (except pain)

Variable	*Orig. table*	*Number of Sex Difference Studies (and Consistency Scores)*						*Overview*	*Countries*	*Time range*	*Non-human*
		I/T	*Child*	*Adol*	*Adult*	*WAR*	*All ages*				
1. Estimating/ perceiving environmental hazards	9.1. 5.2	–	6F 1x	3F	23F (100.0)	2F	34F 1x (97.1)	F: 1Rev	6	1982–2007	–
2. Right-left confusion	9.1. 2.2	–	–	–	20F 1x (95.2)	3F	23F 1x (95.8)	–	8	1973–2018	–

Table 25.9.6 Pain perceptions and tolerance

Variable	*Orig. table*	*Number of Sex Difference Studies (and Consistency Scores)*						*Overview*	*Countries*	*Time range*	*Non-human*
		I/T	*Child*	*Adol*	*Adult*	*WAR*	*All ages*				
1. Pain perception/ detection sensitivity in general	9.2. 1.1	1F	4F 2x	–	62F 16x 5M (70.5)	18F 4x (81.8)	85F 22x 5M (72.7)	F: 4Rev	10	1940–2004	9F Rodents
2. Perception of pain of a specific type	9.2. 1.2	1M	1x	–	26F 13x 2M (60.5)	6F	32F 13x 4M (60.4)	–	4	1954–2006	Mixed
3. Pain catastrophizing	9.2. 2.1	–	–	1F	15F 1x (93.8)	2F	18F 1x (94.7)	–	5	1994–2012	–

levels negatively correlated with pain perception (review: Sorge and Totsch 2017). This would lead one to expect females to perceive pain more and to tolerate it less.

In row 3, sex differences in pain catastrophizing are summarized. One can see that all but 1 of the 19 studies concluded that females catastrophize more than do males.

25.9.7 Basic Reflexes and Repetitive Motor Coordination

Table 25.9.7 turns attention away from perceptions and toward various types of motor functioning. Row 1 shows that 22 studies of motor coordination in general concluded that males surpass females, although four other studies reported no differences and six additional studies concluded that females surpassed males.

In the case of overall athletic ability, row 2 shows that findings are mixed for children. However, for adolescents and adults, all studies agree that males surpass females.

Findings from studies of sex differences in performance on the finger tapping test (FTT) are summarized in row 3. The studies have reached inconsistent conclusions regarding this test of basic digit coordination.

The last two rows have to do with throwing accuracy and throwing velocity. One can see that most studies have concluded that males are more accurate when throwing objects at specific targets. Regarding throwing velocity, all studies agree that, on average, males surpass females.

25.9.8 Coordination of the Lower Extremities

Sex differences in various abilities of the legs to function in a coordinated fashion are summarized in Table 25.9.8. Row 1 has to do with basic balancing ability. Ten out of the 11 studies have concluded that females surpass males in this regard. However, row 2 indicates that extended balance maintenance does not seem to differ according to sex. It should be noted that the difference between these two aspects of balancing appears to be subtle, especially in terms of measurement.

Two aspects of jumping, i.e., jumping distance and jumping height are shown in rows 3 and 4 regarding sex differences. One can see that all studies agree that both aspects of jumping are greater for males than for females, even in studies that statistically controlled for sex differences in height.

The last row in this table has to do with the frequency with which people experience falls, usually when some significant degree of injury is involved. One can see that findings have been mixed, although the

Table 25.9.7 Basic reflexes and repetitive motor coordination

Variable	*Orig. table*	*Number of Sex Difference Studies (and Consistency Scores)*						*Overview*	*Countries*	*Time range*	*Non-human*
		I/T	*Child*	*Adol*	*Adult*	*WAR*	*All ages*				
1. Motor coordination in general	9.3. 2.1	3M 1x 2F	8M 1x 2F (61.5)	6M 1F	5M 2x 1F	–	22M 4x 6F(57.9)	M: 1Met	7	1912–2007	2F Rodent
2. Athletic ability/ performance in general	9.3. 2.4	–	25M 10x(71.4)	18M (100.0)	12M (100.0)	4M	59M 10x(85.5)	M: 5Met	15 (4)	1936–2015	1M Horse
3. Finger-tapping test performance	9.3. 2.9	–	3F	1F	15M 2x 4F (60.0)	2M	17M 2x 8F(48.6)	M: 1Rev M: 1Met	6	1941–2005	–
4. Throwing accuracy (targeting)	9.3. 2.11	–	11M 1x(91.7)	3M	20M 1x 1F (87.0)	1x	34M 3x 1F(87.2)	M: 1Met	8	1929–2017	1x Primate
5. Throwing velocity or distance	9.3. 2.12	2M	11M(100.0)	8M	5M	3M	29M (100.0)	M: 1Met	8	1947–2019	–

Table 25.9.8 Coordination of the lower extremities

Variable	*Orig. table*	*Number of Sex Difference Studies (and Consistency Scores)*						*Overview*	*Countries*	*Time range*	*Non-human*
		I/T	*Child*	*Adol*	*Adult*	*WAR*	*All ages*				
1. Balancing ability	9.3.3.1	–	7F 1x	2F	1F	–	10F 1x(90.9)	M: 1Met; x: 1Met; F: 1Met	6	1970–2017	–
2. Balance maintenance	9.3.3.2	–	1F 2x	–	4M 3x 1F	–	4M 5x 2F(30.8)	M: 1Met; x: 1Met	3	1969–2005	–
3. Jumping distance	9.3.3.6	–	8M	5M	3M	–	16M (100.0)	–	7 (1)	1940–2018	–
4. Jumping height	9.3.3.7	–	–	3M	21M (100.0)	–	24M (100.0)	M: 2Rev	12	1978–2019	–
5. Frequency of falls	9.3.3.9	–	–	–	45F 19x 2M (66.2)	3M 1F	46F 19x 5M (61.3)	F:1Rev x: 1Met	24 (2)	1977–2019	–

majority of studies have found females experiencing more falls than males, although quite a few studies also reported no significant sex differences.

25.9.9 Fine-Motor Skills of the Upper Extremities

Table 25.9.9 pertains to various fine-motor skills involving the hands and fingers. In row 1, studies of sex differences in manual dexterity in general are examined. One can see that the majority of studies have concluded that females surpass males in terms of overall manual dexterity.

A considerable number of specialized studies of fine-motor skills have relied on what is known as the *pegboard test*. Row 2 shows that nearly all studies using this test have concluded that females are better than males.

Seventeen studies of penmanship have compared males and females using judges who are not aware of whether the writing sample was provided by a male or a female. In row 3, one can see that the penmanship of females was considered superior to that of males.

Sex differences in handwriting speed have been assessed in 14 studies. One can see in row 4 that all but one of these studies concluded that females were able to write faster than were males.

25.9.10 Athletic Skills

Just one type of athletic skill was located for which ten or more studies of sex differences were found. It had to do with running speed. One can see in Table 25.9.10 that all 29 studies agreed that males ran faster on average than did females.

25.9.11 Sidedness

Sidedness refers to a tendency to favor one side of the body over the other when one is engaged in various tasks. As shown in Table 25.9.11, four categories of sidedness have been investigated for which at least ten pertinent studies of sex differences were located.

Findings from 258 studies of sex differences in handedness were located. Row 1 shows that 187 of these studies concluded that males were more likely to have non-right-handedness than females. For a number of other non-right sidedness preferences other than handedness, one can see in row 2 that the evidence of any sex differences has been mixed.

The last two rows in this table have to do with left-sidedness in carrying styles. Row 3 involves 20 studies of the tendency to carry (or cradle) infants on the left side of the body. All but three of the located studies concluded that females were more likely to do so than males. Of 11 studies of carrying other types of objects, such as books and bags,

Table 25.9.9 Fine-motor skills of the upper extremities

Variable	*Orig. table*	*Number of Sex Difference Studies (and Consistency Scores)*						*Overview*	*Countries*	*Time range*	*Non-human*
		I/T	*Child*	*Adol*	*Adult*	*WAR*	*All ages*				
1. Manual digital dexterity/fine-motor skills (except pegboard test)	9.3. 4.1	5F	20F 1x 5M (64.5)	7F 2x 1M (63.6)	13F 3x 3M (59.1)	2F	47F 6x 9M (66.2)	F: 2Met	12	1911–2021	–
2. Fine-motor skills using the pegboard test	9.3. 4.2	–	2F	1F	20F 2x(90.9)	3F	26F 2x(92.9)	–	5	1985–2009	–
3. Handwriting legibility (penmanship)	9.3. 4.3	–	8F	2F	3F	4F	17F (100.0)	–	3 (2)	1931–2016	–
4. Handwriting speed	9.3. 4.4	–	5F	3F	1F	4F 1x	13F 1x(92.9)	–	5	1933–2015	–

Table 25.9.10 Athletic skills

Variable	*Orig. table*	*Number of Sex Difference Studies (and Consistency Scores)*						*Overview*	*Countries*	*Time range*	*Non-human*
		I/T	*Child*	*Adol*	*Adult*	*WAR*	*All ages*				
Running speed	9.3. 5.1	–	7M	2M	20M (100.0)	–	29M (100.0)	M: 1Met	34 (5)	1947–2018	–

Table 25.9.11 Sidedness

Variable	*Orig. table*	*Number of Sex Difference Studies (and Consistency Scores)*						*Overview*	*Countries*	*Time range*	*Non-human*
		I/T	*Child*	*Adol*	*Adult*	*WAR*	*All ages*				
1. Non-right-handedness	9.3.6.1	1M 1F	21M 6x(77.8)	15M 5x(75.0)	116M 46x 4F (68.2)	34M 9x(79.1)	187M 66x 5F (77.1)	M: 3Rev2Met	34 (5)	1931–2021	complex
2. Non-right-earedness, eyedness, facedness, footedness, and mouthedness	9.3.6.3	1M	1M	–	6M 2x 3F(42.8)	1M 1x 2F	9M 3x 5F(40.9)	–	7	1972–2019	1M Rodent
3. Left-sidedness in infant cradling/ carrying infants	9.3.6.6	–	–	3F	14F 3x(82.4)	–	17F 3x(85.0)	F: 1Met	7 (3)	1976–2021	–
4. Left-sidedness in carrying objects other than infants	9.3.6.7	–	–	–	5F 6x(45.5)	–	5F 6x(45.5)	–	3	1981–2019	–

six studies found no significant sex difference while the remaining five concluded that females tend to favor the left side.

25.9.12 Voice

Table 25.9.12 contains a summary of findings on sex differences in five traits having to do with voice and basic speech. Row 1 has to do with vocalization made from infancy through late childhood. One can see that all 11 studies of these vocalizations (such as cooing and babbling) have reported significant sex differences favoring baby girls. However, thereafter, findings of sex differences have been mixed.

The concept of *fundamental frequency* refers to the opposite of what is meant by having a low deep voice, i.e., as fundamental frequency goes up, the depth of one's voice becomes less. As shown in row 2, while there appear to be no significant sex differences before puberty, afterward, all findings point toward greater fundamental frequency among females.

In row 3, findings of sex differences in laryngeal airway resistance are summarized. Half of the 12 pertinent studies reported more resistance among males, while most of the remaining half have found no significant sex difference.

Rows 4 has to do with nasality of one's voice during speech. All 11 studies indicated that females exhibit more nasality than do males.

25.10 Emotional Factors: Condensed Findings

This portion provides a summary of research findings having to do with sex differences in a broad spectrum of emotional feelings, as covered more extensively in Chapter 10. The feeling covered include negative emotions such as anger, disgust, fear, jealousy, sadness, anxiety, and stress to positive emotions such as humor appreciation and empathy. Studies of sex differences in emotional *expressions* are also reviewed.

25.10.1 Emotionality in General

Table 25.10.1 has to do with sex differences in overall tendencies to react to experiences in emotional terms. As one can see, the majority of studies have concluded that females are more likely than males to react to most emotion-provoking situations emotionally. Most of the exceptions have to do with anger.

25.10.2 Anger and Disgust

Sex differences in tendencies to feel anger are discussed and summarized in Table 25.10.2. As shown in row 1, 34 studies of anger in general have

Table 25.9.12 Voice

Variable	*Orig. table*	*Number of Sex Difference Studies (and Consistency Scores)*						*Overview*	*Countries*	*Time range*	*Non-human*
		I/T	*Child*	*Adol*	*Adult*	*WAR*	*All ages*				
1. Vocalizing behavior	9.3.7.1	11F (100.0)	9F 10x (47.4)	6F 1x	–	2F 6x 1M	28F 17x 1M (70.0)	F: 1Rev	3	1930–1999	–
2. Fundamental frequency when vocalizing (the opposite of having a low deep voice)	9.3.7.4	–	7x	6F	41F(100.0)	–	47F 7x(87.0)	F: 1Rev	11	1950–2014	–
3. Laryngeal airway resistance	9.3.7.5	–	–	–	6M 4x 2F(42.9)	–	6M 4x 2F(42.9)	–	7	1971–2010	–
4. Nasality of voice during speech	9.3.7.6	1F	2F	–	2F	6F	11F (100.0)	–	2	1970–1996	–

Table 25.10.1 Emotionality in general

Variable	*Orig. table*	*Number of Sex Difference Studies (and Consistency Scores)*						*Overview*	*Countries*	*Time range*	*Non-human*
		I/T	*Child*	*Adol*	*Adult*	*WAR*	*All ages*				
Emotionality in general	10.1.1.1	4F 1x 2M	8F 1M	3F	48F 2x 4M (82.8)	5F	68F 3x 7M (80.0)	F: 4Rev	1 (4)	1910–2020	–

Table 25.10.2 Anger and disgust

Variable	*Orig. table*	*Number of Sex Difference Studies (and Consistency Scores)*						*Overview*	*Countries*	*Time range*	*Non-human*
		I/T	*Child*	*Adol*	*Adult*	*WAR*	*All ages*				
1. Anger in general, feelings of	10.2.1.1	2M	4M 2x 1F	4M 6x 1F(33.3)	12M 21 x 11F (21.8)	1F 5x	22M 34x 14F(26.2)	x: 1Rev	14 (3)	1927–2018	–
2. Being vengeful or spiteful	10.2.1.5	–	1M	–	10M (100.0)	–	11M (100.0)	–	4	1992–2017	–
3. Hostility, feelings of	10.2.1.6	–	1M	1F	7M 3x 7F(29.2)	4M	12M 3x 8F(38.7)	–	9 (3)	1934–2021	–

found no significant sex differences, but when sex differences are reported, 22 studies found males feeling anger more, while the remaining 14 found more females having feelings of anger more.

Being vengeful refers not only to feelings of anger, but with these feeling being accompanied by desires to retaliate. One can see in row 2 that males appear to have vengeful feelings more than do females.

Hostility usually refers to long-term feelings of anger and resentment toward others. Row 3 indicates that there are no consistent sex differences in feelings of hostility, although the types of circumstances typically provoking these feelings may differ between males and females.

25.10.3 Fearfulness

Findings from studies of fearfulness are summarized in Table 25.10.3. Studies of fearfulness in general are provided in row 1. This row leads one to conclude that nearly all studies have indicated that females experience fear more often than do males for all age groupings, with the possible exception of infants and toddlers. In the case of fearing contracting AIDS or HIV infections, however, row 2 shows that most studies have found males being more fearful than females.

Twelve studies of sex differences in fear of animals of various kinds. Row 3 shows that all of these studies reached the conclusion that females have more fear of this kind than do males.

Are there sex differences in fear of blood or to give blood? Thirteen relevant studies were located, 12 of which indicated that females have a greater fear of this type than do males (row 4).

Many studies have assessed people's fear of dying (also known as *death anxiety*). In row 5, one can see that 94 out of a total of 124 studies have concluded that females fear death more, with 27 of the remaining 30 studies reporting no significant sex difference.

Another type of specific type of fear has to do with falling. Row 6 shows that 32 studies of this type of fear have concluded that females have this fear more than do males. The remaining three studies found no significant differences.

Sex differences in fear of being the victim of crime have received a great deal of research attention. One can see in row 7 that all but 2 of the 59 studies concluded that females have greater degrees of fear than do males.

The final row in this table has to do with fear of being negatively evaluated by others. All ten of the pertinent studies concluded that females have more fear of this type than do males.

Table 25.10.3 Fearfulness

Variable	*Orig. table*	*Number of Sex Difference Studies (and Consistency Scores)*						*Overview*	*Countries*	*Time range*	*Non-human*
		I/T	*Child*	*Adol*	*Adult*	*WAR*	*All ages*				
1. Fearfulness in general	10.2.2.2	3F 1M	28F (100.0)	19F (100.0)	32F (100.0)	14F (100.0)	96F 1M(**98.0**)	F: 2Rev	16 (2)	1928 2014	Com-plex
2. Fear of AIDS or HIV infection	10.2.2.3	–	–	–	9M 3x(75.0)	–	9M 3x(75.0)	–	2	1988–2004	–
3. Fear of animals	10.2.2.4	–	2F	–	6F	4F	12F(**100.0**)	F: 1Rev	3 (1)	1983–2012	–
4. Fear of blood or giving blood	10.2.2.5	–	1F	1F	10F 1x(90.9)	–	12F 1x(92.3)	–	3 (1)	1961–2009	–
5. Fear of dying	10.2.2.8	–	–	7F 2x	87F 25x 3M (73.7)	–	94F 27x 3M (74.0)	F: 5Rev x: 1Rev	22 (3)	1961–2021	–
6. Fear of falling	10.2.2.9	–	–	–	32F 3x(91.4)	–	32F 3x(91.4)	F: 3Rev F: 1Met	12 (2)	1994–2019	–
7. Fear of crime victimization	10.2.2.12	–	–	1F	11F 2x(84.6)	45F (100.0)	57F 2x(**96.6**)	–	8 (1)	1974–2020	–
8. Fear of being negatively evaluated by others	10.2.2.17	–	3F	4F	3F	–	10F(**100.0**)	F: 1Rev	1	1935–2005	–

25.10.4 Jealousy and Envy

Jealousy and envy, along with two fairly specific types of jealousy, are given attention in Table 25.10.4 regarding possible sex differences. Concerning jealousy and envy in general, findings have been mixed. Row 1 shows that about twice as many studies indicate that females report having these emotional experiences more than do males.

The two types of jealousy that have received substantial research attention regarding sex differences both have to do with mating behavior. As shown in row 2, nearly all studies that found that jealousy regarding the possibility of a mate falling in love with someone else is greater for females than for males. On the other hand, according to row 3, most studies have indicated that experiencing jealousy after suspecting or knowing that their mate has had sex with someone else is more common among males than among females.

25.10.5 Sadness and Bereavement

Three forms of sadness have been studied sufficiently to be considered as possible universal sex differences. Row 1 of Table 25.10.5 pertains to general feelings of sadness. It shows that all 14 studies concluded that this feeling was more common among females than among males.

Research pertaining two types of bereavement (or grief) is summarized in rows 2 and 3. Regarding the death of a child, most studies indicate that females feel more grief than do males. In the case of the death of a spouse, findings are quite mixed regarding any sex differences.

25.10.6 Stress and Anxiety

Table 25.10.6 has to do with stress and anxiety in general, followed by ten different aspects (or forms) of stress and anxiety. A very large number of studies of stress and anxiety in general have been reported. Row 1 shows that the vast majority of these studies have reported more stress and anxiety among females than males, although quite a few studies failed to detect any significant sex differences.

Regarding stress and anxiety associated with some type of traumatic event, 23 studies were located. Row 2 shows that most of these studies indicate that the associations are usually stronger for females than for males. In rows 3 and 4, one can see that similar findings have been reported for stress and anxiety linked to traumatic sexual experiences and social relationships. Stress associated with providing care to others (usually loved ones) was found to affect females more in all 19 pertinent studies making this a universal sex difference.

Table 25.10.4 Jealousy and envy

Variable	*Orig. table*	*Number of Sex Difference Studies (and Consistency Scores)*						*Overview*	*Countries*	*Time range*	*Non-human*
		I/T	*Child*	*Adol*	*Adult*	*WAR*	*All ages*				
1. Jealousy and envy in general	10.2.3.1	–	4F	1F 1x	4F 4M	–	9F 1x 4M(50.0)	–	2	1927–2017	–
2. Jealousy in response to love/emotional infidelity	10.2.3.2	–	–	–	29F 4x 1M (82.9)	–	29F 4x 1M (82.9)	F:2Rev	5 (3)	1987–2019	–
3. Jealousy in response to sexual infidelity	10.2.3.3	–	–	–	29M 9x 10F (50.0)	–	29M 9x 10F (50.0)	M:2Rev M:1Met	7 (5)	1981–2019	–

Table 25.10.5 Sadness and bereavement

Variable	*Orig. table*	*Number of Sex Difference Studies (and Consistency Scores)*						*Overview*	*Countries*	*Time range*	*Non-human*
		I/T	*Child*	*Adol*	*Adult*	*WAR*	*All ages*				
1. Sadness (dysphoria) in general	10.2.4.1	–	–	6F	8F	–	14F (100.0)	–	2 (2)	1986–2009	–
2. Bereavement/ grief after death of a child	10.2.4.2	–	–	–	8F 3x(72.7)		8F 3x(72.7)	F: 1Rev	4 (1)	1992–2013	–
3. Bereavement/ grief after death of a spouse	10.2.4.3	–	–	–	3M 4x 3F(23.1)	–	3M 4x 3F(23.1)	M: 2Rev F: 1Ref	3	1971–2005	–

Table 25.10.6 Stress and anxiety

Variable	*Orig. table*	*Number of Sex Difference Studies (and Consistency Scores)*						*Overview*	*Countries*	*Time range*	*Non-human*
		I/T	*Child*	*Adol*	*Adult*	*WAR*	*All ages*				
1. Stress/anxiety in general	10.2.5.1	2F	35F 14x 1M (68.6)	65F 4x 1M (91.5)	131F 4x 6M (89.1)	22F 4x(84.6)	255F 45x 8M (80.7)	F: 1Gen 4Rev 2Met	23 (10)	1936–2021	Com-plex
2. Stress/anxiety associated with some type of traumatic event	10.2.5.3	–	6F	2F	7F 1x 3M(50.0)	4F	19F 1x 3M (73.1)	F: 1Met	3	1970–2006	–
3. Stress/anxiety associated with sexuality	10.2.5.9	–	–	–	9F 1M(81.8)	–	9F 1M(81.8)	F: 1Met	2	1961–1998	–
4. Stress/anxiety associated with social relationships	10.2.5.10	–	–	4F	5F 1x 2M	–	9F 1x 2M(64.3)	–	3 (1)	1989–2014	–
5. Stress/anxiety associated with providing care to others	10.2.5.11	–	–	–	19F(**100.0**)	–	19F(**100.0**)	F: 1Met	2	1961–1998	–
6. Stress/anxiety associated with work (outside the home)	10.2.5.12	–	–	–	20F 1x 2M (80.0)	–	20F 1x 2M (80.0)	–	5 (2)	1975–2005	–
7. Stress/anxiety associated with finances or being unemployed	10.2.5.13	–	–	–	6M 4x 2F(42.9)	–	6M 4x 2F(42.9)	–	3	1984–2004	–
8. Stress/anxiety associated with technology	10.2.5.14	–	1F	4F 1x 1M	27F 1x(**96.4**)	1F	33F 2x 1M (89.2)	F: 2Met	5	1985–2006	–
9. Math anxiety	10.2.5.17	–	1F	9F 1x(90.0)	–	–	10F 1x(90.9)	F: 1Met	3 (2)	1978–2019	–
10. Test anxiety	10.2.5.18	–	4F 1x	17F 2x(89.5)	35F 4x(89.7)	–	56F 7x(88.9)	F: 1Met	9 (5)	1959–2016	–
11. Separation anxiety	10.2.5.19	4M	1F 1M	2F	5F	2F	10F 5M(50.0)	–	4	1963–2008	–

Sex differences in stress associated with work (outside the home) and with financial matters (including being unemployed) are summarized in rows 6 and 7. One can see that findings for work-related stress indicate that females usually experience more than males. However, in the case of stress dealing with financial matters and/or being unemployed, a slight majority of studies indicate that males usually have more than females.

Stress linked to technology has been studied substantially. Based on findings from 36 studies, 33 indicate that females experience more stress than do males (row 8).

Two forms of academic-related anxiety are summarized in rows 9 and 10 regarding possible sex differences. One can see that all but one of the 11 studies of math anxiety concluded that females experience it more than males. Similarly, 56 out of 63 studies of test anxiety have indicated that females are more prone to have these feelings than males.

The final row in this table pertains to separation anxiety. Row 11 shows that 10 out of the 15 studies revealed that this form of anxiety is more common among females, although the other five studies reported greater prevalence for males.

25.10.7 Other Negative Emotions

Research findings regarding sex differences for four additional negative emotions are identified in Table 25.10.7. Sex differences in having emotions such as guilt, embarrassment, and shame are summarized in row 1. It indicates that 36 out of 42 studies have concluded that females experience these types of emotions more than do males. Feelings of guilt and regrets specifically having to do with sexual matters are summarized in row 2 rather than row 1. For this type of emotion, 12 studies were located, 7 of which indicated that the experiences are more for females than for males.

Row 3 has to do with feelings of helplessness and discouragement. Of the 22 relevant studies, nine indicate that these are experiences that occur more often for females than for males.

The last row in this table has to do with feelings of loneliness. Nine out of 15 pertinent studies concluded that males have such feelings more often than females, although five of the 15 studies reached the opposite conclusion.

25.10.8 Residual Aspects of Emotions

Table 25.10.8 pertains to two aspects of emotions that are difficult to classify. One is humor appreciation. As shown in row 1, 16 out of 28 studies of sex differences have concluded that males appreciate humor more, while the remaining 12 studies were evenly divided between females being more appreciative and no significant differences.

Table 25.10.7 Other negative emotions

Variable	*Orig. table*	*Number of Sex Difference Studies (and Consistency Scores)*						*Overview*	*Countries*	*Time range*	*Non-human*
		I/T	*Child*	*Adol*	*Adult*	*WAR*	*All ages*				
1. Guilt, embarrassment, and shame in general, feelings of	10.2.6.5	1F	7F 1x 1M	7F	20F 3x 1M (80.0)	1F	36F 4x 2M (81.8)	F: 1Rev 2Met x: 1Met	5 (1)	1963–2005	–
2. Guilt and regrets having to do with sexual experiences	10.2.6.6	–	–	3F	3F 3x 1M	1F 1M	7F 3x 2M(50.0)	–	2	1979–2020	–
3. Helplessness and discouragement, feelings of	10.2.6.7	–	3F	5F	5F 8x(38.5)	1x	13F 9x(59.1)	F: 1Rev	1	1975–1989	–
4. Loneliness, feelings of	10.2.6.-10	–	1M	4F 1M	7M 1F	1x	9M 1x 5F(45.0)	M: 1Met	5 (1)	1978–2012	–

Table 25.10.8 Residual aspects of emotions

Variable	*Orig. table*	*Number of Sex Difference Studies (and Consistency Scores)*						*Overview*	*Countries*	*Time range*	*Non-human*
		I/T	*Child*	*Adol*	*Adult*	*WAR*	*All ages*				
1. Humor appreciation	10.2.8.2	–	2M	1M 1F	13M 6x 5F (44.8)	–	16M 6x 6F (47.1)	M: 1Rev; x: 1Rev	7	1937–2012	–
2. Empathy, feelings of	10.2.8.3	4F	25F 13x(65.8)	19F 1x(95.0)	54F 4x(93.1)	7F	109F 18x(85.8)	F: 1Rev 1Met; x: 1Rev	12 (3)	1936–2019	–

Empathy refers to the ability to virtually feel the emotions experienced by others. Well over 100 studies of sex differences were located. In no case did any of the findings indicate that males were more empathetic than females, but a small minority of studies have failed to identify any significant sex differences.

25.10.9 Emotional Expressions

Table 25.10.9 pertains to emotional expressions rather than emotional feelings. One can see in row 1 that a sizable number of studies have been conducted regarding sex differences in general. Of the 96 located studies, 84 concluded that females are more emotionally expressive than are males, with all of the remaining 12 studies reporting no significant differences.

Row 2 shows that over 100 studies of sex differences in crying have been published. Specifically, of the 107 total findings, 88 concluded that females cried more, with just eight indicating that males do. All of the studies suggesting that crying is more common among males involved prepubertal samples.

Expressions of anger along with overt threats of physical aggression were the focus of study in 60 different studies. Row 3 shows that 31 of these studies found males making these threats more than females, although 19 studies reached the opposite conclusion.

Who laughs more, males or females? According to the 19 studies summarized in row 4, females were found to do so more in 15 studies, with the remaining four studies evenly split between males and no significant differences.

Row 5 has to do with experiments in which both sexes were suddenly exposed to emotion-provoking stimuli, such as photographs of gruesome bodily injuries or happy play among children. Of the 24 pertinent studies, all but one concluded that females were more facially expressive than males.

25.11 Cognitive, Academic, and Intellectual Factors: Condensed Findings

Chapter 11 was devoted to assessing sex differences in terms of cognitive, academic, and intellectual factors. It revealed an especially large number of studies pertaining to these topics. The following 11 tables provide a condensed summary of the findings from these studies.

25.11.1 Academic Performance According to Grades

Academic performance is most often assessed based on grades assigned to students by their teachers. One can see in row 1 of Table 25.11.1 that 421 studies were located having to do with sex differences in the average

Table 25.10.9 Emotional expressions

Variable	*Orig. table*	*Number of Sex Difference Studies (and Consistency Scores)*						*Overview*	*Countries*	*Time range*	*Non-human*
		I/T	*Child*	*Adol*	*Adult*	*WAR*	*All ages*				
1. Emotional expressive-ness in general	10.2.9.1	3F 1x	13F 5x(72.2)	3F	62F 6x(91.2)	3F	84F 12x(87.5)	F: 1Rev 4Met	7 (2)	1972–2007	–
2. Crying	10.2.9.5	7F 10x7M (22.6)	7F 1M	4F	70F 1x(**98.6**)	–	88F 11x 8M (76.5)	F: 4Rev	36 (3)	1934–2013	–
3. Expression of anger/ threats	10.2.9.7	2M 2x 1F	7M 1x	1M	21M 7x 18F (32.8)	–	31M 10x 19F (39.2)	M: 2Rev	6 (1)	1975–2005	–
4. Laughing	10.2.9.12	–	3F	1M	12F 2x 1M (75.0)	–	15F 2x 2M (71.4)	–	3 (1)	1973–2002	M: 1Rodent; x: 1Rodent
5. Facial expressions in response to emotion-provoking stimuli	10.2.9.16	–	1F	2F	19F 1M(90.5)	1F	23F 1M(92.0)	–	3	1943–2000	–

Table 25.11.1 Academic performance according to grades

Variable	*Orig. table*	*Number of Sex Difference Studies (and Consistency Scores)*						*Overview*	*Countries*	*Time range*	*Non-human*
		I/T	*Child*	*Adol*	*Adult*	*WAR*	*All ages*				
1. Grade point average, grades in general	11.1. 1.1	–	46F 3x 2M (86.8)	167F 17x 4M (87.0)	127F 27x 27M (61.1)	1M	340F 47x 34M (74.1)	F: 1Rev 2Met	30 (4)	1910–2016	–
2. Grades in arithmetic and mathematics	11.1. 2.1	–	7F 5x 1M(74.7)	44F 13x 9M (58.7)	13F 5x 8M(38.2)	–	64F 23x 18M (52.0)	F: 2Rev 1Met	10	1911–2016	–
3. Grades in humanities and language-related courses	11.1. 2.3	–	11F (100.0)	32F 1x 1M(91.4)	4F 2x	–	47F 3x 1M(90.4)	F: 1Met	11	1935–2013	–
4. Grades in the physical and biological sciences	11.1. 2.4	–	4F 2x 1M	16F 7x 9M(39.0)	10M 4x 5F(41.7)	–	25F 13x 20M (32.1)	–	9 (1)	1933–2014	–
5. Grades in the social sciences	11.1. 2.5	–	1F	9F 3x 6M(37.5)	9M 4x 2F(52.9)	–	15M 7x 12F (32.6)	–	7 (1)	1935–2009	–
6. Grades in statistics	11.1. 2.6	–	–	–	5F 5x 2M(35.7)	–	5F 5x 2M(35.7)	F: 1Met	1	1978–1992	–

grades given to students. Eighty-one percent of these studies concluded that females have higher grade point averages (GPAs) than do males. Differences are the greatest among children and adolescents, but still substantial among adults (i.e., college students).

The remaining five rows in this table have to do with grades earned in various subject areas. In this regard, females receive higher grades in arithmetic and math (row 2), in humanities and language classes (row 3), in the physical and biological sciences (row 4), and in statistics (row 6). The only area in which males had a slight proportional lead in grades received was in the social sciences (row 5).

25.11.2 Academic Assessment Other Than Grades

In addition to grades given to students by teachers, academic performance can be assessed in terms of scores obtained on standardized tests and a few other ways. Most standardized tests that have been used in comparing males and females are tests given to students as they near graduation from high school, particularly if they intend to go to college. Table 25.11.2 provides a summary of findings on scores obtained on these standardized tests as well as on various other tests that assess academic performance other than simply in terms of grades given to students by teachers.

Row 1 has to do with scores received specifically on multiple choice tests. One can see that 12 of the 13 studies concluded that males outperform females on these tests.

In Rows 2, 3, and 4, summaries are provided specifically for undergraduate college admissions exams, especially the Scholastic Achievement Test (SAT) and the American College Test (ACT). The first of these rows involves overall test scores, the vast majority of which favor males. However, when scores specifically pertaining to the verbal (or language) portion are considered separately, most studies still indicate that males score higher, but not to the degree exhibited for overall test scores. Regarding the math portion of these standardized tests, one finds the strongest evidence of higher scores for males.

Various standardized tests have been given to students from childhood through adulthood. One can see in row 5 that most of the scores on these tests are also higher for males.

In England, some universities award what are considered particularly promising students what are called first-class degrees when they graduate. Row 6 shows that most studies have concluded that males are more likely than females to receive these degrees.

The final row in this table has to do with sex differences in scores on standardized exams used for admission to graduate school. Most of the findings on these exams have also concluded that males score higher on average.

Table 25.11.2 Academic assessment other than grades

Variable	*Orig. table*	*Number of Sex Difference Studies (and Consistency Scores)*						*Overview*	*Countries*	*Time range*	*Non-human*
		I/T	*Child*	*Adol*	*Adult*	*WAR*	*All ages*				
1. Scores obtained on multiple choice tests	11.1.3.1	–	1M	10M 1x(90.9)	1M	–	12M 1x(92.3)	–	5	1981–1997	–
2. Undergrad college admissions exam scores overall	11.1. 3.4	–	–	30M 3x 1F (85.7)	–	–	30M 3x 1F (85.7)	M: 1Rev	3	1971–2012	–
3. Undergrad college admissions exam scores for language	11.1. 3.5	–	–	9M 7x 4F(37.5)	–	–	9M 7x 4F(37.5)	M: 1Rev	2	1957–2012	–
4. Undergrad college admissions exam scores for mathematics	11.1. 3.6	–	–	53M 4x(93.0)	–	–	53M 4x(93.0)	–	3	1957–2016	–
5. Standardized test scores except for undergrad college admissions	11.1 3.8	–	1F 3x	3M 1F	9M	1M	13M 3x 2F (65.0)	–	4 (1)	1928–2019	–
6. Awarded a first-class degree	11.1.3.9	–	–	–	19M 2x 1F (82.6)	–	19M 2x 1F (82.6)	–	2	1984–2006	–
7. Graduate school exam scores	11.1. 3.10	–	–	–	8M 4x(66.7)	–	8M 4x(66.7)	–	1	1972–2012	–

25.11.3 Additional Indicators of Academic Ability

In Table 25.11.3, five more rough indicators of academic ability and/or performance are considered. The first of these indicators, shown in row 1, involves repeating or failing to pass a grade in school. All 11 studies of this phenomenon concluded that males are more likely than females to repeat a grade in school. However, in the case of dropping out of school before graduating, row 2 shows that the findings are mixed.

Row 3 has to do with sex differences in having learning disabilities (or unusual difficulties learning in academic settings). In all 28 pertinent studies, males were found to receive such diagnoses more often than females.

Regarding reading disabilities, row 4 shows that 30 out of 33 studies concluded that males are more affected than females. And, in the case of dyslexia, row 5 indicates that 11 out of 14 studies point toward this specific form of reading disability to be more prevalent in males than in females.

25.11.4 Intelligence in General

Over 400 studies have assessed sex differences in intelligence, some dating back more than a century. As shown in row 1 of Table 25.11.4, most studies have concluded that there are no significant differences, a conclusion also reached by both literature reviews. When sex differences are found, most of them indicate that males are higher, although a substantial minority point toward higher scores for females, especially among toddlers and children.

Within-sex variability in intelligence has been examined by 45 studies. Row 2 shows that most of the evidence indicates that male intelligence is more variable than is female intelligence.

Regarding mental retardation, one can see in row 3 that the majority of findings point toward higher rates among males than among females. Row 4 indicates that being institutionalized with mental retardation is especially more common among males than among females.

25.11.5 Subcomponent Aspects of Intelligence (Except Spatial Reasoning)

In Table 25.11.5, various aspects of intelligence (except spatial reasoning) are summarized regarding possible sex differences. The first aspect is verbal (or linguistic intelligence). Row 1 shows that, out of 276 studies, 123 reported no significant sex differences, while 96 found females performing better and 57 other studies found males performing better.

Table 25.11.3 Additional indicators of academic ability

Variable	*Orig. table*	*Number of Sex Difference Studies (and Consistency Scores)*						*Overview*	*Countries*	*Time range*	*Non-human*
		I/T	*Child*	*Adol*	*Adult*	*WAR*	*All ages*				
1. Repeating (failure to pass) a grade in school	11.1.4.10	–	11M (100.0)	–	–	–	11M (100.0)	–	2	1967–2002	–
2. Dropping out of school	11.1.4.2	–	1F	5M 3x 1F	1M 1F	–	6M 3x 3F(40.0)	–	3	1949–2005	–
3. Learning disabilities or difficulties in general	11.1.4.3	2M	21M (100.0)	2M	2M	1M	28M (100.0)	M: 1Rev	4	1949–2011	–
4. Reading disabilities	11.1.4.4	–	29M 3x(90.6)	–	–	1M	30M 3x(90.9)	M: 3Rev	4 (1)	1933–2007	–
5. Dyslexia	11.1.4.5	–	6M 3x	2M	2M	1M	11M 3x(78.6)	M: 1Rev	2	1959–2017	–

Table 25.11.4 Intelligence in general

Variable	*Orig. table*	*Number of Sex Difference Studies (and Consistency Scores)*						*Overview*	*Countries*	*Time range*	*Non-human*
		I/T	*Child*	*Adol*	*Adult*	*WAR*	*All ages*				
1. Intelligence in general	11.2.1.1	5F 7x(41.7)	40M 59x 33F(24.2)	49M 44x 16F (39.2)	59M 50x 4F(50.4)	19M 18x 4F(42.2)	167M 178x 62F(41.0)	M: 7Gen 1Met x: 30Gen 5Rev 1Met	43 (6)	1891–2021	–
2. Intelligence, intra-sex variability in	11.2.1.2	–	13M 3x 1F(72.2)	15M 1x 1F (83.3)	3M 2x 1F	5M	36M 6x 3F (75.0)	M: 4Gen 4Revx:1Rev	9 (1)	1928–2021	–
3. Mental retardation	11.2.2.1	1M	13M 2x(86.7)	3M	1x	6M	23M 3x(88.5)	–	7	1911–2003	–
4. Mental retardation, institutional-ized	11.2.2.2	–	–	–	–	13M (100.0)	13M(100.0)	–	7	1897–1975	–

Table 25.11.5 Subcomponent aspects of intelligence (except spatial reasoning)

Variable	*Orig. table*	*Number of Sex Difference Studies (and Consistency Scores)*						*Overview*	*Countries*	*Time range*	*Non-human*
		I/T	*Child*	*Adol*	*Adult*	*WAR*	*All ages*				
1. Verbal (linguistic) intelligence	11.2.3.1	5F 3x	34F 67x 16M (25.6)	22F 15x 16M(31.9)	28F 29x 19M(29.5)	7F 9x 6M(25.0)	96F 123x 57M (28.8)	F: 1Met x: 1Met	24 (2)	1916–2021	–
2. Processing speed	11.2.3.2	–	2F 1x	–	8F	7F 1x	17F 2x(89.5)	F: 2Rev	4	1983–2017	–
3. Performance IQ	11.2.3.4	–	1M 7x 1F	27M 6x(81.8)	14M 9x 2F (51.9)	4M 2x	46M 24x 3F (60.5)	–	13 (4)	1953–2021	–
4. Fluid IQ	11.2.3.6	–	1M 1x	4M	5M 5x(50.0)	1M 3x	11M 9x(55.0)	x: 1Rev	10 (2)	1987–2021	–
5. Intellectual/ cognitive decline with age	11.2.3.8	–	–	–	7M 13x 6F (21.9)	3F 1x	9F 14x 7M(24.3)	–	4 (1)	1947–2008	M: 1 primate

Processing speed has to do with how quickly individuals are able to perform identification tasks (e.g., the number of letter *f*'s in the preceding paragraph) within a prescribed amount of time. One can see in row 2 that most studies have indicated that females are faster than males.

Performance IQ has to do with the ability to use information that has already been learned, whereas fluid IQ involves the ability to figure out appropriate and efficient ways of dealing with new situations. Rows 3 and 4 indicate that findings of sex differences are not consistent, although both suggest that, if there are differences, they tend to favor males, especially following puberty.

The final entry in this table has to do with intellectual or cognitive decline with age. One can see that findings have been very inconsistent regarding possible sex differences.

25.11.6 Spatial Reasoning

A great deal of research has focused on sex differences in reasoning having to do with spatial relationships. Spatial reasoning is a rather unique aspect of intellectual reasoning for at least two reasons. First, unlike all other forms of intellectual reasoning, it can be performed with little or no input of a linguistic nature. Second, spatial reasoning can not only be assessed among humans, but also among other animal species. In fact, Table 25.11.6 presents results from findings of sex differences in human spatial reasoning as well as in other animals.

Excluding self-ratings, row 1 shows that 401 studies were located pertaining to sex differences in spatial reasoning. One can see that the vast majority of findings (328) indicated that males were better at such reasoning than were females, with all but nine of the remaining studies pointing toward no significant sex differences. More than ten studies of sex differences in spatial reasoning were located for non-humans of various species, but mainly of rodents and primates. Therefore, findings from these 13 studies are summarized in row 2. All but two of these studies reported that males surpass females in spatial reasoning using various measures.

In the last row of this table, sex differences in self-rated spatial reasoning are assessed. The row indicates that all 17 studies concluded that males on average assessed their abilities to reason in spatial terms higher than did females.

25.11.7 Various Types of (or Tests for) Spatial Reasoning

Table 25.11.7 shows that there are numerous established ways of assessing spatial reasoning. Each one provides somewhat different perspectives on varying spatial reasoning abilities.

Table 25.11.6 Spatial reasoning

Variable	*Orig. table*	*Number of Sex Difference Studies (and Consistency Scores)*						*Overview*	*Countries*	*Time range*	*Non-human*
		I/T	*Child*	*Adol*	*Adult*	*WAR*	*All ages*				
1. Spatial reasoning in general	11.2.4.1	4M 2x	53M 23x 3F(64.6)	75M 5x 2F (89.3)	165M 26x(86.4)	31M 8x 4F (66.0)	328M 64x 9F (80.0)	M: 9Rev 1Met x: 5Rev 1Met	30 (19)	1911–2021	{*See next row*}
2. *Spatial reasoning in general among non-humans*	11.2.4.1e	–	–	2M 1x	8M 1x	1M	11M 2x(84.6)	M:1Rev 1Met	–	1978–2005	–
3. Spatial reasoning ability, self-rated	11.2.4.2	–	1M	3M	9M	4M	17M (100.0)	–	5 (3)	1982–2018	–

Table 25.11.7 Various types of (or tests for) spatial reasoning

Variable	*Orig. table*	*Number of Sex Difference Studies (and Consistency Scores)*						*Overview*	*Countries*	*Time range*	*Non-human*
		I/T	*Child*	*Adol*	*Adult*	*WAR*	*All ages*				
1. Block design ability	11.2.5.1	–	5M 3x	3M 2x	10M 2x(83.3)	3M 7x 2F (21.4)	21M 14x 2F (53.8)	–	13 (1)	1953–2016	–
2. Estimating distances and projectile landings	11.2.5.3	–	2M	1M	14M (100.0)	–	17M (100.0)	–	8	1983–2015	–
3. Aptitude tests of spatial relations	11.2.5.4	–	2x	1M 1x	8M 1x	–	9M 4x(69.2)	–	2	1974–1988	–
4. Embedded figure test of field independence	11.2.5.5	1x	8M 22x 2F(23.5)	10M 9x 4F (37.0)	29M 12x 3F (61.7)	7M 7x 1F (43.8)	54M 51x 10F (43.2)	–	17 (3)	1949–2015	–
5. Rod-and-frame test of field independence	11.2.5.6	–	5M 2x	4M 3x	19M 10x(65.5)	6M 2x	34M 17x(66.7)	–	7	1912–2021	–
6. Mechanical problem solving	11.2.5.7	–	11M (100.0)	14M 1x(93.3)	11M (100.0)	3M	39M 1x(97.5)	–	7	1912–2021	–
7. Mental rotation	11.2.5.8	6M 8x(42.9)	36M 19x (65.5)	36M 11x 2F(36.5)	188M 21x(90.0)	10M 2x(83.3)	276M 61x 2F (81.0)	M: 11Met M: 4Rev	28 (5)	1947–2021	1M primate

8. Mirror image reversal discrimination	11.2.5.17	–	3M	1F	7F	–	8F 3M(72.7)	–	2	1930–2005	–
9. *Morris water maze performance non-human*	11.2.5.1-8b	–	–	2x	22M 9x 2F (62.9)	–	22M 11x 2F (59.5)	M: 1Rev	–	1984–2013	Rodent
10. Piaget's water-level test	11.2.5.21	1x	7M 4x(63.6)	9M 1x(90.0)	36M 3x(92.3)	3M	55M 9x(85.9)	–	8	1964–2012	–
11. Object location memory	11.2.5.23	–	2F 1x	1F 2x 1M	29F 8x 3M (67.4)	3F 1x	35F 12x 4M (63.6)	F: 1Met	6 (4)	1969–2007	3M 3x 2F
12. Spatial navigation ability	11.2.5.24	3M	23M 2x(92.0)	8M 2x(80.0)	61M 16x 1F (77.2)	5M 2x 2F	100M 22x 1F (80.6)	M: 3Met	11 (4)	1918–2019	{*Next row*}
13. *Spatial navigation ability in non-humans*	11.2.5.2-4c	–	1M	6M	26M 1x(96.3)	1M	34M 1x(97.1)	–	–	1915–2008	–

One method for making assessments of sex differences in spatial reasoning is known as a *block design* test. As shown in row 1, most studies have found males scoring higher on this test when compared to females, although quite a few studies have reported no significant differences.

Another type of spatial reasoning test has to do with people's ability to accurately estimate traveling time or distances that various projectiles will travel when shot out at various angles and velocities before hitting ground. Row 2 shows that all 17 studies that were based on these types of measures concluded that males are more accurate than females.

A test developed to measure people's abilities to learn mechanical and other spatially-based reasoning is known as the Differential Aptitude Test (DAT). As shown in row 3, of the 13 studies that were located in which the sexes were compared, nine studies concluded that males do better on this test, while the remaining five found no significant sex difference.

Embedded figure tests involve showing individuals line drawings in which a figure such as a rabbit might be disguised inside a sketch of a complex forest. Findings of sex differences in readily locating features of this type have received the attention of more than 100 studies. Results, summarized in row 4, have not been consistent, although most either indicated that males are better than females or there are no significant differences.

Row 5 has to do with sex differences in what is known as a rod-and-frame test. It indicates that either males are more field independent or there are no significant differences.

Forty studies of mechanical problem solving were located. In row 6, one can see that all but one of these studies concluded that males surpass females in such tasks.

The single most common spatial reasoning test is known as mental rotation; in it, individuals are asked to examine the drawing of either a 2-D or 3-D object, and then asked to imagine it presented from a different physical perspective in order to determine which of three or four other drawings the first drawing would most closely resemble. One can see in row 7 that a substantial majority of studies have concluded that males score higher on mental rotation tasks than do females, although a considerable minority of studies report no significant differences.

Sex differences in the ability to interact with the mirror image of an object has been assessed by 11 studies. Row 8 shows that eight of these studies (all among adolescents and adults) found that females were better than males at such tasks. However, the remaining three studies (all involving children) reached the opposite conclusion.

At least 35 studies of rodents (primarily rats) have utilized the Morris water maze, a maze that has been used to assessed the ability to navigate in a water environment. In row 9, one can see that 22 of these studies concluded that males are quicker than females at traversing this maze, although 11 studies found no significant sex difference, while two reported quicker maze traversing among females.

Piaget's water-level test has been widely used to assess people's ability to judge how the surface level of water would appear in a glass as the glass is being tilted. As shown in row 10, 55 out of 64 studies have concluded that males judge more accurately than do females, with the remaining nine studies reporting no significant differences.

Object memory location tasks involve showing people pictures of various objects arranged in a specific location, then waiting a few minutes and showing them the same objects, some of which have been moved to a different location, and determining if the changes can be identified. Many studies of this type of spatial reasoning task have been conducted with regard to sex differences. Row 11 shows that, out of 51 located studies, 35 concluded that females have more accurate location recall than do males, with 12 of the remaining studies reporting no significant sex differences and four finding that males are more accurate.

The last two rows in this table pertain to spatial navigation, such as in various types of mazes. Row 12 is based on 123 studies of humans, while row 13 is derived from 35 studies of non-humans (mainly rodents). In the case of humans, 100 studies concluded that males were better than females, while all but one of the 23 exceptions indicating that there were no significant differences. Regarding the studies of non-humans, 34 concluded that males were better than females, while the remaining one study reported no significant difference.

25.11.8 Methods Used in Spatial Navigation

Table 25.11.8 has to do with assessing sex differences in the methods used by both sexes when seeking to navigate space. These assessments are based on interviewing research participants, typically following their performing a spatial navigation task. Basically, two methods have been identified. One involves using distance and direction, while the other involves relying primarily on landmarks.

As shown in row 1, 34 out of 35 studies concluded that males use distance plus directional cues more than females, with the remaining study reporting no significant difference. Regarding the use of landmarks, row 2 reveals that this is more common for females, although five studies found no significant sex difference.

Table 25.11.8 Methods used in spatial navigation

Variable	*Orig. table*	*Number of Sex Difference Studies (and Consistency Scores)*						*Overview*	*Countries*	*Time range*	*Non-human*
		I/T	*Child*	*Adol*	*Adult*	*WAR*	*All ages*				
1. Using distance estimates and sense of direction to navigate physical space	11.2.6.5	–	1M	6M	26M 1x(96.3)	1M	34M 1x(97.1)	–	6	1941–2012	4M
2. Using landmarks to navigate physical space	11.2.6.6	–	2F 1x	1F	22F 3x(88.0)	2F 1x	27F 5x(84.4)	–	6	1986–2012	6F

25.11.9 Para-Intellectual Reasoning

Para-intellectual reasoning refers to a variety of reasoning abilities and tendencies that are of an intellectual nature but not normally included in conventional tests of intelligence. Since the 1960s, a number of tests for general reasoning ability have been developed. As shown in row 1 of Table 25.11.9, most of the studies utilizing these methods have concluded that males surpass females regarding scores on these tests.

Creativity is difficult to define and measure, although many scientific efforts have been made to do so. As one can see in row 2, findings have been very inconsistent regarding any possible sex differences in people's tendencies to be creative.

A test called the *digit symbol substitution test* (*DSST*) has been developed to assess early indications of dementia, i.e., the better one can perform on the test, the lower are the risks of dementia. As shown in row 3, most studies indicate that females usually perform better than males.

People's musical abilities have been studied by at least ten studies regarding possible sex differences. Row 4 shows that females were rated better in five of these studies while there were no significant sex differences in the remaining five.

Reaction time studies involve presenting research participants with stimuli (such as a series of tones or flashes of light) and then measuring how quickly they respond (such as by pressing a button). As one can see in row 5, when sex differences are significant, nearly all of them indicate that males exhibit faster reaction times.

Twelve studies of a weight conservation test developed by Piaget were located in which sex differences in performance were assessed. One can see in row 6 that eight of these studies found males performing more accurately with the remaining four studies reporting no significant differences.

The Stroop color-word naming test involves presenting words of specific colors (e.g., red, green) written in letters of another color. Research participants are asked to name the color regardless of the word that is written in that color, and then asked to do the opposite. One can see in row 7 that most studies have found no significant sex difference in performance of this test, although when sex differences are found, they usually favor females.

Systematizing behavior involves organizing words or physical objects into categories according to specific rules and arranging them into hierarchical orders. As shown in row 8, all 18 studies of sex differences in systematizing have concluded that males perform systematizing activities better than do females.

The Wisconsin card sorting test gives research participants in which they must shift strategies several times in order to perform well. Row 9 shows that most studies have found no significant sex differences in this test.

Table 25.11.9 Para-intellectual reasoning

Variable	*Orig. table*	*Number of Sex Difference Studies (and Consistency Scores)*						*Overview*	*Countries*	*Time range*	*Non-human*
		I/T	*Child*	*Adol*	*Adult*	*WAR*	*All ages*				
1. Reasoning abilities in general	11.2.7.1	–	2x	14M 1x 1F (82.3)	8M 4x 4F(40.0)	4M	26M 7x 5F (60.5)	–	13	1967–2011	–
2. Creativity in general	11.2.7.3	–	10M 7x 10F (27.0)	5M 5x 5F(25.0)	7M 4x 6F(30.4)	1M 2x	23M 18x21F (27.7)	–	19	1950–2018	–
3. Digital symbol-coding test	11.2.7.5	–	–	2F	6F 1x 2M	–	8F 1x 2M(61.5)	F: 1Rev	5	1953–2004	–
4. Musical ability	11.2.7.7	–	1F 1x	1x	3F 3x	1F	5F 5x(50.0)	–	6	1931–2005	–
5. Reaction time scores	11.2.7.8	1F	1M 2x	2M	35M 4x (89.7)	4M	42M 6x 1F(84.0)	M: 2Rev 2Met	13 (1)	1931–2018	–
6. Piaget's weight conservation test	11.7.2.11	–	3M	4M 3x	–	1M 1x	8M 4x(66.7)	–	5	1967–2007	–
7. Stroop color-word naming test	11.2.7.17	–	2M 2x 1F	1F 1x	8F 14x(36.4)	–	10F 17x 2M(32.3)	–	7	1932–2010	–
8. Systemizing behavior	11.2.7.19	–	2M	–	17M (100.0)	1M	20M(100.0)	–	8 (1)	2006–2016	–
9. Wisconsin card sorting test	11.2.7.23	–	1M 5x	–	1M 3x 1F	1x	2M 9x 1F(15.4)	–	3	1974–2017	–
10. Relationship between IQ scores and academic performance	11.2.8.5	–	3F 4x	7F 3x 1M (58.3)	10F (100.0)	–	20F 7x 1M(69.0)	–	5	1924–2017	–

In the final row of this table, findings from studies of sex differences in the relationship between IQ scores and academic performance have been assessed. One can see that 20 out of the 28 studies concluded that the relationship is significantly stronger among females than among males, with all but one of the remaining studies reporting no significant sex differences.

25.11.10 Communication-Focused Cognitive Abilities

Sex differences in cognitive abilities having to do primarily with the ability to use language are given attention in Table 25.11.10. The first two rows involve assessments of the overall ability to use language effectively. One can see in row 1 that a large majority of studies have concluded that females have better language skills than do males, especially during childhood. Row 2 shows that 11 out of 12 studies of self-assessed language ability also favor females.

Literacy rates among males and females have been extensively studied. One can see in row 3 that findings of significant sex differences are inconsistent. Regarding the number of words known or used, row 4 shows that they are somewhat more consistent in pointing toward females having the largest vocabularies.

Studies of the ability to accurately name common animals or other objects have been used among adults to screen for signs of dementia. Row 5 shows that the findings regarding sex differences are inconsistent with just a slight leaning toward female superiority.

Thirteen studies of sex differences in the ability to learn a second language were located. As shown in row 6, nine of these studies indicated that females were better, although three other studies favored males.

Rows 7, 8, and 9 all have to do with average sex differences in reading ability and reading comprehension. One can see that over 200 studies of objectively measured reading ability were located, a substantial majority of which have concluded that females surpassed males in reading ability. However, of the 16 studies based on self-rated reading ability, only half of them reported greater ability among females. In the case of reading comprehension, 15 of the 33 studies concluded that females were better, along with 15 other studies suggesting there are no significant sex differences.

Thirty-two studies of sex differences in spelling ability were located. As shown in row 10, 24 of these studies indicated that females surpassed males, five found no significant sex differences, and the remaining three reported greater ability among males.

One hundred and twenty-four studies of verbal linguistic ability in general were located. One can see in row 11 that all but nine of these

Table 25.11.10 Communication-focused cognitive abilities

Variable	*Orig. table*	*Number of Sex Difference Studies (and Consistency Scores)*						*Overview*	*Countries*	*Time range*	*Non-human*
		I/T	*Child*	*Adol*	*Adult*	*WAR*	*All ages*				
1. Language ability	11.3.2.1	12F 2x(85.7)	37F **1x(97.4)**	21F 3x(87.5)	18F 1x 4M (66.7)	3F 2x	91F 9x 4M(84.3)	F: 1Met x: 1Gen	16	1891–2021	–
2. Language ability, self-assessed	11.3.2.2	–	4F	7F	1x	–	11F 1x(91.7)	–	5	1984–2008	–
3. Literacy	11.3.2.3	–	4F	13F 1x(92.9)	16M 2x 2F (72.7)	3M 1x 1F	20F 4x 19M(32.3)	–	23 (3)	1973–2006	–
4. Number of words known or used	11.3.2.6	–	1F	5F	3F 2M	–	9F 2M (81.8)	M: 1Rev	3	1968–1991	–
5. Picture naming test	11.3.2.9	–	–	–	6M 2x 2F(50.0)	–	6M 2x 2F(50.0)	–	3	1986–2007	–
6. Proficiency in a second language	11.3.2.10	–	1F	7F 1x	3M 1F	–	9F 1x 3M(56.3)	–	9	1965–2016	–
7. Reading ability	11.3.2.11	–	67F 15x 8M (68.4)	97F 9x 4M (85.1)	8F 5x 3M(42.1)	10F 2x (83.3)	182F 31x 15M (74.9)	F: 3Met 2Rev x: 1Gen	63 (20)	1912–2021	–
8. Reading ability, self-rated	11.3.2.12	–	6F 6X 1M(42.9)	1F 1M	–	1F	8F 6x 2M(44.4)	x: 1Met	3	1979–1997	–
9. Reading comprehension	11.3.2.14	–	4F 8x(33.3)	10F 6x 1M (55.6)	2M 1x 1F	–	15F 15x 3M(41.7)	F: 1Rev	4 (6)	1937–2011	–
10. Spelling ability	11.3.2.15	–	7F 2x 3M(46.7)	12F 3x(80.0)	1F	4F	24F 5x 3M(68.6)	–	6(2)	1933–2020	–

11. Verbal ability in general	11.3.2.17	8F 2x 1M (66.7)	28F 34x 3M(41.2)	6F 9x 1M(35.3)	15F 10x 3M (48.4)	3F 1M	60F 55x 9M(45.1)	F: 5Met 1Rev; x: 1Met 1Rev	8 (2)	1907–2018	–
12. Verbal fluency	11.3.2.19	1F	9F	10F 1x(90.9)	34F 20x 3M (56.7)	6F 2x	60F 23x 3M(67.4)	F: 1Met	13 (3)	1925–2021	–
13. Verbal reasoning among persons with Alzheimer's disease	11.3.2.20	–	–	–	10M 3x(76.9)	–	10M 3x(76.9)	M: 1Met	2	1994–2008	–
14. Vocabulary comprehension	11.3.2.23	–	12M 11x 2F(44.4)	2M 5x	2F 10x(16.7)	1M	15M 26x 4F(30.6)	x: 1Met	4	1953–2009	–
15. Vocabulary, extensiveness of	11.3.2.24	8F 2x(80.0)	11F 16x 6M(28.2)	7F 5x 6M(29.2)	12F 8x 4M(42.9)	1F 1x	39F 32x 16M(37.9)	F: 1Met 1Rev	7 (2)	19113–20-21	–
16. Writing ability	11.3.2.27	–	7F 1x	14F 1x 1M (82.4)	6F 1x	4F	31F 3x 1M (86.1)	F: 1Met 1Rev	4 (1)	1911–2019	–

studies were more or less evenly split between those indicating that females were more capable of verbal communication or that there were no significant sex differences. The overview column shows that literature reviews and meta-analyses were also fairly equally divided between these two options. A closely related phenomenon – that of verbal fluency – has also been extensively studied. Row 12 shows that most of these studies have indicated that females surpassed males. However, 13 studies of verbal reasoning among persons who appear to have Alzheimer's disease were located. Row 13 shows that most of their results pointed toward males being better than females at retaining verbal reasoning.

Forty-five studies of sex differences in vocabulary comprehension were found. Row 14 shows that most of these studies found no significant sex differences. When significant differences were identified, most of them favored males. However, row 15 indicates that most studies of the actual use of a broader vocabular favors females rather than males.

The ability to write coherently has been investigated by at least 35 studies. The vast majority of these studies indicate that females do so to a greater degree than males.

25.12 Learning, Memory, Knowledge, and Cognitive States: Condensed Findings

In Chapter 12, research findings regarding sex differences in learning, memory, knowledge and various cognitive states were reviewed. The tables in this chapter provide further condensation of the findings.

25.12.1 Learning Ability

Just one table was located in Chapter 12 having to do with sex differences in learning ability. As shown in Table 25.12.1, it indicated that eight out of ten studies concluded that females were better at language-based learning than were males.

25.12.2 Memory and Recall

Table 25.12.2 provides a summary of findings having to do with sex differences in people's memory and recall abilities. Regarding memory and recall abilities in general, row 1 shows that 110 studies were located, and that 60 of these indicated that females were better than males, compared to only 16 studies favoring males. If one focuses simply on what is known as short-term memory, row 2 indicates that most studies have reported no significant sex differences. In the case of working memory, however, 10 out of 13 studies favored males (row 3).

Table 25.12.1 Learning ability

Variable	*Orig. table*	*Number of sex difference findings (and consistency scores)*						*Overview*	*Countries*	*Time range*	*Non-human*
		I/T	*Child*	*Adol*	*Adult*	*WAR*	*All ages*				
1. Linguistic learning	12.1.1.9	–	–	–	6F 1x	2F 1x	8F 2x (80.0)	–	3	1986–2006	–

Table 25.12.2 Memory and recall

Variable	*Orig. table*	*Number of sex difference findings (and consistency scores)*						*Overview*	*Countries*	*Time range*	*Non-human*
		I/T	*Child*	*Adol*	*Adult*	*WAR*	*All ages*				
1. Memory and recall abilities in general	12.1.4.1	1x	13F 10x 6M (37.1)	12F 4x(75.0)	32F 17x 9M (47.8)	3F 2x 1M	60F 34x 16M (47.6)	–	14 (1)	1910–2009	–
2. Short-term memory	12.1.4.3	–	1F 4x	1x	1F 1M	3x	2F 8x 1M(20.0)	x: 2Rev	3	1970–2009	Complex
3. Working memory	12.1.4.4	–	–	1M 1x	7M 1F	2M 1x	10M 2x 1F(71.4)	–	5 (1)	1998–2007	–
4. Facial recognition and recall	12.1.4.10	1F 1x	2F 3x 1M	5F	29F 16x 1M (61.7)	1x	37F 21x 2M (59.7)	F: 1Met	5 (1)	1916–2020	–
5. Recalling emotional experiences	12.1.4.13	–	1F	–	8F 2x 2M(57.1)	–	9F 2x 2M(60.0)	–	3	1985–2004	–
6. Episodic memory	12.1.4.14	–	2F	7F 2x	42F 7x(85.7)	13F 1x(92.9)	64F 10x(86.5)	F: 1Rev; F: 3Met	5 (3)	1982–2019	–
7. Recall visual-spatial stimuli (other than faces)	12.4.4.18	–	–	2M	6M 6x 4F(30.0)	2M	10M 6x 4F (41.7)	–	6	1937–2002	–
8. Recalling word lists, object names, number lists	12.1.4.19	3x	9F 18x 1M (31.0)	5F 1x	33F 10x 1M (73.3)	7F	54F 32x 2M (60.0)	–	8	1913–2021	–

Findings for sex differences in facial recognition and recall are summarized in row 4. One can see that 37 of the 50 studies indicated that females surpassed males in this regard.

Another type of recall ability has to do with past experiences. At least in the case of emotional experiences, row 5 indicates that nine out of 13 studies concluded that females were more accurate that males. Similarly, 64 out of 74 studies reported that females are better than males at the ability to accurately remember episodes in the sequence in which they actually occurred (row 6).

The only type of recall for which most studies favored males had to do with visual-spatial stimuli (excluding faces). Examples of visual-spatial memory tasks have to do with distances between objects and how objects are arranged in a two- or three-dimentional space. As shown in row 7, 10 out of the 20 located studies of visual-spatial memory concluded that males surpassed females.

In the final row of this table, research findings on people's abilities to recall word lists, object names, and sometimes number lists are summarized. Of the 88 studies that were located, 54 indicated that females surpassed males in recalling these types of memory challenges.

25.12.3 General Knowledge and Health Knowledge

Table 25.12.3 pertains to sex differences in overall knowledge along with one type of health knowledge. Regarding knowledge in general, 54 studies of sex differences were located. As shown in row 1, 48 of these studies concluded that males had more general knowledge than females.

The only aspect of health knowledge for which at least ten findings were located was knowledge having to do with AIDS. In this regard, 10 of the 21 studies found no significant sex difference. However, when differences were significant, most of the studies indicated that females knew more than males.

25.12.4 Math Knowledge and Reasoning

A major area of study in the fields of both psychology and education has involved sex differences in math knowledge and reasoning. This knowledge or reasoning is typically measured in terms of performance on standardized tests. The findings are summarized in Table 25.12.4.

Row 1 has to do with performance on various math exams in general. Out of 446 studies dating back over a century, 292 concluded that male performance surpassed that of female. Of the remaining 154 findings, 88 indicating that there were no significant sex differences and 66 reported that females out-scored males. In this row's cell for Overview, one can see

Table 25.12.3 General knowledge and health knowledge

Variable	*Orig. table*	*Number of sex difference findings (and consistency scores)*						*Overview*	*Countries*	*Time range*	*Non-human*
		I/T	*Child*	*Adol*	*Adult*	*WAR*	*All ages*				
1. Knowledge in general	12.2. 1.1.	–	9M 2x 1F(69.2)	13M 3x (81.3)	18M (100.0)	8M	48M 5x 1F (87.3)	–	12 (2)	1910–2021	–
2. Knowledge about AIDS	12.2. 2.2	–	–	4F 1M	4F 10x 2M(22.2)	–	8F 10x 3M (33.3)	–	8	1987–2004	–

Table 25.12.4 Math knowledge and reasoning

Variable	*Orig. table*	*Number of sex difference findings (and consistency scores)*						*Overview*	*Countries*	*Time range*	*Non-human*
		I/T	*Child*	*Adol*	*Adult*	*WAR*	*All ages*				
1. Performance on mathematic exams in general	12.2.3.1	–	41M 30x 24F (34.5)	196M 47x 40F (60.7)	43M 9x 1F (79.6)	12M 2x 1F (75.0)	292M 88x 66F (57.0)	M: 12Rev, 7Met; x: 5Rev, 6Met	58 (11)	1918–2021	–
2. Mathematical ability, self-rated	12.2.3.2	–	20M 14x (58.8)	81M 6x 2F	32M 3x 1F	4M	137M 23x 3F (84.0)	M:1Rev M:2Met	49 (5)	1976–2019	–
3. Math ability, intra-sex variability in	12.2.3.3	–	5M	8M	–	1M	14M (100.0)	–	4 (3)	1980–2020	–
4. Exceptionally high math scores	12.2.3.4	–	6M	30M (100.0)	1M	–	37M (100.0)	M:2Rev M:3Met	7 (4)	1980–2020	–
5. Math ability, trends in sex differences	12.2.3.5	–	–	7 Decrease 6x (53.8)	–	–	7 Decrease 6x (53.8)	Decrease: 1Ref, 1Met	1 (1)	1990–2015	–

that there have been quite a number of literature reviews and meta-analyses on sex differences in math performance, and that these too have reached inconsistent conclusions. While a slight majority of Overview have concluded that males outperform females, several others indicated that the differences are insignificant.

In row 2, findings regarding sex differences in self-rated math ability are summarized. Of the 163 pertinent findings, 137 indicated that males rated their ability higher, while there were no significant differences for 23 studies and three others in which females self-rated their abilities higher.

Sex differences in intra-sex variability in math ability is the focus of row 3. One can see that all 14 studies that reported on this variability concluded that it was greater among males than among females. This fact makes the results of row 4 rather predictable, since it pertains to sex differences in the proportions of individuals with exceptionally high math scores, i.e., usually scores that are at least two standard deviations above the mean. Of 37 studies, every one indicated that more males scored in the exceptionally high range.

The last row in this table is a special one in that it deals with *trends in sex differences in math ability*. In other words, since national data first began to be published on sex differences in math performance (i.e., roughly the mid-20th century), do the findings suggest that the average differences in scores between males and females seem to be increasing, decreasing, or remaining the same? One can see that the answer is mixed. While six studies found no significant change, seven studies found significant *decreases* in the average sex differences.

25.12.5 Specific Realms of Math Knowledge or Reasoning Ability

Three tables were located in Chapter 12 that provided information from ten or more studies on sex differences in specific aspects of math knowledge (or math reasoning ability). Row 1 of Table 25.12.5 shows that 123 studies were located regarding sex differences in the ability to perform basic arithmetic. Sixty-three of these studies concluded that males were better, 38 concluded that females were better, and the remaining 22 studies reported no significant difference.

In the case of geometry, 18 studies were located. In row 2, one can see that 16 of these studies concluded that males were better, while the remaining two indicated there were no significant sex difference.

The third row has to do with algebra and calculus. Knowledge or reasoning ability in these subject areas were assessed by 19 studies. Eight studies concluded that males were better, two indicated that females were, and the remaining nine studies reported no significant sex difference.

Table 25.12.5 Specific realms of math knowledge or reasoning ability

Variable	*Orig. table*	*Number of sex difference findings (and consistency scores)*						*Overview*	*Countries*	*Time range*	*Non-human*
		I/T	*Child*	*Adol*	*Adult*	*WAR*	*All ages*				
1. Arithmetic (numeracy) ability	12.2.4.1	1F	20F 8x 17M (32.3)	21M 10x 13F (36.8)	20M 3x 3F (69.0)	5M 1x 1F	63M 22x 38F (39.1)	M: 1Rev	16 (7)	1891–2021	–
2. Geometric (or spatial) knowledge or reasoning	12.2.4.2	–	2M	14M 2x(87.5)	–	–	16M 2x(88.9)	–	6 (1)	1913–1999	–
3. Algebra and calculus knowledge or reasoning	12.2.4.3	–	–	7M 8x 1F(41.2)	1M 1x 1F	—	8M 9x 2F(38.1)	–	6 (1)	1913–2014	–

25.12.6 Science Knowledge

Sex differences in science knowledge has received a great deal of research attention. Findings in this regard are summarized in Table 25.12.6.

One can see in row 1 that 183 findings were located covering comparisons made in 58 specific countries (plus 20 additional studies of multiple countries). Conclusions regarding sex differences are mixed, although the majority (i.e., 129) of the findings point toward greater knowledge among males than among females.

In row 2, self-rated science knowledge is covered. Of the 38 studies that were located on sex differences, 25 indicated that males perceive themselves as having more science knowledge than females.

Rows 3 through 6 pertaining to knowledge in the fields of biology, chemistry, computer science, and physical science. Evidence of sex differences in the first two of these areas of science are mixed. However, the vast majority of studies of knowledge in computer science and physical science support concluding that males surpass females.

25.12.7 Knowledge of Social Science and the Humanities

Although there is obviously no sharp distinction between the sciences, typically social science is treated separately from the other sciences. This separation is partly due to the fact that the social sciences often overlap with what are known as the *humanities*, such as history and the study of politics. Table 25.12.7 summarizes findings of sex differences in four categories of social science and the humanities.

Row 1 shows that studies have strongly indicated that males are more knowledgeable regarding geography. To a slightly less degree, rows 2 and 3 indicate that males also surpass females in knowledge of history and politics (or world affairs). In the case of language-related knowledge, however, row 4 suggests that females have greater knowledge than do males.

25.12.8 Sleep

Sleep appears to be widespread in the animal kingdom. Among humans, Table 25.12.8 shows that four different aspects of sleep were assessed by at least ten studies with regard to possible sex differences.

The first aspect of sleep that is listed in the table pertains to a phenomenon known as *morningness vs. eveningness*. This refers to evidence that some people prefer getting up fairly early in the morning and then going to sleep earlier at night, while others prefer staying up fairly late at night and then waking up later in the morning. As shown in row 1, all 14 studies that have found sex differences have concluded that females are

Table 25.12.6 Science knowledge

Variable	*Orig. table*	*Number of sex difference findings (and consistency scores)*						*Overview*	*Countries*	*Time range*	*Non-human*
		I/T	*Child*	*Adol*	*Adult*	*WAR*	*All ages*				
1. Science knowledge in general	12.2.5.1	–	13M 4x 1F (68.4)	99M 18x 30F (55.9)	12M (100.0)	5M 1x	129M 23x 31F (57.9)	M: 4Rev 2Met F: 1Rev	58 (20)	1928–2020	–
2. Science knowledge or ability, self-rated	12.2.5.3	–	4M	19M 3x 9F (47.5)	1M	1M	25M 3x 9F (54.4)	M: 1Met	29	1988–2019	–
3. Biological science knowledge	12.2.5.4	–	2M 1x	10M 9x 7F (30.3)	4M	–	16M 10x 7F (40.0)	–	8	1928–2007	–
4. Chemistry knowledge	12.2.5.5	–	1M	7M 1F	1x	–	8M 1x 1F(72.7)	–	3	1969–2014	–
5. Computer science knowledge/ computer literacy	12.2.5.6	–	7M	16M 1x(94.1)	12M 1x(92.3)	–	35M 2x(94.6)	–	7	1979–2014	–
6. Physical science knowledge	12.2.5.7	–	1M	31M 1x(96.9)	3M	1M	36M 1x(97.3)	–	6 (4)	1934–2014	–

Table 25.12.7 Knowledge of social science and the humanities

Variable	*Table*	*Number of sex difference findings (and consistency scores)*						*Overview*	*Countries*	*Time range*	*Non-human*
		I/T	*Child*	*Adol*	*Adult*	*WAR*	*All ages*				
1. Geography knowledge	12.2.6.3	–	1M 1x	6M	14M (100.0)	–	21M 1x(95.5)	–	4	1913–2004	–
2. History knowledge	12.2.6.4	–	2M 1F	6M	4M	–	12M 1F(85.7)	–	5 (1)	1913–2004	–
3. Politics/world affairs knowledge	12.2.6.6	–	4M	5M 1x	16M 2x(88.9)	3M	28M 3x(90.3)	M: 1RevM: 1Met	5	1913–2004	–
4. Language-related knowledge	12.2.7.2	–	4F	5F 1x	1x	1F	10F 2x(83.3)	–	6	1933–2018	–

Table 25.12.8 Sleep

Variable	*Table*	*Number of sex difference findings (and consistency scores)*						*Overview*	*Countries*	*Time range*	*Non-human*
		I/T	*Child*	*Adol*	*Adult*	*WAR*	*All ages*				
1. Morningness (vs. eveningness)	12.3.1.1	–	–	1F 4x	12F 6x(66.7)	1F	14F 10x(58.3)	–	9 (1)	1987–2013	–
2. Time spent sleeping	12.3.1.2	6F 5x (54.5)	1x	1F 2x	13F 6x 3M (52.0)	1F	21F 14x 3M (51.2)	–	11	1961–2020	F: 3 Birds M: 3 Rodents
3. Sleep quality	12.3.1.4	–	–	5M 4x 4F (29.4)	–	–	5M 4x 4F(29.4)	–	8	2004–2020	–
4. Deep, slow-wave sleep	12.3.1.10	–	–	–	15F 2x(88.2)	–	15F 2x(88.2)	F: 1Met	3	1982–2005	–

more likely than males to be early morning risers (while men are more likely to prefer being "late owls"). However, ten additional studies found no significant sex differences. While the consistency score for this variable is not close to attaining a 95.0 level, it is worth mentioning that one study found circulating levels of testosterone positively correlated with people going to bed relatively late at night (Wittert 2014).

Row 2 summarizes findings regarding sex differences in time spent sleeping. One can see that findings have been mixed although 21 out of 38 studies reported females sleep more than males, while three studies concluded the opposite, and the remaining 14 studies reported no significant difference.

Thirteen studies of sleep quality were located. Row 3 shows that findings are quite mixed regarding any possible sex difference.

The last row in this table is concerned with the amount of time adult males and females spend in what is known as deep sleep. One can see that 15 studies indicated that females spend more sleep time in deep sleep than do males, while the remaining two studies reported no significant sex difference.

25.12.9 Dreams

Interesting studies of sex differences in dreaming have been published. Table 25.12.9 shows that ten or more studies were located for sex differences in dream recall and five more for dream content were found.

The first row of this table indicates that 28 out of 30 studies of sex differences in dream recall concluded that females are more likely to recall their dreams once awake than are males. The remaining two studies found no significant difference.

In terms of dream content, row 2 reveals that 12 out of 14 studies concluded that males remember having dreams about aggression more. When asked to report the sex of the most prominent character in their dreams, male characters were more commonly recalled by males than by females in 18 out of 24 findings (row 3).

Row 4 summarizes findings regarding physical objects (other than clothing) being a part of dreams. In this regard all 11 studies agreed that males were more likely than females to have objects (such as cars and weapons) in their dreams.

Another distinct sex difference in dream content involved sexual themes. As shown in row 5, all 13 studies concluded that these themes were more common for males than for females.

The last row has to do with dreams containing social themes (other than aggression and sexuality). One can see that seven of the 12 studies found these types of social themes more common in the dreams of females than in those for males.

Table 25.12.9 Dreams

Variable	*Orig. table*	*Number of sex difference findings (and consistency scores)*						*Overview*	*Countries*	*Time range*	*Non-human*
		I/T	*Child*	*Adol*	*Adult*	*WAR*	*All ages*				
1. Dreaming, frequency of recall	12.3.2.1	–	1F 1x	9F	18F (100.0)	6F 1x	28F 2x(93.3)	F: 1Met	5	1907–2005	–
2. Dream content: Aggression themes	12.3.2.2	–	–	1M 1F	11M 1F(91.7)	–	12M 2F(85.7)	–	3 (1)	1958–2010	–
3. Dream content: Male characters	12.3.2.4	–	–	2M 1F	12M 2x 2F (66.7)	4M 1x	18M 3x 3F (66.7)	–	7	1963–2003	–
4. Dream content: Physical objects and artifacts other than clothing	12.3.2.6	–	–	1M	7M	3M	11M (100.0)	–	3 (1)	1958–2005	–
5. Dream content: Sexual themes	12.3.2.7	–	–	1M	9M	3M	13M (100.0)	–	4 (1)	1958–2010	–
6. Dream content: Social themes (other than aggression or sexuality)	12.3.2.8	–	–	2F	3F 4x 1M(33.3)	2F	7F 4x 1M(53.8)	–	2 (1)	1935–2002	–

25.12.10 Fantasies

Two tables in Chapter 12 having to do with sex differences in fantasizing contained ten or more studies. In row 1 of Table 25.12.10, findings regarding fantasizing about aggression and violence are summarized. One can see that 17 of the 20 studies concluded that males have these types of fantasies more than females.

Row 2 has to do with fantasizing about sexual behavior. Of the 41 located studies, 39 indicated that males experience these types of fantasies more than do females.

25.13 Self-Assessment and States of Mind: Condensed Findings

This section provides a condensed summary of research findings regarding sex differences in how people assess themselves regarding various traits and how people vary regarding happiness and other states of mind. Citations to the actual studies for this section are contained in Chapter 13, which is part of Volume II. The traits for which at least ten relevant findings were located are condensed into four summary tables.

25.13.1 Self-Reflections

The first section of Table 25.13.1 has to do with various types of self-reflections. Row 1 pertains to self-consciousness. It shows that 22 out of a total of 30 studies concluded that females are more self-conscious than males.

In the case of self-esteem, a very large number of studies of sex differences were located. One can see in row 2 that out of the 333 located studies, 192 concluded that males had higher self-esteem than females. This pattern is consistent with two meta-analyses that reported self-esteem in males to be higher, although two literature reviews concluded that there were no significant sex differences, a conclusion consistent with 113 of the remaining studies that were located.

Two types of self-confidence have been studied with respect to possible sex differences, one consisting of overall self-confidence and the other dealing with confidence in terms of a specific type of skill or activity. Row 3 shows that overall (or general) self-confidence is predominantly found among males. As shown in row 4, studies of domain specific self-confidence also largely indicate that males express more confidence than do females, at least regarding the skills and activities specified in most of the pertinent studies.

Seventeen studies were located having to do with sex differences in self-compassion. Row 5 shows that 12 of these studies concluded that males are more self-compassionate than are females.

Table 25.12.10 Fantasies

Variable	*Orig. table*	*Number of sex difference findings (and consistency scores)*						*Overview*	*Countries*	*Time range*	*Non-human*
		I/T	*Child*	*Adol*	*Adult*	*WAR*	*All ages*				
1. Fantasizing about aggression and violence in general	12.4.1.2	–	9M 2x(81.8)	–	8M 1x	–	17M 3x(85.0)	–	3	1967–2008	–
2. Fantasizing about sexual behavior in general	12.4.1.3	–	–	1M	35M 2x(94.6)	3M	39M 2x(95.1)	M: 1Rev	9	2958–2003	–

Table 25.13.1 Self-reflections

Variable	*Orig. table*	*Number of sex difference findings (and consistency scores)*						*Overview*	*Countries*	*Time range*	*Non-human*
		I/T	*Child*	*Adol*	*Adult*	*WAR*	*All ages*				
1. Self-consciousness	13.1.1.1	–	–	3F	11F 2x (84.6)	8F 4x 2M (50.0)	22F 6x 2M (73.3)	–	7 (4)	1957–2018	–
2. Self-esteem in general	13.1.1.8	–	26M 58x 8F (26.0)	82M 29x 10F (62.6)	76M 26x 10F (62.3)	8M 2F (80.0)	192M 113x 28F (53.2)	M: 2Met; x: 2Rev	27 (2)	1957–2019	–
3. Overall self-confidence	13.1.2.1	–	1M 1x	1M	19M (100.0)	1M	22M 1x (95.6)	M: 1Rev	8	1932–2012	–
4. Self-confi-dence, domain specific	13.1.2.2	–	2M 1x	6M	13M 1x (92.8)	–	21M 2x (91.3)	–	5	1969–2016	–
5. Self-compassion	13.1.2.3	–	–	4M	8M 5x (61.5)	–	12M 5x (70.6)	M: 1Met	9	2003–2019	–
6. Relationship between self-esteem and body satisfaction	13.1.2.1	–	2F	5F	5F 1x	1F	13F 1x (92.9)	–	5	1974–2005	–

The final row in this table reports findings on sex differences in the relationship between self-esteem and body satisfaction. One can see that 13 of the 14 located studies concluded that these two variables are more strongly correlated among females than among males.

25.13.2 Self-Assessments

Sex differences in self-assessments having to do with judgments that people make regarding themselves are summarized in Table 25.13.2 if at least ten studies could be located. Row 1 has to do with self-ratings of physical attractiveness. One can see that the findings have been mixed with 12 studies reporting that females provided higher average ratings while ten other studies reported males did so.

Body satisfaction has been widely studied regarding possible sex differences. One can see in row 2 that nearly all studies have found males expressing greater satisfaction with their bodies than do females. At least one of the reasons for this sex difference could involve sex differences in considering themselves to be overweight. Row 3 shows that 94 out of 95 studies bearing on weight-loss issues have concluded that greater proportions of females than males considered themselves to be overweight (or express a desire to lose weight).

Rows 4 and 5 pertain to related issues: academic ability and intelligence. Studies of sex differences in self-ratings for both of these variables are substantial. Regarding academic ability, most studies have indicated that males give themselves higher ratings than females, although a considerable minority of studies have found no significant sex difference. In the case of intelligence, the tendency for males to give themselves higher average ratings is even greater, although still not to the degree that this can be considered a universal sex difference.

Regarding assessment of ability in general, row 6 shows that males usually give themselves higher ratings, although a couple of studies found no significant sex differences. In several studies, both sexes were provided with specific experimental tasks. After doing so, they were asked to self-rate their performance. As shown in row 7, while 11 studies found no sex differences, when there were differences, males perceived their performance to be higher in ten studies, while females did so in just two studies. Other experimental studies ensured that research participants failed several times in a row, and then asked them to assess their own ability. Row 8 shows that this experimental procedure virtually eliminated sex differences in self-assessed ability.

The last two rows in this table have to do with two self-ratings, one for computer competency and the other for risk-taking tendencies. One can see in row 9 that all 12 studies found males giving themselves higher

Table 25.13.2 Self-assessments

Variable	*Orig. table*	*Number of sex difference findings (and consistency scores)*						*Overview*	*Countries*	*Time range*	*Non-human*
		I/T	*Child*	*Adol*	*Adult*	*WAR*	*All ages*				
1. Self-assessed physical attractiveness	13.1.3.1	–	1M	6F 6M (33.3)	6F 1x 3M (46.2)	1M	12F 1x 10M (36.4)	M: 2Met	6	1959–2012	–
2. Body satisfaction	13.1.3.4	–	4M 1F	34M **(100.0)**	38M 2x (90.5)	2M	78M 2x 1F **(95.1)**	M: 1Rev 1Met	11	1966–2010	–
3. Self-assessment of being overweight	13.1.3.6	–	3F 1x	23F **(100.0)**	58F **(100.0)**	10F **(100.0)**	94F 1x **(98.9)**	–	12	1965–2010	–
4. Self-assessed academic ability	13.1.4.1	–	13M 31x 6F (23.2)	4M 3x	32M 1x 2F (86.5)	3M	52M 35x 8F (50.5)	M: 1Met	14	1959–2005	–
5. Self-assessed intelligence	13.1.4.2	–	2M 1x	10M 1F (83.3)	61M 8x 1F (85.9)	3M 1x	76M 10x 1F (86.4)	M: 1Met	20	1959–2018	–
6. Assessment of one's ability in general	13.1.5.1	–	1M 2x	2M	13M **(100.0)**	1M	17M 2x (89.5)	M: 1Rev	1	1972–2000	–
7. Assessment of one's ability to perform various experimental tasks	13.1.5.2	–	7M 11x 2F (31.8)	–	3M	–	10M 11x 2F (40.0)	–	2	1971–1998	–
8. Assessment of one's ability to perform experimental tasks after induced failure	13.1.5.3	–	1M 10x (9.1)	–	1M	–	2M 10x (16.7)	–	1	1975–1989	–
9. Self-assessed computer competency	13.1.5.5	–	1M	6M	12M **(100.0)**	–	19M **(100.0)**	–	5	1985–2013	–
10. Self-assessed tendencies to make risky decisions	13.1.5.12	–	–	–	10M **(100.0)**	–	10M **(100.0)**	–	1	1977–2007	–

ratings in computer competency. All studies report a greater male tendency to make risky decisions.

25.13.3 Locus of Control

Locus of control basically pertains to the forces that people perceive determining the lives. Table 25.13.3 provides a summary of findings regarding sex differences in aspects of locus of control for which at least ten studies were located.

Row 1 shows that 27 of the 44 studies of locus of control in general indicate that males are more likely than females to perceive their lives being internally controlled to a greater degree (as opposed to being externally controlled). Similarly, row 2 shows that, when asked to explain any particular successes they have had in life, most research indicates that males are more likely than females to mention internal factors such as effort and ability. Nonetheless, one can see that, often depending on the nature of the success being explained, there are many studies in which there are no significant sex differences, or where females are more likely to consider internal factors more important.

In row 3, one can see that females appear to be more likely than males to attribute successes they have had to luck or other external factors. As shown in row 4, most studies have found that females are also more likely than males to attribute failures they have had to external (rather than internal) forces.

Finally, row 5 provides a summary of experimental studies in which research participants were repeatedly "made to fail" in order to discover if they seemed to perform worse or better. One can see that most of the research found no significant sex differences in this regard, although when differences were found, performance by females seems to be more adversely affected.

25.13.4 States of Mind and Mood

Sex differences in states of mind such as happiness and satisfaction are summarized in Table 25.13.4. Row 1 has to do with feelings of emotional well-being and stability. Out of the 20 pertinent studies, 13 indicated that males exhibited more well-being and stability than females, with six of the remaining studies suggesting no significant sex differences.

Studies of sex differences in personal happiness or satisfaction with life appear in row 2. Of the very large number of findings, the majority found no significant sex difference. About two-thirds of the remainder concluded that males were happier, with the remainder pointing toward greater happiness among females.

Table 25.13.3 Locus of control

Variable	*Orig. table*	*Number of sex difference findings (and consistency scores)*						*Overview*	*Countries*	*Time range*	*Non-human*
		I/T	*Child*	*Adol*	*Adult*	*WAR*	*All ages*				
1. Internal (vs. external) locus of control in general	13.1.7.1	–	1M 5x 3F	4M 2x 1F	22M 4x 2F (73.3)	–	27M 11x 6F (54.0)	–	10 (3)	1969–2013	–
2. Attribute success to effort, ability, and other internal factors	13.1.7.5	–	14M 23x 1F (35.9)	5M 3F	14M 1x 9F (42.4)	–	33M 24x 10F (42.9)	M: 3Rev	6	1969–1997	–
3. Attribute success to luck or other external factors	13.1.7.6	–	5F	3F	20F 7x (74.1)	1F	29F 7x (80.6)	F: 1Rev 1Met	2	1962–1999	–
4. Attribute failures to external (not internal) factors	13.1.7.7	–	–	5F	9F 1M (81.8)	–	14F 1M (87.5)	–	3	1973–2000	–
5. Diminished performance following past failures	13.1.7.10	–	5F 11x (31.2)	–	–	–	5F 11x (31.2)	–	2	1973–1992	–

Table 25.13.4 States of mind and mood

Variable	*Orig. table*	*Number of sex difference findings (and consistency scores)*						*Overview*	*Countries*	*Time range*	*Non-human*
		I/T	*Child*	*Adol*	*Adult*	*WAR*	*All ages*				
1. Emotional well-being and stability	13.2.1.1	1x	2M	3M 1F	4M 4x	4M 1x	13M 6x 1F (61.9)	M: 2Met; x: 1Met	6 (1)	1923–2016	–
2. Personal happiness (or satisfaction with life)	13.2.1.2	–	2M	4M 4x	56M 69x 44F (33.1)	3M 10x 1F (20.0)	64M 83x 45F (33.3)	M:1Rev 2Met; x: 1Gen 1Rev 2Met; F: 3Rev	36 (8)	1949–2019	–
3. Marital satisfaction	13.2.1.5	–	–	–	19M 12x 8F (40.4)	–	19M 12x 8F (40.4)	–	2	1960–2009	–
4. Job satisfaction	13.2.1.11	–	–	–	27F 26x 8M (39.1)	6F 20x 5F (16.7)	31F 46x 13M (30.1)	–	10 (2)	1951- 2014	–
5. Optimism in general	13.2.2.1	–	–	1x	9M 1F (81.8)	–	9M 1x 1F (75.0)	–	7	1972–2010	–

Row 3 has to do with marital satisfaction. One can see that about half of findings point toward males being more satisfied, while the remainder are split between greater satisfaction among females and no significant differences.

Sex differences in job satisfaction has received considerable research attention. As shown in row 4, most findings indicate that there is no significant sex difference. When there are differences, most studies indicate that females are more satisfied with their jobs than are males.

The last row pertains to sex differences in people's tendencies to be optimistic about the future. Nine of the 11 studies point toward males being more optimistic than females.

25.14 Mental Health and Illness: 10+ Condensed Findings

Who's mentally healthier, males or females? Few would be surprised to learn that this question has no single scientific answer. Nonetheless, there are numerous studies using various indicators of mental health as well as measures of specific mental illnesses that have addressed this issue. Chapter 14 (located in Volume II) presented citations to all of the studies that were located on sex differences in mental health and illness, and the present section will provide a condensed summary of what these studies have found regarding variables for which at least ten studies were located.

25.14.1 Mental Illness and Mental Disorders in General

Mental illnesses and disorders can be organized into many categories, some quite specific while others are rather general. Studies of sex differences in the most general grouping of mental illness and disorders are summarized in Table 25.14.1. The broadest category of all appears in row 1. It indicates that 77 of the 96 studies of mental illness in general, from childhood to adulthood, have concluded that females are afflicted more. Nevertheless, 15 of the remaining studies concluded that males are affected more.

Thirteen studies of sex differences in self-assessed mental health (instead of mental illness) were located. Row 2 shows that ten of these indicated that males rated their mental health as being significantly better than females, although two of the remaining three found the opposite.

Another general indicator of the prevalence of mental illness is the proportion of each sex that has actively sought to obtain mental health services. As shown in row 3, of the 41 located studies, 36 concluded that females seek such services more.

Turning to somewhat more specific forms of mental illnesses, albeit still fairly broad, row 4 summarizes what has been found regarding the

Table 25.14.1 Mental illness and mental disorders in general

Variable	*Orig. table*	*Number of sex difference findings (and consistency scores)*						*Overview*	*Countries*	*Time range*	*Non-human*
		I/T	*Child*	*Adol*	*Adult*	*WAR*	*All ages*				
1. Mental illness in general	14.1.1.1	–	5M 1F	12F 2x 1M (75.0)	32F 1x 6M (71.1)	33F 1x 3M (82.5)	77F 4x 15M (80.2)	F: 2Rev; M: 1Rev	15 (2)	1940–2010	–
2. Self-assessed mental health	14.1.1.2	–	1F	1M	9M 1x 1F(75.0)	–	10M 1x 2F(66.7)	–	6	1961–2016	–
3. Seeking or obtaining mental health services in general	14.1.1.3	–	1F	–	34F 1x 4M (79.1)	1F	36F 1x 4M(80.0)	F: 1Rev	6 (1)	1972–2006	–
4. Seeking help for drug dependency	14.1.1.4	–	–	2M	5M 5F(33.3)	2F	7M 7F(33.3)	–	4	1986–2007	–
5. Behavioral disorders and disturbances in general	14.1.1.5	–	39M 1x(**97.5**)	7M	2M	16M(100.0)	64M 1x(**98.5**)	–	11	1941–1997	–
6. Psychoneurosis and psychosomatic disorders	14.1.1.6	–	1M	4F 1x 1M	19F1x 3M(73.1)	1F 1M	24F 1x 6M(64.9)	–	11 (1)	1925–2014	–
7. Emotional problems and minor psychiatric problems	14.1.1.7	–	2F	3F	6F	1F 1x	12F 1x(92.3)	–	8	1944–2006	–
8. Internalizing behavior problems	14.1.1.8	–	3F 5x	22F (100.0)	3F	1F 1x	29F 1x(**96.7**)	F: 1Met	9 (3)	1986–2020	–
9. Externalizing behavior problems	14.1.1.9	4M	20M 8x(71.4)	38M 1x 2F (88.4)	9M	7M	78M 9x 2F(85.7)	M: 1Rev M: 2Met	17 (5)	1941–2020	–

proportion of each sex that has sought help for drug dependence, both in the case of prescription and non-prescription drugs. One can see that, for this type of mental disturbance, the proportion of males and females are equal.

Research findings on sex differences in the prevalence of behavioral disorders and disturbances are summarized in row 5. This row reveals that all but one of the 65 studies have concluded that males are more affected than females.

Sex differences in the prevalence of psychoneurosis and psychosomatic disorders are summarized in row 6. Of the 31 located studies, 24 concluded that females have these types of disorders more than do males. Similarly, rows 7 and 8 indicate that emotional problems and minor psychiatric problems, along with internalizing behavior, all appear to be more prevalent among females.

The final row in this table has to do with behavior problems of an externalizing nature (most commonly including behaving aggressively toward others). Of the 89 findings, 78 concluded that males are affected more than females.

25.14.2 Mood Disorders

Table 25.14.2 pertains to multiple aspects of depression for which sex differences have been investigated. The first entry deals with depression in general as self-diagnosed by research participants. One can see that 619 studies have been published, 467 of which have concluded that females are more likely to self-report feelings of depression, with 134 reporting no significant sex difference and the remaining 18 indicating that males self-reported more depression than females. Among children, however, no significant differences are typically reported. In the case of depression symptoms (without being either self-diagnosed or clinically diagnosed), the second row shows that findings have been mixed, although predominantly favoring females.

The table's third row has to do with major depression, i.e., depression that has been clinically diagnosed. One can see that a total of 446 pertinent studies were located, of which 378 concluded that females are more often diagnosed with major depression than males. Nevertheless, 68 exceptional studies were found, 11 of which concluded that males received such a diagnosis more than did females. Only in the case of diagnoses of adolescents is the consistency score higher than 95.0.

In row 4, sex differences in depression following one or more stressful experiences are summarized. Of the 15 relevant studies, ten concluded that such depressive experiences are more common among females. Similarly,

Table 25.14.2 Mood disorders

Variable	*Orig. table*	*Number of sex difference findings (and consistency scores)*						*Overview*	*Countries*	*Time range*	*Non-human*
		I/T	*Child*	*Adol*	*Adult*	*WAR*	*All ages*				
1. General depression (affective disorders), self-diagnosed	14.2.1.1	–	20F 59x 14M (18.7)	108F 24x 1M (80.6)	286F 46x 3M (84.6)	71F 2x(**95.4**)	467F 134x 18M (73.3)	F: 10Rev 7Met x: 1Rev 2Met	48 (14)	1953–2020	–
2. Depression symptoms	14.2.1.2	–	1F 1x	11F 5x 5M (42.3)	12F 1x(84.6)	1F	25F 7x 5M (59.5)	F: 1Met	5 (2)	1979–2017	F: 1 for rodents
3. Major depression, clinical	14.2.1.3	1x	5F 25x 5M (12.5)	71F 1x 1M (**95.9**)	218F 22x 4M (87.9)	84F 8x 1M(89.4)	378F 57x 11M (82.7)	F: 8Rev F: 4Met	46 (5)	1942–2020	–
4. Depressive feelings after stress	14,2,1,5	–	–	10F 5x(66.7)	–	–	10F 5x(66.7)	F: 1Rev	2	1988–2003	–
5. Multiple episodes of depression	14.2.1.7	–	–	–	–	8F 5x(61.5)	8F 5x(61.5)	–	3	1984–2000	–
6. Depression, early age of onset	14.2.1.8	–	–	–	–	13F 14x(48.1)	13F 14x(48.1)	–	4	1952–2000	–

(*Continued*)

Table 25.14.2 (Continued)

Variable	*Orig. table*	*Number of sex difference findings (and consistency scores)*						*Overview*	*Countries*	*Time range*	*Non-human*
		I/T	*Child*	*Adol*	*Adult*	*WAR*	*All ages*				
7. Severity of depression	14.2.1.9	–	–	–	8F 4x(66.7)	2F 4x(33.3)	10F 8x(55.6)	–	4	1988- 2000	–
8. Bipolar (manic) depression	14.2.1.11	–	1x	3F	14F 24x 4M (30.4)	5F 7x 2M(31.3)	22F 32x 6M (33.3)	–	15	1925–2016	–
9. Bipolar (manic) depression, age of onset	14.2.1.12	–	–	–	–	9M 9x 3F(37.5)	9M 9x 3F(37.5)	–	5	1919–2005	–
10. Ruminating over negative experiences	14.2.1.15	–	1F	10F 2x(83.3)	50F 8x(86.2)	2F	63F 10x(86.3)	F: 2Met	13 (2)	1982–2013	–
11. Relationship between depression and suicidal behavior	14.2.2.14	–	–	9F 4M(52.9)	–	1M	9F 5M(47.4)	–	4	1988–2005	–

eight out of 13 studies of individuals who have experienced multiple depressive episodes report females having greater frequencies.

Regarding age of onset of an individual's first experience of depression, 27 studies were found. Row 5 shows that 13 of these studies concluded that females had an earlier age of onset, while the remaining 14 reported no significant sex difference.

Eighteen studies were located that sought to assess sex differences in the severity of depression that people experience. Row 6 shows that ten studies concluded that females experienced more severe depression than males, while the remaining eight indicated that sex differences were not statistically significant.

Rows 8 and 9 pertained specifically to bipolar (manic) depression. One can see that the evidence is quite mixed regarding any sex differences in the prevalence of this type of depression and regarding an early age of onset for those with this ailment.

Rumination refers to repeatedly rethinking an unpleasant event that happened in one's past. As one can see in row 10, the majority of studies of ruminating over past negative experiences and feelings is more common among females than among males.

Fourteen studies were located that assessed the apparent relationship between feelings of depression and committing suicide or having thoughts of committing suicide. Row 11 shows that the evidence is quite mixed regarding any possible sex differences.

25.14.3 Schizophrenia

Schizophrenia is a mental disease associated with delusions, hallucinations, and disordered thinking. Symptoms typically begin in early adulthood. The first row of Table 25.14.3 provides a summary of 71 findings regarding sex differences in the prevalence of this disease. Large number of findings have assessed sex differences in the prevalence of schizophrenia. One can see that the vast majority of studies have indicated that males are more likely to be affected by this disease than females, a conclusion that comports with most of the literature reviews and meta-analyses. The most notable exception came from studies of developing countries, where one meta-analysis reported that most studies actually indicate there were no significant sex differences (Aleman, Kahn, and Selten 2003).

Regarding age of onset, the table's second entry shows that, when significant sex differences are found, most studies have concluded that males have an earlier age of onset than do females. This conclusion is consistent with findings from five literature reviews.

Findings from studies of sex differences in recovery from schizophrenia are summarized in the third row. It indicated that the majority of studies

Table 25.14.3 Schizophrenia

Variable	*Orig. table*	*Number of sex difference findings (and consistency scores)*						*Overview*	*Countries*	*Time range*	*Non-human*
		I/T	*Child*	*Adol*	*Adult*	*WAR*	*All ages*				
1. Schizophrenia	14.3.1.1	–	–	–	28M 1x(**96.6**)	29M 11F2F (87.5)	57M 12x 2F (78.1)	M: 1Gen 10Rev 1Met x:1Met (developing countries)	15 (3)	1951–2010	–
2. Schizophrenia, age of onset	14.3.1.4	–	–	–	–	109M 25x 2F (79.0)	109M 25x2F (79.0)	M: 5Rev 1Gen	18 (5)	1896–2008	–
3. Recovering from schizophrenia	14.3.1.5	–	–	–	10F 1x(90.9)	–	10F 1x(90.9)	F: 1Rev	5	1988–2006	–
4. Social competence of schizophrenics	14.3.1.8	–	–	–	11F (**100.0**)	1F	12F(**100.0**)	F: 1Rev	4	1990–2007	–

have concluded that females are more likely than males to exhibit at least some significant degree of recovery. Along similar lines, the table's final entry indicates that females with schizophrenia exhibit a greater degree of social competence than do males with the disease.

25.14.4 Behavioral Disorders

Everyone exhibits behavior that is troublesome to others, at least occasionally. However, some individuals do so to such an extent that they are said to exhibit *behavioral disorders*. Table 25.14.4 summarizes findings from a substantial number of findings of sex differences in three broadly defined behavioral disorders.

Row 1 of this table shows that all but one of 17 studies concluded that males exhibit behavioral disorders in general more than do females. In the case of *psychological problems in general*, row 2 indicates all 15 studies reported that females exhibit these traits more than males. The third entry in this table pertains to *borderline personality disorder*. One can see that the evidence for sex differences has been mixed, although when differences are reported, most studies point toward females having this condition more than males.

25.14.5 Drug Abuse Disorders

Findings of sex differences in drug abuse disorders are summarized in Table 25.14.5. Most forms of drug abuse refer to the abuse of alcohol or other so-called *recreational drugs*; however, the abuse of prescription medicines such as opioids are also considered in this table.

The table indicates that findings of any overall sex differences in drug addiction or substance abuse (row 1) are inconsistent, albeit with a substantial leaning toward males exhibiting higher rates than females. In the case of both alcohol abuse and alcoholism (rows 2 and 3), however, virtually all of the research indicates that males exhibit higher rates than do females.

Most studies also indicate that more males seek treatment for alcohol-related disorders (row 4). Among persons afflicted with alcoholism, all studies agree that males also exhibit an earlier age of onset (row 5).

25.14.6 Antisocial Behavioral Disorders

Antisocial behavior basically encompasses all behavior that observers consider unpleasant, sometimes to the point of punishing it. Such behavior is found both in children – usually termed *conduct disorders* – as well as among adolescents and adults, when it is usually referred to as *antisocial personality disorder* or *psychopathy*. Sex differences in the prevalence of both types of antisocial behavior have been widely reported. One can see in Table 25.14.6 that the vast majority of studies have found that males

Table 25.14.4 Behavioral disorders

Variable	*Orig. table*	*Number of sex difference findings (and consistency scores)*						*Overview*	*Countries*	*Time range*	*Non-human*
		I/T	*Child*	*Adol*	*Adult*	*WAR*	*All ages*				
1. Behavioral problems/disorders in general	14.4.1.1	–	12M 1F (85.7)	3M	1M	–	16M 1F (88.9)	–	2	1923–2006	–
2. Psychological problems in general	14.4.1.2	–	1F	2F	12F(100.0)	–	15F(100.0)	–	5	1979–2015	–
3. Borderline personality disorder	14.4.1.5	–	–	–	10F 8x 1M (50.0)	–	10F 8x 1M (50.0)	–	3	1987–2006	–

Table 25.14.5 Drug abuse disorders

Variable	Orig. table	Number of sex difference findings (and consistency scores)						Overview	Countries	Time range	Non-human
		I/T	Child	Adol	Adult	WAR	All ages				
1. Drug addiction/ substance abuse disorders in general	14.4.2.2	–	–	2M 1x	26M 1x 12F (51.0)	7M 1x	35M 3x 12F (56.5)	F: 1Rev	9	1950–2020	F: 4Rod; F: 2Pri; F: 1Rev
2. Problem drinking/alcohol abuse	14.4.2.4	–	–	4M	20M 1x(95.2)	1M	25M 1x(96.2)	–	5	1953–2012	–
3. Alcoholism (alcohol dependence)	14.4.2.5	–	–	1M	57M (100.0)	20M (100.0)	78M (100.0)	M: 2Rev	13 (2)	1965–2012	–
4. Seeking or obtaining treatment for alcoholism	14.4.2.6	–	–	–	15M 1x(93.8)	–	15M 1x(93.8)	–	3	1953–2004	–
5. Alcoholism, age of onset	14.4.2.7	–	–	–	–	19M (100.0)	19M (100.0)	–	1	1948–2004	–

Table 25.14.6 Antisocial behavioral disorders

Variable	Orig. table	Number of sex difference findings (and consistency scores)						Overview	Countries	Time range	Non-human
		I/T	Child	Adol	Adult	WAR	All ages				
1. Conduct disorder	14.4.3.1	–	126M 5x (96.2)	18M 4x(81.8)	–	3M 2x	147M 11x(93.0)	M: 8Rev	19 (1)	1957–2017	–
2. Antisocial personality disorder and psychopathy	14.4.3.2	–	–	36M 4x(90.0)	82M 9x1F (88.2)	34M (100.0)	152M 14x(91.6)	M: 5Rev	14 (1)	1973–2020	–

outnumber females in these two disorders. However, because in both cases, several studies failed to detect any significant sex differences, neither condition reached consistency scores of 95.0.

25.14.7 Acting Out and Compulsive Behavioral Disorders

Table 25.14.7 cites evidence of sex differences for disorders that can be regarded as pertaining to acting out behavior and compulsive behavior. Regarding narcissistic personality disorder, most studies indicate that males exhibit the disorder more than females; the couple of exceptions simply failed to detect significant differences.

Findings pertaining to oppositional defiant disorder (ODD) are summarized in the second row. If there are significant sex differences, males exhibit ODD more than do females. Nonetheless, one can see that quite a few studies failed to identify significant sex differences.

In the case of obsessive-compulsive disorder (OCD), research findings of sex differences are quite mixed (row 4). However, regarding studies of the age of onset, a distinct majority of studies of affected individuals suggest that OCD begins to manifest itself earlier among males than among females.

Rows 5 and 6 have to do with compulsive gambling to such an extent that it jeopardizes an individual's ability to be self-supporting. Over 50 studies indicate that males exhibit this type of disorder more than females, although four studies failed to find any significant sex difference. Regarding age of onset of compulsive gambling, all ten studies reported earlier ages for affected males than for affected females.

The table's last row in this table has to do with Tourette's syndrome. One can see that all but one of the 15 studies reported this syndrome to be more prevalent among males, with the exception simply failing to identify a significant sex difference.

25.14.8 Eating Disorders

Eating disorders have to do with a variety of eating practices that are not healthy over the long term. They include what is known as binge eating as well as anorexia and bulimia, both of which are also considered individually as two additional entries. As one can see in Table 25.14.8, all studies agree that females are more likely than males to exhibit eating disorders.

25.14.9 Language-Related Disorders

Two language-related disorders had sufficient numbers of studies to warrant consideration as universal sex differences. One can see in Table 25.14.9 that all 11 studies of speech and language disorders as well

Table 25.14.7 Acting out and compulsive behavioral disorders

Variable	*Orig. table*	*Number of sex difference findings (and consistency scores)*						*Overview*	*Countries*	*Time range*	*Non-human*
		I/T	*Child*	*Adol*	*Adult*	*WAR*	*All ages*				
1. Narcissism (including narcissistic personality disorder)	14.4.3.4	–	–	–	15M 2x(88.2)	3M	18M 2x(90.0)	M: 1Met	4 (2)	1972–2020	–
2. Oppositional defiant disorder (ODD)	14.4.3.5	–	8M 4x(66.7)	5M 2x 1F	–	10M 5x(66.7)	23M 11x 1F (63.9)	M: 1Met	13	1993–2019	–
3. Obsessive compulsive-disorder (OCD) in general	14.4.4.1	–	3x	3F 1M	3F 1x 3M	10M 5x 9F (30.3)	15F 9x 14M (28.8)	–	13	1978–2014	–
4. Obsessive compulsive-disorder, age of onset	14.4.4.2	–	–	–	–	16M 1x 3F (69.6)	16M 1x 3F (69.6)	–	6	1989–2004	–
5. Pathological/ compulsive gambling	14.4.4.4	–	–	2M	37M 3x(92.5)	16M 1x(94.1)	55M 4x(93.2)	M: 1Rev	8	1986–2013	–
6. Pathological/ compulsive gambling, age of onset	14.4.4.5	–	–	–	–	10M (100.0)	10M(100.0)	–	3	1992–2012	–
7. Tourette's syndrome	14.4.4.8	–	2M	1M	–	11M 1x(91.7)	14M 1x(93.3)	–	4 (1)	1980–2009	–

Table 25.14.8 Eating disorders

Variable	*Orig. table*	*Number of sex difference findings (and consistency scores)*						*Overview*	*Countries*	*Time range*	*Non-human*
		I/T	*Child*	*Adol*	*Adult*	*WAR*	*All ages*				
1. Eating disorders in general	14.4.5.1	–	–	4F	23F (100.0)	5F	32F (100.0)	F: 3Rev	10	1981–2019	–
2. Anorexia nervosa	14.4.5.4	–	2F	5F	12F (100.0)	24F 1x(96.0)	43F 1x(97.7)	F: 3Rev	9	1976–2015	–
3. Bulimia	14.4.5.6	–	1F	7F	25F (100.0)	14F(100.0)	47F (100.0)	F: 2Rev	12	1981–2001	–

Table 25.14.9 Language-related disorders

Variable	*Orig. table*	*Number of sex difference findings (and consistency scores)*						*Overview*	*Countries*	*Time range*	*Non-human*
		I/T	*Child*	*Adol*	*Adult*	*WAR*	*All ages*				
1. Speech and language disorders in general	14.4.6.1	–	9M	–	2M	–	11M (100.0)	–	2	1916–1997	–
2. Stuttering	14.4.6.3	–	8M	2M	3M	8M	21M (100.0)	M: 1Rev M: 1Met	8	1931–2020	–

as all 21 studies of stuttering have concluded that males are affected by these disorders more than females.

25.14.10 Stress- and Fear-Related Disorders

Stress and anxiety disorders have been studied extensively with respect to possible sex differences, with findings summarized in in the first two rows of Table 25.14.10. One can see that, in both cases, most research has indicated that females exhibit these types of disorders more than males. In the case of post-traumatic stress disorder (PTSD), however, the findings are likely to be confounded somewhat by the fact that males have more stressful experiences that are conducive to PTSD (such as during wars).

All of the last three rows in this table primarily have to do with fear. In the case of panic disorder, all 27 studies agree that females are affected more than males. Regarding phobias, the last two rows indicated that, while findings are mixed, the vast majority also indicate that phobias are more common among females.

25.14.11 Attention-Related Disorders

Three disorders having to do with people failing to pay attention to their social surroundings (often in combination with other disturbing behavior) are covered in Table 25.14.11. In all three cases, males are substantially more likely than females to be affected. This is the case for all studies of attention deficit disorder (ADD) and nearly all studies of attention deficit disorder with hyperactivity (ADHD) and autism spectrum disorder (ASD). The few exceptional findings for the latter two of these disorders simply failed to find statistically significant sex differences.

25.14.12 Residual Mental Disorders

The last two mental disorders for which at least ten relevant studies of sex differences were located are presented in Table 25.14.12. One disorder known as dissociative disorder (as well as dissociative thinking) involves mental states that effectively remove affected individuals from most aspects of reality. These conditions are often associated with past traumatic experiences that cause extremely unpleasant emotions. The table shows that research findings of any significant sex differences in dissociative thinking or dissociative disorders are mixed, although most studies indicated that, when there are sex differences, females experience these types of thought processes more than do males.

Ten studies of sex differences in gender identity disorder (also known as gender dysphoria) were located. The second entry in the table shows that

Table 25.14.10 Stress- and fear-related disorders

Variable	*Orig. table*	*Number of sex difference findings (and consistency scores)*						*Overview*	*Countries*	*Time range*	*Non-human*
		I/T	*Child*	*Adol*	*Adult*	*WAR*	*All ages*				
1. Post-traumatic stress disorder (PTSD)	14.4.7.1	–	4F	13F (100.0)	33F 8x 9M(55.9)	9F 1x(90.0)	49F 9x 9M(64.5)	F: 2Met	13 (1)	1987–2019	–
2. Anxiety disorder in general	14.4.7.4	–	3M 2x 1F	6F	36F 3x(92.3)	8F 1x	51F 6x 3M(81.0)	–	15 (4)	1970–2013	–
3. Panic disorder	14.4.7.7	–	–	2F	13F (100.0)	12F(100.0)	27F (100.0)	–	5 (1)	1984–2012	–
4. Phobias in general	14.4.7.9	–	2x	1F	15F 2x(88.2)	20F 1x(95.2)	34F 5x(87.8)	F: 1Gen	10 (1)	1966–2012	–
5. Specific phobias	14.4.7.10	–	2F	1F 1x	15F 1x(93.8)	3F	21F 2x(91.3)	–	8 (1)	1969–2010	–

Table 25.14.11 Attention-related disorders

Variable	*Orig. table*	*Number of sex difference findings (and consistency scores)*						*Overview*	*Countries*	*Time range*	*Non-human*
		I/T	*Child*	*Adol*	*Adult*	*WAR*	*All ages*				
1. Attention deficit hyperactivity disorder (ADHD)	14.4.8.1	–	92M (100.0)	15M 1x(93.8)	6M	11M 1x (91.7)	124M 2x(98.4)	M: 2Rev 3Met	18	1962–2015	–
2. Attention deficit disorder (ADD)	14.4.8.2	IM	6M	2M	1M	–	10M (100.0)	M: 1Rev	4	1975–2015	–
3. Autism spectrum disorder (ASD)	14.4.8.3	5M	44M (100.0)	1M	2M 1x	16M (100.0)	68M 1x(98.6)	M: 9Rev 3Met	16	1962–2018	–

Table 25.14.12 Residual mental disorders

Variable	*Orig. table*	*Number of sex difference findings (and consistency scores)*						*Overview*	*Countries*	*Time range*	*Non-human*
		I/T	*Child*	*Adol*	*Adult*	*WAR*	*All ages*				
1. Dissociative disorder/ dissociative thinking	14.4.8.4	–	1x	–	10F 5x 1M (58.8)	–	10F 6x 1M (55.6)	x: 1Met	4	1996–2003	–
2. Gender identity disorder	14.4.8.7	–	3M	1M	3M	3M	10M (100.0)	–	3	1937–2018	–

all studies have found males being more likely than females to experience this type of disorder.

25.14.13 Suicidal Behavior

Although suicidal behavior should not be considered a mental disorder per se, it is often associated with intense feelings of depression. Very larger numbers of studies of suicide, including thoughts of suicide and suicide attempts, have been conducted. Findings are summarized in Table 25.14.13. In the case of suicide ideation, one can see that findings have been quite mixed with a distinct leaning toward more females than males reporting having had such thoughts. Similarly, the second row of the table shows that most studies have concluded that more females than males report actually attempting suicide.

In the case of actually succeeding in ending one's life, most research indicates that males do so more than females. However, one may notice that there are a substantial number of findings that reported higher rates of completed suicide among females. Only among adolescents were all 34 studies in agreement that males commit suicide at significantly higher rates than females.

A number of studies have focused on suicidal behavior specifically among persons with manic (bipolar) depression. We tallied results from these findings separately. In the fourth row of the summary table, one can see that most of these studies indicate that greater proportions of females with manic depression attempt or complete suicide than males with this particular mental disorder, although there are several exceptions.

The table's final entry has to do with the lethality of the methods used to commit suicide. This basically means the extent to which individuals use suicide methods that have very high probability of "success," such as firearms and falls from great distances, in contrast to overdosing on sedatives or arm slashing. While exceptions have been reported, most studies indicate that male attempts to kill themselves involve more lethal methods than the methods used by females.

25.15 Attitudes, Beliefs, Interests, and Preferences: Condensed Findings

People's attitudes and beliefs, along with their interests and preferences, are delightfully diverse (at least to those with an open mind). Chapter 15 documented findings regarding average sex differences in the diversity of these ideas. In this section, all of the sex differences in Chapter 15 for which ten or more pertinent findings were located are summarized in a condensed form.

Table 25.14.13 Suicidal behavior

Variable	*Orig. table*	*Number of sex difference findings (and consistency scores)*						*Overview*	*Countries*	*Time range*	*Non-human*
		I/T	*Child*	*Adol*	*Adult*	*WAR*	*All ages*				
1. Suicide ideation	14.4.9.1	–	–	33F 4x(89.2)	10F 14x 3M(33.3)	8F 1x 3M(53.3)	51F 19x 6M(62.2)	F: 2Rev	18 (3)	1972–2019	–
2. Attempted suicide	14.4.9.2	–	1M	40F 6x 5M (71.4)	17F 3x 2M(70.8)	56F 5x 6M(76.7)	113F 14x 14M (72.9)	F: 3Rev	14 (5)	19152–20-10	–
3. Completed suicide	14.4.9,3	–	–	34M (100.0)	42M 2x 5F (77.8)	176M 7x 18F (80.4)	252M 9x 23F(82.1)	M: 3Rev 2Gen F: 1Gen	34 (44)	1881–2020	–
4. Suicide among manic depressives	14.4.9.4	–	–	–	11F 2x 4M(52.4)	–	11F 2x 4M(52.4)	–	6	1921–1988	–
5. Lethality of suicide methods	14.4.9.6	–	–	2M	4M	15M 4F (65.2)	21M 4F(72.4)	–	10	1986–2014	–

25.15.1 Attitudes and Values

The first table having to do with sex differences in attitudes has to do with broad categories. As shown in row 1 of Table 25.15.1, most studies have indicated that males are more opinionated than females. However, most studies have found no significant differences in people's susceptibility to persuasion, although when differences are found females appear to be more susceptible than males (row 2). In the final row, 11 studies having to do with sex differences in valuing power were located, all of which indicated that males value this trait more than do females.

25.15.2 Moral Reasoning and Moral Beliefs

Sex differences in moral beliefs and reasoning have been extensively studied over the years, with some of the earliest studies dating back a century. Eight categories of such attitudes are summarized in Table 25.15.2.

Substantial evidence has assessed sex differences in what are known as levels of (or stages in) moral reasoning. The most basic levels involve equating right and wrong with personal pleasure and pain, while the highest levels only consider the principles that are being promoted or violated by a moral action. Row 1 shows that the vast majority of studies have found no significant sex differences in moral reasoning levels.

Even though most studies have found no sex differences in moral reasoning levels, there may be differences in how the sexes arrive at roughly equal levels. Row 2 shows that most studies have concluded that females are more likely to arrive at their highest levels of moral reasoning by *caring* a great deal about others. Most studies have indicated that males seem to arrive at their highest levels of moral reasoning by adopting a *justice* approach to morality, as shown in row 3.

Substantial research has compared the sexes regarding their attitudes concerning ethical business practices. Row 4 shows that, while many studies have found no significant sex differences, when differences are found, they all indicate that females are more ethical in business dealings.

Row 5 shows that considerable research has assessed sex differences in people's willingness to accept physical aggression as an acceptable response to interpersonal conflict. All of the findings agree that males consider such responses more acceptable than do females.

When all is said and done, is the world just or is it just about as likely to punish those who do good and reward those who don't (epitomized by the expression "No good deed goes unpunished")? People's beliefs in this regard are known as *just world beliefs* (*JWB*), and various questionnaire items have been developed for measuring their opinions. Among the questions addressed by JWB researchers has been whether there are identifiable sex differences. In row 6, one can see that most studies have

Table 25.15.1 Attitudes and values

Variable	*Orig. table*	*Number of sex difference findings (and consistency scores)*						*Overview*	*Countries*	*Time range*	*Non-human*
		I/T	*Child*	*Adol*	*Adult*	*WAR*	*All ages*				
1. Being opinionated	15.1.1.1	–	1M	4M	4M 1F	3M 1F	12M 2F(75.0)	–	2	1958–2000	–
2. Susceptibility to persuasion	15.1.1.4	–	–	4F 3M	19F 50x 2M (26.0)	1F	24F 50x 5M(28.6)	F:1Met; x:1Rev 1Met; M:1Rev	10	1925–2005	–
3. Valuing power	15.1.2.5	–	1M	–	9M	1M	11M (100.0)	–	4 (2)	1987–2011	–

Table 25.15.2 Moral reasoning and moral beliefs

Variable	*Table Orig.*	*Number of sex difference findings (and consistency scores)*						*Overview*	*Countries*	*Time range*	*Non-human*
		I/T	*Child*	*Adol*	*Adult*	*WAR*	*All ages*				
1. Level of (or stage in) moral reasoning	15.2.1.1	–	3M 16x 2F (13.0)	3M 27x 10F (6.0)	10M 19x 3F (28.6)	2M 9x 1F (15.4)	18M 71x 16F (14.9)	M: 1Met; x: 3Rev; F: 3Met	18	1924–2014	–
2. Adopting a caring approach to moral reasoning	15.2.1.2	–	6F 8x 1M(37.5)	18F 2x(90.0)	36F 11x 3M (67.9)	4F 3x	64F 24x 4M (66.7)	F: 1Met	9(3)	1979–2020	–
3. Adopting a justice approach to moral reasoning	15.2.1.3	–	3M	1M 1F	15M 6x 2F (60.0)	–	19M 6x 3F(61.3)	M: 1Met	6(1)	1983–2020	–
4. Ethical reasoning in business dealings	15.2.1.8	–	–	–	21F 17x(55.3)	–	21F 17x(55.3)	–	4	1978–2012	–
5. Preference for, or acceptance of, aggressive responses to conflicts	15.2.1.10	–	9M	1M	17M (100.0)	1M	28M(100.0)	–	4	1927–2002	–
6. Belief in a just world	15.2.2.3	–	–	2M 1x	13M 20x 6F (28.9)	1x2F	15M 22x 8F (28.3)	M: 1Met	9	1980–2016	–
7. Importance of humane treatment of animals	15.2.2.5	–	–	1F	9F	16F(100.0)	26F(100.0)	F: 1Rev	4	1980–2001	–
8. General concern for the feelings and welfare of others	15.2.2.6	–	–	3F	6F 1x	–	9F 1x(90.0)	–	1	1976–2018	–

found no significant sex differences in such beliefs, although when there are differences, males appear to subscribe to JWBs more than do females.

Twenty-six studies have assessed sex differences in the importance people place on behaving humanely toward other animals. Row 7 shows that all of these studies have concluded that females subscribe to this principle more than is the case for males, making this a strong candidate for being a universal sex difference.

This table's last row has to do with the importance of being concerned with the feelings and welfare of other human beings. It shows that nine out of ten of the available studies concluded that females consider this type of concern to be more important than do males.

25.15.3 Attitudes and Beliefs Regarding the Paranormal

Ten studies of sex differences were found regarding people's beliefs in astrology or palm reading. As shown in row 1 of Table 25.15.3, all ten studies concluded that females have favorable or supportive beliefs of this nature to a greater degree than males.

Twenty-seven studies of sex differences in beliefs in extrasensory perception and paranormal phenomena were located. Row 2 shows that 19 of these studies indicated that females held such belief to a greater degree than males.

25.15.4 Supernatural and Religious Beliefs and Attitudes

Beliefs and attitudes of a religious nature have been widely studied in terms of possible sex differences. Table 25.15.4 provides a summary of the results for the six beliefs and attitudes for which at least ten findings were located. Row 1 has to do with people's overall religiosity, usually based on people's self-assessments. One can see that the overwhelming majority of studies have concluded that females are more religious than males. Nonetheless, there are a sufficient number of exceptions so as to remove this variable from being considered a possible universal sex difference.

In rows 2 and 3, findings from studies of the intensity of people's religious faith and what is known as religious saliency are summarized. For both of these variables, females have been found to have stronger faith and consider religion more central to their lives, but there are several exceptional findings.

Belief in God is at the core of most religions. Studies of sex differences in this belief largely indicate that females are more likely to be confident in the existence of God than are males, although, as shown in row 4, exceptions have also been found.

Table 25.15.3 Attitudes and beliefs regarding the paranormal

Variable	*Orig. table*	*Number of sex difference findings (and consistency scores)*						*Overview*	*Countries*	*Time range*	*Non-human*
		I/T	*Child*	*Adol*	*Adult*	*WAR*	*All ages*				
1. Belief in astrology or palm reading	15.3.2.1	–	–	–	10F(100.0)	–	10F(100.0)	–	5	1972–1990	–
2. Belief in (or experience with) extrasensory perception and the paranormal	15.3.2.3	–	–	–	19F 4x 3M (65.5)	1x	19F 5x 3M (63.3)	–	7	1971–2017	–

Table 25.15.4 Supernatural and religious beliefs and attitudes

Variable	*Orig. table*	*Number of sex difference findings (and consistency scores)*						*Overview*	*Countries*	*Time range*	*Non-human*
		I/T	*Child*	*Adol*	*Adult*	*WAR*	*All ages*				
1. Religiosity in general	15.3.3.1	–	8F 1M	17F 1M(89.5)	77F 7x 3M(85.6)	10F (100.0)	112F 7x 5M(86.8)	F: 7Rev M: 1Rev	26 (1)	1931–2019	–
2. Religious faith, intensity of	15.3.3.2	–	–	2F	8F 5x(61.5)	6F	16F 5x(76.2)	–	9 (1)	1943–2020	–
3. Importance of religion to one's life (religious saliency)	15.3.3.3	–	1F	11F 1x 1M (78.6)	29F 4x(87.9)	8F	49F 5x 1M(87.5)	–	11 (3)	1953–2013	–
4. Belief in God or a supreme being	15.3.3.5	–	–	3F 1x 1M	12F 4x (75.0)	2F	17F 5x 1M(70.8)	–	6 (5)	1971–2013	–
5. Religious fundamentalism	15.3.3.6	–	–	–	8F 5x(61.5)	–	8F 5x(61.5)	–	6 (1)	2001–2016	–
6. Belief in life after death	15.3.3.10	–	–	1F	18F 3x 3M(66.7)	–	19F 3x 3M(67.9)	–	6 (2)	1985–2017	–

Religious fundamentalism refers to tendencies to believe whatever is asserted in a particular religion's sacred writings reflect the will of God. Row 5 reveals that, when significant sex differences are found, females are more likely to consider themselves to be fundamentalists than are males.

The last entry in this table has to do with belief in the existence of a conscious life following death. As shown in row 6, most studies have concluded that greater proportions of females than males hold such beliefs.

25.15.5 Sensory Preferences

Table 25.15.5 has to do with sensory preferences, particularly regarding color. The initial entry (row 1) pertains to sex differences in preferences for sweet tastes. Nineteen out of 26 findings point toward females having stronger preferences for sweet tastes than do males. It is interesting to note that 14 studies of sex differences among rodents have all found that females prefer sweet tastes more than do males.

Regarding preferences for colors, four colors were located for which at least ten studies of sex differences were found. One can see in rows 2 through 5 that most findings suggest that males prefer blue and green more and that females usually prefer pink and red more. Nonetheless, none of the sex differences in color preferences were consistent enough to be considered culturally universal.

25.15.6 Residual Aspects of Sensory Preferences

Forty-eight studies were found to have assessed sex differences in the speed with which individuals habituate (i.e., stop paying attention) to stimuli. While the type of stimuli may be relevant, one can see in row 1 of Table 25.15.6 that most studies have found no significant sex differences in habituation tendencies.

The other row in this brief table has to do with sex differences in the tendency to give attention to faces. One can see that all but one of the studies concluded that females spend more time looking at faces (nearly always of humans) than do males. However, the one exception reported males looking longer at faces than did females.

25.15.7 Attitudes Pertaining to Sex Differences

Findings regarding three attitudes having to do with sex roles and gender equality are presented in Table 25.15.7. Row 1 has to do with sex differences in people's attitudes toward maintaining distinct sex roles for males and for females. One can see that 30 out of 35 findings indicated that males preferred maintaining distinct roles more than did females.

Table 25.15.5 Sensory preferences

Variable	*Orig. table*	*Number of sex difference findings (and consistency scores)*						*Overview*	*Countries*	*Time range*	*Non-human*
		I/T	*Child*	*Adol*	*Adult*	*WAR*	*All ages*				
1. Presence for sweet tastes	15.4.1.5	1F 2x	1M	4F 3M	4F 1M	10F(100.0)	19F 2x 5M(61.3)	–	3 (1)	1958–2014	14F rodent
2. Preference for blue	15.4.3.4	1x	10M 3x 1F(66.7)	1M 1F	10M (100.0)	3M	24M 4x 1F(80.0)	–	11	1927–2018	–
3. Preference for green	15.4.3.6	–	6M 1F	3M 1x 1F	–	1x	9M 2x 2F(60.0)	–	8 (1)	1941–2019	–
4. Preference for pink	15.4.3.8	1x	10F 4x(71.4)	1F	11F	–	22F 5x(81.5)	–	14 (1)	1981–2021	–
5. Preference for red	15.4.3.10	1x	5F	1M	4F 4x 2M(33.3)	–	9F 5x 3M(45.0)	–	8 (1)	1931–2019	–

Table 25.15.6 Residual aspects of sensory preferences

Variable	*Orig. table*	*Number of sex difference findings (and consistency scores)*							*Overview*	*Countries*	*Time range*	*Non-human*
		Fetal	*I/T*	*Child*	*Adol*	*Adult*	*WAR*	*All ages*				
1. Habituation to stimuli in general	15.4.4.3	1M	7F 35x 5M (13.5)	–	–	–	–	7F 35x 6M(13.0)	–	2	1964–1991	–
2. Preference for/ attention given to faces (usually as opposed to inanimate objects)	15.4.4.4	–	12F 1M(85.7)	–	–	1F	–	13F 1M(86.7)	–	3	1966–2009	–

Table 25.15.7 Attitudes pertaining to sex differences

Variable	*Orig. table*	*Number of sex difference findings (and consistency scores)*						*Overview*	*Countries*	*Time range*	*Non-human*
		I/T	*Child*	*Adol*	*Adult*	*WAR*	*All ages*				
1. Attitudes toward maintaining distinct sex roles	15.5.1.1	–	–	4M	26M 5F(72.2)	–	30M 5F(75.0)	–	6 (1)	1975–2007	–
2. Attitude toward gender equality/ women's rights in general	15.5.2.2	–	–	7F	38F 3x(92.7)	4F 3x	49F 6x(89.1)	F: 1Met	16 (3)	1971–2006	–
3. Attitude toward gender equality in political participation	15.5.2.7	–	–	2F	5F 2x 1M(55.6)	–	7F 2x 1M(63.3)	–	2	1973–1997	–

In row 2, studies are reviewed regarding sex differences in attitudes toward gender equality (or women's rights) in general. It indicates that of the 55 findings located, 49 indicated that females expressed more support for gender equality than did males.

Attitudes toward allowing women to participate equally in the political arena, including running for political offices, have been assessed in ten studies. One can see that seven of these studies concluded that females are more supportive of these types of attitudes than are males.

25.15.8 Strength of Sex Drive

Two types of measures of sex differences in the strength of sex drives were found to have ten or more pertinent findings. These are summarized in Table 25.15.8. As shown in row 1, nearly all studies have concluded that males express stronger average desires for sex than do females. Similarly, row 2 reveals that over 50 studies have found males reporting that they have stronger desires for (or interests in) promiscuous sexual relationships than do females.

25.15.9 Sexual Orientation

Sexual orientation has to do with which sex one finds attractive. Table 25.15.9 shows that five different variables pertaining to these attractions (sometimes including actual sexual experiences) are presented. As a point of reference, one may keep in mind that most studies have found over 90% of both males and females reporting being more or less exclusively attracted to members of the opposite sex. These individuals would be considered heterosexuals. The issue addressed in this table has to do with possible sex differences in sexual orientation.

In row 1, one can see that the evidence of any significant sex differences in heterosexual preferences is quite inconsistent. Out of 41 studies, 23 concluded that greater proportions of males were heterosexual, while 11 studies indicated females were. Many of the inconsistencies are likely due to differences in how questions are worded or on how exclusive one's heterosexual preferences have to be in order to meet the criteria set in various studies.

Homosexual preferences regarding sexual feelings have been extensively studied. Row 2 shows that 20 out of 24 studies concluded that homosexuality is more common among males, with the remaining four studies finding no significant differences between males and females.

Bisexual preferences pertain to individuals who experience being sexually attracted to both males and females. As shown in row 3, all but one of the 20 pertinent studies have concluded that females are more likely to be

Table 25.15.8 Strength of sex drive

Variable	Orig. table	Number of sex difference findings (and consistency scores)						Overview	Countries	Time range	Non-human
		I/T	Child	Adol	Adult	WAR	All ages				
1. Sex drive, strength of (desire for sex)	15.6.1.1	–	–	3M	50M 1F(**96.2**)	1M	54M 1F(**96.4**)	M: 2Rev; M: 1Met; x: 1Gen	11 (3)	1967–2018	–
2. Desire for promiscuous sexual relationships	15.6.1.6	–	–	4M	46M 1x(**97.9**)	3M	53M 1x(**98.1**)	M: 1Gen 1Met	8 (4)	1973–2021	–

Table 25.15.9 Sexual orientation

Variable	Orig. table	Number of sex difference findings (and consistency scores)						Overview	Countries	Time range	Non-human
		I/T	Child	Adol	Adult	WAR	All ages				
1. Heterosexuality (exclusive or predominant attraction to the opposite sex)	15.6.2.1	–	1F	1M 1x	22M 5x 9F (48.9)	1F 1x	23M 7x 11F (44.2)	–	12 (2)	1930–2016	–
2. Homosexuality (exclusive or predominant attraction to one's own sex)	15.6.2.2	–	–	1M 1x	15M 3x(83.3)	4M	20M 4x(83.3)	M: 1Met	8 (2)	1988–2019	–
3. Bisexual preference (substantial sexual attraction to both sexes)	15.6.2.3	–	–	4F	14F 1x(93.3)	1F	19F 1x(**95.0**)	F: 1Rev	6 (2)	1992–2019	–
4. Ambivalent sexual orientation	15.6.2.4	–	–	–	17F(**100.0**)	–	17F(**100.0**)	F: 1Rev	4	1934–2013	–
5. Sexual experience with same sex (homosexual behavior)	15.6.2.6	–	–	3M 2F	13M 3x 5F (50.0)	4M 4x	20M 7x 7F (48.8)	–	6 (2)	1934–2016	–

sexually attracted to both sexes. Row 4 shows that similar conclusions for a concept closely related to bisexuality, known as ambivalent sexual orientation, i.e., tendencies to be at least mildly attracted to both males and females. In this case, all 17 studies concluded that females had higher probabilities of exhibiting ambivalent sexual orientations than was true for males.

The overall pattern in terms of sex differences in sexual orientation (as reported in rows 1 through 4) point toward the following conclusions: First, roughly equal proportions of both sexes are more or less exclusively heterosexual. Second, when there are significant sex differences, more males are more or less exclusively homosexual, while most females are either bisexual or ambiguous in their sexual orientation.

The final entry in this table had to do with the extent to which the sexes have had actual non-heterosexual sexual experiences. One can see in row 5 that results are mixed, although most suggest that males report having had these experiences more than females.

25.15.10 Attitudes toward Sexual Orientation

As noted above, the vast majority of both sexes are heterosexual in terms of their sexual attractions. Negative attitudes toward people whose sexual orientation is not heterosexual, especially if their preferences are made known to others, are common in most, and possibly all, human societies. Since the mid-1900s, social scientists have sought to assess sex differences in people's attitudes toward homosexuality (or non-heterosexuality). Findings for variables that have been studied by at least ten studies in this regard will now be summarized.

Row 1 of Table 25.15.10 shows that 55 of the 98 findings having to do with unfavorable attitudes toward homosexuality in either sex indicate that males have more unfavorable attitudes than do females. The remaining studies are roughly equally divided between those indicating that females have more unfavorable attitudes and those reporting no significant differences.

A large number of studies have asked people how willing they are to accept homosexuals into their social sphere or otherwise have favorable feelings toward homosexuals or their behavior. One can see in row 2 that 86 out of 89 studies have concluded that females are more inclined to have favorable or accepting feelings toward homosexuals (or their behavior) than are males.

The last two rows in this table summarize findings of studies that have asked people the degree to which they have unfavorable attitudes toward either *male* homosexuality or *female* homosexuality (rather than homosexuality regardless of sex). Findings in row 3 show that most studies have

Table 25.15.10 Attitudes toward sexual orientation

Variable	*Orig. table*	*Number of sex difference findings (and consistency scores)*						*Overview*	*Countries*	*Time range*	*Non-human*
		I/T	*Child*	*Adol*	*Adult*	*WAR*	*All ages*				
1. Unfavorable attitudes toward homosexuality in either sex	15.6.3.1	–	–	3M	44M 21x2F (63.8)	8M 1x	55M 22x 21F (46.2)	M: 2Met x: 1Met	9	1965–2013	–
2. Favorable/ accepting attitudes toward homosexuals or homosexual behavior	15.6.3.2	–	–	3F 1x	74F 2x(**97.4**)	9F	86F 3x(**96.6**)	F: 1Met	14 (1)	1965–2006	–
3. Unfavorable attitude toward male homosexuality	15.6.3.3	–	–	1M	40M 6x 2F (80.0)	3M	44M 6x 2F (81.5)	M: 3Met	9	1974–2006	–
4. Unfavorable attitude toward female homosexuality	15.6.3.4	–	–	1F	21M 5x 10F (45.7)	1M 2x 4F	22M 7x 15F (37.3)	–	7	1974–2006	–

found that males are more likely than females to express negative attitudes toward male homosexuals (i.e., gays) than toward female homosexuality (i.e., lesbians). However, one can see in row 4 that the sex differences in unfavorable attitudes toward female homosexuals (or their behavior) are considerably more evenly divided, although there is still a tendency for most studies to report males having more unfavorable attitudes than do females.

25.15.11 Aspects of Sexual Attractions and Preferences Other Than Sexual Orientation

Table 25.15.11 provides a summary of findings regarding sex differences in various aspects of sexual attractions and preferences except for those involving sexual orientation.

Row 1 summarizes sex differences in the tendency to attribute sexual meaning to interactions between the sexes. It shows that all 27 studies concluded that males are more likely to do so than are females.

In row 2, sex differences in tendencies to fall in love soon after meeting a romantic partner are assessed. The results indicate that 11 out or 15 studies concluded that males have a tendency to fall in love more rapidly than females.

Eleven studies were found in which samples of men and women were asked how much time should lapse between a couple being mutually sexually attracted to one another and the time they should first have sexual relationships. Row 3 shows that all studies concluded that males prefer or consider a shorter time more acceptable.

Sexual arousal from visual erotica and pornography has been extensively investigated with respect to possible sex differences. In row 4, one can see that if there are sex differences, they favor males being more aroused than females, although several studies found no significant sex difference.

There are physiological measures of sexual arousal that can be used with both sexes. In ten studies, these measures were compared to self-reported sexual arousal. Row 5 shows that all of these studies concluded that the two types of measures correlated more strongly for males than for females.

Which sex is choosier when it comes to selecting sex partners? As shown in row 5, nine out of ten studies concluded that females are choosier than males.

Twenty-eight studies assessed sex differences in the degree to which individuals find sexually oriented photographs, movies, and videos enjoyable. Row 7 shows that 21 studies concluded that males did, with the remaining seven indicating that there were no significant sex differences.

Table 25.15.11 Aspects of sexual attractions and preferences other than sexual orientation

Variable	*Orig. table*	*Number of sex difference findings (and consistency scores)*						*Overview*	*Countries*	*Time range*	*Non-human*
		I/T	*Child*	*Adol*	*Adult*	*WAR*	*All ages*				
1. Attribution of sexual meaning/motivation to male-female interactions	15.6.4.5	–	–	–	27M (100.0)	–	27M(100.0)	-	2	1974–2006	–
2. Falling in love, speed of	15.6.4.6	–	1F	1x	11M 2x(84.6)	–	11M 3x 1F (68.8)	–	2	1930–2005	–
3. Preference for a shorter time between sexual attraction and sexual intimacy	15.6.4.9	–	–	–	11M (100.0)	–	11M(100.0)	–	1	1977–2001	–
4. Sexual arousal from visual erotica and pornography	15.6.4.15	–	–	–	12M 7x (63.2)		12M 7x (63.2)	–	4	1970–2016	–
5. Relationship between self-reported sexual arousal and physiological measures	15.6.4.16	–	–	–	10M (100.0)	–	10M(100.0)	M: 1Met	2	1977–2007	–
6. Choosiness in selecting sex partners	15.6.4.18	–	–	–	9F 1x(90.0)	–	9f 1x(90.0)	–	1 (1)	1979–2006	–
7. Enjoyment of sexually oriented pictures, movies, and videos	15.6.4.22	–	–	–	21M 7x (75.0)	–	21M 7x (75.0)	M: 1Met	7	1970–2015	–
8. Interest in (aroused by) visual aspects of sexuality	15.6.4.28	–	–	1M	14M 1x 2F (73.7)	–	15M 1x 2F (75.0)	M: 1Met	2	1965–2005	–

Along similar lines, row 8 shows a summary of results from studies undertaken to assess the degree to which people are specifically interested in, or aroused by, visual aspects of sexuality. One can see that most, but not all, research has indicated that males are more drawn to visual elements of sex than are females.

25.15.12 Mate Preferences

Research having to do with sex differences in mate preferences is substantial. It is numerically summarized in Table 25.15.12. One can see that, except for eight studies reporting no significant differences, the vast majority of studies (i.e., 137) indicate that males express stronger preferences for mates who are physically attractive than do females. However, the consistency score for this preference is just slightly below the 95.0 cutoff point, i.e., it is 94.8.

Rows 2 and 3 indicate that there are universal sex differences in height preferences. Specifically, all but one of 27 studies found females preferring mates who were taller than themselves (the exception finding no significant difference). In the case of males, all 16 studies indicate that they express a preference for mates who are shorter than themselves.

Regarding age of one's preferred mate, rows 4 and 5 indicate that these too are universal sex differences. Whereas all 57 studies of preferences for younger mates found this preference stronger among males, all 26 studies of preferences for older mates found this preference stronger among females.

One more apparent universal sex difference in mate preferences is identified in row 6. It indicates that females are more likely to prefer mates with access to resources (typically in the form of wealth or income). Of the 105 studies, 102 reported this to be statistically significant, while the remaining three studies found no significant sex difference.

As shown in rows 7 and 8, preferences for mates who are intelligent or highly educated as well as ambitious or industrious were found to be more common among females than among males with just a few exceptions. In the case of preferring mates who are chaste or who have never before had sex, row 9 shows that findings have been somewhat mixed, but most indicated that males have such preferences more than females.

25.15.13 Attitudes Surrounding Sex, Reproduction, and Health

Four types of findings are listed in Table 25.15.13 regarding sex differences in people's attitudes toward sex, reproduction, and health. Regarding sex, row 1 shows that all ten relevant studies agree that females want to be in love before having sexual relationships more than is the case for males.

Table 25.15.12 Mate preferences

Variable	*Orig. table*	*Number of sex difference findings (and consistency scores)*						*Overview*	*Countries*	*Time range*	*Non-human*
		I/T	*Child*	*Adol*	*Adult*	*WAR*	*All ages*				
1. Preference for a physically attractive mate	15.6.5.2	–	–	–	131M 8x (94.2)	14M **(100.0)**	145M 8x(94.8)	M: 1Gen 4Rev 1Met	23 (15)	1936–2022	–
2. Preference for a mate who is taller than oneself	15.6.5.4	–	–	1F	25F 1x**(96.2)**	–	26F 1x**(96.3)**	–	10 (1)	1977–2020	–
3. Preference for a mate who is shorter than oneself	15.6.5.5	–	–	–	16M **(100.0)**	–	16M **(100.0)**	M: 1Met	8	1977–2020	–
4. Preference for a mate who is young or younger than oneself	15.6.5.6	–	–	–	48M **(100.0)**	9M	57M **(100.0)**	M: 1Gen	17 (7)	1976–2020	–
5. Preference for a mate who is old or older than oneself	15.6.5.7	–	–	3F	21F **(100.0)**	2F	26F **(100.0)**	–	10 (4)	1989–2020	–
6. Preference for a mate with resources (wealth, high income)	15.6.5.9	–	–	3F	93F 3x**(96.8)**	9F	105F 3x**(97.2)**	F: 1Gen2Rev 1Met	27 (13)	1958–2022	–
7. Preference for a mate with high intelligence or highly education	15.6.5.13	–	–	–	21F 2x(91.3)	–	21F 2x(91.3)	F: 1Met	8 (3)	1955–2015	–
8. Preference for a mate who is ambitious/ industrious	15.6.5.14	–	–	–	15F 1x 1M (83.3)	1F	16F 1x 1M (84.2)	F: 1Met	4 (4)	1945–1955	–
9. Preference for chastity or virginity in one's mate	15.6.5.19	–	–	–	11M 5x 3F (50.0)	–	11M 5x 3F (50.0)	–	8 (5)	1989–2019	–

Table 25.15.13 Attitudes surrounding sex, reproduction, and health

Variable	*Orig. table*	*Number of sex difference findings (and consistency scores)*						*Overvie-w*	*Countries*	*Time range*	*Non-human*
		I/T	*Child*	*Adol*	*Adult*	*WAR*	*All ages*				
1. Wanting to be in love before having sex	15.7.1.3	–	–	1F	9F	–	10F (100.0)	–	3	1976–2001	–
2. Permissive attitude toward premarital/casual sexual intercourse	15.7.1.4	–	–	9M	51M (100.0)	5M 1F	65M 1F(97.0)	M: 3Met	12 (1)	1975–2013	–
3. Attitude toward nudity and sexually explicit material	15.7.4.3	–	1M	1M	8M	–	10M (100.0)	–	2	1973–2000	–
4. Being health conscious	15.8.1.1	–	–	1F	12F (100.0)	3F	16F (100.0)	–	6 (2)	1974–2005	–

In the case of having permissive attitudes toward premarital sex (or casual sexual relationships), 66 studies were located regarding sex differences. As summarized in row 2, all but one of these studies found males expressing more positive (or less negative) attitudes toward premarital sex than was the case for females.

In ten studies of people's attitudes toward nudity and access to sexually explicit maternal (usually in the form of pornography). Row 3 shows that all of these studies concluded that males were more tolerant (or less intolerant) than were females.

The final row in this table has to do with tendencies to be health conscious. All 16 studies reached the conclusion that such tendencies are greater for females than for males. Overall, one can see that all four variables listed in this table are highly likely to be universal sex differences.

25.15.14 Academic and Intellectual Interests

Many studies have compared the sexes in terms of their academic and intellectual interests. Findings from these studies for which at least ten findings were located are summarized in Table 25.15.14.

The table's first row shows that 39 studies of sex differences in people's (primarily children and adolescents) enjoyment of school were located. Without exception, all of these studies concluded that females express greater enjoyment of school than do males, making this an almost certain universal sex difference.

In row 2, the studies of sex differences in ratings given to teaching quality by college students are provided. One can see that ten of the 12 pertinent studies concluded that females provide higher average ratings than do males, with the remaining two studies reporting no significant sex difference.

Row 3 has to do with sex differences in the number of years of education individuals seek to acquire. Of the 13 located studies, 12 of them concluded that males sought to obtain more education than did females, with the remaining study reporting no significant difference. Obviously, this seems somewhat at odds with evidence presented in row 1 that females report enjoying school more.

The remaining five rows in this table have to do with differences in interests in various academic subject areas. The only area for which all of the evidence is in agreement involved the biological and health sciences (row 7), where females consistently expressed a greater interest than did males. Findings do not attain 95.0 consistency scores for any of the other subject areas. Nevertheless, most research suggests that females have greater interest in language-focused subjects (row 4), while males have more interest in math (row 5), science in general (row 6), and in the social sciences (row 8).

Table 25.15.14 Academic and intellectual interests

Variable	*Orig. table*	*Number of sex difference findings (and consistency scores)*						*Overview*	*Countries*	*Time range*	*Non-human*
		I/T	*Child*	*Adol*	*Adult*	*WAR*	*All ages*				
1. Enjoyment of (liking) school in general	15.9.1.1	–	22F (100.0)	16F (100.0)	1F	–	39F (100.0)	–	10 (2)	1956–2013	–
2. Ratings given to teaching quality	15.9.1.2	–	–	–	10F 2x(83.3)	–	10F 2x(83.3)	–	2	1955–2006	–
3. Desired years of education	15.9.1.8	–	–	12M 1x(92.3)	–	–	12M 1x(92.3)	–	1	1955–1988	–
4. Interest in/ enjoyment of language-focused academic subjects	15.9.2.2	–	5F 1x	11F 1x(91.7)	–	–	16F 2x(88.9)	–	4	1922–2011	–
5. Interest in/ enjoyment of mathematics	15.9.2.3	–	8M 2x 1F(66.7)	66M 15x 10F (65.3)	2M	1M	77M 17x 11F (66.4)	M: 3Met	48 (3)	1924–2019	–
6. Interest in/ enjoyment of science in general	15.9.2.5	–	10M (100.0)	96M 11F(81.4)	8M	2M	116M 11F(84.1)	M: 2Rev5Met	34 (7)	1922–2020	–
7. Interest in/ enjoyment of the biological and health sciences	15.9.2.6	–	4F	8F	–	–	12F (100.0)	–	5	1979–2006	–
8. Interest in/ enjoyment of social science/ social studies	15.9.2.7	–	2M 1F	5M 2F	–	–	7M 3F(70.0)	–	4	1922–1996	–

25.15.15 Family-Related Attachments and Attitudes

Studies of sex differences in family-related attachments and values are summarized in Table 25.15.15. Row 1 of the table indicates that females have more pro-family attitudes than do males in all 14 studies that were located. These pro-family attitudes have to do with the degree to which individuals like spending time with family members and take an active interest in maintaining their health and well-being.

In row 2, sex differences in interest in spending time with infants and toddlers are summarized. One can see that, while four studies report no significant difference, the evidence from 45 studies point toward females having more interests of this nature than do males. Only in the case of adults is the consistency score higher than 95.0.

Row 3 has to do with sex differences in being more attached to one's mother than to one's father. Out of 12 pertinent studies, nine indicated that sons report having a greater attachment to their mother than to their father when compared to daughters.

25.15.16 Social Interests and Attachments

Table 25.15.16 provides a summary of findings regarding interests and attachments of a social nature. The first two rows have to do with two primary components of romantic attachments, the anxiety component and the avoidance component. While neither component exhibits anything close to consistent sex differences, one can see that most studies indicated that slightly more females than males exhibited the anxiety component. Also, more males exhibited more of the avoidance component than did females. While the former is consistent with three meta-analyses, the latter is inconsistent with these same meta-analyses.

Research findings having to do with preferences for intimate same-sex relationships (of a non-sexual nature) are summarized in row 3. Of the 12 studies, ten indicated that females had a stronger preference than did males, with the remaining two studies reporting no significant sex difference. Along similar lines, row 4 shows that all ten studies agreed that females are more interested in forming and maintaining close interpersonal relationships with either sex than is the case for males.

Sex differences in being interested in people more than in things has been studied by at least ten studies, covering the entire spectrum of age groupings. Row 5 shows that all of these studies agreed that females have such interests to a greater degree than males. It is worth noting that essentially the same sex difference will be noted when occupational interests are examined.

Finally, for this particular table, sex differences in interest in politics are summarized. One can see in row 6 that findings were mixed, although the

Table 25.15.15 Family-related attachments and attitudes

Variable	*Orig. table*	*Number of sex difference findings (and consistency scores)*						*Overview*	*Countries*	*Time range*	*Non-human*
		I/T	*Child*	*Adol*	*Adult*	*WAR*	*All ages*				
1. Pro-family attitudes and values	15.10.1.1	–	–	3F	11F (100.0)	–	14F (100.0)	–	4 (1)	1963–2006	–
2. Interest in spending time with infants and toddlers	15.10.1.4	2F	16F 2x (88.9)	6F 1x	23F 1x(95.8)	4F	51F 5x(91.1)	–	3	1957–2002	–
3. Greater attachment to mother than father	15.10.1.6	2M	2M 1x	5M 1x 1F	–	–	9M 2x 1F(69.2)	–	1	1935–2007	–

Table 25.15.16 Social interests and attachments

Variable	*Orig. table*	*Number of sex difference findings (and consistency scores)*						*Overview*	*Countries*	*Time range*	*Non-human*
		I/T	*Child*	*Adol*	*Adult*	*WAR*	*All ages*				
1. Romantic attachments, anxiety component	15.10.2.1a	–	–	1x 1F	19F 15x 12M (32.8)	–	19F 16x 13M (31.1)	F: 3Met	13 (1)	2002–2019	–
2. Romantic attachments, avoidance component	15.10.2.1b	–	–	1M 1x	23M 13x 10F (41.1)	–	24M 14x 10F (41.4)	F: 3Met	14 (1)	2002–2019	–
3. Preference for intimate (non-sexual) same-sex friendships	15.10.2.2	–	1F	–	8F 2x(80.0)	1F	10F 2x(83.3)	–	1	1977–1996	–
4. Interests in forming and maintaining close interpersonal relationships	15.10.2.4	–	–	4F	10F(100.0)	2F	16F(100.0)	–	4 (1)	1962–2016	–
5. Interest in people more than things	15.10.2.5	2F	2F	1F	3F	2F	10F(100.0)	F: 1Met x: 1Gen	4 (1)	1914–2012	–
6. Interest in politics	15.10.2.10	–	12M 3F (66.7)	3M 2F	18M 2x 2F (75.0)	3M	36M 2x 7F (69.2)	M: 1Gen	7 (5)	1944–2006	–

majority of studies indicate that males have a greater interest than do females, especially when national and international politics are separated from local politics. Then, essentially all studies agree that males are more interested in the former, while females are more interested in the latter.

25.15.17 Preferences Regarding Social Interactions

Four different preferences regarding social interactions are summarized in Table 25.15.17 regarding possible sex differences. Row 1 has to do with general preferences for socializing with others. One can see that all 20 studies concluded that females expressed stronger preferences in this regard than did males, making this a very probable universal sex difference.

Eleven studies of sex differences in the tendency to desire to be popular or liked by others. Ten of the 11 studies found females expressing stronger desires of this nature than males.

The tendencies that people have for preferring friendly and accommodating social interactions were explored in 16 studies. One can see in row 3 that the evidence was quite mixed regarding sex differences, with most studies reporting nonsignificant differences.

In 20 studies, both males and females were asked if they preferred to socialize with their own sex or with the opposite sex. Row 4 shows that 16 of these studies concluded that females preferred same-sex socializing more than did males.

25.15.18 Aesthetic and Recreational Preferences and Interests

Research findings having to do with sex differences in average preferences and interests of an aesthetic and recreational nature are summarized in Table 25.15.18. Row 1 is concerned with differences in overall aesthetic appreciation. One can see that, when sex differences have been identified, they indicate that females seem to have a greater aesthetic sense than do males.

The second row indicates that findings have been inconsistent regarding sex differences in enjoyment of shopping. To provide an explanation, all 14 studies that reported females enjoy shopping more had to do with brick-and-mortar shops, while the three studies reporting high enjoyment by males involved online shopping.

In the table's last row, research findings regarding sex differences in enjoyment of a wide variety of competitive activities are summarized. Of the 42 located studies, all but one concluded that males enjoyed competitive types of activities more than females, making this a probable cultural universal sex difference. This conclusion is consistent with two published overviews.

Table 25.15.17 Preferences regarding social interactions

Variable	*Orig. table*	*Number of sex difference findings (and consistency scores)*						*Overview*	*Countries*	*Time range*	*Non-human*
		I/T	*Child*	*Adol*	*Adult*	*WAR*	*All ages*				
1. Preference for socializing in general	15.10.3.1	2F	4F	2F	11F(100.0)	1F	20F(100.0)	–	4	1932–2007	–
2. Wanting to be popular or liked by others	15.10.3.2	–	1F	9F 1x(90.0)	–	–	10F 1x(90.9)	–	2	1961–1992	–
3. Preference for friendly and accommodating social interactions	15.10.3.3	–	1F 5x 1M	1F 9x(10.0)	3F 4x	1F 1x	6F 19x 1M (22.2)	–	5	1911–2004	–
4. Preferring to socialize with one's own sex (rather than the opposite sex)	15.10.3.7	1M	6F	1F	9F 3M(60.0)	–	16F 4M(66.7)	–	4	1978–1996	–

Table 25.15.18 Aesthetic and recreational preferences and interests

Variable	*Orig. table*	*Number of sex difference findings (and consistency scores)*						*Overview*	*Countries*	*Time range*	*Non-human*
		I/T	*Child*	*Adol*	*Adult*	*WAR*	*All ages*				
1. Aesthetic judgment and aesthetic appreciation	15.10.4.1	–	–	3F 1x	5F 3x	–	8F 4x(66.7)	–	6 (1)	1950–2005	–
2. Interest in, or enjoyment of, shopping	15.10.4.3	–	–	14F 3M(70.0)	–	–	14F 3M(70.0)	–	4	1985–2008	–
3. Interest in, or enjoyment of, competitive activities	15.10.4.4	–	8M	14M 1F(87.5)	11M (100.0)	8M	41M 1F(95.3)	M: 1Gen 1Rev	10 (4)	1977–2021	–

25.15.19 Interests in Sports

Considerable research has investigated sex differences in people's interests in sports. Findings are summarized in Table 25.15.19 in three separate rows. Results consistently point toward greater interest among males when compared to females. This is true for 17 studies of interests in sports in general (row 1), as well as interests in being a sports spectator (row 2) and in being an actual participant in sports (row 3).

25.15.20 Play and Leisure Time Preferences

Table 25.15.20 provides a summary of research findings regarding sex differences in six types of play and leisure time preferences. Row 1 shows that all 17 studies of the breadth of play and reading interests are broader for males than for females.

Regarding preferences for playing with (including collecting) dolls, 18 studies all concluded that females exhibit stronger preferences than do males (row 2). While there are no consistent sex differences in preferences for playing computer video games (row 3), for preferences for mechanical and construction objects of play, there were, with all 28 studies showing that males prefer these types of play objects more than females (row 4). The same consistent pattern was found for playing with vehicles, or objects resembling vehicles (row 5). It is worth adding that parallel studies of preferences for dolls and for vehicles as toys have been conducted with non-human primates, and the patterns are entirely consistent with what has been reported for human sex differences.

Finally, for this table, row 6 pertains to sex differences in preferences for adventure stories (or stories away from home). All ten relevant studies reported that males preferred these types of stories more than females.

25.15.21 Work-Related Preferences and Interests

A substantial body of research findings have accumulated regarding sex differences in work-related preferences and interests. As shown in rows 1 through 10 of Table 25.15.21, more than half of these sex differences appear to be culturally universal.

Row 1 has to do with attitudes toward females participating in the paid workforce. It shows that females expressed more support than did males in all 16 pertinent studies.

Thirty-seven studies provided research participants with lists of various male-typical occupations, and asked respondents to rate their degrees of interest in such lines of work. In every study, from childhood onward, males expressed greater degrees of interest in these types of occupations than did females (row 2).

Table 25.15.19 Interest in sports

Variable	Orig. table	Number of sex difference findings (and consistency scores)						Overview	Countries	Time range	Non-human
		I/T	Child	Adol	Adult	WAR	All ages				
1. Interest in sports in general	15.10.5.1	–	4M	8M	5M	–	17M (100.0)	M: 1Rev	3 (2)	1925–2005	–
2. Interest in spectator sporting activities	15.10.5.2	–	–	8M	13M (100.0)	3M	24M (100.0)	–	5	1970–2010	–
3. Interest in sports participation	15.10.5.3	–	8M	22M (100.0)	4M	3M	37M (100.0)	–	8 (1)	1930–2010	–

Table 25.15.20 Play and leisure time preference

Variable	Orig. table	Number of sex difference findings (and consistency scores)						Overview	Countries	Time range	Non-human
		I/T	Child	Adol	Adult	WAR	All ages				
1. Range in play and readings interest	15.10.7.4	–	4M	8M	5M	–	17M(100.0)	–	3	1933–2010	–
2. Preference for playing with dolls	15.10.7.8	12F (100.0)	14F (100.0)	1F	1F	–	28F(100.0)	F: 4Rev 1Met	4	1925–2019	F: 6Primat-es
3. Preference for playing computer video games	15.10.8.1	–	2M 3F	5M 1F	–	–	7M 4F(46.7)	–	2	1994–2000	–
4. Preference for mechanical and building objects of play	15.10.8.6	9M	16M (100.0)	2M	1M	–	28M (100.0)	M: 1Rev	7	1913–2015	–
5. Preference for vehicles as objects of play	15.10.8.7	12M (100.0)	9M	1M	1M	–	23M(100.0)	M: 3Rev 1Met	5	1931–2019	M: 5Primat-es
6. Preference for adventure (and away from home) stories	15.10.9.4	–	4M	6M	–	–	10M(100.0)	–	3	1926–1958	

Table 25.15.21 Work-related preference and interest

Variable	*Orig. table*	*Number of sex difference findings (and consistency scores)*						*Overview*	*Countries*	*Time range*	*Non-human*
		I/T	*Child*	*Adol*	*Adult*	*WAR*	*All ages*				
1. Attitudes toward female participation in the paid workforce	15.11.1.1	–	–	–	16F (100.0)	–	16F (100.0)	–	1	1975–1998	–
2. Preference for male-typical occupations	15.11.2.1	–	10M (100.0)	11M (100.0)	16M (100.0)	–	37M (100.0)	–	7 (2)	1936–2016	–
3. Preference for people-oriented occupations	15.11.2.5	–	6F 1x	9F	28F (100.0)	7F	50F 1x(98.0)	F: 2Rev 4Met	9	1922–2020	–
4. Preference for things-oriented occupations	15.11.2.6	–	2M	10M (100.0)	18M (100.0)	1M	31M (100.0)	M: 3Met	8	1933–2020	–
5. Interest in helping-oriented occupations	15.11.2.7	–	–	16F 2x (88.9)	11F (100.0)	2F	29F 2x (93.5)	F: 1Met	4	1931–2014	–
6. Preference for career advancement	15.11.2.11	–	–	5M 1x	3M 3x	–	8M 4x(66.7)	–	1 (1)	1982–1997	–
7. Preference regarding flexible hours	15.11.2.15	–	–	–	10F (100.0)	–	10F (100.0)	–	1 (1)	1981–2006	–
8. Breadth of occupational interests	15.11.2.25	–	4M	6M	2M	–	12M (100.0)	–	1	1957–1978	–
9. Preference regarding monetary compensation of jobs	15.11.2.28	–	1M	22M (100.0)	39M 9x(81.3)	8M 2x (80.0)	70M 11x(86.4)	M: 1Met	4 (1)	1922–2015	–
10. Preference regarding supervisory/power-oriented jobs	15.11.2.29	–	1M	11M 3x(78.6)	9M 4x 1F (60.0)	–	21M 7x 1F (70.0)	M: 1Met	3	1922–2009	–

Many occupations are people-oriented (such as teaching and counseling), while many other occupations are distinctly things-oriented (such as construction and engineering). Rows 3 and 4 indicate that, with the exception of one study, that found no significant difference, females express stronger preferences for being involved in the former. Males, however, had stronger average preferences for the latter.

Thirty-one studies of sex differences in people's interests in helping-oriented occupations (such as providing social services and health care to those in need) were located. As shown in row 5, all but two of these studies concluded that females considered this work feature more appealing than did males.

In 12 studies, both sexes were asked to rate their preference for career advancement as a feature of whatever job they want. Row 6 shows that eight of these studies concluded that males have stronger preferences for career advancement, while the remaining four indicated there were no significant sex differences. On the other hand, row 7 shows that all ten studies of preferences for flexible work schedules indicated that this was more important to females than to males.

Twelve studies sought to determine if the sexes differ in terms of the breadth of their occupational interests. Row 8 reveals that all 12 of these studies point toward males having broader interests than do females.

Eighty-one studies investigated sex differences in the importance given to the monetary compensation associated with a job as a criterion for making a career choice. Findings are summarized in row 9. While all 22 studies of adolescents indicated that males consider this to be more important than for females, by adulthood, several studies have reported no significant sex difference in this regard.

The final entry in this table shows that 29 studies were located regarding preferences for occupations that involve supervisory and managerial components activities. In 21 of these studies, males expressed a greater preference, but seven of the findings reported no significant sex difference and one found females reporting a greater preference.

25.15.22 Interests in Specific Types of Occupations

Table 25.15.22 summarizes the results of studies pertaining to four types of occupations regarding sex differences. The first line of work is that of artistic or expressive occupations such as artists and actors. Row 1 shows that 19 out of 20 studies concluded that females expressed more interest in such occupations than did males.

Sex differences in people's interests in computer programming are summarized in row 2. It shows that all 14 studies that were located

Table 25.15.22 Interests in specific types of occupations

Variable	*Orig. table*	*Number of sex difference findings (and consistency scores)*						*Overview*	*Countries*	*Time range*	*Non-human*
		I/T	*Child*	*Adol*	*Adult*	*WAR*	*All ages*				
1. Interest in artistic/creative/ expressive occupations	15.11.4.3	–	–	9F	8F	2F 1x	19F 1x(95.0)	F: 2Met x: 1Met	5	1935–2017	–
2. Interest in being computer programmers	15.11.4.17	–	–	10M (100.0)	3M	1M	14M (100.0)	–	5	1933–2005	–
3. Interest in being scientist	15.11.4.51	–	–	19M 2x (90.5)	3M	–	22M 2x(91.7)	–	4	1922–2017	–
4. Interest in being teachers	15.11.4.56	–	2F	5F	2F	1F	10F (100.0)	–	3	1922–2010	–

concluded that these interests are greater among males than among females.

In row 3, the 24 studies of people's interests in being scientists are summarized. One can see that all but two of these studies concluded that greater proportions of males than females expressed interests in these types of occupations.

The fourth row of this table summarizes findings regarding sex differences in interests that people have in being teachers. It indicates that, from childhood onward, females seem to have more inclinations to be teachers.

25.15.23 Political Participation and Voting

Summaries of studies of three aspects of sex differences in political participation appear in Table 25.15.23. Row 1 simply has to do with sex differences in the proportion who voted in elections. It shows that 24 out of 33 studies concluded that males were more likely to vote than were females.

People's varying political perspectives can be roughly positioned along a left-wing vs. right-wing continuum. Accordingly, many studies have sought to determine how males and females position themselves. One can see in row 2 that 95 out of 110 studies found that females are more left-wing, while row 3 indicated that 11 out of 13 studies reported males being more right-wing in most of their political views.

25.15.24 Attitudes Pertaining to Political or Quasi-Political Issues

Attitudes toward several specific political or quasi-political issues have been assessed in regard to sex differences. Results for attitudes based on ten or more findings are summarized in Table 25.15.24. One can see that only two attitudes attained consistency scores higher than 95.0, although two other attitudes came close to this cutoff.

Row 3 shows that ten studies all found that males are more likely than females to support the idea of social dominance. This idea refers to belief that some human populations should dominate over others. In row 7, findings from ten studies also indicate that males had more favorable or accepting attitudes toward nuclear power than did females.

In the case of an attitude for which the "all ages" consistency scores came close to 95.0, row 4 indicates that males have more favorable attitudes toward war. The other high consistency score was for attitudes toward providing welfare to the disadvantaged, in which case 12 out of 13 studies found females expressing more favorable attitudes (row 5).

Table 25.15.23 Political participation and voting

Variable	*Orig. table*	*Number of sex difference findings (and consistency scores)*						*Overview*	*Countries*	*Time range*	*Non-human*
		I/T	*Child*	*Adol*	*Adult*	*WAR*	*All ages*				
1. Voting in elections	15.12.1.2	–	–	–	17M 5x 2F (65.4)	7M 2F	24M 5x 4F (64.9)	–	5	1952–2002	–
2. Voting for or favoring left-wing (liberal) political views	15.12.1.3	–	–	3F 1M	64F 11x 6M (73.6)	28F 2M(87.5)	95F 11x 9M (70.9)	–	10 (9)	1925–2012	–
3. Voting for or favoring right-wing (conservative) political views	15.12.1.4	–	–	–	11M 2x(84.6)	–	11M 2x(84.6)	M: 1Gen	4	1982–2014	–

Table 25.15.24 Attitudes pertaining to political or quasi-political issues

Variable	*Orig. table*	*Number of sex difference findings (and consistency scores)*						*Overview*	*Countries*	*Time range*	*Non-human*
		I/T	*Child*	*Adol*	*Adult*	*WAR*	*All ages*				
1. Attitude toward environmental conservation	15.12.2.8	–	1F 1x	1F	50F 3x 4M (82.0)	2F 1x	54F 5x 4M(80.6)	F: 3Rev	9 (7)	1980–2016	–
2. Racial prejudice	15.12.2.13	–	1M	5M	8M 2x	1M	15M 2x(88.2)	–	4 (1)	1939–2009	–
3. Attitude toward social dominance	15.12.2.15	–	1M	1M	7M	1M	10M (100.0)	–	3 (1)	1979–2004	–
4. Attitude toward war and the use of military power	15.12.2.17	–	3M	5M	41M 2F(91.1)	11M (100.-0)	60M 2F(93.8)	–	8 (3)	1954–2011	–
5. Attitude toward providing welfare for the disadvantaged	15.12.2.18	–	–	–	9F 1x(90.0)	3F	12F 1x(92.3)	–	3	1949–2003	–
6. Attitude toward abortion legalization	15.12.3.3	–	–	1F	7M 12x 3F (25.0)	1M 1x	8M 13x 4F(27.6)	–	3 (2)	1976–2004	–
7. Attitude toward nuclear power	15.12.3.10	–	–	–	10M (100.0)	–	10M (100.0)	–	2 (1)	1976–1989	–

25.15.25 Attitudes toward Sexually Offensive Behavior

Table 25.15.25 summarizes findings regarding sex differences in five different types of attitudes regarding sexually offensive behavior. Row 1 pertains to what is known as *rape myth acceptance*. To measure this concept, research participants are presented with various largely untrue statements about rape (e.g., "No sometimes means yes.") in order to assess variations in the degree of agreement. It should be noted that the originally developed scale contains a number of statements that can actually be interpreted as empirically true (Hahnel-Peeters and Goetz 2022), thus calling for revision. One can see that, while ten studies report that males score significantly higher on these scales than do females, seven other studies found no significant sex differences.

For all four of the remaining attitudes in this table, consistency scores were 100.0, suggesting that they represent universal sex differences. The first two indicate that males are more likely to consider rape more acceptable (row 2) and are more likely to attribute some of the responsibility for rape to behavior by the victim (row 3). Regarding the last two attitudes, males are more tolerant of sexual harassment (row 4) and females are more inclusive in describing what constitutes sexual harassment (row 5).

25.15.26 Attitudes toward Non-Sexual Offensive Behavior

Three types of attitudes having to do with offensive or criminal behavior of a non-sexual nature are summarized in Table 25.15.26. The first of these involve people's attitudes toward aggression and violence under various circumstances. Row 1 indicates that in all 12 studies, males were more tolerant of the display of aggression and violence than were females, making this a possible universal sex difference. This conclusion would be more certain except for the limited number of countries sampled.

In rows 2 and 3, people's attitudes toward the harshness of punishment for crimes committed were assessed. One can see that sex differences in these opinions were quite inconsistent regarding the severity of punishment in general. In the case of using the death penalty, most studies reported males having more favorable attitudes, but there were several exceptions.

25.15.27 Attitudes toward Technology and Its Use

The last table dealing with attitudes is concerned with technology. In this case, it was impossible to separate having favorable attitudes toward technology and actual use of technology; thus, they are considered in combination. Table 25.15.27 shows that most studies have indicated that

Table 25.15.25 Attitudes toward sexually offensive behavior

Variable	*Orig. table*	*Number of sex difference findings (and consistency scores)*						*Overview*	*Countries*	*Time range*	*Non-human*
		I/T	*Child*	*Adol*	*Adult*	*WAR*	*All ages*				
1. Rape myth acceptance	15.13.1.2	–	–	1M	9M 4x(69.2)	3x	10M 7x(58.8)	–	3	1981–2008	–
2. Excusing/ tolerating rape	15.13.1.4	–	–	–	12M(100.0)	–	12M(100.0)	–	3	1980–2002	–
3. Attribute more of the responsibility for rape to the victim	15.13.1.5	–	–	–	32M(100.0)	4M	36M(100.0)	M: 1Rev	3	1973–2006	–
4. Tolerance of sexual harassment	15.13.1.7	–	–	6M	6M	–	12M(100.0)	–	1	1986–2004	–
5. Inclusiveness in assessing what constitutes sexual harassment	15.13.1.9	–	–	–	26F(100.0)	–	26F(100.0)	F: 1Met	2	1986–2004	–

Table 25.15.26 Attitudes toward non-sexual offensive behavior

Variable	*Orig. table*	*Number of sex difference findings (and consistency scores)*						*Overview*	*Countries*	*Time range*	*Non-human*
		I/T	*Child*	*Adol*	*Adult*	*WAR*	*All ages*				
1. Attitude toward aggression and violence	15.13.2.4	–	2M	3M	7M	–	12M (100.0)	–	1 (1)	1984–2013	–
2. Attitude toward harsh criminal punishment (except the death penalty)	15.13.2.7	–	–	–	10M 9x 6F(32.3)	–	10M 9x 6F (32.3)	–	5	1977–2016	–
3. Attitude toward the death penalty	15.13.2.8	–	–	–	21M 5x 2F(70.0)	–	21M 5x 2F (70.0)	–	7	1984–2014	–

Table 25.15.27 Attitudes toward technology and its use

Variable	*Orig. table*	*Number of sex difference findings (and consistency scores)*						*Overview*	*Countries*	*Time range*	*Non-human*
		I/T	*Child*	*Adol*	*Adult*	*WAR*	*All ages*				
1. Using (or enjoying use of) computers in general	15.14.2.4	–	6M 2x 3F(42.9)	35M 9x 3F (70.0)	40M 9x 2F (75.5)	3M 2x	84M 22x 8F (68.9)	M: 1Met	15	1983–2013	–
2. Using (or enjoying use of) the internet in general	15.14.2.5	–	1F	–	9M 2x 3F(52.9)	–	9M 2x 4F(47.4)	–	3	1997–2007	–

males express more favorable attitudes toward computers and the internet (and also use them more) than do females (rows 1 and 2). However, it should be noted that as computers have become easier to use and as the internet has become broader in coverage, the sex differences in this regard seem to have become less noticeable.

25.16 Personality and Behavioral Tendencies: Condensed Findings

Studies of personality as well as so-called *general behavior patterns* have been part of scientific efforts to identify sex differences for well over a century. Chapter 16 contains a great deal of research bearing on the possibility of universals in these differences. In the summary below, findings from this chapter for which at least ten studies were located are divided into eight summary tables.

25.16.1 The Big Five Personality Traits

Table 25.16.1 has to do with what are known as *the Big Five personality traits*. Based on factor analysis, these five traits have been shown to be more or less universally exhibited by humans and to roughly encapsulated all of the "lesser" personality traits. In alphabetical order, the Big Five personality traits are agreeableness, conscientiousness, extraversion, neuroticism, and openness to new experiences.

The table's first row identifies 53 findings having to do with average sex differences in agreeableness. One can see that 45 of these findings concluded that females are higher in agreeableness than is the case for males.

In the case of conscientiousness, 42 findings were located. Of these, row 2 shows that 28 studies concluded that females scored significantly higher than males.

Row 3 has to do with extraversion, for which a total of 66 findings were located. One can see that that research has been very inconsistent in finding sex differences in this trait, with roughly equal proportions reporting females being more, males being more, and no significant sex differences.

Sex differences in neuroticism has been investigated by at least 105 different studies. Of these, row 4 shows that 90 studies indicated that females score higher, while the remaining 15 reported no significant sex difference.

Thirty-nine studies were located having to do with sex differences in openness to new experiences. Row 5 shows that the findings regarding possible sex differences are not consistent, with just a slight tendency for females scoring higher than males on this personality trait.

Table 25.16.1 The Big Five personality traits

Variable	*Orig. table*	*Number of sex difference findings (and consistency scores)*						*Overview*	*Countries*	*Time range*	*Non-human*
		I/T	*Child*	*Adol*	*Adult*	*WAR*	*All ages*				
1. Agreeableness	16.1.1.1	1x	4F 1x	5F 1x	30F 3x 2M (81.1)	6F 1x	45F 6x 2M (81.8)	F: 1Met	14 (5)	1973–2021	F: 4Chimp F: 1Orang
2. Conscientious-ness	16.1.1.2	–	5F	2F 2x	14F 11x 1M (51.8)	7F	28F 13x 1M (84.8)	F: 1Met	17 (7)	1924–2018	–
3. Extraversion	16.1.1.3	–	1F 1x 1M	6F 4x 1M (50.0)	16F 18x 24M (–32.4)	4F	27F 23x 26M (26.5)	F: 1Met	20 (15)	1925–2021	M: 1Chimp F: 1Orang
4. Neuroticism	16.1.1.4	–--	2F 1x	19F 3x	66F 10x (86.8)	4F 1x	90F 15x (85.7)	F: 2Rev 4Met	18 (14)	1934–2021	M: 1Chimp F:1Orang
5. Openness to new experiences	16.1.1.5	–	–	4F 2x 2M	12F 7x 9M (32.4)	1F 1x 1M	17F 10x 12M (41.5)	M: 1Met	11 (12)	1966–2021	x: 1Chimp

Before leaving this table, it is worth noting that none of these five personality traits were associated with consistency scores of 95.0 or higher. Thus, even though meta-analyses have concluded that females score higher on the first four traits and that males score higher on the fifth trait, our analysis casts doubt on the possibility of universal sex differences in all five traits.

25.16.2 Additional Personality Traits

Turning to various personality (or personality-like) traits other than the Big Five, Table 25.16.2 shows a summary of findings for seven traits regarding possible sex differences. Of these, the first two had consistency scores of 95.0 or higher. Specifically, females appear to be more gregarious (or friendly), while males are more prone toward acting out when stressed or emotionally upset.

Of the remaining five personality traits shown in this table, none yielded consistency scores of 95.0 or higher. Thus, while not to the point of being culturally universal, males appear to exhibited more aggressive and hostile behavior than females (rows 3 and 4). Males also have been found to be more Machiavellian and to display psychoticism (i.e., antisociality) than females (rows 6 and 7). Females, on the other hand, were found in most studies to be significantly more introverted (row 5).

25.16.3 Honesty and Deception

Who's more honest, males or females? Examination of the five traits dealing with honesty and deceptive behavior in Table 25.16.3 reveals that consistency scores were quite low. Therefore, it does not appear that there are any noteworthy universal sex differences in honesty-related traits.

25.16.4 Self-Regulating Behavior

Table 25.16.4 summarizes findings regarding sex differences in seven traits of a self-regulatory nature. Row 1 of the table shows that boredom proneness is more common among males than females according to 25 out of a total of 27 research findings. In the case of impulsiveness (or disinhibition), row 2 shows that a slight majority (i.e., 37 of the 69 findings) point toward greater prevalence among males.

Sex differences in tendencies to resist temptation have been studied in 15 studies, seven of which indicate that females are better than males, while six found no significant sex difference and two suggested that males were better. For studies of self-control, 29 of the 37 studies found that females surpassed males. Six of the eight exceptions reported no significant sex difference.

Table 25.16.2 Additional personality traits

Variable	*Orig. table*	*Number of sex difference findings (and consistency scores)*						*Overview*	*Countries*	*Time range*	*Non-human*
		I/T	*Child*	*Adol*	*Adult*	*WAR*	*All ages*				
1. Gregariousness (friendliness)	16.1.2.10	–	2F	2F	19F 1x (**95.0**)	4F	27F 1x (**96.4**)	–	4 (7)	1957–2018	–
2. Acting out behavior	16.1.3.1	–	9M	1M	1M	–	11M (**100.0**)	–	3	1964–2001	–
3. Aggressive personality	16.1.3.2	–	5M	2M	6M	4M	15M 1x (93.7)	–	7	1986–2006	1x: Mammal
4. Hostile/conflict-oriented behavior	16.1.3.11	1M	7M 1x	5M	10M 3x (76.9)	1M	24M 2x (84.6)	–	5	1952–2013	2M: Primates
5. Introversion (shyness)	16.1.3.12	–	2F	4F 1x	12F 4x	–	21F 5x (77.8)	–	4	1927–2007	–
6. Machiavellianism (exploiting others)	16.1.3.13	–	2M 1x	3M	7M	1M:1-x	12M 2x (80.0)	–	2 (1)	1964–2020	–
7. Psychoticism	16.1.3.16	–	4M 1x	2M	8M 2x (80.0)	–	14M 2x (87.5)	–	6 (2)	1971–2014	–

Table 25.16.3 Honesty and deception

Variable	*Orig. table*	*Number of sex difference findings (and consistency scores)*						*Overview*	*Countries*	*Time range*	*Non-human*
		I/T	*Child*	*Adol*	*Adult*	*WAR*	*All ages*				
1. Honesty	16.2.1.2	–	1M	–	3M 3x 3F	1M	5M 3x 3F (41.7)	–	3	1925–2019	–
2. Lying and deception	16.2.1.3	1M 1F	1M 2x 1F	5M 1x 1F	1M 2x 2F	–	8M 5x 4F (38.1)	M: 1Met	3	1936–2012	–
3. Lie scale scores	16.2.1.4	–	2M	3F 1x	4F 2x 1M	–	7F 3x 3M (43.7)	–	3	1936–2012	–
4. Cheating in general	16.2.1.6	1x	2M 8x 5F	5M 4x	17M 13x 2F (50.0)	–	24M 25x 7F (38.1)	M: 1Met	5	1928–2014	–
5. Cheating on tests or assignments	16.2.1.7	–	4x 1M	6F 4x	7F 5x 5M (31.8)	1M	13F 13x 7M (34.2)	M: 1Met x: 1Met	3	1943–2015	–

Table 25.16.4 Self-regulating behavior

Variable	*Orig. table*	*Number of sex difference findings (and consistency scores)*						*Overview*	*Countries*	*Time range*	*Non-human*
		I/T	*Child*	*Adol*	*Adult*	*WAR*	*All ages*				
1. Boredom proneness	16.2.2.1	1M	1M	4M 1x	17M 1x 1F (85.0)	–	25M 1x 1F (89.3)	–	6 (1)	1970–2017	–
2. Impulsiveness/ disinhibition	16.2.2.3	1M	17M 2x 1F	4M 2x 1F	19M 15x 1F (51.4)	7M 1x 11F	37M 18x 14F (44.6)	M: 1Rev 4Met	11 (2)	1972–2017	1M: Chimp
3. Resisting temptation	16.2.2.4	1F	4F 6x 2M	–	–	2F	7F 6x 2M (41.2)	F: 1Rev 1Met	1	1963–1997	
4. Self-control (self-regulatory behavior)	16.2.2.6	5F	8F	12F 2x 1M (75.0)	4F 4x 1M	–	29F 6x 2M (74.3)	F: 2Met x: 1Rev M: 1Met	8 (1)	1979–2014	–
5. Self-discipline	16.2.2.7	–	1x 3M	4F 3M	3F	–	7F 1x 6M (35.0)	F: 1Met	4 (3)	1984–2015	–
6. Self-sufficiency/ resourceful-ness/ independence	16.2.2.10	–	2M:1x	8M 1x	15M 1x (93.7)	–	25M 3x (89.2)	–	5 (1)	1941–2013	–
7. Reward dependency	16.2.2.11	–	–	–	28F 2x (93.3)	–	28F 2x (93.3)	F: 1Met	18	1991–2015	–

The last three entries in this table pertain to self-discipline, self-sufficiency, and reward dependence. In the case of self-discipline, findings regarding sex differences are quite inconsistent, even though one meta-analysis concluded that females exhibit more than do males. Regarding self-sufficiency, 25 out of 28 studies concluded that males are more so than females. On the other hand, 28 out of 30 studies of reward dependency point toward this trait being greater for females than for males. None of these differences warrant consideration as being universal sex differences, although the latter two came fairly close.

25.16.5 Risky Behavior

Sex differences in risk-taking have been widely investigated; relevant findings are summarized in Table 25.16.5. Regarding risk-taking (including reckless) in general, row 1 shows that 165 studies were located, 151 of which reported that males surpassed females to a significant degree. However, only in the case of adults was the consistency score high enough to be considered a likely universal sex difference. One can also see that two literature reviews and two meta-analyses all agreed that males were more prone to take risks in general than were females. Also, one study of chimpanzees reported the same conclusion.

A specific form of risk-taking is gambling. Row 2 shows that 100 studies of sex differences in gambling were located (not including compulsive or addictive gambling, see Table 14.4.4.5). While 78 studies concluded that males gambled more than females, nine studies reached the opposite conclusion, and 13 studies found no sex differences. Most of the exceptional findings were limited to certain types of gambling, such as purchasing lottery tickets or playing bingo.

Cautiousness (or risk-aversion) refer to tendencies that are the opposite of risk-taking. Therefore, it is not surprising to find that most studies have reported females are more risk-aversive than males. Nevertheless, as shown in row 3, the consistence scores for this trait fall substantially below the 95.0 cutoff needed to consider this a universal sex difference.

The last two rows in this table pertain to two fairly specific types of risk-taking, one involving health and the other involving finances. For health-related risk-taking, 26 of the 31 studies reported that males do so more. For risk-taking of a monetary nature (not including actual gambling), most studies pertain to making stock-market investments and other business-related risks. One can see that of the 36 studies, 34 concluded that males took more monetary risks than did females.

Table 25.16.5 Risk-taking types of behavior

Variable	*Orig. table*	*Number of sex difference findings (and consistency scores)*						*Overview*	*Countries*	*Time range*	*Non-human*
		I/T	*Child*	*Adol*	*Adult*	*WAR*	*All ages*				
1. Risk-taking/ recklessness	16.2.3.1	4M	18M 4x 1F (75.0)	28M 3x	95M 5x **(95.0)**	6M:1x	151M 13x 1F (91.0)	M: 3Rev 2Met	14 (2)	1927–2020	1M: Chimp
2. Gambling	16.2.3.6	–	4M 3x 1F	22M 5x 1F (75.9)	34M 6x 7F (63.0)	18M 2x (90.0)	78M 13x 9F (71.6)	M: 2Rev 1Met	11	1964–2019	–
3. Cautiousness/ risk-aversion	16.2.3.3	1F	–	2F	8F 7x 1M (47.1)	–	11F 7x 1M (57.9)	F: 1Gen 1Rev lMet	3	1959–2015	–
4. Taking health risks	16.2.3.10	–	2M	3M 1x 1F	18M 1x 2F (78.3)	3M	26M 2x 3F (76.5)	–	9 (1)	1982–2009	–
5. Taking monetary risks	16.2.3.11	–	–	–	34M 2x (94.4)	–	34M 2x (94.4)	M: 1Met	4 (1)	1978–2009	–

25.16.6 Health-Related Aspects of Personality and Behavior

Considerable research has compared males and females regarding their involvement in various health-related behavior patterns. One can see in Table 25.16.6 that this research has resulted in consistency scores surpassing 95.0 for two of the six traits: dieting and exercise.

Regarding eating healthy foods, row 1 reveals that 15 of the 17 studies concluded that females do so more than males, with the remainder reporting no significant sex difference.

Seventy-one studies of dieting (or trying to lose weight) were located. Row 2 shows that all but one of these studies concluded that females do so more frequently than males.

Fifteen studies of sex differences in the tendency to eat high proportions of fruits and vegetables were found. In row 3, one can see that all but one of these studies concluded that females surpassed males in this regard.

In terms of physical exercise, 149 studies were located, 142 of which found males exercising more than females. The remaining seven studies simply failed to detect significant sex differences. (Parenthetically, evidence will be reviewed shortly regarding sex differences in overall physical activity. It too will show that males usually surpass females, especially during adolescence.)

The final entry in this table has to do with the use of seatbelts when driving (and, sometimes, when riding in) a motor vehicle. One can see that 20 of the 29 studies concluded that females were more likely than males to do so.

25.16.7 Persistence and Competitive Behavior

Persistence and competitiveness have been extensively studied regarding possible sex differences. Findings are summarized in Table 25.16.7.

As shown in rows 1 and 2, most studies have indicated that males are more task-persistent than females, both regarding task-persistence in general and regarding task-persistence in tasks that are self-selected, especially the latter.

In the case of assertiveness, row 3 shows that 51 studies were located, 35 of which indicated that males were more assertive than females. Most of the exceptions indicated that there are no significant sex differences.

Ninety studies of sex differences in competitiveness were found. Of these, 84 concluded that males were more so than females, with the remaining six all indicating there were no significant sex differences. The overall pattern comports with conclusions reached by one literature review and three meta-analyses.

Table 25.16.6 Health-related aspects of personality and behavior

Variable	*Orig. table*	*Number of sex difference findings (and consistency scores)*						*Overview*	*Countries*	*Time range*	*Non-human*
		I/T	*Child*	*Adol*	*Adult*	*WAR*	*All ages*				
1. Eating healthy foods	16.3.1.1	–	1x	7F	8F:1x	–	15F:2x (88.2)	–	9 (1)	1985–2012	–
2. Dieting (trying to lose weight)	16.3.1.3	–	3F	38F **(100.0)**	29F 1x **(96.7)**	–	70F 1x **(98.6)**	–	12 (2)	1950–2010	–
3. Eating fruits and vegetables	16.3.1.9	–	–	–	14F 1x (93.3)	–	14F 1x (93.3)	–	10 (1)	1994–2012	–
4. Taking preventive health precautions	16.3.2.1	–	–	4F 2x	22F 3x 2M (78.6)	12F 1x (92.3)	38F 6x 2M (79.2)	–	7	1966–2007	–
5. Physical exercise	16.3.2.2	4M	41M 3x **(97.6)**	46M 1x **(97.9)**	40M 3x (93.0)	11M **(100.0)**	142M 7x **(95.3)**	M: 2Rev	18 (4)	1934–2012	–
6. Seatbelt usage	16.3.2.9	–	–	2F 3x	4F 1x 2M	14F 3M (70.0)	20F 4x 5M (58.8)	–	5	1969–2002	–

Table 25.16.7 Persistence and competitive behavior

Variable	*Orig. table*	*Number of sex difference findings (and consistency scores)*						*Overview*	*Countries*	*Time range*	*Non-human*
		I/T	*Child*	*Adol*	*Adult*	*WAR*	*All ages*				
1. Task-persistence in general	16.4.1.1	–	1x	1M	17M 13x 10F (34.0)	1M	20M 13x 10F (40.0)	x: 1Met	17	1974–2005	–
2. Task-persistence in self-selected tasks	16.4.1.2	1M	3M 1x	2M	10M 2x 1F (71.4)	2M	18M 3x 1F (78.3)	–	2	1951–2001	–
3. Assertiveness	16.4.2.1	1x	10M 5x 1F (58.8)	1M 1x 1F	23M 4x 4F (65.7)	3M	35M 10x 6F (61.4)	M: 1Rev 1Met	7 (7)	1952–2018	–
4. Competitiveness	16.4.2.2	1M	33M 4x (89.2)	10M 1x (90.9)	40M 1x (97.6)	6M	84M 6x (93.3)	M: 1Rev 3Met	12 (2)	1953–2020	1M: Chimp
5. Competitiveness by females depending on their opponent's sex	16.4.2.4	–	4F	2M 1x	1F 6x	–	5F 6x 2M (33.3)	x: 1Rev	1	1968–1982	–

The last row in this table was concerned with a specific type of competitiveness. Based on various experimental games, the relevant studies sought to determine if the tendency for females to be competitive can be influenced by the sex of their opponent. One can see that findings have been quite mixed, with six studies concluding that sex of the opponent made no significant sex difference, five studies suggesting they are more competitive toward other females, and the remaining two studies indicating they are more competitive when competing against males.

25.16.8 Curiosity and Achievement Aspects of Personality and Behavior

As an extension of traits reviewed above regarding task-persistence and competitiveness, Table 25.16.8 identifies other traits that have some bearing on work, broadly defined. According to row 1 of this table, there are no consistent patterns regarding sex differences in achievement orientation or ambitiousness, although roughly half of the studies concluded that males are more ambitious than females.

Sex differences in physical activity levels have been studied extensively since the 1930s, some of which even involved fetuses. One can see in row 2 that the vast majority of studies have concluded that males are more physically active than females, especially during adolescence. This generalization is consistent with two literature reviews and three meta-analyses, although one literature review reported no significant sex difference.

Regarding curiosity, row 3 shows that 12 findings indicated that males surpassed females, while one study reported the opposite. In the case of engaging in exploratory behavior, row 4 shows that 40 studies were located, all of which concluded more exploratory behavior is exhibited by males than by females.

The last entry for this table pertains to what is known as *Type A personality*. Individuals with this type of personality tend to be highly driven to achieve goals they or others have established for them and are usually never satisfied what they may have already accomplished. One can see that seven out of 12 studies of this type of behavior failed to identify any significant sex differences.

25.16.9 Behavior Involving the Upper Body

Research findings regarding sex differences in behavior involving the use of the upper body to carry objects are summarized in Table 25.16.9. Two such behavior patterns were located with ten or more relevant findings. The first had to do with studies of sex differences in the use of the cradling-like book carrying style. Carrying objects similar to books, such as bundles

Table 25.16.8 Curiosity and achievement aspects of personality and behavior

Variable	*Orig. table*	*Number of sex difference findings (and consistency scores)*							*Overview*	*Countries*	*Time range*	*Non-human*
		Fetal	*I/T*	*Child*	*Adol*	*Adult*	*WAR*	*All ages*				
1. Achievement oriented, ambitiousness	16.4.3.1	–	–	3M 2x	9M 2F (69.2)	14M 12x 8F (33.3)	1M	27M 14x 10F (44.3)	M: 1Met x: 1Rev	10 (2)	1969–2018	–
2. Activity levels	16.4.3.2	3M 10x (23.1)	32M 14x 2F (64.0)	66M 4x (94.3)	63M 3x (95.5)	38M 10x 1F (76.0)	4M	206M 41x 1F (82.7)	M: 2Rev 3Met x: 1Rev	20 (7)	1930–2020	Com-plex
3. Curiosity and inquisitiveness	16.4.3.6	–	3M	4M 1F	1M	2M	2M	12M 1F (85.7)	–	4 (1)	1962–2003	–
4. Exploratory behavior in general	16.4.3.9	–	6M	22M (100.0)	2M	9M	1M	40M (100.0)	M: 2Rev	9 (1)	1937–1998	Com-plex
5. Type A personality	16.4.3.19	–	–	–	–	1M 6x 2F	2M 1F	3M 7x 2F (21.4)	–	5	1976–1993	–

Table 25.16.9 Behavior involving the upper body

Variable	*Orig. table*	*Number of sex difference findings (and consistency scores)*						*Overview*	*Countries*	*Time range*	*Non-human*
		I/T	*Child*	*Adol*	*Adult*	*WAR*	*All ages*				
1. Using a cradling-like book carrying style	16.5.2.1	–	2F	1F	11F (100.0)	–	14F (100.0)	–	2	1976–2013	–
2. Carrying weapons	16.5.2.1	–	–	28M (100.0)	1M	1M	30M (100.0)	–	5	1994–2019	–

of paper and satchels were also included in these studies, but carrying infants were not. One can see that all 14 studies agree that females are more likely than males to use a cradling-like book carrying style than were men.

Regarding the second entry, 30 studies of sex differences in weapons carrying were found, primarily by adolescents. Usually, these weapons are concealed, such as in one's pocket or waste-band covered by a jacket. The most commonly reported weapons were knives and guns. All of the studies have concluded that males are more likely to carry these types of weapons than are females.

25.16.10 Religious Behavior

Table 25.16.10 condenses findings for which at least ten findings having to do with four types of religious behavior. Row 1 shows that all 26 studies of sex differences in participating in religious activities have concluded that females surpass males. Also, in row 2, one can see that all but one of the 23 studies of prayer concluded that females do so more than males.

In row 3, studies having to do with attending religious services are summarized. One can see that of the 92 studies, 81 concluded that females attend more than males. It is worth adding that most of the exceptions have to do with attending religious services by members of two religions: Islam and Judaism.

When it comes to religious membership, as shown in row 4 of this table, 18 of the 22 findings pointed toward more females than males considering themselves to be members of a particular religion. The four exceptions all found no significant sex differences in this regard. (Note that findings having to do with religious beliefs are presented in Table 25.15.4)

25.17 Social Behavior: Condensed Findings

Chapter 17 was devoted to research findings having to do with sex differences in social behavior, broadly defined. Even sex differences in various forms of communication, including mass communication, and sexuality were covered in this chapter. Here, all of the tables located in Chapter 17 containing citations to ten or more studies are condensed into cellular rows regarding the nature of what the original tables for this chapter indicates.

25.17.1 Social Behavior

Table 25.17.1 summarizes findings from eight aspects of social behavior, starting with what can be thought of as *socializing ability*. Row 1 indicates that 12 out of 15 studies concluded that the sexes do not significantly

Table 25.16.10 Religious behavior

Variable	*Orig. table*	*Number of sex difference findings (and consistency scores)*						*Overview*	*Countries*	*Time range*	*Non-human*
		I/T	*Child*	*Adol*	*Adult*	*WAR*	*All ages*				
1. Involvement in religious activities	16.6.1.1	–	–	4F	13F (100.0)	9F	26F (100.0)	–	5	1949–2002	–
2. Prayer	16.6.1.4	–	–	–	21F 1M (95.5)	1F	22F 1M (95.7)	–	5 (3)	1983–2014	–
3. Attending religious services	16.6.1.5	–	2F	16F 1x (94.1)	51F 5x 5M (77.3)	12F (100.0)	81F 6x 5M (84.4)	–	16 (3)	1949–2017	–
4. Religious membership	16.6.1.6	–	–	1F	11F 4x (73.3)	6F	18F 4x (81.8)	–	3 (1)	1975–2009	–

Table 25.17.1 Social behavior

Variable	*Orig. table*	*Number of sex difference findings (and consistency scores)*						*Overview*	*Countries*	*Time range*	*Non-human*
		I/T	*Child*	*Adol*	*Adult*	*WAR*	*All ages*				
1. Socializing ability	17.1.1.1	–	1F 12x 1M (6.7)	1F	–	–	2F 12x 1F (12.5)	–	3	1981–1999	–
2. Socializing tendency	17.1.1.2	–	1F	9F	–	3F	13F (100.0)	–	2	1975–1999	–
3. Time spent socializing	17.1.1.5	4F	2F 1x	8F 1M	22F 2x (91.7)	–	36F 3x 1M (87.8)	–	4 (1)	1942–2020	F: most species
4. Having intimate friendships	17.1.2.1	–	10F 1x (90.9)	21F 1x (95.5)	40F 3M (87.0)	7F	71F 2x 3M (89.9)	F: 2Rev 1Met	6 (1)	1963–2005	–
5. Gang membership	17.1.2.7	–	–	40M (100.0)	–	10M (100.0)	50M (100.0)	–	6 (1)	1974–2014	–
6. Number of friends in general	17.1.3.3	–	1F 1M	4F 1M	14F 2x 3M (63.6)	–	19F 2x 4M (65.5)	–	5	1972–2010	M: 1 chimp
7. Same-sex social interactions	17.1.3.4	–	12M 6x (66.7)	5M	3M	2M	22M 6x (78.6)	–	2 (1)	1973–2003	M: 3 various species
8. Number of same-sex friends	17.1.3.5	–	2F 3x 3M	2F	5F	1F	10F 3x 3M (52.6)	–	1	1933–1995	–

differ regarding their abilities to socialize. In the case of socializing tendencies (or tendencies to seek out socializing opportunities), however, one can see in row 2 that all 13 studies concluded that females surpassed males. Similarly, row 3 shows that 36 out of 40 studies of time spent socializing concluded that females do so more than males.

Seventy-six studies were found that assessed sex differences in tendencies to have intimate friendships. The vast majority of these studies, especially those involving adolescents, concluded that females are more likely to have close intimate friendships.

Fifty studies of sex differences in being members of a gang or involved in gang activities were examined. This behavior is primarily found among adolescents and young adults. One can see in row 5 that these studies unanimously concluded that gang members were primarily males.

In 25 studies, sex differences were assessed in the number of friends that individuals have. Row 6 shows that 19 of these studies reported more friends for females than for males.

Another set of studies investigated the frequency of same-sex social interactions. As shown in row 7, 22 of the 28 studies concluded that males have more of these types of social interactions than do females. However, in the case of having same-sex friends, row 8 shows that, while findings were mixed, most studies concluded that females have more than males.

25.17.2 Social Interactions

Sex differences in various forms of social interaction have received considerable attention over the years. Findings from original tables containing ten or more findings are summarized in Table 25.17.2.

Fifteen studies of sex differences in the amount of conflict with friends have been located. Row 1 shows that findings have been mixed, albeit most research has reported no significant sex difference.

Male-female differences in general tendencies to conform has been addressed by 11 studies. In row 2, one can see that findings have been quite mixed. Two types of experimental studies of conforming behavior have helped to refine scientific understanding of possible sex differences. In the first type, where a research participant must state his/her conforming decision (either for or against) the statement by another person. Row 3 shows that most of these studies find no significant differences, although when sex differences are found, the majority of studies have concluded that females conform more than do males. Nearly all results from conforming experiments in which research participants express their judgments in the absence of a person who expressed a contrary view, shown in row 4, indicate that there are no significant sex differences.

Table 25.17.2 Social interactions

Variable	*Orig. table*	*Number of sex difference findings (and consistency scores)*						*Overview*	*Countries*	*Time range*	*Non-human*
		I/T	*Child*	*Adol*	*Adult*	*WAR*	*All ages*				
1. Conflict with friends	17.1.3.7	–	4x 1M	3F 5x 1M	–	1x	3F 10x 2M (17.6)	–	3	1984–2006	–
2. Conformity in general	17.1.4.1	–	–	1F	4F 5x 1M (36.4)	–	5F 5x 1M (41.7)	–	4 (1)	1982–2009	–
3. Experimental conformity in the presence of others	17.1.4.2	–	2F 2x	4F 7x 1M (30.8)	28F 38x 3M (38.9)	–	34F 47x 4M (38.2)	F: 3Met	5	1955–1988	–
4. Experimental conformity in the absence of others	17.1.4.3	–	2x	1F 3x	1F 13x 1M	–	2F 18x 1M (9.1)	x: 1Met	2	1960–1975	–
5. Compromising behavior	17.1.4.5	–	4F 1x	2M	3F 1x	–	7F 2x 2M (53.8)	–	2	1986–2003	–
6. Using competitive strategies when mediating conflict	17.1.4.6	–	–	–	8M 6x (57.1)	–	8M 6x (57.1)	–	1	1971–1991	–
7. Touching others in general	17.1.5.1	2F	2x	1M	3F 1M	1F 1M	6F 2x 3M (42.9)	–	3 (1)	1968–1980	–
8. Opposite-sex touching	17.1.5.3	–	–	2M	2M	4M 3F	8M 3F (57.1)	x: 1Rev	1 (1)	1971–2004	–
9. Parent-child conflict	17.1.6.7	–	–	10F (100.0)	–	–	10F (100.0)	–	3	1987–2018	–

Possible sex differences in tendencies to compromise have been investigated in at least 11 studies. Row 5 shows that seven of these studies found females compromising more than males, with two of the remaining studies reporting no significant sex difference and two others concluding that males compromise more.

Several studies have compared males and females who are mediating conflicts regarding the type of strategies used. As shown in row 6, most research has indicated that males use various types of competitive strategies more than do females.

In 11 studies, sex differences in people's tendencies to touch others were reported. One can see in row 7 that the evidence is inconsistent, with a slight tendency for females to touch others more. Along related lines, row 8 pertains to 11 studies of opposite-sex touching. In this case, the findings have also been mixed, although most evidence points toward males touching females more than vice versa.

The last row in this table has to do with conflicts between parents and their children. Usually based on reports by the offspring, all ten studies among adolescents have indicated that females have more conflicts with one or both of their parents than is the case for adolescent males.

25.17.3 Providing Care to Others

Four variables have been given research attention having to do with sex differences in tendencies to provide care to others are listed in Table 25.17.3. In row 1, studies of sex differences in providing care to one's own offspring is considered. One can see that all 92 studies have concluded that females are the primary caregivers to their children than are males. This conclusion is consistent with nine other literature reviews on the topic. Studies of most other species have reached the same conclusion, although there are several exceptions.

Row 2 has to do with sex differences in tendencies to provide caregiving to the offspring of others (including siblings). It shows that 49 out of 57 pertinent studies have concluded that females provide greater care than do males.

In the case of providing care to family members other than offspring (such as offspring caring for their aging parents), row 3 reveals that all {71} studies that were located have indicated that females surpass males in this regard. Furthermore, row 4 suggests that all but one of 22 studies concluded that females are more likely to provide care and comfort to others in general.

Table 25.17.3 Providing care to others

Variable	*Orig. table*	*Number of sex difference findings (and consistency scores)*						*Overview*	*Countries*	*Time range*	*Non-human*
		I/T	*Child*	*Adol*	*Adult*	*WAR*	*All ages*				
1. Providing care to one's own offspring (parenting behavior)	17.1.6.10	–	–	–	92F (100.0)	–	92F (100.0)	F: 1Gen 9Rev	14 (10)	1967–2014	Mixed results
2. Exhibiting caregiving to the offspring of others	17.1.6.12	–	19F 3x (86.4)	10F 1x (90.9)	20F 6x 1M (71.4)	–	49F 7x 1M (84.5)	–	6	1975–2013	Mixed results
3. Providing care to family members other than offspring	17.1.6.15	–	–	–	71F (100.0)	–	71F (100.0)	F: 4Rev	10 (2)	1981–2017	–
4. Providing care and comfort to others in general	17.1.6.17	–	4F 1x	–	16F (100.0)	1F	21F 1x (95.5)	F: 2Met	5 (2)	1960–2016	–

25.17.4 Proximate Aspects of Social Interactions

In Table 25.17.4, findings having to do with variations in people's proximity to others are reviewed regarding sex differences. Studies of toddlers and children having to do with the amount of time offspring spend close to their mothers are summarized in row 1. It shows that 12 out of 14 studies concluded that females stay close to their mothers more than do males.

Research having to do with the amount of time individuals of either sex spend time socially interacting with others is presented in row 2. One can see that 28 out of a total of 32 studies indicated that females surpass males in this regard.

Fifteen studies were located in which researchers determined group size during social interactions. As shown in row 3, 13 of the studies concluded that during social interactions, the size of the groups tends to be larger for males than for females, albeit two studies reported no significant sex differences. Along similar lines, row 4 summarizes findings from 27 studies of the amount of time individuals spend socially interacting in small groups (i.e., usually groups of two to four). It indicates that all of the studies concluded that females spend more time in small group interactions than do males.

25.17.5 Living Arrangements

Sex differences in three types of living arrangements were located for which at least ten relevant studies were located. These are summarized in Table 25.17.5.

Row 1 has to do with sex differences among adults who live alone. It indicates that eight out of ten studies indicated that this is more common among females than among males. However, row 2 shows that all ten studies of being homeless have all concluded that this is more prevalent among males than females.

The third entry in this table has to do with which parent is more likely to have primary custody of children following divorce by the parents. One can see that all 19 findings were consistent in indicating that mothers are more likely than fathers to have primary custody.

25.17.6 Altruism and Helpfulness

Table 25.17.6 pertains to sex differences in people's varying tendencies to be altruistic and helpful toward others. In row 1, studies are reviewed regarding various general measures of altruism and prosocial behavior. It indicates that most studies have concluded that females surpass males, although several studies have reported no significant differences. Similarly, row 2 reveals that ten studies of people's tendencies to give attention to the

Table 25.17.4 Proximate aspects of social interactions

Variable	*Orig. table*	*Number of sex difference findings (and consistency scores)*						*Overview*	*Countries*	*Time range*	*Non-human*
		I/T	*Child*	*Adol*	*Adult*	*WAR*	*All ages*				
1. Time spent close to mother	17.1.6.20	9F 1x 1M (75.0)	3F	–	–	–	12F 1x 1M (80.0)	–	2	1925–1997	F: 13 1x Mammals
2. Time spent socially interacting in general	17.1.7.1	1F 1x	7F	2F 2M	13F 1M (86.7)	5F	28F 1x 3M (80.0)	x: 1Rev	6	1972–2007	F: 6; M: 1 Mostly mammals
3. Group size during social interactions	17.1.7.3	1M	12M 2x (85.7)	–	–	–	13M 2x (86.7)	M: 1Rev	2	1973–2004	M: 9 Primates
4. Social interactions in small groups	17.1.7.4	–	19F (100.0)	4F	3F	1F	27F (100.0)	–	4	1902–2004	F: 3 Chimps

Table 25.17.5 Living arrangements

Variable	*Orig. table*	*Number of sex difference findings (and consistency scores)*						*Overview*	*Countries*	*Time range*	*Non-human*
		I/T	*Child*	*Adol*	*Adult*	*WAR*	*All ages*				
1. Living along	17.1.8.1	–	–	–	8F 2x (80.0)	–	8F 2x (80.0)	–	7 (3)	1998–2010	–
2. Being homeless	17.1.8.3	–	–	–	6M	4M	10M (100.0)	M: 3Rev 1Met	5	1987–2016	–
3. Having primary custody of children after a divorce	17.1.8.4	–	–	–	8F	11F (100.0)	19F (100.0)	F: 1Rev	3	1975–2001	–

Table 25.17.6 Altruism and helpfulness

Variable	*Orig. table*	*Number of sex difference findings (and consistency scores)*						*Overview*	*Countries*	*Time range*	*Non-human*
		I/T	*Child*	*Adol*	*Adult*	*WAR*	*All ages*				
1. Exhibiting altruistic or prosocial behavior	17.2.1.1	–	22F 9x (71.0)	10F 2x (83.3)	28F 1M (93.3)	2F 1x	62F 12x 1M (81.6)	–	10 (6)	1929–2019	–
2. Attending to the needs of others	17.2.1.2	–	–	–	5F 1x	5F	10F 1x (90.9)	–	4	1989–2003	–
3. Forgiving others	17.2.1.3	–	–	1F	9F 5x 3M (45.0)	1F	11F 5x 3M (50.0)	F: 1Rev 1Met	9	2002–2018	–
4. Helping behavior	17.2.1.5	–	8F 3x (72.7)	1x 2F	55M 17x 36F (38.2)	–	55M 17x 46F (33.5)	M: 2Met	6	1943–2004	–
5. Distributing rewards equitably	17.2.1.7	–	6F 2x 4M (33.3)	–	–	–	6F 2x 4M (33.3)	–	1	1977–2007	–
6. Donating to charity	17.2.1.-10	–	–	–	13F 6x 10M (33.3)	–	13F 6x 10M (33.3)	–	3	1981–2011	–
7. Being an organ donor	17.2.2.2	–	–	–	15F 1x 1M (83.3)	1F	16F 1x 1M (84.2)	–	3	1977–2016	–

needs of others concluded that females did so more than males, although one study reported no significant difference.

Nineteen studies of sex differences in individual tendencies to forgive others for what are perceived as their misdeeds. Row 3 shows that 11 of these studies concluded that females were more likely than males to do so.

Tendencies to help others have been frequently studied, primarily by conducting experiments in which someone exhibits symptoms of distress or being in the midst of being harmed. As shown in row 4, findings have been very inconsistent regarding possible sex differences.

A number of experiments have been conducted in which researchers have assessed potential sex differences in how rewards are distributed. These experiments usually involve looking to see if rewards are distributed by research participants according to the amount of work others may have performed or equitably, regardless of effort. One can see in row 5 that the findings have not been consistent.

Twenty-nine studies were located regarding sex differences in the tendency to make charitable donations. Findings reviewed in row 6 suggest that there are no consistent differences.

The last row in this table has to do with willingness to be organ donors. Out of 18 pertinent studies, 16 revealed that females were more likely to donate organs, particularly one of their kidneys, than were males.

25.17.7 Help Seeking and Receiving

It is usually difficult to separate acts of seeking help and actually receiving the help being sought. Therefore, Table 25.17.7 provides a summary of sex differences regarding either one or both of these social activities.

Row 1 indicates that, in general, females seek and/or receive help from others more than do males according to 34 studies, with all of the remaining five studies concluding that there were no significant sex differences.

Findings of sex differences in seeking or receiving help for health-related matters are summarized in row 2 of this table. They suggest that there are no consistent differences.

In the case of seeking or receiving help when under stress, row 3 shows that 36 studies have found females doing so more than males. The remaining four studies simply found no differences. Roughly the same pattern appears to be the case for those seeking or receiving help when interpersonal disputes arise. As shown in row 4, 22 studies reported higher proportions of females doing so, while the remaining three failed to find significant differences.

Table 25.17.7 Help seeking and receiving

Variable	*Orig. table*	*Number of sex difference findings (and consistency scores)*						*Overview*	*Countries*	*Time range*	*Non-human*
		I/T	*Child*	*Adol*	*Adult*	*WAR*	*All ages*				
1. Seeking or receiving help in general	17.2.3.1	1x	4F 1x	7F	23F 3x (88.5)	–	34F 5x (87.2)	F: 2Met	6 (1)	1974–2015	–
2. Seeking or receiving help for health-related issues	17.2.3.2	–	–	–	8M 5x 5F (34.8)	–	8M 5x 5F (34.8)	–	6	1984–2006	–
3. Seeking or receiving help when under stress	17.2.3.4	–	3x	15F 1x (93.7)	21F (100.0)	–	36F 4x (90.0)	–	3 (1)	1987–2005	–
4. Help seeking when interpersonal dispute arises	17.2.3.6	–	9F 3x (75.0)	12F (100.0)	–	1F	22F 3x (88.0)	–	3 (1)	1961–2004	–

25.17.8 Nonlinguistic Communication

Table 25.17.8 have to do with sex differences in nonlinguistic forms of communication. Original tables for four types of such behavior were found to contain information from ten or more studies.

Row 1 pertains to gazing behavior. While this is not always a form of communication, it often is. One can see that 13 out of 16 studies indicated that females gazed longer at others than did males, with the remaining three studies concluding the opposite. Similarly, row 2 shows that most studies of establishing or maintaining eye contact point toward females doing so more than males. It is worth adding that the only two exceptions both involved sex differences maintaining eye contact when engaged in public speaking. Under this condition, males were found to maintain eye contact more than females.

In row 3, studies of ten studies of sex differences in making eye contact with the opposite sex were observed. Eight of these studies indicated that females did so more, while the remaining two studies reached the opposite conclusion.

The last row in this table pertained to smiling under social conditions (including smiling when being photographed). One can see that 115 pertinent studies were located, 94 of which indicated that females smile more than males, with all but two of the remining studies reporting no significant sex differences. It is worth adding that the 11 studies of adolescents all concluded that females smile more than males.

25.17.9 Linguistic Communication

As shown in Table 25.17.9, substantial research has assessed sex differences in linguistic communication. Row 1 has to do with the amount of talking (or number of words) in a typical day. One can see that the evidence of sex differences is quite mixed for all age groupings identified. Even the meta-analyses were inconsistent, possibly due to age differences. Specifically, before puberty, when sex differences have been found, most studies point toward females being more talkative, while most studies of adults point toward males being more talkative.

Seventeen studies of sex differences in story telling were located. One can see in row 2 that findings were not consistent regarding sex differences in such behavior.

In the case of average sentence length, row 3 summarizes results from 17 studies. All but two of these indicated that the average sentence length by females tends to be longer than those by males.

Who swears and curses more, males or females? According to 41 studies reported in row 4, males do so. The two exceptional findings both

Table 25.17.8 Nonlinguistic communication

Variable	*Orig. table*	*Number of sex difference findings (and consistency scores)*						*Overview*	*Countries*	*Time range*	*Non-human*
		I/T	*Child*	*Adol*	*Adult*	*WAR*	*All ages*				
1. Gazing at others	17.3.1.1	1F	2F	–	10F 3M	–	13F 3M (68.4)	F: 1Met	3	1963–2010	F: 1 Primate
2. Establishing or maintaining eye contact	17.3.1.3	5F	5F	–	10F 2M (71.4)	3F	23F 2M (85.2)	F: 1Rev 2Met	2	1963–1999	–
3. Making eye contact with the opposite sex	17.3.1.4	1F	1F	2M	3F	3F	8F 2M (80.0)	–	2	1976–1998	–
4. Smiling	17.3.2.4	8F 1x 2M (61.5)	8F 6x (57.1)	11F (100.0)	56F 11x (83.6)	11F 1x (91.7)	94F 19x 2M (80.3)	F: 2Met	7 (2)	1966–2018	–

Table 25.17.9 Linguistic communication

Variable	*Orig. table*	*Number of sex difference findings (and consistency scores)*						*Overview*	*Countries*	*Time range*	*Non-human*
		I/T	*Child*	*Adol*	*Adult*	*WAR*	*All ages*				
1. Extensiveness of linguistic communication in general (talkativeness)	17.4.1.1	10F 14x 1M (38.5)	15F 18x 6M (33.3)	6F 2x 2M (50.0)	9F 5x 11M (25.0)	2F 3x 1M	42F 42x 21M (33.3)	M: 1Rev 2Met F: 1Met	5	1951–2007	–
2. Storytelling	17.4.1.3	–	–	–	9F 2x 5M (42.9)	1M	9F 2x 6M (39.1)	–	10	1960–2018	–
3. Average sentence length	17.4.2.2	1F	7F 2M	–	7F	–	15F 2M (78.9)	–	1	1960–2004	–
4. Swearing and cursing	17.4.3.2	–	3M	4M	24M 1x **(96.0)**	8M 1x	39M 2x **(95.1)**	–	6	1935–2009	–
5. Using proper grammar	17.4.3.7	–	1F 2x	–	–	9F	10F 2x (83.3)	–	3 (1)	1925–2002	–
6. Using tentative language	17.4.3.8	–	–	3F	24F 8x 2M (66.7)	2F	29F 8x 2M (70.7)	F: 1Met	6	1972–2014	–
7. Using intensifying adverbs	17.4.3.10	–	–	1F	9F	–	10F **(100.0)**	–	1	1977–1995	–
8. Using pronouns	17.4.3.13	–	2F	1M	7F 2M	–	9F 3M (60.0)	–	1 (1)	1959–2008	–

had to do with using swear words in the presence of the opposite sex. Basically, in mixed company, males and females appear to swear and curse to equal degrees. But, when they are in same-sex groups, males use these words more often according to all of the evidence.

Row 5 reports findings from 12 studies of sex differences in people's tendencies to use proper grammar. Ten of these studies reported that females did so more, while the remaining two found no significant differences.

Thirty-nine studies of sex differences in the use of tentative language were located. As shown in row 6, 29 of these studies reported greater use by females, with eight of the remaining ten reporting no significant differences. Along similar lines, row 7 shows that all ten studies of sex differences the use of intensifying adverbs concluded that females do so more than males.

The last row of this table has to do with the use of pronouns. Of the 12 located studies, nine concluded that females surpass males in this regard, while the remaining three studies reached the opposite conclusion.

25.17.10 Information Communicated

In Table 25.17.10, research findings regarding sex differences in terms of 13 types of information that people communicate are summarized. Row 1 has to do with disclosing personal or private information. It shows that the majority of studies indicate that females do so more than males.

Affirming speech is that which a listener uses to confirm what someone else is saying. One can see in row 2 that roughly half of the 29 located studies indicated that females used this type of speech more, while most of the remaining studies found no significant sex differences.

Sex differences in studies of communication in mixed-sex (vs. same-sex) company are reviewed in row 3. Nearly all of the studies indicate that males are more likely than females to address both sexes when speaking.

Thirteen studies of sex differences in the use of written communication were located. Row 4 reveals that 11 of these studies concluded that females use written communication more than males.

Male-female differences in arguing or quarreling has been assessed in at least 15 studies. As shown in row 5, all but two of these studies indicated that males use this form of communication more than females.

Which sex asks more questions? According to 13 out of the 16 pertinent studies, females do (row 6). So, who brags more? Regarding sex differences in bragging (or emphasizing one's accomplishments or abilities) more, 20 studies were found. Eighteen of these studies indicated that males brag more, while the other two studies reported the opposite (row 7).

Table 25.17.10 Information communicated

Variable	*Orig. table*	*Number of sex difference findings (and consistency scores)*						*Overview*	*Countries*	*Time range*	*Non-human*
		I/T	*Child*	*Adol*	*Adult*	*WAR*	*All ages*				
1. Disclosing personal or private information	17.4.1.5	–	3F	2F	18F 1x 1M (85.7)	2F 1x	25F 2x 1M (86.2)	F: 1Met	2	1958–2000	–
2. Affirming speech or responses	17.4.6.5	2F	6F 9x 1M (35.3)	6F 3x	–	1F 1x	15F 13x 1M (50.0)	F: 1Met	3	1958–2000	–
3. Mixed-sex (vs. same-sex) linguistic communication	17.4.6.10	1M	–	4M	13M 1x (92.9)	6M	24M 1x (**96.0**)	–	4	1951–2002	–
4. Written communication	17.4.6.12	–	2F	1F	8F 1x 1M (72.7)	–	11F 1x 1M (78.6)	–	1	1990–2012	–
5. Arguing or quarreling	17.4.7.1	1M	6M 1x	–	4M 1x	2M	13M 2x (86.7)	–	1	1933–2006	–
6. Asking questions	17.4.7.2	–	1F	1x	11F 1x 1M (84.6)	1F	13F 2x 1M (76.5)	–	3 (1)	1977–2008	–
7. Bragging or emphasizing one's accomplishments	17.4.7.3	–	4M	4M	10M 2F	–	18M 2F (90.0)	–	2	1938–1995	–
8. Confiding and sharing secrets	17.5.7.5	–	7F 3x (70.0)	18F (**100.0**)	23F (**100.0**)	8F	56F 3x (94.9)	F: 1Rev 1Met	4	1958–2005	–
9. Gossiping	17.4.7.6	–	2F	2F 2x	9F 1x (90.0)	–	13F 3x (86.7)	F: 1Rev	6	1985–2019	–
10. Insulting others	17.4.7.7	–	–	3M 1x	1M	5M 1x	9M 2x (81.8)	–	3	1980–2003	–
11. Interrupting others as they are speaking	17.4.7.8	–	1M 1x	–	27M 11x 2F	2M 1x	30M 13x 2F (63.8)	M: 1Gen 1Rev x: 1Rev	3	1955–2004	–
12. Issuing commands and instructions or assertive speech in general	17.4.7.9	1M 1x	11M 4x 4F (47.8)	1M 2x	10M (100.0)	–	23M 7x 4F (60.5)	M: 1Met	3	1976–2004	–
13. Supportive and inclusive communication	17.4.7.11	–	4F	3F	6F	–	13F (**100.0**)	–	2	1990–2010	–

In the case of confiding and sharing secrets with others, 59 studies were found. All but three of these studies concluded that females do so more, with the remainder reporting no significant sex difference (row 8). Although the research is less extensive for gossiping, all but three of the 16 studies have concluded that females do so more than males (row 9).

Eleven studies of people's tendencies to insult others were located. Row ten shows that nine of these studies concluded that males do so more than females, with the remaining three studies reporting no significant differences.

Sex differences in the tendency to interrupt others as they are speaking has been assessed in 45 studies. In row 11, one can see that 30 of these studies concluded that males do so more, 13 found no significant sex difference, and the remaining two reported that this was more common among females.

Row 12 has to do with studies of sex differences in issuing commands, giving instructions, and using assertive speech in general. Of the 34 studies located, 23 found males doing so more, seven found no significant differences, and four reported that females did so more.

The last row in this table pertained to sex differences in providing supportive or inclusive communication. It shows that all 13 studies bearing on this matter agreed that females do so more than males.

25.17.11 Conversational Focus

As shown in Table 25.17.11, two of the original tables with at least ten pertinent studies in Chapter 17 had to sex differences in conversational focus. In row 1, one can see that conversations focused on emotions and feelings are more common among females according to 44 out of the 47 pertinent studies. On the other hand, row 2 indicates that all 15 studies of sex differences in conversations about work or leisure were found to be more common among males.

25.17.12 Play

Play is hard to define, but rather easy to recognize. It appears to be a type of behavior that is unique to mammals, including humans, especially among animals before reaching puberty. Research findings having to do with sex differences in play behavior are summarized in Table 25.17.12. The first row has to do with time spent playing in general. One can see that nine out of ten studies concluded that males play more than females. This pattern is quite similar to studies of mammals generally, where 15 out of 18 studies found males playing more than females.

Table 25.17.11 Conversational focus

Variable	*Orig. table*	*Number of sex difference findings (and consistency scores)*						*Overview*	*Countries*	*Time range*	*Non-human*
		I/T	*Child*	*Adol*	*Adult*	*WAR*	*All ages*				
1. Conversations focus on emotions and feelings	17.4.8.5	1F	9F	6F 1x	22F 2M	6F	44F 1x 2M	F: 1Gen	2	1932–2004	–
2. Conversational focus on work and/or leisure	17.4.8.10	–	–	3M	7M	5M	15M (100.0)	–	2	1922–1993	–

Table 25.17.12 Play

Variable	*Orig. table*	*Number of sex difference findings (and consistency scores)*						*Overview*	*Countries*	*Time range*	*Non-human*
		I/T	*Child*	*Adol*	*Adult*	*WAR*	*All ages*				
1. Play in general	17.5.1.1	–	3M 1x	4M	2M	–	9M 1x (90.0)	–	3	1977–2004	15M 3x mammals
2. Play or recreating outside (instead of indoors)	17.5.1.8	1M	11M (100.0)	2M	1M	–	15M (100.0)	–	3	1930–2006	–
3. Competitive social play	17.5.2.1	–	14M 1x (93.3)	2M	3M	2M	21M 2x (95.5)	–	4	1961–2010	3M mammals
4. Rough-and-tumble play	17.5.2.2	13M 1x (92.9)	16M 4x (80.0)	4M	–	4M	37M 5x (88.1)	M: 1Rev	7 (2)	1910–2010	85M 9x 5F mammals
5. Cooperative social play	17.5.3.1	–	10F (100.0)	–	–	–	10F (100.0)	–	1	1976–2002	–
6. Playing house	17.5.3.3	–	19F (100.0)	–	–	–	19F (100.0)	–	3	1910–2008	–
7. Pretend or fantasy play	17.5.3.6	–	11F 4M (57.9)	–	–	–	11F 4M (57.9)	–	3	1895–2014	–
8. Playing with mechanical or construction objects	17.5.3.7	8M	19M (100.0)	4M	–	–	31M (100.0)	–	3	1910–2004	–

Fifteen studies of the amount of time spent playing (or recreating) outside, as opposed to doing so indoors. Row 2 reveals that all of these studies concluded that males do so more than females.

Competitive social play comes in many forms, all of which have to do with two or more individuals attempting to outperform others regarding some type of task. In row 3, one can see that all but one of 23 studies concluded that males engage in more social play of a competitive nature than do females.

Rough and tumble play has been extensively investigated regarding possible sex differences. Row 4 shows that 37 out of a total of 42 concluded that males engage in this type of scuffling more than do females.

Sex differences in social play in which playmates cooperate (rather than compete) have been assessed in ten studies. One can see in row 5 that all of these studies concluded that females exhibit this type of play more than do males. The same conclusion has been reached regarding time spent playing house, where row 6 shows that 19 different studies indicated that females do so more.

Fifteen studies of sex differences in what is known as pretend or fantasy play. Row 7 shows that 11 of these studies reported more play of this nature among females, while the remaining four reported that males did so more.

The final row in this table has to do with findings from 31 studies of sex differences in tendencies to play with objects of a mechanical or construction nature. In all of these studies, males spent more time in such play than did females.

25.17.13 Toy Choices and Preferences

Table 25.17.13 has to do with sex differences in choices of toys (or sometimes expressed preferences for one type of toy over another). In the first two rows, toy choices and preferences were simply designated feminine or masculine. Obviously, this type of toy classification could bias researchers somewhat when making their assessments of sex differences. Nevertheless, as one would probably expect, clear patterns emerged: Overall, in 95 studies of feminine toys, more girls chose them, or expressed preferences for them to a greater degree than did boys. Conversely, all 112 studies of masculine toys revealed that boys were significantly more likely than girls to choose or express preferences for these toys.

In rows 3 and 4, findings from more specific studies were summarized regarding the types of toys boys and girls choose or prefer as objects of play. One can see that all 42 studies of sex differences in playing with (or preferring to play with) dolls reported that girls selected this type of toy more than did boys. On the other hand, all 19 studies of choosing or

Table 25.17.13 Toy choices and preferences

Variable	*Orig. table*	*Number of sex difference findings (and consistency scores)*						*Overview*	*Countries*	*Time range*	*Non-human*
		I/T	*Child*	*Adol*	*Adult*	*WAR*	*All ages*				
1. Feminine toy choices and preferences	17.5.4.1	36F (100.0)	59F (100.0)	–	–	–	95F (100.0)	F: 2Met	10 (1)	1932–2011	–
2. Masculine toy choices and preferences	17.5.4.2	38M (100.0)	74M (100.0)	–	–	–	112M (100.0)	M: 2Met	13 (1)	1890–2015	–
3. Choosing or preferring dolls	17.5.4.3	12F (100.0)	28F (100.0)	2F	–	–	42F (100.0)	–	8	1910–2015	2F primates
4. Choosing or preferring toys with moved parts or that can be used for building	17.5.4.4	10M (100.0)	9M	–	–	–	19M (100.0)	–	5	1975–2017	2M primates
5. Choosing gender ambiguous toys	17.5.4.5	4M 1F	7M 1x 2F (70.0)	2M	–	–	13M 1x 2F (72.2)	–	3	1910–2008	1M and 1x cats (ball)

preferring toys with moving parts (such as mobiles or wheels) or that could be used to build something were selected by boys more often than by girls. It is interesting to add that two studies of monkeys reached the same conclusion about sex differences in toy choices.

The last row of this table reported results from studies that compared males and females (usually under the age of puberty) regarding their preferences for gender ambiguous toys. These include such toys as balls, paint brushes, and puzzles. The row reveals that 13 out of the 16 studies concluded that males chose these types of toys more often than did females. Regarding studies of non-humans, two studies of domestic cats were found regarding sex differences in time spent playing with balls. One reported that male kittens did so more, while the other study found no significant sex differences.

25.17.14 Sports Involvement and Preferences

Table 25.17.14 reviews evidence having to do with sex differences in involvement and preferences for being involved in sports. It shows that three of the book's original tables contained ten or more studies bearing on sports involvement and preferences. Furthermore, findings compiled in all three of these tables concluded that males were more involved or had stronger preferences for being involved in sports. Specifically, greater proportions of males than females were (or expressed interest in being) involved in sports and athletic activities in general (row 1), in team sports (row 2), and in competitive sporting activities (row 3).

25.17.15 Audience-Based Entertainment Choices and Preferences

Research having to do with sex differences in audience-based entertainment and media is summarized in Table 25.17.15. The first row pertains to sex differences in time spent watching television. One can see that the majority of studies indicate that males do so more than females, although there are quite a few studies that have found no significant sex difference.

Studies having to do with time spent watching or enjoying comedic performances has received attention regarding sex differences. Out of the 21 studies located, row 2 shows that a substantial majority reported no significant sex differences.

Row 3 pertains to sex differences in time spent watching, or reported enjoyment of, pornography. All 44 studies reported greater involvement by males in this type of activity. Regarding time spent watching or reported degree of enjoyment of romantic mass media programs (mainly in the form of what are known as soap operas), all 23 studies concluded that females do so more than males (row 4).

Table 25.17.14 Sports involvement and preferences

Variable	*Orig. table*	*Number of sex difference findings (and consistency scores)*						*Overview*	*Countries*	*Time range*	*Non-human*
		I/T	*Child*	*Adol*	*Adult*	*WAR*	*All ages*				
1. Playing sports and athletics in general	17.5.5.1	1M	8M	16M (100.0)	15M (100.0)	8M	48 (100.0)	–	8 (4)	1985–2015	–
2. Playing team sports	17.5.5.2	–	4M	9M	1M	4M	18 (100.0)	–	5 (1)	1989–2012	–
3. Playing competitive sports	17.5.5.6	–	19M (100.0)	12M (100.0)	5M	1M	37 (100.0)	M: 1Rev	6 (1)	1930–2013	–

Table 25.17.15 Audience-based entertainment choices and preferences

Variable	*Orig. table*	*Number of sex difference findings (and consistency scores)*						*Overview*	*Countries*	*Time range*	*Non-human*
		I/T	*Child*	*Adol*	*Adult*	*WAR*	*All ages*				
1. Watching TV	17.5.7.2	–	7M 4x (63.6)	8M 5x 1F (53.3)	6M 1x 1F	2M	23M 10x 2F (62.2)	–	4	1970–2018	–
2. Watching or enjoying comedic performances	17.5.8.3	–	1F 1M	1M	3F 15x	–	4F 15x 2M (17.4)	–	3	1927–2005	–
3. Watching pornography	17.5.8.4	–	–	25M (100.0)	30M (100.0)	6M	62M (100.0)	M: 1Rev 2Met	15 (1)	1960–2020	–
4. Watching or enjoying romantic programs	17.5.8.6	–	1F	6F	15F (100.0)	1F	23F (100.0)	–	2 (2)	1929–2011	–
5. Watching or enjoying sports programs	17.5.8.9	–	2M	4M	8M	–	14M (100.0)	–	4 (1)	1932–2011	–
6. Using computers in general	17.5.9.1	–	–	15M 2x (88.2)	25M 3x 4F (69.4)	2M	42M 5x 4F (76.4)	–	10 (2)	1984–2007	–
7. Accessing the internet	17.5.9.2	–	–	17M 1F (94.4)	6M 1x 2F	–	23M 1x 3F (76.7)	–	7 (4)	1999–2015	–
8. Playing electronic or video games	17.5.9.4	–	11M (100.0)	18M (100.0)	16M 1x (94.1)	15M (100.0)	60M 1x (98.4)	–	4 (1)	1983–2017	–

Regarding sports programs, the available evidence is summarized in row 5. All 14 of these studies lead to the conclusion that males devote more time, or report enjoying, these types of programs more than do females.

Fifty-one studies of sex differences in time spent using computers were located. Row 6 shows that 42 of these studies indicated that males do so more, although five other studies reported no sex differences and four reported greater use by females. Sex differences in accessing the internet have also been extensively investigated. In row 7, one can see that most of the relevant studies have concluded that males do so more.

The final sex difference considered in this table has to do with playing video and other types of electronic games. One can see that all but one of the 61 pertinent studies concluded that males do so more, with the exception reporting no significant sex difference. The one exceptional study deserves a special qualifying comment. It was based on a sample of adult engineering students rather than on a more general sample of males and females.

25.17.16 Physical Aggression

Many studies have sought to determine if males and females differ in terms of physical aggression of various types. In fact, sex differences in physical aggression are one of the most widely researched variables we located. Findings from the available studies are summarized in row 1 of Table 25.17.16. Nearly all of these studies were based on observing behavior in naturalistic settings or on self-reports (rather than on experiments or studies conducted in institutionalized settings). One can see that most studies have indicated that males are more physically aggressive than females. This pattern conforms with conclusions reached by seven literature reviews and 13 meta-analyses. Even so, because we set the consistency score cutoff for potential inclusion as a universal sex difference so high (i.e., 95.0), only one indicator of such aggression will be considered in this regard. Specifically, as discussed more below, males are more involved in combative warfare in all located studies.

We were frankly surprised that most measures of sex differences in physical aggression did not attain or surpass the 95.0 cutoff. Nevertheless, it is relevant to note that physical aggression can occur under many circumstances, including retaliatory aggression or aggression in defense of others. Also, if one examines the last column of this table, one will see that only 147 out of 176 studies of sex differences among other animals concluded that males were more physically aggressive than females. Using the same formula as for human studies, the data for non-humans yielded a consistency score 78.6, which is even lower than the 89.0 score for humans.

Table 25.17.16 Physical aggression

Variable	*Orig. table*	*Number of sex difference findings (and consistency scores)*						*Overview*	*Countries*	*Time range*	*Non-human*
		I/T	*Child*	*Adol*	*Adult*	*WAR*	*All ages*				
1. Physical aggression in general	17.6.1.1	17M 5x 1F (70.8)	163M 18x 1F (89.1)	104M 5x 2F (92.0)	143M 15x 1F (89.4)	18M 5x 1F (72.0)	445M 43x 6F (89.0)	M: 7Rev 13Met	33 (10)	1933–2021	147M 18x 11F (78.6)
2. Physical aggression among the mentally ill	17.6.1.4	–	2M 1x	–	–	8M 13x 2F (32.0)	10M 14x 2F (35.7)	–	7 (1)	1982–2003	–
3. Experimentally provoked physical aggression	17.6.1.5	–	25M 2x 3F (75.8)	–	41M 10x 3F (71.9)	–	66M 10x 6F (75.0)	M: 1Met	3	1951–2007	7F
4. Inflicting pain on others	17.6.1.6	–	3M 1x	4M 1x	5M 4x	–	12M 6x (66.7)	–	1	1962–1980	–
5. Imitating physical aggression by others	17.6.1.9	–	9M 2x (81.8)	–	1M	–	10M 2x (83.3)	–	1	1966–1978	–
6. Direct involvement in warfare (combat)	17.6.1.10	–	–	–	14M (100.0)	3M	17M (100.0)	M: 2Rev	6 (6)	1978–2016	2M Chimps

Several studies of physical aggression have been collected for persons with various forms of mental illness, primarily in hospitals. One can see in row 2 that ten of these studies found males being more physically aggressive, while 14 reported no significant sex differences and two reported more physical aggression among females.

Since the 1950s, many experiments have been conducted in which research participants are provoked (usually by being insulted) and then asked to inflict electric shock on another "research participant" (who is actually part of the experiment and not actually being shocked). Row 3 reveals that while most studies have found males inflicting more electric shock, there are a substantial number of exceptions. Additional number of studies in which individuals are allowed to inflict pain (or supposed pain) on others are summarized in row 4. Most of these studies also report such behavior to be more common among males, although several studies found no significant sex differences.

Twelve studies of physical aggression under experimental circumstances in which research participants were presented with aggression by others and then allowed to retaliate or displace their aggression elsewhere were tabulated as a separate category. In row 5, one can see that ten of these studies concluded that males were more likely to imitate physical aggression by others, while the remaining two studies found no significant sex differences.

The last row in this table on physical aggression pertains to studies of people who become directly involved in warfare (not including those who performing support roles in warfare such as medical personnel). All 17 studies concluded that males were more likely than females to be combatants in wars. It is also worth adding that two studies of warfare-like behavior between tribal groups of chimpanzees both concluded that the combatants were all males.

25.17.17 Non-Physical Aggression

Three forms of non-physical aggression were located for which at least ten studies of sex differences were found. Their findings are summarized in Table 25.17.17. Row 1 has to do with verbal aggression. Over 150 relevant studies were located. Findings have been mixed, although most of the evidence points toward males exhibiting such aggression to greater degrees than females.

Indirect aggression refers to such things as slamming doors and punching inanimate objects when one is upset. As shown in row 2, about twice as many studies point toward this type of aggression being more common among males than among females.

Table 25.17.17 Non-physical aggression

Variable	*Orig. table*	*Number of sex difference findings (and consistency scores)*						*Overview*	*Countries*	*Time range*	*Non-human*
		I/T	*Child*	*Adol*	*Adult*	*WAR*	*All ages*				
1. Verbal aggression	17.6.2.1	2x	24M 4x 8F (54.5)	10M 4x 4F (45.5)	62M 35x 10F (53.0)	1x	96M 46x 22F (51.6)	M: 1Met x: 1Met	17 (1)	1934–2014	–
2. Indirect aggression	17.6.2.2	–	9M 1x 7F (37.5)	15M 3F (71.4)	7F 2x 3M (46.7)	1M	28M 3x 17F (43.1)	F: 1Rev x: 1Met	8 (1)	1930–2012	–
3. Relational aggression	17.6.2.4	–	20F 5x 2M (69.0)	1F 1x 1M	3F 3M	–	24F 6x 4M (63,2)	F: 1Gen 1Met x: 1Met	8	1986–2007	–

When someone tells you that they will no longer be their friend, this is an example of behavior termed *relational aggression*. Row 3 show that most pertinent studies have indicated that this type of aggression is more common among females than among males.

25.17.18 Sexuality

Studies of sex differences in sexual behavior for which at least ten findings were located are summarized in Table 25.17.18. While most forms of sexual behavior are social in nature, masturbation can be considered an exception. As shown in row 1, all but one of the 50 studies of sex differences in such behavior have concluded that it is more frequent among males than females. This pattern comports with two meta-analyses and with five studies of non-human mammals.

Twenty-nine studies were located regarding sex differences in tendencies to initiate or at least trying to initiate sexual intimacy. One can see in row 2 that 27 of these studies concluded that males do so more than females. Along similar lines, row 3 shows that 12 out of 13 studies of exhibiting pushy forms of sexual overtures concluded that this too was more common among males.

Sex differences in individuals having had premarital sexual intercourse have been widely studied, with nearly all of the evidence coming from self-reports. Row 4 shows that a substantially greater proportion of males have reported having had premarital sex than is the case for females, i.e., according to 80 out of the 113 of the relevant studies.

25.17.19 Bonding Aspects of Sexuality

Sex differences in the bonding aspects of sexuality for which ten or more pertinent findings were located are summarized in Table 25.17.19. The first row has to do with the age of marriage or the relative age of those while dating or engaged in courtship. It shows that nearly all studies indicate that males are older on average when they marry (and are older than their dating partner when courting). It may be pertinent to note that the four studies of the relative age of non-human animals during courtship also concluded that, on average, males were older than females.

Many studies have assessed sex differences among adults in terms of them either being married or ever have been married. One can see in row 2 that 84 out of 94 of the located studies have concluded that adult males are more likely than adult females to have been married at least once.

Table 25.17.18 Sexuality

Variable	*Orig. table*	*Number of sex difference findings (and consistency scores)*						*Overview*	*Countries*	*Time range*	*Non-human*
		I/T	*Child*	*Adol*	*Adult*	*WAR*	*All ages*				
1. Masturbation	17.7.1.1	–	2M	10M (100.0)	32M 1x (97.0)	5M	49M 1x (98.0)	M: 2Met	13 (2)	1937–2020	5M mammals
2. Initiating or attempting to initiate sexual intimacy	17.7.1.11	–	–	3M	21M 1x 1F (87.5)	3M	27M 1x 1F (90.0)	–	5	1976–2011	–
3. Pushy sexual overtures	17.7.1.13	–	–	1M 1x	11M (100.0)	–	12M 1x (92.3)	–	2	1984–1996	–
4. Premarital sexual intercourse	17.7.1.19	–	–	34M 9x 7F (59.6)	31M 13x 3F (62.0)	15M 1x (93.7)	80M 23x 10F (65.0)	M: 1Met	21 (2)	1966–2016	–
5. Age of first sexual intercourse	17.7.1.21	–	–	14F 1x 12M (35.9)	18M 7x 4F (54.5)	1M 4x	31M 12x 18F	M: 2Met	20 (3)	1970–2005	3F 1x
6. Number of sex partners	17.7.1.24	–	–	7M	38M 10x 4F (67.9)	15M (100.0)	60M 10x 4F (76.9)	M: 2Met	16 (2)	1982–2020	2M rodents 6F non-mammals

Table 25.17.19 Bonding aspects of sexuality

Variable	*Orig. table*	*Number of sex difference findings (and consistency scores)*						*Overview*	*Countries*	*Time range*	*Non-human*
		I/T	*Child*	*Adol*	*Adult*	*WAR*	*All ages*				
1. Age at marriage or relative age when courting	17.7.4.1	–	–	4M	39M 1x 1F (92.8)	83M (100.0)	126M 1x 1F (97.7)	–	27 (10)	1955–2013	4M during courtship
2. Ever married	17.7.4.2	–	–	–	84M 3x 6F (84.8)	1F	84M 3x 7F (83.2)	–	8	1959–2021	–
3. Divorce initiation	17.7.4.11	–	–	–	23F 1x 1M (88.5)	–	23F 1x 1M (88.5)	–	3 (1)	1925–2002	–
4. Remarriage after divorce	17.7.4.14	–	–	–	11M (100.0)	–	11M (100.0)	–	1 (1)	1959–2021	–

Which sex is most likely to initiate divorce? Row 3 shows that 23 out of the 25 studies addressing this question have concluded that females do so more often than males.

Eleven studies examined people's likelihood of remarrying after having gotten divorced. As shown in row 4, all of these studies concluded that divorced males remarry at higher rates than do divorced females.

25.18 Acquiring, Selling, and Consuming Behavior: Condensed Findings

In the present section, a condensation of research findings regarding sex differences in the acquisition, selling, and consumption of a wide spectrum of goods and services is provided. The information is based on evidence provided in tables contained throughout Chapter 18.

25.18.1 Acquiring and Consuming (Except Drug Consumption)

Setting aside drug consumption, three variables were found for which ten or more relevant studies were located regarding possible sex differences. As shown in Table 25.18.1, the first variable was that of shopping. One can see in row 1 that nearly all studies have concluded that females spend more time shopping than do males.

In row 2, findings from studies of sex differences in the consumption of fruits and vegetables are summarized. It shows that 17 out of 19 studies concluded that females do so more than males. Incidentally, seven studies of non-human primates indicated that eating of fruits and vegetables is also more common among females than males.

The last row in this table pertains to the consumption of meat and fats. All 16 located studies of sex differences concluded that meat and fats were a more prevalent part of the diet for males than for females. It may be worth adding that nine studies of non-human primates also concluded that males consume more meat and fats than do females.

25.18.2 Consuming Therapeutic Drugs

Substantial research has been undertaken to assess sex differences in the consumption of drugs for therapeutic purposes. As shown in the first row of Table 25.18.2, nearly all research has concluded that females take more therapeutic drugs than do males, especially following childhood.

Rows 2 and 3 both focus on physician-prescribed drugs. One can see that the majority of studies have found females using prescription drugs

Table 25.18.1 Acquiring and consuming (except drug consumption)

Variable	*Orig. table*	*Number of sex difference findings (and consistency scores)*						*Overview*	*Countries*	*Time range*	*Non-human*
		I/T	*Child*	*Adol*	*Adult*	*WAR*	*All ages*				
1. Shopping in general (except online)	18.1.3.3	–	–	5F 1x	26F (100.0)	3F	34F 1x (**97.1**)	–	12	1982–2016	–
2. Eating fruits and vegetables	18.1.3.3	–	–	4F	10F 1x 1M (76.9)	3F	17F 1x 1M (85.0)	F: 2Rev	8 (1)	1984–2019	7F primates
3. Eating meat and fats	18.1.4.2	–	–	–	13M (100.0)	3M	16M (100.0)	M: 2Rev	6 (2)	1984–2015	9M primates

Table 25.18.2 Consuming therapeutic drugs

Variable	*Orig. table*	*Number of sex difference findings (and consistency scores)*						*Overview*	*Countries*	*Time range*	*Non-human*
		I/T	*Child*	*Adol*	*Adult*	*WAR*	*All ages*				
1. Consuming therapeutic medications in general	18.1.4.2	–	1x	1F	18F (100.0)	8F	27F 1x (**96.4**)	–	3 (2)	1973–2004	–
2. Consuming prescription medications in general	18.1.4.5	–	–	1x	24F 2x (92.3)	4F	28F 3x (90.3)	–	7	1970–2004	–
3. Consuming prescription psychotropic medications	18.1.4.6	–	–	–	24F 4x 1M (80.0)	18F (100.0)	42F 4x 1M (87.5)	–	8 (3)	1960–2007	–

more than males. This is true in general and specifically regarding psychotropic drugs, particularly for treatment of depression.

25.18.3 Consuming Alcohol

The distinction between drugs taken for medical purposes and those taken for recreation, relaxation, and fun is fairly clear. One of the primary distinctions is that recreational drugs nearly always affect one bodily organ, i.e., the brain, whereas medications are taken to treat or prevent discomfort anywhere in the body. Table 25.18.3 summarizes sex differences in the use of the single most widely used recreational drug: alcohol.

Row 1 pertains to sex differences in the consumption of alcohol in general. This is followed by a summary of the findings having to do with information on the actual amount of alcohol consumed in row 2. One can see that the vast majority of studies in both rows indicate that males drink more alcohol than do females, especially regarding the amount of alcohol consumed.

In row 3, research findings having to do with sex differences in abstinence from alcohol consumption are presented. Nearly all of these studies suggest that females are more likely than males to be abstainers.

Twenty-two studies sought to determine if the sexes differ in terms of when they first consumed alcohol. Row 4 shows that all but two of these studies concluded that males begin consuming alcohol at younger ages than do females.

In 51 studies, researchers assessed the extent to which males and females might differ in terms of experiencing any alcohol-related problems. As shown in row 5, all but two of these studies found more problems among males than among females.

Of the 11 studies of sex differences in tendencies to drink light to moderate amounts of alcohol (as opposed to some form of heavy drinking), row 6 shows that all but one study concluded there were no significant sex differences.

Row 7 has to do with alcohol abuse. While definitions vary, nearly all studies agree that such abuse is more common among males than females. In the case of binge drinking, row 8 shows that all findings agree that this type of drinking is most common among males.

The last row in this table has to do with sex differences in preferring beer over other forms of alcoholic beverages. It shows that 11 out of the 12 pertinent studies concluded that this preference is stronger for males than for females.

Table 25.18.3 Consuming alcohol

Variable	Orig. table	Number of sex difference findings (and consistency scores)						Overview	Countries	Time range	Non-human
		I/T	Child	Adol	Adult	WAR	All ages				
1. Alcohol consumption in general	18.1.15.1	–	–	70M 8x 3F (83.3)	176M 8x 4F (91.7)	42M 2x (95.5)	288M 18x 7F (90.0)	M: 3Rev	34 (16)	1952–2012	mixed
2. Amount of alcohol consumed	18.1.5.3	–	–	6M	66M 1F (97.0)	4M	76M 1F (97.4)	–	12 (3)	1953–2010	5F 1x mammals
3. Abstinence from alcohol consumption	18.1.5.4	–	–	2F 1x	22F (100.0)	15F (100.0)	39F 1x (97.5)	–	11 (1)	1947–2020	–
4. Alcohol consumption, age of onset	18.1.5.6	–	–	1M	–	19M 2x (90.5)	20M 2x (90.9)	–	5	1951–2005	–
5. Experience alcohol-related problems	18,1,5,7	–	–	1M 1F	26M 1x (96.3)	22M (100.0)	49M 1x 1F (94.2)	–	7 (1)	1947–2005	–
6. Light to moderate alcohol consumption	18.1.5.8	–	–	–	1M 10x (9.1)	–	1M 10x (9.1)	–	10	2000–2005	–
7. Alcohol abuse	18.1.5.9	–	–	8M 3x (72.7)	65M 1x (98.5)	11M (100.0)	84M 4x (95.5)	–	16 (2)	1968–2010	–
8. Binge drinking	18.1.5.10	–	–	1M	25M (100.0)	4M	30M (100.0)	–	9	1971–2007	–
9. Preferring beer over other forms of alcohol	18.1.5.15	–	–	–	11M 1x (91.7)	–	11M 1x (91.7)	–	3	1953–1985	–

Table 25.18.4 Consuming nicotine

Variable	*Orig. table*	*Number of sex difference findings (and consistency scores)*						*Overview*	*Countries*	*Time range*	*Non-human*
		I/T	*Child*	*Adol*	*Adult*	*WAR*	*All ages*				
1. Smoking cigarettes in general	18.1.6.1	–	–	44M 18x 28F (37.3)	125M 14x 15F (74.0)	53M 14x 8F (63.9)	222M 28x 51F (62.9)	M: 1Rev	33 (1)	1925–2016	–
2. Abstinence from smoking (never smoked)	18.1.6.3	–	–	–	13F 1M (86.7)	–	13F 1M (86.7)	–	8	1981–2020	–
3. Age of smoking onset	18.1.6.4	–	–	–	–	6M 5x (54.5)	6M 5x (54.5)	–	6	1972–2010	–
4. Quit smoking once established	18.1.6.5	–	–	1x	37M 2x (97.4)	14M 3x (82.4)	51M 5x (91.1)	–	12 (1)	1966–2015	–
5. Severity of smoking withdrawal symptoms	18.1.6.7	–	–	–	2F 3x	10x	2F 13x (13.3)	–	2	1967–2004	–

25.18.4 Consuming Nicotine

The consumption of nicotine comes primarily from smoking cigarettes. As shown in the first row of Table 25.18.4, over 300 studies of sex differences in the consumption of cigarettes have been published, with the first one dating back to 1925. Overall, a substantial majority of findings have concluded that males are more likely than females to smoke cigarettes.

Fourteen studies of smoking abstainers, persons who have never smoked, were located. Row 2 shows that all but one of these studies indicated that females are more likely to be abstainers than are males.

Is there a sex difference among established smokers regarding the age when they first began to smoke? According to row 3, roughly half of the 11 studies reported males doing so at a younger age while the remainder reported no significant sex difference.

Over 50 studies sought to determine if male and female established smokers differ in terms of their success in quitting. One can see in row 4 that the vast majority of studies indicate that males are more likely to succeed than females. Additional research has asked both sexes how difficult it was to quit smoking (or how severe were their withdrawal symptoms). Row 5 shows that all but two of the 15 studies found no significant sex differences in this regard.

25.18.5 Consuming Other Recreational Drugs

Sex differences in the consumption of largely recreational drug use not yet covered for which at least ten pertinent studies were located are examined in Table 25.18.5. The specific drugs are cocaine, marijuana, and opiates (without a prescription).

In row 1, sex differences in the consumption of cocaine are summarized. Out of 20 studies, 12 found males being more likely to use cocaine, with the remaining eight studies evenly split between those reporting no significant differences and those indicating that females use more than males.

Over 70 studies have assessed sex differences in the consumption of marijuana. One can see in row 2 that findings have been mixed, although a substantial majority have concluded that males are more likely to use this drug than females.

Opiate drugs include both natural opiates such as heroin and a variety of synthetic opiates, some of which are prescribed. Of the 23 located studies on the use of these drugs, 13 concluded that males use them more than females.

Table 25.18.5 Consuming other non-prescription drugs

Variable	*Orig. table*	*Number of sex difference findings (and consistency scores)*						*Overview*	*Countries*	*Time range*	*Non-human*
		I/T	*Child*	*Adol*	*Adult*	*WAR*	*All ages*				
1. Cocaine consumption	18.1.7.3	–	–	5M 2x 1F	7M 2x 2F (53.8)	3M 1F	12M 4x 4F (50.0)	–	4 (2)	1988–2005	10F 4x rodents
2. Marijuana consumption	18.1.7.7	–	–	22M 7x 1F (71.0)	32M 6x 1F (80.0)	2M	56M 13x 2F (76.7)	–	16 (1)	1970–2016	1M rodents
3. Opiate consumption	18.1.7.10	–	–	1M	6F 2x 5M (21.4)	7M 1x 1F	13M 3x 7F (43.3)	–	7	1989–2005	3F rodents

Table 25.18.6 Controlling machines or tools

Variable	*Orig. table*	*Number of sex difference findings (and consistency scores)*						*Overview*	*Countries*	*Time range*	*Non-human*
		I/T	*Child*	*Adol*	*Adult*	*WAR*	*All ages*				
1. Driving motor vehicles aggressively	18.2.1.5	–	–	1M	2M	3M 4x	6M 4x (60.0)	–	3	1980–2015	–
2. Driving while under the influence of alcohol	18.2.1.7	–	–	–	3M	11M (100.0)	14M (100.0)	–	3	1968–2008	–
3. Owning weapons	18.2.2.2	–	–	–	12M (100.0)	–	12M (100.0)	M: 1Rev	1	1991–2015	–

25.18.6 Controlling Machines or Tools

As a final residual category in this chapter, Table 25.18.6 provides a summary of sex differences having to do with the possession or use of certain machines or tools. Only three traits contained ten or more relevant findings.

In row 1, one can see that ten studies were located having to do with sex differences in tendencies to drive motor vehicles aggressively, sometime known as *road rage*. Six of these studies reported that males do so more, while the remaining four studies concluded that there were no significant sex differences.

Row 2 summarizes findings of sex differences in driving while under the influence of alcohol. According to all 14 studies, males do so more than females.

In the last row for this table, findings regarding sex differences in the ownership of weapons, most often firearms, are summarized. One can see that all 12 studies reached the conclusion that males are more likely than females to own weapons.

25.19 Criminality, Near-Criminality, and Victimization: Condensed Findings

Criminality is obviously a legal concept, and, because criminal statutes vary from country to country as well as over time, criminal behavior is difficult to study in a scientific context. Despite these limitations, many studies of sex differences in criminal behavior have been published over the years. A summary of the results of these studies appeared in Chapter 19, and condensed summaries of all the tables in this chapter containing ten or more findings, appear in the tables below.

25.19.1 Official Criminal and Delinquent Behavior

Official crimes are acts that violate the laws governing where they occur, and delinquent acts are those committed by a juvenile that would be crimes if committed by adults. Row 1 of Table 25.19.1 shows the results of the large number of studies of sex differences in commission of criminal acts in general. These studies indicate a likely universal sex difference because the studies have taken place in a large number of countries and nearly all of them have found most crimes are committed by males. The same conclusion can be reached regarding studies limited to violent offenses other than homicide summarized in row 2, where all of the studies found males committing most of these crimes. Studies of homicide, listed in row 3, also show that males commit this crime more, as do nearly all of the studies of

Table 25.19.1 Official criminal and delinquent behavior

Variable	*Orig. table*	*Number of Sex Difference Studies (and Consistency Scores)*						*Overviews*	*Countries*	*Time range*	*Non-human*
		I/T	*Child*	*Adol*	*Adult*	*WAR*	*All ages*				
1. Officially determined offending in general	19.1.1.1	–	–	100M 3x 1F (98.0)	36M (100.0)	53M (100.0)	189M 3x 1F (99.0)	M: 2Rev 1Met	30 (+10)	1842–2016	–
2. Officially determined violent offending (except homicide)	19.1.1.2	–	–	15M (100.0)	14M (100.0)	68M (100.0)	97M (100.0)	M: 6Rev	12 (+11)	1963–2017	–
3. Officially determined homicide perpetration (excluding genetic relative victims)	19.1.1.3	–	–	3M	13M (100.0)	85M (100.0)	101M (100.0)	–	25 (+22)	1950–2020	–
4. Officially determined homicide followed by suicide	19.1.1.4	–	–	–	28M 1x (96.6)	–	28M 1x (96.6)	M: 1Rev	12 (+1)	1965–2011	–
5. Infanticide among non-humans (scientifically documented)	*19.1.1.5b*	–	–	*9F 2x 4M (47.4)*	*13M 6x 8F (37.1)*	*1M 1x*	*18M 9x 17F (29.5)*	–	–	*1961–2000*	*various species*
6. Officially determined property offending	19.1.1.8	–	1M	8M	1M	32M 5F (76.2)	42M 5F (80.8)	–	11 (+2)	1931–2016	–
7. Officially determined sex offenses in general (except prostitution)	19.1.1.10	–	–	–	7M	11M (100.0)	18M (100.0)	M: 1Rev 1Met	9 (+1)	1994–2020	–
8. Officially determined illegal drug use/ possession/sale	19.1.1.11	–	–	6M	2M	15M (100.0)	23M (100.0)	–	10	1973–2018	–

homicide followed by suicide. In contrast to the above findings, row 4 indicates that there is not a clear difference among non-humans regarding which sex commits infanticide.

Row 6 shows that 42 out of 47 studies found property crimes to be committed primarily by males, but the other five studies found more property offenses being committed by females. Consistent with findings discussed earlier, sexual offenses (other than prostitution) were found to be committed more by males than by females in all of the 18 studies presented in row 7. Finally, all of the studies of crimes involving illegal drugs found males to commit this crime more frequently than females.

25.19.2 Self-Reported Victimizing Offenses

Studies of the crimes reported by victims include crimes that may or may not have been officially recorded as crimes by law enforcement. Table 25.19.2 indicates that the many studies of this type have found that individuals report being victimized by males more often than by females, with three of the four rows surpassing the threshold of probably being a universal sex difference, and the exception, concerning property offending, coming very close (94.1). At the other extreme, all of the studies of rape or sexual assault found males were more often reported than females as committing these offenses.

25.19.3 Victimless Offending

Both of the questions regarding possible sex differences in the self-reporting of victimless crime where ten or more studies were examined involved illegal drug use. Row 1 of Table 25.19.3 shows that illegal drug use was found to be self-reported more often by males, although many studies found no sex difference. Row 2 indicates that all of the dozen studies on the age of drug use onset found that males reported starting illegal drug use at a younger age than females.

25.19.4 Residual Criminality and Variables Related to Criminality

Table 25.19.4 shows the results of questions regarding a variety of variables related to criminality that were examined by ten or more studies. Row 1 shows that all of the studies based on victim reports of violent crimes, including over a century of research, found males to be reported more often than females. Row 2 indicates that recidivism is higher among male criminals, but ten of the 49 studies found no sex difference. The specific crime of tax evasion, examined in row 3, shows that 13 out of 15 studies found that males commit tax evasion more than females. Finally, row 4 indicates that nine studies found that the greater

Table 25.19.2 Self-reported victimizing offenses

Variable	*Orig. table*	*Number of Sex Difference Studies (and Consistency Scores)*						*Overviews*	*Countries*	*Time range*	*Non-human*
		I/T	*Child*	*Adol*	*Adult*	*WAR*	*All ages*				
1. Self-reported offending in general	19.1.2.1	–	–	196M 5x (97.5)	16M 1x (94.1)	15M (100.0)	227M 6x (97.4)	–	46 (+70	1947–2021	–
2. Self-reported violent offending (excluding sexual assault)	19.1.2.2	–	–	46M 1x (97.9)	7M	10M (100.0)	63M 1x (98.4)	–	9 (+4)	1979–2019	–
3. Self-reported property offending	19.1.2.3	–	–	35M 2x (94.6)	21M (100.0)	8M 2x (80.0)	64M 4x (94.1)	M: 1Rev	11 (+7)	1926–2019	–
4.Self-reported (or victim-reported when indicated) rape or sexual assault	19.1.2.4	–	–	2M	14M (100.0)	2M	18M (100.0)	M: 1Rev	7	1982–2021	–

Table 25.19.3 Victimless offending

Variable	*Orig. table*	*Number of Sex Difference Studies (and Consistency Scores)*						*Overviews*	*Countries*	*Time range*	*Non-human*
		I/T	*Child*	*Adol*	*Adult*	*WAR*	*All ages*				
1. Self-reported illegal drug use or possession	19.1.3.1	–	–	77M 18x 3F (76.2)	82M 16x 3F (78.8)	21M 2x (91.3)	180M 36x 6F (79.0)	–	30 (1)	1969–2020	–
2. Illegal drug use, age of onset	19.1.3.3	–	–	–	–	12M (100.0)	12M (100.0)	–	1	1985–2001	–

Table 25.19.4 Residual evidence of criminality and variables related to criminality

Variable	*Orig. table*	*Number of Sex Difference Studies (and Consistency Scorses)*						*Overviews*	*Countries*	*Time range*	*Non-human*
		I/T	*Child*	*Adol*	*Adult*	*WAR*	*All ages*				
1. Perpetration of violent crime based on victim reports	19.1.4.1	–	–	1M	–	10M (100.0)	11M (100.0)	M: 1Met	4	1911–2011	–
2. Criminal recidivism	19.1.4.8	–	–	1M	38M 10x (79.2)	–	39M 10x (79.6)	M: 1Met	8	1969–2014	–
3. Tax evasion	19.1.4.9	–	–	–	13M 1x 1F (81.3)	–	13M 1x 1F (81.3)	–	7	1978–2010	–
4. Trends in sex differences in criminal behavior	14.9.4.10	–	–	2F 1x	–	7F 3x (70.0)	9F 4x (69.2)	–	2	1981–2014	–

commission of crimes by males was decreasing, and the remaining four studies found no change.

25.19.5 Responsible for Near-Criminality

Many offensive actions that are often not prosecuted as official crimes, are still strongly disapproved of and studied to identify ways to eliminate them or at least lessen their frequency. Bullying is one such action, and most of the studies of bullying behavior have found that males most often committed this behavior. Row 1 of Table 25.19.5. shows that this was particularly true when the bullying did not involve the internet, but it was true to a lesser extent when cyberbullying was studied (row 2). Studies of sexual assertiveness or pushiness, summarized in row 3 found that 15 out of 16 studies found that males most frequently behaved in this manner, with the remaining study finding no sex difference.

An interesting pattern was found in regard to violence among couples resulting in serious injury. Row 5 reveals that there was no clear sex difference in such violence among courting and intimate partners. However, in row 6, one can see that all 13 of the studies specifically concerned with violence among spouses concluded that, when the injuries were serious (usually assessed in terms of requiring medical treatment), the offender was significantly more likely to be a male.

25.19.6 Responsibility for Disruptive or Reckless Behavior

Sex differences in two types of disruptive or reckless behavior were the subjects of the required number of studies to be presented in this chapter. Rows 1 and 2 in Table 25.19.6 found that all of the studies on the recipients of school discipline (row 1) and suspension or expulsion from school (row 2) found males are the recipients more often than females. Males were also more likely to engage in unsafe driving in all of the studies on that topic (row 4), and in all but one of the 16 studies concerning sex differences in traffic violations and suspensions (row 3).

25.19.7 Crime Victimization

Table 25.19.7 presents the condensed results of studies on sex differences in the likelihood of being a victim of crime in general, and of being a victim of a variety of specific crimes. Row 1 indicates that nearly all (21 out of 22) of the studies of victims of crime in general found males to be the most frequent victims, with the remaining study finding no significant sex difference. This sex difference exists despite the great variety in sex differences found when specific crimes are studied. The male dominance in being victims of crimes in general appears to be largely due to the

Table 25.19.5 Responsible for near-criminality.

Variable	*Orig. table*	*Number of Sex Difference Studies (and Consistency Scores)*						*Overviews*	*Countries*	*Time range*	*Non-human*
		I/T	*Child*	*Adol*	*Adult*	*WAR*	*All ages*				
1. Bullying behavior in general (except cyberbullying)	19.2.1.1	–	15M 1x (93.8)	38M 6x (86.4)	–	2M 1x	55M 8x (87.3)	M: 1Rev x: 1Rev	17	1982–2021	–
2. Cyberbullying	19.2.1.2	–	–	11M 6x 1F (57.9)	5M 1F	–	16M 6x 2F (61.5)	M: 1Rev 2Met	12 (1)	2008–2021	–
3. Cruelty and sadistic behavior	19.2.1.3	–	3M	3M	4M	3M	13M (100.0)	M: 1Rev	4	1971–2020	–
4. Sexual assertiveness/ pushiness	19.1.2.4	–	–	7M	6M 1x	2M	15 M 1x (93.8)	M: 1Rev	3 (1)	1982–2011	1M primate
5. Commission of courtship and intimate partner violence with significant injury	19.1.2.6a	–	–	10F 4x 2M (55.6)	26F 26x 21M (27.7)	7F 15x 6M (20.6)	43F 45x 29M (29.5)	M: 1Gen x: 1Rev	15 (1)	1974–2019	–
6. Commission of domestic violence (spousal violence) with serious injury	19.1.2.7	–	–	–	13M (100.0)	–	13M (100.0)	M: 1Met	2	1980–2007	–

Table 25.19.6 Responsibility for disruptive or reckless behavior

Variable	*Orig. table*	*Number of Sex Difference Studies (and Consistency Scores)*						*Overviews*	*Countries*	*Time range*	*Non-human*
		I/T	*Child*	*Adol*	*Adult*	*WAR*	*All ages*				
1. Recipient of teacher discipline or school discipline	19.2.2.2	–	11M (100.0)	10M (100.0)	–	2M	23M (100.0)	–	2	1952–2002	–
2. Suspension/ expulsion from school	19.2.2.3	–	–	10M (100.0)	–	–	10M (100.0)	–	2	1986–2014	–
3. Traffic violation and suspension	19.2.2.4	–	1M	1M	3M	10M 1x (90.9)	15M 1x (93.8)	–	3	1970–2007	–
4. Unsafe (reckless) driving in general	19.2.2.5	–	–	16M (100.0)	3M	3M	22M (100.0)	M: 1Met	11	1973–2019	–

Table 25.19.7 Crime victimization

Variable	*Orig. table*	*Number of Sex Difference Studies (and Consistency Scores)*						*Overviews*	*Countries*	*Time range*	*Non-human*
		I/T	*Child*	*Adol*	*Adult*	*WAR*	*All ages*				
1. Victim of crime in general	19.3.1.1	–	–	2M	2M 1x	17M	21M 1x **(95.5)**	–	5 (1)	1975–2021	–
2. Victim of infanticide, investigative evidence	19.3.1.2a	4M 3x 3F (37.8)	–	–	–	–	4M 3x 3F (37.8)	–	3 (1)	1984–2006	–
3. Victim of murder/ homicide (except infanticide)	19.3.1.3	–	1x	1M	5M	74M 1x 2F (93.7)	80M 2x 2F (93.0)	–	23 (6)	1932–2016	–
4. Victim of violent offenses other than murder (non-sexual)	19.3.1.4a	–	–	24M 1x 2F (82.8)	4M 1F	13M 3x 5F (50.0)	41M 4x 8F (67.2)	M: 1Met	8 (3)	1967–2017	1M chimp
5. Victim of child abuse/ neglect (non-sexual)	19.3.1.6	2F	3M 1x	2M 1F	–	1M 1x 1F	6M 2x 4F (37.5)	x: 1Rev	6	1980–2020	–
6. Victim of coerced or forced sex	19.3.1.9	–	1F	4F 1x	20F 1x **(95.2)**	3F	28F 2x (93.3)	–	8 (1)	1985–2021	–
7. Victim of rape or sexual assault	19.3.1.10	–	3F	14F (100.0)	51F **(100.0)**	23F **(100.0)**	91F **(100.0)**	–	11 (1)	1964–2019	–
8. Victim of sexual abuse in general (excluding violent forms)	19.3.1.11	–	10F 1x (90.9)	18F **(100.0)**	7F 2x	12F **(100.0)**	47F 3x (94.0)	F: 2Rev	10	1979–2020	–
9. Victim of spouse abuse/ domestic violence	19.3.1.12	–	–	–	23F 3x 1M (82.1)	3F 3M	26F 3x 4M (70.3)	–	3	1971–2005	–

high proportion of male victims of murder/homicide (other than infanticide), row 3, and male victims of other non-sexual violent crimes, row 4. The greater number of male victims of crime in general also exists despite females being the most likely victims of sex crimes (rows 6, 7, and 8), as well as the most likely victims of spouse abuse and domestic violence (row 9). The two specific crimes found to have the least sex difference are infanticide (row 2) and non-sexual child neglect and abuse (row 5).

25.19.8 Victimization from Bullying and Non-Criminal Antisocial Behavior

Of the eight research categories concerning victimization from bullying and non-criminal antisocial behavior in which at least ten studies were located, Table 25.19.8 shows that only "being pressured to have sex" (row 8) had a consistency score above 95.0, and this only involved adults. However, it is worth noting that all of the pertinent studies were conducted in the United States.

25.20 Education, Work, Social Status, and Territorial Behavior: Condensed Findings

The resources that are used to sustain and enjoy life are distributed in complex ways, many of which have been investigated with regard to possible sex differences. The tables contained in the present section provides a summary of findings regarding sex differences in how people prepare for and actually earn a living. Also, sex differences in earnings and in other aspects of social stratification, including the establishment and control over territories, are examined in this section.

25.20.1 Academic Effort

Given the importance of education in the social stratification process, at least in literate societies, the first table examines how males and females appear to differ in terms of academic effort. Two variables are particularly relevant to this issue. The first is shown in row 1 of Table 25.20.1. It indicates that most studies have found males more likely to be truant from school than is true for females.

Row 2 pertains to sex differences in the amount of time spent studying and doing homework. One can see that all 29 pertinent studies concluded that females spend more time studying than do males.

Table 25.19.8 Victim of bullying and non-criminal antisocial behavior

Variable	*Orig. table*	*Number of Sex Difference Studies (and Consistency Scores)*						*Overviews*	*Countries*	*Time range*	*Non-human*
		I/T	*Child*	*Adol*	*Adult*	*WAR*	*All ages*				
1. Being bullied in general	19.3.2.1	–	5M	8M 3x (72.7)	–	–	13M 3x (81.3)	–	8	1982–2021	–
2. Being physically bullied	19.3.2.2	–	12M 2x (85.7)	20M 4x (83.3)	–	2M	34M 6x (85.0)	–	13	1982–2018	–
3. Being verbally bullied	19.3.2.3	–	5M 3x 2F (41.7)	4M 4x 2F (33.3)	–	–	9M 7x 4F (37.5)	–	6	1988–2018	–
4. Victim of physical aggression in general (except criminal)	19.3.2.6	–	11M (100.0)	8M 3x (72.7)	11M 1x 1F (78.6)	–	30M 4x 1F (83.3)	–	5	1966–2019	3F 2M
5. Victim of relational aggression	19.3.2.7	1F	9F 3x (75.0)	2x	1x	–	10F 6x (62.5)	–	4	1996–2007	–
6. Victim of dating violence or domestic partner violence	19.3.2.8	–	–	3F 1x 1M	10F 6x 3M (45.5)	3F	16F 7x 4M (51.6)	F: 2Rev x: 1Met	5	1983–2019	–
7. Victim of sexual harassment (or unwelcomed sexual overtures)	19.3.2.9	–	–	6F 1M	27F 2M (77.1)	–	33F 3M (84.6)	–	7	1976–2018	–
8. Being pressured to have sex	19.3.2.11	–	–	2F 1x	15F (100.0)	–	17F 1x (94.4)	–	1	1987–2019	–

Table 25.20.1 Academic effort

Variable	*Orig. table*	*Number of Sex Difference Studies (and Consistency Scores)*						*Overviews*	*Countries*	*Time range*	*Non-human*
		I/T	*Child*	*Adol*	*Adult*	*WAR*	*All ages*				
1. Truancy	20.1.1.3	–	1M	8M 3x (72.7)	–	–	9M 3x (75.0)	–	4 (1)	1927–2013	–
2. Time spent studying and doing homework	20.1.1.4	–	–	21F (100.0)	8F	–	29F (100.0)	–	8	1929–2013	–

25.20.2 Years of Education

Sex differences in the number of years spent as a student has been widely studied over the past half-century, and nine research questions had been the subject of at least ten studies. Their findings are presented in Table 25.20.2. None of the sex differences found approached the threshold required to be considered a likely universal difference.

25.20.3 Areas of Educational Training

Substantial research has assessed sex differences in specialized areas of educational training, and there are numerous striking sex differences. Table 25.20.3 reveals that, out of the 15 categories where ten or more studies were located, all of the studies of four areas of education found more female than male students, and all of the studies of three other areas found more male than female students. All of the relevant studies found females to take more training in the fields of education (row 6), fine arts (row 8), humanities (row 9), and language (row 10). Studies of taking more course training by males were limited to computer science (row 4), economics (row 5), and engineering (row 7).

25.20.4 Education and Financial Rewards

Considerable research has sought to determine if there are sex differences in the tendency to financially benefit from obtaining additional years of education. As one can see in Table 25.20.4, the evidence is mixed, although most studies have concluded that females benefit more than males do for each additional year of education they receive.

25.20.5 Participation in Paid Workforce

Table 25.20.5 contains rows pertaining to three variables involving people's work. Row 1 shows that nearly all studies have found males are more likely than females to participate in work, particularly among adults.

Twelve studies were located regarding trends in sex differences in participating in the paid workforce, row 2 shows that all of these studies have concluded that, at least since the 1960s or 1970s, that the sex differences have diminished. In other words, the proportion of females in the paid workforce has grown over roughly the past half century, even though more males than females are still full-time workers.

The last row in this table has to do with job absences, whether due to sickness or other factors. One can see that all but one of 18 studies concluded that females have more absences than do males.

Table 25.20.2 Years of education

Variable	*Orig. table*	*Number of Sex Difference Studies (and Consistency Scores)*						*Overviews*	*Countries*	*Time range*	*Non-human*
		I/T	*Child*	*Adol*	*Adult*	*WAR*	*All ages*				
1. Average years of education in general	20.1.2.1	–	2M	1F	54M 24x 9F (56.3)	12M 7x 2F (52.3)	68M 31x 12F (55.3)	–	32 (18)	1967–2019	–
2. Attending primary school	20.1.2.3	–	5M 5x 1F (41.7)	–	–	–	5M 5x 1F (41.7)	–	10 (1)	1993–2001	–
3. Attending secondary school	20.1.2.5	–	–	9F 3x 3M (50.0)	–	–	9F 3x 3M (50.0)	–	9 (2)	1973–2004	–
4. Graduating high school	20.1.2.6	–	–	20F 11x 7M (44.4)	13M 7x 1F (59.1)	3M 1F	23M 18x 22F (27.1)	–	20 (2)	1971–2012	–
5. Attending college	20.1.2.7	–	–	–	41F 7x 33M (36.0)	–	41F 7x 33M (36.0)	–	17 (8)	1966–2020	–
6. Graduating from college/awarded a bachelor's degree	20.1.2.9	–	–	–	29F 22x 28M (27.1)	–	29F 22x 28M (27.1)	–	22	1983–2015	–
7. Enrollment in an advanced degree program	20.1.2.10	–	–	–	10M 1F (83.3)	–	10M 1F (83.3)	–	4	1971–2010	–
8. Receiving an advanced/graduate degree in general	20.1.2.11	–	–	–	19M 2x 2F (76.0)	–	19M 2x 2F (76.0)	–	5 (1)	1970–2012	–
9. Receiving an advanced/graduate degree in a specific field	20.1.2.12	–	–	–	12M 1x 2F (70.6)	–	12M 1x 2F (70.6)	–	2	1966–2017	–

Table 25.20.3 Areas of educational training

Variable	*Orig. table*	*Number of Sex Difference Studies (and Consistency Scores)*						*Overviews*	*Countries*	*Time range*	*Non-human*
		I/T	*Child*	*Adol*	*Adult*	*WAR*	*All ages*				
1. Math courses taken	20.1.3.2	–	–	13M 2x (86.7)	2M	–	15M 2x (88.2)	–	3	1976–2006	–
2. Majoring (or taking advanced courses) in biology	20.1.4.6	–	–	3F 1M	6M 3x 4F (35.3)	–	7M 3x 7F (29.2)	–	6 (2)	1992–2017	–
3. Majoring (or taking advanced courses) in business	20.1.4.7	–	–	1M	7M 3x (70.0)	–	8M 3x (72.7)	–	3	1984–2017	–
4. Majoring (or taking advanced courses) in computer science	20.1.4.9	–	–	3M	35M (100.0)	–	38M (100.0)	–	5 (3)	1983–2017	–
5. Majoring (or taking advanced courses) in economics	20.1.4.13	–	–	–	11M (100.0)	–	11M (100.0)	–	4	1986–2007	–
6. Majoring (or taking advanced courses) in education	20.1.4.14	–	–	1F	11F (100.0)	–	12F (100.0)	–	2 (1)	1971–2017	–
7. Majoring (or taking advanced courses) in engineering	20.1.4.15	–	–	1M	43M (100.0)	–	44M (100.0)	–	6 (6)	1981–2017	–
8. Majoring (or taking advanced courses) in fine arts	20.1.4.16	–	–	–	10F (100.0)	–	10F (100.0)	–	3	1980–2007	–

9. Majoring (or taking advanced courses) in the humanities	20.1.4.19	–	–	–	18F (100.0)	–	18F (100.0)	–	7 (2)	1980–2016	–
10. Majoring (or taking advanced courses) in language	20.1.4.20	–	–	4F	9F	–	13F (100.0)	–	5	1975–2007	–
11. Majoring (or taking advanced courses) in mathematics	20.1.4.23	–	–	33M 5x (86.8)	42M (100.0)	–	75M 5x (93.8)	M: 1Rev	9 (5)	1972–2017	–
12. Majoring (or taking advanced courses) in medicine or health sciences	20.1.4.24	–	–	–	8M 5x 6F (32.0)	–	8M 5x 6F (32.0)	–	3 (1)	1985–2017	–
13. Majoring (or taking advanced courses) in physics, chemistry, or other physical science and technology	20.1.4.26	–	–	30M 1x 1F (90.9)	68M 2x (97.1)	–	98M 3x 1F (95.1)	–	11 (5)	1963–2016	–
14. Majoring (or taking advanced courses) in psychology	20.1.4.29	–	–	–	4M 11x (26.7)	–	4M 11x (26.7)	–	2	1986–2015	–
15. Majoring (or taking advanced courses) in social science	20.1.4.31	–	–	1F	9M 3x 3F (50.0)	–	9M 3x 4F (45.0)	–	4 (1)	1980–2017	–

Table 25.20.4 Education and financial rewards

Variable	*Orig. table*	*Number of Sex Difference Studies (and Consistency Scores)*						*Overviews*	*Countries*	*Time range*	*Non-human*
		I/T	*Child*	*Adol*	*Adult*	*WAR*	*All ages*				
Financial return from education	20.1.5.5	–	–	–	18F 4x 7M (50.0)	–	18F 4x 7M (50.0)	–	11 (3)	1969–2011	–

Table 25.20.5 Participation in paid workforce

Variable	*Orig. table*	*Number of Sex Difference Studies (and Consistency Scores)*						*Overviews*	*Countries*	*Time range*	*Non-human*
		I/T	*Child*	*Adol*	*Adult*	*WAR*	*All ages*				
1. Participation in paid workforce in general	20.2.1.1	–	3M 1x 2F	5M 1x 3F	112M 4x (96.6)	7M	127M 6x 5F (88.8)	–	40 (20)	1962–2020	–
2. Trends in sex differences in paid workforce participation	20.2.1.2	–	–	–	12 Diminis-hing (100.0)	–	12 Diminish-ing (100.0)	–	11 (1)	1985–2011	–
3. Job absences/ missing work	20.2.1.4	–	–	–	17F 1x (94.4)	–	17F 1x (94.4)	–	4	1978–2012	–

25.20.6 Employment History

Three variables are cited in Table 25.20.6 having to do with sex differences in employment histories. One can see in row 1 that nearly all studies of adults have concluded that males are more likely to be employed full time than are females. Row 2 documents that female employment is more likely than that of males to be part time. Finally, for this table, row 3 indicates that males are more commonly employed in high-paying businesses than is the case for females.

25.20.7 Importance of Work

Three dimensions of sex differences in the importance of work were the subject of at least ten studies, and their findings are presented in Table 25.20.7. Of these, the only likely universal sex difference was in regard to the centrality of one's job to one's life as a whole. Row 3 shows that all 18 studies of this dimension found that males considered work outside the home more central to their lives than was the case among females.

25.20.8 Time Devoted to Work

Table 25.20.8 presents the condensed findings of studies regarding sex differences in the amount of time devoted to work, and two of the research questions indicate possible universal sex differences. Row 1 shows that all 20 of the studies concerning the duration of lifetime spent in the paid labor market found a greater amount of time spent by males. Nearly all of the 46 studies of the number of hours worked outside the home (row 3) found a higher number of hours worked by males, and the two other studies found no sex difference.

25.20.9 Basic Occupational Characteristics

A large number of studies during the past half-century have focused on sex differences in the basic kinds of occupations of males and females, and if those differences are changing over time. Table 25.20.9 presents the eight specific research questions asked by at least ten studies. Not surprisingly, all of the relevant studies found that males were more likely to have masculine occupations (row 2), and females were more likely to have female occupations (row 3). Also, all of the 15 studies of manual/blue collar jobs found that they were more often held by males than females. However, 15 of the 17 studies shown in row 4, found that such sex segregation in occupations was decreasing, and the remaining two studies found no change. In regard to the long-studied division of labor in regard

Table 25.20.6 Employment history

Variable	*Orig. table*	*Number of Sex Difference Studies (and Consistency Scores)*						*Overviews*	*Countries*	*Time range*	*Non-human*
		I/T	*Child*	*Adol*	*Adult*	*WAR*	*All ages*				
1. Full-time paid employment	20.2.2.1	–		–	79M 1x 1F (96.3)	1M	80M 1x 1F (96.4)	–	25 (12)	1941–2020	–
2. Employed part time	20.2.2.2	–	–	–	10F (100.0)	3F	13F (100.0)	–	3 (2)	1987–2018	–
3. Employment in high-paying establishments/ businesses	20.2.2.3	–	–	–	13M (100.0)	–	13M (100.0)	–	4 (2)	1980–2004	–

Table 25.20.7 Importance of work

Variable	*Orig. table*	*Number of Sex Difference Studies (and Consistency Scores)*						*Overviews*	*Countries*	*Time range*	*Non-human*
		I/T	*Child*	*Adol*	*Adult*	*WAR*	*All ages*				
1. Work ethic/ work motivation	20.2.3.1	–	–	–	3M 7x 3F (18.7)	–	3M 7x 3F (18.7)	–	7	1973–2000	–
2. Workaholism/ careerism	20.2.3.2	–	–	–	3M 3x	4M 1x	7M 4x (63.6)	–	3	1969–1999	–
3. Centrality of one's job or career (outside the home) to one's life as a whole	20.2.3.3	–	1M	1M	11M (100.0)	5M	18M (100.0)	M: 1Met	3	1966–1997	–

Table 25.20.8 Time devoted to work

Variable	*Orig. table*	*Number of Sex Difference Studies (and Consistency Scores)*						*Overviews*	*Countries*	*Time range*	*Non-human*
		I/T	*Child*	*Adol*	*Adult*	*WAR*	*All ages*				
1. Duration of lifetime spent in paid labor market	20.2.4.1	–	–	–	20M (100.0)	–	20M (100.0)	–	5 (2)	1970–2014	–
2. Job-quitting frequency	20.2.4.2	–	–	–	9F 2x (81.8)	–	9F 2x (81.8)	–	3	1974–2005	–
3. Average hours worked outside the home	20.2.4.5	–	–	–	44M 2x (95.7)	–	44M 2x (95.7)	–	4 (3)	1980–2014	–
4. Leisure time	20.2.4.8	–	–	–	6M 6x (50.0)	–	6M 6x (50.0)	–	4 (1)	1985–2003	–

Table 25.20.9 Basic occupational characteristics

Variable	*Orig. table*	*Number of Sex Difference Studies (and Consistency Scores)*						*Overviews*	*Countries*	*Time range*	*Non-human*
		I/T	*Child*	*Adol*	*Adult*	*WAR*	*All ages*				
1. Manual/blue-collar occupations	20.2.5.3	–	–	–	15M (100.0)	–	15M (100.0)	–	6	1971–2006	–
2. Masculine occupations/male-typical occupations	20.2.6.1	–	–	1M	94M **(100.0)**	–	95M **(100.0)**	–	11 (10)	1973–2012	–
3. Feminine occupations/ female-typical occupations	20.2.6.2	–	–	–	42F (100.0)	3F	45F (100.0)	–	9 (7)	1970–2003	–
4. Trends in sex segregation of occupations	20.2.6.6	–	–	–	–	15Dec 2x (88.2)	15Dec 2x (88.2)	–	2 (2)	1981–2013	–
5. Occupational level/prestige	20.2.6.9	–	–	–	28M 14x 4F (56.0)	–	28M 14x 4F (56.0)	M: 3Gen	8 (2)	1964–2016	–
6. Jobs involving management/ supervisory responsibility	20.2.6.13	–	–	–	139M 2x 2F **(95.9)**	–	139M 2x 2F **(95.9)**	–	19 (10)	1967–2010	–
7. Hunting	20.2.6.21	–	–	–	20M 1x 2F (80.0)	–	20M 1x 2F (80.0)	M: 1Rev	7 (6)	1971–2017	17M pri-mates
8. Foraging and gathering	20.2.6.23	–	–	–	–	13F 4x 6M (44.8)	13F 4x 6M (44.8)	–	7 (12)	1973–2003	–

to hunting and foraging, the majority of studies of humans found that hunting was more likely to be performed by males, a finding also found in all of the 17 studies of primates. On the other hand, foraging was more likely to be done by females. However, neither of these findings appear to qualify as a universal sex difference.

25.20.10 Work in Specific Occupational Areas

Table 25.20.10 presents the findings concerning sex differences in specific occupations that have been the subject of at least ten studies. Overall, this table reveals a high degree of sex difference at this level of analysis as all of the studies of 15 of the 20 specific occupations found one sex was more likely than the other to have that occupation. That list included 11 occupations where all of the studies found males dominated the occupation, and four occupations where all of the studies found females more likely to be engaged in that particular form of work. Further, row 1 shows that all of the ten studies of the number of different jobs held by an individual during their career found that number to be higher among males. However, the specific nature of the occupations examined in Table 24.20.10 means that the occupations are only found in a small number of cultures. This reduces the likelihood that any of the specific sex differences are universal.

25.20.11 Wealth, Income, and Promotions

Table 25.20.11 examines sex differences in the wealth and status resulting from work and other sources. All of the variables examined show that wealth, income, and promotion indicators are higher for males than for females. For example, rows 1–5 are all related to wealth resulting from work, and all of these measurements are higher for males. The possibility that some of these general level findings represent potential universal sex differences is increased by the fact that the findings are not restricted to studies in North America (row 1).

25.20.12 Domestic Work

Work comes not only in the form of the activities people are paid to perform, but also in activities for which people usually receive no direct payment. Such work is especially common in and around the home, such as cleaning, repairing, and preparing meals.

A very large number of studies were located having to do with performing household cleaning and maintenance chores. Row 1 of Table 25.20.12 shows that all but three of the 337 studies concluded that females were significantly more likely than males to routinely do

Table 25.20.10 Work in specific occupational areas

Variable	*Orig. table*	*Number of Sex Difference Studies (and Consistency Scores)*						*Overviews*	*Countries*	*Time range*	*Non-human*
		I/T	*Child*	*Adol*	*Adult*	*WAR*	*All ages*				
1. Diversity (variability) in occupations	20.2.7.1	–	–	–	10M (100.0)	–	10M (100.0)	M:2Gen	2 (1)	1976–2001	–
2. Clerical/service occupations	20.2.7.13	–	–	–	16F (100.0)	2F	18F (100.0)	–	8	1976–2006	–
3. Computer-related occupations	20.2.7.14	–	–	–	11M (100.0)	–	11M (100.0)	–	1	1990–2007	–
4. Construction/ carpentry occupations	20.2.7.15	–	–	–	12M 3x 2F (63.2)	–	12M 3x 2F (63.2)	–	5 (9)	1973–2005	–
5. Corporate executive official (CEO/manager)	20.2.7.17	–	–	–	18M (100.0)	–	18M (100.0)	–	3 (2)	1977–2001	–
6. Engineer	20.2.7.27	–	–	–	36M (100.0)	4M	40M (100.0)	M: 2Rev	6	1969–2012	–
7. Geographical mapping scientist	20.2.7.30	–	–	–	17M (100.0)	–	17M (100.0)	–	4	1973–2005	–
8. Janitor and housekeeper	20.2.7.33	–	–	–	8M 3F (57.1)	–	8M 3F (57.1)	–	2 (1)	1984–2004	–
9. Law enforcement/ police officers	20.2.7.35	–	–	–	29M (100.0)	–	29M (100.0)	–	4	1953–2006	
10. Lawyer	20.2.7.36	–	–	–	28M (100.0)	–	28M (100.0)	–	5	1965–2004	–
11. Manufacturing/ production work	20.2.7.39	–	–	–	10M 7x 7F (24.4)	–	10M 7x 7F (24.4)	–	2 (21)	1973–2001	–

12. Mechanic or machine operator	20.2.7.41	–	–	–	10M (100.0)	–	10M (100.0)	–	3	1984–2006	–
13. Nursing	20.2.7.44	–	–	–	18F (100.0)	1F	19F (100.0)	–	7	1969–2008	–
14. Physical scientist	20.2.7.46	–	–	–	19F (100.0)	–	19F (100.0)	–	3 (1)	1969–2004	–
15. Physician	20.2.7.47	–	–	–	33M 8F (67.3)	–	33M 8F (67.3)	–	8 (1)	1955–2003	–
16. Politician: Appointed office holder	20.2.7.49	–	–	–	10M (100.0)	–	10M (100.0)	–	4 (4)	1955–2003	–
17. Politician: Candidate for elected office	20.2.7.50	–	–	–	10M (100.0)	–	10M (100.0)	–	3	1976–2004	–
18. Politician: Holder of elected office	20.2.7.51	–	–	–	127M 2x (98.5)	–	127M 2x (98.5)	–	22 (22)	1930–2019	–
19. Social/welfare worker	20.2.7.60	–	–	–	9F	2F	11F (100.0)	–	4	1969–2003	–
20. Teacher: College level	20.2.7.64	–	–	–	29M (100.0)	–	29M (100.0)	–	8 (4)	1969–2012	–
21. Teacher: Primary and secondary teacher	20.2.7.66	–	–	–	25F 2x 1M (86.2)	–	25F 2x 1M (86.2)	–	10	1971–2009	–

Table 25.20.11 Wealth, income, and promotions

Variable	*Orig. table*	*Number of Sex Difference Studies (and Consistency Scores)*						*Overviews*	*Countries*	*Time range*	*Non-human*
		I/T	*Child*	*Adol*	*Adult*	*WAR*	*All ages*				
1. Earnings/ salaries for workers excluding North America	20.3.1.1a	–	–	–	–	244M 3F **(97.6)**	244M 3F **(97.6)**	M: 3Met	45 (58)	1967–2016	–
2. Earnings/ salaries for North American workers	20.3.1.1b	–	–	–	–	335M 6x **(98.2)**	335M 6x **(98.2)**	–	2	1961–2021	–
3. Earnings/ salaries (all countries combined)	20.3.1.1	–	–	–	–	579M 6x 3F **(98.0)**	579M 6x 3F **(98.0)**	M: 3Met	47 (58)	1961–2021	–
4. Earnings/ salaries for specific occupational areas	20.3.1.2	–	–	–	161M 8x 1F (94.2)	–	161M 8x 1F (94.2)	–	11 (2)	1968–2016	–
5. Retirement income	20.3.1.5	–	–	–	10M 1x (90.9)	–	10M 1x (90.9)	–	3 (1)	1986–2004	–
6. Wealth inheritance	20.3.1.12	–	–	–	–	9M 4x 4F (42.9)	9M 4x 4F (42.9)	–	11	1979–2010	–
7. Parental social status	20.3.1.13	–	–	1M 6x	1M 3x	2x	2M 11x (15.4)	–	3	1980–2019	–
8. Promotion rates	20.3.1.14	–	–	–	–	116M 7x (94.3)	116M 7x (94.3)	–	12 (3)	1964–2012	–

9. Entrepreneur (profit-making) business owner	20.3.1.15	–	–	–	–	15M 3x (83.3)	15M 3x (83.3)	–	6	1979–2004	–
10. Effects of worker sex ratio on an occupation's average salary	20.3.1.16	–	–	–	50F decrease 1x (98.0)	–	50F decrease 1x (98.0)	–	5 (1)	1979–2004	–
11. Financial well-being	20.3.1.18	–	–	–	10M (100.0)	–	10M (100.0)	M: 1Gen	7	1992–2003	–
12. Relationship between social status and mortality	20.3.2.4	–	–	–	–	14M 2x (87.5)	14M 2x (87.5)	–	11	1994–2006	–

Table 25.20.12 Domestic work

Variable	*Orig. table*	*Number of Sex Difference Studies (and Consistency Scores)*						*Overviews*	*Countries*	*Time range*	*Non-human*
		I/T	*Child*	*Adol*	*Adult*	*WAR*	*All ages*				
1. Performing indoor household maintenance and chores in general	20.3.3.1	–	31F 3x (91.2)	26F (100.0)	22F (100.0)	255F (100.0)	334F 3x (99.1)	F: 11Rev	31 (41)	1967–2018	–
2. Performing outdoor or construction household chores	20.3.3.3	–	2M	7M	1M	–	10M (100.0)	–	4	1981–2006	–
3. Food preparation	20.3.3.4	–	1F	2F	2F	5F 5x 4M (27.8)	10F 5x 4M (43.5)	–	2 (14)	1971–2015	–
4. Knitting, sewing, or making clothes	20.3.3.6	–	1F	1F	–	4F 3x 2M (36.4)	6F 3x 2M (46.2)	–	3 (8)	1971–2000	–

housework and indoor household chores. However, while there were far fewer studies, row 2 shows that all of the studies located regarding outdoor chores (such as mowing the lawn and raking leaves) concluded that males performed these chores more. Males also seemed to be involved in household construction.

According to rows 3 and 4, findings were mixed regarding sex differences in both food preparation and in such tasks as knitting and sewing.

25.20.13 Work Productivity

Table 25.20.13 shows that three variables having to do with sex differences in work productivity primarily among academic researchers had at least ten studies. All three of these studies reported that males exhibit higher productivity than do females, but none to the extent of these variables being candidates for universality.

25.20.14 Dominance and Dominance-Related Behavior

Dominance and dominance-related behavior primarily has to do with interpersonal relationships, rather than with how individuals are viewed in large societal contexts. The latter is primarily captured by the term *social status*, even though the terms *dominance* and *social status* are sometimes used interchangeably. Five rows pertaining to dominance are presented in Table 25.20.14, the first four of which pertain to humans.

Row 1 has to do with sex differences in people's dominance or tendencies to strive for dominance. Using various types of measures, one can see that most of the evidence indicates that males exhibit greater tendencies in this regard than do females. In a related vein, 13 studies were located pertaining to dominance specifically between the sexes. One can see in row 2 that the evidence was roughly split between studies concluding that there was no sex difference in tendencies to dominate the opposite sex and studies indicating that males dominate over females, or seek to do so, more than females dominate over males, or seek to do so.

Status striving, often in the form of career aspirations, was investigated by 16 studies. As shown in row 3, 13 of these studies concluded that males strive for status more than females.

Having a social dominance orientation involves considering one's own social group as being dominant (or should be dominant) over one or more other social groups. One can see in row 4 that the vast majority of studies have concluded that this type of sentiment is more prevalent among males than among females.

Table 25.20.13 Work productivity

Variable	*Orig. table*	*Number of Sex Difference Studies (and Consistency Scores)*						*Overviews*	*Countries*	*Time range*	*Non-human*
		I/T	*Child*	*Adol*	*Adult*	*WAR*	*All ages*				
1. Publication rates and scholarly productivity	20.3.4.1	–	–	–	81M 5x 4F (86.2)	–	81M 5x 4F (86.2)	M: 3Rev	12 (10)	1914–2020	–
2. Having a published work cited by others	20.3.4.3	–	–	–	10M 5x 3F (47.6)	–	10M 5x 3F (47.6)	–	5 (3)	1978–2011	–
3. Obtaining a research grant	20.3.4.5	–	–	–	8M 8x (50.0)	–	8M 8x (50.0)	M: 1Met x: 1Met	6	1983–2012	–

Table 25.20.14 Dominance and dominance-related behavior

Variable	*Orig. table*	*Number of Sex Difference Studies (and Consistency Scores)*						*Overviews*	*Countries*	*Time range*	*Non-human*
		I/T	*Child*	*Adol*	*Adult*	*WAR*	*All ages*				
1. Dominance or dominance-striving in general	20.4.1.1a	–	6M 2x	8M	38M 3x 1F (88.4)	6M 1F	58M 5x 2F (86.6)	M: 7Gen	12 (1)	1933–2012	–
2. Inter-sex dominance or status	20.4.1.4	–	–	–	3M 7x (30.0)	3M	6M 7x (46.2)	M: 1Gen	3 (2)	1956–2008	–
3. Status striving (career aspirations)	20.4.1.5	–	2M	7M 1x 1F	4M 1x	–	13M 2x 1F (76.5)	–	1	1959–2007	–
4. Social dominance orientation	20.4.1.6	–	–	3M	50M 4x (92.6)	–	53M 4x (93.0)	M: 1Met	10 (1)	1992–2011	–
5. *Inter-generational similarity in dominance among non-humans*	20.4.1.7	–	–	1F	2F	9F	12F (100.0)	–	N.A.	1958–1986	X

In the final row of this table, a variable relevant to dominance among non-humans is summarized regarding sex differences. It has to do with determining if male or female offspring are more likely to assume the same dominance stature as their mothers. Twelve studies, all but one involving species of non-human primates, reported that female offspring more closely resembled the same dominance position as their mothers than did males. Basically, this implies that males are more likely to "earn" their own dominance position than is the case for females.

25.20.15 Leadership and Social Mobility

Six variables having to do with leadership and social mobility were found for which there were ten or more studies of sex differences. Their findings are summarized in Table 25.20.15. Regarding being leaders or assuming leadership roles, row 1 shows that ten out of 12 studies concluded that males do so more than females, with the remaining two studies reporting no significant difference.

In row 2, findings of sex differences in being effective leaders are assessed. One can see that findings from 13 studies were mixed, although most of the findings point toward males scoring higher. Obviously, the precise criteria used to assess leadership effectiveness could account for some of the inconsistent findings.

Rows 3 and 4 have to do with styles of leadership (or management). One can see that most studies have indicated that males appear to be more prone to be authoritative leaders, while most research suggests that females are more democratic (or collegial) in their leadership style.

The last two rows in this table have to do with mobility in social status of one form or another. According to row 5, all but two of 138 studies concluded that males were more likely to achieve high job status, authority, or eminence than were females. Similarly, row 6 shows that 98 out of 99 studies of upward social mobility (usually assessed in generational terms) have concluded that males surpass females.

25.20.16 Territorial Behavior

Territorial behavior refers to the tendency to either establish or at least lay claim to ownership of a territory, including defending such claims when ownership of the territory is being sought by others. The only row in Table 25.20.16 having to do with humans is row 3. It indicates that most studies have concluded that males claim or maintain larger territories or home ranges than do females.

Regarding territorial behavior among non-humans, most studies indicate that males are more likely to scent mark than do females, at least in

Table 25.20.15 Leadership and social mobility

Variable	*Orig. table*	*Number of Sex Difference Studies (and Consistency Scores)*						*Overviews*	*Countries*	*Time range*	*Non-human*
		I/T	*Child*	*Adol*	*Adult*	*WAR*	*All ages*				
1. Being leaders or assuming leadership	20.4.3.1	–	1M	–	9M 2x (81.8)	–	10M 2x (83.3)	F: 2Rev	4	1974–2005	–
2. Leadership effectiveness	20.4.3.2	–	–	2M	7M 2x 2F (53.9)	–	9M 2x 2F (60.0)	M: 2Met x: 1Met F: 1Met	2	1976–2007	–
3. Authoritative leadership/ management styles	20.4.3.3a	–	–	–	8M 3x (72.7)	–	8M 3x (72.7)	M: 3Met	3 (1)	1983–2012	–
4. Democratic/ collegial leadership/ management styles	20.4.3.3b	–	–	–	11F 1x 1M (78.6)	–	11F 1x 1M (78.6)	F: 3Met	2 (1)	1961–2012	–
5. Achieving high job status/ authority/ eminence	20.4.3.4	–	–	–	136M 2x (98.6)	–	136M 2x (98.6)	M: 1Met	22 (20)	1930–2006	–
6. Upward social status mobility	20.4.3.5	–	–	–	98M 1x (99.0)	–	98M 1x (99.0)	–	12 (4)	1964–2016	–

Table 25.20.16 Territorial behavior

Variable	*Orig. table*	*Number of Sex Difference Studies (and Consistency Scores)*						*Overviews*	*Countries*	*Time range*	*Non-human*
		I/T	*Child*	*Adol*	*Adult*	*WAR*	*All ages*				
1. *Scent marking among non-humans*	20.5.1.1	–	–	1M	10M 1x 4F (52.6)	6M 3x	17M 4x 4F (58.6)	–	N.A.	1944–2017	–
2. *Territorial/ resource procurement or defense among non-humans*	20.5.1.2	–	–	–	8M 3F (57.1)	8M	16M 3F (72.7)	–	N.A.	1943–2017	–
3. Size of territory or home range	20.5.1.3a	–	3M	–	7M 1x	–	10M 1x (90.9)	–	7	1971–2016	–
4. *Size of territory or home range among non-humans*	20.5.1.3b	–	–	–	40M 16x 14F (47.6)	1M	41M 16x 14F (48.2)	–	N.A.	1949–1998	X

most species (row 1). *Scent marking* is considered an indirect measure of claiming ownership or control over a territory in many species. As shown in row 2, most studies have also concluded that more direct forms of territorial or resource procurement and defense, such as making threats or engaging in combat against other claimants, is more common among males, at least for most species. Regarding possible sex differences in the size of the territories or home ranges an individual controls, one can see in row 4, that the evidence of sex differences is quite mixed, although most findings point toward these variables being greater for males than females in most species.

25.20.17 Migration or Dispersal Behavior

Even though migration and dispersal behavior are often not indicative of establishing new territories, they are being considered here due to their sometimes having such effects. As shown in the first row of Table 25.20.17, the majority of studies among humans have indicated that males migrate or disperse away from their homes more than do females.

The remaining research findings summarized in this table pertain to studies of non-humans. The amount of evidence in this regard is substantial. One can see in all four of these rows (rows 2 through 5) that males in most, if not all, species tend to migrate and disperse more than do females.

25.21 Sex Stereotypes: Condensed Findings

Stereotypes are generalization that people make about one another based on personal experiences, rather than being based on specific empirical evidence (such as would come from an actual population survey). *Sex stereotypes*, of course, are ones having to do with male-female differences.

Readers will see that this section is briefer than all other section in this chapter. This is because, while large number of studies of sex stereotypes have been published over the years, less than a dozen specific stereotypes have received the attention of ten or more studies.

25.21.1 Sex Stereotyping Trends

Regarding overall aspects of sex stereotyping, only one table was located containing ten or more citations. This table had to do with trends in people's tendencies to stereotype traits associated with sex. In other words, do people in recent years subscribe to sex stereotypes more (or less) than people in the past (with the earliest studies going back to the 1950s)? It should be noted that the vast majority of stereotyping studies have been conducted in the United States along with other Western countries.

Table 25.20.17 Migration or dispersal behavior

Variable	Orig. table	Number of Sex Difference Studies (and Consistency Scores)						Overviews	Countries	Time range	Non-human
		I/T	Child	Adol	Adult	WAR	All ages				
1. Migration and dispersal	20.5.2.1a	–	2M	1M	14M 1x (93.3)	1M 1x	18M 2x (90.0)	M: 1Rev	11 (4)	1981–2016	–
2. *Migration and dispersal among mammals except humans*	20.5.2.1c	–	–	97M 10x 9F (77.6)	11M (100.0)	16M (100.0)	124M 10x 9F (81.6)	M: 2Rev	N.A.	1937–2008	–
3. *Dispersal before puberty among birds and fish*	20.5.2.1d	–	33F 18x 9M (47.8)	4F 2x	20F 10x 1M (62.5)	13F 6x 3M (52.0)	70F 35x 13M (53.4)	–	N.A.	1958–2004	–
4. *Peripheralization among non-humans*	20.5.2.3	–	–	7M	1M	23M (100.0)	31M (100.0)	–	N.A.	1963–1990	–
5. *Moving away from where one grew up among non-humans*	20.5.4b	–	–	33M 4F (80.5)	2M	1F	35M 5F (77.8)	–	N.A.	1965–2001	–

Table 25.21.1 Sex stereotyping trends

Variable	Orig. table	Number of sex difference findings (and consistency scores)						Overview	Countries	Time range	Non-human
		I/T	Child	Adol	Adult	WAR	All ages				
Trends in stereotyping	21.1.2.3	–	–	1x	9x 1 Increase	4x	14x 1 Increase (06.7 Increase)	x: 1Rev	3	1978–2016	–

Given the social (and sometime even legal) pressure on people to treat males and females more equally now than in the past, most people would probably expect to find less stereotyping of the sexes today than in the past. Contrary to this expectation, however, one can see in Table 25.21.1 that 14 of the 15 studies of trends in stereotypes have concluded that there has been no significant change in people's tendencies to concur with stereotypes regarding male-female differences. The one exception was a study that found significantly *more* sex stereotyping in recent years than in the mid-20th century.

25.21.2 Specific Sex Stereotypes

A substantial number of stereotypes have been investigated to the extent of at least ten individual studies having been published. Table 25.21.2 provides a summary of the findings from these studies, all of which indicate a high degree of consensus among research participants. For example, if one examines the "All Ages" category, one can see that all but one of the stereotypes (that for friendliness) achieved a consistency score higher than 95.0, and most actually attained scores of 100.0. This suggests that there are several apparently universal sex stereotypes. Nonetheless, if one returns to the source tables in Volume III, one will see that the samples are largely limited to Western countries, particularly the United States.

25.22 Attitudes and Actions toward Others According to Their Sex: Condensed Findings

The summary of findings in the preceding section pertained to stereotypes. In the present section, attention will be given to how people think about or act toward others based on the sex of the recipient. Collectively, we will refer to these as *sex-of-recipient* variables. For example, in many countries, people (of both sexes) are more likely to view prenatal sex as being more undesirable for females than for males. This variable, along with many other sex-or-recipient variables, will be summarily documented in this chapter.

25.22.1 Assessments of Others Based on Their Sex (or Presumed Sex)

Table 25.22.1 summarizes four variables having to do with people's assessments of others based on the sex of those being assessed. Row 1 has to do with people's ability to correctly guess the sex of a writer from his or her style of writing. In other words, how accurately are people able to assess the writer of an essay simply on the basis of the writing style (and sometimes on the subject matter being described)? All 15 studies concluded that the sex of a writer could be correctly identified well above chance levels.

Table 25.21.2 Specific sex stereotypes

Variable	*Orig. table*	*Number of sex difference findings (and consistency scores)*						*Overview*	*Countries*	*Time range*	*Non-human*
		I/T	*Child*	*Adol*	*Adult*	*WAR*	*All ages*				
1. Sex stereotyping of math ability	21.1.2.3	–	–	5M	13M (100.0)	3M	21M (100.0)	M: 1Gen 1Met	8	1976–2012	–
2. Sex stereotyping of emotionality	21.2.5.1	1F	2F	3F	21F (100.0)	–	27F (100.0)	–	2 (2)	1968–2002	–
3. Sex stereotyping of emotional expressiveness	21.2.6.2	–	–	–	12F (100.0)	–	12F (100.0)	–	2	1972–2003	–
4. Sex stereotyping of fearfulness	21.2.6.14	–	4F	–	6F	–	10F (100.0)	–	1 (1)	1957–2003	–
5. Sex stereotyping of adventurousness	21.2. 7.3	–	–	–	10M (100.0)	–	10M (100.0)	–	4 (2)	1957–2001	–
6. Sex stereotyping of physical aggressiveness	21.2.7.5	1M	5M	2M	18M (100.0)	–	26M (100.0)	–	4 (1)	1957–2015	–
7. Sex stereotyping of friendliness	21.2.7.32	–	–	2F	9F 1M (81.8)	–	11F 1M (84.6)	–	4	1985–2008	–

(*Continued*)

Table 25.21.2 (Continued)

Variable	*Orig. table*	*Number of sex difference findings (and consistency scores)*						*Overview*	*Countries*	*Time range*	*Non-human*
		I/T	*Child*	*Adol*	*Adult*	*WAR*	*All ages*				
8. Sex stereotyping of being independent	21.2.7.39	–	1M	3M	13M (100.0)	–	17M (100.0)	–	3 (2)	1957–2003	–
9. Sex stereotyping of being nurturing	21.2.9.10	–	1F	–	9F	–	10M (100.0)	–	2	1976–2015	–
10. Sex stereotyping of dominance and leadership	21.2.11.2	–	1M	–	29M 1x (96.7)	–	30M 1x (96.8)	M: 1Met	3	1957–2008	–
11. Sex stereotyping of earnings	21.2.11.5	–	–	1M	9M	–	10M (100.0)	–	3	1982–2014	–

Table 25.22.1 Assessments of others based on their sex (or presumed sex)

Variable	*Orig. table*	*Number of sex difference findings (and consistency scores)*						*Overview*	*Countries*	*Time range*	*Non-human*
		I/T	*Child*	*Adol*	*Adult*	*WAR*	*All ages*				
1. Correctly judging sex based on writing style	22.1.1.1	–	–	–	5 Above chance	10 Above chance (100.0)	15 Above chance **(100.0)**	–	4	1910–2005	–
2. Evaluation of another's facial attractiveness	22.1.2.4	–	–	3F	6F 1x	–	9F 1x (90.0)	–	3	1983–2011	–
3. Evaluation of employees by supervisors	22.1.2.9	–	–	–	11F 13x 4M (34.4)	–	11F 13x 4M (34.4)	F: 1Met x: 1Met	3	1976–2005	–
4. Evaluation of writing quality according to the sex of the presumed writer	22.1.2.13	–	–	1M 1x 1F	7M 3x 1F (58.3)		8M 4x 2F (50.0)	–	1	1968–1985	–

In row 2, findings from ten studies of people's evaluations of facial attractiveness. Nine of these studies concluded that female faces were considered more attractive on average, while one study – involving adults assessing the attractiveness of babies – concluded that there were no significant sex differences in attractiveness ratings.

Twenty-eight studies were located on ratings given by supervisors to their employees relative to the sex of the employees. Row 3 shows that, while most studies have concluded that female employees receive higher ratings than males. However, many studies failed to find any significant sex difference and a few reported higher ratings given to males.

The last row in this table has to do with findings from experiments in which research participants were given samples of people's writings, and then assess the *quality of the writing*. In these experiments, the participants are indirectly led to believe that the writer was either a male or a female (at random) based on whether the name appearing on the top of the report (such as with names such as Mary or Suzie or John or Bill). One can see in the table that the findings from these studies are quite mixed, often depending on whether the subject of the written report involved something that males would normally be better at or more interested in (e.g., adventure, football) or something that females would normally be more likely to report on (e.g., clothing styles, child care). In other words, if the report was about sports, reports that were supposedly written by males usually received higher ratings, but, if the report was about taking care of a child, the reports by female writers usually got higher ratings.

25.22.2 Judgments and Preferences Based on the Sex of Another

Table 25.22.2 summarizes findings from studies of the preferences people have when it comes to the sex of others. In the first entry, approval of premarital sexual intercourse is considered. In this regard, the so-called *double standard* refers to the tendency to be more approving (or less disapproving) of premarital sex by males than by females. One can see that 15 studies have found research participants (usually of both sexes) endorse this double standard more, while five studies found approval (or disapproval) of premarital sex by males and females to statistically equal degrees.

Row 2 pertains to parental reports of their preferences for giving birth to a boy or a girl. One can see that boys were preferred (usually by prospective mothers and fathers) according to 73 studies, although there are a substantial number of exceptions. Specifically, eight studies found no significant difference in sex-of-offspring preferences, and eight additional studies found prospective parents expressing preferences for girls.

Substantial research has investigated the sex of persons who are most often chosen as social interactants. In other words, are boys or girls more

Table 25.22.2 Judgments and preferences based on the sex of another

Variable	*Orig. table*	*Number of sex difference findings (and consistency scores)*						*Overview*	*Countries*	*Time range*	*Non-human*
		I/T	*Child*	*Adol*	*Adult*	*WAR*	*All ages*				
1. Approval of double standard for premarital sex	22.1.4.3	–	–	3M	12M 5x (85.7)	1M	15M 5x (75.0)	M: 1Met	9	1983–2009	–
2. Parental preference regarding sex of offspring	22.1.5.1	–	–	–	9M 1x (90.0)	64M 7x 8F (73.6)	73M 8x 8F (75.3)	–	22 (3)	1931–2010	
3. Choice of social interactants, own sex	22.1.5.2	3M	21M	–	–	–	24M (100.0)	–	5	1927–1999	1M primate
4. Preference for sex of one's boss or leader	22.1.5.9	–	–	–	7M 2F	4M	11M 2F (73.3)	–	1	1975–2011	

often chosen as social interactants? All of the located studies involved toddlers and children. One can see that 24 out of 24 studies agreed that males were chosen (usually as playmates) more often than are females, making this an apparent universal sex difference. Incidentally, one study of rhesus monkeys reached the same conclusion.

The last variable summarized in this table has to do with the preferences that are expressed regarding the sex one would prefer as a boss or leader. Responses have been mixed, although 11 out of 13 studies indicated that males were more often preferred than females.

25.22.3 Assessments Made Based on Sex

In Table 25.22.3, a summary of findings from sex differences in judgments of three empirical phenomena are presented: emotional states of others, assessment of risk associated with various events, and salary expectations. Beginning in the 1930s, many studies began assessing the accuracy with which males and females could assess the emotional states of others. Row 1 shows that most studies of children have found no significant sex differences in this regard. However, studies of adults have indicated that females provide more accurate assessments. Some of the inconsistent results seem to be the result of whether spontaneous facial expressions are being assessed as opposed to posed facial expressions. In the overview cell, one can see that inconsistent conclusions have also been reached by meta-analyses.

Research that was located pertaining to sex differences in risk assessment consisted of 14 studies. One can see in row 2 that all these studies agree that females assess risk for adverse outcomes of various events as being higher than do males for the same outcome.

The final set of research studies pertaining to sex differences in judgments and assessments for which at least ten studies were located involved salary expectations. Of the 20 pertinent findings, 19 concluded that males expected to receive higher salaries than do females, while one study reported the opposite.

25.22.4 Responses by Others Relative to the Recipient's Sex

Table 25.22.4 has to do with how people respond to others depending on whether the recipient is a male or a female. The first entry in this table involves how one or both parents (sometimes including guardians) respond to baby boys and baby girls in terms of providing basic nurturing and care. One can see that, while findings are mixed, most studies indicate that greater nurturing is provided to boys than to girls. It is noteworthy that five studies of non-humans also reached the same conclusion.

Table 25.22.3 Assessments made based on sex

Variable	*Orig. table*	*Number of sex difference findings (and consistency scores)*						*Overview*	*Countries*	*Time range*	*Non-human*
		I/T	*Child*	*Adol*	*Adult*	*WAR*	*All ages*				
1. Accuracy in assessing the emotional states of others	22.1.6.1	3x	12F 23x (34.3)	2F 1x	64F 15x 3M (75.3)	11F 1x (91.7)	89F 42x 3M (67.3)	F: 4Rev 5Met x: 2Met	13 (1)	1936 2013	–
2. Risk assessment	22.1.6.8	–	5F	–	9F	–	14F (100.0)	–	6	1984–2014	–
3. Salary expectations	22.1.6.19	–	–	1M	18M 1F	–	19M 1F (90.5)	–	2	1979–2009	–

Table 25.22.4 Responses by others relative to the recipient's sex

Variable	*Orig. table*	*Number of sex difference findings (and consistency scores)*						*Overview*	*Countries*	*Time range*	*Non-human*
		I/T	*Child*	*Adol*	*Adult*	*WAR*	*All ages*				
1. Receipt of parental nurturing in general	22.2.1.1	14M 2x 6F (50.0)	–	–	–	–	14M 2x 6F (50.0)	–	7	1967–2002	5M various species
2. Being breastfed by mother	22.2.1.3	15M 12x 2F (48.4)	–	–	–	–	15M 12x 2F (48.4)		14	1981–2011	6M 1x 3F various species
3. Receiving food other than breast milk	22.2. 1.4	6M	5M 2x 2F	–	–	–	11M 2x 2F (61.1)	–	6 (2)	1981–2007	1M chimp

In the case of being breast-fed, studies of mothers have indicated that, if a sex difference exists, baby boys are breast-fed more (or for greater amounts of time) than is the case for females. Nevertheless, a substantial minority of studies found no significant sex difference in this regard. Ten studies of breastfeeding among non-human mammals were located. Most of them also indicated that infant males were breast-fed more or over longer durations than was the case for infant females.

Fifteen studies of sex differences in infants receiving food other than breast milk have been reported. One can see that most of these studies indicate that boys are given more food in general than is the case for girls. In one study of chimpanzees, the same tendency was reported.

25.22.5 Responses by Parents Relative to the Offspring's Sex

In Table 25.22.5, research findings regarding how parents treat their male and female offspring differently are summarily examined. Row 1 has to do with physical punishment. It shows that 36 studies have concluded that male offspring are physically punished more often (or more severely) than are female offspring. Just one study failed to document a significant difference in this regard.

Parental supervision (or monitoring) of their children has been the focus of 18 studies. One can see in row 2 that all of these studies concluded that female offspring are supervised more than male offspring.

The third row has to do with parent-offspring discussions in general. Of the 14 located studies, nine indicated that parent-offspring discussions are more frequent with female offspring than with male offspring.

Row 4 has to do with parent-offspring discussions of emotions specifically. A total of 38 relevant studies were found. One can see that the findings were mixed, although the majority (i.e., 30) found that parents have more discussions of emotions with daughters. However, it is worth adding that the type of emotions discussed seems to make a difference. Parents appear to discuss sadness and disappointments with daughters more, but their discussions with boys usually involve anger.

25.22.6 Health Factors in Relationship to the Recipient's Sex

Table 25.22.6 identifies three aspects of health that have been frequently studied in relationship to sex. The first factor has to do with an individual receiving medical services in general. One can see that findings have been mixed although roughly twice as many studies indicate that males receive more medical attention than do females. The excess of attention given to males appears to occur primarily before puberty.

In the case of sex differences in receiving surgery for coronary heart disease, row 2 shows that 49 studies have concluded that males receive

Table 25.22.5 Responses by parents relative to the offspring's sex

Variable	*Orig. table*	*Number of sex difference findings (and consistency scores)*						*Overview*	*Countries*	*Time range*	*Non-human*
		I/T	*Child*	*Adol*	*Adult*	*WAR*	*All ages*				
1. Physical punishment of offspring by parents	22.2.1.12	4M 1x	28M (100.0)	4M	–	–	36M 1x (97.3)	M: 1Met	5	1961–2006	3M 3F various primates
2. Parental supervision of offspring	22.2.1.23	5F	6F	7F	–	–	18F (100.0)	–	4 (1)	1968–2013	–
3. Parent-offspring discussions in general	22.2.2.1	7F 1x 4M (43.7)	2F	–	–	–	9F 1x 4M (50.0)	–	2	1969–2013	–
4. Parental-offspring discussions of emotions	22.2.2.2	11F 2M	19F 2x 4M (65.5)	–	–	–	30F 2x 6M (68.2)	–	1	1979–2003	–

Table 25.22.6 Health factors in relationship to the recipient's sex

Variable	*Orig. table*	*Number of sex difference findings (and consistency scores)*						*Overview*	*Countries*	*Time range*	*Non-human*
		I/T	*Child*	*Adol*	*Adult*	*WAR*	*All ages*				
1. Accessing medical care in general	22.2.6.1	1M	7M 1x	–	7M 1x 7F (29.2)	–	15M 2x 7F (48.4)	–	6 (2)	1978–2003	–
2. Accessing surgery for coronary heart disease	22.2.6.4	–	–	–	49M 7x (87.5)	–	49M 7x (87.5)	–	5	1987–2006	–
3. Survival following surgery	22.2.6.12	–	–	–	5F 2x 4M (33.3)	–	5F 2x 4M (33.3)	–	7	1987–2013	–

more surgery than do females. Seven additional studies found no significant sex difference.

The table's third entry has to do with survival following surgery. One can see that findings are inconsistent regarding any sex differences in this regard.

25.22.7 Social Treatment by Others Relative to the Recipient's Sex

Various forms of social treatment have been studied in relationship to whether the recipients were males or females. Findings are covered in Table 25.22.7. Most of these studies are based on experiments in which either a male or a female exhibits some sort of need for assistance.

As one can see in row 1, most studies indicate that females are more often helped by others than are males. However, the only results that yielded a high consistency score in this regard involved the ten studies in which adolescents were exhibiting a need for some type of help.

Sex differences in the sentences given to persons convicted of a juvenile or an adult criminal offense are summarized in row 2. It shows that a substantial majority of studies have concluded that males receive more severe penalties than do females for similar types of offenses, although there are exceptions, particularly for adolescent offenders.

Quite a few studies have assessed the extent to which males and females receive encouragement or help from their parents. Row 3 shows that the findings have been very inconsistent.

Regarding studies of sex differences in receiving attention from teachers, row 4 shows that the majority of studies indicate that male students receive more attention from teachers than do female students. It should be added that not all of the attention given to students is positive; a substantial minority is either corrective or even punitive. Research having to do with teacher-student contact outside of class has also been extensively studied. Row 5 shows that findings have been mixed with respect to sex differences, although most research points toward males receiving greater out-of-class attention from teachers than do females.

Finally, in this table, studies are reviewed regarding the evaluations that teachers receive from the students primarily among those attending college. Row 6 shows that the findings have been extremely mixed, with most studies indicating that there are no significant sex differences in the ratings that college teachers receive from their students.

25.22.8 Single-Sex Education versus Co-Education

Both rows in Table 25.22.8 have to do with single-sex education vs. coeducation. The first row involves students attending one or more single-sex

Table 25.22.7 Social treatment by others relative to the recipient's sex

Variable	*Orig. table*	*Number of sex difference findings (and consistency scores)*						*Overview*	*Countries*	*Time range*	*Non-human*
		I/T	*Child*	*Adol*	*Adult*	*WAR*	*All ages*				
1. Being helped by others in general	22.4.1.1	–	4F 3x	10F (100.0)	25F 4x 6M (49.0)	–	35F 4x 6M (68.6)	–	1	1971–2010	–
2. Sentences administered by courts for offenses	22.4.2.2	–	–	4M 5F	20M 2x (90.9)	3M 1x	27M 3x 5F (67.5)	M: 2Rev 1Met	8	1977–2000	–
3. Receiving encouragement or help from parents	22.4.3.4	2F 1M	3F 2M	1x 1M	–	–	5F 1x 5M (31.2)	–	3	1971–1999	–
4. Receiving attention from teachers	22.4.4.1	–	20M 2x 1F (83.3)	19M 1F (90.5)	8M	3M	50M 2x 2F (89.3)	M: 1Met	2	1956–2004	–
5. Teacher-student instructional contact outside of class	22.4.4.2	–	7M 2x 3F (46.7)	13M 1x (92.9)	–	–	20M 3x 3F (69.0)	–	2	1956–2000	–
6. Evaluation of teacher by students	22.4.4.6	–	–	1M	17M 27x 17F (21.8)	–	18M 27x 17F (22.8)	–	3	1932–2003	–

Table 25.22.8 Single-sex education versus co-education

Variable	*Orig. table*	*Number of sex difference findings (and consistency scores)*						*Overview*	*Countries*	*Time range*	*Non-human*
		I/T	*Child*	*Adol*	*Adult*	*WAR*	*All ages*				
1. Single-sex classes	22.4.5.1	–	–	7 Enhanced; 6x 2Diminished (41.2)	–	–	7 Enhanced 6x; 2 Diminished (41.2)	Varied: 1Met	6	1999–2018	–
2. Single-sex schools	22.4.5.2		1x	13 Enhanced; 6x; 3 Diminished (52.0)	–	–	13Enhanced; 7x; 3 Diminished (50.0)	x: 1Rev	6	1974–2012	–

classes, while the second row is concerned with students attending entire single-sex schools. For the studies in these tables, the issue is whether or not either sex benefits from single-sex education (when compared to co-educated students) with regard to learning.

Assessments of learning usually involves performance on standardized tests. Regarding attendance of single-sex classes, the first entry shows that most studies indicate enhanced performance. This seems to be particularly true for females attending single-sex math and science classes. It is worth mentioning that one meta-analysis examined the effects of single-sex education and concluded that the benefit to females was limited to math and science classes, but only among studies that did not use randomized experimental designs. The studies that did use randomized assignment methods (considered more scientifically rigorous than non-random designs), detected no significant benefits.

In the case of attending entire single-sex schools (relative to co-education schools), 13 studies found that both sexes seem to benefit in terms of performance on standardized tests, while seven other studies found no significant differences and three more actually found poorer performance in the single-sex schools. However, the one literature review of this line of research reported no significant beneficial effects of single-sex schools.

25.22.9 Social Prominence

Two measures of social prominence, broadly defined, were located with respect to sex differences. Table 25.22.9 shows that neither of them attained consistency scores high enough to be candidates for universality. Row 1 indicates that 14 studies have found males being more likely to receive prominent obituaries or funerals than was the case for females, but three studies reported no significant sex difference. The other set of studies had to do with measures of popularity at various stages in life. These studies reached mixed conclusions, although most indicated a tendency for more females than males to be considered popular by their peers.

25.22.10 Characteristics Portrayals in the Mass Media

Another way to think of sex differences in how others are treated has to do with their portrayed in various forms of mass media. Especially now with the availability of the internet, the types of mass media are vast, ranging from written descriptions in books, newspapers, and magazines, to presentations in movies and television programs, along with a wide range of internet sources.

The first entry in Table 25.22.10 has to do with the proportions of males and females who appear in the mass media in general. One can see

Table 25.22.9 Social prominence

Variable	*Orig. table*	*Number of sex difference findings (and consistency scores)*						*Overview*	*Countries*	*Time range*	*Non-human*
		I/T	*Child*	*Adol*	*Adult*	*WAR*	*All ages*				
1. Recipient of obituary or prominent funeral	22.4.7.7	–	–	–	–	14M 3x (82.4)	14M 3x (82.4)	–	1 (2)	1977–1999	–
2. Popularity	22.4.7.9	–	6F 2x 1M	1x	1F 1M	–	7F 3x 2M (50.0)	–	3	1939–2007	–

Table 25.22.10 Characteristics portrayed in the mass media

Variable	*Orig. table*	*Number of sex difference findings (and consistency scores)*						*Overview*	*Countries*	*Time range*	*Non-human*
		I/T	*Child*	*Adol*	*Adult*	*WAR*	*All ages*				
1. Featured in mass media in general (except commercials)	22.5.1.1	–	17M 1x 2F (77.3)	1M	6M	14M 2x 6F (50.0)	38M 3x 8F (66.7)	M: 2Rev	4 (1)	1972–2011	–
2. Featured in commercials	22.5.1.3	–	4M	–	4F 3M	6F 2x 4M (37.5)	11M 2x 10F (33.3)	–	9 (1)	1993–2011	–
3. Commercial voice-overs	22.5.1.5	–	–	–	9M	17M 5F (62.9)	26M 5F (72.2)	–	10 (2)	1975 2003	–
4. Age of persons featured in mass media	22.5.2.1	–	–	–	11M (100.0)	8M	19M (100.0)	–	5 (2)	1972–2005	–

that, out of a total of 49 studies, 38 concluded that males are more prominent in the mass media than females. Of the remaining 11 studies, eight concluded that females are more prominent and three found no significant difference.

Row 2 of this table focuses only on being features in commercials (usually on television or in magazines). Out of 23 studies, 11 concluded that males are featured more often, while ten concluded females were and two reported no significant sex difference. Some of the research indicates that sex differences vary considerably, depending on the types of products being advertised.

In the third row, information regarding sex differences in so-called *voice-overs* are assessed. These are people's voices that used primarily in commercials without those producing the voices ever being seen in the commercial. One can see that 26 of the 31 studies of sex differences in voice-overs concluded that more males than females are used. However, the remaining five studies all reached the opposite conclusion. Again, the type of product being advertised is likely to be responsible for some of these inconsistent findings.

The last entry in this table has to do with the age of persons who appear in mass media. In this case, one can see that all 19 studies agree that the males who appear in the mass media are significantly older than are the females that appear.

25.22.11 Behavior Portrayed in the Mass Media

In Table 25.22.11, attention is focused on the types of behavior males and females are portrayed performing in the mass media. The table's first entry has to do with portrayals of aggression and violence. One can see that all studies agree that males are portrayed being more aggressive and violent than is the case for females. Along similar lines, row 2 shows that 17 out of 18 studies of individuals being dominant or of high social status were males rather than females.

Findings of sex differences in mass media portrayals of persons performing work inside the home are summarized in row 3. One can see that all but 1 of the 50 pertinent studies indicated that females were portrayed as doing more housework than were males. Row 4 has to do with portrayals of specific types of jobs being performed by each sex. While the consistency scores fell short of 95.0 this table basically indicated that males are more often depicted as performing male-typical jobs and females performing female-typical jobs. In the final row, of this table, mass media portrayals of sex differences in persons working outside the home are reported. All but 1 of the 23 studies concluded that males are depicted as working outside the home more than are females.

Table 25.22.11 Behavior portrayed in the mass media

Variable	Orig. table	Number of sex difference findings (and consistency scores)						Overview	Countries	Time range	Non-human
		I/T	Child	Adol	Adult	WAR	All ages				
1. Portrayals of aggression and violence	22.5.3.5	–	1M	2M	1M	16M (100.0)	20M (100.0)	–	2 (1)	1974–2004	–
2. Portrayals of dominance and high status	22.5.3.7	–	–	–	5M 1x	12M (100.0)	17M 1x (94.4)	–	4	1979–2006	–
3. Portrayals of working inside the home	22.5.6.1	–	1F	1F	36F 1x (97.3)	11F (100.0)	49F 1x (98.0)	–	15 (1)	1971–2008	–
4. Portrayals of specific jobs or occupations being performed	22.5.6.2	–	–	–	18M 1F (90.0)	12M 1F (85.7)	30M 2F (88.2)	–	8 (1)	1975–2009	–
5. Portrayals of working outside the home	22.5.6.3	–	–	–	7M	15M 1x (93.7)	22M 1x (95.7)	–	6 (1)	1971–2003	–

25.23 Ecologically Based Sex Differences: Condensed Findings

Studies of sex differences associated with ecological factors have a distinct design structure when compared to nearly all of the other studies examined throughout the rest of this book. In particular, the units of analysis are not individual males or females. Instead, countries (or sometimes U.S. states) are the units of analysis. The most common design of these ecological studies involves comparing samples of countries regarding average sex differences in a particular trait (such as math ability) with some other variable in those same countries, such as how industrialized each country happens to be.

Because most of the ecological research published on sex differences in traits and national variations in variables such as gender equality/inequality, in all of Chapter 23, only two tables contained ten or more citations. For this reason, the present summary of this chapter is brief.

It is also worth noting that the structure of the tables that will be used to provide a summary of the two tables with ten or more citations in Chapter 23 differs from all of the tables used throughout the preceding 22 sections of this condensed summary chapter. Specifically, the first column presents a description of the nature of each relationship, and the second column identifies the table in which each relevant study is cited. In the third column, the direction of the findings is described, along with the calculated consistency scores. To calculate each consistency score, the number for finding with the greatest support was divided by the number of studies finding no relationship plus a doubling of any studies actually finding the opposite of the most frequent finding.

In the last three columns, the following information is presented: First, an overview column reports conclusions reached by any literature reviews or meta-analyses that were located. Second, the number of pertinent studies is identified, and, third, the dates for the publication of the first and the most recent studies bearing on each relationship is provided.

25.23.1 National Gender Equality and Sex Differences in Math Ability

Twelve studies were found that assessed the relationship between sex differences in average math performance on objective exams and national variations in gender equality. These studies were inspired by speculation that the well-documented gender gap in mathematics is likely due to differential treatment of the sexes along with the prevalence of sex stereotypes (Hyde and Mertz 2009; Else-Quest et al. 2010; Kane and Mertz 2012). In other words, countries with the greatest sex differences in math scores should be those in which one finds less gender equality regarding how males and females are treated in their particular countries. For instance, countries such as Norway and Sweden should exhibit fewer sex differences in math performance than countries such as Saudi Arabia and Sudan.

The results of the 12 studies are summarized in Table 25.23.1. One can see that findings were not consistent. Four studies found gender equality positively correlated with sex equality in math performance, while five studies found the opposite, and three others reported no significant relationship.

Table 25.23.1 National gender equality and sex differences in math ability

Nature of Relationship	*Table*	*Findings (Consistency Score)*	*Overviews*	*# Studies*	*Time range*
National gender equality ratings and sex differences in math ability	23.1.4.2	4Pos 3x 5Neg (31.2)	–	12	1993–2020

25.23.2 Ecological Percent of Males and Crime Rates

Males are much more involved in nearly all types of crime than are females, especially in the case of violent offenses (Ellis, Farrington, and Hoskin 2019: 102). Primarily for this reason, many studies have sought to determine if crime rates are higher in populations where the percentage of males are the highest. Table 25.23.2 shows that findings are quite mixed, with almost equal numbers of studies reporting positive correlations (i.e., 15) as those reporting the reverse (i.e., 13). Furthermore, nine additional studies (plus one literature review and two meta-analyses) all concluded that there are no significant associations between the percent of males in a population and the population's crime rate.

Table 25.23.2 Ecological percent of males and crime rates

Nature of Relationship	*Table*	*Findings (Consistency Score)*	*Overviews*	*# Studies*	*Time range*
Association between the percent male in a population and the population's crime rate	23.3.3.1	15Pos 9x 13Neg (3.0)	x: 1Rev 2Met	37	1991–2012

26 Identifying and Theoretically Explaining Each Potential Universal Sex Difference

As the review presented in Chapter 25 has shown, hundreds of traits were found in Chapters 1 through 23 to have consistency scores of 95.0 or higher. For some of the variables, these high consistency scores were true only for certain age ranges, while, for others, all ages exhibited high scores.

In this chapter, attention will be given exclusively to variables for which consistency scores of 95.0 were found for at least one age group. These variables will be considered one-by-one with a goal being to decide if each variable should be considered a *universal sex difference* (*USD*). If the answer is affirmative, each USD will be designated as one in which males exhibit the trait more (a *USD-M*) or females do so (a *USD-F*).

Decisions regarding the USD status of each variable will largely depend on their consistency scores, but also sometimes on the number of countries sampled. Additional consideration will occasionally be given to whether or not a particular sex difference seems to comport with any theoretical reasoning.

Structure of Each Précis

The word *précis* refers to a boiled-down summary of essential evidence. We will use this term throughout this chapter (instead of the word *table*) since each précis will pertain to a single variable, rather than a grouping of variables. The headers for each précis are as follows:

In the first cell of each précis, the age grouping, and the number of studied according to sex are summarily shown for each age grouping containing at least ten findings. Regarding sex, *M* meaning males were greater, *x* meaning no significant sex differences were found, and *F* meaning females were greater. For instance, if there were 40 findings for a particular variable and 37 of these findings reported that males exhibited the trait to a greater degree, while two findings concluded that there were no significant sex differences, and one remaining finding indicated that females exhibited the trait more, the first column will contain the following: 37M 2x 1F.

DOI: 10.4324/9781003405290-3

Age categories, number of studies, and sex	Consist. Score	Overviews	Countries	Time Range	Non-Human	Social Role Theory			Evolutionary Theory		Theological
						Basic	Found	Biosoc	Select	ENA	

The second column reports the consistency score for each variable with at least one age grouping containing a score of 95.0 or higher. Findings from age groupings with fewer than ten findings are not individually reported, although they are included in the "All Ages" total. In the case of variables for which there were age groupings with ten or more findings *but* the consistency score was below 95.0, the scores were reported, albeit simply in parentheses.

The third column, entitled *Overviews*, indicates the nature of conclusions about sex differences based on any published literature reviews and meta-analyses. Following this column is one labeled *Countries*. It reports the number of countries sampled along with the number of studies that were based on samples from two or more countries (the latter being provided in parentheses).

In the fifth column, the publication time range is reported. The sixth column reports the nature and number of studies that were located for studies of sex differences in species other than humans.

This brings us to the last six headers in each précis, all of which are lightly shaded. Whereas the first six headers characterize the nature of the empirical evidence, the six shaded headers pertain to potential *explanations* for sex differences. The names given to each of these headings pertain to each of the five scientific theories of sex differences that were specifically described in Chapter 24. In the last header, theological explanations, should they exist, are listed. The header abbreviations are as follows:

Origin – original social role theory
Found – foundation effect social role theory
Biosoc – biosocial role theory
Select – sexual selection theory
ENA – evolutionary neuroandrogenic theory
Theological – theological/religious scriptural explanations

Under each of the six shaded headers, one of six abbreviated descriptive terms is offered.

a *Expl* (explained). This means that a particular sex difference can be reasonably explained by a particular theory along with there being some supportive evidence.
b *Poss* (possible). If a particular theory (or theological explanation) offers some way of predicting a particular sex difference but evidence for or against the prediction is still minimal, a theory is designated as a possible explanation.

c *Circ* (circular). This means that the explanation is essentially circular. In other words, if Variable A is used to explain Variable B, then it would be circular to argue that Variable B also explains Variable A.
d *Contra* (contradicted). If a theory leads one to expect a particular sex difference (e.g., cultural exceptions to exist but no exceptions were found) the theory is designated as having been refuted.
e *NoTest*. This term refers to theoretical arguments that are impossible to either confirm or refute.
f *NoEvid*. In some cases, a theory makes a prediction for which evidence is possible to collect but could not be located.
g *Silent*. When a theory does not specifically address the possibility of a sex difference in a variable, it will be described as *silent* in this regard.

Each of the variables identified in this chapter were selected because information summarized in the preceding chapter indicated that (a) at least ten relevant findings were located and (b) the consistency scores were 95.0 or higher. Of course, all of the findings summarized in Chapter 25 can be traced back to tables cited in Volumes I through III. To allow readers to refer back to the actual studies, the present chapter also reports the original table numbers as part of the title for each précis.

All sex difference variables for which ten or more findings were located *and* for which consistency scores of 95.0 or higher were obtained will now be individually listed and briefly discussed. Each one of these variables will have its own précis as well as a brief three-part discussion, consisting of a description of the evidence, a consideration of how best to explain the evidence, and a final assessment.

Because the present chapter is unusually lengthy, it is divided into three parts, each one corresponding to the book's first three volumes. Thus, the first part pertains to variables covered in Volume I (i.e., Chapters 1 through 8). These variables were primarily of a biological, non-cognitive, and non-behavior nature. The next section pertains to variables first covered in Volume II (i.e., Chapters 9 through 15). They pertain primarily to perceptual, emotional, cognitive, and attitudinal sex differences. In the third part of this chapter, variables identified in Volume III (i.e., Chapters 16 through 23) are reported. These variables mainly involve sex differences in personality, behavior, social interactions, stereotypes, and mass media characterizations.

26.1 High Consistency Scores in Volume I

As noted earlier in this book, *biology* is the science of life. All sex-related traits are about features of living things, animals in particular. Therefore, broadly speaking, this entire book is about biological variables. Even so, sex differences in cognition and behavior are often thought of as being separate from "mere biology." This conceptual distinction reflects the fact that cognitive and behavioral factors usually involves learning, and, in the case of humans, most learning occurs in sociocultural context. Nevertheless, when making distinctions between biology, cognition, and behavior, one should keep in mind that the latter two phenomena are in fact largely the result of functioning by a biological organ, the brain.

We have tried to organize all of the findings cited in this book in accordance with the somewhat fuzzy distinction made between basic biology, cognition, and behavior. In a number of cases, the differences between these three broad categories of variables were ambiguous, and therefore rather arbitrary. This chapter is divided into three sections, each section corresponding to Volumes I, II, and III. The first section deals primarily with basic biology; the second section focuses on cognition; and the third section primarily deals with behavior.

26.1.1 Reproduction, Development, and Morphology

Six variables of a reproductive, developmental, or morphological nature were identified in Chapter one that had consistency scores of 95.0 or higher. Each one of these variables will be discussed regarding the possibility of it being a USD. This discussion will focus on a summary table accompanying each variable's narrative.

26.1.1.1 Stillbirth or Early Infant Death

Evidence. Based on samples derived from ten countries (plus three multi-country samples), published over a 73-year time span, Précis 26.1.1.1 shows that 31 studies have all found stillbirths and early infant deaths to be more common in males. Two studies of non-human mammals also concluded that males are more likely to die before or soon after birth.

Explanation. As shown in the shaded region of this précis, the only theory that seems to offer an explanation for this particular sex difference is ENA theory. It does so by suggesting that high prenatal exposure to androgens, as occurs in males, seems to affect development in ways that retard maturation, especially of the nervous system. This explanation has considerable support, both from studies of humans (Emery, Krous et al. 2005; DiPietro & Voegtline 2017), as well as other primates (Bachevalier & Hagger 1991). Another study noted that smoking by mothers during pregnancy is associated with an increased risk of their baby dying unexpectedly soon after birth (Anderson, Lavista Ferres et al. 2019). This finding is relevant here because of evidence that nicotine exposure tends to increase testosterone levels (LM Smith, Cloak et al. 2003).

Assessment. Overall, the greater probability of males being stillborn or dying in infancy constitutes a USD-M. The most reasonable theoretical

Précis 26.1.1.1 Stillbirth and early infant death (original Table 1.1.2.4)

Age categories, number of studies, and sex	*Consist. score*	*Overviews*	*Countries*	*Time range*	*Non-human*	*Social role theory*			*Evolutionary theory*		*Theological*
						Origin	*Found*	*Biosoc*	*Select*	*ENA*	
Fetal or Infancy 31M	100.0	M: 1Rev	10 (3)	1945–2018	2M mammals	Silent	Silent	Silent	Silent	Expl-M	Silent

explanation is provided ENA theory. Additional information on this matter will be presented later specifically regarding a phenomenon known as *sudden infant death syndrome* (which is also more common among males). As research continues to accumulate on this matter around the world, it is reasonable to expect that infant boys will continue to be less viable than infant girls.

26.1.1.2 Age of Onset of Puberty

Evidence. Which sex reaches puberty at a younger age? As shown in Précis 26.1.1.2, a total of 39 studies were located based on samples obtained from 11 different countries with their publications covering a 71-year time span. Without exception, these studies concluded that females reach puberty at significantly younger average ages than do males. Further bolstering this conclusion are findings from five studies of non-human mammals, which also reported early ages of puberty onset for females.

Explanation. The précis's shaded region shows that none of the social role theories offer an explanation for this sex difference, and evolutionary theory by natural selection does not seem to do so either. Also, as far as we could determine, no religious explanation has been offered either. Neuroendocrine factors, however, appear to play a role in controlling the onset of puberty (Foster, Olton et al. 2004). Therefore, it is possible that ENA theory can contribute to an understanding of this sex difference, although nothing in the theory currently constituted would lead to a specific explanation.

Assessment. Overall, the age of pubertal onset can be considered a well-established USD-F. Nevertheless, more work is needed to provide a theoretical explanation for the difference.

26.1.1.3 Physical Maturation Rate

Evidence. Sex differences in the rate at which individuals mature has been investigated by 41 studies based on samples drawn from eight different countries that were published over a 96-year time frame. One can see in Précis 26.1.1.3 that all studies agreed that females mature more rapidly than do males, a conclusion reached by four studies of non-human mammals as well. This sex difference also seems to be in accordance with the sex difference in age of pubertal onset (discussed above), with females experiencing puberty earlier than males.

Explanation. In terms of explanations, the shaded region of the précis shows the same pattern as was indicated above for age of pubertal onset. Generally, ENA theory seems to come closest to providing an explanation for why females would mature more rapidly than males but it is not specific in this regard. It implies that androgens must be playing a significant role in physical maturation. One study found that, in within both sexes,

Précis 26.1.1.2 Age of onset of puberty (original Table 1.1.6.2)

Age categories, number of studies, and sex	*Consist. score*	*Overviews*	*Countries*	*Time range*	*Non-human*	*Social role theory*			*Evolutionary theory*		*Theological*
						Origin	*Found*	*Biosoc*	*Select*	*ENA*	
Adolescence-Adult 39F	100.0	-0-	11	1935–2006	5F mammals	Silent	Silent	Silent	Silent	PossF	Silent

Précis 26.1.1.3 Physical maturation rate (original Table 1.1.6.1)

Age categories, number of studies, and sex	*Consist. score*	*Overviews*	*Countries*	*Time range*	*Non-human*	*Social role theory*			*Evolutionary theory*		*Theological*
						Origin	*Found*	*Biosoc*	*Select*	*ENA*	
All Ages 41F	100.0F	-0-	8	1913–2009	4F mammals	Silent	Silent	Silent	Silent	Poss-F	Silent

testosterone levels correlated with self-reported measures of physical maturation (Shirtcliff, Dahl & Pollak 2009). However, because pubertal levels of testosterone among males vastly surpasses that of females (see Précis 26.3.2), leaves one to wonder why females would mature more rapidly than males.

Assessment. The consistency of the studies and sizable number of countries sampled, along with the same findings reported in other mammals, all support the view that sex differences in physical maturation is a USD-F. However, no theory at the present time seems to precisely explain this sex difference.

26.1.1.4 Height

Evidence. One can see in Précis 26.1.1.4 that, at least by the age of 15, all studies agree that males are taller than females. However, prior to mid-adolescence, the evidence of consistent sex differences is quite mixed.

Fifty-eight different countries were sampled by these studies, plus 12 additional multi-national studies. The time frame separating the oldest and the most recent studies was 148 years. It is worth noting that several studies of the skeletal remains of extinct hominids (e.g., Neandertals and Australopithecus) have also concluded that males were taller than their female counterparts.

Explanation. None of the three social role theories offer an explanation for sex differences in height, but both evolution-based theories provide some insight into the process. Specifically, sexual selection theory asserts that males have been favored for being taller in part because females bias their mate choices toward males who are taller than themselves (see Précis 26.2.7.15). Some of this female mate preference likely reflects the fact that large body size is associated with dominance and control of territory among many species of mammals (Ellis 1994b). Furthermore, male mating frequency is also positively correlated with dominance and control of territory (Ellis 2005).

Among humans, similar patterns have been documented. In particular, male height is positively correlated with social status, especially income (Ellis 1994a; Deaton & Arora 2009), and with number of sex partners males report having had (Frederick & Jenkins 2015).

Because ENA theory has both evolutionary and neurohormonal components, it goes beyond sexual selection theory to predict that androgens are contributing to sex differences in height and body size. In this regard, considerable evidence points toward testosterone promoting bone growth (Thompson et al. 1972; Wiren 2005; Clarke & Khosla 2009; Giri, Patil et al. 2017; Chen, Lin et al. 2019), thereby enhancing height (Tremblay, Schaal et al. 1998) and overall body size (Moisey, Swinburne & Orme 2008; Cox, Stenquist, & Calsbeek 2009). The process by which testosterone has these effects is complex, however. This is partly due to the fact that testosterone

Précis 26.1.1.4 Height (original Tables 1.2.1.1)

Age categories, number of studies, and sex	*Consist. score*	*Overviews*	*Countries*	*Time range*	*Non-human*	*Social role theory*			*Evolutionary theory*		*Theological*
						Origin	*Found*	*Biosoc*	*Select*	*ENA*	
Child 17M 21x 4F Young adol. 21F 17x Late adol. 54M Adult 167M All Ages 244M 40x 26F	(36.9) (55.3) 100.0 100.0 (72.6)	M: 1Rev (adults)	58 (12)	1872–2020	15M extinct pre-humans (adults)	Silent	Silent	Silent	Expl-M	Expl-M	Poss-M

can be metabolized into estradiol (usually considered a "female hormone"). Some evidence suggests that estradiol may be playing a more direct role in promoting bone growth than testosterone (Vanderschueren, Gaytant et al. 2008). There appear to be additional hormonal complexities surrounding the sexual differentiation in height that are not fully explained by ENA theory (Maione, Pala et al. 2020).

Regarding theological perspectives on sex differences, we are aware of no religious teachings that explain why males are taller than females. However, because various biblical passages indicate that males should be leaders and females should be their helpers (Corinthians 11:3; Genesis 2:18), some may argue that one way God helps to ensure that this leadership edict is adhered to is by making males physically larger.

Assessment. The evidence is extremely strong regarding adult height being a USD-M. Also, evolutionary theories appear to be especially strong in helping to explain why males in all societies are taller than females, on average.

26.1.1.5 Body Weight

Evidence. Among humans, height and body weight are strongly correlated (Nakamura, Hoshino et al. 1999; Elgar, Roberts et al. 2005). Knowing this allows one to conclude that, because males are taller than females, at least beyond the age of 15, it is to be expected that they would also be heaver on average. A total of 144 studies of sex differences in body weight were found. In the case of adults (but none of the other age groupings), all 72 studies concluded that, without exception, males were taller. The number of countries sampled was 26 plus four other studies of multiple countries. These studies were published over a 136-year time frame. Additional confidence in this conclusion comes from 37 studies of various species of apes, all of which also found adult males being heavier than adult females.

Explanation. At least among adults, body weight is a well-established USD-M. The only theoretical explanations we located for explaining average sex differences in body weight were evolutionary in nature (Kirchengast & Marosi 2008; Askovic & Kirchengast 2012). Much of the evolutionary reasoning centers around noting that body size helps to promote dominance and status for males (Ellis 1994b). Substantial evidence indicates that testosterone is a major contributor to body size, thereby helping to promote dominance and status (Mazur & Booth 1998).

Assessment. Among adults, body weight is a USD-M. Evolutionary theory appears to offer the best explanation (Précis 26.1.1.5).

26.1.1.6 Percent Body Fat

Evidence. As shown in Précis 26.1.1.6, 78 studies were found pertaining to sex differences in the percent body fat among adolescents and adults, all

Précis 26.1.1.5 Body weight (or body size) (original Tables 1.2.1.4)

Age categories, number of studies, and sex	*Consist. score*	*Overviews*	*Countries*	*Time range*	*Non-human*	*Social role theory*			*Evolutionary theory*		*Theological*
						Origin	*Found*	*Biosoc*	*Select*	*ENA*	
Infant/Toddler 10M 2x Child 18M 8x 5F Young Adol. 8M 5x 7F Older Adol. 13M 1x Adult 72M All Ages 116M 16F 12F	(83.3) (50.0) (29.6) (92.9) 100.0 (73.9)	-0-	26 (4)	1885–2021	37M apes	Silent	Silent	Silent	Expl-M	Expl-M	Silent

Précis 26.1.1.6 Percent body fat (original Table 1.2.2.13)

Age categories, number of studies, and sex	*Consist. score*	*Overviews*	*Countries*	*Time range*	*Non-human*	*Social role theory*			*Evolutionary theory*		*Theological*
						Origin	*Found*	*Biosoc*	*Select*	*ENA*	
Adol & Adult 77F 1x	98.7	-0-	14 (3)	1963–2015	--	Silent	Silent	Silent	Silent	Poss-F	Silent

but one of which concluded that females surpass males, with the exception simply failing to find a significant sex difference. The number of countries sampled was 14, along with three studies of multiple countries, covering a 52-year time period.

Explanation. To explain sex difference in the percent body fat, the best support seems to be provided by ENA theory. Specifically, testosterone appears to diminish body fat (Katznelson, Finkelstein et al. 1996: 4360; Bhasin, Storer et al. 1997; Bhasin, Parker et al. 2007). It does so at least in part by lowering the levels of leptin in the body (Blum, Englaro et al. 1997). As will be noted later in this chapter, natural levels of leptin are consistently higher in females than in males (see Précis 26.1.3.3).

Assessment. The findings are very supporting of the conclusion that percent body fat being a USD-F. The best explanation appears to involve the bodily influences of testosterone on leptin levels.

26.1.2 Anatomical and Physiological Factors

Chapter 2 provided a summary of anatomical and physiological factors (other than the most general factors such as body size and growth covered in Chapter 1). Twenty-two such variables were found to have consistency scores of 95.0 or higher. These variables are individually examined below regarding their status as USDs.

26.1.2.1 Bone Density (Bone Mass)

Evidence. Précis 26.1.2.1 shows that 53 studies of sex differences in bone density (or bone mass) were located. All but one of these studies concluded that bone density is greater for males (the exception simply failing to find the difference statistically significant). Ten countries (plus two multi-country studies), covering a 51-year time frame, were part of this assessment. Furthermore, five studies of non-human primates also indicated that males have denser bones than females of the same species.

Explanation. In the case of one of the three social role theories (i.e., biosocial role theory), one is led to expect that males would have denser bones. This is because this theory *assumes* that males will be larger and stronger than females. However, the theory does not specifically offer an explanation for why such a sex difference exists.

Both evolutionary theories, on the other hand, offer explanations for sex differences in physical size, of which bone density is a part. According to sexual selection theory, males have been selected for being physically larger, stronger, more combative, and more prone to take risks than females in order to better compete with other males for mates (Sell, Hone & Pound 2012; Glaudas, Rice et al. 2020). In other words, at least in

Précis 26.1.2.1 Bone density (bone mass) (original Table 2.1.4.1)

Age categories, number of studies, and sex	*Consist. score*	*Overviews*	*Countries*	*Time range*	*Non-human*	*Social role theory*			*Evolutionary theory*		*Theological*
						Origin	*Found*	*Biosoc*	*Select*	*ENA*	
Adult 41M 1x Wide Age Range 11M All Ages 75M 10x	97.6 100.0 (88.2)	-0-	10 (2)	1966–2014	5M: primates	Silent	Silent	Poss-M	Poss-M	Expl-M	Silent

ancestral times, males with these types of traits have tended to reproduce more prolifically than other males, while, for females, these same types of traits would do little to promote reproduction, and might even inhibit it.

In addition to the above evolutionary arguments, ENA theory predicts that exposure to high levels of androgens should promote bone density. Correlative empirical evidence has supported this prediction (Vanderschueren & Bouillon 1995; Katznelson, Finkelstein et al. 1996: 4360; Kenny, Prestwood et al. 2000). Additional evidence comes from research in which daily administration of a testosterone nasal gel has been found to help increase bone density among hypogonadal males (Rogol, Tkachenko & Bryson 2016).

Assessment. Overall, we conclude that bone density is a USD-M. Evolved hormonal factors appear to be major contributors to the sex difference in bone density.

26.1.2.2 Finger Length or Hand Size

Evidence. As shown in Précis 26.1.2.2, 14 studies of sex differences in average finger length (or hand size) were found. Fourteen studies, based on samples drawn for eight countries (plus one multi-country study) and covering a 75-year time span, all indicated that males have longer fingers or hands than do females. This sex difference was also reported in two studies of non-human primates.

Explanation. In as much as finger or hand lengths reflect overall growth and strength of the body, both evolutionary theories predict longer fingers and larger hands among males. ENA theory also predicts that androgens must be contributing to the growth of these extremities. Supporting this deduction, at least one study has indicated that prenatal exposure to testosterone promotes the growth of fingers (DM Moffit & Swanik 2011).

Assessment. Overall, finger length and hand size can be jointly considered USD-Ms. ENA theory seems to offer the most comprehensive explanation for this sex difference.

26.1.2.3 2D:4D Finger Length Ratio

Evidence. Précis 26.1.2.3 shows that a total of 250 studies were located having to do with human sex differences in what is known as the *2D:4D finger length ratio*. This variable refers to the relative length of the second (pointing) finger and the fourth (ring) fingers. Most studies have found that males have longer ring fingers, whereas females have slightly longer index (or pointing) fingers (or the length of these two fingers are indistinguishable).

Despite the large number of studies based on samples of children, adults, and the Wide Age Range, only for the latter of these three groups was the consistency score above 95.0. In addition, one literature review

Précis 26.1.2.2 Finger length or hand size (original Table 2.1.3.3)

Age categories, number of studies, and sex	*Consist. score*	*Overviews*	*Countries*	*Time range*	*Non-human*	*Social role theory*			*Evolutionary theory*		*Theological*
						Origin	*Found*	*Biosoc*	*Select*	*ENA*	
All Ages 14M	100.0	-0-	8 (1)	1943–2018	2M: primates	Silent	Silent	Silent	Expl-M	Expl-M	Silent

Précis 26.1.2.3 2D:4D finger length ratio (Original Table 2.1.3.5)

Age categories, number of studies, and sex	*Consist. score*	*Overviews*	*Countries*	*Time range*	*Non-human*	*Social role theory*			*Evolutionary theory*		*Theological*
						Origin	*Found*	*Biosoc*	*Select*	*ENA*	
Child 13F 2x Adult 180F 18x 2M Wide Age Range 15F All Ages 224F 24x 2M	(86.7) (90.0) 100.0 (88.9)	F: 1Rev F: 3Met	7 (4)	1998–2021	Mixed	Silent	Silent	Silent	Silent	Expl-F	Silent

and three meta-analyses all concluded that females have a higher 2D:4D ratio than do males, i.e., females have a shorter ring finger relative to their pointing finger.

Explanation. The main contributor to this sex difference appears to be the degree to which fetuses are exposed to testosterone. During this time, fingers grow at slightly different rates, and testosterone appears to contribute to this variation. When exposure is high, the ring finger grow elongates slightly more than does the pointing finger (Knickmeyer, Woolson et al. 2011; Abbott, Colman et al. 2012; Ventura, Gomes et al. 2013; Lofeu, Brandt, & Kohlsdorf 2017).

Testosterone exposure following birth may contribute at least slightly to differences produced during fetal development (McIntyre, Ellison et al. 2005; Trivers, Manning & Jacobson 2006; Wong & Hines 2016). Also, studies have indicated that the sex differences seem to be more apparent on the right hand than on the left (Lippa 2003:186; Xu & Zheng 2015).

Assessment. Overall, the 2D:4D ratio can be considered a USD-F, although more research is needed to explain the exceptions, the vast majority of which simply reported no significant sex difference (especially when measuring the left, rather than the right, hand). Even though the overall consistency score was below the 95.0 cutoff, the evidence surrounding the ratio and testosterone is strong. Many of the failures to replicate the main findings appear to be due to small sample size and imprecision in measurement.

26.1.2.4 Loss of Bone Density with Age

Evidence. Two related variables will be considered jointly. As shown in Précis 26.1.2.4, ten studies have all concluded that females lose a greater percentage of bone density with age than is the case for males. The number of countries sampled was three, published over a 23-year time frame.

Explanation. As noted earlier (see Précis 26.1.2.1), females have less dense bones than do males throughout adulthood. This is likely due to sexual selection for greater physical strength among males, given males' heavier physical labor throughout most of human history and prehistory. Regarding ENA theory, one can note that bodily exposure to testosterone appears to be a major contributor to the development of denser bones (Katznelson, Finkelstein et al. 1996). Oddly enough, much of the effect that testosterone has on bone growth (in both sexes) is accomplished by converting into a closely related so-called *female hormone* known as estradiol. This conversion is made possible by an enzyme known as *aromatase* (Carani, Qin et al. 1997; Sinnesael, Boonen et al. 2011). Females produce much less testosterone than males do throughout adolescence and adulthood (see Précis 26.1.3.2), and the estradiol that they produce

Précis 26.1.2.4 Loss of bone density with age (original Table 2.1.4.10)

Age categories, number of studies, and sex	*Consist. score*	*Overviews*	*Countries*	*Time range*	*Non-human*	*Social role theory*			*Evolutionary theory*		*Theological*
						Origin	*Found*	*Biosoc*	*Select*	*ENA*	
Adults 10F	100.0	-0-	3	1981–2004	-0-	Silent	Silent	Silent	Poss-M	Expl-M	Silent

without first being testosterone begins to precipitously decline after menopause. As a result, despite their having less bone density throughout adulthood, females tend to lose even more than males do after around age 50 (Ebeling, Atley et al. 1996; Ahlborg, Johnell et al. 2003). Once it is noted that androgens not only appear to affect the brain, but also the body more generally, one can see that ENA theory is the best of the five theories for explaining why females experience greater proportional bone loss with age than do males.

Assessment. Overall, bone density can be considered a well-established USD-M. Nevertheless, research involving more countries would strengthen this conclusion.

26.1.2.5 Joint Laxity

Evidence. The degree of flexibility in the joints, particularly of the elbows, hips, knees, and ankles is referred to as *joint laxity*. Studies have indicated that joint laxity is associated with ligament injuries following strenuous athletic activities (Denegar, Hertel & Fonseca 2002; Myer, Ford et al. 2008). As shown in Précis 26.1.2.5, row 5, studies from 20 studies indicate that, at least following puberty, females have greater joint laxity than do males. The evidence in this regard was found in samples drawn from 11 countries (plus one multi-country study), spanning 34 years in publication dates. This sex difference was also reported in one literature review.

Explanation. At least one study indicated that joint laxity was associated with low testosterone exposure (Stijak, Kadija et al. 2015). This would support ENA theory's assertion that most sex difference traits are the result of differential exposure to testosterone and other androgens. The only other theory that might predict greater joint laxity in females would be basic evolutionary theory, since it argues that males have been naturally selected over females for involvement in hunting and other forms of strenuous physical activities (Eshed, Gopher et al. 2004; O'Keefe, Vogel et al. 2011).

Assessment. At least for adolescents and adults, the evidence points toward joint laxity being a USD-F. The best explanation appears to be that provided by ENA theory.

26.1.2.6 Anogenital Distance

Evidence. The distance between the anus and the genitals of either sex is known as the *anogenital distance* (*AGD*), and is also called the *perineum*). Nearly all studies of the AGD are carried out among infants. As shown in Précis 26.1.2.6, all 19 located studies agreed that the AGD is greater for males. The number of countries sampled was 12, but the time frame between publication of the first and the most recent research report was

Précis 26.1.2.5 Joint laxity (original Table 2.1.4.12)

Age categories, number of studies, and sex	*Consist. score*	*Overviews*	*Countries*	*Time range*	*Non-human*	*Social role theory*			*Evolutionary theory*		*Theological*
						Origin	*Found*	*Biosoc*	*Select*	*ENA*	
Adolescent 8F Adult 12F	100.0 100.0	F: 1Rev	11 (1)	1978–2012	-0-	Silent	Silent	Silent	PossF	Expl-F	Silent

Précis 26.1.2.6 Anogenital distance (original Table 2.1.5.2)

Age categories, number of studies, and sex	*Consist. score*	*Overviews*	*Countries*	*Time range*	*Non-human*	*Social role theory*			*Evolutionary theory*		*Theological*
						Origin	*Found*	*Biosoc*	*Select*	*ENA*	
Infant 19M All Ages 24M	100.0	-0-	12 (1)	2009–2021	26M: multiple species	Silent	Silent	Silent	Silent	Expl-M	Silent

just 12 years. It is also worth noting that 26 studies of other species were located, all of which agreed that the AGD is greater for males.

Explanation. In terms of theoretically understanding this sex difference, only ENA theory makes a specific prediction. It predicts that prenatal exposure to testosterone or other androgens causes the AGD to elongate. Several studies, both of humans (Avidime et al. 2011; Jain, Goyal et al. 2018) as well as other animals (Hotchkiss, Lambright et al. 2007; Manno 2008) support this explanation.

Assessment. The evidence is very supportive of the conclusion that anogenital distance is a USD-M.

26.1.2.7 Muscularity

Evidence. Findings of sex differences in muscularity are summarized in Précis 26.1.2.7. A total of 68 pertinent studies were located. One can see that all studies concluded that males were more muscular than females. The number of countries samples regarding muscularity was nine (plus one multi-country study) and 22 countries were sampled for the studies of physical strength (plus three multi-country studies). The conclusion that post-pubertal males are more muscular and physically stronger than females is very solid (with the possible exception of the elderly). This is based on scientific publications spanning 47 years and 119 years, respectively. It is also in accordance with conclusions reached from one literature review and one meta-analysis regarding physical strength. Also, studies of several non-human animals also concluded that males appear to be more muscular and stronger than females.

Explanation. Regarding theoretical explanations, the only theories that seem to provide accounts of why sex differences in muscularity and strength would be found throughout the human species are the two evolutionary theories. In this regard, the sexual selection version argues that males who reproduce the best tend to be those who are stable suppliers of resources to themselves and to their spouse and offspring (EM Miller 1994; Gangestad & Simpson 2000). Accordingly, research has shown that females usually prefer relatively strong males over their weaker rivals (Frederick & Haselton 2007; Lidborg, Cross, & Boothroyd 2020).

ENA theory adds to this basic evolutionary reasoning the notion that testosterone and other androgens are the hormonal mechanisms that make muscularity and strength possible. Supporting this reasoning, several studies have shown that testosterone promotes musculature (Bhasin, Storer et al. 1996; Hansen, Bangsbo et al. 1999; Snyder, Peachey et al. 1999; Iannuzzi-Sucich, Prestwood, & Kenny 2002; Auyeung, Lee et al. 2011; Folland, McCauley et al. 2012).

Précis 26.1.2.7 Muscularity (original Table 2.1.9.1)

Age categories, number of studies, and sex	*Consist. score*	*Overviews*	*Countries*	*Time range*	*Non-human*	*Social role theory*			*Evolutionary theory*		*Theological*
						Origin	*Found*	*Biosoc*	*Select*	*ENA*	
Adult 58M All Ages 68M	100.0 100.0	-0-	9 (1)	1974–2021	18M	Silent	Silent	Silent	Expl-M	Expl-M	Silent

Assessment. From adolescents onward, it appears safe to conclude that muscularity and physical strength are both USD-Ms, particularly among adults.

26.1.2.8 Physical Strength

Evidence. Précis 26.1.2.8 provides a summary of sex differences in physical strength, a trait that is obviously closely related to muscularity (as discussed directly above). Except in the case of two studies of the elderly (where no sex differences were found), all studies agree that males are physically stronger than females, at least following puberty. The number of separate countries sampled was 22, plus three studies of multiple countries, and the studies were published over a 119-year time frame. One literature review and one meta-analysis also agreed that males are physically stronger than females.

Explanation. It is worth noting that one of the three social role theories, i.e., the biosocial version, *assumes* that males are physically stronger (Wood & Eagly 2012: 56). However, neither this theory or the other two versions offer a specific explanation for this sex difference.

In the case of the two evolutionary theories, both lead one to expect that males would be physically stronger than males. An argument put forth by sexual selection theorists has been that males who are physically stronger tend to be more likely to successfully compete for resources (Archer 2009; CW Miller & Svensson 2014).

The ENA version of the theory asserts that males are made to be physically stronger than females primarily by exposing their bodies to high levels of testosterone. This view is supported by studies showing a substantial correlation between circulating testosterone and physical strength (Storer, Woodhouse et al. 2008; KY Chin, Soelaiman et al. 2012; Handelsman, Hirschberg & Bermon 2018).

Assessment. With the possible exception of the elderly, there is very strong evidence that physical strength is a USD-M. Evolutionary theories offer the best theoretical explanations.

26.1.2.9 Skin Color (Skin Darkness)

Evidence. Fifty-six studies were located having to do with sex differences in skin color. As shown in Précis 26.1.2.9, all but two of these studies found that males have darker skin than do females, with the two exceptions simply findings no significant difference. The relevant studies were conducted in 22 different countries, along with eight other studies involving multiple countries, and were published over a 49-year time frame.

Explanation. The only theory that seems to offer an explanation for sex differences in skin color is ENA theory. It does so by stipulating that

Précis 26.1.2.8 Physical strength (original Table 2.1.9.2)

Age categories, number of studies, and sex	*Consist. score*	*Overviews*	*Countries*	*Time range*	*Non-human*	*Social role theory*			*Evolutionary theory*		*Theological*
						Origin	*Found*	*Biosoc*	*Select*	*ENA*	
Strength: Adol 14M Adult 42M 2x All 56M 2x	100.0 97.2 97.7	M: 1Rev 1Met	22 (3)	1902–2021	2M: mammals	Silent	Silent	Silent	Expl-M	Expl-M	Silent

Précis 26.1.2.9 Skin color (skin darkness) (original Table 2.1.10.1)

Age categories, number of studies, and sex	*Consist. score*	*Overviews*	*Countries*	*Time range*	*Non-human*	*Social role theory*			*Evolutionary theory*		*Theological*
						Origin	*Found*	*Biosoc*	*Select*	*ENA*	
Adult 42M 2x Wide Age Range 11M All 54M 2x	95.5 100.0 96.4	-0-	22 (8)	1958–2007	-0-	Silent	Silent	Silent	Silent	Expl-M	Silent

exposure to testosterone (both prenatally and post-pubertally) is higher for males than for females. Studies have shown that testosterone enhances skin pigmentation (Edwards, Hamilton et al. 1941; Kupperman 1944). While the biochemistry underlying this process is not fully understood, at least some evidence points toward testosterone being able to enhance the body's production of melanin (Wilson & Spaziani 1976). This enhancement even appears to affect the sclera (the whitish portion of the human eye), which one study found to be slightly darker among males than females (Kramer & Russell 2022).

Assessment. Despite two studies reporting no significant sex difference and the lack of any overview studies, the evidence of males having darker skin than females overwhelming supports the conclusion that skin color is a USD-M.

26.1.2.10 Larynx (Vocal Cords) Size

Evidence. The larynx is the primary organ controlling pitch variations when vocalizing. Individuals with large larynxes usually produce deeper (baritone) sounds than those with smaller larynxes. As shown in Précis 26.1.2.10, the evidence from 25 studies all indicate that, from adolescence onward, males have larger larynxes than do females. Findings in this regard come from studies of three different countries published over a 76-year time frame.

Explanation. The most viable theoretical explanations for these sex differences involve evolution. In particular, larynx size has been found to be positively correlated with body size and strength (Riede & Brown 2013). Therefore, it is likely that all of these traits have helped males physically compete with one another for access to resources and to attract mates (Fink 1963; Tecumseh Fitch & Reby 2001).

In the case of ENA theory, research findings strongly support the notion that testosterone promotes enlargement of the larynx (Dabbs & Mallinger 1999; Gugatschka, Kiesler et al. 2010; Kim, Shin, et al. 2020). Therefore, it not only offers an evolutionary explanation, but also a specific biochemical explanation.

As an alternative to evolutionary theory, one could offer a theological explanation for sex difference in larynx size. Specifically, as with males being physically larger than females, perhaps, God has helped to ensure that males would be more likely to lead females (rather than vice versa) by giving males more authoritative-sounding voices (see Corinthians 11:3; Genesis 2:18).

Assessment. Overall, both empirically and theoretically, there is strong support for the conclusion that larynx size is a USD-M.

Précis 26.1.2.10 Larynx (vocal cords) size (original Table 2.1.13.4)

Age categories, number of studies, and sex	*Consist. score*	*Overviews*	*Countries*	*Time range*	*Non-human*	*Social role theory*			*Evolutionary theory*		*Theological*
						Origin	*Found*	*Biosoc*	*Select*	*ENA*	
Adol 11M Adult 14M	100.0 100.0	-0-	3	1925–2001	-0-	Silent	Silent	Silent	Expl-M	Expl-M	Poss-M

26.1.2.11 QT Interval

Evidence. The QT interval refers to the span of time (measured in milliseconds) between the start of what is known as the Q wave and the end of the T wave in heart beats. As shown in Précis 26.1.2.11, all 20 studies located on this matter have concluded that females have longer QT intervals than do males. These studies were conducted in five different countries and were published over a 42-year time span. Among nonhuman mammals, however, only six of the eight studies identified females as having QT intervals greater than males.

Explanation. The only theory that predicts longer QT intervals among females is ENA theory. Offering confirmation for this prediction, two literature reviews have both concluded that testosterone tends to shorten QT intervals (Sedlak, Shufelt et al. 2012; Gutierrez, Wamboldt & Baranchuk 2021).

Assessment. Overall, at least for humans, all of the evidence is consistent with the QT interval being greater among females. However, because not all animal research has confirmed this pattern, we will not declare the QT interval to be a USD-F.

26.1.3 Biochemical Factors

Chapter three was devoted to summarizing findings regarding sex differences in biochemical factors. Just four such factors were found to have consistency scores of 95.0 or higher. Findings from these studies regarding them possibly being USDs are reviewed.

26.1.3.1 DHEA-S

Evidence. DHEA-S (dehydroepiandrosterone-sulfate) is a naturally occurring precursor to most of the androgens, including testosterone. As shown in Précis 26.1.3.1, all 12 studies of adults, derived from samples obtained from five different countries published over a 36-year time period, concluded that males have higher DHEA-S levels than do females. It seems relevant to mention, however, that one study of adolescents (see the original table) found no significant sex difference in DHEA-S levels.

Explanation. The only theory represented in the shaded region of this précis that makes any predictions regarding DHEA-S being a USD is ENA theory. It does so by simply noting that the biochemical structure of DHEA-S is such as to be a precursor to the production of testosterone and other androgens. Supporting this theoretical reasoning, at least one experiment indicated that oral administration of DHEA-S caused a significant increase in the production of free testosterone (Liu, Lin et al. 2013). Nevertheless, at least one study found no evidence that elevated DHEA-S

Précis 26.1.2.11 QT interval (original Table 2.3.1.17)

Age categories, number of studies, and sex	*Consist. score*	*Overviews*	*Countries*	*Time range*	*Non-human*	*Social role theory*			*Evolutionary theory*		*Theological*
						Origin	*Found*	*Biosoc*	*Select*	*ENA*	
Adult 20F	100.0	F: 1Rev adults	5	1961–2003	6F 1x 1M: mammals	Silent	Silent	Silent	Silent	Expl-M	Silent

Précis 26.1.3.1 DHEA-S (Original Table 3.1.1.7)

Age categories, number of studies, and sex	*Consist. score*	*Overviews*	*Countries*	*Time range*	*Non-human*	*Social role theory*			*Evolutionary theory*		*Theological*
						Origin	*Found*	*Biosoc*	*Select*	*ENA*	
Adult 12M All Ages 12M 1x	100.0 (92.3)	-0-	5	1966–2002	-0-	Silent	Silent	Silent	Silent	Expl-M	Silent

produced a rise in testosterone, although another androgen, that of androstenedione, was enhanced after a few weeks of DHEA-S treatment (Brown, Vukovich et al. 1999).

Assessment. Overall, while DHEA-S may be a USD, the evidence and the theory are not entirely clear in this regard, so we will not make a declaration at this point in time.

26.1.3.2 Testosterone

Evidence. Testosterone is classified as a male sex hormone. As shown in Précis 26.1.3.2, it is more abundant in males than in females, especially prenatally, and following puberty. The pertinent research has been conducted in 16 different countries plus one multi-country study, published over a 55-year time frame. This conclusion has also been reached by three literature reviews. Many studies of non-humans have also largely confirmed this age-dependent pattern.

Explanation. Testosterone plays a key role in sexual reproduction of all known vertebrates (Husak, Fuxjager et al. 2021). In humans, as illustrated in Figure 24.1, males have higher average levels of this hormone than females during fetal development (known as the *organizational phase* of sexual differentiation), soon after birth (sometimes known as *mini-puberty*), and following the onset of puberty (during the so-called *activational phase*). Animal experiments have shown that testosterone plays key roles in sexually differentiating basic physiology, neurology, as well as various forms of behavior (Hines 2006; Hau 2007).

Of the five theories of human sex differences, only the evolutionary neuroandrogenic (ENA) theory (Ellis 2011a, 2011b; Ellis & Ratnasingam 2015) hypothesizes that testosterone is directly involved. It should be mentioned that sexual selection theory is certainly compatible with the hypothesis that testosterone and other androgens help to orchestrate sex differences. In fact, several proponents of sexual selection theory incorporate androgens and other hormonal elements into their explanations of sex differences (e.g., Ligon, Thornhill et al. 1990; Owens & Short 1995).

Also worth noting is that one of the three social role theories of sex differences has argued for sex hormones playing a role in producing sex differences. Specifically, Wood and Eagly (2012:Figure 2.1) include "sex hormones" in their theoretical model (also see Eagly & Wood 2012: Figure 49.1). However, both of these theoretical models show sex hormones *resulting from* "gender role beliefs." This does not appear to be the case.

While social environmental experiences can interactively influence sex hormone levels, the effects are complex and usually temporary. For example, when subjected to stress, testosterone levels will usually rise, but when the stress persists, they normally decline, at least over the next few

Précis 26.1.3.2 Testosterone (original Table 3.1.1.14)

Age categories, number of studies, and sex	*Consist. score*	*Overviews*	*Countries*	*Time range*	*Non-human*	*Social role theory*			*Evolutionary theory*		*Theological*
						Origin	*Found*	*Biosoc*	*Select*	*ENA*	
Fetus 36M Infant/Toddler 17M 2F Child 4M 3x 2F Adol 16M Adult 53M Wide Age Range 3M 1x All Ages 129M 1x 4F	100.0 (80.9) (44.4) 100.0 100.0 (75.0) (93.5)	M: 3Gen 3Rev	16 (1)	1966–2021	31M 5x	Silent	Silent	Contra	Poss-M	Expl-M	Silent

hours (W Daly, Seegers et al. 2005; Chichinadze & Chichinadze 2008). However, sex hormones primarily *impact* bodily functioning (including brain functioning), rather than reacting to it (Hiller-Sturmhöfel & Bartke 1998). This is true for circulating levels of sex hormones (Barth, Villringer & Sacher 2015), and especially for prenatal sex hormones.

Assessment. Overall, the evidence is strong that testosterone is a USD-M. This is true during fetal development and following the onset of puberty. The evidence having to do with infancy through childhood varies regarding sex differences.

26.1.3.3 Leptin

Evidence. Leptin is a protein-based hormone that helps to regulate appetite and fat storage. One can see in Précis 26.1.3.3 that 34 studies of adults have all concluded that greater levels of leptin are found in females. Eight different countries were sampled by these studies covering, although the time spanning the publication of these studies was just 12 years.

Explanation. As shown in the shaded portion, ENA theory is the only theory that predicts higher leptin levels among females than males. It does so by asserting that circulating testosterone suppresses leptin production. Support for this hypothesis comes from studies showing that leptin levels are inversely correlated with circulating testosterone levels, at least among men (Behre, Simoni & Nieschlag 1997; Blum, Englaro et al. 1997; Vettor, De Pergola et al. 1997; Luukkaa, Pesonen et al. 1998).

Assessment. Given the consistency of a larger number of studies located, we consider leptin to be a USD-F, at least among human adults. More research is needed to fully understand how testosterone affects leptin levels.

26.1.4 The Brain: Structure and Functioning

Chapter 4 had to do with findings of sex differences in the structure and functioning of the brain. Three variables associated with the brain had consistency scores of 95.0 or higher regarding sex differences. They are summarized below.

26.1.4.1 Brain Volume (Brain Size)

Evidence. Many studies have reported on sex differences in the size of the brain, most of which are based on samples of adults. The first of these were brain volume (or brain size). As shown in Précis 26.1.4.1, among adults, 120 of the 122 relevant studies of adults concluded that male brains were larger than were the brains of females, with the remaining two studies failing to report significant differences, making the consistency score 98.4. It is worth adding that, when studies of all age groupings were

Précis 26.1.3.3 Leptin (Original Table 3.4.1.11)

Age categories, number of studies, and sex	*Consist. score*	*Overviews*	*Countries*	*Time range*	*Non-human*	*Social role theory*			*Evolutionary theory*		*Theological*
						Origin	*Found*	*Biosoc*	*Select*	*ENA*	
Adult 34F	100.0	-0-	8	1995–2007	-0-	Silent	Silent	Silent	Silent	Expl-F	Silent

Précis 26.1.4.1 Brain volume (brain size) (Original Table 4.1.1.1)

Age categories, number of studies, and sex	*Consist. score*	*Overviews*	*Countries*	*Time range*	*Non-human*	*Social role theory*			*Evolutionary theory*		*Theological*
						Origin	*Found*	*Biosoc*	*Select*	*ENA*	
Infant/Toddler 11M 4x Child Adult 120M 2x Wide Age Range 35M 2x All Ages 186M 10x	(73.3) (92.3) 98.4 (94.6) (94.9)	M: 2Rev 3Met; x: 4Gen	23 (1)	1908–2021	11M 5x mammals	Silent	Silent	Silent	Expl-M	Expl-M	Silent

combined, 186 studies reported males having larger brains while ten studies found no significant sex differences. As shown in parentheses, this sex difference variable has a consistency score that is exceedingly close to the 95.0 threshold, i.e., it was 94.9.

The number of countries represented in the samples of these studies was 23, with one more study drawing its sample from multiple countries, and the number of years between the publication of the first and the most recent evidence on this matter was 113. Additional evidence comes from two literature reviews and three meta-analyses, all of which concluded that male brains were larger, although four published generalizations have concluded that there are no significant sex differences. One final line of evidence has come from studies of various non-human mammals. Eleven of these found larger brains in mammals and five reported no significant sex differences.

Explanation. In terms of theoretical explanations, both evolutionary theories would reason as follows: Enlarged brains are likely to promote male behavior patterns that are females find attractive, such as status-striving and competitiveness. In other words, an enlarged brain could promote both the ability and motivation to compete for resources. Among the evidence supporting this reasoning comes from studies of brain size and intelligence (the latter being a general measure of reasoning and learning ability). Brain size has been found to be positively associated with scores on standardized IQ tests, typically with correlation coefficients in the range of r = .40–.50 (Andreasen, Flaum et al. 1993; Tramo, Loftus et al. 1998; Gur, Turetsky et al. 1999; Pennington, Filipek et al. 2000; Witelson, Beresh & Kigar 2006). This may be especially true for spatial reasoning (Geary 1996).

Regarding ENA theory, substantial evidence indicates that exposing the brain prenatally to testosterone causes it to enlarge. This is true both in humans (Perrin, Hervé et al. 2008; Peper, Brouwer et al. 2009: Figure 1; Paus, Nawaz-Khan et al. 2010; Whitehouse, Maybery et al. 2010; Lombardo, Ashwin et al. 2012; Heany, van Honk et al. 2016 males only; Savic, Frisen et al. 2017) and in other species (Parolini, Romano et al. 2017). It seems reasonable to infer that, by increasing the size of the brain, testosterone contributed to various cognitive traits that enhance male attractiveness to females. One possibility in this regard is the ability to reason in spatial terms (Sherry & Hampson 1997).

Assessment. Overall, evidence predominantly indicates that brain size is a USD-M, particularly among adults. This pattern can be explained by both of the evolutionary theories.

26.1.4.2 Size of Brain Ventricles

Evidence. The ventricles of the brain carry cerebrospinal fluid throughout the brain, helping to flush out toxins and waste products. Précis 26.1.4.2

Précis 26.1.4.2 Size of brain ventricles (original Table 4.1.1.8)

Age categories, number of studies, and sex	*Consist. score*	*Overviews*	*Countries*	*Time range*	*Non-human*	*Social role theory*			*Evolutionary theory*		*Theological*
						Origin	*Found*	*Biosoc*	*Select*	*ENA*	
All Ages 12M	100.0	-0-	6	1995–2011	Mixed	Silent	Silent	Silent	Silent	Silent	Silent

shows that all 12 studies of sex differences that were located concluded that males have larger ventricles than do females. Six countries were represented in these studies spanning 16 years between publication of the oldest and the newest studies. No overview studies were located and findings for non-humans were inconsistent regarding sex differences.

Explanation. None of the theories specifically predicts a sex difference in the size of the brain ventricles, although ENA theory predicts that, if there are sex differences, they would be the result of differential exposure to androgens. No specific evidence was located regarding the possibility that androgens affect brain ventricle enlargement.

Assessment. Because the number of studies reporting sex differences was fairly small (i.e., 12) and the findings from studies of non-humans were mixed, and no specific evidence was found to support any theoretical arguments that have been made regarding sex differences in the size of the brain ventricles, we reserve judgment about the USD status of this variable.

26.1.4.3 Size of Neocortex

Evidence. The neocortex (sometimes simply called the *cortex*) refers to the outer layer of the brain. In humans, it is unusually large when compared to other mammals, comprising roughly 40% of the entire brain. This portion of the brain is the most active area as your read and make sense of these sentences! As shown in Précis 26.1.4.3, 37 studies of the size of the neocortex were located, all but one indicated that this brain area is larger among males than among females. Relevant studies were based on samples from eight different countries (plus one multi-country study).

Explanation. ENA theory seems to uniquely predict that males would have larger neocortices than females. This prediction comes in part from noting that testosterone appears to promote cell proliferation of many types, including the proliferation of nerve cells (Rasika, Nottebohm & Alvarez-Buylla 1994; Louissaint, Rao et al. 2002; Fowler, Freeman & Wang 2003). Research pertaining to the possibility that exposure to testosterone promotes greater growth of the neocortex has been supported in at least one study of non-human animals (Ryzhavskii 2002).

Assessment. Overall, we conclude that the size of the neocortex is a USD-M. The best theoretical explanation is offered by the neurohormonal component of ENA theory.

26.1.5 Physical Health and Illness

Twenty-five variables having to do with sex differences in physical health and/or physical illness were located with consistency scores of 95.0 or higher. (Research having to do with sex differences in mental health are the focus of a separate chapter.) Findings for these variables were initially

Précis 26.1.4.3 Size of neocortex (original Table 4.1.2.1)

Age categories, number of studies, and sex	*Consist. score*	*Overviews*	*Countries*	*Time range*	*Non-human*	*Social role theory*			*Evolutionary theory*		*Theological*
						Origin	*Found*	*Biosoc*	*Select*	*ENA*	
Wide Age Range 12M All Ages 36M 1x	100.0 97.3	M: 1Met	8 (1)	1979–2011	3M 1x rodents	Silent	Silent	Silent	Silent	Expl-M	Silent

summarized in Chapter 5. Here, these health or illness variables are reviewed regarding the possibility of their being USDs are reviewed below.

26.1.5.1 Seeking or Utilizing Healthcare Services Including Ob/Gyn

Evidence. Seventy-five studies were found in which males and females were compared regarding the seeking of, or utilizing of, healthcare services (including obstetric and gynecological services). One can see in Précis 26.1.5.1 that all but three of these studies concluded that females seek or obtain more services than do males. The studies involved came from nine countries plus one multi-country study, with 48 years of publications covered.

Explanation. Although some have reported no significant sex differences, the vast majority of studies indicate that males tolerate pain to a greater degree than females, both in humans (Table 4.2.1.3a) and in various species of rodents (Table 4.2.1.3b). If so, one can expect that women would seek to obtain medical care to a greater degree than males.

As to why females would be less prone to tolerant pain, ENA theory implies that brain exposure to testosterone helps to diminish sensitivity to nearly all forms of stimuli. Evidence in this regard is supportive, both among males (Apkhazava, Kvachadze et al. 2018) and among females (Bartley, Palit et al. 2015). More broadly, one could also explain greater pain tolerance among males as having been sexually selected in order for males to better compete for dominance and status (Vigil & Strenth 2014; Hill, Bailey & Puts 2017).

Assessment. While the evidence of sex differences in the seeking or utilizing of healthcare services is not entirely clear when ob/gyn services are excluded (see Table 5.1.2.1), when included, this variable is a well-documented USD-F.

26.1.5.2 Sudden Infant Death Syndrome

Evidence. As its name implies, *sudden infant death syndrome* (*SIDS*) refers to instances in which an infant is found dead, usually after sleeping and no prior symptoms of ill health. Précis 26.1.5.2 shows that all 13 studies all concluded that males suffer more than do females from this syndrome. Three countries (plus one multi-country study) served as samples, with 36 years separating publication of the first and the most recent studies.

Explanation. The précis's shaded region indicates that the only theory that specifically predicts higher rates of SIDS among males than females is ENA theory. Supporting this prediction, there are some studies that have found testosterone higher among babies who are SIDS victims (Emery, Krous et al. 2005). Especially suspect in this regard, is a phenomenon sometimes known as "mini-puberty," an event involving a

Précis 26.1.5.1 Seeking or utilizing healthcare services including ob/gyn (original Table 5.1.2.2)

Age categories, number of studies, and sex	*Consist. score*	*Overviews*	*Countries*	*Time range*	*Non-human*	*Social role theory*			*Evolutionary theory*		*Theological*
						Origin	*Found*	*Biosoc*	*Select*	*ENA*	
Adolescent 40F Adult 24F All Ages 72F 3x	100.0 100.0 96.0	-0-	9 (1)	1957–2005	-0-	Silent	Silent	Silent	Poss-F	Expl-F	Silent

Précis 26.1.5.2 Sudden infant death syndrome (Original Table 5.1.5.2)

Age categories, number of studies, and sex	*Consist. score*	*Overviews*	*Countries*	*Time range*	*Non-human*	*Social role theory*			*Evolutionary theory*		*Theological*
						Origin	*Found*	*Biosoc*	*Select*	*ENA*	
Infant 13M	100.0	-0-	3 (1)	1968–2004	-0-	Silent	Silent	Silent	Silent	Expl-M	Silent

sudden short-term spike in testosterone levels that occurs in nearly all males a few months following birth (see Figure 24.1). One study reported that this event was associated with an elevated male risk of SIDS (Hadziselimovic, Verkauskas et al. 2019).

Assessment. SIDS, like stillbirths and death soon after birth (reviewed earlier in this chapter), all support the conclusion that males are at greater risk. The evidence is sufficiently strong to consider SIDS a USD-M.

26.1.5.3 Mortality Following Puberty

Evidence. Eighteen studies were located having to do with sex differences in the probability of people dying following the age of puberty. One can see in Précis 26.1.5.3 that all of these studies agree that males are more likely to die after reaching puberty than are females. This conclusion was based on samples drawn from five countries (plus one multi-country study) covering a publication span of 47 years. It also comports with one literature review and one study of a non-human primate species.

Explanation. Both evolutionary theories lead one to expect higher death rates among males than females. Specifically, because males have a much higher potential "reproductive ceiling" than do females, early in their reproductive lives, they are favored for greater risk-taking in order to realize as much of this high potential as possible (Kruger & Nesse 2006; BJ Ellis, Del Giudice et al. 2012; Apicella, Carré & Dreber 2015).

In the case of ENA theory, it also predicts that exposure to high levels of testosterone enhances the probability of death. The evidence in this regard does not appear to be entirely supportive (Assari, Caldwell & Zimmerman 2014). Also, as males reach old age, and their reproductive potential is minimal, males with high testosterone are actually less likely to die than those with low testosterone (*review*: Muraleedharan & Hugh Jones 2014).

Assessment. Overall, we will not declare mortality following puberty to be a USD-M. Although the evidence is consistent with this conclusion, the ability of any theory to fully account for why males would have greater post-pubertal mortality needs clarification.

26.1.5.4 Accidental Injuries and Fatalities in General

Evidence. A very large number of studies of sex differences in accidental injuries (including fatalities) have been published. Précis 26.1.5.4 shows that, of the 134 studies that were located, all but four concluded that males suffered more injuries than did females, making the consistency score 97.0. One literature review along with one meta-analysis also reached this conclusion. The number of individual countries sampled was 17 (plus 11 studies of more than one country), and the years during which these studies were

Précis 26.1.5.3 Mortality following puberty (original Table 5.1.5.4)

Age categories, number of studies, and sex	*Consist. score*	*Overviews*	*Countries*	*Time range*	*Non-human*	*Social role theory*			*Evolutionary theory*		*Theological*
						Origin	*Found*	*Biosoc*	*Select*	*ENA*	
All Ages (after puberty) 18M	100.0	M: 1Rev	5 (1)	1957–2004	1M primates	Silent	Silent	Silent	Expl-M	Expl-M	Silent

Précis 26.1.5.4 Accidental injuries and fatalities in general (original Table 5.2.1.1)

Age categories, number of studies, and sex	*Consist. score*	*Overviews*	*Countries*	*Time range*	*Non-human*	*Social role theory*			*Evolutionary theory*		*Theological*
						Origin	*Found*	*Biosoc*	*Select*	*ENA*	
All Ages 130M 4x	97.0	M: 1Rev 1Met	17 (11)	1964–2011	4M primates	Silent	Silent	Silent	Expl-M	Expl-M	Silent

published was 47. Furthermore, four studies of non-human primates also reached the conclusion that males experience more than females do regarding accidental injuries.

Explanation. None of the six explanations identified in the shaded region of this Précis specifically pertained to sex differences in accidental injuries. However, if one assumes that traits such as risk-taking and activity levels contribute substantially to accidental injuries, both evolutionary theories lead one to expect more such injuries among males. Regarding ENA theory specifically, studies have indicated that neurological exposure to androgens appears to promote risk-taking (circulating T in humans: Vermeersch, T'sjoenet al. 2008; Apicella, Carre & Dreber 2015; Mehta, Welker et al. 2015) and to enhance overall activities levels (prenatal T in humans: Alexander & Saenz 2012; circulating T in mice: Klomberg, Garland et al. 2002).

Assessment. Overall, the evidence is strong that males suffer more accidental injuries (including fatal ones) than do females. While the theoretical explanation for this sex difference is somewhat indirect, it can be considered a USD-M.

26.1.5.5 Motor Vehicle Injuries

Evidence. Précis 26.1.5.5 shows that ten studies of sex differences in adolescent motor vehicle injuries all concluded that males have these experiences more than females. Samples were obtained from three different countries over a 52-year time frame.

Explanation. It is likely that sex difference in sustaining injuries from motor vehicles can be at least partly explained by noting that males are more prone to take risks (see Précis 26.3.1.2). If so, evidence that risk-taking has been sexually selected among males (BJ Ellis, Del Guidice et al. 2012; Engqvist, Cordes & Reinhold 2015) and promoted by exposing the brain to testosterone (*review*: Apicella, Carré & Dreber 2015; *meta-analysis*: Kurath & Mata 2018) could help to account for sex differences in motor vehicle injuries.

Assessment. Given that only the minimum of ten pertinent studies were located, with just three countries sampled, and the fact that some unknown factors could be confounding efforts to theoretically explain these findings, we will not deem this variable to be a USD.

26.1.5.6 Bicycle Injuries

Evidence. Substantial research has sought to determine if there are sex differences in the incidence of bicycle injuries. At least regarding studies of wide age range research participants, Précis 26.1.5.6 shows that all 16 studies concluded that males have more bicycle-related injuries than do

Précis 26.1.5.5 Motor vehicle injuries (original Table 5.2.1.2)

Age categories, number of studies, and sex	*Consist. score*	*Overviews*	*Countries*	*Time range*	*Non-human*	*Social role theory*			*Evolutionary theory*		*Theological*
						Origin	*Found*	*Biosoc*	*Select*	*ENA*	
Adolescent 10M All Ages 50M 5x 1F	100.0 (87.7)	-0-	3	1961–2013	-0-	Silent	Silent	Silent	Poss-M	Poss-M	Silent

Précis 26.1.5.6 Bicycle injuries (original Table 5.2.1.4)

Age categories, number of studies, and sex	*Consist. score*	*Overviews*	*Countries*	*Time range*	*Non-human*	*Social role theory*			*Evolutionary theory*		*Theological*
						Origin	*Found*	*Biosoc*	*Select*	*ENA*	
Wide Age Range 16M All Ages 17M 1x 1F	100.0 (85.0)	M: 1Rev	7 (1)	1985–2015	-0-	Silent	Silent	Silent	Poss-M	Poss-M	Silent

females. As noted above, one factor that confounds studies of all specific types of accidental injuries has to do with exposure differentials. In other words, the higher rate of bicycle injuries for males could be simply due to their riding bicycles more.

Explanation. Aside from the possibility that males may ride bicycles more than females (a possibility for which we found no evidence), the most reasonable explanation is that males ride bicycles in riskier ways than do females. If the latter is the case, both of the evolutionary theories would help to explain why males experience more bicycle injuries than do females due to evidence that males are more prone to take risks (see Précis 26.3.1.2).

Assessment. Even though we did not declare sex differences in motor vehicle injuries to be a USD, we will do so for bicycle injuries. Specifically, the greater number of studies and the greater number of countries sampled, all indicate that the greater male risk of bicycle injuries is a USD-M.

26.1.5.7 Accidental Drownings

Evidence. Drownings nearly always occur in large bodies of water, such as swimming pools, lakes, and the like, and are usually fatal. As shown in Précis 26.1.5.7, the evidence of sex differences came from 47 studies obtained from nine different countries (plus seven multi-country studies). In agreement with seven literature reviews that males suffer more accidental drownings than do females, all but one of the 47 studies reached this conclusion.

Explanation. Later in this chapter, evidence will be reviewed regarding sex differences in risk-taking (Précis 26.3.1.2). The greater tendency for males to take risks could help to account for sex differences in accidental drowning rates. If so, research findings having to do with risk-taking tendences would be relevant to accidental drownings (see Précis 26.3.1.2). In this regard, greater tendencies by males to take risks appears to have been sexually selected (BJ Ellis, Del Guidice et al. 2012; Engqvist, Cordes & Reinhold 2015). Also, as predicted by ENA theory, brain exposure to testosterone appears to promote risk-taking (*review*: Apicella, Carré & Dreber 2015; *meta-analysis*: Kurath & Mata 2018).

Assessment. Because nearly all of the located research points toward more males being victims of accidental drownings, as do seven literature reviews, we will declare this variable to be a USD-M.

26.1.5.8 Accidental Lightning Strike Injuries or Fatalities

Evidence. Précis 26.1.5.8 shows that all 25 studies of sex differences in accidental lightning strike injuries (including fatalities) indicated that males are the greatest victims. Fourteen different countries (plus three multi-country studies) were involved in providing these results.

Précis 26.1.5.7 Accidental drownings (original Table 5.2.1.10)

Age categories, number of studies, and sex	*Consist. score*	*Overviews*	*Countries*	*Time range*	*Non-human*	*Social role theory*			*Evolutionary theory*		*Theological*
						Origin	*Found*	*Biosoc*	*Select*	*ENA*	
All Ages 46M 1F	95.8	M: 7Rev	9 (7)	1974–2014	-0-	Silent	Silent	Silent	Expl-M	Expl-M	Silent

Précis 26.1.5.8 Accidental lightning strike injuries or fatalities (original Table 5.2.1.14)

Age categories, number of studies, and sex	*Consist. score*	*Overviews*	*Countries*	*Time range*	*Non-human*	*Social role theory*			*Evolutionary theory*		*Theological*
						Origin	*Found*	*Biosoc*	*Select*	*ENA*	
Wide Age Range 25M	100.0	-0-	14 (3)	1990–2018	-0-	Silent	Silent	Silent	Poss-M	Poss-M	Silent

Explanation. Theoretically explaining these sex differences can only be done indirectly. The most reasonable way of doing so would involve assuming that individuals who spend more time outdoors, where risk of being struck by lightning is greater than when indoors. Evidence that males spend more time outdoors than do females is substantial (original Table 20.3.3.3). This tendency could be partly attributable to males being more physically active (original Table 16.3.2.2), which evolutionary theorists have attributed to males having been selected for engaging in more physically active work (such as hunting) away from home (Joseph 2000: 57).

Assessment. Overall, while theoretical explanation is indirect, the strength of the evidence is sufficiently strong and culturally diverse as to confidently declare accidental injuries from lightning strikes a USD-M.

26.1.5.9 Breast Cancer

Evidence. While breast cancer occasionally afflicts men (Giordano 2018), the majority of those suffering from this form of cancer are women. In Précis 26.1.5.9, one can see that 27 studies reached this conclusion, with samples drawn from 11 different countries (plus two multi-country studies), and their publications covering a 63-year time frame.

Explanation. The only sex difference theory that seems to bear on breast cancer risk is ENA theory. In this regard, high levels of both testosterone and especially its metabolite, estradiol, have been found positively correlated with breast cancer, at least among women (Cauley, Lucas et al. 1999; Hankinson & Eliassen 2007; Zhang, Tworoger et al. 2013). Also, one experimental study indicated that administering testosterone in conjunction with an aromatase inhibitor (which prevents testosterone from metabolizing into estradiol) appears to inhibit the development of breast cancer (Glaser & Dimitrakakis 2015).

Assessment. Overall, even though the theoretical understanding of breast cancer is still sketchy, the consistency of the evidence that it primarily affects females is strong. Therefore, it is considered a USD-F.

26.1.5.10 Leukemia and Non-Hodgkin's Lymphoma

Evidence. Both leukemia and non-Hodgkin's lymphoma involve excessive production of lymphocytes. As shown in Précis 26.1.5.10, males appear to be affected more than females. Pertinent research, covering a 42-year publication time span, has come from six countries as well as one multi-country study.

Explanation. Regarding theoretical explanations, only ENA theory offers a possible partial explanation, i.e., that it is promoted by exposure to androgens. While no research was found of a causal nature, several

Précis 26.1.5.9 Breast cancer (original Table 5.3.3.5)

Age categories, number of studies, and sex	*Consist. score*	*Overviews*	*Countries*	*Time range*	*Non-human*	*Social role theory*			*Evolutionary theory*		*Theological*
						Origin	*Found*	*Biosoc*	*Select*	*ENA*	
All Ages 27F	100.0	F: 1Rev	11 (2)	1943–2006	-0-	Silent	Silent	Silent	Silent	Poss-F	Silent

Précis 26.1.5.10 Leukemia and non-Hodgkin's lymphoma (original Table 5.3.3.13)

Age categories, number of studies, and sex	*Consist. score*	*Overviews*	*Countries*	*Time range*	*Non-human*	*Social role theory*			*Evolutionary theory*		*Theological*
						Origin	*Found*	*Biosoc*	*Select*	*ENA*	
All Ages 18M	100.0	-0-	6 (1)	1961–2003	-0-	Silent	Silent	Silent	Silent	Expl-M	Silent

studies have actually found males with non-Hodgkin lymphoma to have low levels of testosterone levels compared to males generally (Olsson 1984; *review*: Arden-Close, Eiser & Pacey 2011).

Assessment. Overall, the evidence that males suffer from both leukemia and non-Hodgkin's lymphoma is substantial, as is the number of countries sampled. While a theoretical explanation needs to be more fully developed, we deem Leukemia and non-Hodgkin's lymphoma to be a USD-M.

26.1.5.11 Insomnia

Evidence. Insomnia refers to consistent difficulties falling or staying asleep. Précis 26.1.5.11 shows that a total of 51 studies of sex differences in this ailment were located. Of the 48 involving adults. all but two concluded that females experience insomnia to a great degree than do males. The studies of this condition drew samples from 16 different countries (plus one multi-country study), and were published over a 42-year time frame.

Explanation. No attempts to theoretically explain sex differences in insomnia were found. Nonetheless, one study reported results regarding the possibility of 2D:4D being correlated with insomnia among a sample of college students. No significant relationships were found for either hand (Verster, Mackus et al. 2017). This would be contrary to predictions by ENA theory that essentially all sex differences are the result of brain exposure to testosterone.

Assessment. At least for adults, insomnia appears to be a USD-F. However, no theoretical explanations of sex differences have yet been developed.

26.1.5.12 Sleep Apnea

Evidence. Sleep apnea refers to repeatedly stopping breathing during sleep. It is also often associated with loud snoring and can gradually cause serious heart problems due to lack of oxygen. As shown in Précis 26.1.5.12, all 34 studies – drawn from eight countries (plus one multi-country study) – concluded that this disorder is more prevalent among males. The studies were published over a 53-year time frame. Two literature reviews also concluded that males suffer from sleep apnea more than do females.

Explanation. The only theory that purports to explain sex differences in sleep apnea is ENA theory. It does so by asserting that essentially all cognitive and behavioral sex differences are at least partly the result of the brain being exposed to testosterone. The evidence is generally supportive. For example, a study of males with this disorder concluded that their circulating levels of testosterone were higher than for males without the disorder (Kirbas, Abakay et al. 2007). In addition, two studies in which

Précis 26.1.5.11 Insomnia (original Table 5.4.3.2)

Age categories, number of studies, and sex	*Consist. score*	*Overviews*	*Countries*	*Time range*	*Non-human*	*Social role theory*			*Evolutionary theory*		*Theological*
						Origin	*Found*	*Biosoc*	*Select*	*ENA*	
Adults 46F 2x All Ages 48F 2x 1M	95.8 (92.8)	F: 1Rev 1Met	16 (1)	1976–2018	-0-	Silent	Silent	Silent	Silent	Contra	Silent

Précis 26.1.5.12 Sleep apnea (original Table 5.4.3.5)

Age categories, number of studies, and sex	*Consist. score*	*Overviews*	*Countries*	*Time range*	*Non-human*	*Social role theory*			*Evolutionary theory*		*Theological*
						Origin	*Found*	*Biosoc*	*Select*	*ENA*	
All Ages 34M	100.0	M: 2Rev	8 (1)	1955–2008	-0-	Silent	Silent	Silent	Silent	Poss-M	Silent

synthetic testosterone were administered both reported that this increased sleep apnea symptoms. One study involved male patients (Wittert 2014), and an earlier study involved females (Johnson, Anch & Remmers 1984). However, in a literature review of clinical data, one researcher argued that testosterone therapy has only a minor risk of causing sleep apnea at most (Hanafy 2007).

Assessment. Sleep apnea is all but certainly a USD-M. Theoretically explaining this sex difference is so far poorly developed.

26.1.5.13 Nightmares

Evidence. The last health-related variable to be identified with having ten or more findings and a consistency score of 95.0 or higher is nightmares. While probably everyone has experienced nightmares at least sometime in their lives, Précis 26.1.5.13 shows that these experiences are more common for females than for males. Twelve pertinent studies were located, based on samples from eight different countries, published over a 32-year time frame. One literature review also supported the conclusion that females have more nightmares than do males.

Explanation. Of the five theories of sex differences, only ENA theory made a specific prediction regarding sex differences in the prevalence of nightmares. This is because of its blanket assertion that essentially all cognitive sex differences are influenced by exposing the brain to testosterone. No evidence either confirming or refuting this prediction were located.

Assessment. With a high degree of confidence, it appears that the frequency with which individuals have nightmares is a USD-F. However, no theory has yet offered a specific explanation for this sex difference.

26.2 High Consistency Scores in Volume II

The variables now to be examined regarding their potential for being USDs are those first identified in Volume II for which high consistency scores were found (i.e., 95.0 or higher). These variables are primarily of a cognitive nature, as covered in Chapters 9 through 15 . The first group of cognitive variables have to do with perceptual and motor factors. Following this group of variables, attention is given to variables of an emotional, intellectual, mental health, and attitudinal nature.

26.2.1 Perceptual and Motor Factors

Research pertaining to sex differences in a wide range of perceptual and basic motor skills were covered in the book's Chapter 9. Here, the variables for which high consistency scores were found are given individual attention.

Précis 26.1.5.13 Nightmares (original Table 5.4.3.7)

Age categories, number of studies, and sex	*Consist. score*	*Overviews*	*Countries*	*Time range*	*Non-human*	*Social role theory*			*Evolutionary theory*		*Theological*
						Origin	*Found*	*Biosoc*	*Select*	*ENA*	
All Ages 12F	100.0	F: 1Rev	8	1984–2016	-0-	Silent	Silent	Silent	Silent	NoTest	Silent

26.2.1.1 Auditory Acuity

Evidence. The 13 studies on sex differences in hearing ability (auditory acuity) among adults located were performed over an 80-year period but were limited to only two countries. Précis 26.2.1.1. shows that all of the studies found females to have greater auditory acuity.

Explanation. The only theory that seems to directly bear on sex differences in auditory acuity is ENA theory. While it offers no specific evolutionary explanation, it suggests that testosterone inhibits auditory acuity. Some evidence points toward auditory acuity being diminished by prolonged testosterone exposure and enhanced by estrogen exposure, both prenatally (McFadden 1998) as well as postpubertally (Williamson, Zhu et al. 2020).

Assessment. Although the number of countries sampled needs to be expanded and just 13 relevant studies were located, we consider auditory acuity a USD-F. Among the reasons for this decision is that (a) the time range for the relevant studies was substantial (i.e., 70 years), (b) the sex difference can be theoretically explained (by ENA theory), and (c) the sex difference has been confirmed by at least one study of rats.

26.2.1.2 Color Discrimination/Recognition (Naming Colors)

Evidence. Précis 26.2.1.2 summarizes findings from 35 studies of people's abilities to distinguish different colors. All of the 35 studies, reported over a 101-year time frame, concluded that females are better able to discriminate colors and accurately report their fine gradations in shade. These results were found among research participants of different ages and in seven countries plus one multi-national study.

Explanation. Sex differences in the ability to discriminate between colors that are close to one another on the color spectrum has obviously received a great deal of research attention. To our knowledge, only one of the theories identified in the shaded region of the précis has any implications regarding such sex differences, i.e., ENA theory. It implies that sex hormones must be operating on brain functioning in ways that promote female color discrimination and/or inhibits male color discrimination. In this regard, Abramov, Gordon et al. (2012) found that females require slightly longer wave lengths in order to detect changes in color hues, and proposed that genes located on the X chromosome could be involved. However, no evidence for or against the idea that androgens affect color discrimination ability was found.

Assessment. Despite precise theoretical understanding for why females would be better than males at color discrimination, the number of studies and their breadth in terms of countries sampled and the time frame involved gives us confidence in declaring this ability to be a USD-F.

Précis 26.2.1.1 Auditory acuity (original Table 9.1.1.1)

Age categories, number of studies, and sex	*Consist. score*	*Overviews*	*Countries*	*Time range*	*Non-human*	*Social role theory*			*Evolutionary theory*		*Theological*
						Origin	*Found*	*Biosoc*	*Select*	*ENA*	
Adult 13F	100.0	-0-	2	1931–2001	1F rat	Silent	Silent	Silent	Silent	Expl-F	Silent

Précis 26.2.1.2 Color discrimination/recognition (naming colors) (original Table 9.1.4.5)

Age categories, number of studies, and sex	*Consist. score*	*Overviews*	*Countries*	*Time range*	*Non-human*	*Social role theory*			*Evolutionary theory*		*Theological*
						Origin	*Found*	*Biosoc*	*Select*	*ENA*	
Adult 21F All Ages 35F	100.0 100.0	-0-	7 (1)	1903–2004	-0-	Contra	Silent	Silent	Silent	NoTest	Silent

26.2.1.3 Estimating/Perceiving Environmental Hazards

Evidence. Most people are aware that the world faces climate change and other environmental problems, but is there a sex difference in how dangerous environmental hazards are perceived to be? As shown in Précis 26.2.1.3, females perceived environmental hazards to be more dangerous in 34 out of 35 studies, a conclusion also reached by one literature review. These studies were conducted in six different countries that were published over a 25-year time frame. The only study not reaching this result simply failed to find a significant sex difference.

Explanation. Since it is reasonable to assume that sex differences in perceiving environmental hazards involve substantial learning, the original social role theory would predict that these perceptions would vary from one culture to another. This prediction does not appear to be the case.

Elsewhere, we will discuss sex differences in risk-taking (which males tend to do more of than females), a sex difference that can be theoretically explained in evolutionary terms. Here it is worth mentioning that one of the reasons males take greater risks is that they simply perceive fewer environmental hazards than do females.

If ENA theory has bearing on perceptions of environmental hazards, one would expect to find brain exposure to testosterone inhibiting the perception of hazards. None was found.

Assessment. Overall, we conclude that the evidence is sufficiently strong to consider estimating or perceiving more hazards in one's environment to be a USD-F. This conclusion is primarily based on the number and consistency of studies, but also on the substantial number of countries sampled. Theories that have the greatest potential to explain this sex difference need further development and testing.

26.2.1.4 Left/Right Confusion

Evidence. Many people occasionally confuse their left and right when indicating where something is or which way one should go, but is there a sex difference in how often such mistakes are made? As shown in Précis 26.2.1.4, 23 out of 24 studies concluded that females exhibit left/right confusion more often than males. The remaining study failed to find a statistically significant sex difference. Studies on left/right confusion were conducted in eight different countries over a 45-year time frame.

Explanation. The fact that left/right confusion appears to be pan-cultural is contrary to the original social role theory. However, it is consistent with the hypothesis that males have been sexually selected for engaging in more hunting than females (Geary 1995; Joseph 2000: 44). One aspect of hunting often involves searching for and tracking prey over long distances, sometimes for days, and then finding one's way back home. If one assumes that

Précis 26.2.1.3 Estimating/perceiving environmental hazards (original Table 9.1.5.2)

Age categories, number of studies, and sex	*Consist. score*	*Overviews*	*Countries*	*Time range*	*Non-human*	*Social role theory*			*Evolutionary theory*		*Theological*
						Origin	*Found*	*Biosoc*	*Select*	*ENA*	
Adult 23F All Ages 34F 1x	100.0 97.1	F 1Rev	6	1982–2007	-0-	Contra	Silent	Silent	Silent	Silent	Silent

Précis 26.2.1.4 Left/right confusion (original Table 9.1.5.3)

Age categories, number of studies, and sex	*Consist. score*	*Overviews*	*Countries*	*Time range*	*Non-human*	*Social role theory*			*Evolutionary theory*		*Theological*
						Origin	*Found*	*Biosoc*	*Select*	*ENA*	
Adult 20F 1x All Ages 23F 1x	95.2 95.8	-0-	8	1973–2018	-0-	Contra	Silent	Silent	Poss-F	Poss-F	Silent

the ability to navigate has been sexually selected, and that left/right confusion might be a significant component in successful navigation, both evolutionary theories predict that males would be *less* likely to exhibit this type of confusion than females.

The ENA version of evolutionary theory would also lead one to expect neuroandrogenic exposure to help reduce right/left confusion. Evidence bearing on this possibility so far appears to be inconsistent (Hjelmervik, Westerhausen et al. 2015:204–205).

Assessment. Despite the lack of well-grounded theory responsible for sex differences in left/right confusion, the evidence that females do so more than males is strong. The only exception was a study of 12 males and 12 females which reported no significant sex difference. Overall, 23 studies derived from eight countries lead us to conclude that left/right confusion is a USD-F.

26.2.1.5 Athletic Ability/Performance in General

Evidence. Précis 26.2.1.5 shows the results of studies measuring general athletic ability and performance. All of the studies, and all five of the meta-analysis performed on this topic, found males have greater athletic ability. The relevant research was performed over eight decades and includes studies conducted in 15 different countries and four multi-national studies. An additional study of horses reached the same result.

Explanation. The sex difference in general athletic ability and performance is surely related to sex differences in muscularity and physical strength (see Précis 26.2.7 and Précis 26.2.8). The only theories that seem to provide an account for why the sex differences would differ in muscularity and strength are the two evolutionary theories. In this regard, both of these theories argue that males who reproduce most prolifically are those capable of competing for status that is attractive to the opposite sex. While successful competition obviously requires more than physical strength, it is often a significant component. This is consistent with studies that have found that females usually find physically strong males more attractive than their weaker rivals (Frederick & Haselton 2007; Lidborg, Cross, & Boothroyd 2020).

ENA theory adds to this basic evolutionary reasoning by hypothesizing that testosterone and other androgens help to promote muscularity, strength, and coordination, thereby boosting athletic potential. Supportive evidence is substantial, especially regarding prenatal testosterone exposure (Golby & Meggs 2011; DM Moffit & Swanik 2011; Schorer, Rienhoff et al. 2013; Trivers, Hopp & Manning 2013 running speed; Sudhakar, Majumdar et al. 2014; Ranson, Stratton & Taylor 2015 in both sexes; Meggs, Chen et al. 2019 in females). In the case of circulating testosterone, two studies also

Précis 26.2.1.5 Athletic ability/performance in general (original Table 9.3.2.4)

Age categories, number of studies, and sex	*Consist. score*	*Overviews*	*Countries*	*Time range*	*Non-human*	*Social role theory*			*Evolutionary theory*		*Theological*
						Origin	*Found*	*Biosoc*	*Select*	*ENA*	
Adolescent 18M Adult 12M	100.0 100.0	M: 5Met	15 (4)	1936–2015	1M horse	Silent	Silent	Silent	Expl-M	Expl-M	Silent

reported positive correlations with athletic ability (L Hansen, Bangsbo et al. 1999; Folland, McCauley et al. 2012).

Assessment. The consistency of the results for studies of sex differences in athletic ability and performance have been conducted in a large number of countries over many years, and is supported by five meta-analyses. This, plus the fact that it can be theoretically predicted, given us considerable confident in concluding that greater general athletic ability and performance is a USD-M.

26.2.1.6 Throwing Velocity or Distance

Evidence. Twenty-nine studies of throwing velocity or throwing distance (also known as *explosive throwing speed*) were located. All of the studies found males to be capable of throwing objects greater distances and/or with greater velocity. These studies, presented in Précis 26.2.1.6, were conducted in eight different countries, and the one meta-analysis on this topic reached the same conclusion.

Explanation. Assuming that this variable essentially reflects sex differences in physical strength (which favors males), it is possible to predict that males would be able to throw longer distances and/or at greater speed. As noted earlier in this chapter (Précis 26.1.2.7 and Précis 26.1.2.8), males have been shown to be physically stronger than females. Therefore, it is reasonable to expect that they will be able to throw projectiles at higher velocities. Especially in pre-agrarian societies, throwing ability could have been sexually selected because it enabled males to kill wild animals for food and clothing to a greater degree (Geary 1995; Joseph 2000: 44). It could also have enabled males to more effectively compete with rivals for access to mates (Daly 2017).

Regarding ENA theory, throwing velocity seem to be promoted by exposing the brain to androgens. This appears to be true, at least regarding postpubertal testosterone exposure (Gorostiaga, Izquierdo et al. 1999; Jäncke 2018).

Assessment. The consistency with evolutionary theories, number of countries where the trait has been studied and the support of a meta-analysis makes us confident in declaring the sex differences in throwing ability a USD-M.

26.2.1.7 Jumping Distance

Evidence. Can one sex jump further than the other? Précis 26.2.1.7 shows that males jumped greater distances in 16 studies conducted in seven different countries. A multi-national study reached the same conclusion. The studies were published over a span of nearly 80 years.

Précis 26.2.1.6 Throwing velocity or distance (original Table 9.3.2.12)

Age categories, number of studies, and sex	*Consist. score*	*Overviews*	*Countries*	*Time range*	*Non-human*	*Social role theory*			*Evolutionary theory*		*Theological*
						Origin	*Found*	*Biosoc*	*Select*	*ENA*	
Child 11M All Ages 29M	100.0 100.0	M: 1Met	8	1947–2019	-0-	Silent	Silent	Silent	Expl-M	Expl-M	Silent

Précis 26.2.1.7 Jumping distance (original Table 9.3.3.6)

Age categories, number of studies, and sex	*Consist. score*	*Overviews*	*Countries*	*Time range*	*Non-human*	*Social role theory*			*Evolutionary theory*		*Theological*
						Origin	*Found*	*Biosoc*	*Select*	*ENA*	
All Ages 16M	100.0	-0-	7 (1)	1940–2018	-0-	Silent	Silent	Silent	Expl-M	Expl-M	Silent

Explanation. The sex difference in jumping distance is surely related to sex differences in muscularity and physical strength (Précis 26.2.7–26.2.8). The only theories that seem to provide accounts of why the sex differences in muscularity and strength responsible for the sex difference in jumping distance would be the two evolutionary theories. In this regard, both of these theories argue that males who reproduce the best tend to be those capable of competing for sufficient status that will ensure fairly stable supplies of resources. This position is consistent with studies that have found that females usually prefer relatively strong males over their weaker rivals (Frederick & Haselton 2007; Lidborg, Cross, & Boothroyd 2020).

ENA theory adds to this basic evolutionary reasoning the notion that testosterone and other androgens are the hormonal mechanisms that make muscularity and strength possible. Supporting this reasoning, several studies have shown that male sex hormones appear to promote both of these physical traits (Hansen, Bangsbo et al. 1999; Folland, McCauley et al. 2012).

Assessment. Given the consistency with the evolutionary theories, the number of countries that have been studied, and the length of time covered by those studies, it appears safe to conclude that the ability to jump greater distances is a USD-M.

26.2.1.8 Jumping Height

Evidence. As shown in Précis 26.2.1.8, all 24 of the studies, and two literature reviews, on sex differences in jumping height reached the conclusion that males can jump higher. This sex difference was found in 12 countries and published over a span of almost 80 years.

Explanation. Like the previously discussed sex difference in jumping distance, the ability of males to jump higher is surely related to sex differences in muscularity and physical strength (Précis 26.2.7–26.2.8). The only theories that seem to provide accounts of why the sex differences in muscularity and strength responsible for the sex difference in jumping distance would be the two evolutionary theories. In this regard, both of these theories argue that males who reproduce the best tend to be those capable of competing for sufficient status that will ensure fairly stable supplies of resources. This position is consistent with studies that have found that females usually prefer relatively strong males over their weaker rivals (Frederick & Haselton 2007; Lidborg, Cross, & Boothroyd 2020).

Regarding the possible involvement of testosterone, as hypothesized by ENA theory, the evidence is also supportive, provided one assumes that muscularity is a major contributor to jumping height. In this regard, circulating levels of testosterone has been shown to promote muscular enlargement and physical strength (Bhasin, Storer et al. 1996; Iannuzzi-Sucich,

Précis 26.2.1.8 Jumping height (original Table 9.3.3.7)

Age categories, number of studies, and sex	*Consist. score*	*Overviews*	*Countries*	*Time range*	*Non-human*	*Social role theory*			*Evolutionary theory*		*Theological*
						Origin	*Found*	*Biosoc*	*Select*	*ENA*	
Adult 21M All Ages 24M	100.0 100.0	M: 2Rev	12	1978–2019	-0-	Silent	Silent	Silent	Expl-M	Expl-M	Silent

Prestwood, & Kenny 2002; Auyeung, Lee et al. 2011). Even experimental studies among elderly males have indicated that testosterone patches in which the hormone is steadily released into the body promote muscular growth and strength (Urban, Bodenburg et al. 1995; Snyder, Peachey et al. 1999).

Assessment. Given the consistency with the evolutionary theories, the number of countries that have been studied, and the length of time covered by those studies, it is safe to conclude that the ability to jump greater distances is a USD-M.

26.2.1.9 Handwriting Legibility/Penmanship

Evidence. Penmanship can also be an indication of fine-motor skills. Is there a sex difference in the legibility of penmanship? Seventeen studies on this topic are summarized in Précis 26.2.1.9. These studies were conducted in four different countries and there were also two multi-national studies. The studies were published over 85 years. Sixteen out of the 17 studies found that females had more legible handwriting. Although the consistency score of 94.12 just missed the 95.0 cutoff point, we include this variable because the one exception was a study of only 56 research participants that also found that females had better penmanship, but the difference was not statistically significant.

Explanation. The two evolutionary explanations are the only ones that appear to be consistent with this explanation. Specifically, keeping in mind the difference between gross-motor skills and fine-motor skills, many evolutionists have recognized that males have evolved gross-motor skills in order to fascinate intra-male aggression and hunting as well as other strength-dependent activities (Puts 2010; Hill, Hunt et al. 2013). Females, on the other hand, may have developed better fine-motor skills, such as those involving the care of infants and the harvesting and cleaning of edible plants (Sanders 2013). These basic sex differences in motor coordination could help to explain sex differences in handwriting legibility, which is obviously a fine-motor skill.

Regarding the ENA version of evolutionary theory, one study reported a positive correlation between 2D:4D ratio (a rough indicator of *low* exposure to prenatal testosterone) and fine-motor skills (Wang, Wang et al. 2016). While greater strength in leg and arm muscles are likely to be associated with hunting and warfare, the fine-motor skills associated with penmanship may be better in females as a result of females performing more subsistence tasks involving fine-motor skills, such as gathering and preparing food for consumption (Sanders 2013).

Assessment. Overall, we assess handwriting legibility to be a USD-F. Both evolutionary theories offer reasonable explanations.

Précis 26.2.1.9 Handwriting legibility/penmanship (original Table 9.3.4.3)

Age categories, number of studies, and sex	*Consist. score*	*Overviews*	*Countries*	*Time range*	*Non-human*	*Social role theory*			*Evolutionary theory*		*Theological*
						Origin	*Found*	*Biosoc*	*Select*	ENA	
All Ages 16F 1x	94.12	-0-	4 (2)	1931–2016	-0-	Silent	Silent	Silent	Poss-F	Expl-F	Silent

26.2.1.10 Running Speed

Evidence. In addition to five multi-national studies, sex differences in running speed have been measured in 34 countries. Précis 26.2.1.10 reveals all of these studies, and two literature reviews, have found males to be faster runners.

Explanation. The ability of males to run faster is substantially associated with sex differences in muscularity and physical strength (Précis 26.2.7 and Précis 26.2.8). Only the two evolutionary explanations would appear to account for difference in running speed. Further, there would have been obvious evolutionary advantages to higher running speeds in the predominantly male activities of warfare and large animal hunting.

In the case of ENA theory, which asserts that higher testosterone among males contributes to their ability to run faster, especially over long distances, one supportive finding was located. It indicated that prenatal testosterone exposure (as inferred from 2D:4D data) was positively correlated with endurance running speed (JT Manning, Morris & Caswell 2007).

Assessment. Given the consistency with the evolutionary theories, the very large number of countries that have been studied, and the length of time covered by those studies, it appears safe to conclude that faster running speed is distances is a USD-M.

26.2.1.11 Vocalizing Behavior

Evidence. Précis 26.2.1.11 summarizes findings from 46 studies of sex differences in vocalizing behavior. Among infants and toddlers, vocalizations often take the form of babbling, but with age, they are primarily in the form of spoken language. One can see that only among the 11 studies of infants and toddlers was the consistency score beyond 95.0. All of these studies found that females vocalized more than do males.

Explanation. Of the five theories of sex differences, only the two evolutionary theories seem to bear on why females would vocalize more as infants and toddlers than do males. Both theories recognize that language usage has been one of the main traits that has made humans so numerically successful when compared to any other large mammal (Pinker & Bloom 1990; Nowak & Krakauer 1999; Gillespie-Lynch, Greenfield et al. 2014). But, why would girls begin vocalizing and eventually acquiring language sooner than boys? At least part of the reason is that girls have been shown to mature more rapidly than boys (see Précis 26.1.1.3).

Regarding the possible involvement of testosterone, which ENA theory asserts should retard vocalizing tendencies, as well as early language development, the evidence is supportive. Specifically, three studies have found testosterone levels in the saliva of 2- to 4-month-olds negatively correlated with various measures of language development (e.g., number

Précis 26.2.1.10 Running speed (original Table 9.3.5.1)

Age categories, number of studies, and sex	*Consist. score*	*Overviews*	*Countries*	*Time range*	*Non-human*	*Social role theory*			*Evolutionary theory*		*Theological*
						Origin	*Found*	*Biosoc*	*Select*	*ENA*	
Adult 20M All Ages 29M	100.0 100.0	M: 1Met	8 (11)	1947–2018	-0-	Silent	Silent	Silent	Expl-M	Expl-M	Silent

Précis 26.2.1.11 Vocalizing behavior (original Table 9.3.7.1)

Age categories, number of studies, and sex	*Consist. score*	*Overviews*	*Countries*	*Time range*	*Non-human*	*Social role theory*			*Evolutionary theory*		*Theological*
						Origin	*Found*	*Biosoc*	*Select*	*ENA*	
Infant/Toddler 11F All Ages 28F 17x 1M	100.0 (70.0)	F: 1Rev	3	1930–1999	-0-	Silent	Silent	Silent	Poss-M	Poss-M	Silent

of words spoken) by the time they were toddlers (Saenz & Alexander 2013; Schaadt, Hesse & Friederici 2015; Kung, Browne et al. 2016).

Assessment. The consistency with evolutionary explanations and the high consistency score makes it likely that earlier vocalization is a USD-F.

26.2.1.12 Fundamental Frequency (F_0) *When Vocalizing*

Evidence. Fundamental frequency (F_0) refers to the depth of one voice. As shown in Précis 26.2.1.12, all 41 studies of adults indicate that males have a lower tone of voice than do females. This sex difference was found in 11 countries and was consistent with a literature review.

Explanation. Only the two evolutionary theories address sex differences in fundamental frequency. From the standpoint of sexual selection, males appear to have been favor for exhibiting competence at resource procurement and retention. This includes signs of dominance and intimidation, especially when confronting other males (Evans, Neave & Wakelin 2006; Hodges-Simeon, Gaulin, & Puts 2011; Hodges-Simeon, Gurven & Gaulin 2015). Females have been shown to prefer mating with males with relatively low deep voices (Feinberg, Jones et al. 2005; Re, O'Connor et al. 2012). Also, fundamental frequency appears to be positively correlated with social status (Pittam 1990). One study reported that it was also positively correlated with the number of offspring fathered (Rosenfeld, Sorokowska et al. 2020).

Consistent with evolutionary reasoning, research has found that, throughout the primate order, males have lower fundamental frequencies than do females, especially among species who tend to mate polygamously (rather than monogamously). In this regard, human sex differences in F_0 are among the most extreme of all hominids (Puts, Hill et al. 2016:4).

Evidence pertaining to the ENA version of evolutionary theory comes from the same research report just cited. It indicated that, at least for males, testosterone was inversely correlated with fundamental frequency, suggesting that testosterone helps to make male voices more attractive to females (Hodges-Simeon, Gurven & Gaulin 2015; Puts, Hill et al. 2016:6). As far as sexual selection is concerned, this increased attractiveness to females, like other traits such as height and muscularity, may be due to relatively low deep voices being indicative of dominance. At least two studies have indicated that circulating testosterone is associated with lower F_0 for both sexes (Evans, Neave et al. 2008; Raj, Gupta et al. 2010).

Assessment. At least among adults, having a high-pitched voice can be considered a USD-F. This difference is presumably due to the tendency for females to prefer mating with males with relatively low-pitched voices.

Précis 26.2.1.12 Fundamental frequency (F_0) when vocalizing (original Table 9.3.7.4)

Age categories, number of studies, and sex	*Consist. score*	*Overviews*	*Countries*	*Time range*	*Non-human*	*Social role theory*			*Evolutionary theory*		*Theological*
						Origin	*Found*	*Biosoc*	*Select*	*ENA*	
Adult 41F	100.0	F: 1Rev	11	1950–2014	-0-	Silent	Silent	Silent	Expl-M	Expl-M	Silent

26.2.1.13 Nasality of Voice During Speech

Evidence. When air travels through the nose during speech, it is said to produce a relatively high-pitched nasal sound. Précis 26.2.1.13 shows that the 11 available studies all found females to have more of this nasal sound when they talk than do males. However, these studies were conducted in only two countries and published over a relatively short period of time.

Explanation. None of the three social role theories seem to have any bearing on evidence that females have more nasality of voice than males. Both evolutionary theories, on the other hand, lead one to expect that the sexes should differ in this regard. This prediction comes from noted that males have been favored for exhibiting dominance to a greater degree than females (Buss & Kenrick 1998: 983; Weisfeld & Dillon 2012: 25). Voice nasality appears to be one of the indicators of *non*-dominance. For evidence of a related nature, see Précis 26.1.2.10 on vocal cord size and Précis 26.2.1.12 on fundamental frequency.

In addition to the above sexual selection arguments for less nasality in male voices, ENA theory also asserts that androgens should be involved. No evidence directly linking androgen exposure and voice nasality was found.

Assessment. While the number of countries sampled needs to be enlarged, and the theoretical explanations are indirect, the consistency of the findings warrants categorization nasality of voice during speech a USD-F.

26.2.2 Emotional Factors

Chapter 10 of this book had to do with sex differences in emotions, including emotional expressions. In the section below, findings of sex difference variables with high consistency scores are identified with regard to the nature of the evidence and the ability of various theories to explain why those sex differences might be found in virtually all human societies.

26.2.2.1 Being Vengeful or Spiteful

Evidence. Are the members of one sex more likely to be vengeful or spiteful than the other sex? Précis 26.2.2.1 shows that evidence from 11 studies from four different countries all agreed that such tendencies are more common or intense among males than among females.

Explanation. The greater vengeful or spiteful behavior exhibited by males does not appear to be explained by any of the five scientific theories nor by any theological passages with which we are familiar. The only possible theoretical explanation would involve assuming that these behaviors are associated with competitiveness, a trait that appears to be more common among males than among females (see Précis 26.3.1.5). In other words, perhaps feelings of vengefulness and spite are part of the emotional reactions

Précis 26.2.1.13 Nasality of voice during speech (original Table 9.3.7.6)

Age categories, number of studies, and sex	*Consist. score*	*Overviews*	*Countries*	*Time range*	*Non-human*	*Social role theory*			*Evolutionary theory*		*Theological*
						Origin	*Found*	*Biosoc*	*Select*	*ENA*	
All Ages 11F	100.0	-0-	2	1970–1996	-0-	Silent	Silent	Silent	Expl-M	NoEvid	Silent

Précis 26.2.2.1 Being vengeful or spiteful (original Table 10.2.1.5)

Age categories, number of studies, and sex	*Consist. score*	*Overviews*	*Countries*	*Time range*	*Non-human*	*Social role theory*			*Evolutionary theory*		*Theological*
						Origin	*Found*	*Biosoc*	*Select*	*ENA*	
All Ages 11M	100.0	-0-	4	1992–2017	-0-	Silent	Silent	Silent	Silent	Silent	Silent

that motivate competitive behavior. If this assumption is true, then it may be possible to eventually explain greater vengefulness and spitefulness among males in evolutionary terms.

Assessment. Evidence strongly suggests that vengefulness and spitefulness are USD-M. Therefore, even though little in the way of theoretically understanding this sex difference is currently available, we will consider such emotion responses to be universal.

26.2.2.2 Fearfulness in General

Evidence. All but one of 97 studies of sex differences in fearfulness concluded that females were more fearful than males. These studies drew samples from 16 countries (along with two multiple-country studies) and were published over an 86-year time period. Précis 26.2.2.2 shows that this sex difference was also reported by two literature reviews. The lone exception was a study of toddlers that reported boys being more fearful of strangers than were girls.

Explanation. The best explanations of sex differences in fearfulness seem to be those that are evolutionary in nature. Specifically, fearfulness can obviously be both beneficially and detrimental from a reproductive standpoint, depending on the severity of the threat and possibly on which sex is being threatened.

Research has shown that, within both sexes, physical strength is inversely correlated with fearfulness, i.e., individuals who are physically stronger tend to be less fearful and anxious than those with less physical strength (Kerry & Murray 2021; Manson, Chua et al. 2022). This suggests that individuals who are relatively strong physically will be less fearful of being harmed, all of which could promote risk-taking and competitiveness, thereby benefiting males in reproductive terms. It is also worth noting that testosterone exposure has been shown to promote physical strength (see Précis 26.1.7.8). Furthermore, one study reported that administration of one dose of testosterone reduced various symptoms of fearfulness (Hermans, Putman et al. 2007). Overall, ENA theory appears helpful for understanding sex differences in fearfulness.

Assessment. Overall, numerous studies conducted in many countries, spanning over eight decades, all point toward fearfulness being a USD-F. Evolutionary theories seem to offer the best explanations for this difference.

26.2.2.3 Fear of Animals

Evidence. Turning to studies of fairly specific types of fear, Précis 26.2.2.3 shows that all 12 studies of the fear of animals found this type of fear to be higher among females. The pertinent studies were conducted in three

Précis 26.2.2.2 Fearfulness in general (original Table 10.2.2.2)

Age categories, number of studies, and sex	*Consist. score*	*Overviews*	*Countries*	*Time range*	*Non-human*	*Social role theory*			*Evolutionary theory*		*Theological*
						Origin	*Found*	*Biosoc*	*Select*	*ENA*	
All Ages 96F 1M	98.0F	F: 2Rev	16 (2)	1928–2014	Complex	Silent	Silent	Silent	Poss-F	Poss-F	Silent

Précis 26.2.2.3 Fear of animals (original Table 10.2.2.4)

Age categories, number of studies, and sex	*Consist. score*	*Overviews*	*Countries*	*Time range*	*Non-human*	*Social role theory*			*Evolutionary theory*		*Theological*
						Origin	*Found*	*Biosoc*	*Select*	*ENA*	
All Ages 12F	100.0	F: 1Rev	3 (1)	1983–2012	-0-	Silent	Silent	Silent	Poss-F	Poss-F	Silent

countries along with one multi-national study, all published over a 31-year time frame. One literature review also reported that females were more fearful of animals than were males.

Explanation. The lesser size and strength of females is likely related to females having a greater fear of animals. Thus, both evolutionary explanations are consistent with this finding. However, studies finding the same sex difference in additional countries is required before confidently claiming the greater fear of animals is a USD-F.

Assessment. Studies in more countries are required before confidently concluding that greater fear of animals is a USD-F.

26.2.2.4 Fear of Crime Victimization

Evidence. Over 100 studies were located regarding sex differences in the fear of becoming a victim of crime. Two studies found no sex difference but the remaining studies found females were more fearful of crime victimization. These studies were conducted in eight countries and there was one multi-national study. As shown in Précis 26.2.2.4, the studies were published over a 46-year time frame.

Explanation. An evolutionary explanation for females exhibiting greater fear of crime victimization has been proposed (Fetchenhauer & Buunk 2005). These authors proposed that "gender differences may be the result of sexual selection that favored risk-taking and status fights among males, and being cautious and protecting one's offspring among females" (p. 95). Basically, this explanation fits with evidence reviewed above (see Précis 26.2.2.2), indicating that females have been sexually selected for higher fearfulness than males.

If ENA theory is true, research should find high brain exposure to testosterone positively correlated with fear of crime victimization. No specific test of this hypothesis was located.

Assessment. The large number of studies, the number of countries where those studies have taken place, and the consistency with the evolutionary theories lead us to conclude that the fear of crime victimization is a USD-F.

26.2.2.5 Fearing Being Negatively Evaluated by Others

Evidence. There is probably no one who likes being negatively evaluated by others, but is there a sex difference in the amount of fear experienced by the possibility of such an evaluation? Précis 26.2.2.5 shows that females had a greater fear of being negatively evaluated by others in all ten studies located on this topic. The studies were published over a 70-year time frame. One one literature review on the topic also concluded that females have such fear to a greater degree than males.

Précis 26.2.2.4 Fear of crime victimization (original Table 10.2.2.12)

Age categories, number of studies, and sex	*Consist. score*	*Overviews*	*Countries*	*Time range*	*Non-human*	*Social role theory*			*Evolutionary theory*		*Theological*
						Origin	*Found*	*Biosoc*	*Select*	*ENA*	
WAR 45F All Ages 57F 2x	100.0 96.6	-0-	8 (1)	1974–2020	-0-	Silent	Silent	Silent	Poss-F	Poss-F	Silent

Précis 26.2.2.5 Fearing being negatively evaluated by others (original Table 10.2.2.17)

Age categories, number of studies, and sex	*Consist. score*	*Overviews*	*Countries*	*Time range*	*Non-human*	*Social role theory*			*Evolutionary theory*		*Theological*
						Origin	*Found*	*Biosoc*	*Select*	*ENA*	
All Ages 10F	100.0	F: 1Rev	1	1935–2005	-0-	Silent	Silent	Silent	Poss-F	Poss-F	Silent

Explanation. Tendencies for females to fear being negatively evaluated by others to a greater degree than males could have something to do with females being more socially oriented (see Précis 26.3.2.1) and having a stronger desire to cooperate with others (see Précis 26.3.2.23). A social orientation, in turn, could have been sexually selected for its tendency to promote prolonged child care. If so, it can be explained in terms of sexual selection.

Assessment. The relatively small number of studies, and their limitation to only one country, and the lack of a clear connection to an explanation, does not give us great confidence in fear of being negatively evaluated by others as being a USD. Also, the pertinence of any of the five theories for explaining why females would exhibit this trait more warrants more development.

26.2.2.6 Sadness (Dysphoria) in General

Evidence. Does one sex experience greater sadness than the other? Précis 26.2.2.6 shows that the 14 studies located on this topic all found that females experiencing greater sadness than males. Most of these studies were conducted in one of two countries, but there were also two multinational studies. Additional evidence on this matter can be gleaned from studies of depression (see Précis 26.2.6.3).

Explanation. All three social role theories can explain sex differences in sadness by simply noting that eight studies have reported that females are *stereotyped* as being sad and depressed more than males (see Table 21.2.6.11). However, none of these theories offer an explanation for *why* females would be stereotyped this way (unless, of course, one was to assume that the stereotype reflects reality). And, if the stereotype is true, one is brought back to the original question: Why are females are more often sad or depressed than are males.

Various evolutionary explanations for sadness and depression being more common in females have been put forth. Most of these explanations center around asserting that females have evolved tendencies to respond to unpleasant experiences less energetically than is the case for males (Welling 2003: 149–152; Hagen & Rosenström 2016). In other words, whereas males tend to respond to environmental challenges in assertive and aggressive ways, females do so by acceptance and withdrawal. These differing types of responses presumably reflect the fact that males are more risk-prone in their approach to reproduction than are females (MD Baker & Maner 2008; Pawlowski, Atwal & Dunbar 2008). Of course, the consequences to aggressive types of responses to environmental challenges are risks of retaliation and escalating dangers, while the repercussions for accepting and withdrawal responses are sadness and depression.

Précis 26.2.2.6 Sadness (dysphoria) in general (original Table 10.2.4.1)

Age categories, number of studies, and sex	*Consist. score*	*Overviews*	*Countries*	*Time range*	*Non-human*	*Social role theory*			*Evolutionary theory*		*Theological*
						Origin	*Found*	*Biosoc*	*Select*	*ENA*	
All Ages 14F	100.0	F: one Rev	2 (2)	1986–2009	-0-	Circ-F	Circ-F	Circ-F	Expl-F	Expl-F	Silent

Regarding the possible involvement of androgens, as would by hypothesized by ENA theory, the evidence is mixed. Indirect support comes from studies showing that physical strength (which is promoted by muscular exposure to testosterone) has been repeatedly found to be negatively correlated with depression, even within each sex (van Milligen, Lamers et al. 2011; Suija, Timonen et al. 2013; Fukumori, Yamamoto et al. 2015; Hagen & Rosenström 2016: Figure 2). Also, a study of hypogonadal males in which they were intermittently administered synthetic testosterone in a nasal gel indicated that the drug resulted in a significantly elevated mood (Rogol, Tkachenko & Bryson 2016). However, other evidence of testosterone being able to reduce negative affect has been equivocal (Rohr 2002; Seney & Sibille 2014).

Assessment. It is interesting to note that despite sadness and depression being more prevalent among females than males, studies of sex differences in self-reported happiness are quite inconsistent (see Table 13.2.1.2). Why nearly all studies have concluded that females are sad and depressed more while the sexes appear to be nearly equal in terms of happiness is certainly a conundrum.

Overall, the sex difference in sadness is a USD-F. This difference is accentuated by noting that general and clinical depression are also higher for females, at least among adults (see Tables 14.2.1.1 and 14.2.1.3).

26.2.2.7 Stress/Anxiety Associated with Providing Care to Others

Evidence. Précis 26.2.2.7 provides a summary of 19 studies having to do with the amount of stress and anxiety resulting from providing care to others. All of the studies found females experiencing more stress and anxiety from this cause. These studies were conducted in two different studies, and were published over a 37-year time period. One meta-analysis also concluded that females surpassed males in this regard.

Explanation. Several theorists have proposed that the greater tendencies for females to feel stress and anxiety have evolutionary roots (Nesse 1999; Palanza 2001; Troisi 2001). The reasoning behind this suggestion is based in part on the fact that males have been favored for being more competitive and prone to take risks, both traits that feelings of stress and anxiety would often discourage. Nevertheless, no specific application of the sexual selection version of evolutionary theory was found applied specifically to stress associated with providing care to others.

ENA theory specifically predicts that any trait for which there are universal sex differences results from brain exposure to androgens. It does so in part because cortisol, a stress hormone, tends to be associated with lower testosterone (Bedgood, Boggiano & Turan 2014: 114). Only a limited amount of evidence was located that specifically tested the hypothesis that

Précis 26.2.2.7 Stress/anxiety associated with providing care to others (original Table 10.2.5.11)

Age categories, number of studies, and sex	*Consist. score*	*Overviews*	*Countries*	*Time range*	*Non-human*	*Social role theory*			*Evolutionary theory*		*Theological*
						Origin	*Found*	*Biosoc*	*Select*	*ENA*	
Adult 19F All Ages 19F	100.0 100.0	F: 1Met	2	1961–1998	-0-	Silent	Silent	Silent	Poss-F	Poss-F	Silent

that stress and anxiety are inversely correlated with testosterone, none of which directly pertaining to providing care to others. Specifically, one study reported that males with low exposure to prenatal testosterone (as assessed using the 2D:4D ratio) reported more feelings of anxiety than did males with low exposure, while no such relationship was found among females (Evardone & Alexander 2009). In another study, circulating levels of testosterone were found to be negatively correlated with self-reported feelings of anxiety among females, although there was no significant relationship for males (Giltay, Enter et al. 2012).

Assessment. Overall, sex differences in feeling of stress and anxiety associated while providing care to others has a reasonable theoretical explanation and is well documented, although collecting data from more than two countries would be desirable. Currently, all indications are that this is a USD-F.

26.2.2.8 Stress/Anxiety Associated with Technology

Evidence. Many people experience anxiety and stress when they encounter some unfamiliar form of technology with which they need to interact. A particularly common example involves computers. Sex differences in this regard are summarized in Précis 26.2.2.8. One can see that all but one of the 28 studies on this topic found that females experienced a higher level of stress or anxiety when attempting to utilize an unfamiliar form of technological device than males. The remaining study found no significant sex difference. The studies were based on samples of adults (often college students) residing in five different countries, and were published over a 21-year time span. Two meta-analyses also found females had a higher level of such stress than did males.

Explanation. Why would the sexes differ regarding the amount of stress they experience dealing with technology, particular computers? The best proposals have been along evolutionary lines. Regarding sexual selection theory, it appears that males have been favored for spending time interacting with physical, non-social, environmental objects more than has been the case for females (see Précis 26.2.7.44). Such a focus is likely to promote learning about mechanical and technological devices more by males than by females. As a result, males are likely to prefer "things-oriented" educational and occupational fields that will females, including computer-related occupations (see Précis 26.3.5.23).

Turning to ENA theory, it hypothesizes that brain exposure to testosterone should help to prevent feelings of stress when confronting and interacting with complex technology. At least one study has provided support for this hypothesis, particularly regarding prenatal testosterone exposure. Specifically, as measured by 2D:4D, increased testosterone exposure was

Précis 26.2.2.8 Stress/anxiety associated with technology (original Table 10.2.5.14)

Age categories, number of studies, and sex	*Consist. score*	*Overviews*	*Countries*	*Time range*	*Non-human*	*Social role theory*			*Evolutionary theory*		*Theological*
						Origin	*Found*	*Biosoc*	*Select*	*ENA*	
Adult 27F 1x	96.4	F: 2Met	5	1985–2006	-0-	Silent	Silent	Silent	Poss-F	Poss-F	Silent

found inversely correlated with computer anxiety among males, but not significantly so among females (Evardone & Alexander 2009).

Assessment. Studies over a greater time-period and more types of technology are required before concluding that greater stress associated with technology is a USD-F. Theoretical efforts to explain these differences are minimally developed.

26.2.2.9 Feelings of Empathy

Evidence. People differ in the extent to which they are able to feel the experiences of others; tendencies that are known as empathy. Well over 100 studies of sex differences in empathy have been published. As shown in Précis 26.2.2.9, the findings have been somewhat mixed, although, out of the 127 studies located, 18 reported no significant sex difference, but not a single study reported that males surpassed females to a significant degree.

The available studies of sex differences in empathy were conducted in 12 different countries (plus three additional studies of multiple countries). A literature review and a meta-analysis also found females to be more empathetic, while another literature review failed to find a sex difference.

Explanation. While no research specifically pertaining to *empathy* as a sex stereotype was found, females do appear to be stereotyped as more soft-hearted (Table 21.2.6.23) and sympathetic (Table 21.2.6.24), both concepts close to that of empathy. According to all three versions of social role theory, females learn to conform to empathy-like stereotypes as they grow up. The original social role theory implies that this should not be true in all societies, while the founder effect version and the biosocial version both imply that sex differences in this regard could be universal. Regarding the founder effect version, there is no way of testing it without any empirical access to the very first human societies. In the case of the biosocial version, it is difficult to see the relevance of body size or strength (in males) or the fact that females alone can bear offspring.

Turing to the evolutionary perspective, sexual selection theory implies that empathy promotes female reproduction more than male reproduction (Josephs & Shimberg 2010). In this regard, a general tendency to empathize could promote caregiving, a possibility that has sometimes been termed the *primary caretaker hypothesis* (Hames, Babchuk & Thomason 1985; Christov-Moore, Simpson et al. 2014).

According to the ENA version of evolutionary theory, not only should females have evolved greater tendencies to be empathetic than males, but neuroandrogenic factors should be contributing to this sex difference. Findings have provided support for this conclusion by showing that testosterone diminishes tendencies to be empathetic. This was found in several studies of prenatal testosterone exposure (Auyeung, Baron-Cohen

Précis 26.2.2.9 Feelings of empathy (original Table 10.2.8.3)

Age categories, number of studies, and sex	*Consist. score*	*Overviews*	*Countries*	*Time range*	*Non-human*	*Social role theory*			*Evolutionary theory*		*Theological*
						Origin	*Found*	*Biosoc*	*Select*	*ENA*	
Adolescent 19F 1x All ages 109F 18x	95.0 (85.8)	F: 1Rev F: 1Met; x: 1Rev	12 (3)	1936–2019	-0-	Contra	NoTest	Silent	Expl-F	Expl-F	Silent

et al. 2006; Chapman, Baron-Cohen et al. 2006; Wakabayash & Nakazawa 2010; Terburg & van Honk 2013: Figure 3; Nitschke & Bartz 2020), although in one 2D:4D study, the pattern only held for males, not for females (Pickering, Anger et al. 2022: Figure 3). Regarding post-pubertal circulating exposure, high testosterone was also found associated with diminished empathy (Harris, Rushton et al. 1996; Hermans, Putman & van Honk 2006; Turan, Guo et al. 2014; Nitschke & Bartz 2020). Research has even indicated that empathy can be experimentally reduced by exposing research participants to elevated levels of testosterone (*review*: Bos, Panksepp et al. 2012). Experimental testosterone injections were also found to reduce people's tendencies to be generous (Ou, Wu et al. 2021).

In terms of qualifying the above findings, some studies point toward testosterone being more inversely correlated with empathy among females than among males. One experiment with adult women found that administration of testosterone caused a reduction in empathy (Hermans, Putman & van Honk 2006). However, a similar experiment with adult males found no significant effects (Nadler, Camerer et al. 2019). Likewise, among children, testosterone levels were inversely correlated with empathy among females but not correlated among males (Pascual-Sagastizabal, Azurmendi et al. 2013; Pascual-Sagastizabal, Del Puerto et al. 2019).

Assessment. The majority of studies have concluded that females are more empathetic than are males. All of the exceptions have simply failed to find statistically significant sex differences, possibly due in part to difficulties in precisely measuring empathy. While the only consistency score attaining the 95.0 cutoff was for the 20 studies of adolescents, we declare empathy a USD-F in light of the general consistency of the evidence for all age groups combined, the large number of countries that have been sampled over extended periods of time, and the strength of the supportive theoretical arguments along evolutionary lines. Also, all but one of the overviews concluded that females are more empathetic than females.

26.2.2.10 Crying

Evidence. Crying appears to be one of the ways that emotions are expressed by humans throughout the world. The emotions expressed by crying are nearly always unpleasant ones among infants and children. Among adolescents and adults, however, pleasant emotions, such as those associated with listening to inspiring music or a heart-warming story, are sometimes accompanied by crying.

Sex differences in crying have been studied over many years and in many different cultures, and a sex difference in the amount of crying has been found. Précis 26.2.2.10 shows that 70 out of 71 studies conducted among adults found that females cry more than males, with the remaining

Précis 26.2.2.10 Crying (original Table 10.2.9.5)

Age categories, number of studies, and sex	*Consist. score*	*Overviews*	*Countries*	*Time range*	*Non-human*	*Social role theory*			*Evolutionary theory*		*Theological*
						Origin	*Found*	*Biosoc*	*Select*	*ENA*	
Adult 70F 1x All Ages 88F 11x 8M	98.6 (76.5)	F: 4Rev	36 (3)	1934–2013	-0-	Contra	NoTest	Poss-F	Poss-F	Poss-F	Silent

study failing to find a sex difference. It is also worth noting that 36 countries were sampled in these studies plus three multi-country studies. Four literature reviews also concluded that females cry more often.

Explanation. Several researchers have noted that some, but not all, crying can be considered as a way of influencing others, such as when children cry when hungry or under other forms of stress (Vingerhoets & Scheirs 2000: 143). However, some crying is not vocally expressed and can occur simply in response to both unpleasant as well as pleasant experiences. In other words, the circumstances under which humans cry appear to be quite varied, especially in the case of adults (Vingerhoets & Bylsma 2016).

At least three studies have found that females are stereotyped as being more prone to cry than males (see Table 21.2.6.10). This being the case, one can say that all three social role theories can explain sex differences in crying as being due to sex stereotypes. This prediction comes from social role theories considering cultural norms, which are often reflected by stereotypes, are responsible for variations in people's behavior.

Regarding the two evolutionary theories, it is obvious that crying could have been naturally selected as a way for infants to draw attention to themselves when hungry or in danger (Lummaa, Vuorisalo et al. 1998; CJ Lane 2006). One might call this the cry-for-help hypothesis, and one would expect no sex differences in this regard. Examination of Table 10.2.9.5 reveals that the evidence for sex differences in crying among infants and toddlers is quite mixed. Studies of adolescents and especially adults, however, have strongly indicated that females cry more than do males (Hastrup, Kraemer et al. 2001). So, why would such a change occur?

Perhaps, due to the greater size and strength of males following puberty, their ability to use crying in order to satisfy their wants and needs is likely to take second place to the use of intimidation and physical power. Therefore, their tendencies to cry for help will be lower than for females, particularly after reaching puberty. Put another way, individuals who want something that is not readily available to them can either threaten to use force or seek it through pity. Especially after puberty, males may gravitate toward the use of force (dominance) while females persist in using non-dominant methods when stressed such as crying.

If the above reasoning is correct, one can surmise that neurohormonal factors are likely to be involved, as hypothesized by ENA theory. The evidence pertaining to this hypothesis is currently weak. Specifically, while considerable research pertaining to how infant crying affects sex hormones levels among adults (especially parents) have been conducted (Bos, Hechler et al. 2018), no evidence was found bearing on the possibility of post-pubertal levels of androgens being associated with the probability of crying.

Assessment. Among adults, females have been overwhelmingly shown to cry more than males. Thus, the generalization that "it is a well-stablished fact that adult women cry more frequently than adult men" (Van Tilburg, Unterberg & Vingerhoets 2002:77) is beyond reasonable doubt. However, it should be emphasized that the evidence among pre-adults is unsettled. Especially among infants, there appears to be no discernable sex differences. Overall, crying can be considered a USD-F, at least among adults.

26.2.3 Academic and Intellectual Factors

Sex differences in academic and intellectual factors have received a great deal of research attention. Findings were summarized in Chapter 11 (in Volume II). In the section below, the sex difference variables with consistency scores of 95.0 or higher are given individual attention regarding the possibility of being USDs.

26.2.3.1 Repeating (Failing to Pass) a Grade in School

Evidence. Précis 26.2.3.1 highlights the evidence of a sex difference in individuals repeating a grade in school (i.e., failing to pass all grades the first time). All 11 studies located on this topic found that males repeated grades more frequently. However, the studies were limited just two countries, and to a 35-year time frame in terms of publication dates.

Explanation. The only theory that seems to have a bearing on this variable is ENA theory, since it predicts that females mature more rapidly than do males due to testosterone's tendency to retard brain maturation (Styne & Grumbach 2002; Del Giudice, Angeleri & Manera 2009; Llaurens, Raymond & Faurie 2009).

Theoretically, by slowing neurological maturation, androgens should be positively correlated with the tendency to repeat grade and poor school performance even though, by full-adulthood, males should have caught up with females in this regard. Unfortunately, no specific evidence was located regarding either prenatal or postpubertal exposure to androgens being directly linked to repeating grades or doing poorly in school.

Assessment. Due to the fact that only 11 studies indicated that males repeat grades in school more than do females and only two countries were represented among these studies, declaring this variable a USD could be strengthened in empirical terms. However, the theoretical evidence is strong. First, the evidence is strong that males are exposure to higher levels of testosterone, both prenatally and post-pubertally (see Précis 26.1.3.2). Second, links between this hormone and slow brain development is substantial, especially regarding the left hemisphere (which is primarily responsible for language) is also very supportive (Gouchie & Kimura 1991; Tan 1994; Grimshaw, Bryden & Finegan 1995; Nguyen 2018). Third, males exhibit learning disabilities more than do females (see Précis 26.2.3.2). Fourth, females have stronger language skills than males,

Précis 26.2.3.1 Repeating (failing to pass) a grade in school (original Table 11.1.4.1)

Age categories, number of studies, and sex	*Consist. score*	*Overviews*	*Countries*	*Time range*	*Non-human*	*Social roles*			*Evolutionary*		*Theological*
						Origin	*Found*	*Biosoc*	*Select*	*ENA*	
All Ages 11M	100.0M	-0-	2	1967–2002	-0-	Silent	Silent	Silent	Silent	Expl-M	Silent

Précis 26.2.3.2 Learning disabilities or difficulties in general (Original Table 11.1.4.3)

Age categories, number of studies, and sex	*Consist. score*	*Overviews*	*Countries*	*Time range*	*Non-human*	*Social roles*			*Evolutionary*		*Theological*
						Origin	*Found*	*Biosoc*	*Select*	*ENA*	
Child 21M All Ages 28M	100.0 100.0	M: 1Rev	4	1949–2011	-0-	Silent	Silent	Silent	Silent	Expl-M	Silent

especially in childhood (when most academic learning is focused on language skills) (see Précis 26.2.3.10). Fifth, females enjoy school more (see Précis 26.2.7.24) and devote more time to studying and doing homework (see Précis 26.3.5.1). All of these universal sex differences reinforce the conclusion that repeating at least one grade in school is a USD-M.

26.2.3.2 Learning Disabilities or Difficulties in General

Evidence. Twenty-eight studies of sex differences in learning disabilities were located based on samples drawn from four different countries, published over a 62-year time frame. As shown in Précis 26.2.3.2, the studies unanimously indicated that males had more learning disabilities and difficulties than did females. The one literature review on the subject reached the same conclusion.

Explanation. Sex differences in learning disabilities is likely to help explain sex differences in tendencies to repeat grades (discussed in the preceding narrative). Accordingly, ENA theory provides a strong explanation for both sex differences. Furthermore, it is worth mentioning that, before puberty, both male and females with learning disabilities had higher circulating testosterone levels than did peers without learning disabilities (Kirkpatrick, Campbell et al. 1993). Also, to the extent that learning disabilities are closely related to slow reading, a literature review concluded that difficulties learning to read are positively correlated with prenatal testosterone exposure (WH James 2008).

Assessment. Overall, learning disabilities can be deemed a USD-M. The best explanation appears to involve neurohormonal factors.

26.2.3.3 Institutionalized for Mental Retardation or Mental Disabilities

Evidence. Thirteen studies of the institutionalization of individuals for mental disabilities or mental retardation are presented in Précis 26.2.3.3, all of which indicated that higher proportions of males than females are affected by these conditions. These studies were conducted in seven different countries over a 78-year time period.

Explanation. Theoretically explaining why more males than females would be mentally disabled in all cultures is difficult with any of the six available theories. The most promising possibility seems to involve ENA theory regarding the possibility of brain exposure to testosterone being involved. In this regard, one study did find a relationship between prepubertal levels of testosterone among a sample of 7- to 9-year-olds and their scores on IQ tests, but the pattern was a bit complicated (Ostatnikova, Celec et al. 2007). Among males, those with the highest testosterone levels were most likely to have average IQ scores, while males who were in the lower testosterone ranges had both the lowest and the highest IQs. In other words,

Précis 26.2.3.3 Institutionalized for low intelligence or mental disabilities (original Table 11.2.2.2)

Age categories, number of studies, and sex	*Consist. score*	*Overviews*	*Countries*	*Time range*	*Non-human*	*Social roles*			*Evolutionary*		*Theological*
						Origin	*Found*	*Biosoc*	*Select*	*ENA*	
All Ages 13M	100.0M	-0-	7	1897–1975	-0-	Silent	Silent	Silent	Silent	Poss-M	Silent

they were most likely to be either in the retarded or in the genius ranges. For the females in this study, no relationship was identified regarding testosterone levels and IQ scores.

Assessment. While just 13 relevant studies were located, they were based on samples drawn from seven different countries compiled from three-quarters of a century. Therefore, while theoretical explanations are tenuous, we consider the greater institutionalization of males for mental retardation to be a USD-M.

26.2.3.4 Spatial Reasoning Ability, Self-Rated

Evidence. Very large numbers of studies have been conducted in which sex differences in spatial reasoning has been assessed. As summarized in Précis 26.2.3.4, the vast majority of these studies have concluded that males surpass females in such reasoning. However, because several studies found no significant sex differences, the general studies of sex differences in spatial reasoning did not achieve 95.0+ consistency scores.

However, in studies of self-rated spatial reasoning ability, males rated themselves higher than did females. Précis 26.11.4 shows that this was true in all of the 17 studies located on this subject. The précis also shows that the same result was found in studies of eight different countries and three multi-national research projects.

Explanation. As one can see in Table 21.2.5.15 (in Volume II), three studies were found in which males were stereotyped as being better than females in spatial reasoning, with all three studies having been conducted in European countries. Based on this evidence, all three social role theories could be invoked to explain why empirical studies have confirmed that males are better than females, at least in terms of self-ratings. However, only the latter two social role theories would predict that the relationship might be found to be universal. In order for the founder effect version to predict universality, males would have had to be better in primordial human societies, which, of course, is unknown. Regarding the biosocial role theory to predict universality, spatial reasoning would have had to be associated with the larger size and strength of males and/or with the tendency for females to give birth.

Many have argued that evolutionary forces are likely to have contributed to sex differences in spatial reasoning. The gist of these arguments has been that, throughout nearly all of human prehistory, males were sexually selected for their ability to track down and kill wild game, thereby providing animal protein to themselves and their kin, while females devoted most of their time gestating, nursing, and caring for offspring (Geary 1995; Wynn, Tierson & Palmer 1996; Joseph 2000; Silverman, Choi et al. 2000; Jones, Braithwaite & Healy 2003; Ecuyer-Dab & Robert 2004; Geary 2016). Supporting this theoretical reasoning, studies of several species of

Précis 26.2.3.4 Spatial reasoning ability, self-rated (original Table 11.2.4.2)

Age categories, number of studies, and sex	*Consist. score*	*Overviews*	*Countries*	*Time range*	*Non-human*	*Social role theory*			*Evolutionary theory*		*Theological*
						Origin	*Found*	*Biosoc*	*Select*	*ENA*	
All Ages 17M	100.0M	-0-	5 (3)	1982–2018	-0-	Contra	NoTest	Poss-M	Expl-M	Expl-M	Silent

non-human animals in which spatial reasoning helps to promote territorial behavior, wayfinding, and resource procurement have concluded that males, on average, surpass females in spatial reasoning tasks (Owen-Smith 1977; Baylis 1981; Geary 1995).

Going beyond this basic evolutionary logic, ENA theory predicts that spatial reasoning will be promoted by exposing the brain to high levels of androgens (Ellis 2011a). The evidence in this regard is mixed, perhaps depending in part on the timing of brain exposure. In the case of prenatal testosterone, a review concluded that, at least regarding mental rotation ability, there seems to be no significant relationship (M Hines 2006). In a subsequent study based on 2D:4D, another research team found no evidence of a significant relationship with spatial reasoning in either sex (Hampson, Ellis & Tenk 2008).

Regarding postpubertal exposure, the evidence supporting the hypothesis that testosterone is positively correlated with spatial reasoning is very consistent, not only among humans (Gouchie & Kimura 1991; Janowsky, Oviatt & Orwol 1994; Berenbaum, Korman & Leveroni 1995; Grimshaw, Sitarenios & Finegan 1995; Aleman, Bronk et al. 2004; Hooven, Chabris et al. 2004; Zitzmann 2006; Valla & Ceci 2011) but also among rodents (Roof & Havens 1992). However, there is an interesting caveat: There appears to be limits to the effects that testosterone can have on spatial reasoning. Specifically, at least two studies indicate that *moderate* male-typical levels of testosterone (not *extremely* high male-typical levels) promote spatial reasoning to the greatest degree (Moffat & Hampson 1996; Ostatníková, Laznibatová & Dohnányiová 1996).

To provide more evidence on sex differences in spatial reasoning, an experiment with rats damaged the hippocampus of both sexes to assess the effects that it had on spatial reasoning. The study revealed that this damage curtailed the ability of males (but not females) to learn how to traverse mazes, suggesting that females utilize some other part of the brain to engage in spatial reasoning (Roof, Zhang et al. 1993). Perhaps related to this rat experiment, brain imaging studies of humans have indicated that females utilize the frontal portion of the neocortex when performing spatial reasoning tasks more than do males (E Weiss, Siedentopf et al. 2003). Therefore, if brain exposure to testosterone is contributing to variations in spatial reasoning, it may do so by causing males and females to utilize different parts of the brain when they are confronted with spatial reasoning tasks.

Assessment. The evidence is strong that self-rated spatial reasoning is a USD-M. It is also worth adding that a substantial majority of studies of objectively measured spatial reasoning ability have also indicated that males surpass females according to studies of humans as well as other species (see Table 25.11.6). The best explanations appear to be of an evolutionary nature.

26.2.3.5 Estimating Distances and Projectile Landings

Evidence. Can one sex better estimate distances and predict projectile landings? As shown in Précis 26.2.3.5, all of the 17 studies of this topic found males to be better at these tasks. These studies were based on samples from eight different countries and were published over a 32-year time frame.

Explanation. As noted above regarding self-rated spatial reasoning, sexual selection theory predicts that, due to the long prehistoric period in which the human species has lived – roughly 98% (Orians 1980) – males were favored for their ability to track down and kill wild game, while females focused more on gestating, nursing, and caring for offspring (Joseph 2000; Silverman, Choi et al. 2000; Jones, Braithwaite & Healy 2003; Geary 2016). This has come to be known as the *savanna hypothesis*, given that humans appear to have branched off of an ape tree-dwelling linage similar to modern-day chimpanzees and then primarily occupied grassy savannas rather than heavy woodlands thereafter. According to this hypothesis, while males specialized in hunting, females evolved a bias toward mating with males who were the most capable hunters, allowing them to focus on child care (Kruger & Byker 2009; Cashdan, Marlowe et al. 2012). At least in pre-agrarian times, these complementary sex roles presumably gave rise to males who are better at making judgments regarding spatial skills such as wayfinding and the throwing of projectiles. Following the rise of agriculture, these spatial skills allowed males to take up many occupations involving engineering and science.

If the above evolutionary reasoning is correct, one would expect the ability to estimate distances and projectile landing to be positively correlated with exposure to androgens. No evidence specifically having to do with androgenic effects on brain functioning and the ability to estimate distances was located. However, there is evidence of androgenic effects on spatial reasoning indicating that neurological exposure to androgens promotes such reasoning (see Précis 26.11.4).

Assessment. The empirical evidence as well as the theoretical arguments provide substantial support for considering the ability to accurately estimate distances and targets a USD-M. Nonetheless, more work on theoretically explaining this sex difference in neurohormonal terms is needed.

26.2.3.6 Mechanical Problem Solving

Evidence. Dating back to the early 1900s, over a century of studies of sex differences in mechanical problem solving were located. These studies were based on samples drawn from seven different countries. Précis 26.2.3.6 shows that 39 of the 40 studies found males were better at solving mechanical problems, with the remaining study reporting no significant sex difference.

Précis 26.2.3.5 Estimating distances and projectile landings (original Table 11.5.2.3)

Age categories, number of studies, and sex	*Consist. score*	*Overviews*	*Countries*	*Time range*	*Non-human*	*Social role theory*			*Evolutionary theory*		*Theological*
						Origin	*Found*	*Biosoc*	*Select*	*ENA*	
Adult 14M All Ages 17M	100.0 100.0	-0-	8	1983–2015	-0-	Silent	Silent	Silent	Poss-M	Poss-M	Silent

Précis 26.2.3.6 Mechanical problem solving (original Table 11.2.5.7)

Age categories, number of studies, and sex	*Consist. score*	*Overviews*	*Countries*	*Time range*	*Non-human*	*Social role theory*			*Evolutionary theory*		*Theological*
						Origin	*Found*	*Biosoc*	*Select*	*ENA*	
Child 11M Adult 11M All Ages 39M 1x	100.0 100.0 97.5	-0-	7	1912–2021	-0-	Silent	Silent	Silent	Expl-M	Expl-M	Silent

Explanation. From a theoretical standpoint, the social role theories would all attribute these sex differences to stereotypes or to social role expectations and training. While these social factors could serve to exaggerate (or dampen) such sex differences, they would be hard-pressed to explain their apparent universality.

If mechanical problem solving is largely a reflection of spatial reasoning interests and abilities, both evolutionary theories offer explanations based on the assumption that males have been selected for their ability to reason in spatial terms. Specifically, from the time that males were first selected for spatial reasoning in order to successfully hunt to modern times, they have been reproductively favored over females for their abilities to (and interests in) manipulating physical objects (see Précis 26.2.3.5).

In addition, ENA theory asserts that brain exposure to androgens play a major role in enhancing spatial reasoning. The evidence supporting this neuroandrogenic hypothesis is substantial, especially regarding prenatal exposure (Wai, Lubinski & Benbow 2009; Browne 2013: 780; Casey & Ganley 2021).

Assessment. The weight of the evidence as well as theoretical reasoning both strongly supports the conclusion that mechanical problem solving is a USD-M.

26.2.3.7 Using Distance Estimates to Navigate

Evidence. Thirty-five studies were located having to do with sex differences in tendencies to use distance estimates (usually as opposed to landmarks) when navigating. As shown in Précis 26.2.3.7, all but one of these studies concluded that males use this type of navigating strategy more than do females. Samples involved in these studies came form six different countries, and the studies were published over a 71-year time frame.

Explanation. None of the three social role theories seem to have specifically proposed any explanations for sex differences in the use of distance estimates for navigating. Both evolutionary theories do so.

The essence of sexual selection theory involves noting that, at least up to the advent of agriculture, males were favored for their ability to hunt, a set of skills that often took them far from home. For this reason, male skills at navigating are likely to have been important for survival. In accordance with this line of reasoning, a substantial majority of studies have indicated that males are better at spatial reasoning (see Table 11.2.4.1). Presumably, this ability has included the ability to navigate using a feel for the distances traveled in various directions.

Regarding the effects of brain exposure to testosterone contributing to sex differences in navigational skills when distance estimates are important,

Précis 26.2.3.7 Using distance estimates to navigate (original Table 11.2.5.7)

Age categories, number of studies, and sex	*Consist. score*	*Overviews*	*Countries*	*Time range*	*Non-human*	*Social role theory*			*Evolutionary theory*		*Theological*
						Origin	*Found*	*Biosoc*	*Select*	*ENA*	
Adult 26M 1x All Ages 34M 1x	96.3 97.1	-0-	6	1941–2012	4M	Silent	Silent	Silent	Expl-M	Expl-M	Silent

ENA theory leads to the expectation that the effects will be positive. While evidence is fragmentary, it is largely supportive. After documenting that males perform better than females in a virtual water maze task (where distance estimating is important), the research team found correlated circulating testosterone levels with maze performance among both sexes. This revealed that, among females, those with the highest circulating levels of testosterone performed considerably better than females with the lowest levels; however, for the males sampled, testosterone levels and maze performance did not correlate (Burkitt, Widman & Saucier 2007).

Another study was conducted only with women who received a single dose injection with testosterone. Results from this study revealed that virtual maze performance was significantly elevated following the injection (Pintzka, Evensmoen et al. 2016).

Assessment. Research strongly suggests that the tendency to use distance estimates more when navigating is a USD-M. The strongest explanations appear to be evolutionary in nature.

26.2.3.8 Systemizing Thinking

Evidence. Systemizing thinking refers to the tendency to try to think about events as if they were part of an interactive system and then seek to control events as part of a system. Put another way, it is "the drive to analyze a system in terms of the rules that govern the system, in order to predict the behavior of the system" (Baron-Cohen, Knickmeyer & Belmonte 2005: 820).

Although the studies of systemizing behavior only reach back to 2006, 20 were located. Précis 26.2.3.8 shows that males have been found to be better at exhibiting such behavior than females in all of these studies. Eight countries sampled in these studies plus one multi-national study.

Explanation. From an evolutionary perspective, systemizing thinking is associated with the ability to analyze and manipulate system components and sometimes entire systems. Such analyses and manipulations appear to be part of many occupational pursuits, such as those involving science, engineering, mathematics, and even business and government. As such, researchers have proposed that systemizing thinking has been favored among males in order to compete with other males in stable resource procurement (Charlton & Rosenkranz 2016).

Some have argued that systemizing thinking is favored among males by sexual selection because it promotes cognitive skills helpful in competing for resources with which to attract mates (Del Giudice, Angeleri et al. 2010; Doll, Cárdenas et al. 2016). However, when systemizing thinking becomes extreme, it often manifests itself in the form of what is known as the *autism spectrum disorder* (Baron-Cohen 2006; Spikins, Wright & Scott 2018). This trait appears to have no reproductive advantage. In this

Précis 26.2.3.8 Systemizing thinking (original Table 11.2.7.19)

Age categories, number of studies, and sex	*Consist. score*	*Overviews*	*Countries*	*Time range*	*Non-human*	*Social role theory*			*Evolutionary theory*		*Theological*
						Origin	*Found*	*Biosoc*	*Select*	*ENA*	
All Ages 20M	100.0	-0-	8 (1)	2006–2016	-0-	Silent	Silent	Silent	Expl-M	Expl-M	Silent

regard, the evidence of males being more often diagnosed with autism is very strong autism (see Précis 26.2.6.24).

According to ENA theory, not only should males have evolved greater tendencies to systemize than females, but that neuroandrogenic factors contribute to the sex difference. Findings have supported this conclusion, at least regarding prenatal testosterone exposure (Auyeung, Baron-Cohen et al. 2006; Manning, Reimers et al. 2010; Wakabayash & Nakazawa 2010).

Assessment. The conclusion that systemizing thinking is greater among males than among females is strong even though the time frame during which this trait has been studied is quite brief. We declare that such thinking is a USD-M and that evolution-based explanations appear to be the strongest.

26.2.3.9 Relationship between IQ Scores and Academic Performance

Evidence. It is of little surprise that IQ scores are positively correlated with academic performance (Imre 1963; Duckworth & Seligman 2005). But, is this relationship stronger for males or for females? Ten studies were located pertaining to this question. As shown in Précis 26.2.3.9, all studies agree that the correlation between these two variables is stronger among females. The studies were conducted in five different countries and took place over a 103-year time frame.

Explanation. None of the five theories of sex differences seems to offer a specific explanation for why IQ scores and academic performance would correlate more strongly among females than among males. One might begin to better understand sex differences in the connection between these two variables by seeking to determine if sex differences in the verbal aspects of intelligence IQ (i.e., VIQ, see Table 11.2.3.1) correlate better with academic performance than do other aspects of intelligence.

Assessment. Despite the lack of any specific theoretical explanation, the consistency of findings and the number of countries and time frame covered gives us considerable confidence that the relationship between IQ and academic performance is a USD-F.

26.2.3.10 Language Ability

Evidence. Sex differences in language ability have been studied over the past 130 years in 16 different countries. Précis 26.2.3.10 shows that, among children, 37 out of 38 studies found that females had superior language ability. While considerable research has assessed language ability among other age groupings and concluded that females were better, the consistency scores were all below the 95.0 cutoff. One meta-analysis also supported the conclusion that females have greater language abilities than males on average.

Précis 26.2.3.9 Relationship between IQ scores and academic performance (original Table 11.2.8.5)

Age categories, number of studies, and sex	*Consist. score*	*Overviews*	*Countries*	*Time range*	*Non-human*	*Social role theory*			*Evolutionary theory*		*Theological*
						Origin	*Found*	*Biosoc*	*Select*	*ENA*	
Adult 10F	100.0	-0-	5	1924–2017	-0-	Silent	Silent	Silent	Silent	Silent	Silent

Précis 26.2.3.10 Language ability (original Table 11.3.2.1)

Age categories, number of studies, and sex	*Consist. score*	*Overviews*	*Countries*	*Time range*	*Non-human*	*Social role theory*			*Evolutionary theory*		*Theological*
						Origin	*Found*	*Biosoc*	*Select*	*ENA*	
Infant/Toddler 12F 2x Child 37F 1x Adolescent 21F 3x Adult 18F 1x 4M All Ages 91F 9x 4M	(85.7) 97.4 (87.4) (66.7) (84.3)	F: 1Met x: 1Gen	16	1891–2021	-0-	Silent	NoTest	Poss-F	Poss-F	Poss-F	Silent

Explanation. The ability of humans to use language is one of the most unique characteristics of our species (Oller & Griebel 2014). Because females are stereotyped as being better at the use of language (see Table 21.2.5.9), one can use all three versions of social role theory to explain why, at least among children, females are universally better. However, the original version of the theory would lead one to expect exceptions, which, at least for this age group, there appears to be none. The founder effect version would argue that this is due to the primordial human society exhibiting this sex difference (for which there is no evidence). According to the biosocial version of social role theory, most sex difference traits are the result of sex differences in size and strength or the ability to bear offspring. In this regard, males may be able to use greater size and strength to accomplish their goals, while females may have to rely on language. The merit of such reasoning seems questionable given that the highest consistency score was for children, when size and strength sex differences are minimal and pregnancy is all but impossible.

Regarding the use of sexual selection theory for explaining sex differences in the development of proficient language skills, it is difficult to argue that language would benefit one sex more than the other in reproductive terms. Instead, it may be necessary to recognize that language ability, at which females tend to be better and spatial reasoning and systemizing reasoning, at which males tend to be better (see Précis 26.2.3.4 and Précis 26.2.3.8, respectively) are somewhat competing ways of learning. If so, because males are more disposed to think and learn in spatial-systemizing terms, their tendency to acquire language skills is delayed relative to that of females. In other words, due to males have been sexually selected for spatial-systemizing reasoning in order to compete for resources as adults, the uniquely human linguistic abilities that both sexes possess are usually slower to develop in males.

Turning to ENA theory, evidence based on amniotic fluid samples has suggested that prenatal brain exposure to testosterone delays childhood development of language skills, at least in terms of vocabulary size (Lutchmaya, Baron-Cohen & Ragatt 2002b). Another study reported that the amount of blood in the umbilical cord was inversely correlated with various language development skills, at least among boys (Whitehouse, Mattes et al. 2012). These findings suggest that, even though brain exposure to androgens can promote reason in spatial and systemizing terms, traits for which males tend to be higher (see Précis 26.2.3 and Précis 26.2.3.8), language skills are slower to development in males. Nonetheless, Précis 26.2.3.10 clearly suggests that beyond childhood, males appear to slowly catch up with females in their language skills.

At least one study suggested that the slowing of language development is due to prenatal testosterone slowing down neocortical maturation, especially in the left hemisphere (Halpern, Benbow et al. 2007). Complicating the matter, one study found no significant relationship between amniotic testosterone levels and language comprehension at age 4 for boys, while there an inverted U-shape pattern was found for young girls (Finegan, Nicols & Sitarenois 1992).

Assessment. The amount of evidence that females develop language skills more rapidly than do males is substantial, at least among children. Therefore, the development of language skills among children can be considered a USD-F. More work is needed to account for these sex differences in theoretical terms.

26.2.4 Learning, Memory, Knowledge, and Cognitive States

The focus of Chapter 12 was on a number of cognitive variables. These variables primarily have to do with styles of learning, the ability to remember, the amount of knowledge that individuals possess, and various cognitive states, such as sleeping and dreaming.

26.2.4.1 Knowledge in General

Evidence. Précis 26.2.4.1 shows that, among adults, 18 studies have all found that males have more knowledge. However, it should be noted that when studies for all ages are combined, the consistency score is substantially under 95.0. The research included samples drawn for 12 different countries plus two multi-national studies and the dates of publication for these studies covered 110 years.

Explanation. This is an interesting and controversial sex difference. Not only is it likely to be politically incorrect to say that males have more knowledge than females (at least by the time they reach adulthood), but this conclusion seems contrary to most evidence indicating that females receive higher grades in school (see Table 11.1.1.1). Also, females enjoy school more (Précis 26.2.7.24), spend more time studying (Précis 16.3.20.1), and, in recent decades, are even more likely to attend and even graduate from college (Table 25.20.2).

Part of the answer involves noting that nearly all measures of people's general knowledge are based on scores derived from multiple-choice tests, where answers are either correct or incorrect (Gayef, Oner & Telatar 2014; Vlazneva & Androsova 2021). In this regard, nearly all studies have concluded that males perform better on multiple-choice tests than do females (see Table 11.1.3.1 in Volume II), while females appear to perform better than males on essay-type tests (Table 11.1.3.2 in Volume II).

A second component involves noting that males exhibit more systematizing thinking than do females (Précis 26.2.3.8). This type of thinking is

Précis 26.2.4.1 Knowledge in general (original Table 12.2.1.1)

Age categories, number of studies, and sex	*Consist. score*	*Overviews*	*Countries*	*Time range*	*Non-human*	*Social role theory*			*Evolutionary theory*		*Theological*
						Origin	*Found*	*Biosoc*	*Select*	*ENA*	
Child 9M 2x 1F Adult 18M All Ages 48M 5x 1F	(69.2) 100.0 (87.3)	-0-	12 (2)	1910–2021	-0-	Silent	Silent	Silent	Silent	Expl-M	Silent

most likely to focus on objective information, devoid of all emotional connotations (Golan & Baron-Cohen 2006; Tracy, Robins et al. 2011). Such information is emphasized in tests of general knowledge, thereby leading to the prediction that males would score higher than females on tests of general knowledge.

The theory that makes the most testable prediction about sex differences in general knowledge is ENA theory. It asserts that exposing the brain to high levels of testosterone causes thinking to be more systemizing. Since the average male brain receives greater brain exposure to testosterone than the average female brain, one would expect that males would perform better on tests of general knowledge.

Assessment. Even though more work is needed to better explain sex differences in overall knowledge, this variable appears to be a USD-M.

26.2.4.2 Intra-Sex Variability in Math Ability

Evidence. Evidence regarding sex differences in math ability did not attain consistency scores of 95.0 or higher (see Table 25.12.4). However, there are two aspects of math ability that did attain a 95.0+ consistency score. One involves variability in math ability measures. As shown in Précis 26.2.4.2, all of the 14 studies found on this topic indicated that males achieved greater variability than did females. The relevant studies drew samples from four different countries plus three multi-national studies.

Explanation. The three social role theories are all silent to the possibility of males exhibiting more *variability* in their math abilities than females. Both of the evolutionary theories, however, recognize that the ways males can pass their genes onto future generations is inherently more varied than in the case of (Feingold 1992).

In fact, in many species, males (but not females) come in two recognizable forms, often called *alternative reproductive strategies* (Gross 1996; Sih & Bell 2008; Sinervo & Zamudio 2001). One type of male strategy involves males who gain voluntary sexual access to females by complying with female desires for reliable resource procuring mates. The other strategy usually entails males who use some form of deception or force to obtain sexual access to mates. An informative list of these evolved male strategies, and the species in which they are found, is provided by Gross (1996:Table 1).

Some have proposed that alternative reproductive strategies may have evolved among humans (Ellis & Walsh 1997; Belsky 2012; Yao, Långström et al. 2014). If so, it is reasonable to hypothesize that some of the greater variability in math-related cognitive skills could be an expression of this variability, thereby helping to promote greater male diversity in how resources are obtained in order to obtain mating opportunities (Geary 1996: 238). Given that females are more likely than males to choose mates on the

Précis 26.2.4.2 Intra-sex variability in math ability (original Table 12.2.3.3)

Age categories, number of studies, and sex	*Consist. score*	*Overviews*	*Countries*	*Time range*	*Non-human*	*Social role theory*			*Evolutionary theory*		*Theological*
						Origin	*Found*	*Biosoc*	*Select*	*ENA*	
All Ages 14M	100.0	-0-	4 (3)	1980–2020	-0-	Silent	Silent	Silent	Poss-M	Poss-M	Silent

basis of their abilities to make a living wage (see Précis 26.2.7.19), one would expect males to have a wider diversity of occupational interests and skills than females. Later in this chapter, specific evidence will be provided that this is the case: i.e., males throughout the world do work in a wider range of occupations than is the case for females (see Précis 26.3.5.21).

Regarding the ENA version of evolutionary theory, there is evidence that testosterone influences alternative reproductive strategies in a variety of ways (Goetz, Weisfeld & Zilioli 2019), including exposing the brain to varying regimens of testosterone (Geary 1996: 245). In this regard, one study indicated that, based on 2D:4D measurement, prenatal testosterone appeared to subtly alter the right ("non-language") hemisphere in ways that influence interests in mathematics (Brosnan 2006).

Assessment. There is a need for more theoretical developments on how variability in math abilities is produced. Nonetheless, the evidence is pointing toward intra-sex variability in math ability being a USD-M.

26.2.4.3 Exceptionally High Math Scores

Evidence. As discussed directly above, males appear to exhibit greater variability in math scores than is the case for females. In other words, while females are more heavily concentrated in the middle of the roughly normal distribution curve, males tend to be more spread out, with greater proportions scoring extremely low and extremely high. From this, one can actually deduce that a greater proportion of males would be at the extremely high end of the distribution curve. Support for this deduction is found in Précis 26.2.4.3. There, one can see that all 37 studies have concluded that exceptionally high scores on tests of math ability (i.e., usually two to three standard deviations above the mean) are attained by males.

The relevant evidence came from seven different countries along with four multi-country studies that were published over a 40-year time frame. One can also see that two literature reviews as well as three meta-analyses all reached the conclusion that greater proportions of males than females are at the extremely high end of the distribution curve regarding math ability measures.

Explanation. Because math scores are more or less normally distributed for both sexes (Hedges & Friedman 1993) and males are more widely varied in their math scores than are females (see Précis 26.2.4.2), one can infer that a greater proportion of males will score in the exceptionally high *and* exceptionally low ranges of the distribution in math ability. Findings of sex differences in scores on math tests are surprisingly mixed (Table 12.2.3.1). Only among adults do the majority (i.e., about 2/3) indicate that male attain higher average scores. Because the proportion of males attaining exceptionally high scores appear to be largely explainable

Précis 26.2.4.3 Exceptionally high math scores (original Table 12.2.3.4)

Age categories, number of studies, and sex	*Consist. score*	*Overviews*	*Countries*	*Time range*	*Non-human*	*Social role theory*			*Evolutionary theory*		*Theological*
						Origin	*Found*	*Biosoc*	*Select*	*ENA*	
Adolescent 30M All Ages 37M	100.0M 100.0M	M: 2Rev M:3Met	7 (4)	1980–2020	-0-	Silent	Silent	Silent	Silent	Silent	Silent

in statistical terms, we will make not effort to explain the findings with any of the five theories.

Assessment. Attaining exceptionally high math scores is a USD-M. However, the explanation appears to have more to do with statistical features of two normally distributed variables with substantially different standard deviations than with any empirical theory.

26.2.4.4 Science Knowledge in General

Evidence. Précis 26.2.4.4 shows that 183 studies have reported on sex differences in objectively measured science knowledge. These studies published over a 90-year period in 58 different countries (plus 20 multi-national studies). Both of the two meta-analyses concluded that males have higher science knowledge, along with four of the five literature reviews, but one literature review actually concluded that females have greater science knowledge. Only in the case of the 12 studies of adults was the consistency score higher than the 95.0 cutoff.

Explanation. For two reasons, we will not attempt to explain this variable in theoretical terms. First, the findings are very inconsistent regarding all age groupings other than adults. Second, many of the findings of science knowledge in general are likely to have included biology as well as the physical sciences. As shown in the upcoming précis, it appears that sex differences in science knowledge is only consistently greater for males when it comes to the so-called physical sciences, not the biological sciences.

Assessment. We will not declare science knowledge a USD, for the two reasons provided above.

26.2.4.5 Physical Science Knowledge

Evidence. Précis 26.2.4.5 shows that all but one of the 37 studies of sex differences in physical science knowledge have concluded that males surpass females. These studies were conducted in six different countries as well as four additional studies of multiple countries. The publication of these studies spanned 80 years.

Explanation. No stereotype studies of sex differences in physical science knowledge were found. However, males do appear to be stereotyped as enjoying science and engineering more (Table 21.2.4.3). From this stereotype, social role theorists may argue that, because cultures teach their boys to enjoy physical science more, they are thereby inclined to learn more about physical science. The original version of the theory implies that this type of sex difference will only hold for some cultures, not all (which does not appear to be the case). In the case of the founder effect social role theory, this sex difference in interests would have to have

Précis 26.2.4.4 Science knowledge in general (original Table 12.2.5.1)

Age categories, number of studies, and sex	*Consist. score*	*Overviews*	*Countries*	*Time range*	*Non-human*	*Social role theory*			*Evolutionary theory*		*Theological*
						Origin	*Found*	*Biosoc*	*Select*	*ENA*	
Child 13M 4x 1F Adolescent 99M 18x 30F Adult 12M All Ages 129M 23x 31F	(68.4) (55.9) 100.0 (57.9)	M: 4Rev M: 2Met F: 1Rev	58 (20)	1928–2020	-0-	Silent	Silent	Silent	Poss-M	Poss-M	Silent

Précis 26.2.4.5 Physical science knowledge (original Table 12.2.5.7)

Age categories, number of studies, and sex	*Consist. score*	*Overviews*	*Countries*	*Time range*	*Non-human*	*Social role theory*			*Evolutionary theory*		*Theological*
						Origin	*Found*	*Biosoc*	*Select*	*ENA*	
Adolescent 31M 1x All Ages 36M 1x	96.9 97.3	-0-	6 (4)	1934–2014	-0-	Contra	NoTest	Poss-M	Poss-M	Poss-M	Silent

originated by accident in the first primordial human cultures, a proposal for which there appears to be no evidence. The biosocial version of social role theory would attribute sex differences in physical science to biological factors, such as physical size and strength or the ability to bear offspring. Neither of these latter two social role theories seem very likely, but we will list them both as *possible*.

Using evolutionary theory to help explain sex differences in physical science knowledge would come from noting that people's knowledge tends to be positively correlated with their interests (Hyland, Hoff & Rounds 2022: 16). In this regard, research has consistently shown that, from infancy onward, boys are more interested in things than in people (see Table 15.10.2.6 in Volume II). These differences appear to have been sexually selected (Byrd-Craven, Massey et al. 2015).

ENA theory extends beyond sexual selection theory by asserting that brain exposure to androgens contributes to sex differences in interest and subsequent knowledge in the physical science. While the evidence is still fragmentary, various researchers have argued that brain exposure to testosterone appears to promote both interest in and ability in the physical sciences (Ethington & Wolfle 1986; Benbow & Lubinski 2007; Penner & Cadwallader Olsker 2012).

Assessment. Overall, we consider knowledge of physical sciences to be a USD-M. More work is needed to theoretically account for this difference.

26.2.4.6 Geography Knowledge

Evidence. When it comes to knowing the location of places, nearly all of the studies found that males have more geography knowledge. As shown in Précis 26.2.4.6, the only exception was one study that did not find a sex difference. The studies that were located were based on samples from four different countries, and were published over a 91-year time frame.

Explanation. Some research has indicated that males are stereotyped as having better spatial reasoning than is the case for females (Table 21.2.5.15). If so, social role theorists could argue that one of the ways males are taught to express this type of reasoning is by learning more about geography. Of course, the original version of this theory would have difficulty explaining why the sex difference appears to be pan-cultural. The founder effect version of the theory might argue that this stereotype was present in the first human cultures, a plausible but impossible-to-test supposition. Proponents of the biosocial version of social role theory would argue that the greater size and strength of males (plus their inability to get pregnant) promotes geography knowledge, a line of reasoning that seems difficult to defend.

Regarding an evolutionary perspective, a number of researchers have argued that males have been sexually selected for spatial reasoning, particularly in ancestral times when males spent hours and even days stalking prey to bring back to their families (Gaulin 1993; Browne 2006; Giery & Layman 2019). In contemporary times, the navigational skills that successful hunters acquired has resulted in males learning of geography more easily than females (Self & Golledge 2018).

ENA theory asserts that brain exposure to testosterone and other androgens are responsible for producing sexually selected traits. In this regard, there is considerable evidence that brain exposure to testosterone enhances spatial reasoning (see Précis 26.2.3.4).

Assessment. We conclude that geographic knowledge is a USD-M. The best supported theories seem to be evolutionary in nature.

26.2.4.7 Dreaming, Frequency of Recall

Evidence. Fourteen studies conducted over a 98-year time frame carried out in five different countries all reached the same conclusion: Females either dream more or at least remember their dreams better than do males. The nature of the evidence in this regard is summarized in Précis 26.2.4.7.

Explanation. No evidence was found that females are stereotyped as dreaming more, or remembering their dreams more, than males. Therefore, none of the three social role theories seems to offer an account for such a sex difference.

In the case of the evolutionary reasoning, Nielsen (2012) suggested that females recall dreams more than males do because they have better episodic memory. While episodic memory fell slightly short of attaining a 95.0 consistency score (see Table 12.1.4.14), the vast majority of studies do indicate that females surpass females in this regard. The apparent greater episodic memory of females may partially result from their keener verbal ability (see Précis 26.2.3.10) and their greater interests in social relationships (see Précis 26.2.7.31). Such skills and interests may have evolved among females more than among males in order to enhance the caregiving tendencies of females (Archer 2019; Asperholm, Högman et al. 2019).

A particularly novel argument along evolutionary lines was put forth by McNamara (1996). She noted that dreams occur in a limited time during sleep when the eyes tend to flutter, thus given its name: *rapid eye movements* (or *REM sleep*). After noting that dreams are largely limited to just two senses, sight and sound (i.e., rarely are odor, taste, or touch sensations a part of dreaming). This and evidence that most dreaming involves the limbic-frontal lobe inter-connections lead McNamara to hypothesize that REM sleep primarily helps to promote social attachments. If so, this could help to explain why dreaming – at least in terms of what individuals are able to consciously recall – tends to be greater for females, and conforms

Précis 26.2.4.6 Geography knowledge (Original Table 12.2.6.3)

Age categories, number of studies, and sex	*Consist. score*	*Overviews*	*Countries*	*Time range*	*Non-human*	*Social role theory*			*Evolutionary theory*		*Theological*
						Origin	*Found*	*Biosoc*	*Select*	*ENA*	
Adult 14M All Ages 21M 1x	100.0 95.5	-0-	4	1913–2004	-0-	Contra	NoTest	Poss-M	Expl-M	Expl-M	Silent

with evidence that social attachments are more pronounced among females (see Précis 26.2.7.29).

Regarding the possibility of neuroandrogenic involvement, as ENA theory would suggest, it is worth noting that two of the three studies that failed to identify significant sex differences in dreaming were conducted among pre-pubertal samples. This suggests that post-pubertal hormones may play a crucial role in diminishing dream recall among males. However, no specific evidence was located on this line of reasoning.

Assessment. Overall, evidence substantially supports the conclusion that the frequency of dream recall is a USD-M, particularly for adults. Nevertheless, more work is needed to better understand this sex difference.

26.2.4.8 Dream Content: Physical Objects and Artifacts (Other Than Clothing)

Evidence. Are their sex differences in the content of dreams? Précis 26.2.4.8 shows that 11 studies found males were more likely to dream about physical objects (other than clothing). Those studies were conducted in three countries, and there was one multi-national study. The studies covered most of the second half of the 20th century.

Explanation. No attempts to use any of the three social role theories to explain sex differences in dream content. Also, no studies of stereotypes regarding dream content were found.

The evolutionary theories seem to offer some leads toward explaining why males would dream more about physical objects than do females. Specifically, studies have indicated that males have greater interests in inanimate objects than do females. This is true in terms of their educational training (see Précis 26.2.7.42), their occupational interests (see Précis 26.2.7.44) and the actual occupations they pursue (see Précis 26.3.5.19). In the narrative surrounding these sex differences, one can find evidence suggesting that these relatively non-social interests may have been sexually selected for as well as influenced by brain exposure to testosterone.

Assessment. While more theoretical development would be desirable, the number of studies and the number of countries that have been sampled seems sufficient to consider dreaming about physical objects and artifacts (excluding clothing) is a USD-M.

26.2.4.9 Dream Content: Sexual Themes

Evidence. Précis 26.2.4.9 shows that all 13 studies of how often dreams included sexual content found that males reported such content more often. The relevant research was conducted over a 52-year time frame in four countries along with one multi-national study.

Précis 26.2.4.7 Dreaming, frequency of recall (original Table 12.3.2.1)

Age categories, number of studies, and sex	*Consist. score*	*Overviews*	*Countries*	*Time range*	*Non-human*	*Social role theory*			*Evolutionary theory*		*Theological*
						Origin	*Found*	*Biosoc*	*Select*	*ENA*	
Adult 18F All Ages 34F 3x	100.0 (91.9)	-0-	5	1907–2005	-0-	Silent	Silent	Silent	Poss-F	Poss-F	Silent

Explanation. As noted earlier in this chapter, males have been shown to have stronger sex drives on average than do females (see Précis 26.2.7.6). Therefore, it is rather predictable that sexual themes would be more common in the dreams of males. Also, the narrative surrounding the research on strength of sex drive noted that the best explanations seem to involve (a) sexual selection and (b) neuroandrogenic factors.

Assessment. Considering the number of consistent studies, the number of countries assessed, and the theoretical explanations that have been offered, we declare dream contents with sexual themes to be a USD-M.

26.2.4.10 *Fantasizing about Sexual Behavior*

Evidence. Fantasizing in the present context refers to imaginary scenarios of a sexual nature that usually involve oneself and at least one other person. According to Précis 26.12.2.10, 39 out of 41 studies found that males fantasize more about sexual behavior, with the last two studies simply finding no significant sex difference. One can see that the pertinent studies were based on samples drawn from nine different countries. The literature review located on this topic concurred that males fantasize more about sexual behavior than do females.

Explanation. No evidence of stereotypes regarding sex differences in fantasizing about sex were found. This makes the three social role theories difficult to consider relevant to such sex differences.

In the case of the two evolutionary theories, both seem to apply if one assumes that sexual fantasies largely reflect the influence of an individual's sex drive. As discussed earlier, using various methods, a large number of studies have concluded that males have a stronger sex drive than do females (see Précis 26.2.7.6). This difference can be explained in terms of sexual selection, given that male investment in offspring tends to be less than that of females (Ellis & Symons 1990; Wilson 1997). Regarding the possible influence of androgens on the sex drive, most research points toward an affirmative conclusion (Van Goozen, Cohen-Kettenis et al. 1995; Demers 2010).

Assessment. The large number of studies along with the sizable number of countries sampled leaves little doubt that fantasizing about sexual behavior is a USD-M. Evolutionary reasoning seems to offer the best explanations for this sex difference.

26.2.5 *Self-Assessments and States of Mind*

Research findings having to do with sex differences in people's self-assessments of various kinds were presented in Chapter 13. Variables for which consistency scores were 95.0 or higher in at least one age grouping will now be summarized.

Précis 26.2.4.8 Dream content: Physical objects and artifacts (other than clothing) (original Table 12.3.2.6)

Age categories, number of studies, and sex	*Consist. score*	*Overviews*	*Countries*	*Time range*	*Non-human*	*Social role theory*			*Evolutionary theory*		*Theological*
						Origin	*Found*	*Biosoc*	*Select*	*ENA*	
All Ages 11M	100.0	-0-	3 (1)	1958–2005	-0-	Silent	Silent	Silent	Poss-F	Poss-F	Silent

Précis 26.2.4.9 Dream content: Sexual themes (original Table 12.3.2.7)

Age categories, number of studies, and sex	*Consist. score*	*Overviews*	*Countries*	*Time range*	*Non-human*	*Social role theory*			*Evolutionary theory*		*Theological*
						Origin	*Found*	*Biosoc*	*Select*	*ENA*	
All Ages 13M	100.0	-0-	4 (1)	1958–2010	-0-	Silent	Silent	Silent	Poss-F	Poss-F	Silent

26.2.5.1 Overall Self-Confidence

Evidence. Are the members of one sex more self-confident? Although one study failed to find a sex difference in self-confidence, the remaining 22 studies determined that males were more self-confident. The studies were spread over eight decades of research in eight different countries. Précis 26.2.5.1 shows that a literature review also concluded that males were more self-confident.

Explanation. Seven studies of stereotypes all indicated that males are believed to be more self-confident or self-assured than is the case for females (see Table 21.2.7.56). In light of this stereotype (all based on samples drawn from Western countries), it is reasonable to infer that, at least in Western countries, stereotyping may be responsible for sex differences in overall self-confidence. Table 13.1.2.1 shows that all of the eight countries sampled were essentially Western (with the possible exception of Israel). Without a more culturally diverse sample of countries, it is not possible to specifically test the original version of the social role theory. However, both of the other two versions, which lead to predictions of possible universal conformity to stereotypes, more evidence is needed. Obtaining such evidence in the case of the founder effect version appears impossible, but it could be argued that the fact that males tend to be stronger and/or do not have to worry about getting pregnant, could give them more self-confidence.

From an evolutionary perspective, greater male risk-taking has been favored since at least moderately high risk-prone males are likely to compete and acquire resources with which to attract mates. As noted elsewhere, most evidence supports this prediction (see Précis 26.3.1.2). Among the factors that may help to promote risk-taking are diminished fearfulness and reduced sensitivity to pain. Both fearfulness (see Précis 26.2.2.2) and pain sensitivity (Table 5.4.2.1 in Volume I) appear to be lower among males than females. Another trait that could help to promote risk-taking is greater confidence in being successful when facing risks (Browne 2006).

Regarding the ENA version of evolutionary theory, some evidence was found pertaining to prenatal testosterone exposure possibly promoting self-confidence under competitive circumstances. Two studies reported that, among males (but not females), high prenatal testosterone exposure (inferred based on 2D:4D) was positively correlated with at least modest degrees of self-confidence (but not with "overconfidence") in one's ability to successfully complete an experimental task involving a monetary reward (Neyse, Bosworth et al. 2016; Dalton & Ghosal 2018).

Assessment. Given the numbers of studies and the countries sampled, we deem overall self-confidence to be a USD-M. Nonetheless, theoretically explaining the sex differences appears to be poorly developed.

Précis 26.2.4.10 Fantasizing about sexual behavior (original Table 12.4.1.3)

Age categories, number of studies, and sex	*Consist. score*	*Overviews*	*Countries*	*Time range*	*Non-human*	*Social role theory*			*Evolutionary theory*		*Theological*
						Origin	*Found*	*Biosoc*	*Select*	*ENA*	
All Ages 39M 2x	95.1	M: 1Rev	9	1958–2003	-0-	Silent	Silent	Silent	Expl-F	Expl-F	Silent

26.2.5.2 Body Satisfaction

Evidence. The past five decades have seen considerable research on how satisfied people are with their bodies. Précis 26.2.5.2 reveals that 78 studies found that males were more satisfied with their bodies, while the remaining two studies failed to find a sex difference. The studies were conducted 11 countries, and both a literature review and a meta-analysis also concluded that males have greater body satisfaction than do females.

Explanation. No evidence could be located regarding sex stereotypes having to do with body satisfaction. This implies that social role theories are largely silent to the possibility of sex difference in body satisfaction.

Applying either of the two evolutionary theories to sex differences in body satisfaction is also problematic. One line of reasoning in this regard has involved noting that males are more likely than females to emphasize physical attractiveness as an important criterion for mate selection (see Précis 26.2.7.14). Among the results is sexual selection pressure on females to be as attractive as possible and probably to be more negatively impacted when their appearance falls short of their hopes (Ålgars, Santtila et al. 2009). Perhaps, therefore, even minor "flaws" in bodily appearance could be more consequential for females than for males. Another component in this argument involves noting that males appear to focus more on musculature in assessing body satisfaction, while females put a stronger emphasis on maintaining a thin body weight (Petrie, Greenleaf & Martin 2010; da Silva, Marôco & Campos 2021). Both musculature and body weight are substantially influenced by various hormones, which could suggest that neuroandrogenic factors could be contributing to sex differences in body satisfaction to some significant degree.

Assessment. The number of pertinent studies and the diversity of countries sampled give us substantial confidence in declaring body satisfaction a USD-M. Nevertheless, more clarification is needed in order to offer a clear theoretical explanation of the sex difference.

26.2.5.3 Self-Assessment of Being Overweight (or Wanting to Lose Weight)

Evidence. Over the past-half century, 95 studies have been published on sex differences in considering oneself to be overweight (or expressing the desire to lose weight). Précis 26.2.5.3 shows that only one of these studies failed to find significant tendencies for females to express such opinions. Twelve different countries were sampled in these studies.

Explanation. One research team proposed an evolutionary explanation for sex differences in desires to lose weight (Kirchengast & Marosi 2008). Their arguments involved noting that females have greater percent body

Précis 26.2.5.1 Overall self-confidence (Original Table 13.1.2.1)

Age categories, number of studies, and sex	*Consist. score*	*Overviews*	*Countries*	*Time range*	*Non-human*	*Social role theory*			*Evolutionary theory*		*Theological*
						Origin	*Found*	*Biosoc*	*Select*	*ENA*	
Adult 19M All Ages 22M 1x	100.0 95.6	M:1Rev	8	1932–2012	-0-	NoTest	NoTest	Poss-M	Expl-M	Expl-M	Silent

fat (see Précis 26.1.1.6), which has been favored for successfully carrying maintaining pregnancy (Thong, McLean & Graham 2000).

Another evolutionary proposal centers around males being sexually attracted to females with relatively thin waste (Jackson & McGill 1996; Singh & Luis 1995). This attraction could have been evolutionarily favored as a way of minimizing the risk to males of being attracted to mates who are already pregnant (L Ellis 2011b:559).

Assessment. Even though theoretical explanations are poorly developed, we conclude that this variable is a USD-F in light of the very large number of pertinent studies and the diversity of countries sampled.

26.2.5.4 Assessment of One's Ability in General

Evidence. Is there a sex difference in how people assess their overall ability to get things accomplished? Précis 26.2.5.4 shows that, at least among adults, all of the 13 located studies found that males appear to make more positive assessments of their abilities. This was also the conclusion reached by one literature review. However, all studies were conducted in just one country (the United States).

Explanation. According to seven different studies, males are stereotyped as being more self-confident than are females (Table 21.2.7.56). It seems reasonable to assume that self-confidence is associated with high ratings to one's overall abilities. If so, all three of the social role theories would lead one to expect males to surpass females in ratings given to their overall abilities. The latter two of these theories would predict that the sex differences could be universal. However, the founder effect theory appears to be impossible to test (given the absence of any data from the earliest human societies). In the case of the biosocial version of social role theory, it could be argued that, larger size and greater strength gives individuals more confidence in their overall ability to perform most tasks.

Regarding the evolutionary theories, they too could correctly predict that males would be more prone to give higher average ratings to their abilities. This would come from evidence that males are more prone to take risks (Précis 26.2.5.6), be less fearful (see Précis 26.2.2.2), and to have more self-confidence (see Précis 26.2.5.1). These types of male-typical traits are likely to promote resource provisioning tendencies, there by attracting mates. Furthermore, as noted in the narratives surrounding these traits, brain exposure to testosterone appears to be involved. If so, it would be reasonable to hypothesize that such brain exposure would also promote how highly one assesses his or her abilities. No evidence for or against this hypothesis was located.

Assessment. Because just one country was sampled and the theoretical arguments are somewhat indirect and underdeveloped, we will not make a declaration regarding the USD status of assessing one's own ability.

Précis 26.2.5.2 Body satisfaction (Original Table 13.1.3.4)

Age categories, number of studies, and sex	*Consist. score*	*Overviews*	*Countries*	*Time range*	*Non-human*	*Social role theory*			*Evolutionary theory*		*Theological*
						Origin	*Found*	*Biosoc*	*Select*	*ENA*	
Adolescent 34M All Ages 78M 2x	100.0 95.1	M: 1Rev M: 1Met	11	1966–2010	-0-	Silent	Silent	Silent	Poss-M	Poss-M	Silent

Précis 26.2.5.3 Self-assessment of being overweight (or wanting to lose weight) (original Table 13.1.3.6)

Age categories, number of studies, and sex	*Consist. score*	*Overviews*	*Countries*	*Time range*	*Non-human*	*Social role theory*			*Evolutionary theory*		*Theological*
						Origin	*Found*	*Biosoc*	*Select*	*ENA*	
Adolescent 23F Adult 57F Wide Age Range 10F All Ages 94F 1x	100.0 100.0 100.0 98.9	-0-	12	1965–2010	-0-	Silent	Silent	Silent	Poss-F	Poss-F	Silent

26.2.5.5 Self-Assessed Computer Competency

Evidence. Précis 26.2.5.5 indicates that all of the studies about how people self-assessed their computer competency found that the self-assessments by males were more positive. This research was conducted in five different countries. No overviews of research on this topic were located.

Explanation. Sex stereotype studies have found that males are perceived as being more logical (Table 21.2.5.10), better at math (Table 21.2.5.11), and better at spatial reasoning (Table 2.2.5.15). If one assumes these traits all contribute to computer competency, it is reasonable to argue that sex stereotypes are responsible for sex differences in such competency. However, because this sex difference appears to be universal, the original social role theory would be nullified by the evidence. Since there were no computers when the first human societies emerged, the founder effect version of social role theory would also be unable to explain this sex difference. In the case of the biosocial version of social role theory, one would have difficulty explaining why any connection would exist between greater male size and strength or the ability of females to bear offspring with computer competency.

Turning to the evolutionary theories, it seems safe to assume that self-assessed computer competency at least roughly reflecting actual competency. Sexual selection theory would attribute greater male competency with regard to computer technology to traits such as being more interested in things than in people (see Table 15.10.2.6 in Volume II) and tendencies to be more systemizing in thinking patterns (see Précis 26.2.3.8). These traits, in turn, have been attributed to selection for mates who spend less time interacting with offspring and more time competing for resources.

ENA theory asserts that computer competency should be promoted by exposing the brain to androgens. Substantial evidence supports the hypothesis that neuroandrogenic factors contribute to variations in interest in things (vs. people) (Beltz, Swanson & Berenbaum 2011; Hell & Pabler 2011) and in tendencies to think in systemizing ways (rather than emotion-laden ways) (Auyeung, Baron-Cohen, Chapman et al. 2006; Durdiaková, Celec et al. 2015).

Assessment. We conclude with considerable confidence that self-assessed computer competency, first, basically reflects reality, and, second, that it is a USD-M. The strongest explanations appear to be evolutionary in nature.

26.2.6 Mental Health and Illness

Sex differences in mental health and illness have received a great deal of research attention. Findings are summarized below regarding mental health and illness variables with consistency scores of 95.0 or higher.

Précis 26.2.5.4 Assessment of one's ability in general (Original Table 13.1.5.1)

Age categories, number of studies, and sex	*Consist. score*	*Overviews*	*Countries*	*Time range*	*Non-human*	*Social role theory*			*Evolutionary theory*		*Theological*
						Origin	*Found*	*Biosoc*	*Select*	*ENA*	
Adult 13M	100.0	M:1Rev	1	1972–2000	-0-	Contra	NoTest	Poss-M	Poss-M	Poss-M	Silent

26.2.6.1 Behavioral Disorders and Externalizing Behavior

Evidence. Many studies have been performed on sex differences in behavioral disorders, or externalizing behavior problems. Précis 26.2.6.1 shows that although no overviews were located on this issue, studies have taken place in nearly a dozen different countries. The vast majority of these studies have found more behavioral disorders and externalizing behavior are among males, with just one exception, it finding no significant sex difference.

Explanation. Studies have shown that males are stereotyped as being more aggressive (21.2.7.5) and obnoxious (Table 21.2.8.6), while females are stereotyped as being more polite and well-mannered (Table 21.2.7.55) and obedient (Table 21.2.9.11). Together, these stereotyped sex differences could be considered indicative of males exhibiting behavioral disorders and externalizing behavior more than females. Thereby, one can argue that social role theory predicts such a sex difference. However, the original version of the theory would not predict universality in this regard.

Regarding the founder effect version of social role theory, one could argue this type of sex difference existed in the first human societies and has simply persisted ever sense. Unfortunately, of course, there is no way to actually know what sex differences actually existed in the earliest human societies.

The biosocial version of the theory would attribute the difference to the fact that, as adults, males are larger and stronger and/or the fact that only females bear offspring. This type of explanation does not seem to apply to behavioral disorders and externalizing behavior.

Turning to evolutionary lines of reasoning, one can begin by arguing that males have been sexually selected for being relatively more competitive, aggressive, and prone to take risks, as evidence suggests. If one thinks of externalizing behavior as being rather "crude" (early age) expressions of competitiveness, aggression, and risky behavior, it is possible to expect that males will have been sexually selected for exhibiting externalizing behavior more than females (Martel 2013).

Studies undertaken to assess the possible role of neuroandrogenic factors (as hypothesized by ENA theory) have been largely supportive. Specifically, most studies have found both prenatal and post-pubertal testosterone exposure to be positively correlated with aggression (*meta-analysis*: Geniole, Bird et al. 2020), conduct disorder and antisocial behavior (*reviews*: Yildirim & Derksen 2012a; Yildirim & Derksen 2012b). Regarding externalizing behavior per se, two studies have reported that it was positively correlated with prenatal testosterone exposure (Martel 2013), especially among males (Booth, Johnson et al. 2003).

Assessment. Overall, the evidence strongly supports concluding that behavioral disorders (particularly in the form of externalizing behavior) is

Précis 26.2.5.5 Self-assessed computer competency (Original Table 13.1.5.5)

Age categories, number of studies, and sex	*Consist. score*	*Overviews*	*Countries*	*Time range*	*Non-human*	*Social role theory*			*Evolutionary theory*		*Theo-logical*
						Origin	*Found*	*Biosoc*	*Select*	*ENA*	
Adult 12M All Ages 19M	100.0 100.0	-0-	5	1985–2013	-0-	Contra	NoTest	Contra	Expl-M	Expl-M	Silent

a USD-M. We will refer back to this set of disorders several times as we consider closely rated traits, such as conduct disorder and antisocial personality disorder.

26.2.6.2 Internalizing Behavior

Evidence. Internalizing behavior essentially refers to the tendency to turn inward to such a degree that one is often sad and seems to dislike oneself. Rather than acting out (as in the case of externalizing behavior), symptoms of internalizing behavior can include prolonged social isolation and even causing harm to oneself.

Findings from studies of internalizing behavior problems are summarized in Précis 26.2.6.2. One can see that 29 of the 30 studies concluded that females exhibit internalizing behavior problems more. The exception did not find a significant sex difference. The studies of this variable were conducted in nine different countries along with three multi-national studies. A meta-analysis also concluded that internalizing behavior problems are more common among females.

Explanation. If one assumes that traits such as sad and depressed are fairly similar to internalizing behavior problems, one could argue that the social role theories predict that females would exhibit internalizing behavior more. Research has indicated that, not only are females stereotyped as being sad and depressed more often than males (Table 21.2.6.11), but they are also considered more often shy and timid (Table 21.2.8.11). The original version of social role theory would not predict universality in this sex difference, contrary to the actual evidence. Regarding the founder effect version of the theory, there is no way of testing it. In the case of the biosocial version of social role theory, it would be difficult to argue that body size, strength, or capability of giving birth would have any bearing on exhibiting internalizing behavior.

Turning to the evolutionary perspective, we found no attempt to apply sexual selection theory to such behavior. However, we did locate one study that correlated internalizing behavior to prenatal testosterone exposure. As ENA theory would predict, the correlation was negative (Martel 2013). It can also be noted that internalizing behavior seems to be essentially the opposite of externalizing behavior. And, as noted in the preceding précis, the evidence points toward males exhibiting more externalizing behavior is very strong. Furthermore, in the surrounding narrative, considerable evidence points toward brain exposure to testosterone as promoting externalizing behavior. A reasonable inference would be that internalizing behavior would be inhibited by high exposure to testosterone, thus accounting for why more females than males exhibit such behavior.

Précis 26.2.6.1 Behavioral disorders and externalizing behavior (Original Table 14.1.1.5)

Age categories, number of studies, and sex	*Consist. score*	*Overviews*	*Countries*	*Time range*	*Non-human*	*Social role theory*			*Evolutionary theory*		*Theological*
						Origin	*Found*	*Biosoc*	*Select*	*ENA*	
Child 39M 1x Wide Age Range 16M All Ages 64M 1x	97.5 100.0 98.5	-0-	11	1941–1997	-0-	Contra	NoTest	Contra	Expl-M	Expl-M	Silent

Assessment. Overall, internalizing behavior can be confidently declared a USD-F. The best theoretical explanation seems to along neurohormonal lines.

26.2.6.3 Self-Diagnosed General Depression

Evidence. Sex differences in general depression is one of the most extensively studied topics that was identified in this book. Two categories (or levels) of depression were recognized, one based on self-reports (or self-diagnosis) and the other based on clinical diagnoses. Attention will first focus on self-reported general depression.

One can see in Précis 26.2.6.3 that a total of 634 studies were located, but their findings had to be broken down according to age categories because sex differences in general depression seem to vary a great deal in this regard. In particular, studies of children reveal no consistent sex differences. However, following puberty, the pattern substantially shifts, with most studies pointing toward depression being more common among females. Nonetheless, the post-pubertal differences only attain a consistency score of 95.0 in the case of the Wide Age Range group.

Forty-eight different countries (plus 14 multi-national studies) were sampled regarding sex differences in self-diagnosed depression. Also noteworthy is that 11 different literature reviews and nine meta-analyses were identified. For the ten reviews and seven meta-analyses based on adolescent and adult samples, all concluded that females self-report having more depression than is the case for males. However, the one literature review and two meta-analyses that separated out studies of children all concluded that the sex differences were not discernible.

Readers wishing to explore the cross-cultural nature of sex differences in depression may want to examine Table 14.2.1.1. This table reveals no particular nations in which one finds exceptions to the generalization that females experience more depression than males, at least among adolescents and adults. Therefore, our assessment of why quite a few exceptions were found regarding sex differences in self-reported depression following puberty is that many of the assessment measures for such depression are rather poor at discriminating degrees of depression, a conclusion also reached by another research team (Wilhelm, Parker & Hadzi-Pavlovic 1997).

Explanation. Studies have indicated that females are stereotyped as being depressed or sad more than males (Table 21.2.6.11). From this evidence, the original social role theorists might argue that societal expectations and other sex role training encourages depression more in females than in males. However, this version of the theory would have difficulty with the fact that sex differences in depression only seem to be present following the onset of puberty.

Précis 26.2.6.2 Internalizing behavior (Original Table 14.1.1.8)

Age categories, number of studies, and sex	*Consist. score*	*Overviews*	*Countries*	*Time range*	*Non-human*	*Social role theory*			*Evolutionary theory*		*Theological*
						Origin	*Found*	*Biosoc*	*Select*	*ENA*	
Adolescent 22F All Ages 29F 1x	100.0 96.7	F: 1Met	9 (3)	1986–2020	-0-	Contra	NoTest	Silent	Silent	Poss-F	Silent

Turning to the two social role theories that predict universal tendencies for sex differences in traits, Fausto-Sterling's (1992) founder effect theory would argue that female depression was present in the first human societies and has persisted ever since. Unfortunately, this assertion is not really testable. Wood and Eagly's (2012) biosocial version of social role theory traces all sex differences to the fact that males are larger and stronger, while only females can bear offspring. Some support for this line of reasoning has been found. Specifically, several studies have reported that upper body strength is inversely correlated with depression, even within each sex (van Milligen, Lamers et al. 2011; van Milligen, Vogelzangs et al. 2012; Suija, Timonen et al. 2013; Fukumori, Yamamoto et al. 2015; Hagen & Rosenstrom 2016:128). However, as will be described below, this particular finding can also be interpreted from an evolutionary perspective (Hagen 2011).

Several proposals along evolutionary lines have been offered for variations in human depression (Nettle 2004; Almeida 2011; Varga 2012). However, only a few proposals have been made to account for sex differences. One proposal has been that, due to the greater physical strength of males, they can more often deal with social frustration and conflict through aggression (or treats of aggression) than can females. As a result, females are more likely to feel frustrated, anxious, and depressed (Hagen 2011; Hagen & Rosenstrom 2016:119). It should be noted that this particular version of evolutionary theory fails to specify how female depression affects reproduction, which is essential to evolutionary reasoning in terms of sexual selection.

The other evolutionary proposal asserts that females have been sexually selected for being highly sensitive to interpersonal relationships with others (McGuire, Troisi & Raleigh 1997; Martel 2013). Some of this sensitivity is likely to be promoted by depressive symptoms whenever relationships begin to deteriorate. Such deterioration causes females to withdrawal from all but the most conflict-free social interactions. We would add to this proposal by noting sex differences in competitiveness, which appears to have been favored among males more than females (see Précis 26.3.1.5), while females are more empathetic (see Précis 26.2.2.9). In light of these differences, females are likely to be drawn toward providing care to others, including offspring, thereby enhancing their reproduction. One way to test this theoretical proposal would be to determine if women with at least modest degrees of depression tend to have more offspring than do women with little or no depression.

ENA theory leads to the hypothesis that brain exposure to testosterone helps to prevent depression. Research evidence in this regard is substantial and largely supportive, at least regarding circulating testosterone levels, particularly among males (*males*: Barrett-Connor, von Mühlen & Kritz-

Précis 26.2.6.3 Self-diagnosed general depression (Original Table 14.2.1.1)

Age categories, number of studies, and sex	*Consist. score*	*Overviews*	*Countries*	*Time range*	*Non-human*	*Social role theory*			*Evolutionary theory*		*Theological*
						Origin	*Found*	*Biosoc*	*Select*	*ENA*	
Child 20F 59x 14M	(18.7)	F: 10Rev	48 (14)	1953–2020	-0-	Contra	NoTest	Poss-F	Poss-F	Expl-F	Silent
Adolescent 108F 24x 1M	(80.6)	F: 7Met									
Adult 286F 46x 3M	(84.6)	x:1 Rev*									
Wide Age Range 71F 2x	95.4	x: 2Met*									
All Ages 485F 131x 18M	(73.3)										

Note
*Before puberty.

Silverstein 1999; A Booth, Johnson & Granger 1999; Seidman 2003; Rhoden & Morgentaler 2004; Shores, Moceri et al. 2005; Almeida, Yeap et al. 2008; Hintikka, Niskanen et al. 2009; Jockenhovel, Minnemann et al. 2009; *females*: Giltay, Enter et al. 2012; *review*: McHenry, Carrier et al. 2014; *meta-analysis*: Zarrouf, Artz et al. 2009). This inverse relationship between testosterone and depression could provide a better explanation for why upper-body strength is also inversely correlated with depression than does the biosocial social role theory (as discussed above).

Furthermore, one research team reviewed the literature regarding studies that used synthetic testosterone as a form of treatment for depression in placebo-controlled trials (Ebinger, Sievers et al. 2009:Table 1). The review indicated that 16 of the 25 relevant studies found no significant effects, but all of the remaining nine studies concluded that testosterone administration significantly diminished depression symptoms.

One study was located having to do with 2D:4D and depression. Consistent with ENA theory, the study indicated that low exposure to prenatal testosterone was positively correlated with depressive symptoms later in life among females, but not among males (Smedley, McKain & McKain 2014).

A tangential factor to consider with regard to sex differences in depression involves sex differences in meat consumption. As will be documented later in this chapter, research has strongly indicated that males eat meat (animal-based protein) more than do females (Précis 26.3.3.2). One research team suggested that, meat consumption appears to promote the production of an amino acid known as *creatine*, and that creatine seems to have anti-depressive properties (Roitman, Green et al. 2007; also see Kious, Kondo & Renshaw 2019). If this line of reasoning is correct, one could argue that both prenatal and post-pubertal testosterone appear to promote meat consumption (Wingfield, Jacobs & Hillgarth 1997; Asarian & Geary 2013), which could in turn at least partially reduce tendencies toward depression among males.

Assessment. Overall, following the onset of puberty, most evidence suggests that females suffer from depression to a greater degree than males, although the consistency scores fall below the 95.0 cutoff except in the case of the Wide Age Range samples. Note that additional evidence appears directly below with regard to clinical depression. We declare general depression to be a USD-F for adolescents and adults but not for children.

26.2.6.4 Major (Clinical) Depression

Evidence. Précis 26.2.6.4 shows that clinical (unipolar) depression appears to be more common among females. However, the only age grouping for which a consistency score was attained 95.0 was for adolescents, and especially among children, most studies report no significant sex differences. The located studies were published over a 78-year time frame. These studies were derived from 46 different countries, plus five multi-country studies. One can see that an overall sex difference has been reported by eight different reviews and four different meta-analyses.

Explanation. Theoretically explaining sex differences in clinical depression would resemble the explanations offered for general depression (see the précis directly above). However, it may also be noted that experiments in which synthetic testosterone has been administered to males have indicated that these injections are usually followed by an elevation in mood (Wang, Alexander et al. 1996; Nieschlag, Behre et al. 2004). Since females have much lower levels of testosterone than do males, especially during adolescence, one would be led to expect their mood to be lower.

Assessment. As with self-reported depression, the vast majority of studies have concluded that, following puberty, females are more likely than males to suffer from clinical depression. At least in the case of adolescents, major depression appears to be a USD-F.

26.2.6.5 Schizophrenia

Evidence. Considerable research on sex differences in schizophrenia has been reported, the vast majority of which indicates that this type of mental illness is more prevalent among males. Nevertheless, as shown in Précis 26.2.6.5, only among studies restricted to samples of adults was the consistency score above the 95.0 cutoff.

The number of countries sampled totaled 15 in addition to three multi-country studies. The studies were published over a 59-year time range. In the overview column, one can see that all ten literature reviews agreed that males suffer from schizophrenia more than do females. However, in the case of the only located meta-analysis, analyses were performed separately for samples drawn from developed (industrialized) countries and developing countries. Using this approach, schizophrenia rates were more pronounced in males living in developed countries, but no sex differences were apparent among persons residing in developing countries.

Explanation. No evidence of sex stereotypes associated with schizophrenia was found. Thus, the social role theories seem to be silent regarding any sex differences in this mental illness.

Schizophrenia has been shown to be a highly heritable disease, with genetics accounting for 85–90% of its variation within most populations (Tandon, Keshavan & Nasrallah 2008). So, how could it be sexually selected in favor of males? Some interesting proposals have been offered in this regard. These proposals basically note that families with high rates to schizophrenia tend to have unusually high proportions of highly creative

Précis 26.2.6.4 Major (clinical) depression (Original Table 14.2.1.3)

Age categories, number of studies, and sex	*Consist. score*	*Overviews*	*Countries*	*Time range*	*Non-human*	*Social role theory*			*Evolutionary theory*		*Theological*
						Origin	*Found*	*Biosoc*	*Select*	*ENA*	
Child 5F 25x 5M	(12.5)	F: 8Rev	46 (5)	1942–20-20	-0-	Contra	NoTest	Poss-F	Poss-F	Expl-F	Silent
Adolescent 71F 1x 1M	95.9	F: 4Met									
Adult 218F 22x 4M	(87.9)										
WAR 84F 8x 1M	(89.4)										
All Ages 378F 57x 11M	(82.7)										

persons, e.g., artists, musicians, and even creative scientists (GF Miller 2001; Nettle 2001). They also argue that, while fully manifested schizophrenia is detrimental to reproduction among both sexes, modest forms (often known as *schizoid-typicality*) may help to promote creative thinking, especially for males, and, by so doing, often promote their attractiveness to the opposite sex (Nettle & Clegg 2006; Del Giudice 2010).

Regarding the hypothesis derived from ENA theory, that neuroandrogenic factors contribute to schizophrenia, evidence has been mixed. Some research has been inconclusive (Elias & Kumar 2007) regarding any contribution that testosterone makes to schizophrenia, while other research seems supportive, albeit in complex ways (Goyal, Sagar et al. 2004; Moore, Kyaw et al. 2013; Misiak, Frydecka et al. 2018).

Assessment. Overall, we conclude that schizophrenia is a USD-M, at least among adults. Considerably more theoretical work is needed to understand this sex difference.

26.2.6.6 *Social Competence of Schizophrenics*

Evidence. Studies of the social competence of schizophrenia sufferers have found that females tend to have higher levels of social competence. Précis 26.2.6.6 shows that the relatively recent studies in four countries, and one literature review, have all reached this conclusion.

Explanation. The apparent greater social competence of females may large reflect one or both of two apparent sex differences: First, schizophrenia symptoms are usually less severe for females than for males (Versola-Russo 2005; also see Table 14.3.1.3 in Volume II). Second, females tend to be more socially adept than males (see Précis 26.3.1.1 and Précis 26.3.2.1). Beyond noting these two factors, no explanation for the apparent greater social competence of female schizophrenics appears possible. In particular, none of the five theories of sex differences seems to offer insight into this sex difference.

Assessment. While no specifically theories seem to pertain to social competence of schizophrenics, the number of studies, their consistency, and the number of countries sampled points toward this variable being a USD-F.

26.2.6.7 *Psychological Problems in General*

Evidence. Psychological problems refer to a variety of obviously ill-defined difficulties that people experience of a mental nature. Most of these problems seem to involve either depression or anxiety. Précis 26.2.6.7 reveals that all 12 studies located on psychological problems in general concluded that females experience them to a greater degree than males.

Précis 26.2.6.5 Schizophrenia (Original Table 14.3.1.1)

Age categories, number of studies, and sex	*Consist. score*	*Overview*	*Countries*	*Time range*	*Non-human*	*Social role theory*			*Evolutionary theory*		*Theological*
						Origin	*Found*	*Biosoc*	*Select*	*ENA*	
Adult 28M 1x WAR 29M 11F 2F All Ages 57M 12x 2F	96.6 (87.5) (78.1)	M: 10Rev M: 1Met x: 1Met	15 (3)	1951–2010	-0-	Silent	Silent	Silent	Poss-M	Poss-M	Silent

The relevant studies were based on samples from five different countries, all published within a 17-year time frame.

Explanation. Given the vagueness of this concept, and the fact that it seems to primarily reflect varying degrees and forms of depression, readers are referred to Précis 26.2.6.3 for theoretical explanations. For this reason, we have assigned the same codes under the shaded region of the present précis as was used in Précis 26.2.6.3.

Assessment. With the understanding that psychological problems primarily have to do with what is known as general depression and anxiety, we conclude with a fair degree of confidence that psychological problems in general is a USD-F.

26.2.6.8 Conduct Disorder

Evidence. Conduct disorder primarily refers to behavior patterns that are frequently irritates, and sometimes even injurious, to others. Behavior captured by the term varies from chromic disobedience, not respecting the property of others, and even serious forms of aggression toward parents, teachers, and peers. The condition is usually assessed during childhood, and, if it persists through adolescence and adulthood, it is often referred to as *antisocial personality disorder* (see below).

One can see in Précis 26.2.6.8 that, among children, all but five of 131 pertinent studies concluded that conduct disorder is more common in males than in females, with the five exceptions simply failing to report significant differences. The pertinent studies came from 19 different countries in addition to one multi-national study covering a 60-year publication time frame.

Explanation. No stereotype findings regarding sex differences in exhibiting conduct disorder were located. Therefore, we will list the social role theories as being essentially silent to the possibility of sex differences in possessing this disorder.

The basic evolutionary explanation for sex differences in the prevalence of conduct disorders is essentially the same as for behavioral disorders in general (see Précis 26.2.6.1). It seems that conduct disorder has been sexually selected as a way of laying the foundation for behavior associated with resource procurement. For additional arguments in this regard, see the section directly below having to do with antisocial personality disorder (which is sometimes considered the adolescent and adult expression of childhood conduct disorder).

ENA theory hypothesizes that conduct disorder is promoted by exposing the brain to testosterone. Specifically, at least two studies have found circulating levels of testosterone positively correlated with conduct disorder, at least among males (R Rowe, Maughan et al. 2004; Kirillova, Vanyukov et al. 2008:Table 3). Another line of evidence in this regard

Précis 26.2.6.6 Social competence of schizophrenics (Original Table 14.3.1.8)

Age categories, number of studies, and sex	*Consist. score*	*Overviews*	*Countries*	*Time range*	*Non-human*	*Social role theory*			*Evolutionary theory*		*Theological*
						Origin	*Found*	*Biosoc*	*Select*	*ENA*	
All Ages 12F	100.0	F: 1Rev	4	1990–2007	-0-	Silent	Silent	Silent	Silent	Silent	Silent

Précis 26.2.6.7 Psychological problems in general (Original Table 14.4.1.2)

Age categories, number of studies, and sex	*Consist. score*	*Overviews*	*Countries*	*Time range*	*Non-human*	*Social role theory*			*Evolutionary theory*		*Theological*
						Origin	*Found*	*Biosoc*	*Select*	*ENA*	
Adult 12F	100.0	-0-	5	1990–2007	-0-	Contra	NoTest	Poss-F	Poss-F	Expl-F	Silent
All Ages 15F	100.0										

involves a gene located on the X chromosome that influences cellular receptivity to testosterone. This gene comes in a variety of forms, known as androgen *receptors CAG repeats*. Basically, the more AR CAG repeats individuals have, the more resistant their bodies (including their brains) are to the presence of testosterone (Ellis & Hoskins 2022). Consistent with the prediction of ENA theory, one study reported an inverse correlation between AR CAG repeats and diagnosed conduct disorder (Comings, Chen et al. 1999).

Assessment. Given the large number of studies, the numerous countries sampled, and the concurrence of several prior literature reviews, we confidently declare conduct disorders to be a USD-M.

26.2.6.9 Antisocial Personality Disorder, Including Psychopathy

Evidence. *Antisocial behavior* basically encompasses all behavior that those observing such behavior consider irritating and unacceptable in a polite social setting. When such behavior is found in children, it is usually termed *conduct disorders* (see above). When such behavior is chronically displayed by adolescents and especially by adults, it is usually referred to as *antisocial personality disorder* or *psychopathy*. Sex differences in the prevalence of such behavioral tendencies have been widely reported.

One can see in Précis 26.2.6.9 that the vast majority of studies have found that males outnumber females in these disorders. However, only in the case of 34 studies recorded under the Wide Age Range category was the consistency scores above 95.0.

Explanation. No sex stereotypes pertaining to antisocial personality disorder (or psychopathy) were located. Thus, none of the three social role theories seem to have offered specific explanations for such disordered behavior patterns.

Both evolutionary theories of sex differences offer explanations for why males tend to exhibit antisocial personality traits more than do females. Much of the explanation centers around evidence that males have been sexually selected for being more competitive and victimizing in their approach to reproduction (Ellis 2005; Mededovic, Wertag & Sokic 2018; Delgado, Maya-Rosero et al. 2020). Research comparing males with competitive and victimizing tendencies have indicated that they have more offspring than males generally, although they tend to spend less time and resources on caring for their offspring (Gladden, Figueredo et al. 2009; Eme 2016; Brazil & Volk 2022). Consequently, the term *cads* (as opposed to being *dads*) has sometimes been used to refer to psychopaths (Ellis & Walsh 1997; Brazil & Volk 2022).

In the case of ENA theory, studies of antisocial personality disorder and psychopathy are substantial. As theoretically predicted, most (but not all) studies point toward both prenatal and post-pubertal brain exposure to

Précis 26.2.6.8 Conduct disorder (Original Table 14.4.3.1)

Age categories, number of studies, and sex	*Consist. score*	*Overviews*	*Countries*	*Time range*	*Non-human*	*Social role theory*			*Evolutionary theory*		*Theological*
						Origin	*Found*	*Biosoc*	*Select*	*ENA*	
Child 126M 5x Adolescent 18M 4x All Ages 147M 11x	96.2 (81.8) (93.0)	M: 8Rev	19 (1)	1957–2017	-0-	Silent	Silent	Silent	Poss-M	Expl-M	Silent

androgens contributing to antisocial and psychopathic behavior within each sex (*review*: Yildirim & Derksen 2012; Ellis, Farrington & Hoskin 2019:351–357). It is worth adding that one study confirmed this association while also indicating that androgen exposure promoted prosocial behavior as well (Dreher, Dunne et al. 2016).

Assessment. We declare that antisocial personality disorder (including psychopathy) is a USD-M. The most viable theoretical explanations for this sex differences appears to be evolutionary in nature.

26.2.6.10 Problem Drinking/Alcohol Abuse

Evidence. As shown in Précis 26.2.6.10, 26 studies of problem drinking (including alcohol abuse) were located, 25 of which indicated that males were more likely than females to exhibit these problems. Studies of what are known as problem drinking or alcohol abuse have taken place in five different countries since the middle of the 20th century. While one study found no significant sex difference, the remaining 25 studies found these problems to be more common among males.

Explanation. From adolescence through adulthood, the vast majority of studies have concluded that males drink more alcohol than do females (Table 18.1.5.1). From this, one can assume that the greater tendency for males to consume alcohol generally could be responsible for at least some of a greater male tendency to exhibit problem drinking – as well as alcoholism (see directly below).

We found no evidence of stereotyping specifically pertaining to alcohol consumption nor evidence of any attempt to use sex role theory to explain such consumption, either in general or drinking to excess. Furthermore, no attempt to use sexual selection theory to account for sex differences in alcohol consumption was located.

Regarding the neurohormonal component of ENA theory, however, some research has sought to explain sex differences in problem drinking in terms of higher brain exposure to testosterone. In this regard, one study reported that problem drinking was positively correlated with circulating testosterone (especially if this exposure is accompanied by low exposure to the stress hormone cortisol) (Witt 2007). In addition, some evidence suggests that alcohol consumption and testosterone levels interact, where both can have positive influences on one another depending on dosages (Forquer, Hashimoto et al. 2011).

In the case of prenatal testosterone exposure, most of the research has focused on alcohol dependence rather than mere problem drinking. As shown in Précis 26.2.6.11 (directly below), on will see that substantial evidence points toward high testosterone exposure during fetal development contributing to alcohol dependence.

Précis 26.2.6.9 Antisocial personality disorder, including psychopathy (Original Table 14.4.3.2)

Age categories, number of studies, and sex	*Consist. score*	*Overviews*	*Countries*	*Time range*	*Non-human*	*Social role theory*			*Evolutionary theory*		*Theological*
						Origin	*Found*	*Biosoc*	*Select*	*ENA*	
Adolescent 36M 4x Adult 82M 9x Wide Age Range 34M All Ages 152M 14x	(90.0) (88.2) 100.0 (91.6)	M: 5Rev	14 (1)	1973–2020	–	Silent	Silent	Silent	Expl-M	Expl-M	Silent

Précis 26.2.6.10 Problem drinking/alcohol abuse (Original Table 14.4.2.4)

Age categories, number of studies, and sex	*Consist. score*	*Overviews*	*Countries*	*Time range*	*Non-human*	*Social role theory*			*Evolutionary theory*		*Theological*
						Origin	*Found*	*Biosoc*	*Select*	*ENA*	
Adult 20M 1x All Ages 25M 1x	95.2 96.2	-0-	5	1953–2012	-0-	Silent	Silent	Silent	Silent	Poss-M	Silent

Overall, if evolutionary forces have helped to produce sex differences in excessive alcohol consumption, it is likely to be *coincidental* effects, such as would occur when high testosterone is favored for having other favorable effects on male reproduction. Perhaps some knowledge can be gained along these lines by determining why in rodent species, females appear to voluntarily consume more alcohol more than do males (see Table 18.1.5.1c).

Assessment. We conclude that problem drinking is a USD-M. As shown below, this conclusion is reinforced by noting that alcoholism is also a USD-M. Nevertheless, theoretically explaining this sex differences calls for more work.

26.2.6.11 Alcoholism (Alcohol Dependence)

Evidence. Alcohol dependence, usually referred to as alcoholism, has been studied in 13 countries. It has also been the subject of two multi-national studies and two literature reviews. As shown in Précis 26.2.6.11, all of this research has found alcoholism to be more common among males.

Explanation. To explain sex differences in the prevalence of alcoholism, one can refer back to the explanation offered for problem drinking (see Précis 26.2.6.10 directly above). In addition, multiple studies have indicated that, at least among alcohol dependent males, the 2D:4D finger length ratios have indicated that alcoholics were exposed to more prenatal testosterone than was the case for non-alcoholic controls (Kornhuber, Erhard et al. 2011; JT Manning & Fink 2011; Han, Bae et al. 2016; Lenz, Mühle et al. 2017; Gegerhuber, Weinland et al. 2018; Lenz, Bouna-Pyrrou et al. 2018). This, of course, is consistent with the neurohormonal component of ENA theory, which argues that alcoholism is at least partly promoted by exposing the brain to testosterone and other androgens.

Assessment. The evidence that alcoholism is more prevalent among males than among females is beyond dispute. Therefore, alcoholism is a USD-M, but more theory development is needed.

26.2.6.12 Age of Onset Alcoholism

Evidence. Précis 26.2.6.12 provides a summary of research findings having to do with sex differences in the age of the onset of alcoholism. All 19 of the studies of alcoholics found the age of onset for alcoholism to be younger for males than for females. Unfortunately, from a research standpoint, all of the studies were conducted in just one country.

Explanation. No attempt to theoretically explain age differences in the onset of alcoholism will be made beyond those pertaining to alcoholism and problem drinking (see above).

Assessment. While all of the studies of age of alcoholism onset were based on samples drawn from just one county (the United States), in light of the

Précis 26.2.6.11 Alcoholism (alcohol dependence) (Original Table 14.4.2.5)

Age categories, number of studies, and sex	*Consist. score*	*Overviews*	*Countries*	*Time range*	*Non-human*	*Social role theory*			*Evolutionary theory*		*Theological*
						Origin	*Found*	*Biosoc*	*Select*	*ENA*	
Adult 57M Wide Age Range 20M All Ages 78M	100.0 100.0 100.0	M: 2Rev	13 (2)	1965–2012	-0-	Silent	Silent	Silent	Silent	Poss-M	Silent

Précis 26.2.6.12 Age of onset alcoholism (Original Table 14.4.2.7)

Age categories, number of studies, and sex	*Consist. score*	*Overviews*	*Countries*	*Time range*	*Non-human*	*Social roles*			*Evolutionary*		*Theological*
						Origin	*Found*	*Biosoc*	*Select*	*ENA*	
All Ages 19M	100.0	-0-	1	1948–2004	-0-	Silent	Silent	Silent	Silent	Poss-M	Silent

consistency of the findings and the fact the evidence extends over a substantial time frame (i.e., 56 years), we will tentatively declare this variable to be a USD-M. As with alcoholism and problem drinking more broadly, more theoretical work is needed to explain why such a sex difference would exist.

26.2.6.13 Age of Onset of Compulsive Gambling

Evidence. As a sex difference variable, compulsive (or pathological) gambling came extremely close to being a potential USD. With a consistency score of 93.2, the vast majority of studies indicated that males are more likely than females to be diagnosed with this disorder (see Table 25.14.7).

Précis 26.2.6.13 pertains to the findings of the age at which those affected by compulsive gambling are first diagnosed. One can see that all ten pertinent studies concluded that males are diagnosed at an earlier average age than are females. This research took place in three countries and was published over a 20-year time frame.

Explanation. We found no theoretical attempts to use any of the three social role theories to account for sex differences in the age of onset for compulsive gambling. In the case of evolutionary theory, one can predict that males would be more likely to gamble, sometimes compulsively at a relatively early age, as follows: First, gambling can be considered a form of risk-taking. As noted elsewhere, risk-taking appears to be more common (or intense) among males relative to females (see Précis 26.3.1.2). Second, this sex difference in risk-taking (along with differences in competitiveness) is likely due to the usefulness of these traits for obtaining resources (Mishra, Barclay & Sparks 2017), which, in turn, is due to females preferring mates with fairly high risk-taking tendencies (Wilke, Hutchinson et al. 2006).

Furthermore, brain exposure to testosterone seems to promote risk-taking behavior (see Précis 26.3.1.2). From such reasoning, one might argue that, not only should males gamble more (when freely allowed to do so), but that they would probably start gambling at a younger average age. Presumably, this younger average age for gambling among males would also apply to the males with compulsive gambling tendencies as well as those without these tendencies. These arguments are obviously not developed well enough to fit well into any specific theory of sex differences.

Assessment. Given that just ten pertinent studies were located, and no specific theoretical explanation for sex differences in age of onset of compulsive gambling, we can only declare it to be a USD-M with caution.

26.2.6.14 Eating Disorders in General

Evidence. The results of studies of eating disorders in general are presented in Précis 26.2.6.14. All 32 of these studies found that females suffer from

Précis 26.2.6.13 Age of onset of compulsive gambling (Original Table 14.4.4.5)

Age categories, number of studies, and sex	*Consist. score*	*Overviews*	*Countries*	*Time range*	*Non-human*	*Social role theory*			*Evolutionary theory*		*Theological*
						Origin	*Found*	*Biosoc*	*Select*	*ENA*	
All Ages 10M	100.0	-0-	3	1992–2012	-0-	Silent	Silent	Silent	Silent	Silent	Silent

Précis 26.2.6.14 Eating disorders in general (original Table 14.4.5.1)

Age categories, number of studies, and sex	*Consist. score*	*Overviews*	*Countries*	*Time range*	*Non-human*	*Social role theory*			*Evolutionary theory*		*Theological*
						Origin	*Found*	*Biosoc*	*Select*	*ENA*	
Adult 23F All Ages 32F	100.0 100.0	F: 3Rev	10	1981–2019	-0-	Silent	Silent	Silent	Poss-F	Poss-F	Silent

eating disorders more than do males. This research has been conducted in ten different countries over a 38-year time span.

Explanation. No evidence that females are stereotyped as being more inclined to be develop eating disorders were found. The closest we could come to such evidence had to do with females being stereotyped as more concerned about their personal appearance (Table 21.2.4.4). In any case, no attempt by social role theorists to offer an explanation for sex differences in eating disorders was located.

Regarding the application of evolutionary reasoning, several lines of evidence are pertinent. First, twin studies have provided evidence of substantial genetic influences on people's tendencies to become extremely concerned about their bodily appearance following the onset of puberty, especially regarding the accumulation of fat (Reichborn-Kjennerud, Bulik et al. 2004; Keski-Rahkonen A, Bulik et al. 2005; Baker, Maes et al. 2009; Culbert, Burt et al. 2009). Specifically, while genetic influences appear to be close to zero prior to the onset of puberty, they usually surpass 50% in both sexes following the puberty (Klump, Perkins 2007; Culbert, Burt, Klump 2017; SM O'Connor, Culbert et al. 2020). Evidence that genes have little to no influence before puberty, but a substantial influence afterwards, leads one to suspect that sex hormones operate to trigger changes in people's attitudes toward the importance of fat reduction and weight suppression. It is also the time that females in particular begin to accumulate fat more than do males (see Précis 26.1.1.6).

The basic prediction of ENA theory is that exposure to testosterone should reduce the likelihood of developing eating disorders. Evidence in this regard is not entirely consistent. In the case of prenatal testosterone (inferred based on 2D:4D finger length), studies have indicated that prenatal testosterone reduces the risk of eating disorders (Klump, Gobrogge et al. 2006; AR Smith, Hawkeswood & Joiner 2010; LA Martin & Ter-Petrosyan 2020). This is consistent with ENA theory. Likewise, research inferring testosterone exposure from same-sex vs. opposite-sex twin data has indicated that high prenatal testosterone seems to reduce the risk of developing eating disorders (Culbert, Breedlove et al. 2008; Culbert, Breedlove et al. 2013).

In terms of non-support, two studies found no evidence that prenatal testosterone is associated with eating disorders (JH Baker, Lichtenstein & Kendler 2009; Lydecker, Pisetsky et al. 2012). No evidence for or against the possibility of post-pubertal testosterone being related to eating disorders were found. Additional findings being on ENA theory having to do with anorexia nervosa and bulimia nervosa, the main forms of eating disorders, are presented in the narratives surrounding the next two précises.

Assessment. The evidence is very strong that eating disorders can be collectively considered a USD-F. However, theories for understanding why females exhibit these disorders more are not well developed.

26.2.6.15 *Anorexia Nervosa*

Evidence. The severe restriction of food intake to avoid being fat is known as anorexia nervosa. Précis 26.2.6.15 shows that the vast majority of studies have found this condition to be more common among females. These results have been found in nine different countries and are supported by three literature reviews. The only exception was a single study that did not find a sex difference.

Explanation. Attempts to theoretically explain sex differences in anorexia nervosa, as with eating disorders in general, do not appear to be well developed. All of the social role theories seem to be silent, as is the sexual selection version of evolutionary theory.

ENA theory essentially predicts that brain exposure to androgens will inhibit development of anorexia nervosa. Evidence in this regard is mixed. Regarding circulating testosterone levels, one study of women concluded that those who were anorexic had *higher* levels than control women (Monteleone, Luisi et al. 2001:Table 2). This is contrary to what ENA theory would predict.

One study based on the 2D:4D measure provided evidence based on same-sex versus opposite-sex twins that prenatal testosterone exposure was *positively* correlated with anorexia nervosa (Quinton, Smith & Joiner 2011). This is the opposite of what ENA theory predicts.

Assessment.

26.2.6.16 *Bulimia Nervosa*

Evidence. Bulimia involves binge eating followed by acts such as vomiting or the use of laxatives to avoid weight gain. Fourth-seven studies were located having to do with sex differences in this disorder. As shown in Précis 26.2.6.16, all of them concluded that bulimia was more common among females. Samples for these studies came from 12 different countries, and were published over the course of 23 years. One can see that two literature reviews also concluded that bulimia was more prevalent among females than males.

Explanation. Of the five theories of possible universal sex differences, only ENA theory offers a testable prediction. It suggests that, this eating disorder should be associated with low brain exposure to testosterone and other androgens. One study of prenatal testosterone (based on the 2D:4D finger length measure) and bulimia was reported. It provided support for the view that exposure to testosterone prior to birth helps to prevent bulimia later in life (Oinonen & Bird 2012).

Précis 26.2.6.15 Anorexia nervosa (original Table 14.4.5.4)

Age categories, number of studies, and sex	*Consist. score*	*Overviews*	*Countries*	*Time range*	*Non-human*	*Social role theory*			*Evolutionary theory*		*Theological*
						Origin	*Found*	*Biosoc*	*Select*	*ENA*	
Adult 12F Wide Age Range 24F 1x All Ages 53F 1x	100.0 96.0 97.7	F: 3Rev	9	1976–2015	-0-	Silent	Silent	Silent	Silent	Contra	Silent

Précis 26.2.6.16 Bulimia nervosa (Original Table 14.4.5.6)

Age categories, number of studies, and sex	*Consist. score*	*Overviews*	*Countries*	*Time range*	*Non-human*	*Social role theory*			*Evolutionary theory*		*Theological*
						Origin	*Found*	*Biosoc*	*Select*	*ENA*	
Adult 25F Wide Age Range 14F All Ages 47F	100.0 100.0 100.0	F: 2Rev	12	1981–2001	-0-	Silent	Silent	Silent	Silent	Poss-F	Silent

Regarding post-pubertal testosterone, a study of women with and without bulimia found no significant difference in testosterone levels (Monteleone, Luisi et al. 2001:Table 2), while an earlier study reported a positive correlation, which would be directly contrary to ENA theory (Sundblad, Bergman & Eriksson 1994). On the other hand, a series of experiments with various drugs designed to combat bulimic symptoms concluded that drugs that block androgen receptors seemed to reduce these symptoms (Sundblad, Landén et al. 2005).

Assessment. Bulimia nervosa is an all but certain USD-F. ENA theory shows some promise in helping to understand sex differences in this disorder.

26.2.6.17 *Speech and Language Disorders in General*

Evidence. Précis 26.2.6.17 shows that when all kinds of speech and language disorders are considered together, they occur more frequently among males. The 11 studies on this topic span most of the 20th century, but they were all conducted in just two countries.

Explanation. *None of the social role theories seem to address the possibility of sex differences in speech and language disorders. In the case of the 2 evolutionary theories, they two are somewhat vague, although ENA theory implies that brain exposure to testosterone may interfere with the development of language skills.*

Various studies have indicated that brain exposure to testosterone delays language development (Lutchmaya, Baron-Cohen & Raggatt 2001; Hollier, Mattes et al. 2013; Kung, Browne et al. 2016) and is positively correlated with language disorders (Mouridsen & Hauschild 2010). It is also relevant to note that a key feature of the autism spectrum disorder also involves poor language skills, and that several studies have linked high fetal exposure to testosterone to this male-dominant disorder (see Précis 26.2.6.24).

As to how this testosterone has its effects on the development of speech and language skills, some studies suggest that when the brain is exposed to high levels of testosterone during fetal development, the tendency to process linguistic information shifts away from predominantly involving the left hemisphere to being shared more with the right hemisphere (Hassler, Gupta & Wollmann 1992). One of the results is a slowing of the processing of linguistic information. (This point is also made with respect to stuttering, as discussed below.)

Despite all of this evidence pointing toward prenatal testosterone exposure being positively correlated with slow language development along with speech and language disorders, one study reported contrary evidence. It compared 15 boys with various language disorders with 16 boys with normal language development. Based on 2D:4D finger length data, it

Précis 26.2.6.17 Speech and language disorders in general (Original Table 14.4.6.1)

Age categories, number of studies, and sex	*Consist. score*	*Overviews*	*Countries*	*Time range*	*Non-human*	*Social role theory*			*Evolutionary theory*		*Theological*
						Origin	*Found*	*Biosoc*	*Select*	*ENA*	
All Ages 11M	100.0	-0-	2	1916–1997	-0-	Silent	Silent	Silent	Silent	Poss-M	Silent

concluded that the control boys actually received higher testosterone exposure than was the case for those with language disorders (Font-Jordà, Gamundí et al. 2018). Given the imprecision of the 2D:4D measure of prenatal testosterone exposure, this study's findings need to be viewed with some degree of skepticism.

Assessment. Overall, speech and language disorders in general constitute a USD-M. The best explanations appear to be evolutionary in nature, particularly when neurohormonal factors are included.

26.2.6.18 Stuttering

Evidence. All 21 studies of sex differences in stuttering, summarized in Précis 25.2.14.18, concluded that this disorder is more common among males. These studies were published over an 89-year time frame and were based on samples drawn from eight different countries. A literature review as well as a meta-analysis also reached this same conclusion.

Explanation. No evidence of any sex stereotypes having to do with stuttering were located. Thus, it is not particularly surprising to find that all of the social role theories are silent regarding possible sex differences in stuttering.

Specific assertions that sexual selection might be responsible for sex differences in tendency to stutter were not found. Overall, it appears that both sexes have been equally selected for their linguistic abilities (Lange, Hennighausen et al. 2016), so the relevance of sexual selection is in doubt.

Regarding the possible involvement of neuroandrogenic factors, as asserted by ENA theory, most of the evidence is supportive. Specifically, prenatal testosterone (as inferred from 2D:4D), several studies have found significant relationships for one or both hands (Dönmez, Özcan et al. 2019; Yuksel, Sizer & Durak 2019) while another study reported no significant relationship (Montag, Bleek et al. 2015). Also, circulating levels of testosterone have been found positively correlated with stuttering tendencies, even among children (Selcuk, Erbay et al. 2015).

Perhaps, rather than any sex differences in stuttering being sexually selected, stuttering is a coincidental side-effect that accompanies one or more other testosterone-influenced traits that are being sexually selected. In this regard, one study indicated that when the brain is exposed to high levels of testosterone, it tends to shift the processing of linguistic information from the left to the right hemisphere (Hassler, Gupta, & Wollmann 1991).

Assessment. Overall, theoretically explaining sex differences in stuttering is still in need of development. Nonetheless, given the consistency of the findings and the number of countries sampled, we confidently declare stuttering to be a USD-M.

Précis 26.2.6.18 Stuttering (Original Table 14.4.6.3)

Age categories, number of studies, and sex	*Consist. score*	*Overviews*	*Countries*	*Time range*	*Non-human*	*Social role theory*			*Evolutionary theory*		*Theological*
						Origin	*Found*	*Biosoc*	*Select*	*ENA*	
All Ages 21M	100.0	M: 1Rev M: 1Met	8	1931–2020	-0-	Silent	Silent	Silent	Contra	Poss-M	Silent

26.2.6.19 Post-Traumatic Stress Disorder (PTSD)

Evidence. A substantial proportion of people who experience an extremely traumatic event will have a great deal of difficulty "putting the experience behind them". These people are said to be suffering from post-traumatic stress disorder (PTSD). Précis 26.2.6.19 shows that, among adolescents, females are more likely to suffer from the mental condition known as PTSD than are females. However, among other age groups, evidence for sex differences is mixed, although still strongly leaning toward females being the primary sufferers. Samples for these studies were drawn from 13 different countries, plus one multi-national study. The publication of the located studies spanned 32 years, and two meta-analyses concurred that female sufferers outnumbered male sufferers.

Explanation. According to ENA theory, PTSD symptoms should be inversely correlated with brain exposure to testosterone. The evidence located on this matter offers no support for this prediction, at least regarding circulating levels of testosterone among males. When males with and without PTSD exhibit little to no differences in circulating testosterone (JW Mason, Giller et al. 1990; Karlović, Serretti et al. 2012; Reijnen, Geuze & Vermetten 2015).

Assessment. At least among adolescents, PTSD appears to be a USD-F. No adequate theoretical explanation for this sex difference was found.

26.2.6.20 Panic Disorder

Evidence. Panic disorder refers to repeated, unexpected panic attacks involving intense fear, often associated with a racing heart and shortness of breath. Précis 26.2.6.20 shows that panic attacks have been found to be more common among females, according to all 27 studies that were located. The findings were confirmed in five different countries, plus one multi-national study. Publication of the findings spanned 28 years.

Explanation. The most specific hypothesis regarding sex differences in panic disorder comes from ENA theory. Given that panicking is usually based on fear, and fear appears to be promoted by low exposure to testosterone, the theory would lead to the hypothesis that testosterone will be inversely correlated with panic disorder. Evidence in this regard is mixed. Regarding circulating testosterone, one study found males who had this disorder had lower testosterone levels than control males, but among females, no significant difference was found (Masdrakis, Papageorgiou & Markianos 2019). However, an earlier study of males concluded that those with panic disorder did not differ from non-suffering males regarding circulating testosterone (Bandelow, Sengos et al. 1997).

Précis 26.2.6.19 Post-traumatic stress disorder (PTSD) (Original Table 14.4.7.1)

Age categories, number of studies, and sex	*Consist. score*	*Overviews*	*Countries*	*Time range*	*Non-human*	*Social role theory*			*Evolutionary theory*		*Theological*
						Origin	*Found*	*Biosoc*	*Select*	*ENA*	
Adolescent 13F All Ages 56F 9x 9M	100.0 (35.1)	F: 2Met	13 (1)	1987–2019	-0-	Silent	Silent	Silent	Silent	Contra	Silent

Précis 26.2.6.20 Panic disorder (Original Table 14.4.7.7)

Age categories, number of studies, and sex	*Consist. score*	*Overviews*	*Countries*	*Time range*	*Non-human*	*Social role theory*			*Evolutionary theory*		*Theological*
						Origin	*Found*	*Biosoc*	*Select*	*ENA*	
Adult 13F Wide Age Range 12F All Ages 27F	100.0 100.0 100.0	-0-	5 (1)	1984–2012	-0-	Silent	Silent	Silent	Silent	Mixed	Silent

It may also be relevant to note that general tendencies to experience anxiety may have relevance to panic disorder. In this regard, one study reported that males with low exposure to prenatal testosterone (as assessed using the 2D:4D ratio) reported more feelings of anxiety, but no correlation was found among females (Evardone & Alexander 2009).

Assessment. Overall, the evidence is strong that panic disorder is a USD-F. Nonetheless, theoretically explaining this sex differences is poorly developed.

26.2.6.21 Phobias in General

Evidence. Phobias refer to objects or environmental situations that some people find irrationally frightening. Précis 26.2.6.21 reveals that 34 out of 39 studies concluded that females experience more (or stronger) phobias than do males, with all of the exceptions simply reporting no significant sex difference. The consistency score for all of these studies considered together was below the 95.0 threshold, but 20 out of 21 of the studies listed in the wide age range category concluded that females experience more phobias than do males. The only exception found no significant sex difference. The number of countries sampled was ten, along with one multi-national study. The topic of phobias in general has been the subject of one multi-national study and studies in ten different countries.

Explanation. If one assumes that phobias are essentially an expression of fearfulness, both of the evolutionary theories predict that females will be more likely to have phobias than males. In terms of sexual selection theory, the prediction of greater female phobias comes from noting that males have been favored for engaging in more risk-taking activities, primarily as a way of acquiring resources with which to attract mates (see Précis 26.3.1.2).

ENA theory asserts that brain exposure to testosterone should be a major factor contributing to sex differences in phobias. Specifically, the greater the exposure the less prone toward phobias individuals should be. Evidence for this assertion is fragmentary but supportive. Among humans, testosterone appears to prevent various forms of avoidance behavior (Terburg & van Honk 2013) as well as feeling of anxiety (Enter, Spinhoven & Roelofs 2016), fear (Derntl, Windischberger et al. 2009), and phobias (Giltay, Enter et al. 2012; Zitzmann 2020:Figure 1). Similarly, among rats, males who exhibited anxiety and frequent cowering behavior had lower levels of circulating testosterone levels than did males in general (Khonicheva, Livanova et al. 2008).

In an experiment with women who were diagnosed with social anxiety disorder, half were injected with testosterone and the other half received a placebo injection. Those receiving testosterone reported significantly reduced feelings of anxiety a few hours later than did the placebo group

Précis 26.2.6.21 Phobias in general (Original Table 14.4.7.9)

Age categories, number of studies, and sex	*Consist. score*	*Overviews*	*Countries*	*Time range*	*Non-human*	*Social role theory*			*Evolutionary theory*		*Theological*
						Origin	*Found*	*Biosoc*	*Select*	*ENA*	
Adult 15F 2x Wide Age Range 20F 1x All Ages 34F 5x	(88.2) 95.2 (87.8)	-0-	10 (1)	1966–2012	-0-	Silent	Silent	Silent	Poss-F	Expl-F	Silent

(Enter, Spinhoven & Roelofs 2016). Another experiment with women found that four hours after being injected with testosterone, they expressed less fear of potentially receiving an electrical shock (Hermans, Putman et al. 2006).

Assessment. The tendency to have phobias appears to be a USD-F. Evolutionary reasoning seems to offer the best explanations, especially when this reasoning is linked to neuroandrogenic elements.

26.2.6.22 Attention Deficit Hyperactivity Disorder (ADHD)

Evidence. The last 60 years have seen a very large number of studies on attention deficit hyperactivity disorder (ADHD). Although two studies failed to find a significant sex difference, the remaining 124 studies found ADHD to be more common among males. Those studies were conducted in 18 different countries. Précis 26.2.6.22 also shows that two literature reviews and two meta-analyses reached the same conclusion as the vast majority of studies.

Explanation. None of the three social role theories appear to have ever been used to help explain sex differences in ADHD. In the case of one published effort to offer an evolutionary theory of ADHD, some mention was made to the greater prevalence of this diagnosed condition among males than among females, but few assertions were made that males with this disorder would have any reproductive advantages over those without this disorder (Shelley-Tremblay & Rosen 1996).

Regarding the possible involvement of brain exposure to testosterone, as would be hypothesized by ENA theory, there is substantial support for the connection between this type of behavior and brain exposure to testosterone. This is especially true for prenatal testosterone (Williams, Greenhalgh & Manning 2003; Baron-Cohen, Knickmeyer & Belmonte 2005; de Bruin, Verheij et al. 2006; WH James 2008). In one additional study, the pattern was found for males, but not for females (Martel, Gobragge et al. 2008).

Addition evidence of testosterone contributing to ADHD comes from at least one study of a gene located on the X chromosome that influences cellular receptivity to testosterone. This gene comes in multiple forms, known as *androgen receptors CAG repeats*. Basically, the more AR CAG repeats individuals have, the more resistant their bodies (including their brains) are to the presence of testosterone (Ellis & Hoskins 2022). Consistent with the prediction of ENA theory, one study reported an inverse correlation between AR CAG repeats and diagnosed ADHD (Comings, Chen et al. 1999).

Assessment. Even though the theoretical understanding of why ADHD would be more common among males than females is weak, the very large

Précis 26.2.6.22 Attention deficit hyperactivity disorder (ADHD) (Original Table 14.4.8.1)

Age categories, number of studies, and sex	*Consist. score*	*Overviews*	*Countries*	*Time range*	*Non-human*	*Social role theory*			*Evolutionary theory*		*Theological*
						Origin	*Found*	*Biosoc*	*Select*	*ENA*	
Child 92M All Ages 124M 2x	100.0 98.4	M: 2Rev M: 2Met	18	1962–2015	-0-	Silent	Silent	Silent	Poss-M	Expl-M	Silent

number of studies along with the number of countries studied, leaves little doubt that ADHD is a USD-M.

26.2.6.23 Attention Deficit Disorder (ADD)

Evidence. As the name implies, attention deficit disorder (ADD) is similar to ADHD except that affected individuals simply have difficulty paying attention without signs of being hyperactive. Précis 26.2.6.23 reveals that all ten studies of ADD found the disorder to be more common in males. That was also the conclusion reached by one literature review on this disorder. The studies were conducted in four countries and were published over a 40-year time frame.

Explanation. As with ADHD, none of the five theories of sex differences seem to offer much in the way of explanations for sex differences in ADD. However, ADD could be related to slow development in language skills, which, as noted in Précis 26.2.3.10, characterizes males more than females.

Assessment. ADD appears to be a USD-M. However, none of the five theories seems to offer a clear explanation for why.

26.2.6.24 Autism Spectrum Disorder (ASD)

Evidence. Autism spectrum disorder (ASD) has been studied in 16 different countries over the past 60 years. As shown in Précis 26.2.6.24, out of 69 pertinent studies, all but one concluded that males have this condition more than do females. No fewer than nine literature reviews and three meta-analyses also concluded that more males than females exhibit the autism spectrum disorder.

Explanation. No attempts to explain for sex differences in the autism spectrum disorder with social role theory were located. However, evolutionary explanations have been proposed. In particular, researchers have argued that systemizing thinking is favored at moderate levels, particularly among males, because such thinking is associated with high performance in many occupations in science, engineering, and mathematics (Kanazawa & Vandermassen 2005; Focquaert, Steven et al. 2007). However, when systemizing thinking goes to an extreme, it often manifests itself in the form of autism or at least the autism spectrum disorder (Baron-Cohen 2006; Spikins, Wright & Scott 2018). Thus, there appear to be evolutionarily advantageous for males in terms of exhibiting systemizing thinking, since it often promotes thought processes that are linked to various high-demand technical occupations, thereby usually garnering high pay and mating opportunities (Baron-Cohen 2008). But, extreme forms of systemizing thought results in a tendency to become fixated on and preoccupied by shapes and movement of inanimate objects.

Précis 26.2.6.23 Attention deficit disorder (ADD) (Original Table 14.4.8.2)

Age categories, number of studies, and sex	*Consist. score*	*Overviews*	*Countries*	*Time range*	*Non-human*	*Social role theory*			*Evolutionary theory*		*Theological*
						Origin	*Found*	*Biosoc*	*Select*	*ENA*	
All Ages 10M	100.0	M: 1Rev	4	1975–2015	-0-	Silent	Silent	Silent	Silent	Silent	Silent

Précis 26.2.6.24 Autism spectrum disorder (ASD) (Original Table 14.4.8.3)

Age categories, number of studies, and sex	*Consist. score*	*Overviews*	*Countries*	*Time range*	*Non-human*	*Social role theory*			*Evolutionary theory*		*Theological*
						Origin	*Found*	*Biosoc*	*Select*	*ENA*	
Child 44M Wide Age Range 16M All Ages 68M 1x	100.0 100.0 98.6	M: 9Rev M: 3Met	16	1962–2018	-0-	Silent	Silent	Silent	Expl-M	Expl-M	Silent

According to ENA theory, traits such as systemizing thought are promoted by neuroandrogenic factors, thereby helping to explain why males exhibit both modest and extreme forms of this type of thought. Findings have found substantial evidence that autism spectrum disorder is promoted by prenatal testosterone exposure (Manning, Baron-Cohen et al. 2001; Baron-Cohen, Knickmeyer et al. 2005; Auyeung, Baron-Cohen et al. 2006; Knickmeyer & Baron-Cohen 2006; Auyeung, Baron-Cohen et al. 2008; WH James 2008; Auyeung, Baron-Cohen et al. 2009b; Wakabayash & Nakazawa 2010). Based on a 2D:4D measure, another study concluded that autism was associated with high prenatal testosterone exposure among males, but not among females (De Bruin, De Nijs et al. 2009). Finally, one study found that saliva testosterone levels at 3–4 months of age were positively correlated with toddlers exhibiting at least mild autism symptoms (Saenz & Alexander 2013).

Assessment. The autism spectrum disorder can be confidently declared a USD-M. The social role theories appear to be silent to this sex difference, while both of the evolutionary theories predict that males would exhibit autism symptomology more than females.

26.2.6.25 Gender Identity Disorder (GID)

Evidence. Gender Identity Disorder (GID), where individuals do not feel that their gender identity is consistent with their anatomical gender. Précis 26.2.6.25 shows that all ten located regarding this disorder have concluded that it is more common among males than females. The studies were based on samples drawn from three different countries with their publication having spanned an 81-year time frame.

Explanation. Various efforts have been made to explain GID in evolutionary terms (Roughgarden 2017; Wren, Launer et al. 2019). However, to our knowledge, only the ENA version of the theory predicts high rates of GID among males than females. It does so by asserting that, in mammals (including humans), the basic sex is female, with males being a variant made possible by genes located on the Y chromosome (Ellis 2011a:554). Of course, if this theory is correct, brain exposure to androgens should contribute substantially to variations in GID. Substantial evidence supports this hypothesis (*review*: M Hines 2010).

Assessment. Evidence points toward gender identity disorder being a USD-M. The best theory for explaining this sex differences appears to be ENA theory.

26.2.6.26 Completed Suicide

Evidence. Hundreds of studies have compared the sexes regarding the contemplation of suicide, attempting suicide, and actually completing

Précis 26.2.6.25 Gender identity disorder (GID) (Original Table 14.4.8.7)

Age categories, number of studies, and sex	*Consist. score*	*Overviews*	*Countries*	*Time range*	*Non-human*	*Social role theory*			*Evolutionary theory*		*Theological*
						Origin	*Found*	*Biosoc*	*Select*	*ENA*	
All Ages 10M	100	-0-	3	1937–2018	-0-	Silent	Silent	Silent	Silent	Expl-M	Silent

suicide. Most of these studies indicate that females contemplate and attempt suicide more, while males actually complete suicide more, partly because they tend to use more lethal methods (see Table 25.14.13). Nevertheless, as shown in Précis 26.2.6.26, only in the case of completed suicides among adolescents was the consistency score beyond the 95.0 cutoff.

Explanation. We found no attempts by proponents of social role theory to explain sex differences in the completion of suicide. Also, no stereotypes were found regarding sex differences in this regard either.

Since suicide brings an end to life and evolutionary theory is about perpetuating life, there are some inherent challenges to efforts to understanding suicide from an evolutionary perspective. Nevertheless, there are ways of fitting suicidal behavior within an evolutionary framework (Kruger & Nesse 2004; Aubin, Berlin & Kornreich 2013). Most applications of evolutionary theory to suicidal behavior recognize that an individual's own life is not as important over the long term as the genes that one is carrying and seeking to transmit to future generations. Nevertheless, we found not attempt to explain why males would be universally more likely to commit suicide than females strictly based on sexual selection theory.

Regarding the possibility of brain exposure to testosterone contributing to sex differences in suicide, some proposals have been made (TR Rice & Sher 2017). One Brazilian study of males who had attempted (but not completed) suicide had higher circulating levels of testosterone than control males, albeit not to a significant degree (Perez-Rodriguez, Lopez-Castroman et al. 2011). A Chinese study concluded that circulating testosterone levels among suicide attempters were significantly higher for both males and females when compared to controls, although it was statistically significant only in the case of males (Zhang, Jia & Wang 2015). In the case of prenatal testosterone, a German study of the 2D:4D finger lengths of male cadavers of suicide victims indicated that they had higher exposure than control males, while no significant difference was reported in the case of females (Lenz, Thiem et al. 2016). The latter research team also compared average 2D:4D finger lengths among multiple countries and each country's male suicide rates. This study indicated that national increases in prenatal testosterone among males were positively correlated with male suicide rates (Lenz & Kornhuber 2018: Figure 1). Finally, a recent literature review concluded that age may be a factor in explaining intra-male variations in the association between testosterone and suicide. It asserted that among adolescents and young adults, the association may be positive, while testosterone may actually be inversely correlated with suicide among elderly males (Sher 2013; also see Rice & Sher 2017).

Assessment. Among adolescents, males appear to commit suicide at higher rates than females, making completed suicide a USD-M, at least for

Précis 26.2.6.26 Completed suicide (Original Table 14.4.9.3)

Age categories, number of studies, and sex	*Consist. score*	*Overviews*	*Countries*	*Time range*	*Non-human*	*Social role theory*			*Evolutionary theory*		*Theological*
						Origin	*Found*	*Biosoc*	*Select*	*ENA*	
Adolescent 34M Adult 42M 2x 5F WAR 176M 7x 18F All Ages 252M 9x 23F	100.0 (70.8) (76.7) (82.1)	M: 3Rev	34 (44)	1881–2020	–	Silent	Silent	Silent	Silent	Poss-M	Silent

this age group. The neurohormonal component of ENA theory appears to help to account for this sex difference.

26.2.7 Attitudes, Beliefs, Interests, and Preferences

A large number of studies have sought to determine if males and females agree or disagree regarding all manner of attitudes, beliefs, interests, preferences. Findings from these studies for which consistency scores of 95.0 or higher are summarized below.

26.2.7.1 Valuing Power

Evidence. The tendency to value power, especially in social relationships, is usually assessed with questionnaire items that were first developed and refined by Shalom Schwartz (Schwartz & Bilsky 1987). Précis 26.2.7.1 shows that all 11 studies of sex differences in valuing power have concluded that males do so more than females. The number of countries sampled was four (along with two multi-country studies), covering a 27-year publication time frame.

Explanation. The original social role theory argues that sex differences in the value placed on power can be attributed to sex differences in social training, with males being trained to consider power more important than do females (Lips & Lawson 2009; Pietraszkiewicz, Kaufmann & Formanowicz 2017). However, if all that is involved is social training, one would expect to find some cultures in which there were no sex differences and even some where females would value power more than males. Both the founder effect version and the biosocial version of the social role theory stipulate conditions that might explain the failure to identify sex differences in this trait.

Both evolutionary theories envision sex differences in the tendencies to value power over others as a way of promoting dominance and status, thereby enhancing the potential for reproduction. In other words, males with tendencies to value power are likely to attain relatively high status, thereby leaving more offspring in future generations than males without such values. This prediction is consistent with evidence from a twin study that valuing power is genetically influenced (Knafo & Spinath 2010:728).

In the case of ENA theory, the evolutionary reasoning is augmented with predicts that brain exposure to androgens promotes the tendency to value power. No specific evidence in this regard was found. However, some indirect evidence has been published. In this regard, a literature review indicated that valuing power was strongly correlated with psychopathy (Palmen, Kolthoff & Derksen 2021), a trait that appears to be more common among males (see Précis 26.2.6.9). Psychopathy also appears to be positively correlated with circulating testosterone (*review*: Ellis, Farrington & Hoskin 2019:357).

Précis 26.2.7.1 Valuing power (Original Table 15.1.2.5)

Age categories, number of studies, and sex	*Consist. score*	*Overviews*	*Countries*	*Time range*	*Non-human*	*Social role theory*			*Evolutionary theory*		*Theological*
						Origin	*Found*	*Biosoc*	*Select*	*ENA*	
All Ages 11M	100.0M	-0-	4 (2)	1987–2011	-0-	Silent	NoTest	Poss-M	Poss-M	Expl-M	Yes

Assessment. Overall, while more studies would be worth obtaining, the fact that the sex difference in tendencies to value power can be theoretically explained leads us to declare valuing power a USD-M.

26.2.7.2 *Preference for, or Acceptance of, Aggressive Responses to Conflict*

Evidence. People vary in their tendencies to prefer or accept the "necessity" of aggressive responses to interpersonal, group, or national conflicts. As shown in Précis 26.2.7.2, findings from 28 different studies, based on samples obtained from four different countries and covering a 75-year publication time span, all point toward males being more likely than females to prefer (or accept) the use of aggressive responses to conflicts.

Explanation. It seems reasonable to theoretically explain sex differences in such attitudes and preferences as similar to theoretically explaining sex differences in the *actual* use of violence and aggression when violence arise. As will be explained more later in this chapter, violence and aggression have usually been found to be more common among males than females. Social role theories and evolutionary theories have all offered explanations for these sex differences.

In the case of the social role theories, the explanations are that sex differences are not "innate", but due to social learning. If we assume these sex differences are *entirely* learned, one should not expect the differences to be universal, as the original social role theory asserts. However, the other two versions of social role theory – i.e., the founder effect theory and the biosocial role theory – are able to account for why all societies might have some universal sex differences of a behavioral nature. In the case of Fausto-Sterling's (1992) founder effect theory, there could be universal sex differences simply as the result of "momentum" established in the first human societies that arose. This particular explanation is difficult to test, since no data are available for sex differences in the first human societies.

Regarding Wood and Eagly's (2002) biosocial role theory, one might argue that the greater acceptance of aggression by males is attributable to their larger size and strength compared to females. In other words, size and strength differences could elevate one's confidence in being successful when engaged in violent conflicts. If the theory espoused by Wood and Eagly is true, sex differences in these attitudes would be eliminated by statistically controlling for body size and strength. No evidence bearing on this hypothesis was located.

Both of the evolutionary theories assert that aggressive males are likely to have passed their genes on to future generations than were their less aggressive male counterparts, at least historically (Georgiev, Klimczuk et al. 2013). This assertion has been offered with regard to interpersonal same-sex conflict (Lindenfors & Tullberg 2011) and conflicts between

Précis 26.2.7.2 Preference for, or acceptance of, aggressive responses to conflicts (Original Table 15.2.1.10)

Age categories, number of studies, and sex	*Consist. score*	*Overviews*	*Countries*	*Time range*	*Non-human*	*Social role theory*			*Evolutionary theory*		*Theological*
						Origin	*Found*	*Biosoc*	*Select*	*ENA*	
Adult 17M	100.0M	-0-	4	1927–2002	-0-	Silent	NoTest	Poss-M	Poss-M	Expl-M	Silent
All Ages 28M	100.0M										

groups of individuals, the latter having been termed the *male warrior hypothesis* (Van Vugt 2009).

To the above evolutionary arguments, ENA theory adds the premise that neuroandrogenic factors are involved. Specifically, it asserts that brain exposure to testosterone promotes aggressive tendencies, and, because males are exposed to more of this hormone than females, they should be more aggressive. There is evidence to support this hypothesis regarding both aggression and attitudes toward aggression (Baucom, Besch & Callahan 1985; Johnson, McDermott et al. 2006).

Assessment. The number of studies and the time frame during which they were conducted are sufficient to consider preference for, or acceptance of, aggressive responses to conflict a USD-M.

26.2.7.3 Importance of Humane Treatment of Animals

Evidence. People place different degrees of importance on treating animals properly, with proper treatment often based loosely on how they think humans should be treated in similar situations. Précis 26.2.7.3 reveals that all 26 studies, taking place over a 20-year period in four different countries in Europe or North America found females place greater importance on the humane treatment of animals. The one literature review also reached this conclusion.

Explanation. As to why females would be more supportive of humane treatment of animals in all countries studied, four theories may provide at least a partial answer, provided one assumes that humane behavior is akin to empathy and kindness, and that these latter concepts are generally the opposite of violence and cruelty. Evidence pertaining to these assumptions are largely supportive (*review*: McPhedran 2009). As indicated elsewhere in this chapter, females tend to be more empathetic (see Précis 26.2.2.9) and less prone toward violence and cruelty (see Précis 26.3.4.13).

No specific social role theories were found to offer explanations for the universal (or near universal) tendency for females to be more empathetic than males. However, numerous theorists have offered evolutionary explanations for sex differences in empathy (e.g., Hampson, van Anders & Mullin 2006; De Waal 2008; Vongas & Al Hajj 2015). Most of these proposals have centered around what is known as the *primary caretaker hypothesis.* It asserts that empathy is especially important for developing skills in competently caring for offspring or other close relatives and that females have been favored for providing such care to a greater degree than males (Hampson, van Anders & Mullin 2006; Christov-Moore, Simpson et al. 2014). Others have proposed that the greater empathy of females helps to explain why are usually less aggressive and violent than males (Carlo, Raffaelli et al. 1999).

Précis 26.2.7.3 Importance of humane treatment of animals (Original Table 15.2.2.5)

Age categories, number of studies, and sex	*Consist. score*	*Overviews*	*Countries*	*Time range*	*Non-human*	*Social role theory*			*Evolutionary theory*		*Theological*
						Origin	*Found*	*Biosoc*	*Select*	*ENA*	
Wide Age Range 16F	100.0F	F: 1Rev	4	1980–2001	-0-	Silent	Silent	Silent	Poss-F	Expl-F	Silent
All Ages 26F	100.0F										

Regarding ENA theory, quite a few studies of the relationship between empathy and testosterone have been conducted. Most of these studies have provided support for the conclusion that brain exposure to testosterone diminishes feelings of empathy, particularly in the case of females (see the narrative around Précis 26.2.2.9 for a review).

Assessment. Overall, females consider humane treatment of animals to be more important than males, making this a USD-F. The best explanations seem to be of an evolutionary nature if one is willing to accept that humane treatment of animals is substantially motivated by empathy.

26.2.7.4 *Belief in Astrology or Palm Reading*

Evidence. Ten studies, taking place in five countries in three different parts of the world between 1972 and 1990, found that more females than males self-reported belief in astrology or palm reading. As shown in Précis 26.2.7.4, all ten of these studies used adult research participants.

Explanation. We found nothing in the scientific literature offering a theoretical explanation for sex differences in belief in astrology or palm reading. The closest that was found involved a study by Randall and Desrosiers (1980) in which they reported women being more prone to believe in the supernatural, and that supernaturalism was positively correlated with external locus of control. This finding suggests that, if one were to theoretically explain sex differences in subscribing to external locus of control, advances might be made in explaining sex differences in belief in astrology or palm reading.

Assessment. Because only ten studies were located regarding sex differences in belief in astrology and palm reading, and little work was located regarding theoretically explaining such sex differences, we will withhold making a declaration regarding the USD status of these two related belief variables.

26.2.7.5 *Preference for the Color Blue*

Evidence. As shown in Précis 26.2.7.5, studies taking place over nine decades in 11 countries have found a stronger preference for the color blue among males than among females. However, this difference was only found among adults.

Explanation. Little in the way of theoretical explanations were found for any color preferences might vary by sex. However, a couple of studies have indicated that female retinas appear to be somewhat more sensitive than the retinas of males to colors in the red-yellow light spectrum (McGuinness & Lewis 1976; Hoyenga & Wallace 1979).

Assessment. Because just ten studies of sex differences in preference for one color (blue) were found for one age group (adults) but not for other

Précis 26.2.7.4 Belief in astrology or palm reading (Original Table 15.3.2.1)

Age categories, number of studies, and sex	*Consist. score*	*Overviews*	*Countries*	*Time range*	*Non-human*	*Social role theory*			*Evolutionary theory*		*Theological*
						Origin	*Found*	*Biosoc*	*Select*	*ENA*	
Adult 10F All Ages 10F	100.0F 100.0F	-0-	5	1972–1990	-0-	Silent	Silent	Silent	Silent	Silent	Silent

Précis 26.2.7.5 Preference for the color blue (Original Table 15.4.3.4)

Age categories, number of studies, and sex	*Consist. score*	*Overviews*	*Countries*	*Time range*	*Non-human*	*Social role theory*			*Evolutionary theory*		*Theological*
						Origin	*Found*	*Biosoc*	*Select*	*ENA*	
Adult 10M	100.0M	-0-	5	1933–2014	-0-	Silent	Silent	Silent	Silent	Silent	Silent

age groups, and because no specific theoretical explanation could be located, preference for the color blue will not be considered a USD.

26.2.7.6 *Strength of Sex Drive (Desire for Sex)*

Evidence. Précis 26.2.7.6 shows that over half of a century of studies conducted in a large number of countries, and several overviews, have found males have a stronger sex drive. The evidence often consists of answers to questions about such topics as wanting sex more often, or the frequency of thinking about sex.

Explanation. Males have been found to be stereotyped as having stronger sex drives than females (Table 21.2.10.2). From this stereotype, social role theorists might argue that, were it not for this stereotype, the sexes would not differ regarding their sex drives. The original version of social role theory would not predict that such a stereotype would exist throughout the world (as appears to be the case). In order to the founder effect version of the theory to explain why males have a stronger sex drive, one would have to assume that such this sex difference would have been present since the earliest human societies existed, an obviously untestable assumption. The biosocial version of social role theory would contend that males have a stronger sex drive than females because they are taller, physically stronger or are unable to bear offspring. The connections drawn in this regard seem rather difficult to follow unless one links all of these sex differences to differential exposure to testosterone, including prenatal testosterone. Such links would obviously undercut the basic theme of social role theory, since it attributes all sex differences of a cognitive and behavioral nature to learning within a sociocultural context.

Both evolutionary theories lead one to expect males to have a stronger sex drive. Specifically, sexual selection theory notes that males have a higher potential reproductive ceiling than do females, provided that they can have sexual relationships with multiple sex partners. In a review of pertinent literature available at the end of the 20th century, the following conclusion was reached: "It would be premature to declare that a substantial part of the gender difference in sex drive is biologically innate, but we think the evidence is pointing in that direction" (Baumeister, Catanese & Vohs 2001:269). Since then, evidence supporting this conclusion has continued to grow (Lippa 2009; Holloway & Wylie 2015; J Archer 2019; Frankenbach, J., Weber et al. 2022).

In the case of ENA theory, it asserts that neuroandrogenic factors help to promote stronger sex drives among males than among females. Supportive evidence comes from noting that, circulating levels of testosterone have been found positively correlated with people's sex drive (*among males*: Vignozzi, Corona et al. 2005; Brock, Heiselman et al. 2016; Wu, Zitzmann et al. 2016;

Précis 26.2.7.6 Strength of sex drive (desire for sex) (Original Table 15.6.1.1)

Age categories, number of studies, and sex	*Consist. score*	*Overviews*	*Countries*	*Time range*	*Non-human*	*Social role theory*			*Evolutionary theory*		*Theological*
						Origin	*Found*	*Biosoc*	*Select*	*ENA*	
Adult 50M 1F All Ages 54M 1F	96.2F 96.4F	M: 2Rev; 1Met	11 (3)	1967–2018	-0-	Contra	NoTest	Contra	Expl-M	Expl-M	Silent

among females: Purifoy, Koopmans & Tatum 1980; *both sexes*: Holloway & Wylie 2015; Arnocky, Carre et al. 2018:Figure 3; Raisanen, Chadwick et al. 2018). Circulating testosterone also appears to promote "mate-seeking behavior" (Geniole & Carre 2018). Actual sexual behavior has also been found positively correlated with circulating testosterone, at least among males (Halpern, Udry & Suchindran 1998). Furthermore, several studies have administered synthetic testosterone to hypogonadal males, nearly all of which have resulted in a significant increase in various forms of sexual activity (Jockenhovel, Minnemann et al. 2009; *meta-analysis*: Corona, Isidori et al. 2014). Additional supportive evidence involving both sexes has also been reported (*meta-analysis*: Achilli, Pundir et al. 2017). All of these lines of evidence point toward circulating levels of testosterone substantially promoting the sex drive of both sexes.

Assessment. While cultural factors may serve to both accentuate and dampen sex differences in the sex drive, evidence points toward both evolutionary and neurohormonal factors as contributing to this trait. Overall, the sex drive of males appears to be strong that than of females, making this variable a well-established USD-M. The two evolutionary theories offer specific explanations for this sex difference, while none of the social role theories seem to do so in ways that fit with the pertinent evidence.

26.2.7.7 Desire for Promiscuous Sexual Relationship

Evidence. Fifty-four studies, conducted in numerous countries, have investigated sex differences in tendencies to desire promiscuous sexual relationships. Précis 26.2.7.7 shows that all but one study concluded that males have a greater desire for promiscuous sexual relationships. The only exception found no significant sex difference. These studies were conducted in eight different countries (plus four multicultural studies), and were published over a time span of 48 years. Furthermore, one meta-analysis also concluded that males surpass females in having these desires.

Explanation. A distinction can be made between a desire for promiscuous sexual relationships and having a strong sex drive (the latter having been discussed above). Also, desiring promiscuous sex and actually engaging in it are obviously not the same. The one thing all three of these variables appear to have in common is that males surpass females. Therefore, from a causal standpoint, they are likely to share factors in common.

In evolutionary terms, promiscuous sexual behavior would be sexually selected in males more than in females due to the much shorter amount of time needed to produce offspring by males than by females (Geher & Kaufman 2013). Also, the only way males can capitalize on their higher reproductive potential is by having multiple sex partners (Berger 1989; Mysterud, Langvatn & Stenseth 2004).

Précis 26.2.7.7 Desire for promiscuous sexual relationship (Original Table 15.6.1.6)

Age categories, number of studies, and sex	*Consist. score*	*Overviews*	*Countries*	*Time range*	*Non-human*	*Social role theory*			*Evolutionary theory*		*Theological*
						Origin	*Found*	*Biosoc*	*Select*	*ENA*	
Adults 46M 1x All Ages 53M 1x	97.9M 98.1M	M: 1Met	8 (4)	1973–2021	-0-	Silent	Silent	Silent	Expl-M	Expl-M	Silent

ENA theory predicts that brain exposure to androgens contributes to sex differences in the desire for promiscuous sexual relationships. Evidence from three studies all indicated that desired number of sex partners were positively correlate with circulating testosterone (Bogaert & Fisher 1995; Hirschenhauser 2012; Klimas, Ehlert et al. 2019).

Assessment. As with the strength of an individual's sex drive, we conclude that males have greater desires for promiscuous sexual relationships than do females, making it a USD-M. Only the two evolutionary theories specifically predict such sex differences.

26.2.7.8 *Bisexual Preference (Sexual Attraction to Both Sexes)*

Evidence. Bisexuality refers to the tendency to find both sexes sexually attractive. As shown in Précis 26.2.7.8, 19 out of 20 studies along with one literature review, found that females are more likely to have a clear sexual attraction to both sexes. The remaining study did not find a difference between the sexes.

Explanation. The involvement of testosterone and other sex hormones in shaping sexual orientation has considerable empirical support, both in humans (Ellis, Lykins et al. 2015; Bogaert & Skorska 2020) and in other animals (Wallen & Parson 1997). Regarding having bisexual preferences specifically, however, the evidence is still rather sketchy, although some evidence points to neurohormonal influences as well (Olvera-Hernández, Tapia-Rodríguez et al. 2017; Jeffery, Shackelford et al. 2019).

Assessment. Overall, the evidence is largely supportive of the conclusion that bisexuality is more common among females than among males, giving us considerable confidence in deeming bisexual preference a USD-F.

26.2.7.9 *Ambivalent Sexual Orientation*

Evidence. Some individuals have what is known as an *ambivalent sexual orientation*, i.e., one that involves being potentially attracted to either sex. It is similar to bisexuality. All of the 17 studies summarized in Précis 26.2.7.9 concluded that females are more likely to have an ambivalent sexual orientation. These studies have been conducted in four different countries and published over a 49-year time frame.

Explanation. We do not see any theoretical explanation for why females would be more ambivalent regarding sexual orientation than is the case for males. The only possibility involves the neurohormonal component of ENA theory, since it envisions male sex hormones (especially testosterone and dihydrotestosterone) producing stronger sex drives than any of the female sex hormones (Kaufman & T'sjoen 2002; Federman 2006).

Précis 26.2.7.8 Bisexual preference (sexual attraction to both sexes) (Original Table 15.6.2.3)

Age categories, number of studies, and sex	*Consist. score*	*Overviews*	*Countries*	*Time range*	*Non-human*	*Social role theory*			*Evolutionary theory*		*Theological*
						Origin	*Found*	*Biosoc*	*Select*	*ENA*	
All Ages 19F 1x	95.0M	F: 1Rev	4 (2)	1992–2019	-0-	Silent	Silent	Silent	Silent	Expl-M	Silent

Précis 26.2.7.9 Ambivalent sexual orientation (Original Table 15.6.2.4)

Age categories, number of studies, and sex	*Consist. score*	*Overviews*	*Countries*	*Time range*	*Non-human*	*Social role theory*			*Evolutionary theory*		*Theological*
						Origin	*Found*	*Biosoc*	*Select*	*ENA*	
Adult 17F All Ages 17F	100.0 100.0	F: 1Rev	4	1934–2013	-0-	Silent	Silent	Silent	Silent	Poss-F	Silent

Assessment. The evidence that greater proportions of females than males exhibit ambivalence regarding sexual orientation is strong, making this a USD-F. Theoretical explanations for this sex difference need further development.

26.2.7.10 *Favorable or Accepting Attitudes Toward Homosexuals or Homosexual Behavior*

Evidence. Précis 26.2.7.10 displays the results of the many studies indicating that females tend to have more positive or accepting attitudes toward homosexuality. These studies took place in over a dozen countries and cover more than a 40-year period.

Explanation. Depending on how questions are phrased, most studies have indicated that the majority of people are exclusively (or almost exclusively) sexually attracted to members of the opposite sex (Poston & Baumle 2010:520). Most studies have indicated that this nearly exclusive preference is stronger for males than among females (Rieger & Savin-Williams 2012:Table 1; Fergusson, Horwood et al. 2005:975).

The best overall explanation for why most people exhibit predominantly heterosexual preferences is that such preferences are more strongly favored from a reproductive standpoint than any alternative preferences (Ellis & Ames 1987; EM Miller 2000). Regarding people's varying tendencies to acceptance or at least not condemn homosexuality (or homosexuals), it is reasonable to assume that most heterosexuals do not find pleasure in engaging in homosexual behavior themselves. If so, it is reasonable to argue along evolutionary lines that males have been sexually selected for being primarily attracted to females, while females have been sexually selected for mainly preferring males as sex partners (Ellis & Ames 1987:251; Russock 2011). This being the case could result in both sexes having negative emotional feelings toward homosexual behavior. But why would females have less intense negative feelings than males? Neither the sexual selection version or the ENA version of evolutionary theory seems to offer a specific answer to this question.

In the case of attitudes toward homosexuality, it is possible to explain sex differences in religious terms. Specifically, the only scriptural passages condemning homosexuality pertain to such behavior among males, not among females. The most specific passage in this regard is Leviticus (20:13), where it states that "if a man lies with a man as with a woman, both of them have committed an abomination." Nowhere is a woman lying with a woman similarly condemned. Therefore, one might argue that the main reason males are less tolerant of homosexuality than females are (especially regarding male homosexuality) is that it is more explicitly condemned by biblical scriptures.

Précis 26.2.7.10 Favorable or accepting attitudes toward homosexuals or homosexual behavior (Original Table 15.6.3.2)

Age categories, number of studies, and sex	*Consist. score*	*Overviews*	*Countries*	*Time range*	*Non-human*	*Social role theory*			*Evolutionary theory*		*Theological*
						Origin	*Found*	*Biosoc*	*Select*	*ENA*	
Adult 74F 2x All Ages 86F 3x	97.4 96.6	F: 1Met	14 (1)	1965–2006	-0-	Silent	Silent	Silent	Silent	Silent	Poss-F

Assessment. Having favorable or at least less unfavorable attitudes toward homosexuals or homosexual behavior is a USD-F. Explaining this sex difference calls for more scientific investigation.

26.2.7.11 *Attribution of Sexual Meaning/Motivation to Male-Female Interactions*

Evidence. The 27 studies illustrated in Précis 26.2.7.11 all found that males are more likely to attribute sexual meaning or sexual motivation to male-female interactions. These studies were conducted in two countries over a 30-year period.

Explanation. Most studies of the tendency to attribute sexual meaning or motivation to male-female interactions have to do with how research participants think when interacting with a member of the opposite sex. As such, it is reasonable to assume that most of these attributions reflects what is known as *wishful thinking*. If so, one can predict that males would be more likely to make such attributions than females because males are more interested in sexually interacting with others than are females (see Précis 26.2.7.6 and Précis 26.2.7.7). This assumption would lead to the assumption that the causes of attributing sexual meaning or motivation to male-female interactions are the same as the causes of sexual interests generally. As noted for both of these précis, evolutionary theories seem to provide the most persuasive explanations for male's relatively high sexual interests.

Assessment. Theoretically explaining sex differences in the tendency to attribute sexual meaning or motivation to male-female interactions is still tentative and more countries need to be sampled. Nevertheless, the substantial number of studies provides confidence in concluding that this variable is a USD-M.

26.2.7.12 *Preference for a Shorter Time between Sexual Attraction and Sexual Intimacy*

Evidence. The amount of time following initial sexual attraction to a person and sexual intimacy with that person can vary considerably. All of the 11 studies presented in Précis 26.2.7.12 found that males reported a preference for a shorter time period between sexual attraction and sexual intimacy. However, it is worth noting that all of these studies were conducted in just one country, the United States.

Explanation. Both of the evolutionary theories would predict that males would prefer shorter periods of time between being sexually attracted to another person (usually a female) than would females. This prediction emanates from noting that males have been sexually selected for preferring to have multiple mates. In addition, ENA theory would predict that such

Précis 26.2.7.11 Attribution of sexual meaning/motivation to male-female interactions (Original Table 15.6.4.5)

Age categories, number of studies, and sex	*Consist. score*	*Overviews*	*Countries*	*Time range*	*Non-human*	*Social role theory*			*Evolutionary theory*		*Theological*
						Origin	*Found*	*Biosoc*	*Select*	*ENA*	
Adult 27M All Ages 27M	100.0 100.0	–	2	1974–2006	-0-	Silent	Silent	Silent	Poss-M	Poss-M	Silent

Précis 26.2.7.12 Preference for a shorter time between sexual attraction and sexual intimacy (Original Table 15.6.4.9)

Age categories, number of studies, and sex	*Consist. score*	*Overviews*	*Countries*	*Time range*	*Non-human*	*Social role theory*			*Evolutionary theory*		*Theological*
						Origin	*Found*	*Biosoc*	*Select*	*ENA*	
Adult 11M All Ages 11M	100.0M 100.0M	–	1	1977–2001	-0-	Silent	Silent	Silent	Yes	Yes	Silent

preferences would be promoted by exposing the brain to androgens. No specific evidence was found pertaining to this prediction.

Assessment. While the evidence is quite consistent and reasonable theoretical explanations are apparent, we will not declare this to be a USD mainly due to the fact that just one country was sampled. Also, no pertinent overview studies were located.

26.2.7.13 Relationship between Self-Reported Sexual Arousal and Physiological Measures

Evidence. Sexual arousal is often indicated by self-report, but it can also be measured by physiological measurements (enlargement of the penis or the clitoris). When both measurements are used, it is possible to determine if they reach the same conclusion. Précis 26.2.7.13. shows that ten studies all agreed that self-reported sexual arousal and physiological measures of sexual arousal were more strongly correlated among males.

Explanation. No theoretical explanations were found to explain why such a correlation would be stronger in males than in females. Instead, the sex difference could be due to the fact that the penis is a considerably larger organ than is the clitoris, thereby making measurement of its enlargement more accurate.

Assessment. All of the research is in agreement, and this agreement was confirmed by a meta-analysis. Nevertheless, because of the fairly small number of studies involved in assessing this sex difference and the failure to identify any specific theoretical explanation, we will not consider this to be a USD.

26.2.7.14 Preference for a Physically Attractive Mate

Evidence. One hundred and fifty-two studies, conducted in 23 countries (plus 14 multi-country studies), have been conducted over an 80-year time frame to assess sex differences in preferences for mates who were physically attractive as a primary criterion which choosing a sex or marital partner. Précis 26.2.7.14 shows that, when all results were combined, the findings fell slightly below the 95.0 consistency score. Only in the case of the wide age range samples was the consistency score above this threshold.

Explanation. Nothing in any of the three social role theories seem to address this particular sex difference. According to both evolutionary theories, males and females should prefer physically attractive mates inasmuch as attractiveness provides rough indicators of good health and fecundity (BJ Ellis 1992; Barber 1995; Singh 2006; Little, Jones & DeBruine 2011). However, in order to maximize the number of offspring they can have, females should be even more interested in evidence of their prospective mates being good providers than is the case for males

Précis 26.2.7.13 Relationship between self-reported sexual arousal and physiological measures (Original Table 15.6.4.16)

Age categories, number of studies, and sex	*Consist. score*	*Overviews*	*Countries*	*Time range*	*Non-human*	*Social role theory*			*Evolutionary theory*		*Theological*
						Origin	*Found*	*Biosoc*	*Select*	*ENA*	
Adult 10M All Ages 10M	100.0M 100.0M	M: 1Met	2	1977–2007	-0-	Silent	Silent	Silent	Silent	Silent	Silent

Précis 26.2.7.14 Preference for a physically attractive mate (Original Table 15.6.5.2)

Age categories, number of studies, and sex	*Consist. score*	*Overviews*	*Countries*	*Time range*	*Non-human*	*Social role theory*			*Evolutionary theory*		*Theological*
						Origin	*Found*	*Biosoc*	*Select*	*ENA*	
Adult 131M 8x Wide Age Range 13M All Ages 144M 8x	94.2M 100.0M 94.7M	M: 1Gen 4Rev 1Met	23 (14)	1936–2020	-0-	Silent	Silent	Silent	Expl-M	Poss-M	Silent

(Buss & Shackelford 2008; Davis & Amocky 2022). The end result of these competing interests in prospective mates seems to be that females are more willing to "trade off" good-provider indicators for physical attractiveness, especially when looking for a long-term mate (Bereczkei, Voros et al. 1997; Scheib 2001; Buss & Shackelford 2008).

While both evolutionary theories predict greater male preference for mates who are physically attractive (Davis & Amocky 2022), ENA theory also predicts that neurohormonal influences are involved in producing the sex difference. While there is some evidence that sex hormones contribute to sex differences in mate attraction, the findings are not well enough developed to pass judgment on their theoretical relevance at this time (see Hönekopp, Rudolph et al. 2007; Roney, Simmons & Gray 2011).

Assessment. Obviously, a very large number of studies have assessed sex differences in the degree to which physical attractiveness is preferred when choosing mates. While the vast majority of studies indicate that males are more likely to prioritize physical attractiveness when expressing preferences than are females, there are some exceptions when preferences are measured some other ways, such as when speed dating (Eastwick & Finkel 2008; Eastwick, Eagly et al. 2011). Overall, with a desire to error on the side of caution, we will not declare the tendency to strongly prefer a physically attractive mate a USD.

26.2.7.15 Preference for a Mate Who Is Relatively Tall

Evidence. As indicated in Précis 26.2.7.15, numerous studies in over ten countries have found that females have a stronger preference for a mate to be taller than themselves. Specifically, based on samples from ten countries (plus one multi-country sample), 26 out of 27 studies indicated that females prefer mates who are taller than themselves (or than males in general). The study that failed to find a sex difference in this preference was conducted among the Hadza in Tanzania.

Explanation. Only the evolutionary theories predict such a sex difference, provided one assumes that one assumes that height (or body size) is associated with resource provisioning capabilities. In this regard, considerable evidence points toward height being positively correlated with income, wealth, and social status, typically in the range of $r = .25$ to $.35$ (Komlos 1990; Ellis 1994b; Meyer & Selmer 1999; Judge & Cable 2004; Rashad 2008; Batty, Shipley et al. 2009). It is reasonable to infer that height is being used by women as a rough proxy for the ability of men to be reliable resource providers (BJ Ellis 1992; Stulp & Barrett 2016).

ENA theory also predicts that height and body size are being promoted by exposing the body to testosterone along with other male sex hormones. Considerable evidence is consistent with this expectation (McElduff,

Précis 26.2.7.15 Preference for a mate who is relatively tall (Original Table 15.6.5.4)

Age categories, number of studies, and sex	*Consist. score*	*Overviews*	*Countries*	*Time range*	*Non-human*	*Social role theory*			*Evolutionary theory*		*Theological*
						Origin	*Found*	*Biosoc*	*Select*	*ENA*	
Adult 25F 1x All Ages 26F 1x	96.2F 96.3F	-0-	10 (1)	1977–2020	-0-	Silent	Silent	Silent	Expl-F	Expl-F	Silent

Wilkinson, Ward, & Posen, 1988; Ellison, Lipson & Meredith 1989; Tremblay 1998).

Assessment. Overall, the tendency for females to prefer relatively tall mates compared to what males prefer is deemed a USD-F.

26.2.7.16 *Preference for a Mate Who Is Relatively Short*

Evidence. While females prefer mates taller than themselves, all of the studies displayed in Précis 26.2.7.16. concluded that males have a preference for mates who are shorter than themselves. These studies were conducted in a total of eight countries.

Explanation. Explaining why males would prefer shorter mates than themselves is more difficult to theoretically explain than was the case for the tendency for females to prefer relatively tall mates. One possibility is that males with such preferences would be more likely than other males to be drawn to children, thereby making more committed fathers.

Assessment. The theoretical understanding of such a preference is poorly developed. However, the number of studies, their consistency, and the fact that one meta-analysis leads us to consider the male preference for mates who are relatively short to be a USD-M.

26.2.7.17 *Preference for a Mate Who Is Younger Than Oneself*

Evidence. A large number of studies conducted in numerous countries asked research participants if they preferred a young mate or a mate younger than themselves. Précis 26.2.7.17 shows that these studies found that such preferences were stronger among males. No exceptions to this result were found.

Explanation. No instances were located in which social role theory was invoked to explain male tendencies to prefer younger mates than females. However, several examples of evolutionary theory being used in this regard were found (Okami & Shackelford 2001; Kenrick & Luce 2012). From the standpoint of sexual selection, males should gravitate toward relatively young mates given the relatively short time during which females can bear offspring, i.e., roughly between 15 and 40 years of age (Kenrick & Keefe 1992).

Regarding the possibility that neuroandrogenic factors contribute to male tendencies to prefer younger mates, no evidence was found. However, ENA theory clearly predicts that brain exposure to testosterone should make a contribution.

Assessment. Primarily based on the large number of studies, as well as the sizable number of countries sampled, preference for mates who are relatively young is deemed a USD-M.

Précis 26.2.7.16 Preference for a mate who is relatively short (Original Table 15.6.5.5)

Age categories, number of studies, and sex	*Consist. score*	*Overviews*	*Countries*	*Time range*	*Non-human*	*Social role theory*			*Evolutionary theory*		*Theological*
						Origin	*Found*	*Biosoc*	*Select*	*ENA*	
Adult 16M All Ages 16M	100.0M 100.0M	M: 1Met	8	1977–2020	-0-	Silent	Silent	Silent	Poss-M	Poss-M	Silent

Précis 26.2.7.17 Preference for a mate who is younger than oneself (Original Table 15.6.5.6)

Age categories, number of studies, and sex	*Consist. score*	*Overviews*	*Countries*	*Time range*	*Non-human*	*Social role theory*			*Evolutionary theory*		*Theological*
						Origin	*Found*	*Biosoc*	*Select*	*ENA*	
Adults 48M All Ages 57M	100.0M 100.0M	M: 1Gen	17 (7)	1976–2020	-0-	Silent	Silent	Silent	Expl-M	Poss-M	Silent

26.2.7.18 Preference for a Mate Who Is Older Than Oneself

Evidence. Précis 26.2.7.18 summarizes studies of preference for a mate who is old or older than oneself. Although fewer studies have asked about this preference than have asked about a preference for youth, a large number of studies in many different countries have found that females have a stronger preference for old or older mates. All 26 of the located studies concluded that females have such preferences to a greater degree than males.

Explanation. Why would such a sex difference exist? Social role theory does not appear to address this issue, but evolutionary theory does.

Several evolutionary theorists have proposed that females generally prefer mates who are older than themselves primarily because they prefer mates who have or exhibit promise of being able to provide a stable family income (Hansen & Price 1995; Kokko & Lindstrom 1996; Souza, Conroy-Beam & Buss 2016). In other words, females have been sexually selected for choosing mates with physical and behavioral characteristics, such as height, physical strength, and other signs of dominance and power as evidence of being "good providers" (Moore & Moore 1988; Souza, Conroy-Beam & Buss 2016). These traits are usually more apparent with age.

Regarding ENA theory's neurohormonal hypothesis, studies have shown that height, physical strength, other signs of dominance and status are all promoted by bodily exposure to testosterone (L Ellis 1993). Therefore, it is reasonable to attribute female preferences for relatively older mates to the reproductive benefits that derive from choosing mates with these sorts of traits. Of course, given that males usually have shorter life expectancies (see Table 5.1.5.5), females should not be overly extreme in their preferences for older mates.

Assessment. Given that a substantial number of studies derived from numerous countries all reaching the same conclusion and there are reasonable theoretical ways to explain this sex difference, we conclude that preference for a mate who is relative old is a USD-F.

26.2.7.19 Preference for a Mate with Resources (Wealth, High Income)

Evidence. A great many studies have asked research participants if they have a preference for a mate with resources such as wealth or high income (or at least to appear to have earnings potential). Although three studies did not find a sex difference in this preference, Précis 26.2.7.19 shows that 104 studies concluded that females have stronger preferences in this regard than do males. The number of countries sampled was 27, along with 12 additional studies in which two or more countries were sampled. Publication of these studies spanned 62 years. Most of the studies pertinent to this conclusion have been subjected to two literature reviews and one meta-analysis, all of which agreed that this sex difference is well established.

Précis 26.2.7.18 Preference for a mate who is older than oneself (Original Table 15.6.5.7)

Age categories, number of studies, and sex	*Consist. score*	*Overviews*	*Countries*	*Time range*	*Non-human*	*Social role theory*			*Evolutionary theory*		*Theological*
						Origin	*Found*	*Biosoc*	*Select*	*ENA*	
Adult 21F All Ages 26F	100.0F 100.0F	-0-	10 (4)	1989–2020	-0-	Silent	Silent	Silent	Expl-M	Expl-M	Silent

Précis 26.2.7.19 Preference for a mate with resources (wealth, high income) (Original Table 15.6.5.9)

Age categories, number of studies, and sex	*Consist. score*	*Overviews*	*Countries*	*Time range*	*Non-human*	*Social role theory*			*Evolutionary theory*		*Theological*
						Origin	*Found*	*Biosoc*	*Select*	*ENA*	
Adult 93F 3x All Ages 104F 3x	96.8F 97.1F	F: 2Rev 1Met	27 (12)	1958–2020	-0-	Silent	Silent	Silent	Expl-M	Expl-M	Silent

Explanation. We found no scientific evidence of females being stereotypes as preferring mates with resources. Nor did we locate any attempt for social role theory proponents to account for this sex difference.

Both evolutionary theories do, however, specifically predict that females will be more likely than males to prefer mates with traits such as high income or wealth (Buss & Barnes 1986; L Ellis 2011a). This prediction comes from noting that females with such preferences will be likely to bear and rear more offspring than females who give no consideration to earnings potential when making mate choices (Brooks, Blake & Fromhage 2022).

In addition, ENA theory predicts that neurohormonal factors are likely to be promoting male tendencies to comply with this female preference. Consistent with this prediction, at least three studies have found testosterone levels positively correlated with income, at least among males (Gielen, Holmes & Myers 2016; Hughes & Kumari 2019; Harrison, Davies et al. 2021). Secondarily, the theory suggests that female preferences for high-earning males might be associated with relatively low testosterone levels. No evidence regarding this prediction was located.

Assessment. Vast amounts of research point toward females having stronger preferences for mates with resources than is the case for males. The number of countries sampled in this regard is quite large, as is the time frame over which studies have been conducted. Preference for a mate with resources is a USD-F.

26.2.7.20 *Wanting to Be in Love before Having Sex*

Evidence. Précis 26.2.7.20 shows that all ten of the studies examining possible sex difference in wanting to be in love before having sex found females holding such a preference more strongly than males. The studies took place in three different countries, primarily during the last quarter of the 20th century.

Explanation. The most viable explanation for this type of preference would be of an evolutionary nature. Specifically, females with such a preference would be more likely to sustain a relationship with sex partners, thereby being more likely to receive long-term help caring for whatever offspring result from the relationship.

Assessment. Although just ten studies were located and just three countries were sampled over a fairly short publication time frame (i.e., 25 years), we deem this variable a USD-F. This declaration is warranted in part because there is substantial theoretical support for it being strong among females than among males.

Précis 26.2.7.20 Wanting to be in love before having sex (Original Table 15.6.5.9)

Age categories, number of studies, and sex	*Consist. score*	*Overviews*	*Countries*	*Time range*	*Non-human*	*Social role theory*			*Evolutionary theory*		*Theological*
						Origin	*Found*	*Biosoc*	*Select*	*ENA*	
All Ages 10F	100.0F	-0-	3	1976–2001	-0-	Silent	Silent	Silent	Poss-F	Poss-F	Silent

26.2.7.21 *Attitude towards Premarital/Casual Sexual Intercourse*

Evidence. Sixty-six studies of sex differences in people's permissive attitudes toward premarital/casual sexual intercourse were located. Twelve different countries were sampled (plus one multi-country study) within a 38-year publication period. Précis 26.2.7.21 shows that all but one of these studies concluded that males held more permissive attitudes than did females.

Explanation. If one assumes that permissive attitudes toward casual sex, at least to some extent, reflects one's own desires in this regard, it is possible to offer an evolutionary explanation for why males would be more accepting of such behavior than females. In other words, as noted earlier, males appear to have a stronger sex drive than do females (Précis 26.2.7.6), and they also express desires for sex with more partners (Précis 26.2.7.7) and for waiting less time between attraction and actual sexual relationships (Précis 26.2.7.12).

If the ENA version of evolutionary theory is true, one should find that elevated androgen levels contributing to these permissive attitudes. In this regard, research indicates that males with the highest levels of circulating testosterone do express more sexually permissive attitudes (Halpern, Udry et al. 1994; van Anders, Goldey et al. 2012).

Assessment. Overall, the evidence is quite strong that males have more permissive attitudes toward premarital or casual sexual intercourse than do females, making this variable a USD-M.

26.2.7.22 *Attitude towards Nudity and Sexually Explicit Material*

Evidence. Ten ten studies were located having to do with sex differences in attitudes toward the availability of nude and sexually explicit maternal. Précis 26.2.7.22 shows that all studies agree that such attitudes are more common among males, although just two countries were sampled.

Explanation. Both versions of evolutionary theory predict that males would have more accepting attitudes toward nudity and sexually explicit material than females, at least regarding their own abilities to access these types of material (Malamuth 1996; Saad & Gill 2000; Hald 2006).

Regarding possible involvement of androgens in this sex difference, one study found that testosterone levels among males were positively correlated with time spent viewing pornographic images in a controlled research setting (Rupp & Wallen 2007). However, a study of elderly adult males found no relationship between testosterone per se and time spent viewing near-nude female models, although other sex hormones were positively correlated (Palmer-Hague, Wong et al. 2021).

Assessment. While both the number of located studies and the countries sampled were rather small, the pattern of findings fits with other similar permissive sexual attitudes and interests (see directly above). Therefore,

Précis 26.2.7.21 Attitude towards premarital/casual sexual intercourse (Original Table 15.6.5.9)

Age categories, number of studies, and sex	*Consist. score*	*Overviews*	*Countries*	*Time range*	*Non-human*	*Social role theory*			*Evolutionary theory*		*Theological*
						Origin	*Found*	*Biosoc*	*Select*	*ENA*	
Adult 51M All Ages 65M 1F	100.0M 97.0M	M: 3Met	12 (1)	1975–2013	-0-	Silent	Silent	Silent	Expl-M	Expl-M	Silent

Précis 26.2.7.22 Attitude towards nudity and sexually explicit material (Original Table 15.7.4.3)

Age categories, number of studies, and sex	*Consist. score*	*Overviews*	*Countries*	*Time range*	*Non-human*	*Social role theory*			*Evolutionary theory*		*Theological*
						Origin	*Found*	*Biosoc*	*Select*	*ENA*	
All Ages 10M	100.0M	-0-	2	1973–2000	-0-	Silent	Silent	Silent	Expl-M	Expl-M	Silent

attitudes toward nudity and sexually explicit material are considered a USD-M.

26.2.7.23 Being Health Conscious

Evidence. People vary in the extent to which they consider the health consequences of their actions. The results of studies regarding sex differences in being health conscious are presented in Précis 26.2.7.23. These studies were published over a 31-year time frame in six different countries (plus two multi-country studies). All 16 studies reached the same basic conclusion: Females are more health conscious than males.

Explanation. The only theories that seem to address the possibility of sex differences in health consciousness are the evolutionary theories, and they do so only indirectly. As noted elsewhere in this chapter, compared to males, females are more fearful (see Précis 26.2.2.2) and are less inclined to take risks (see Précis 26.3.1.2). Together, these two traits could prompt females to be more health conscious. In the narratives surrounding both of these précises, evidence was presented indicating that evolution could have favored diminished fearfulness and increased risk-taking among males. Likewise, brain exposure to testosterone appears to inhibit fearfulness (Van Honk, Peper & Schutter 2005) and promote risk-taking (*meta-analysis*: Dekkers, van Rentergem et al. 2019).

Assessment. In light of the number of consistent research findings and the fact that they were based on samples drawn from a substantial number of different countries, we declare being health conscious to be a USD-F. The most persuasive theories seem to be those of an evolutionary nature.

26.2.7.24 Enjoyment of School in General

Evidence. Studies of sex differences in the enjoyment of school have been performed for over half of a century in a variety of different countries. The results from all 39 of the studies that were located are summarized in Précis 26.2.7.24. Based on samples drawn from ten different countries, plus two multi-country studies, the findings have been unanimous: Females report greater enjoyment of school than do males.

Explanation. The finding that females enjoy school more than males is puzzling, given the importance of education for attaining high social status. However, some of the sex difference may have to do with male desires for greater physical activity (see Précis 26.3.1.6), which tends to be less compatible with most forms of education. Another sex difference that could promote greater female enjoyment of school is their tendency to develop language skills more rapidly than in the case for males (see Précis 26.2.3.10). As discussed in their respective narratives, both of these traits

Précis 26.2.7.23 Being health conscious (Original Table 15.8.1.1)

Age categories, number of studies, and sex	*Consist. score*	*Overviews*	*Countries*	*Time range*	*Non-human*	*Social role theory*			*Evolutionary theory*		*Theological*
						Origin	*Found*	*Biosoc*	*Select*	*ENA*	
Adult 12F All Ages 16F	100.0F 100.0F	-0-	6 (2)	1974–2005	-0-	Silent	Silent	Silent	Expl-F	Expl-F	Silent

Précis 26.2.7.24 Enjoyment of school in general (Original Table 15.9.1.1)

Age categories, number of studies, and sex	*Consist. score*	*Overviews*	*Countries*	*Time range*	*Non-human*	*Social role theory*			*Evolutionary theory*		*Theological*
						Origin	*Found*	*Biosoc*	*Select*	*ENA*	
Child 22F Adolescent 16F All Ages 39F	100.0F 100.0F 100.0F	-0-	10 (2)	1956–2013	-0-	Silent	Silent	Silent	Expl-F	Expl-F	Silent

appear to have been sexually selected and seem to be influenced by exposing the brain to testosterone.

Assessment. Even though only indirect theoretical explanations for sex differences in enjoyment of school were located, the number of consistent findings and the diversity of countries sampled provide us with a confident conclusion that such enjoyment is a USD-F.

26.2.7.25 *Interest in or Enjoyment of Science in General*

Evidence. One hundred and twenty-seven studies of sex differences in the interest in or enjoyment of science were located. They were conducted in 34 different countries (plus seven studies of multi-country samples), with publications extending over a 98-year time frame. The majority of these studies found that males reported being more interested in science and enjoying it more. However, as indicated in Précis 26.2.7.25, this difference only reached a consistency score over 95.0 among children, where ten out of ten studies found males to be more attracted to science. Among adolescents, 96 studies also reported greater interest among males, but 11 studies actually reported more interest among females. Therefore, for all age groupings, the consistency score was just 84.1.

We should add as a methodological comment that a number of the studies pertaining to the enjoyment of science did not make a clear distinction between various categories of science, e.g., physical vs. biological. As will be documented in the section immediately below this one, females have actually been found to express greater interest in biological and health sciences than do males. Overall, we suspect that several of the 11 studies cited in the précis on science in general may reflect female interests in the life sciences rather than the physical sciences. In a future edition of this book, we hope to revisit this possibility in greater detail.

Explanation. As noted earlier, males seem to be more interested in physical things than are females (see 26.2.7.30). To the existent that science in general primarily pertains to physical, non-living entities, one would expect deals with

Assessment. We will deem interest in or enjoyment of science in general to be a USD-M for children only. More research is in order before generalizing beyond that. In particular, future editions of this book should seek to more clearly separate sex differences in interest in the physical sciences from interest in science generally, which often include the biological sciences (see below).

26.2.7.26 *Interest in or Enjoyment of the Biological and Health Sciences*

Evidence. As shown in Précis 26.2.7.26, 12 studies of sex differences in people's interests in or enjoyment of the biological and health sciences

Précis 26.2.7.25 Interest in or enjoyment of science in general (Original Table 15.9.2.5)

Age categories, number of studies, and sex	*Consist. score*	*Overviews*	*Countries*	*Time range*	*Non-human*	*Social role theory*			*Evolutionary theory*		*Theological*
						Origin	*Found*	*Biosoc*	*Select*	*ENA*	
Child 10M All Ages 116M 11F	100.0M 84.1M	M: 2Rev; 5Met	34 (7)	1922–2020	-0-	Silent	Silent	Silent	Poss-F	Poss-F	Silent

Précis 26.2.7.26 Interest in or enjoyment of the biological and health sciences (Original Table 15.9.2.6)

Age categories, number of studies, and sex	*Consist. score*	*Overviews*	*Countries*	*Time range*	*Non-human*	*Social role theory*			*Evolutionary theory*		*Theological*
						Origin	*Found*	*Biosoc*	*Select*	*ENA*	
All Ages 12F	100.0	-0-	5	1979–2006	-0-	Silent	Silent	Silent	Poss-M	Poss-M	Silent

were located, all of which concluded that females had more interest in these sciences than was true for males. These studies were carried out in five countries over a 27-year publication time frame.

Explanation. None of the three social role theories seem to explain the apparent universal tendency for females to have greater interest in the life sciences than do males. However, both of the evolutionary theories do by assuming that females have been sexually selected for being more socially oriented (see Précis 26.2.7.30) and in providing care to others (see Précis 26.2.7.45). An interest in and an understanding of biology and the health-related sciences obviously helps individuals to achieve these goals.

Regarding the possibility of ENA theory accurately predicting that brain exposure to testosterone would inhibit an interest in the life sciences, one indirect supportive finding was located. It reported an inverse correlation between prenatal testosterone exposure and the amount of eye contact infants made under standard testing conditions (Lutchmaya, Baron-Cohen & Raggatt 2002a). Obviously, more direct evidence is needed to provide strong support for the hypothesis that females are more interested in the life sciences because of low brain exposure to testosterone and possibly other androgens.

Assessment. We declare interest in (or enjoyment of) the biological and health sciences to be a USD-F. Additional strength for this declaration would come from increasing the number and cultural diversity of the countries sampled.

26.2.7.27 Pro-Family Attitudes and Values

Evidence. Studies in four different countries, and one multi-national study, found that females reported more pro-family attitudes and values. Précis 26.2.7.27 shows that the consistency scores on this topic were 100.0.

Explanation. Compared to males, females have been shown to be stereotyped as being more family-oriented (Table 21.2.9.8) and more nurturing (Table 21.2.9.10). These stereotypes would lead proponents of social role theories to believe that pro-family attitudes and values would be greater among females. However, the original version of the theory would not predict these attitudes and values to be universal. In the case of the other two social role theories, it is possible to predict universal sex differences. Founder effect theorists could argue that such values just happened to be present in the earliest human societies and have persisted essentially unchanged ever sense. However, no empirical evidence exists in this regard. In the case of the biosocial version of social role theory, it can be argued that females are expected and taught to hold more pro-family attitudes than males. No evidence in this regard was located.

Précis 26.2.7.27 Pro-family attitudes and values (Original Table 15.10.1.1)

Age categories, number of studies, and sex	*Consist. score*	*Overviews*	*Countries*	*Time range*	*Non-human*	*Social role theory*			*Evolutionary theory*		*Theological*
						Origin	*Found*	*Biosoc*	*Select*	*ENA*	
Adult 11F All Ages 14F	100.0 100.0	-0-	4 (1)	1963–2006	-0-	Contra	NoTest	NoEvid	Poss-F	Poss-F	Silent

According to evolutionary reasoning, when compared to males, females are usually sexually selected for caring for offspring and other close relatives more (Wade & Shuster 2002; Kokko & Jennions 2008). This theoretical deduction is borne out by nearly all of the available evidence (see Précis 26.3.2.5 and Précis 26.3.2.6). Given that females also tend to be more interested in spending time with children (see Précis 26.2.7.28) and in maintaining close interpersonal relationships with others (see Précis 26.2.7.29), it is reasonable to deduce that females would be more likely than males to subscribe to pro-family attitudes and values (Weisfeld, Weisfeld & Goetz 2017).

Regarding the possible involvement of testosterone in inhibiting pro-family attitudes and values, no specific evidence was found. Therefore, this element of ENA theory does not appear to have been tested.

Assessment. We declare pro-family attitudes and values to be a USD-F. More research is needed to better identify the best theoretical explanations for these types of attitudes.

26.2.7.28 Interest in Spending Time with Infants and Toddlers

Evidence. Précis 26.2.7.28 provides a summary of research having to do with sex differences in wanting to spend time with infants and toddlers. At least among adults, one can see that 23 of the 24 studies found that females have a greater interest in doing so than do males. Three countries were sampled over a publication time frame of 45 years.

Explanation. From the perspective of social role theory, one can argue that females are more interested in young children because their societies expect them to have these interests. In this regard, some research suggests that females are stereotyped as being more nurturing (Table 21.2.9.10). However, the original version of this theory would lead one to expect to find some societies in which males are more interested in infants and toddlers, or at least where there are no sex differences (which does not appear to be the case). Perhaps the founder effect version of social role theory is correct – that this difference just happened to be the case in the earliest human societies and has continued unchanged ever since – but, unfortunately, no evidence exists on this point.

In the case of the biosocial version of social role theory, one might reason that females are more interested in spending time with infants and toddlers because societies foresee the need for females to be the primary caregivers to their own infants and toddlers, partly because females have had to traditionally breastfeed each of their infants for a year or so after giving birth. While such an explanation assumes that societies can think and plan, it makes some sense from the biosocial social role perspective.

Thinking in evolutionary terms, females are likely to have been sexually

Précis 26.2.7.28 Interest in spending time with infants and toddlers (Original Table 15.10.1.4)

Age categories, number of studies, and sex	*Consist. score*	*Overviews*	*Countries*	*Time range*	*Non-human*	*Social role theory*			*Evolutionary theory*		*Theological*
						Origin	*Found*	*Biosoc*	*Select*	*ENA*	
Child 16F 2x Adult 23F 1x All Ages 51F 5x	(88,9) 95.8 (91.1)	-0-	3	1957–2002	-0-	Contra	NoTest	Poss-F	Expl-F	Expl-F	Silent

selected for caring for their own offspring to a greater degree than do males. This is because, at least among mammals, females can have greater confidence than males can (whether consciously or not) in identifying their genetic offspring (Buss & Schmitt 1993; KG Anderson, Kaplan & Lancaster 2007). Of course, while both sexes may have largely unlearned interests in young children, the interests appear to be greater for females. Even when it comes to occupational interests, females appear to prefer interacting with children more than do males (see Table 20.2.7.66 in Volume III).

Regarding ENA theory, two studies were located bearing on the possible role of testosterone. As the theory hypothesizes, one study found that nurturing and caregiving tendencies were inversely correlated with circulating testosterone levels (van Anders & Watson 2006; also see van Anders, Goldey & Kuo 2011).

Assessment. At least among adults, interest in spending time with infants and toddlers is a USD-F. The strongest theoretical explanations appear to be evolutionary in nature.

26.2.7.29 Interest in Forming and Maintaining Close Interpersonal Relationships

Evidence. As shown in Précis 26.2.7.29, all 16 studies located on sex differences in forming and maintaining close interpersonal relationships concluded that this interest was greater for females than for males. The relevant studies were conducted in four countries in addition to one multi-national study. Publications pertaining to this sex difference spanned 54 years.

Explanation. Research has indicated that females are stereotyped as being more communal (see Table 21.2.9.3 in Volume III). Such evidence lends support to the social role theories in that sex differences in people's interests in forming close interpersonal relationships could be rooted in sex role training and expectations. Regarding the original version of social role theory, however, not all societies should exhibit the same sex difference. The evidence that they do casts doubt on this version of the theory. Unfortunately, there is no way to directly test the founder effect version of social role theory since no data from the first human societies were obtained. In the case of the biosocial version of social role theory, one would have to demonstrate why sex differences in body size, physical strength, or the ability to bear offspring would impact interest in interpersonal relationships. No such evidence was located in this regard.

Turning to the evolutionary perspective, proponents have argued that females have been sexually selected for forming and maintaining close interpersonal relationships, while males have been favored more for forming hierarchical relationships (Geary, Byrd-Craven et al. 2003:Figure 2; Del

Précis 26.2.7.29 Interest in forming and maintaining close interpersonal relationships (Original Table 15.10.2.4)

Age categories, number of studies, and sex	*Consist. score*	*Overviews*	*Countries*	*Time range*	*Non-human*	*Social role theory*			*Evolutionary theory*		*Theological*
						Origin	*Found*	*Biosoc*	*Select*	*ENA*	
Adult 10F All Ages 16F	100.0 100.0	-0-	4 (1)	1962–2016	-0-	Contra	NoTest	NoTest	Expl-F	Expl-F	Silent

Giudice & Belsky 2010). The primary basis for this sex difference is that male reproduction is more dependent on stable resource procurement whereas female reproduction depends more on socializing and providing care to offspring.

The ENA version of evolutionary theory asserts that brain exposure to testosterone should be inversely correlated with the formation and maintenance of close interpersonal relationships with others. Support for this hypothesis has been obtained in various studies (CA Pedersen 2004; LE Duncan & Peterson 2010).

Assessment. Interest in forming and maintaining close interpersonal relationships with others can be considered a USD-F. The strongest theoretical explanations appear to be evolutionary in nature.

26.2.7.30 Interest in People More Than Things

Evidence. In their widely cited book on cognitive and behavioral sex differences, Maccoby and Jacklin (1974:349) contended that "the two sexes are equally interested in social (as compared with nonsocial) stimuli." This does not appear to be the case. suggests that there actually are sex differences in social interests, with females being more interested in other human beings than is the case for males. Précis 26.2.7.30 shows that all ten studies that were located agreed that females have a greater interest in people than in things.

Incidentally, in the case of interests in things (rather than people), only nine pertinent studies were located, all of which indicated that males surpassed females (see Table 15.10.2.6 in Volume II). Because only nine studies were found, this variable is not listed as a 10+ variable. However, it is also worth mentioning that additional evidence bearing on the interest in people-versus-things distinction will be reviewed later in this chapter, when sex differences in the types of college majors and occupations are considered.

Explanation. As noted in the section above, research has found females being stereotyped as being more communal than are males (Table 21.2.9.3). For this reason, one could assume that such societal stereotyping is responsible for making females more prone to have an interest in people than do males. However, the same limitations cited above regarding sex differences in being interested in closer social relationships, can be applied here as well. Specifically, the original version of social role theory is called into question by the fact that the sex difference appears to be universal. In the case of the founder effect theory, it is essentially untestable.

The biosocial version of the social role theory asserts that *universal* sex differences are the result of males being larger and stronger, or females being the only sex that can give birth. Unfortunately, these sex differences

Précis 26.2.7.30 Interest in people more than things. (Original Table 15.10.2.5)

Age categories, number of studies, and sex	*Consist. score*	*Overviews*	*Countries*	*Time range*	*Non-human*	*Social role theory*			*Evolutionary theory*		*Theological*
						Origin	*Found*	*Biosoc*	*Select*	*ENA*	
All Ages 10F	100.0	F: 1Met x: 1Gen	4 (1)	1914–2012	-0-	Contra	NoTest	Silent	Expl-F	Expl-F	Silent

do not seem to have any bearing on sex differences in being more interested in people (relative to interest in things).

Turning to the evolutionary theory, the sexual selection version would suggest that females have been favored for caring for offspring to a greater degree than males, not only among humans (Trivers 2017) but nearly all other mammals as well (Plavcan 2001). Among the reasons for this is that females can identify their offspring with greater certainty, they produce breast milk, and will have already invested heavily in each offspring prior to birth. The empirical evidence on sex differences in providing care to offspring is very clear for all countries yet sampled (see Précis 26.2.17.5).

Because females have been favored for providing greater care to offspring than is the case for males, females are likely to have also evolved general tendencies to socialize more than males, especially in friendly non-competitive venues. For evidence in this regard, see Précis 26.3.2.23. The greater evolved female interests in people over things is also likely to be responsible for female preferences for playing with dolls (see Précis 26.3.2.28).

Regarding ENA theory, various studies indicate that testosterone inhibits the formation of close interpersonal relationships. This applies to circulating testosterone and parenting/nurturing behavior, in particular (van Anders & Watson 2006; Edelstein, Chin et al. 2019; Gettler 2020). To provide an overall assessment of the association between circulating testosterone and parenting tendencies as well as parenting quality, at least among males, a meta-analysis concluded that both were inversely correlated with testosterone levels, albeit to just a modest degree (Meijer, van IJzendoorn et al. 2019). Also, a study of people's varying expressed interests in spending time with young children concluded that, as theoretically expected, prenatal testosterone levels were negatively correlated with such interests (Knickmeyer, Baron-Cohen et al. 2005). An earlier study of circular testosterone among women indicated that those with the highest levels expressed relatively low levels of interest in having children (Udry, Morris & Kovenock 1995).

The above evidence suggests that, elevated testosterone is favored among males, not because it helps to make them good at caring for their children, but for providing resources to mates in order to produce more and healthier offspring. Accordingly, when males are compared to females, they tend to be more competitive (see Précis 26.3.1.5), better at spatial reasoning (see Précis 26.2.3.4), and at systemizing thought (see Précis 26.2.3.7), all traits that tend to give males an advantage in procuring resources with which to attract mates.

Assessment. We conclude that interest in people more than in things is a USD-F. Evidence accumulated so far seems to be more in line with evolutionary theory than with social role theory.

26.2.7.31 Preference for Socializing in General

Evidence. Findings from research on sex differences in preferences for socializing are summarized in Précis 26.2.7.32. One can see that all 20 located studies agree that females devote more time to socializing than do males. The studies were based on samples drawn from four countries and were published over a 75-year time period.

Explanation. As noted above with reference to Précis 26.2.7.29 as well as Précis 26.2.7.30, when compared to males, females seem to be more interested in forming close social relationships and in people than in things. Therefore, it is predictable that they would also socialize more in general than is the case for males.

The same theoretical explanations that were applied to forming social relationships and preferring people over things can also be applied to socializing. Therefore, we will not reiterate the evidence here.

Assessment. Preferences for socializing in general is a USD-F. Once again, most of the supportive evidence points toward the two evolutionary theories as offering the best overall explanations.

26.2.7.32 Interest in or Enjoyment of Competitive Activities

Evidence. Competition can come in many forms, ranging from overt physical combat to economic entrepreneurship and numerous competitive games. As shown in Précis 26.2.7.32, 42 studies were located having to do with sex differences in liking competitive activities. Of these, all but one concluded that males express greater interests in and/or enjoyment of these activities. Samples for these studies came from ten different countries, plus four multi-country studies. Their dates of publication spanned 44 years.

Later in this chapter, evidence will be summarized regarding sex differences in actual competitiveness as a basic psychological trait (Précis 26.3.1.5). As with interests in (and enjoyment of) competitive activities, nearly all studies have concluded that males exhibit competitiveness to a greater degree than do females.

Explanation. No empirical evidence having to do with sex stereotypes in interests in competition were located. Therefore, we will simply list the three social role theories as silent to this particular sex difference.

From an evolutionary perspective, one can begin to explain sex differences in interest in competing by noting that one finds males of most species to be more overtly competitive than females, and that most of this competitiveness involves intra-sex disputes over breeding rights (Geary 2006; Clutton-Brock 2007). Especially among mammals, because females must gestate each offspring to be born, male competition over breeding rights can be evolutionarily understood by noting that males have a higher

Précis 26.2.7.31 Preference for socializing in general (Original Table 15.10.3.1)

Age categories, number of studies, and sex	*Consist. score*	*Overviews*	*Countries*	*Time range*	*Non-human*	*Social role theory*			*Evolutionary theory*		*Theological*
						Origin	*Found*	*Biosoc*	*Select*	*ENA*	
Adult 11F All Ages 20F	100.0 100.0	-0-	4	1932–2007	-0-	Contra	NoTest	Silent	Expl-F	Expl-F	Silent

Précis 26.2.7.32 Interest in or enjoyment of competitive activities (Original Table 15.10.4.4)

Age categories, number of studies, and sex	*Consist. score*	*Overviews*	*Countries*	*Time range*	*Non-human*	*Social role theory*			*Evolutionary theory*		*Theological*
						Origin	*Found*	*Biosoc*	*Select*	*ENA*	
Adult 11M All Ages 41M 1F	100.0 95.4	M: 1Gen 1Rev	10 (4)	1977–2021	-0-	Silent	Silent	Silent	Expl-F	Poss-F	Silent

reproductive ceiling than do females. But, to attain some of this higher reproductive potential, males must mate with more than one female. As a result, in most species, males have larger bodies, increased musculature, and often special armaments (e.g., antlers) to better compete with rivals over mating opportunities (Glutton-Brock & Vincent 1991). In behavioral terms, male mammals also typically evolve tendencies to be more combative toward one another than is the case for female mammals (Huxley 1938; Clutton-Brock 1985).

Among humans, the same principles seem to apply (Puts 2016). However, because of our special language skills, human males often express their competitiveness in rather subtle rule-based sports and games where physical aggression and the risk of injuries are relatively low. We will discuss sex differences in interests in sports more specifically in the next three sections.

Overall, from an evolutionary perspective, interest in (or enjoyment of) competition is more pronounced among males because competition is more central to males having mating opportunities than is the case for females. In other words, interest in and enjoyment of competition is likely to be a prelude to actual competition, and the latter has been sexually selected primarily among males (Hill, Bailey & Puts 2017). Supporting this reasoning, research has indicated that frequent

ENA theory contends that neurohormonal factors are contributing to interest in or enjoyment of competition. Two studies were located providing support for this hypothesis, at least regarding circulating testosterone levels (Casto & Edwards 2016; Arnocky, Albert et al. 2018).

Assessment. Interest in or enjoyment of competitive activities appears to be a USD-M. More work is needed to understand this sex difference from a theoretical standpoint.

26.2.7.33 *Interest in Sports in General*

Evidence. Précis 26.2.7.33 summarizes the findings of studies concerning sex differences in people's interests in sports in general. Seventeen studies published over an 80-year time span all concluded that males have greater interest in sports than do females.

Explanation. As noted in Précis 26.2.7.32 directly above, males appear to have a greater interest in (or enjoyment of) competitive activities than do females. Given that nearly all sports have substantial competitive elements, we will assume that the basic explanations presented there also applied to interest in sports.

Assessment. While theoretical understanding of variations in people's interests in sports still needs development, the evidence is strong that such interests constitute a USD-M.

Précis 26.2.7.33 Interest in sports in general (Original Table 15.10.5.1)

Age categories, number of studies, and sex	*Consist. score*	*Overviews*	*Countries*	*Time range*	*Non-human*	*Social role theory*			*Evolutionary theory*		*Theological*
						Origin	*Found*	*Biosoc*	*Select*	*ENA*	
All Ages 17M	100.0	M: 1Rev	3 (2)	1925–2005	-0-	Silent	Silent	Silent	Expl-M	Poss-M	Silent

26.2.7.34 Interest in Spectator Sporting Activities

Evidence. Forty years of published studies have all converged on the conclusion that males are more interested in spectator sports than is true for females. The evidence, summarized in Précis 26.2.7.34, is based on 24 studies conducted in five different countries.

Explanation. Readers are referred to the narrative having to do with interests in competitive activities (Précis 26.2.7.32). There, information is provided to help explain why males would be more interested in spectator sports than females. Overall, the only theories along these lines are evolutionary in nature.

Assessment. Interest in spectator sports is a USD-M. More theoretical work is needed to fully understand this sex difference.

26.2.7.35 Interest in Sports Participation

Evidence. As presented in Précis 6.2. 5.35, all 37 studies of sex differences in being interested in playing sports have concluded that males surpass females. The publication of these studies spanned 80 years, and the number of countries sampled in this regard was eight (along with one multi-country study).

Explanation. Readers should consult the narrative surrounding Précis 26.2.7.32 for theoretical attempts to explain why males would express a greater interest in sports participation than females. So far, the best type of explanation seems to be along evolutionary lines.

Assessment. The evidence strongly supports concluding that interest in sports participation is a USD-M.

26.2.7.36 Range in Play and Reading Interest

Evidence. Précis 26.2.7.36 shows that 17 studies were located bearing on the diversity and flexibility of play and reading interests. All of these studies indicated that females expressed or exhibited wider interest than did males. The studies were based on samples drawn from three countries, and were published over a 77-year time frame.

Explanation. None of the five theories seems to address this particular sex difference.

Assessment. Having a wide range in play and reading interests appears to be a USD-F. No theoretical explanation is apparent.

26.2.7.37 Preference for Playing with Dolls

Evidence. For nearly a century, studies have examined sex differences in playing with dolls. Although limited to four countries, Précis 26.2.7.37

Précis 26.2.7.34 Interest in spectator sporting activities (Original Table 15.10.5.2)

Age categories, number of studies, and sex	*Consist. score*	*Overviews*	*Countries*	*Time range*	*Non-human*	*Social role theory*			*Evolutionary theory*		*Theological*
						Origin	*Found*	*Biosoc*	*Select*	*ENA*	
Adult 13M All Ages 24M	100.0 100.0	-0-	5	1970–2010	-0-	Silent	Silent	Silent	Expl-M	Poss-M	Silent

Précis 26.2.7.35 Interest in sports participation (Original Table 15.10.5.3)

Age categories, number of studies, and sex	*Consist. score*	*Overviews*	*Countries*	*Time range*	*Non-human*	*Social role theory*			*Evolutionary theory*		*Theological*
						Origin	*Found*	*Biosoc*	*Select*	*ENA*	
Adolescence 22M All Ages 37M	100.0 100.0	-0-	8 (1)	1930–2010	-0-	Silent	Silent	Silent	Expl-M	Poss-M	Silent

Précis 26.2.7.36 Range in play and reading interest (Original Table 15.10.7.4)

Age categories, number of studies, and sex	*Consist. score*	*Overviews*	*Countries*	*Time range*	*Non-human*	*Social role theory*			*Evolutionary theory*		*Theological*
						Origin	*Found*	*Biosoc*	*Select*	*ENA*	
All Ages 17F	100.0	-0-	3	1933–2010	-0-	Silent	Silent	Silent	Silent	Silent	Silent

shows that all 28 of these studies have concluded that females exhibit a stronger preference than do males. Furthermore, four literature reviews and one meta-analysis reached the same conclusion. It is worth adding that six studies of non-human primates (of childhood age) also indicated that females were more likely than males to choose dolls as objects of play.

Explanation. Table 21.2.3.4 shows that girls are stereotyped as playing with dolls more than boys. Therefore, it is possible to explain these sex differences as a preferential response to females being stereotyped as liking to play with dolls more. However, the original version of the theory would lead one to expect that there should be some societies in which this type of stereotype does not exist; instead, this sex difference appears to be universal. The founder effect version of the theory does not appear to be applicable, since there were probably no dolls to play with in the first human societies. As for the biosocial version of the theory, one might argue that societies encourage females to play with dolls because most of them will one day bear children. However, one might wonder why at least some societies would not encourage boys to also play with dolls in order to become better fathers.

The evolutionary theories assume that sex differences in playing with dolls reflects the fact that dolls resemble babies along with the fact that females appear to be the primary caregivers to babies in all human societies (see Précis 26.3.2.5). This sex bias in providing care to offspring appears to have been sexually selected so that males can make their main parental contribution in the form of provisioning food and other resources to their mates and to their mutually produced offspring (Gangestad & Thornhill 1997; ML Wilson, Miller & Crouse 2017).

If the above evolutionary reasoning is correct, then what are the underlying physiological processes? According to ENA theory, brain exposure to testosterone and other androgens has the effect of diverting males away from providing long-term care to babies and children in order to focus their efforts more on traits such as competitiveness. Support for this line of reasoning has come from a study indicating that prenatal testosterone exposure (as measured in amniotic fluid) was negatively correlated with children expressing interests in caring for babies (Knickmeyer, Baron-Cohen et al. 2005). Another study found that, among 2-year-old boys, testosterone levels in the first few months following birth were negatively correlated the time they spent playing with dolls (Lamminmaki, Hines et al. 2012). To help explain this finding, it is worth mentioning that testosterone levels among males tend to rise to varying degrees during the first few months of life, a phenomenon sometimes referred to as *mini-puberty* (see Figure 24.1).

Assessment. Overall, preference for playing with dolls is a USD-F. The causes for this universal difference still need investigation, but work along evolutionary lines appears to be particularly promising.

Précis 26.2.7.37 Preference for playing with dolls (Original Table 15.10.7.8.)

Age categories, number of studies, and sex	*Consist. score*	*Overviews*	*Countries*	*Time range*	*Non-human*	*Social role theory*			*Evolutionary theory*		*Theological*
						Origin	*Found*	*Biosoc*	*Select*	*ENA*	
Infant/Toddler 12F Child 14F All Ages 28F	100.0 100.0 100.0	F: 4Rev 1Met	4	1925–2019	F: six primates	Contra	Silent	Poss-F	Expl-F	Expl-F	Silent

26.2.7.38 Preference for Mechanical and Building Objects of Play

Evidence. Researchers have examined possible sex differences in the preference for play objects for over a century. Précis 26.2.7.38 shows that all 28 studies having to do with preferences for toys or other objects with moving parts or that can be used for building purposes agree that these are more popular among males than among females. The findings were based on samples drawn from seven different countries in publications spanning 102 years.

Explanation. No research linking sex stereotype of preferences for mechanical and building objects of play were located. Therefore, the three social role theories are listed as silent in terms of predicting such sex differences in preferences.

Evolutionary theorists would likely explain sex differences in preferences for mechanical and building objects as toys by noting that interests in things rather than in people (see Table 15.10.2.6 in Volume II), especially when it comes to occupations (see Précis 26.3.5.19). These things-oriented interests, in turn, are likely to be driven by males being more spatially oriented (see Précis 26.2.3.4) and more prone to think in systemizing terms (see Précis 26.2.3.7) rather than in terms of social relationships.

According to ENA theory, all of these evolved sex differences are likely to be driven by exposing male brains to greater average testosterone than is the case for female brains. Support for this assertion come from studies of both prenatal testosterone and post-pubertal testosterone (see Précis 26.1.3.2). Additional evidence comes from experiments with laboratory animals in which administering testosterone has been shown to promote spatial reasoning (Hodosy, Páleš et al. 2010; Spritzer, Daviau et al. 2011; Spritzer, Fox et al. 2013).

Assessment. Overall, there is no reason to doubt that preference for mechanical and building objects for play is a USD-M. Both evolutionary theories seem to provide explanations for these sex differences.

26.2.7.39 Preference for Vehicles as Objects of Play

Evidence. Précis 26.2.7.39 shows that all 23 studies that were located having to do with sex differences in interests in playing with vehicles concluded that males surpass females in this regard. The studies were based on samples drawn from five different countries and were published over an 88-year time frame. It is also worth adding that three literature reviews and one meta-analysis also reached the conclusion that males surpass females, as did five studies of non-human primate infants.

Explanation. Basically, preferences for playing with vehicles can be seen as providing another example of non-social play, similar to preferring to

Précis 26.2.7.38 Preference for mechanical and building objects of play (Original Table 15.10.8.6)

Age categories, number of studies, and sex	*Consist. score*	*Overviews*	*Countries*	*Time range*	*Non-human*	*Social role theory*			*Evolutionary theory*		*Theological*
						Origin	*Found*	*Biosoc*	*Select*	*ENA*	
Child 16M All Ages 28M	100.0 100.0	M: 1Rev	7	1913–2015	-0-	Silent	Silent	Silent	Expl-M	Expl-M	Silent

Précis 26.2.7.39 Preference for vehicles as objects of play (Original Table 15.10.8.7)

Age categories, number of studies, and sex	*Consist. score*	*Overviews*	*Countries*	*Time range*	*Non-human*	*Social role theory*			*Evolutionary theory*		*Theological*
						Origin	*Found*	*Biosoc*	*Select*	*ENA*	
Infants/Toddlers 12M All Ages 23M	100.0 100.0	M: 3Rev 1Met	5	1931–2019	M: five primates	Silent	Silent	Silent	Expl-M	Expl-M	Silent

play with mechanical and building materials. We found no basic explanation for why males appear to surpass females in these preferences.

Several researchers have attempted to answer this rather question: How can biology possibly explain why boys prefer cars, while girls prefer dolls? (Alexander & Saenz 2012:503). Evolutionary theory seems to offer at least part of the answer while generating testable hypotheses. As noted in regard to other sex differences, males appear to have been sexually selected for their ability to provide resources to their spouse(s) when they become adults. Especially when food was often in short-supply, by being the primary "breadwinner," males are able to make it possible for females to bear and rear more offspring that would be the case without males providing resources (L Ellis 1993; Kaplan, Hill et al. 2000; Bleske-Rechek & Gunseor 2021). Consequently, males seem to have come to specialize in manipulating objects (and sometimes manipulating people *as* objects) in ways that provide fairly stable supplies of resources for their families.

Regarding the underlying physiology that may promote sex differences in preferring play with vehicles and other types of objects (see Précis 26.2.7.38), some studies support ENA theory's assertion that brain exposure to androgens are involved. Specifically, one study indicated that toddlers who played with vehicles the most had higher levels of circulating testosterone during the third and fourth months following birth (GM Alexander & Saenz 2012). This period of time, sometimes known as *mini-puberty*, is when males typically experience a rather limited burst in testosterone production (see Figure 24.1). The role played by this testosterone burst is still not fully understood, but it is likely to contribute to the overall masculinization process.

Assessment. The tendency to prefer playing with vehicles is a well-established USD-M. The fact that the same sex difference has also been found in two different species of non-human primates makes it very likely that biological factors are contributing a great deal to this sex difference.

26.2.7.40 Preference for Adventure Stories

Evidence. Précis 26.2.7.40 indicates that, when compared to females, males are more likely to prefer to read, watch, or listen to stories in which adventure is a prominent theme. While just ten studies were found, they were based on samples from three countries and were published over a 32-year time frame.

Explanation. Two sex stereotypes we identified that seemed to have some bearing on males preferring stories of adventure more than females. One is that females are more likely than males to be concerned with personal safety and security (Table 21.2.4.7), and the other indicated that males are stereotyped as being more daring and prone to take risks (Table 21.2.7.23).

Précis 26.2.7.40 Preference for adventure stories (Original Table 15.10.9.4)

Age categories, number of studies, and sex	*Consist. score*	*Overviews*	*Countries*	*Time range*	*Non-human*	*Social role theory*			*Evolutionary theory*		*Theological*
						Origin	*Found*	*Biosoc*	*Select*	*ENA*	
All Ages 10M	100.0	-0-	3	1926–1958	-0-	Contra	NoTest	Silent	Poss-M	Poss-M	Silent

With the above two stereotypes in mind, social role theory could be used to predict that males would be more likely to enjoy taking risks than females, even vicariously experienced risks. However, the original version of the theory would not predict that this sex difference would be universal (as it appears to be). Unfortunately, it is not possible to test the prediction of founder effect theory that the first human societies exhibited such a sex difference and has been "faithfully passed down thousands of times over" ever since (Fausto-Sterling 1992:199). In the case of the biosocial version of social role theory, it is difficult to envision why males being larger or stronger or why females being able to bear offspring would have any impact on worldwide sex differences in preferences for adventure stories.

Turning to the evolutionary explanations, sexual selection theory leads one to expect that males have not only been favored for greater risk-taking (see Précis 26.3.1.2), but also for being more physically active (see Précis 26.3.1.6), adventurousness (see Table 16.2.4.2 in Volume III), and to reason more in spatial terms (see Précis 26.2.3.4). Together, these traits are likely to contribute to greater willingness to roam further from home later in life, thereby locating more hunting opportunities (see Table 20.2.6.21 in Volume III). While these selection forces would not be selected for until adulthood, it is reasonable to assume that predisposition could be inborn during gestation.

No evidence was found to suggest that brain exposure to testosterone promotes preferences for stories of an adventurous nature. Therefore, this is hypothesis needs to be tested.

Assessment. Overall, preferring adventure stories appears to be a USD-M. Understanding it from any specific theory is still poorly developed.

26.2.7.41 Attitudes toward Female Participation in the Paid Workforce

Evidence. Précis 26.2.7.41 shows that 16 different studies investigated sex differences in people's attitudes toward females working in the paid workforce. All of these studies concluded that females have a more positive attitudes in this regard than do males. Unfortunately, all of these studies were conducted in just one country (i.e., the United States), published over a fairly narrow span of time (i.e., 23 years).

Explanation. Because of the fact that just one country was sampled in all of the studies that were located, we will not attempt to align their conclusion with any of the five theories.

Assessment. The USD status of sex differences in attitudes toward females participating in the paid workforce still needs to be established in more than just one country before seeking to apply any of the five theories to its explanation.

Précis 26.2.7.41 Attitudes toward female participation in the paid workforce (Original Table 15.11.1.1)

Age categories, number of studies, and sex	*Consist. score*	*Overviews*	*Countries*	*Time range*	*Non-human*	*Social role theory*			*Evolutionary theory*		*Theological*
						Origin	*Found*	*Biosoc*	*Select*	*ENA*	
Adult 16F	100.0	-0-	1	1975–1998	-0-	Silent	Silent	Silent	Silent	Silent	Silent

26.2.7.42 Preference for Male-Typical Occupations

Evidence. Male-typical occupations are typically ones that involve working more with things rather than with people (except in the case of managerial occupations, which also tend to be male-typical). Thirty-seven studies of sex differences in having male-typical occupations were located. Based on studies of children, adolescents, and adults, Précis 26.2.7.42 shows that all studies agree that males have stronger preferences for these types of jobs than do females. The pertinent studies came from seven different countries, plus two multi-national studies, and were published over an 80-year time frame.

Explanation. Are there any societies in which what have come to be recognized as "male-typical occupations" (i.e., principally those involving working with things rather than with people) do not exist or are even reversed (i.e., more typical of females than of males)? All of the evidence we could locate points toward concluding that there are no such societies. This tentative conclusion would be all but impossible to explain along the lines of the original social role theory. Regarding the founder effect version of the theory, there is no way of knowing what sex differences might exist in the earliest human society. As for the biosocial version of social role theory, it fails to specify why male size and strength would be linked to male-typical jobs, except in the case of occupations requiring heavy labor. However, there are many male-typical occupations that involve little or no strenuous activity (e.g., engineering, mathematics, management).

The sexual selection version of evolutionary theory focuses attention on what males and females benefit most from in reproductive terms when they choose mates. When they are compared to males, females tend to have a much greater likelihood of looking for mates with an ability to make a living (see Précis 26.2.7.19). In this regard, male-typical occupations tend to yield higher salaries than do female-typical jobs (Leuze & Straub 2016; Mari & Luijkx 2020). As a result, it is reasonable to assume that males in male-typical occupations will sire more offspring than males in female-typical occupations. This prediction is supported by two studies indicating that prenatal testosterone among fully-adult men, using the 2D:4D finger length measure, is positively correlated with the number of offspring they report having had (Voracek, Pum & Dressler 2010; Klimek, Galbarczyk et al. 2014).

The ENA version of evolutionary theory predicts that neurohormonal factors are important in producing preferences for male-typical jobs. In this regard, two studies have reported on the relationship between prenatal testosterone (both inferred from 2D:4D finger length measures) and male-typical occupational interests. Both reported significant correlated with interests in things-oriented jobs, among adult males, but there were

Précis 26.2.7.42 Preference for male-typical occupations (Original Table 15.11.2.1)

Age categories, number of studies, and sex	*Consist. score*	*Overviews*	*Countries*	*Time range*	*Non-human*	*Social role theory*			*Evolutionary theory*		*Theological*
						Origin	*Found*	*Biosoc*	*Select*	*ENA*	
Child 10M	100.0	-0-	7 (2)	1936–2016	-0-	Contra	NoTest	Poss-M	Expl-M	Expl-M	Silent
Adolescent 11M	100.0										
Adult 16M	100.0										
All Ages 37M	100.0										

no significant correlations among adult females (Hell & Päbler 2011). Brain exposure to this hormone has also been found associated with enhanced competitiveness (see Précis 26.3.1.5) and greater risk-taking (see Précis 26.3.1.2), both traits that are more common in male-typical occupations than in female-typical occupations.

Assessment. The evidence that male-typical occupation preferences is a USD-M is very strong. The evolutionary theories seem to provide stronger explanations for this sex difference than do the social role theories.

26.2.7.43 Preference for People-Oriented Occupations

Evidence. As shown in Précis 26.2.7.43, all but one of 51 studies, published over a 98-year time span, have concluded that females are more interested in people-oriented occupations that are males. These occupations include those of nursing, social work, and teaching). In these studies, nine different countries were sampled, and two literature reviews plus four meta-analysis all agreed that this sex difference was substantial. It is also worth noting that studies of sex differences in the interest people express in other people generally (rather than in inanimate things) have also concluded that females surpass males (see Précis 26.2.7.30).

Explanation. Why would nearly all studies in nine different countries all conclude that females are more likely than males to prefer working in occupations that are primarily concerned with people (rather than with things)? There are at least three stereotypes that might direct one toward an answer. In particular, females are thought to be more friendly (Table 21.2.7.32), communal (Table 21.2.9.3), and sociable (Table 21.2.9.13).

According to all three versions of social role theory, growing up amidst stereotypes about sex differences in friendliness and sociability will incline girls and women to prefer people-oriented occupations. However, the original version of this theory would not predict universality in this regard. As for the founder effect version of the theory, it is unreasonable to believe that the first human society had occupations per se, let along ones that were people-oriented (as opposed to things-oriented). In the case of the biosocial version of social role theory, we see no way that males being larger and stronger, or females being the only sex capable of bearing children, would affect sex differences in preferences for people-oriented occupations.

Turning to the evolutionary perspective, sexual selection theory asserts that, because males have been favored more than females for being reliable resource provisioners, their interest in people at an intimate interpersonal level tends to be lower than for females (Reno, Meindl et al. 2003; L Ellis 2011b; Lippa, Preston & Penner 2014). Females, on the other hand, should be more people-oriented, in part, because this orientation would promote patient concern and care for offspring (Lippa 2010).

Précis 26.2.7.43 Preference for people-oriented occupations (Original Table 15.11.2.5)

Age categories, number of studies, and sex	*Consist. score*	*Overviews*	*Countries*	*Time range*	*Non-human*	*Social role theory*			*Evolutionary theory*		*Theological*
						Origin	*Found*	*Biosoc*	*Select*	*ENA*	
Adult 28F All Ages 50F 1x	100.0 98.0	F: 2Rev 4Met	9	1922–2020	-0-	Contra	NoTest	NoTest	Expl-F	Expl-F	Silent

To the above evolutionary premise involving sexual selection, ENA theory adds that relatively low brain exposure to testosterone should promote a people-orientation in general, including people-oriented occupations (DB Stewart-Williams & Halsey 2021). While little direct evidence pertaining to this hypothesis was located, various indirect supportive evidence has been reported (L Ellis & Ratnasingam 2015; Wright, Eaton & Skagerberg 2015).

Assessment. Overall, the evidence is strong that preference for people-oriented occupations is a USD-F. More work is need in order to fully explain this sex difference in theoretical terms, but an evolutionary perspective appears promising.

26.2.7.44 Preference for Things-Oriented Occupations

Evidence. Whereas females are more interested in people-oriented occupations (see directly above), Précis 26.2.7.44 shows that substantial evidence has found males being more interested in things-oriented occupations. Specifically, 31 studies were located, all of which concluded that things-oriented occupations (e.g., engineering, construction, and transportation) were more appealing to males. Samples for these studies were drawn from eight countries, published over an 87-year time period. This conclusion was also reached by three meta-analyses.

Explanation. As noted in connection with preferences for male-typical occupations (Précis 26.2.7.42), males usually express greater interest in jobs that focus on material things rather than on ones that involve establishing close interpersonal relationships with others. We see sufficient similarity between *male-typical occupations* and *things-oriented occupations* to simply refer readers to the former for an explanation of the latter.

As with male-typical jobs, those that are things-oriented appear to be best explained with evolutionary reasoning. It is also worth adding that one study specifically bearing on interest in things-oriented occupations (e.g., engineering, mathematics) rather than people-oriented occupations. As theoretically expected, based on 2D:4D measurement, the study concluded that prenatal testosterone exposure was higher among males with things-oriented occupational interests, but, among females, the differences were not statistically significant (Hell & Pabler 2011). This finding is very similar to that reported on male-typical occupational interests being positively correlated with high testosterone exposure (Manning, Reimers et al. 2010).

Assessment. The evidence is very strong that preference for things-oriented occupations is a USD-M. The best explanations are those with an evolutionary focus.

Précis 26.2.7.44 Preference for things-oriented occupations (Original Table 15.11.2.6)

Age categories, number of studies, and sex	*Consist. score*	*Overviews*	*Countries*	*Time range*	*Non-human*	*Social role theory*			*Evolutionary theory*		*Theological*
						Origin	*Found*	*Biosoc*	*Select*	*ENA*	
Adolescent 10M Adult 18M All Ages 31M	100.0 100.0 100.0	M: 3Met	8	1933–2020	-0-	Contra	NoTest	Silent	Expl-M	Expl-M	Silent

26.2.7.45 *Interest in Helping-Oriented Occupations*

Evidence. Occupations that focus on helping others would include those of social work, health care, and teaching. Even retail sales may often fit this categorization. Helping occupations obviously overlap with people-oriented occupations to a substantial degree. One can see in Précis 26.2.7.45 that all 11 studies bearing on people's interests in helping-oriented occupations found that females express stronger interests than do males, a sex difference also reported based on a meta-analysis. The studies were based on samples drawn from four different countries, published over an 83-year time period.

Explanation. As with people-oriented occupations (see Précis 26.2.7.43), only the evolutionary theories seem to be geared toward accounting for the apparent universal tendency for females to be drawn toward these types of jobs than is the case for males. Nevertheless, more work is needed, particularly regarding the identification of any influence of androgens on such sex differences.

Assessment. Interest in helping-occupations can be considered a USD-F. More development of theoretical accounting of this sex difference is needed.

26.2.7.46 *Preference for Jobs with Flexible Hours*

Evidence. As shown in Précis 26.2.7.46, ten studies have found females expressing greater interest in jobs with flexible hours than is the case for males. Most of the countries involved a single country over a 25-year time frame.

Explanation. According to sexual selection theory, females should have a preference for jobs with flexible hours more than do males in order to manage their paid work with their obligations to care for their offspring and to maintain a functioning household. In other words, in order for females to contribute to reproduction by bearing and nurse them, and provide them with a healthy and learning environment, any earned employment will need to be more flexible than the earned employment for males.

Assessment. Having a preference for jobs with flexible working hours appears to be a USD-F. More detailed theoretical development is in order.

26.2.7.47 *Breadth of Occupational Interest*

Evidence. As shown in Précis 26.2.7.47, 12 studies were located having to do with sex differences in the breadth of people's occupational interests. All of these studies agreed that males have a broader range of interests than do females. Unfortunately, all of these studies were conducted in just one country (i.e., the United States) and the range of time covered by their publication was just 21 years.

Explanation. Regarding evolutionary theory, the sexual selection version rests heavily on the assumption that females are more likely to prefer

Précis 26.2.7.45 Interest in helping-oriented occupations (Original Table 15.11.2.7)

Age categories, number of studies, and sex	*Consist. score*	*Overviews*	*Countries*	*Time range*	*Non-human*	*Social role theory*			*Evolutionary theory*		*Theological*
						Origin	*Found*	*Biosoc*	*Select*	*ENA*	
Adult 11F	100.0	F: 1Met	4	1931–2014	-0-	Silent	Silent	Silent	Expl-F	Expl-F	Silent

Précis 26.2.7.46 Preference for jobs with flexible hours (Original Table 15.11.2.15)

Age categories, number of studies, and sex	*Consist. score*	*Overviews*	*Countries*	*Time range*	*Non-human*	*Social role theory*			*Evolutionary theory*		*Theological*
						Origin	*Found*	*Biosoc*	*Select*	*ENA*	
Adult 10F	100.0	-0-	1 (1)	1981–2006	-0-	Silent	Silent	Silent	Expl-F	NoTest	Silent

Précis 26.2.7.47 Breadth of occupational interest (Original Table 15.11.2.25)

Age categories, number of studies, and sex	*Consist. score*	*Overviews*	*Countries*	*Time range*	*Non-human*	*Social role theory*			*Evolutionary theory*		*Theological*
						Origin	*Found*	*Biosoc*	*Select*	*ENA*	
All Ages 12M	100.0	-0-	1	1957–1978	-0-	Silent	Silent	Silent	Expl-M	Poss-M	Silent

mates who are able to provide stable resources and share them with their spouse and offspring (Bereczkei, Voros et al. 1997; Ellis 2001; Souza, Conroy-Beam & Buss 2016). There is strong empirical support for this assumption (see Précis 26.2.7.19). Among the consequences of this sex difference is that males will be less choosy than females when selecting jobs, as long as they are fairly lucrative. In other words, males should be more likely than females to be interested in jobs that involve substantial risk of injury, are physically strenuous, or are often dirty and odorous. Even jobs that are often confrontational and legally risking should appeal to males more. Females, on the other hand, should be more prone to limit their occupational interests to work that involves helping others, particularly children, i.e., people-oriented occupations (see Précis 26.2.7.43). For evidence that males are not only more *interested* in a wider range of occupations than females, but that they actually *fill* a wider range of jobs, see Précis 26.3.5.21.

If the ENA version of evolutionary theory has merit, not only should males have broader occupational interests than females, but brain exposure to testosterone should be contributing to these broader interests. No evidence for or against this prediction was located.

Assessment. Even though just one country was sampled, evolutionary reasoning provides a persuasive explanation for why males would have broader occupational interests than females. Therefore, we will reservedly declare the breadth of occupational interests to be a USD-M.

26.2.7.48 Preference Regarding Monetary Compensation of Jobs

Evidence. Many studies have sought to determine if there are sex differences in people's interests in how much money they will make in a particular line of work. As shown in Précis 26.2.7.48, at least among adolescents, all studies agreed that this was a more important consideration than was the case for females. When significant sex differences were found for other age groups, all studies also agreed that monetary compensation was considered more important to males than to females. The number of countries sampled in these studies was four in addition to one multi-country study, and the publication time frame was 93 years.

Explanation. Why would males be more interested in how much they will earn from their work than females? The only theory that appears to provide an explanation is along evolutionary lines. As noted in the narrative directly above, when choosing mates, females weigh their potential spouse's ability to "make a living" to a greater degree than do males (see Précis 26.2.7.19).

Regarding the ENA version of evolutionary theory, no evidence was located specifically regarding the possible involvement of neuroandrogenic factors in preferences for monetary compensation. However, it is

Précis 26.2.7.48 Preference regarding monetary compensation of jobs (Original Table 15.11.2.28)

Age categories, number of studies, and sex	*Consist. score*	*Overviews*	*Countries*	*Time range*	*Non-human*	*Social role theory*			*Evolutionary theory*		*Theological*
						Origin	*Found*	*Biosoc*	*Select*	*ENA*	
Adolescents 22M Adult 39M 9x Wide Age Range 8M 2x All Ages 70M 11x	100.0 (81.3) (80.0) (86.4)	M: 1Met	4 (1)	1922–2015	-0-	Silent	Silent	Silent	Expl-M	Poss-M	Silent

reasonable to suspect that male tendencies to be more competitive that females, as evidence strongly suggests and for which there is considerable evidence of androgen influences (see Précis 26.3.1.5) could partly manifest itself in the form of desire for high monetary compensation.

Assessment. The importance of monetary compensation when choosing a job appears to be a USD-M. The strongest theoretical explanation appears to be evolutionary in nature.

26.2.7.49 Interest in Artistic and Creative/Expressive Occupations

Evidence. Précis 26.2.7.49 provides a summary of findings having to do with sex differences in people's interests in occupations that involve artistic and creative expression. One can see that 19 out of 20 of the pertinent studies concluded that these features of occupations appealed to females more than to males. Five countries were sampled and were published over an 82-year time period. Two meta-analyses agreed that females were more drawn to these types of occupations while one meta-analysis indicated that there were no noteworthy sex differences in this regard.

Explanation. Why would females be more interested in occupations in which they are allowed to express themselves artistically and creatively? One evolution-based explanation involves noting that most individuals who pursue these types of occupations rarely receive long term employment; thereby being unlikely to be a stable lifetime earner. Since males are typically chosen as mates with a heavy emphasis put on stable earnings ability, males in artistic and creative occupations are likely to leave fewer offspring in future generations than males in other types of occupations. We found no evidence for or against this prediction.

Another evolutionary proposal was not limited to occupational interests. Instead, it asserted that females who are artistic in general are likely to be interested in enhancing their physical appearance than are females with few artistic inclinations (Varella, Valentova & Fernández 2017). If so, sex differences in people's interests in artistic and creative occupations would be just a side effect of general aesthetic tendencies.

In the case of ENA theory, there should be a positive correlation between brain exposure to testosterone and interest in artistic and creatively expressive occupations. Just one study was located that provided evidence in this regard. While the sample sizes were small, the study found no significant correlations between 2D:4D finger length ratios and being involved in the visual arts for either sex (Crocchiola 2014).

Assessment. Substantial evidence supports concluding that interest in artistic and creative/expressive occupations is a USD-F. Theories explaining this sex difference appear to be poorly developed and tested.

Précis 26.2.7.49 Interest in artistic and creative/expressive occupations (Original Table 15.11.4.3)

Age categories, number of studies, and sex	*Consist. score*	*Overviews*	*Countries*	*Time range*	*Non-human*	*Social role theory*			*Evolutionary theory*		*Theological*
						Origin	*Found*	*Biosoc*	*Select*	*ENA*	
All Ages 19F 1x	95.0	F: 2Met x: 1Met	5	1935–2017	-0-	Silent	Silent	Silent	Poss-F	Contra	Silent

26.2.7.50 Interest in Being a Computer Programmer or Analyst

Evidence. Twelve studies were located on sex differences in people's interests in being a computer programmer or analyst. Précis 26.2.7.50 shows that all of these studies concluded that males have greater average interest in this line of work than do females. Studies came from five different countries that were published over a 72-year time frame.

Explanation. Given that computer programming and computer analysis involve primarily working with things rather than with people, one can return to Précis 26.2.7.43 and Précis 26.2.7.44 for theoretical foundations with which to explain sex differences in the desire to work with computers. It is also worth noting that interests and abilities tend to be positively correlated (Hyland, Hoff & Rounds 2022:16). This leads one to expect males to be more likely to major in computer science (which they do: see Précis 26.3.5.2), to be more likely to work in computer related occupations (as evidence suggests: see Précis 26.3.5.23). Males should also experience less computer anxiety (as nearly all evidence suggests: see Table 10.2.5.14).

No evidence was located having to do with brain exposure to testosterone and interest in working with computer at a programming level. However, one study reported a negative correlation between grades earned in a Java programming course and 2D:4D finger length. This indicates that the best students had been exposed to relatively high prenatal testosterone (Brosnan, Gallop et al. 2011). Assuming that people tend to be more interested in things they are good at, this rather indirect evidence suggests that ENA theory could help to explain some sex differences in interests in computers.

Assessment. Overall, the evidence is strong that interest in being a computer programmer or computer analyst is a USD-M. Evolutionary theory is likely to have some explanatory power in terms of explaining this sex difference.

26.2.7.51 Interest in Being Teachers

Evidence. Ten studies of sex differences in being teachers were located. Précis 26.2.7.51 shows that females express more interests in this profession than do males. Three different countries were sampled; the studies were published over an 88-year period of time.

Explanation. One can predict that females would be more likely to be teachers by noting that they are more interested in people-oriented occupations than are males (see Précis 26.2.7.43). Also, females have been shown to exhibit more interest in spending time with children (see Précis 26.2.7.28). Even as children themselves, females exhibit stronger tendencies to play with dolls than do males (see Précis 26.2.7.37).

Evolutionary theory would envision sexual selection to be at root of all of these sex differences. Specifically, because females have been favored for

Précis 26.2.7.50 Interest in being a computer programmer or analyst (Original Table 15.11.4.17)

Age categories, number of studies, and sex	*Consist. score*	*Overviews*	*Countries*	*Time range*	*Non-human*	*Social role theory*			*Evolutionary theory*		*Theological*
						Origin	*Found*	*Biosoc*	*Select*	*ENA*	
All Ages 12M	100.0	-0-	5	1933–2005	-0-	Silent	Silent	Silent	Poss-M	Poss-M	Silent

Précis 26.2.7.51 Interest in being teachers (Original Table 15.11.4.56)

Age categories, number of studies, and sex	*Consist. score*	*Overviews*	*Countries*	*Time range*	*Non-human*	*Social role theory*			*Evolutionary theory*		*Theological*
						Origin	*Found*	*Biosoc*	*Select*	*ENA*	
All Ages 10F	100.0	-0-	3	1922–2010	-0-	Silent	Silent	Silent	Poss-F	Mixed	Silent

providing care to their own offspring to a greater degree than are males (Westneat & Sherman 1993; Kempenaers & Sheldon 1996), this tendency is likely to be extended to children generally.

ENA theory predicts that testosterone will inhibit interests in children, and thereby reduce the likelihood of choosing to be a teacher, especially at the elementary level. At least regarding circulating levels of testosterone, the evidence reported so far has provided little to no support for this hypothesis, either in humans (Gordon, Pratt et al. 2017; Bos, Hechler et al. 2018) or in other species (O'Neal, Reichard et al. 2008; Zöttl, Vullioud et al. 2018).

Assessment. Interest in being teachers appears to be a USD-F. The strongest explanations appear to be in terms of sexual selection.

26.2.7.52 Attitude toward Social Dominance

Evidence. Attitudes having to do with social dominance involve viewing social groups (such as races or ethnic groups) as existing in a hierarchy, often with one's own group seen as being at the top. Précis 26.2.7.52 indicates that males tend to have more positive views about this topic than do females. The evidence is based on studies from three individual countries (plus three multi-country study) extending over a 25-year time frame.

Explanation. No evidence of sex stereotypes regarding attitudes toward social dominance were found. Therefore, no attempt will be made to apply any of the three social role theories to such attitudes.

In the case of the evolutionary theories, studies have shown that females prefer mates who exhibit dominance and status in general (Cotton, Small & Pomiankowski 2006; Cummins 2015) as well as physical traits associated with status, such as height (see Précis 26.2.7.15) and physical strength (Sell, Lukazsweski & Townsley 2017). This is likely to have favored males who are both competitive and hierarchically oriented. In essence, the less genetically related individuals are to others, the more competitive and condescending they should be toward others, including entire groups of others. Consequently, sexual selection theory would lead one to expect males to be more likely to subscribe to social dominance attitudes.

Regarding ENA theory's prediction that tendencies to favor social dominance over groups of others, or to behave in socially dominant ways should be positively correlated with testosterone. Considerable evidence supports this hypothesis, at least regarding circulating testosterone levels among males (Ehrenkranz, Bliss & Sheard 1974; Tremblay, Schaal et al. 1998; R Rowe, Maughan et al. 2004; Tarter, Kirisci et al. 2007).

Assessment. More work is needed to better understand why attitudes toward social dominance appear to be higher among males. Nonetheless, given the number of countries sampled and the consistency of the findings, we deem this variable to be a USD-M.

Précis 26.2.7.52 Attitude toward social dominance (Original Table 15.12.2.15)

Age categories, number of studies, and sex	*Consist. score*	*Overviews*	*Countries*	*Time range*	*Non-human*	*Social role theory*			*Evolutionary theory*		*Theological*
						Origin	*Found*	*Biosoc*	*Select*	*ENA*	
All Ages 10M	100.0	-0-	3 (1)	1979–2004	-0-	Silent	Silent	Silent	Poss-M	Expl-M	Silent

26.2.7.53 Attitude toward War and the Use of Military Power

Evidence. Sixty-two studies of sex differences in attitudes toward war and/or the use of military power to settle international disputes were located. Précis 26.2.7.53 shows that the vast majority of these studies concluded that males held more favorable attitudes in this regard than did females. However, only for the studies in which the samples were of a wide age range were the findings above the 95.0 cutoff. Overall, finding came from eight countries (plus three multi-country studies), and they were published over a 57-year period of time.

Explanation. There are different ways of explaining sex differences in attitudes toward international conflict. From the perspective of social role theory, one could argue that females are stereotyped as being more fearful than males (see Table 21.2.6.14 in Volume III), while males are stereotyped as being more prone toward aggression (see Précis 26.3.6.6). Both of these stereotypes could lead one to stereotype males as being more likely to support the use of force in international conflicts. However, the theory's original version would not predict universality in this sex difference, and there is no way to test the founder effect version of social role theory. Regarding the biosocial version, it might be argued that males are more aggressive and less fearful because of their greater physical size and strength, relative to females.

Turning to the evolutionary approaches, no evidence was found specifically attempting to explain sex difference in war-related attitudes with sexual selection theory. Likewise, no research seems to have examined neurohormonal factors as contributing to these sex differences. However, if one were to assume that fear is a major inhibitor of people wanting to avoid war and that aggressiveness and competitiveness are contributors to having favorable attitudes toward war, one could make the case for both evolutionary theories being able to explain such attitudes. This is because sexual selection and brain exposure to testosterone both appear to diminish fearfulness (see Précis 26.2.2.2) and increased by competitiveness (see Précis 26.3.1.5). Nonetheless, no direct evidence for or against the hypothesis that these traits affect attitudes toward war.

Assessment. Nearly all research indicates that favorable attitudes toward war and the use of military power to settle international disputes is a USD-M. Theories explaining these differences are not well developed.

26.2.7.54 Attitude toward Nuclear Power

Evidence. Primarily for safety reasons, people vary in their attitudes toward nuclear power. As presented in Précis 26.2.7.54, one can see that ten studies, derived from two separate countries (plus one multi-national study) all found males to have a more positive (or less negative) attitude

Précis 26.2.7.53 Attitude toward war and the use of military power (Original Table 15.12.2.17)

Age categories, number of studies, and sex	*Consist. score*	*Overviews*	*Countries*	*Time range*	*Non-human*	*Social role theory*			*Evolutionary theory*		*Theological*
						Origin	*Found*	*Biosoc*	*Select*	*ENA*	
Adult 41M 2F Wide Age Range 11M All Ages 60M 2F	(91.1) 100.0 (93.8)	-0-	8 (3)	1954–2011	-0-	Contra	NoTest	Poss-M	Poss-M	Poss-M	Silent

Précis 26.2.7.54 Attitude toward nuclear power (Original Table 15.12.3.10)

Age categories, number of studies, and sex	*Consist. score*	*Overviews*	*Countries*	*Time range*	*Non-human*	*Social role theory*			*Evolutionary theory*		*Theological*
						Origin	*Found*	*Biosoc*	*Select*	*ENA*	
All Ages 10M	100.0	-0-	2 (1)	1976–1989	-0-	Silent	Silent	Silent	Poss-M	Poss-M	Silent

toward nuclear power. These studies were all published within a narrow 13-year time period.

Explanation. None of the three social role theories seems to address the possibility of sex differences in attitudes toward nuclear power. To use evolutionary theories to account for why males hold more favorable attitudes toward nuclear power would require recognizing that nuclear power involves some degree of risk. As noted in Précis 26.3.1.2, males appear to have evolved tendencies to take more risks than is the case for females. Therefore, one could argue that, if one were to control for sex differences in risk-taking, sex differences in attitudes toward nuclear power would roughly equalize.

Assessment. Favorable attitudes toward nuclear power appears to be a USD-M. Little is available to offer a theoretical explanation for this sex difference.

26.2.7.55 Excusing or Tolerating Rape

Evidence. Twelve studies were located having to do with people's tolerance of rape. These studies are usually based on describing hypothetical situations to research participants of both sexes in which a potential rape victim behaves in ways that indicate some degree of interest in having sexual relationships with a potential rapist without actually agreeing to a specific relationship. Précis 26.2.7.55 indicates that all 12 of the studies conducted on this topic found that males were more likely to excuse or tolerate rape than were females. These studies were limited to samples drawn from three countries.

Explanation. No studies of sex stereotypes pertaining to the tendency to excuse or tolerate rape were located. Therefore, no effort will be made to draw on any of the three social role theories for a possible explanation.

From an evolutionary perspective, one could argue as follows: When the sexes are compared, males appear to (a) have a stronger sex drive (see Précis 26.2.7.6), (b) are more tolerant of casual sexual encounters (see Précis 26.2.7.21), and are less desirous of being in love before having sex (see Précis 26.2.7.20). All three of these sex differences are likely to have been selected due to the fact that males have a higher reproductive ceiling than do females, provided they can have sex with multiple sex partners (Clutton-Brock 2007; Puts, Jones & DeBruine 2012). All of these sex differences are likely to lead to males being less able to "empathize" with rapists than is true for females.

If the above theoretical reasoning is correct, ENA theory would assert that brain exposure to testosterone is likely to promote greater tendencies to excuse or tolerate rape. No evidence for or against this hypothesis was located.

Assessment. A greater tendency to excuse or tolerate rape appears to be a USD-M. The most pertinent theory appears to surround evolutionary arguments.

Précis 26.2.7.55 Excusing or tolerating rape (Original Table 15.13.1.4)

Age categories, number of studies, and sex	*Consist. score*	*Overviews*	*Countries*	*Time range*	*Non-human*	*Social role theory*			*Evolutionary theory*		*Theological*
						Origin	*Found*	*Biosoc*	*Select*	*ENA*	
Adult 12M	100.0	-0-	3	1980–2002	-0-	Silent	Silent	Silent	Poss-M	Poss-M	Silent

26.2.7.56 *Attribute More of the Responsibility for Rape to the Victim*

Evidence. Précis 26.2.7.56 shows that 36 studies from three different countries, extending over a 33-year time frame, have all concluded that males attribute more of the responsibility for rape to the victim. The majority of these studies involve research participants being presented with hypothetical rape scenarios in which a rape victim was described as behaving in ways regarding having an interest in sexual relationships with a potential perpetrator.

Explanation. As explained in the preceding précis, on average, males appear to have stronger sex drives and are more tolerant of casual sexual encounters than do females, both for evolutionary reasons. Consequently, it is possible to argue that males are more inclined to put themselves in the shoes of a rapist, so to speak, and thereby attribute more responsibility for rape to those who are victimized than are females.

Assessment. We declare the tendency to attribute more of the responsibility for rape to the victim to be a USD-M. Nevertheless, more work is needed to develop and test theoretical explanations for this apparently universal sex difference.

26.2.7.57 *Tolerance of Sexual Harassment*

Evidence. Twelve studies were located in which hypothetical or real instances of potential sexual harassment are presented to samples of males and females. Then, the research participants are asked to report whether they consider the instances to actually constitute sexual harassment in their minds. As Précis 26.2.7.57 shows, all of these studies concluded that males expressed less inclusive opinions than did females. Unfortunately, all 12 studies were conducted in just one country.

Explanation. The only theories that seem to offer an explanation for this sex difference are those rooted in evolutionary reasoning. As noted in the narrative bearing on Précis 26.2.7.55, because males have evolved stronger sex drives than females, they are more likely to tolerate at least subtle sexual overtures than are females.

Assessment. Because all 12 studies were conducted in just one country (i.e., the United States) and the theoretical arguments surrounding this sex difference is rather tenuous, we will not deem tolerance of sex harassment to be a USD.

26.2.7.58 *Inclusiveness in Assessing What Constitutes Sexual Harassment*

Evidence. Précis 26.2.7.58 shows that 26 studies have compared males and females regarding how many different types of actions constitutes

Précis 26.2.7.56 Attribute more of the responsibility for rape to the victim (Original Table 15.13.1.5)

Age categories, number of studies, and sex	*Consist. score*	*Overviews*	*Countries*	*Time range*	*Non-human*	*Social role theory*			*Evolutionary theory*		*Theological*
						Origin	*Found*	*Biosoc*	*Select*	*ENA*	
Adult 32M All Ages 36M	100.0 100.0	M:1Rev	3	1973–2006	-0-	Silent	Silent	Silent	Poss-M	Poss-M	Silent

Précis 26.2.7.57 Tolerance of sexual harassment (Original Table 15.13.1.7)

Age categories, number of studies, and sex	*Consist. score*	*Overviews*	*Countries*	*Time range*	*Non-human*	*Social role theory*			*Evolutionary theory*		*Theological*
						Origin	*Found*	*Biosoc*	*Select*	*ENA*	
All ages 12M	100.0	-0-	1	1986–2004	-0-	Silent	Silent	Silent	Poss-M	Poss-M	Silent

Précis 26.2.7.58 Inclusiveness in assessing what constitutes sexual harassment (Original Table 15.13.1.9)

Age categories, number of studies, and sex	*Consist. score*	*Overviews*	*Countries*	*Time range*	*Non-human*	*Social role theory*			*Evolutionary theory*		*Theological*
						Origin	*Found*	*Biosoc*	*Select*	*ENA*	
All Ages 26F	100.0	F:1Met	2	1986–2004	-0-	Silent	Silent	Silent	Expl-F	Expl-F	Silent

sexual harassment. Every one of these studies concluded that females were more inclusive than are males. In other words, they tend to have more expansive definitions of what it means to engage in sexual harassment than do males. The evidence came from two countries published over an 18-year time period.

Explanation. As with the variable covered in the preceding précis, an evolutionary perspective seems to offer the most reasonable explanation as to why such a sex difference would be universal. In essence, because males have a stronger sex drive (Précis 26.2.7.6), the are likely to be more tolerant with regard to sexual harassment. In other words, the circumstances surrounding the potentially harassment would have to be more blatant and serious before males would concur with females about the actual harassing nature of the behavior.

Assessment. Inclusiveness in assessing what constitutes sexual harassment seems to be a USD-F. Evolutionary theory seems to provide the most reasonable theoretical explanation for its apparent universality.

26.2.7.59 Attitude toward Aggression and Violence

Evidence. Précis 26.2.7.59 shows that all 12 studies regarding attitudes toward aggression and violence as a way to deal with interpersonal disputes have reported that males are more favorable or accepting of such behavior than are females. The available research was based on samples from one country in addition to one multi-cultural study. Publication of these studies spanned 23 years.

Explanation. As shown in an earlier précis (Précis 26.2.7.41), males have more favorable attitudes toward war and the use of military force to deal with international disputes. Therefore, it is rather predictable that males would hold more favorable attitudes toward interpersonal aggression and violence as a way to settle interpersonal disputes as well. Furthermore, both types of attitudes are likely to be similarly explained from a theoretical standpoint.

Regarding social role theory, the original version would not predict that sex differences in these attitudes would be universal. In the case of the founder effect version, it is essentially untestable because of the lack of any evidence from the original human societies. As for the biosocial version, it could be argued that males are universally prone to have more favorable attitudes toward interpersonal violence because of their relatively large size and strength. One way to test this idea further would be to look for correlations between size and strength, on the one hand, and these types of attitudes. The biosocial version of social role theory would predict that, within each sex, positive correlations should be found. It should be mentioned, however, that ENA theory would also predict positive correlations, at least regarding physical strength.

Précis 26.2.7.59 Attitude toward aggression and violence (Original Table 15.13.2.4)

Age categories, number of studies, and sex	*Consist. score*	*Overviews*	*Countries*	*Time range*	*Non-human*	*Social role theory*			*Evolutionary theory*		*Theological*
						Origin	*Found*	*Biosoc*	*Select*	*ENA*	
All Ages 12M	100.0	-0-	1 (1)	1984–2013	-0-	Contra	NoTest	Poss-M	Poss-M	Poss-M	Silent

Some proposals along evolutionary lines have been made for sex differences in attitudes toward the use of violence. In essence, these proposals have involved stipulating that sexually selection has favored males being more prone to violence in their efforts to compete with other males for resources and sexual access along with the assumption that most behavior patterns will nearly always be preceding by favorable attitudes toward those behavior patterns (Fares, Ramirez et al. 2011; Huppin, Malamuth & Linz 2019).

No specific research directly linking neurohormonal factors to sex differences in attitudes toward aggression and violence. However, it seems reasonable to suspect such associations given that both prenatal and post-pubertal testosterone appears to be positively correlated with serious forms of actual aggression and violence (see Précis 26.3.4.2 and Précis 26.3.4.3).

Assessment. While samples need to be drawn from a wider variety of countries that are currently available, positive attitudes toward aggression and violence as a way to deal with interpersonal conflicts can be tentatively considered a USD-M. More research is needed to better understand this sex difference in theoretical terms.

26.3 High Consistency Scores in Volume III

Most of the variables pertaining to sex differences in Volume III have to do with behavior, including social behavior. Nevertheless, a number of behavioral variables are difficult to separate from cognitive traits, which were the main focus of Volume II.

26.3.1 Personality Traits and Behavioral Tendencies

There are often no sharp distinctions between personality traits and so-called behavioral tendencies. In general, the former tends to be a bit more foundational and long-lasting throughout one's life, while behavioral tendencies are usually more specific and can come and go with age.

26.3.1.1 Gregariousness (Friendliness)

Evidence. Précis 26.3.1.1 indicates that, at least among adults, females are more gregarious and friendly than males. The evidence comes from 29 studies conducted in four countries plus seven multinational studies. Findings were published over a 61-year time frame.

Explanation. All three versions of social role theory can predict that females would be more gregarious and friendly. Their prediction in this regard can be derived from studies of stereotypes, indicating that females are believed to be more sociable (see Table 21.2.9.13) and friendly (Table 21.2.7.32). However, the original version of this theory would not

Précis 26.3.1.1 Gregariousness (friendliness) (Original Table 16.1.2.10)

Age categories, number of studies, and sex	*Consist. score*	*Overviews*	*Countries*	*Time range*	*Non-human*	*Social role theory*			*Evolutionary theory*		*Theological*
						Origin	*Found*	*Biosoc*	*Select*	*ENA*	
Adult 19F 1x All Ages 27F 1x 1M	95.0 (90.0)	-0-	4 (7)	1957–2018	-0-	Contra	NoTest	Poss-F	Poss-F	Expl-M	Silent

predict that sex differences in gregariousness or friendliness would be found universally, while the founder effect version and the biosocial version of social role theory at least imply that the differences could be universal. To confirm the founder effect version, one would have to determine if females were more gregarious in the earliest human societies, which of course is impossible to know. Regarding the biosocial version of social role theory, one would have to document that female gregariousness is inversely associated with sex difference in physical size and strength and/or with female capabilities of giving birth. No specific evidence in this regard were found.

In the case of the two evolutionary theories, it seems relevant to note that (a) females appear to be more empathetic (Table 10.2.8.3) and prone to provide care to others (Table 17.1.6.10; Table 17.1.6.15; Table 17.1.6.17). These traits can be considered pertinent to gregariousness and friendliness, and has been explained in terms of sexual selection theory (Preston & de Waal 2002). It also seems relevant to note that sex differences in being socially oriented appear to be present even among newborns. For example, female infants have been found to make more eye contact with their caregivers (Hittelman & Dickers 1979; Leeb & Rejskind 2004; also see Table 17.3.1.3). Also, infant females have been shown to orient more frequently to the appearance of human faces (Connellan, Baron-Cohen et al. 2000) and to human voices (Osofsky & O'Connell 1977). Furthermore, infant females exhibit more of what is known as *emotional contagion*, such as crying when hearing another baby crying (Sagi & Hoffman 1976). Experimental work with non-human infant primates have reported similar tendencies for females to surpass males in being socially oriented (Simpson, Nicolini et al. 2016).

Regarding the possible involvement of neuroandrogenic factors in producing sex differences in people's interests in living things, one study reported an inverse correlation between prenatal testosterone exposure (derived from amniotic fluid) and the amount of eye contact made by infants with adults (Lutchmaya, Baron-Cohen & Raggatt 2002a). Another study of prenatal testosterone (using the 2D:4D finger length ratio) concluded that "relationship satisfaction" with their spouse was lower among males who had the highest exposure (Voracek, Pum & Dressler 2009).

Assessment. Given the uncertainty surrounding the exact nature of gregariousness and how it is best to measure it, the marginality of the evidence of significant sex differences, and the absence of highly probable theoretical explanations, we do not consider gregariousness to be a USD.

26.3.1.2 Risk-Taking (Recklessness) in General

Evidence. As shown in Précis 26.3.1.2, 165 studies of sex differences were located regarding sex differences in risk-taking. One hundred of these studies were conducted among adults (often college students). For the adult samples, the 95.0 threshold was attained, but for the studies involving all of the age groups combined, this threshold was not reached. Samples for these studies came from 14 different countries in addition to two multi-country samples. The publication time range was also substantial, i.e., 93 years.

Explanation. No sex stereotype regarding risk-taking tendencies were found. Accordingly, no attempts to use any of the three social role theories to explain greater risk-taking among males were located either.

From an evolutionary perspective, many researchers have applied sexual selection theory to the study of risk-taking behavior. In doing so, they often begin by noting that risk-taking is an inherent part of life, and that all animals appear to have the ability to at least roughly detect and estimate the degree to which various activities are to being physically injured and balance these estimates against the potential gains associated with their actions (Dugatkin 2013). Of course, human have tendencies to take risks in order to achieve an especially wide range of desirable outcomes (Wilson, Daly & Pound 2002).

As to why males might have evolved greater tendencies to take risks than females, the assumption is that risk-taking increases the chances that males have of successfully competing for resources, which then helps males to attract mates (BJ Ellis, Del Guidice et al. 2012; Greitemeyer, Kastenmüller & Fischer 2013; Engqvist, Cordes & Reinhold 2015). Stated another way, because males have higher reproductive ceilings than do females, they stand to gain more from taking risks than is the case for females (Puts 2016).

Regarding the ENA version of evolutionary theory, the evidence of neurohormonal influences has been largely supportive, especially regarding circulating testosterone levels. Specifically, the higher the levels are, the more prone both sexes tend to be in terms of general risk-taking behavior (Mazur 1995; Goudriaan, Lapauw et al. 2010; Somerville, Jones & Casey 2010; Dariotis, Schen & Garnger 2016; Votinov, Knyazeva et al. 2022; *review*: Apicella, Carré & Dreber 2015; *meta-analysis*: Kurath & Mata 2018). The same has been concluded regarding financial risk-taking (Coates & Herbert 2008; Sapienza, Zingales & Maestripieri 2009 in females but not males; Peper, Koolschijn & Crone 2013), with one exception (Stanton, Welker et al. 2021).

Other research having to do with circulating testosterone suggest that its effects on risk-taking can be suppressed somewhat by cortisol, a stress hormone. In other words, when individuals are under stress, testosterone seems to be less effective in promoting risk-taking and related behaviors (Carre & Mehta 2011; Mehta, Welker et al. 2015; Barel, Shahrabani & Tzischinsky 2017; Alacreu-Crespo, Costa et al. 2019). Social environmental factors may also impact circulating testosterone's effects on risk-taking. For

Précis 26.3.1.2 Risk-taking (recklessness) in general (Original Table 16.2.3.1)

Age categories, number of studies, and sex	*Consist. score*	*Overviews*	*Countries*	*Time range*	*Non-human*	*Social role theory*			*Evolutionary theory*		*Theological*
						Origin	*Found*	*Biosoc*	*Select*	*ENA*	
Adult 104M 5x All Ages 160M 13x 1F	95.4 (91.4)	M: 3Rev 2Met	14 (2)	1927–2020	1M: chimp	Silent	Silent	Silent	Expl-M	Expl-M	Silent

example, one experiment found that the presence of an attractive woman increased both testosterone levels and risk-taking behavior among male research participants (Ronay & von Hippel 2010a).

Five studies investigated the possibility that prenatal testosterone exposure might promote risk-taking later in life. Using the 2D:4D finger length indicator, all of these studies have conclusion that high prenatal testosterone was associated with risk-taking, particularly among males (Manning, Bundred et al. 2003; Vermeersch, T'sjoen et al. 2008; Coates, Gurnel & Rustichini 2009; Ronay & von Hippel 20010b; Stenstrom, Saad et al. 2011; Aycinena, Baltaduonis & Rentschler 2014; Brañas-Garza, Galizzi & Nieboer 2018).

One study assessed both prenatal and circulating testosterone levels among males relative to their risk-taking tendencies in gambling situations. This study concluded that prenatal testosterone (assessed using the 2D:4D ratio) was predictive of risk-taking, but circulating testosterone was not (Evans & Hampson 2014). In another study using the same risk-taking measure, circulating testosterone was predictive of risk-taking in both sexes (Stanton, Liening & Schultheiss 2011). In this latter study, prenatal testosterone was not measured.

Assessment. More research is needed to fully explain sex differences in risk-taking, especially regarding the apparent neurohormonal underpinnings. While there may be circumstances in which the sexes do not differ significantly in risk-taking tendencies, we conclude that this behavior is a USD-M, particularly among adults.

26.3.1.3 Trying to Lose Weight (Dieting)

Evidence. Do members of one sex attempt to lose weight through dieting more often than members of the other sex? As shown in Précis 26.3.1.3, 70 out of 71 studies found that females tried to do so more frequently than do males. The studies were conducted in 12 different countries, along with two multi-national research projects. Findings in this regard were published over a 60-year time frame. (Note: This variable is similar to a variable appearing in Chapter 13 having to do with sex differences in self-assessment of being overweight – Table 13.1.3.6. For both variables, females score higher than males.)

Explanation. We found no sex stereotype evidence pertaining to greater female tendencies to lose weight. Predictably, therefore, no theoretical efforts by social role theorists to explain such a sex difference was found.

In the case of sexual selection theory, some inconsistencies were found. Some researchers have argued that, not including obesity, female body weight appears to be positively correlated with female fertility (Grammer,

Précis 26.3.1.3 Trying to lose weight (dieting) (Original Table 16.3.1.3)

Age categories, number of studies, and sex	*Consist. score*	*Overviews*	*Countries*	*Time range*	*Non-human*	*Social role theory*			*Evolutionary theory*		*Theological*
						Origin	*Found*	*Biosoc*	*Select*	*ENA*	
Adolescent 38F Adult 29F 1x All Ages 70F 1x	100.0 96.7 98.6	-0-	12 (2)	1950–2010	-0-	Silent	Silent	Silent	Poss-F	Expl-F	Silent

Fink et al. 2003:391; Kirchengast & Marosi 2008). This implies that female attempts to lose weight should actually be *dis*favored by sexual selection.

An alternative proposal has been offered. It asserts that females have been sexually selected for maintaining a low body weight, especially in the stomach area, because males have been sexually selected for preferring mates with minimal chances of already being pregnant (Ellis 2011a:559). This view would also be consistent with evidence that most males are sexually attracted to mates with low waist-to-hip (WtH) ratios, while females are not (Singh 1994; Henss 1995; Platek & Singh 2010).

Regarding the possible relevance of sex hormones to the waist-to-hip ratio, studies have indicated that women's WTH ratios tend to be lower for those with high estradiol and low testosterone (Sowers, Beebe et al. 2001; Van Anders & Hampson 2005; Mondragón-Ceballos, García Granados et al. 2015). This pattern would be in accordance with ENA theory.

Assessment. The evidence strongly indicates that trying to lose (or at least not gain) weight is a USD-F. The best theoretical explanations seem to be of an evolutionary nature.

26.3.1.4 Physical Exercise

Evidence. Four multi-national research projects, two literature reviews, and studies in 18 separate countries, found that males engaged in more physical exercise than was the case for females. The only exceptions were seven studies that did not find a sex difference. All of these studies, presented in Précis 26.3.1.4, took place over a period of almost 80 years.

Explanation. Given that physical exercise can be thought of as being a form of physical activity levels, we will deal with these two slightly different variables jointly. Therefore, for theoretical explanations of sex differences in physical exercise, see Précis 26.3.1.6 (having to do with activity levels).

Assessment. We can confidently declare physical exercise to be a USD-M. As discussed in connection with Précis 26.3.1.6 (below), most of the theoretical reasoning points toward the two evolutionary theories as being strongest.

26.3.1.5 Competitiveness

Evidence. Ninety studies were located having to do with sex differences in competitiveness. As shown in Précis 26.3.1.5, all but six of these 84 studies concluded that males are more competitive than females. Four of the six studies that reported no significant sex differences were ones conducted among prepubertal children. One can see that, especially among adults, the consistency score is above 95.0. The patterns shown are consistent with one literature review and three meta-analyses.

Précis 26.3.1.4 Physical exercise (Original Table 16.3.2.2)

Age categories, number of studies, and sex	*Consist. score*	*Overviews*	*Countries*	*Time range*	*Non-human*	*Social role theory*			*Evolutionary theory*		*Theological*
						Origin	*Found*	*Biosoc*	*Select*	*ENA*	
Child 41M 3x Adolescent 46M 1x Wide Age Range 11M All Ages 142M 7x	97.6 97.9 100.0 95.3	M: 2Rev	18 (4)	1934–2012	-0-	Silent	Silent	Silent	Poss-M	Expl-M	Silent

Précis 26.3.1.5 Competitiveness (Original Table 16.4.2.2)

Age categories, number of studies, and sex	*Consist. score*	*Overviews*	*Countries*	*Time range*	*Non-human*	*Social role theory*			*Evolutionary theory*		*Theological*
						Origin	*Found*	*Biosoc*	*Select*	*ENA*	
Adult 40M 1x All Ages 84M 6x	97.6M (93.3)	M: 1Rev 3Met	12 (2)	1953–2020	1M: Chimp	Contra	NoTest	Poss-M	Expl-M	Expl-M	Silent

Explanation. Research has indicated that males are stereotyped as being more competitive than females (see Table 21.2.7.16). Therefore, based on social role theory, one can predict that males will be more competitive than females, at least in the countries where this stereotype is held. However, the original social role theory would not predict that this sex difference would be universal; it implies the opposite.

The founder effect theory would predict universality of sex differences in competitiveness if the earliest human cultures happened to have such a sex difference (which is not knowable). The only way to approximate any information in this regard would be to assess competitiveness among chimpanzees, humans closest living primate relatives. In this case, one study of competitiveness among chimpanzees were found and it indicated that males were more competitive.

Regarding the biosocial version of social role theory, it might be argued that competitiveness is a manifestation of the fact that males are physically stronger than females, although this would imply that only competition in which physical strength was central would exhibit a sex difference. Monetary forms of competitiveness would not be expected to exhibit a sex difference (a prediction that is contrary to most of the research).

From an evolution perspective, greater competitiveness among males is clearly predicted. As the sex with the highest reproductive potential, males are favored for being more overtly competitive for access to mating opportunities and for resources with which to attract mates (Wilson, Daly & Pound 2002; Puts 2016; Dunsworth 2020).

In the case of ENA theory, testosterone should be making a contribution to sex differences in competitiveness. Supporting this prediction, several studies have found positive correlations between circulating testosterone levels and competitiveness (van Anders & Watson 2006; Hahn, Fisher et al. 2016; Eisenegger, Kumsta et al. 2017; Kordsmeyer & Penke 2019; Casto, Arthur et al. 2022) while one study found no significant correlation (Torrance, Hahn et al. 2018). By and large, most of the evidence points toward a positive effect of circulating testosterone on competitiveness (*review*: Geniole & Carre 2018). A study also indicated that whether one happens to win or lose in experimental competition can sometimes confound the relationship between testosterone and competitiveness (Zilioli & Watson 2014).

One investigation sought to determine if prenatal testosterone (as measured by 2D:4D) correlated with competitiveness among a sample of males. Results revealed no significant relationship (Apicella, Dreber et al. 2011).

A final comment regarding theoretical explanations of sex differences in competitiveness has to do with an interesting study that specifically compared predictions by *both* social role theory and evolutionary theory regarding sex differences in competitiveness. In this study, college students were allowed to play competitive games as themselves and as an avatar of the opposite sex. The reasoning was that if individuals play as an opposite-sex avatar, they would adopt any stereotype of how the opposite sex *should* behave. Contrary to what social role theory would predict, males were found to be more competitive than females even when they played as a female avatar and females were less competitive than males even when they played as a male avatar (Deaner, Dunlap & Bleske-Rechek 2022).

Assessment. Overall, we conclude that, at least among adults, competitiveness is a USD-M. The best explanations appear to be evolutionary in nature.

26.3.1.6 Activity Levels

Evidence. A variable related to physical exercise is activity level. Differences in activity levels have been studied for 90 years in humans. Does one sex tend to have a higher activity level than the other sex? Précis 26.3.1.6 reveals that a large number of studies in 20 different countries found that males engaged in higher levels of activity than did females. Seven multi-national studies, two literature reviews, and three meta-analyses reached this same conclusion. However, one other literature review failed to find a sex difference. The results of studies of sex differences in non-human activity levels are complex.

Explanation. No evidence was found regarding sex stereotypes having to do with physical activity (or physical exercise). However, the biosocial version of social role theory asserts that males are physically stronger, and that this could contribute to universal sex differences in various traits. In this regard, research has indicated that muscular strength and physical activity are positively correlated (Madsen, Adams & Van Loan 1998; Leblanc, Taylor et al. 2015).

Among evolutionary proposal that have been offered for sex differences in physical activity is one that is similar to what was just described for the biosocial role theory, i.e., physical activity and physical strength complement one another (also see Kirchengast & Marosi 2008). Another proposal along evolutionary lines involve noting that physical activity may help to promote and refine spatial reasoning, with the latter being important in performing a number of male-typical occupations (Bjorklund & Brown 1998).

Regarding the possible involvement of neuroandrogenic factors in promoting activity levels, several studies have provided supportive evidence. One study reported that high exposure to prenatal testosterone (inferred from 2D:4D ratio data) was positively correlated with activity levels among children (Alexander & Saenz 2012). Another study using amniotic fluid sample reached the same conclusion (Lamminaki, Hines et al. 2012). At a neurological level, prenatal testosterone was found associated with

Précis 26.3.1.6 Activity levels (Original Table 16.4.3.2)

Age categories, number of studies, and sex	*Consist. score*	*Overviews*	*Countries*	*Time range*	*Non-human*	*Social role theory*			*Evolutionary theory*		*Theological*
						Origin	*Found*	*Biosoc*	*Select*	*ENA*	
Fetal 3M 10x I/T 32M 14x 2F Child 66M 4x Adolescent 63M 3x Adult 38M 10x 1F	(23.1) (64.0) (94.3) 95.5 (76.0)	M: 2Rev M: 3Met x: 1Rev	20 (7)	1930–2020	complex	Silent	Silent	Poss-M	Poss-M	Expl-M	Silent

excitability in the hippocampus, a region in the brain's limbic system (Smith, Jones & Wilson 2002; also see Kelava, Chiaradia et al. 2022).

In the case of circulating testosterone, three studies were located. One indicated that circulating levels of testosterone were positively correlated with restlessness (Dabbs, Strong & Milun 1997). A second study reported a significant positive correlated between circulating testosterone levels and activity levels, at least among young adult males (Haring, Volzke et al. 2010:Table 1). The third study was experimental, involving hypogonadal adult males. After receiving a single injection of synthetic testosterone (versus a placebo), these males reported feeling significantly fewer symptoms of fatigue (Jockenhovel, Minnemann et al. 2009).

Assessment. We confidently declare that activity levels constitute a USD-M, at least in the case of adolescents. Also worth noting is that a closely related variable, that of amount of physical exercise, is also a USD-M (see Précis 26.3.1.4). In terms of theoretically explaining sex differences in activity levels, the most persuasive evidence is of an evolutionary neuroandrogenic nature.

26.3.1.7 Exploratory Behavior in General

Evidence. As shown Précis 26.3.1.7, 40 studies of sex differences in exploratory behavior were located. All of which concluded that males engaged in more exploratory behavior than females. These differences were found in nine different countries, one multi-national study, and reported to be the case in two literature reviews. The located studies were published over a 61-year time frame.

Explanation. Exploratory behavior is clearly more common among males than among females. While no stereotype studies were located having to do with sex differences in exploratory behavior per se, one can see in Précis 26.3.6.5 that males are stereotyped as being more *adventurous* than females. This stereotype suggests that the three social role theories are appropriate for helping to explain sex differences in exploratory behavior.

Regarding the original version of social role theory, it would have difficulty accounting for why males appear to be more prone to explore in all countries yet sampled. The founder effect version would assert that this sex difference existed in the first human societies, and then has been "faithfully passed down thousands of times over" ever since (Fausto-Sterling 1992:199), a seemingly impossible-to-test explanation.

In the case of the biosocial version of social role theory, the tendency for males to exhibit more exploratory behavior would be attributed to their being larger or stronger, or to females being uniquely capable of reproduction (Wood & Eagly 2012:56). However, because the sex difference has been documented even among children – when sex differences in

Précis 26.3.1.7 Exploratory behavior in general (Original Table 16.4.3.9)

Age categories, number of studies, and sex	*Consist. score*	*Overviews*	*Countries*	*Time range*	*Non-human*	*Social role theory*			*Evolutionary theory*		*Theological*
						Origin	*Found*	*Biosoc*	*Select*	*ENA*	
Child 22M All Ages 40M	100.0 100.0	M: 2Rev	9 (1)	1937–1998	complex	Contra	NoTest	Contra	Poss-M	Poss-M	Silent

strength are quite small, and girls are not yet able to bear offspring – this sort of theoretical explanation is not supported.

Turning to the two evolutionary theories, no attempts to explain sex differences in exploratory behavior using sexual selection theory. However, research have come fairly close to applying this theory to exploratory behavior if one broadens the conceptualization of this behavior as often including risk-taking. As noted in Précis 26.3.1.1, at least among adults, risk-taking is a USD-M and it has been explained in terms of sexual selection.

Also, in the case of brain exposure to androgens contributing to exploratory behavior, no direct evidence was found. However, some indirect evidence seems worth mentioning: there are studies that have indicated that circulating testosterone levels are positively correlated with financial risk-taking among both sexes (Peper, Koolschijn & Crone 2013) and with novelty seeking among males (Määttänen, Jokela et al. 2013:2245).

Assessment. Exploratory behavior can be confidently considered a USD-M. Nonetheless, more work is in order to fully understand this sex difference in theoretical terms.

26.3.1.8 Using a Cradling-Like Book-Carrying Style

Evidence. One of the ways students tend to carry books is to cradle them using one or both hands and arms. Précis 26.3.1.8 shows that, according to all 14 located studies, females tended to use a cradling book-carrying style more often than did males. This sex difference has been studied in two countries, published over a 37-year time frame.

Explanation. To our knowledge, no social role theoretical explanation for sex differences in book-carrying style has been offered. Nor were we able to locate any study of sex stereotypes having to do with book-carrying style.

From an evolutionary standpoint, one could argue that females appear to have been sexually selected for caring for offspring to a greater degree than males (see Précis 26.3.2.5). Consequently, when compared to males, females may feel greater comfort cradling many types of objects as they would do with babies, given that (Jenni 1976; Ellis 2011a:557). No evidence linking this type of behavior to brain exposure to testosterone was located.

Assessment. Even though only two countries were sampled, the consistency of the findings provide substantial evidence of the use of a cradling-like book-carrying style being a USD-D.

26.3.1.9 Carrying Weapons

Evidence. Précis 26.3.1.9 shows that all 30 studies that were located on sex differences in weapons carrying concluded that males do so more than females. These studies were based on samples drawn from five countries and were published over a 25-year time frame.

Précis 26.3.1.8 Using a cradling-like book-carrying style (Original Table 16.5.2.1)

Age categories, number of studies, and sex	*Consist. score*	*Overviews*	*Countries*	*Time range*	*Non-human*	*Social role theory*			*Evolutionary theory*		*Theological*
						Origin	*Found*	*Biosoc*	*Select*	*ENA*	
Adult 11F All Ages 14F	100.0 100.0	-0-	2	1976–2013	-0-	Silent	Silent	Silent	Poss-F	Poss-F	Silent

Précis 26.3.1.9 Carrying weapons (Original Table 16.5.2.6)

Age categories, number of studies, and sex	*Consist. score*	*Overviews*	*Countries*	*Time range*	*Non-human*	*Social role theory*			*Evolutionary theory*		*Theological*
						Origin	*Found*	*Biosoc*	*Select*	*ENA*	
Adolescent 28M All Ages 30M	100.0 100.0	-0-	5	1994–2019	-0-	Contra	NoTest	Poss-F	Expl-M	Expl-M	Silent

Explanation. Offering a simple self-defense explanation for sex differences in weapons carrying is difficult. On the one hand, males are more likely to be crime victims than females, especially regarding violent offenses (see Table 25.19.7). However, females express greater fear of being the victim of crime (see Précis 26.2.2.4) and are more likely to be sexually assaulted (see Précis 26.3.4.18). Therefore, it would be difficult to argue that males carry weapons more than females do for the purpose of self-protection.

Turning to the social role theories, if one were to assume that carrying weapons is a reflection of potential physical aggression, one could argue that, because males are stereotyped as being more aggressive (Table 21.2.7.5), they are more inclined to carry weapons. However, the original version of sex role theory would not lead one to expect universality in such a sex difference. In the case of the founder effect version, there were no weapons to be carried (except for sticks and stones) in the earliest human societies, so this theory's relevance also seems in doubt. The biosocial social role theory argues that sex differences in body size, strength, and birthing capabilities are responsible for universal sex differences in behavior. Why weapons carrying would be relevant to any of these sex differences is unclear.

In the case of the two evolutionary theories, no specific proposals were found regarding sex differences in weapons carrying. However, if the carrying of weapons is considered a form of preparedness for physical violence and to take risks, then sexual selection may be pertinent. Specifically, some researchers have proposed that, when compared to females, males have been evolutionarily favored for being prepared to confront all types of changes to their honor (Archer 2009; Hartley 2009; BJ Ellis, Del Guidice et al. 2012).

The ENA version of evolutionary theory asserts that brain exposure to testosterone produces numerous sex differences in behavior, including aggression and risk-taking (Ellis & Hoskin 2015). If so, the carrying of weapons may often be an expression of these types of behavioral tendencies. One study of female college students provided support for this line of reasoning. When women who reported using steroids (a form of synthetic testosterone) with women who did not use these drugs, significant differences in weapons carrying during the past 30 days were found. Just 5.6% of the women with no steroid use said they had carried a weapon, while 30.2% of those who had used steroids reported doing so (DL Elliot, Cheong et al. 2007).

Assessment. Weapons carrying appears to be a USD-M. More work is needed to ferret-out the best theoretical explanation for this difference, but both of the evolutionary theories appear promising.

26.3.1.10 Involvement in Religious Activities in General

Evidence. Is one sex more involved in religious activities (other than specifically attending religious services) more than the other sex? Studies that have attempted to answer this question have been conducted in five different countries, one dating back to the middle of the 20th Century. Readers can see in Précis 26.3.1.10 that all of the 26 pertinent studies found females were more involved than males in religious activities.

Explanation. According to stereotypes, females are more religious than males (see Table 21.2.8.8). To the extent that involvement in religious activities is indicative of religiosity in general, social role theories lead one to expect that females would be more involved. However, the original version would not predict that this would be true of all societies (since it envisions all sex differences to be culturally learned). In the case of the founder effect version of the theory, it is reasonable to assume that there were no religious activities in the original human societies. The biosocial version of social role theory attributes universal sex differences to sex differences in size and strength and the ability to bear offspring. None of these traits seem to have any sociocultural relevance to religiosity.

Turning to the evolutionary explanations, researchers have proposed that greater religiosity has evolved among females as a way of promoting long-term mating strategies (CT Palmer & Begley 2015; Van Slyke 2017). In other words, because females are able to successfully rear more offspring if they have the help of a committed mating partner, and nearly all organized religions advocate that their members remain in committed male-female relationships, it would be predicted that females would be more religious.

A different line of evolutionary reasoning has come to the same conclusion, i.e., females should be more religious than males. It is based on noting that males think in systemizing terms more than do females (see Précis 26.2.3.8). Evidence suggests that systemizing thinking is more compatible with scientific reasoning than with religious reasoning (Nettle 2007; Flannelly, Flannelly & Santos 2017). If so, one would expect males to be less religious than females (Lindeman & Svedholm-Hakkinen 2016), and religious people generally to reason in less scientific ways (AE Lawson & Worsnop 1992; RR Reilly 2014; N Murphy 2018).

If either of the above evolutionary perspectives is correct, ENA theory would predict that brain exposure to androgens negatively correlated with religiosity. The evidence in this regard has not been consistent (Stark 2002; Ellis, Hoskin & Ratnasingam 2016; G Richards, Davies et al. 2018).

Assessment. Overall, the theoretical understanding of why females would be more involved in religious activity than males is low. However, the number of studies, as well as the number of countries sampled, is sufficient to consider involvement in religious activities to be a USD-F.

Précis 26.3.1.10 Involvement in religious activities in general (Original Table 16.6.1.1)

Age categories, number of studies, and sex	*Consist. score*	*Overviews*	*Countries*	*Time range*	*Non-human*	*Social role theory*			*Evolutionary theory*		*Theological*
						Origin	*Found*	*Biosoc*	*Select*	*ENA*	
Adult 13F All Ages 26F	100.0 100.0	-0-	5	1949–2002	-0-	Contra	Silent	Silent	Poss-F	Poss-F	Silent

26.3.1.11 Praying

Evidence. As shown in Précis 26.3.1.11, all but one of 23 studies concluded that females engage in more prayer than do males. These results are based on samples from five different countries along with three multinational studies published over a 31-year time frame.

Explanation. If one assumes that praying reflects general tendencies to be religious in the sense of believing in the supernatural (see above), and that religiosity is inhibited by systemizing thinking, tendencies (i.e., males), one can deduce that females would pray more than males (also see above). If these premises are correct, both evolutionary theories would lead one to expect females to pray more than males.

As noted directly above, ENA theory leads to the hypothesis that brain exposure to androgens may result in enhanced systemizing-type thinking, and such thinking could curtain beliefs in forces operating outside the universe controlling its functioning. One result of systemizing thinking could be skepticism about the power of prayer, thereby inhibiting prayer. Another factor could be the tendency for females to be more language-oriented than males (see Précis 26.2.3.8). Given that language is a central element in nearly all forms of prayer, testosterone appears to interfere with, or at least slow the development of, language. It is worth noting that the only study that found males praying more than females was limited to samples of Jews and Muslims, where prayer is performed by rabbis and imams, rather than by individual religious members themselves (see Loewenthal, MacLeod & Cinnirella 2002:Table 1).

Assessment. Based on the substantial number of studies reporting that females pray more than males from a substantial number of studies, we declare praying to be a USD-F.

26.3.2 Social Behavior

By and large, humans are highly social animals. This section examines findings from studies of sex differences in various aspects of social behavior that have high (i.e., 95.0+) consistency scores.

26.3.2.1 Socializing Tendency

Evidence. Précis 26.3.2.1 indicates that all of the 13 studies found that females tended to be more socially connected when compared to males. A study on non-human primates also found females had more social connections than males did. All of the human studies were conducted in only two countries and no overviews were located.

Explanation. No evidence of stereotypes regarding sex differences in socializing were locate. Likewise, no specific attempt by social role theorists to explain sex differences in socializing tendencies were found.

Précis 26.3.1.11 Praying (Original Table 16.6.1.4)

Age categories, number of studies, and sex	*Consist. score*	*Overviews*	*Countries*	*Time range*	*Non-human*	*Social role theory*			*Evolutionary theory*		*Theological*
						Origin	*Found*	*Biosoc*	*Select*	*ENA*	
Adult 21F 1M All Ages 22F 1M	95.5 95.7	-0-	5 (3)	1983–2014	-0-	Silent	Silent	Silent	Poss-F	Expl-F	Silent

Précis 26.3.2.1 Socializing tendency (Original Table 17.1.1.2)

Age categories, number of studies, and sex	*Consist. score*	*Overviews*	*Countries*	*Time range*	*Non-human*	*Social role theory*			*Evolutionary theory*		*Theological*
						Origin	*Found*	*Biosoc*	*Select*	*ENA*	
All Ages 13F	100.0	-0-	2	1975–1999	F: one primate	Silent	Silent	Silent	Poss-F	Expl-F	Silent

From an evolutionary perspective, high levels of testosterone tend to promote what is known as mating effort, while low levels are associated with parenting effort (McGlothlin, Jawor & Ketterson 2007; Eikenaar, Whitham et al. 2011; Perini, Ditzen, Fischvacher & Ehlert 2012). These generalizations lead one to expect that traits such as sexuality and competition should be more common in males than in females. On the other hand, friendly and cooperative forms of social behavior should be more prominent among females.

Based on the above logic, ENA theory would lead one to expect that friendly behavior forms of social behavior should be promoted by reducing brain exposure to testosterone. Considerable evidence supports this prediction (van Anders, Goldey & Kuo 2011; Wingfield, Hegner et al. 1990). Also, research has indicated that circulating testosterone is to be associated with reduced social bonding (*review*: Edelstein & Chin 2018) and with diminished quality of people's opposite-sex relationships (Edelstein, van Anders et al. 2014).

Assessment. Overall, socializing tendencies appears to be a USD-F. An evolutionary perspective seems to offer the best explanations for this sex difference.

26.3.2.2 Having Intimate Friendships

Evidence. A common consequence of socializing is the development of intimate friendships with some of those with whom one interacts. Précis 26.3.2.2 shows that 76 studies of sex differences in the development of close friendships were located, the vast majority of which have concluded that females do so more than males. Nevertheless, only among adolescents did the consistency score exceed the 95.0 cutoff. For the total number of studies conducted, six different countries were represented, along with one multi-country study, and the publication of these studies ranged over 42 years. It is also noteworthy that two literature reviews and one meta-analysis also concluded that females are more likely than males to develop intimate friendships.

Explanation. Social role theory can be used to explain sex differences in tendencies to have close friendships. These explanations derive from evidence that females are stereotyped as being friendlier (Table 21.2.7.32), more nurturing (Table 21.2.9.10) and more sociable (Table 21.2.9.13). Also, as noted directly above, females appear to spend more time actually socializing. However, the original version of this theory would leave one to expect to find exceptions in this regard, and, at least for adolescents, there appear to be none. In the case of the founder effect version, there is no way of knowing whether or not females had more intimate friendships in the earliest human societies. The biosocial version of social role theory predicts

Précis 26.3.2.2 Having intimate friendships (Original Table 17.1.2.1)

Age categories, number of studies, and sex	*Consist. score*	*Overviews*	*Countries*	*Time range*	*Non-human*	*Social role theory*			*Evolutionary theory*		*Theological*
						Origin	*Found*	*Biosoc*	*Select*	*ENA*	
Child 10F 1x Adolescent 21F 1x Adult 40F 3M All Ages 71F 2x 3M	(90.9) 95.5 (87.0) (89.9)	F: 2Rev F: 1Met	6 (1)	1963–2005	-0-	Contra	NoTest	Poss-F	Silent	Silent	Silent

that males being larger or more muscular, or females being the only sex capable of bearing offspring will impact the formation of intimate friendships, variable connections that fail to be obvious, but that might exist.

Turning to the evolutionary perspective, we found no research directly bearing on why females would be favored for having more intimate friendships than males. However, one could argue that it is a reflection of female's greater tendencies to socialize in general (see directly above).

Assessment. At least among adolescents, the tendency to have more intimate friendships appears to be a USD-F. Theoretical development in explaining this sex differences is still needed.

26.3.2.3 Gang Membership

Evidence. Précis 26.3.2.3 shows that the 50 studies on this topic were published over a 40-year period regarding sex differences in being gang members, all of which concluded that such membership was more common for males than for females. The studies were conducted in six countries, in addition to one multi-national study.

Explanation. No evidence was found that males are stereotyped as being gang members to a greater degree than females. Therefore, we will not attempt to apply any of the three social role theories to this sex difference.

From an evolutionary perspective, it has been argued that human gangs are a form of what is known as *peripheralization* among numerous mammalian species (Ellis & Hoskin 2015:69). The species in which peripheralization occurs are primarily non-human primates. Most peripheralizing events unfold as follows: Soon after males reach puberty, they begin to spend more and more time away from their natal social group (usually called a *troop*) and then form bachelor bands of five to ten other adolescent males on the outskirts of their natal troop (Box 1984:99; Lunardini 1989). As these males mature into adults, some of them eventually integrate into a neighboring mixed-sex primate troop and form its leadership structure. This transition to leadership often occurs after the males repeatedly attack and eventually defeat the aging males in the established leadership structure of this neighboring troop (Koyama 1967; Pruetz, Ontl et al. 2017).

Evolutionary theorists have reasoned that the peripheralization process essentially represents how leadership usually transitions between generations of males while simultaneously helping to minimize incestuous mating (Mueller & Thalmann 2000). Ultimately, in most primate species, males in high-ranking positions have more sex partners than do lower-ranking males, and thereby usually pass their genes on to future generations at relatively high rates (Ellis 1995).

In the case of humans, research has indicated that gang membership is positively correlated with the number of sex partners males report having

Précis 26.3.2.3 Gang membership (Original Table 17.1.2.7)

Age categories, number of studies, and sex	*Consist. score*	*Overviews*	*Countries*	*Time range*	*Non-human*	*Social role theory*			*Evolutionary theory*		*Theological*
						Origin	*Found*	*Biosoc*	*Select*	*ENA*	
Adolescent 40M	100.0	-0-	6 (1)	1974–2014	-0-	Silent	Silent	Silent	Expl-M	Expl-M	Silent
Wide Age Range 10M	100.0										
All Ages 50M	100.0										

had (Palmer & Tilley 1995; Voisin, Salazar et al. 2004). Studies also have indicated that females who dated gang members were more likely to have become pregnant than were females who did not do so (Voisin, Salazar et al. 2004; Minnis, Moore et al. 2008). Such research suggests that, even in modern times, gang membership may have positive reproductive consequences for males. In earlier times, before the advent of modern contraception and abortion access, the reproductive advantages to gang members may have been even higher.

Regarding ENA theory's neurohormonal extension of sexual selection theory, no specific evidence was found linking gang membership to androgens. However, substantial evidence has associated criminality (a frequent accompaniment of gang membership) with both prenatal and post-pubertal testosterone levels (Ellis, Farrington & Hoskin 2019:351–355).

Assessment. Overall, sex differences in gang membership have been well established, both in terms of the empirical evidence and regarding theoretical understanding. Therefore, gang membership is a well-established USD-M.

26.3.2.4 Parent-Offspring Conflict

Evidence. Ten studies were located that examined whether there are sex differences in the extent to which offspring have conflicts with their parents. These studies were generally based on reports by the offspring themselves. Précis 26.3.2.4 shows that all of these studies reported that female offspring had more conflicts with parents than did male offspring. These studies took place in three different countries and were published over a 31-year time frame.

Explanation. No explanation seems to be offered by any of the five theories for sex differences in parent-child conflict.

Assessment. Given the minimum number of studies reaching this conclusion and the absence of any known theory to explain them, we will not declare parent-child conflict to be a USD.

26.3.2.5 Providing Care to One's Own Offspring

Evidence. As shown in Précis 26.3.2.5, over 90 studies reached the conclusion that human females tended to provide more care to their offspring than human males did. While the results of studies on other species are mixed, there were no exceptions among the studies on humans. The human studies were conducted in 13 different countries, and there were an additional ten multi-national studies on the subject. Further, nine literature reviews and one generalization reached the same conclusion as the studies.

Précis 26.3.2.4 Parent-offspring conflict (Original Table 17.1.6.7)

Age categories, number of studies, and sex	*Consist. score*	*Overviews*	*Countries*	*Time range*	*Non-human*	*Social role theory*			*Evolutionary theory*		*Theological*
						Origin	*Found*	*Biosoc*	*Select*	*ENA*	
Adolescent 10F All Ages 10F	100.0 100.0	-0-	3	1987–2018	-0-	Silent	Silent	Silent	Silent	Silent	Silent

Précis 26.3.2.5 Providing care to one's own offspring (Original Table 17.1.6.10)

Age categories, number of studies, and sex	*Consist. score*	*Overviews*	*Countries*	*Time range*	*Non-human*	*Social role theory*			*Evolutionary theory*		*Theological*
						Origin	*Found*	*Biosoc*	*Select*	*ENA*	
Adult 92F All Ages 92F	100.0 100.0	F: 1Gen F: 9Rev	13 (10)	1967–2014	mixed results	Contra	Contra	Contra	Expl-F	Expl-F	Silent

Explanation. Consider the following statement: "Women in US society are more likely to be responsible for raising children and more often found in caregiving positions than are men" (Cross & Madson 1997:8). Notice that this quote implies that there may be societies in which women are *not* primarily responsible for raising children. Obviously, if there are such societies, a different theoretical explanation would be called for than if there are no exceptions.

The original social role theory predicts that sex differences in providing care to one's own offspring, like all others sex differences in cognitive and behavioral traits, will have cultural exceptions. One can see in Table 17.1.6.10 (in Volume 3) that there are no cultural exceptions, thus casting doubt on the original social role theory.

The founder effect theory would predict universality in sex differences if the first primordial human societies happened to have had a given sex difference. Unfortunately, this is impossible to know unless one assumes that modern-day human hunter-gathering societies are reflective of those dating back to what Fausto-Sterling (1992:199) called "small pro-genital stock." In this regard, studies of such societies all indicate that females spend more time caring for offspring than do males (Hewlett 1992; MM West & Konner 1976), thus supporting the founder effect theory. However, the fact that the same sex difference is seen is all mammals, not just humans (M Daly & Wilson 1978:140; Goy & McEwen 1980:58; Jaros & White 1983:132) would suggest that some biological factors underlie the sex difference in providing child care.

In the case of Wood and Eagly's (2012) biosocial theory, the expectation would be that females are the primary caregivers to offspring in all societies because they are the only sex that can bear offspring. This proposal would assume that there is a culturally imposed link between bearing offspring and providing care to offspring. However, one might think that some societies might develop traditions in which males would provide care to offspring after they are born to compensate for females having to go through the gestation and birth process.

Both evolutionary theories provide strong explanations for why females are the primary offspring caregivers in all societies. The most important evolutionary explanation of this sex difference involves noting that females must contribute so much more time and energy *producing* each offspring through the gestation process than is the case for males. Because of this greater initial investment, females would be strongly disfavored for terminating their lopsided investment following birth (Kokko & Jennions 2008; Trivers 2017).

According to ENA theory, various emotional and behavioral traits (such as the ability to empathize and care for others) are likely to be influenced by exposing the brain to testosterone. In the case of these particular traits, *low* testosterone exposure should be associated with pro-childcare traits. Evidence supports this prediction (SK Carter 2017; Roellke, Raiss et al. 2019; Almanza-Sepulveda, Fleming & Jonas 2020). It may also be relevant to note that, when males provide comfort to crying infants, their testosterone levels tend to subside (van Anders, Tolman & Volling 2012).

Assessment. Providing care to one's own offspring is a firmly established USD-F. The number of studies and the number of countries involved are substantial. Furthermore, why such a sex difference would be universally documented is well explained, especially in evolutionary terms.

26.3.2.6 *Providing Care to Family Members Other Than Offspring*

Evidence. Care is often provided to family members other than offspring, particularly in the case of offspring caring for elderly parents. Précis 26.3.2.6 reveals that all 70 studies found females gave more care to family members other than offspring than did males. In addition to ten different countries having been sampled, two multi-national studies were also included. The findings are consistent with four literature reviews on this topic as well.

Explanation. It seems reasonable to assume that, because females are more inclined to care for their offspring than are males, females may have a more general tendency to exhibit concern for other family members as well. If so, all of the comments and evidence presented in the narrative directly above would apply here as well.

Assessment. Given the number of studies, the countries in which they have been conducted and the ability to theoretically understand this variable, there is essentially no doubt that providing care to family members other than offspring is a USD-F.

26.3.2.7 *Providing Care and Comfort to Others in General*

Evidence. Précis 26.3.2.7 indicates that 21 out of 22 studies of sex differences in the tendency to provide care and comfort to others concluded that females did so more than males. The studies were conducted in five different countries and there were two multi-national studies, and were published over a 56-year time frame. Two meta-analyses also reached the conclusion that such behavior was more common in females.

Explanation. Social role theorists can cite at least four different sex stereotypes that would support the view that females will provide more care and comfort to others. Specifically, tables in Volume III report that females are perceived as more nurturing (Table 21.2.9.10), more sociable (Table 21.2.9.13), more affectionate (Table 21.2.6.4), and, above all, more likely to be caregivers (Table 21.2.9.2). All four of these stereotypes

Précis 26.3.2.6 Providing care to family members other than offspring (Original Table 17.1.6.15)

Age categories, number of studies, and sex	*Consist. score*	*Overviews*	*Countries*	*Time range*	*Non-human*	*Social role theory*			*Evolutionary theory*		*Theological*
						Origin	*Found*	*Biosoc*	*Select*	*ENA*	
All Ages 70F	100.0	F: 4Rev	10 (2)	1981–2017	-0-	Contra	Poss-F	Poss-F	Expl-F	Expl-F	Silent

Précis 26.3.2.7 Providing care and comfort to others in general (Original Table 17.1.6.17)

Age categories, number of studies, and sex	*Consist. score*	*Overviews*	*Countries*	*Time range*	*Non-human*	*Social role theory*			*Evolutionary theory*		*Theological*
						Origin	*Found*	*Biosoc*	*Select*	*ENA*	
Adult 16F All Ages 21F 1x	100.0 95.5	F: 2Met	5 (2)	1960–2016	-0-	Contra	Poss-F	Poss-F	Expl-F	Expl-F	Silent

may lead people in most societies to teach females that they are expected to provide care and comfort to others more than males are expected to do so. If female's greater likelihood of providing care and comfort to others in universal, as evidence in this précis suggests, it would be inconsistent with the original version of social role theory, but it might fit with either of the latter two versions.

Evolutionary explanations for why females would provide greater care and comfort to others would likely include the view that such behavior is an extension of providing care to offspring (see Précis 26.3.2.5). The tendency to care for others could also be part of a tendency to be more empathetic, a trait that is more pronounced in females (see Précis 26.2.2.9).

Regarding ENA theory, quite a few studies of the relationship between testosterone exposure and empathy have been reported. As predicted, most of these studies have indicated that brain exposure to testosterone diminishes feelings of empathy (for a review, see Précis 26.2.2.9).

Assessment. We confidently declare providing care and comfort to others to be a USD-F. Both social role theory and evolutionary theory offer reasonable explanations for this sex difference, although the latter appears stronger.

26.3.2.8 *Social Interactions in Small Groups*

Evidence. Twenty-seven studies of sex differences in socially interacting in small groups (i.e., usually just two or three individuals) were located. All of them concluded that females form and interact in small groups more, whereas when males socially interact, they more often do so in groups of four or more individuals. Précis 26.3.2.8 shows that four different countries were sampled in these studies, with the relevant research having been published over a 102-year time frame. Three studies also found the same type of sex difference among chimpanzees.

Explanation. We were not able to identify any specific theoretical explanations for this sex difference.

Assessment. Overall, we confidently declare socially interacting in small groups to be a USD-F. Work on theoretically explaining this difference is needed.

26.3.2.9 *Being Homeless*

Evidence. Précis 26.3.2.9 shows that more males than females were homeless. This was the conclusion in ten recent studies taking place in five countries. Three literature reviews and one meta-analysis reached the same conclusion.

Explanation. To the extent that homelessness refers to not having a permanent domicile in which to live, it is obviously a less-than-ideal way

Précis 26.3.2.8 Social interactions in small groups (Original Table 17.1.7.4)

Age categories, number of studies, and sex	*Consist. score*	*Overviews*	*Countries*	*Time range*	*Non-human*	*Social role theory*			*Evolutionary theory*		*Theological*
						Origin	*Found*	*Biosoc*	*Select*	*ENA*	
Child 19F All Ages 27F	100.0 100.0	-0-	4	1902–2004	F: three chimps	Silent	Silent	Silent	Silent	Silent	Silent

Précis 26.3.2.9 Being homeless (Original Table 17.1.8.3)

Age categories, number of studies, and sex	*Consist. score*	*Overviews*	*Countries*	*Time range*	*Non-human*	*Social role theory*			*Evolutionary theory*		*Theological*
						Origin	*Found*	*Biosoc*	*Select*	*ENA*	
All Ages 10M	100.0	M: 3Rev M: 1Met	5	1987–2016	-0-	Contra	Silent	Silent	Silent	Silent	Silent

to live. We found no theoretical attempt to explain why males would be homeless to a greater degree than females. However, if homelessness is more common worldwide among males than among females, as all ten studies suggest, it would cast doubt on the original social role theory.

Assessment. Overall, more research is need on sex differences in homelessness. Even in the absence of a theoretical understanding, we will tentatively declare it to be a USD-F.

26.3.2.10 Having Primary Custody of Children after Divorce

Evidence. Is there a sex difference when it comes to having primary custody of children following a divorce? Précis 26.3.2.10 reveals that all 19 of the studies found females were more likely than males to have primary custody of children following divorce. This sex difference was examined in three countries primarily during the last quarter of the 20th century. Our finding in this regard comports with conclusions reached in a review of the literature.

Explanation. Females are stereotyped as being more likely to be caregivers (Table 21.2.9.2) and to be nurturing (Table 21.2.9.10). From the perspective of social role theory, both of these stereotypes are likely to prompt judges in divorce proceedings to favor grant primary custody of children to the mother rather than the father. However, the original version of the theory would not lead one to expect this sex difference in stereotypes to be universal. Assuming that the proverbial first human society had something equivalent to divorce, it would still be impossible to test the founder effect theory's assertion that the universal tendency for females to have primary custody of children is because the practice was "faithfully passed down thousands of times over" ever since (Fausto-Sterling 1992:199). In the case of the biosocial version of social role theory, one might be able to make the case that, because females gave birth to each of their offspring, their connection to them would be stronger. This, plus the fact that females are stereotyped as being more nurturing, all societies might see them as being the better parent in most cases.

Explaining why females are more likely to have primary custody of children when a couple divorce in evolutionary terms would include the fact that mothers have greater certainty about who their offspring are than is the case for fathers (J Wright & Cotton 1994; Harts & Kokko 2013). Parental certainty, whether consciously registered or not, will favor females being the primary caregivers. Presumably, judicial authorities recognize the reality of this sex bias when they make decisions about child custody.

ENA theory predicts that exposing the brain to androgens should serve to diminish caregiving behavior. Supporting this prediction, at least among males, studies have shown that circulating testosterone levels are inversely correlated with time spent providing care to offspring (Alvergne, Faurie &

Précis 26.3.2.10 Having primary custody of children after divorce (Original Table 17.1.8.4)

Age categories, number of studies, and sex	*Consist. score*	*Overviews*	*Countries*	*Time range*	*Non-human*	*Social role theory*			*Evolutionary theory*		*Theological*
						Origin	*Found*	*Biosoc*	*Select*	*ENA*	
Wide Age Range 11F	100.0	F: 1Rev	3	1975–2001	-0-	Contra	NoTest	Poss-F	Expl-M	Expl-M	Silent
All Ages 19F	100.0										

Raymond 2009:Figure 3; Bos, Hechler et al. 2018; Beijers, Breugelmans et al. 2022).

Perhaps, as an evolved consequence of the connection between testosterone and the tendency to be sexually promiscuous (see Précis 26.2.16.7), male testosterone levels have been found to decline substantially following marriage (PB Gray, Kahlenberg et al. 2002; Kuzawa, Gettler et al. 2009). Testosterone appears to decline even more after males become fathers (Alvergne, Faurie & Raymond 2009:Figure 1) and as the time fathers spend actually caring for offspring (Gettler, McDade et al. 2011; de Vries, van der Pol, et al. 2019).

As a general principle, it appears that exposing the brain to testosterone has the effect of promoting what is known as *mating effort over parenting effort* (Kuzawa, Gettler et al. 2010; Perini, Ditzen, Fischvacher & Ehlert 2012; Roney & Gettler 2015). Because females have much lower testosterone levels than do males (see Précis 26.1.3.2), females will nearly always gravitate more than males (even those who are the most committed fathers) toward child care. Court judges seem to recognize this sex difference and make child custody decisions accordingly.

Assessment. Overall, the evidence is strong that females have (or receive from court authority) primary custody over children in the case of divorced couples making this a well-established USD-F. The best explanation for this sex difference and its universality appears to be evolutionary in nature.

26.3.2.11 Seeking or Receiving Help When under Stress

Evidence. When faced with a stressful situation, some people reach out to others for help. Précis 26.3.2.11 shows that a total of 39 pertinent studies were located, and that their consistency score was just 90.0. However, for the 21 of these studies based on adult samples, females were more likely than males to seek or receive help from others in stressful situations. The main reason for the failure for the consistency score to be over the 95.0 cutoff for All Ages involved three studies of children, all of which concluded that there were no significant sex differences in seeking or receiving help.

Explanation. Seeking or receiving help from others may be seen as a non-dominant social gesture. Offering a specific theoretical explanation of help seeking based on either social role or evolutionary reasoning would be difficult. Therefore, we will list all five of the theories as *silent*. Nevertheless, it is probably relevant to note that, when compared to males, most studies have found females report feeling stress and anxiety more (see Table 25.10.6). They are also more fearful (Table 25.10.3). Therefore, much of the sex differences in help seeking (or help receiving) could be simply due to sex differences in feeling of stress, anxiety, and fear.

Précis 26.3.2.11 Seeking or receiving help when under stress (Original Table 17.2.3.4)

Age categories, number of studies, and sex	*Consist. score*	*Overviews*	*Countries*	*Time range*	*Non-human*	*Social role theory*			*Evolutionary theory*		*Theological*
						Origin	*Found*	*Biosoc*	*Select*	*ENA*	
Adolescent 15F 1x Adult 21FAll Ages 39F 4x	(93.7) 100.0 (90.0)	-0-	3 (1)	1987–2005	-0-	Silent	Silent	Silent	Silent	Silent	Silent

Assessment. Even though no theoretical explanation is readily apparent, we declare the tendency to seek or receive help from others when under stress to be a USD-F, at least for adults.

26.3.2.12 *Help-Seeking When Interpersonal Dispute Arises*

Evidence. Interpersonal disputes are a specific source of stress that prompt people to reach out to a third party for help. Précis 26.3.2.12 shows that, among adolescents, all 12 studies of sex differences in such behavior are undertaken by females more than by males.

Explanation. As with help seeking (and receiving) in general, discussed directly above, the application of both social role theory and evolutionary theory are illusive. Thus, we will list them all as *silent*.

Assessment. While no theoretical explanation is obvious, help-seeking when interpersonal disputes arise is a USD-F, at least for adults.

26.3.2.13 *Smiling*

Evidence. People smile under a variety of circumstances. Since the 1960s, many studies of sex differences in smiling have been published (115 were located). As shown in Précis 26.3.2.13, out of these 115 studies, all but 21 reported that females smiled more than did males. Notice that some of the inconsistencies appear to be associated with age. For example, findings among infants and toddlers only had a consistency score of 61.5, and, for all age groups combined, the consistency score was just 80.3.

Only among studies that involved adolescents was the consistency score higher than 95.0, i.e., it was 100.0. Thus, while two meta-analyses (both based on samples of wide age ranges) concluded that females smile more, our approach, which involved analyzing data with regard to fairly specific age groupings, suggest that caution is in order regarding the universality of sex differences in smiling.

To elaborate briefly on the relevance of age variations in smiling, one study deserves special attention. This study involved rating the degree of smiling exhibited in over 18,000 yearbook photographs of boys and girls from kindergarten through the twelfth grade. It found that males and females were statistically equal in their smiling tendencies until around age 11. From that age onward, the proportion of males who smiled declined substantially (Wondergem & Friedlmeier 2012:407). In other words, sex differences in tendencies to smile among children (roughly ages five through 18) was similar until the onset of puberty, at which time, smiling by males declined noticeably.

Explanation. The original social role theory seems to offer no explanation for female tendencies to smile more than males, especially following the

Précis 26.3.2.12 Help-seeking when interpersonal dispute arises (Original Table 17.2.3.6)

Age categories, number of studies, and sex	*Consist. score*	*Overviews*	*Countries*	*Time range*	*Non-human*	*Social role theory*			*Evolutionary theory*		*Theological*
						Origin	*Found*	*Biosoc*	*Select*	*ENA*	
Child 9F 3x Adolescent 12F All Ages 22F 3x	(75.0) 100.0 (88.0)	-0-	3 (1)	1961–2004	-0-	Silent	Silent	Silent	Silent	Silent	Silent

Précis 26.3.2.13 Smiling (Original Table 17.3.2.4)

Age categories, number of studies, and sex	*Consist. score*	*Overviews*	*Countries*	*Time range*	*Non-human*	*Social role theory*			*Evolutionary theory*		*Theological*
						Origin	*Found*	*Biosoc*	*Select*	*ENA*	
Infant/Tod 8F 1x 2M Child 8F 6x Adolescent 11F Adult 56F 11x Wide Age Range 11F 1x All Ages 94F 19x 2M	(61.5) (57.1) 100.0 (83.6) (91.7) (80.3)	F: 2Met	7 (2)	1966–2018	-0-	Silent	Silent	Poss-F	Expl-F	Expl-F	Silent

onset of puberty. Regarding the two variants of social role theory, the founder effect version is unable to account for why humankind's "original culture" would have exhibited a sex difference in smiling or why it would vary by age and would have persisted in all cultures to this day, at least among adolescents. In the case of the biosocial role theory, the only sex differences that is envisioned as being universal are those associated with males being physically stronger and/or females exhibiting the burden of pregnancy. Within the framework of biosocial role theory, one could argue that, because males become physically stronger than females during adolescence, they want to accentuate their strength by smiling less. Why smiling would be related to these two sex differences has no obvious answer, but calls for investigation by this theory's proponents.

In evolutionary terms, social smiling is widely recognized as being a friendly, cooperative, and largely non-threatening gesture, the signaling of which helps to deter aggression (Hopcroft 2002; Mehu & Dunbar 2008) or keep aggression in a playful form (McCord 1999; Drummond & Bailey 2013). The use of smiling has even been documented among various species of non-human primates under similar social circumstances (Eibl-Eibesfeldt 1989; Tomonaga 2006; Shilling & Brown 2016). Smiling also appears to be linked to how people are rated in terms of their physical attractiveness and their desires for positive social interactions (Mehu, Little & Dunbar 2008).

While circumstances surrounding smiling (both spontaneous and intentionally-produced) are complex, smiling is largely a non-threatening social gesture. Regarding the possibility of explaining smiling at the neurohormonal level, ENA theory has offered some specific proposals that may account for the tendency for sex differences to be most pronounced during adolescents. Specifically, while prenatal exposure is likely to have some depressing effects on smiling, postpubertal testosterone should have even stronger effects (Ellis 2006). Direct evidence in this regard is fragmentary so far but supportive (Cashdan 1995; Dabbs 1997). Worth adding is the fact that ENA theory asserts that friendly gestures such as smiling are inversely associated with efforts to become dominant over others, a trait that helps promote social status and the accumulation of resources. More will be said about dominance and neuroandrogenic factors later in this chapter (see Précis 26.3.5.42).

Assessment. Too many studies have failed to support the assertion that smiling is more common among females than males to consider smiling a USD, at least for all age groups. However, in the case of 11 studies that sampled adolescents, a USD-F designation is justified, especially given that the sex difference can be theoretically predicted.

26.3.2.14 Swearing and Cursing

Evidence. Does one sex use socially objectionable language more than the other sex? Précis 26.3.2.14 shows that almost all of the research on this topic found that males tended to swear and curse more than females. The two exceptions did not find a sex difference. The research has been based on samples drawn from six different countries, with the earliest study having been published in 1935.

Explanation. We found no proposals for how social role theory would explain sex differences in tendencies to swear and curse. Evolutionary theorists, however, have suggested that swearing and cursing appear to be verbal forms of aggressive and dominating behavior (Güvendir 2015; Finkelstein 2018). Furthermore, "involuntary" swearing and cursing is often a symptom of a mental disorder known as *Tourette's syndrome* (Senberg, Munchau et al. 2021). As discussed in Chapter 25, nearly all studies have found this disorder being more common among males than females (see Table 25.14.7). Some research has pointed toward brain exposure to testosterone as contributing to both the frequent use of swear and curse words (Finkelstein 2018) and to Tourette's syndrome (Senberg, Munchau et al. 2021). Although fragmentary, this evidence provides support for ENA theory.

Assessment. Overall, sex differences in tendencies to swear and curse are deemed a USD-M. More research is needed, however, to better understand such behavior from both an evolutionary and neurohormonal perspective.

26.3.2.15 Using Intensifying Adverbs

Evidence. Does one sex use intensifying adverbs more than the other sex? In other word, rather than saying something is "beautiful" or "exciting," are there any sex differences in tendencies to say "very beautiful" or "really exciting"? Results from all ten pertinent studies in this regard are presented in Précis 26.3.2.15. All of them have reported that females use adverbial intensifiers more than do males. However, all of the studies were published over only two decades and based on samples drawn for just one country (i.e., the United States).

Explanation. We found no efforts to explain this sex difference by any of the five theories. One possibility worth pursuing would involve evidence that females appear to be more emotional (see Table 25.10.1) and emotionally expressive (Table 25.10.9) than males.

Assessment. Given that just a minimum of ten studies were located, all of which came from samples drawn from the same country, and no theoretical explanations were located, we will not declare the use of intensifying adverbs to be a USD.

Précis 26.3.2.14 Swearing and cursing (Original Table 17.4.3.2)

Age categories, number of studies, and sex	*Consist. score*	*Overviews*	*Countries*	*Time range*	*Non-human*	*Social role theory*			*Evolutionary theory*		*Theological*
						Origin	*Found*	*Biosoc*	*Select*	*ENA*	
Adult 24M 1x All Ages 39M 2x	96.0 95.1	-0-	6	1935–2009	-0-	Silent	Silent	Silent	Poss-M	Poss-M	Silent

Précis 26.3.2.15 Using intensifying adverbs (Original Table 17.4.3.10)

Age categories, number of studies, and sex	*Consist. score*	*Overviews*	*Countries*	*Time range*	*Non-human*	*Social role theory*			*Evolutionary theory*		*Theological*
						Origin	*Found*	*Biosoc*	*Select*	*ENA*	
All Ages 10F	100.0	-0-	1	1977–1995	-0-	Silent	Silent	Silent	Silent	Silent	Silent

26.3.2.16 Mixed-Sex (vs. Same-Sex) Linguistic Communication

Evidence. Sex differences in speech might also be dependent on the social context in which the speech takes place. A large number of studies have found that males spoke more in mixed-sex settings than was the case for females. The only exception was a study that did not find a sex difference. The studies took place in four countries, primarily during the second half of the 20th century.

Explanation. It is rather difficult to explain why males would communicate linguistically in mixed-sex gatherings more than females unless one assumes that such a sex difference reflects greater status (or status-seeking) among males. Evidence pertaining to sex differences in status will be given coverage later in this chapter (see 26.3.5.42), but, in the present context, the possible link between mixed-sex communication and status seems too poorly established to be given serious attention here.

Assessment. Even though none of the five scientific theories seems to directly pertain to sex differences in mixed-sex linguistic communication, we will declare it to be a USD-M, given the substantial number of studies and the fairly sizable number of countries sampled.

26.3.2.17 Confiding and Sharing Secrets

Evidence. Précis 26.3.2.17 shows that females tended to share secrets more often than did males according to 56 out of 59 studies. The three exceptions simply failing to find significant sex differences and were based on samples of pre-adolescents. Samples for these studies were obtained in four countries and results were published over a 47-year time frame. The sex difference is consistent with one literature and one meta-analysis.

Explanation. We found no specific theoretical arguments to account for sex differences in confiding behavior unless one assumes that confiding behavior is a form of linguistic behavior, in which case one would suspect that females would confide more than males (see 26.2.3.10). It may also be useful to link confiding and sharing secrets with supportive and inclusive communication, which is also more common among females (see Précis 26.3.2.19). Greater female tendencies to gossip (see Table 17.4.7.6 in Volume III) could also be an expression of confiding and sharing secrets.

Assessment. Even though a theoretical explanation for confiding and sharing secrets is needed, at least for studies of adolescents and adults, we confidently declare it to be a USD-F.

26.3.2.18 Issuing Commands and Instructions or Assertive Speech in General

Evidence. Précis 26.3.2.18 presents evidence concerning a possible sex difference in assertive speech, such as issuing commands and instructions.

Précis 26.3.2.16 Mixed-sex (vs. same-sex) linguistic communication (Original Table 17.4.6.10)

Age categories, number of studies, and sex	*Consist. score*	*Overviews*	*Countries*	*Time range*	*Non-human*	*Social role theory*			*Evolutionary theory*		*Theological*
						Origin	*Found*	*Biosoc*	*Select*	*ENA*	
All Ages 23M 1x	95.8M	-0-	4	1951–2002	-0-	Silent	Silent	Silent	Silent	Silent	Silent

Précis 26.3.2.17 Confiding and sharing secrets (Original Table 17.4.7.5)

Age categories, number of studies, and sex	*Consist. score*	*Overviews*	*Countries*	*Time range*	*Non-human*	*Social role theory*			*Evolutionary theory*		*Theological*
						Origin	*Found*	*Biosoc*	*Select*	*ENA*	
Adolescent 18F Adult 23F All Ages 56F 3x	100.0 100.0 (94.9)	F: 1Rev F: 1Met	4	1958–2005	-0-	Silent	Silent	Silent	Silent	Silent	Silent

It shows that all ten studies on this form of speech found that adult males engaged in assertive speech more than was the case for adult females. Although only studied in three countries, this finding is consistent with one meta-analysis.

Explanation. It seems reasonable to assume that issuing commands and other forms of assertive speech reflect status or status-striving. Most studies have shown that such traits are more common among males than females among humans as well as non-humans (see Table 25.20.14 in this volume). Without more detailed information, it is difficult to assess the relevance of any of the five theories to this particular trait.

Assessment. We will withhold making any declaration about the USD status of tendencies to issue commands and make assertive forms of speech.

26.3.2.19 Supportive and Inclusive Communication

Evidence. Does one sex engage in supportive forms of communication more often than the other sex does. According to the 13 studies presented in Précis 26.3.2.19, females engaged in supportive and inclusive communication more often than did males. Although no study reached a different result, the studies were only conducted in two countries.

Explanation. None of the five theories directly explains why females would engage in more supportive and inclusive communication than is the case for males. However, there are possible indirect ways they might be relevant.

Given that females are stereotyped as being more affectionate (Table 21.2.6.4), more cheerful (Table 21.2.6.8), and more friendly (Table 21.2.7.32, all in Volume III), social role theory could be used to argue that females learn to be more supportive and inclusive in their communication styles because this is expected of them by the people with whom they associate. Of course, the goal here is to explain why sex differences in supportive and inclusive communication appears to be universal. Since the original social role theory assumes that cultures vary in sex role expectations, it would have difficulty explaining any universal sex difference. As noted elsewhere, the founder effect version of the theory is essentially untestable since no research from the first human societies exist. The biosocial version of social role theory could predict the universality of sex differences in supportive and inclusive communication, although the relevance of male size and strength or the ability of females to give birth to such communication styles needs to be established.

Regarding evolutionary theorizing, one could argue that greater supportive and inclusive communication among females could have been sexually selected indirectly by simply favoring greater friendliness (see Précis 26.3.1.1) and concern about the well-being of others (see Précis 26.3.2.7) among females. Males, on the other hand, appear to be selected for being more

Précis 26.3.2.18 Issuing commands and instructions or assertive speech in general (Original Table 17.4.7.9)

Age categories, number of studies, and sex	*Consist. score*	*Overviews*	*Countries*	*Time range*	*Non-human*	*Social role theory*			*Evolutionary theory*		*Theological*
						Origin	*Found*	*Biosoc*	*Select*	*ENA*	
Child 11M 4x 4F Adult 10M All Ages 23M 7x 4F	(47.8) 100.0 (60.5)	M: 1Met	3	1976–2004	-0-	Silent	Silent	Silent	Silent	Silent	Silent

Précis 26.3.2.19 Supportive and inclusive communication (Original Table 17.4.7.11)

Age categories, number of studies, and sex	*Consist. score*	*Overviews*	*Countries*	*Time range*	*Non-human*	*Social role theory*			*Evolutionary theory*		*Theological*
						Origin	*Found*	*Biosoc*	*Select*	*ENA*	
All Ages 13F	100.0	-0-	2	1990–2010	-0-	Contra	NoTest	Poss-F	Poss-F	Poss-F	Silent

physically aggressive (see Table 25.17.16) and competitive (see Précis 26.3.1.5). Both of these latter traits are somewhat contrary to communicating in supportive and inclusive ways.

ENA theory leads to hypothesize that brain exposure to testosterone inhibits supportive and inclusive communication, thus explaining why females exhibit such behavior more than do males. Unfortunately, no direct evidence testing this hypothesis was located.

Assessment. Given that indirect theoretical explanations for sex differences in supportive and inclusive communication being found universally, we consider this variable to be a USD-F.

26.3.2.20 Conversational Focus on Work and/or Leisure

Evidence. Does one sex talk more about work and/or leisure more than the other sex does? Précis 26.3.2.20 shows that all 15 studies on this type of communication found that males communicate more about these topics than was the case for females. However, the studies are limited to two countries, although the studies were published over a 71-year time frame.

Explanation. It seems reasonable to assume that conversational focus usually reflects the interests that individuals happen to have, thus suggesting that males are more interested in work and leisure activities than are females. This might be at least partly due to males enjoying competitive activities, a sex difference that appears to have been sexually selected (see Précis 26.3.1.5). As to how competitiveness was produced in neurological terms, this précis also suggests that brain exposure to testosterone appears to be involved.

Assessment. While just two countries were sampled, we will declare conversational focus on work and leisure to be a USD-M. Work is needed to help explain this difference in theoretical terms.

26.3.2.21 Play or Recreating Outside (Instead of Indoors)

Evidence. Does one sex spend more time recreating outside than the other sex does? According to Précis 26.3.2.21, all of the studies found that males spent more time recreating outdoors than did females. Although the studies were limited to three countries, the research on this topic has spanned 76 years.

Explanation. One reason males spend more leisure time outside than do females may involve their greater overall activity levels (see Précis 26.3.1.6). Regarding sex differences in activity levels, there is evidence that male tendencies to roam about to a greater degree than females reflect evolutionary factors that incline them to be more involved in hunting (see Table 25.20.9). In ancestral times, hunting was a major factor involved in making access to animal protein routinely possible (Fogel, Tuross et al. 1997).

Précis 26.3.2.20 Conversational focus on work and/or leisure (Original Table17.4.8.10)

Age categories, number of studies, and sex	*Consist. score*	*Overviews*	*Countries*	*Time range*	*Non-human*	*Social role theory*			*Evolutionary theory*		*Theological*
						Origin	*Found*	*Biosoc*	*Select*	*ENA*	
All Ages 15M	100.0	-0-	2	1922–1993	-0-	Silent	Silent	Silent	Poss-F	Poss-F	Silent

Précis 26.3.2.21 Play or recreating outside (instead of indoors) (Original Table 17.5.1.8)

Age categories, number of studies, and sex	*Consist. score*	*Overviews*	*Countries*	*Time range*	*Non-human*	*Social role theory*			*Evolutionary theory*		*Theological*
						Origin	*Found*	*Biosoc*	*Select*	*ENA*	
Child 11M All Ages 15M	100.0 100.0	-0-	3	1930–2006	-0-	Silent	Silent	Silent	Expl-M	Expl-M	Silent

Regarding the involvement of brain exposure to androgens, studies have indicated that activity levels in general are positively correlated with circulating testosterone has been reported (Niemann, Wegner et al. 2013; Jardí, Laurent et al 2018). All in all, while there is little research available to specifically explain sex differences in play and recreational outdoor activities, substantial research provides support for evolutionary explanations of sex differences in overall activity levels.

Assessment. While the number of studies needs to be expanded (as do the number of countries sampled), the consistency of the findings over an extensive time frame provides substantial support for considering sex differences in time spent playing and recreating outside to be a USD-M. Evolutionary reasoning appears to provide the strongest explanation.

26.3.2.22 Competitive Social Play

Evidence. Most people find play easy to recognize, but hard to describe. Précis 26.3.2.22 presents the results of studies examining if one sex engages in competitive social play more than the other sex does. Although one study failed to find a sex difference, the remaining 21 studies found that males engaged in competitive play (usually with elements of aggression) more than was the case for females. The studies were conducted from childhood to adulthood in four countries over a time span of about 50 years. We even located three studies of sex differences in competitive social play among non-human mammals, all of which also concluded that it was more common among males.

Explanation. None of the five theories offer a direct explanation for sex differences in competitive social play unless one assumes that such play is essentially a precursor to actual competitive behavior. In this case, one can refer back to Précis 26.3.1.5, which has to do with competitive behavior itself, and overwhelmingly documents that males surpass females in this regard.

Given that males are stereotyped as being more competitive than females (see Table 21.2.7.16 in Volume III), all three social role theories predict that males would begin to exhibit playful forms of competitive behavior soon after they learn this particular sex stereotype. However, the founder effect version would argue that this sex difference was present in the earliest human societies and has then been perpetuated in all societies ever since. Unfortunately, this assertion is not testable.

In the case of the biosocial version of the social role theory proposed by Wood and Eagly (2012), male competitive play would have to be associated with the greater size and strength of males or with the tendency for females to become pregnant. Since many of the examples of competitive play were among children, these criteria would not seem particularly relevant.

Précis 26.3.2.22 Competitive social play (Original Table 17.5.2.1)

Age categories, number of studies, and sex	*Consist. score*	*Overviews*	*Countries*	*Time range*	*Non-human*	*Social role theory*			*Evolutionary theory*		*Theological*
						Origin	*Found*	*Biosoc*	*Select*	*ENA*	
All Ages 21M 1x	95.5	-0-	4	1961–2010	M: three mammals	Contra	NoTest	Poss-M	Poss-M	Poss-M	Silent

Turning to the evolutionary theories, greater overt competitiveness among males is thought to be a part of how males seek to obtain resources, which in turn, tend to attract mates (Wilson, Daly & Pound 2002; Puts 2016; Dunsworth 2020). However, because many of the studies of competitive play involved children, attraction of mates would seem to have little relevance. The way sexual selection theorists have dealt with this qualification is to assume that the male brain tends to be "pre-programmed" to exhibit play behavior that allows them to learn behavior patterns that will benefit them reproductively later in life (Bönte, Procher et al. 2017; Nepomuceno, de Aguiar Pastore & Stenstrom 2021; Moraes, Valentova & Varella 2022).

Regarding the possible involvement of neuroandrogenic factors (as hypothesized by ENA theory), substantial support was found. Specifically, one study of playful aggression among preschoolers of both sexes, not only found such behavior more common among males, but also found it positively correlated with circulating levels of testosterone even before puberty (Sánchez-Martın, Fano et al. 2000). This study is rather surprising given that, during childhood, there tend to be no significant average sex differences in testosterone (see Table 3.1.1.14a).

In the case of prenatal testosterone exposure, the evidence has been mixed. One study found that male-typical play behavior was positively correlated with amniotic fluid levels of testosterone in both sexes (Auyeung, Baron-Cohen et al. 2009a), while an earlier study found no significant relationship (Knickmeyer, Wheelwright et al. 2005). In a comprehensive meta-analysis, Richards and Browne (2022), found very modestly supportive evidence that male-typical play (which includes competitive forms of play) was promoted by prenatal testosterone exposure.

Assessment. Overall, we conclude that competitive social play is a USD-M. More research is needed to more thoroughly test the various theoretical explanations.

26.3.2.23 *Cooperative Social Play*

Evidence. Turning to forms of play that emphasize cooperation, Précis 26.3.2.23 presents the results of studies examining if one sex engages in cooperative social play more than the other sex does. All ten of the studies found that females engaged in cooperative forms of play more than did males. As a limitation, the studies were all conducted in just one country, i.e., the United States, albeit over a 69-year time frame.

Explanation. Since cooperative social play can be essentially considered the opposite of competitive social play, one can refer back to the preceding narrative as providing a backdrop for assessing sex differences in cooperative social play. Cooperative social play can also be considered similar

Précis 26.3.2.23 Cooperative social play (Original Table 17.5.3.1)

Age categories, number of studies, and sex	*Consist. score*	*Overviews*	*Countries*	*Time range*	*Non-human*	*Social role theory*			*Evolutionary theory*		*Theological*
						Origin	*Found*	*Biosoc*	*Select*	*ENA*	
Child 10F	100.0	-0-	1	1933–2002	-0-	Silent	Silent	Silent	Poss-F	Expl-F	Silent

to general sociability, which is also more prevalent among females than males (see Précis 26.3.2.23). Engaging in cooperative social play also usually involves less physically active than engaging in competitive social play (see directly above).

Regarding theoretical explanations, sexual selection theory would lead one to expect that, at least in adulthood, females tend to gain more through social cooperation, while males gain more by social competition. Because play often lays a foundation for the learning of all types of behavioral skills that are helpful throughout life (SL Brown 2009), one would hypothesize that females would exhibit more cooperative forms of play than males, even in childhood.

As far as ENA theory is concerned, cooperative social play, especially among young children, should be associated with low prenatal exposure to testosterone. No evidence was located regarding children, but one study did indicate that, based on 2D:4D finger lengths among adults, low prenatal testosterone exposure was associated with greater social cooperative behavior (Terburg & van Honk 2013:Figure 3).

Assessment. Cooperative social play among children appears to be a USD-F. However, additional research is needed to verify that this pattern exists in more than one country and to further confirm that testosterone is a contributing factor.

26.3.2.24 Playing House

Evidence. "Playing house" is the name of a form of play in which children usually pretend to be adults engaged in arranging and living in an imaginary home, sometimes with elaborate props that are either life-size or miniaturized (such as doll houses). According to Précis 26.3.2.24, all of the 19 studies on this subject found that females engaged in playing house more often than did males. Although the studies only examined three countries, the findings span nearly a century.

Explanation. Females appear to be stereotyped as playing house more than males (Table 21.2.8.3). How can this be theoretically explained? The original version of social role theory would argue that it is due to sex role training that would be found in some cultures, but not others. Contrary to this prediction, the sex difference appears to be universal.

According to the founder effect theory, this type of sex difference began far back in prehistory when the first human cultures just happened to develop a tradition of females being the main caregivers to their offspring, a tradition that has been sustained even to modern times (Fausto-Sterling 1992). For this reason, from a very young age, girls may be encouraged to devote time playing house and playing with dolls.

Précis 26.3.2.24 Playing house (Original Table 17.5.3.3)

Age categories, number of studies, and sex	*Consist. score*	*Overviews*	*Countries*	*Time range*	*Non-human*	*Social role theory*			*Evolutionary theory*		*Theological*
						Origin	*Found*	*Biosoc*	*Select*	*ENA*	
Child 19F All Ages 19F	100.0 100.0	-0-	3	1910–2008	-0-	Silent	NoTest	Silent	Expl-F	Expl-F	Silent

As far as Wood and Eagly's (2012) biosocial theory is concerned, it does not really address the issue of sex differences in the tendency to play house. This is because such behavior has nothing to do with sex differences in size and strength or the fact that only females can become pregnant.

Evolutionary approaches to the study of sex difference in child's play, including the well-documented tendency for girls to play house to a greater degree than boys do, have been proposed (Bjorklund & Junger 2001; Byrd-Craven & Geary 2013). These approaches are derived from indications that adult females have been sexually selected for focusing much of their attention on providing child care and maintaining a healthy and low-stress home environment for children.

According to the ENA version of evolutionary theory, fairly specific neurohormonal mechanisms have been favored for promoting sex differences in child caregiving tendencies, and these mechanisms would be different from those utilized by males to hone skills needed for attracting mates by acquiring resources. In other words, the neurohormonal processes whereby females are inclined to have interests in child care and homemaking are contrary to the male emphasis on mating effort and more supportive of the female emphasis on parenting effort (Zilioli & Bird 2017; PB Gray, Straftis et al. 2020). Since the greater tendency for females to play house occurs in childhood, ENA theory leads to the prediction that prenatal testosterone may be more important than post-pubertal testosterone. While mating effort does appear to be positively correlated with post-pubertal (circulating) testosterone (Perini, Ditzen, Fischvacher & Ehlert 2012), no pertinent studies were located regarding prenatal testosterone.

Assessment. While studies in more countries would be desirable, we conclude that sex differences in playing house is a USD-F.

26.3.2.25 Playing with Mechanical or Construction Objects

Evidence. Another form of play involves building things or using mechanical or construction objects (the later including building blocks and erector sets). Précis 26.3.2.25 shows that all of the 31 studies of this type of play, conducted in three countries and published between 1910 and 2004, found males engaged in this form of play more than did females. The age groups sampled involved eight studies of infants and toddlers, 19 studies of children, and the remaining four studies of adolescents.

Explanation. We found no stereotypes bearing on the possibility of sex differences in tendencies to play with mechanical or construction objects (including toys). Therefore, the three social role theories will be listed as *silent* in this regard.

In the case of the evolutionary proposals, several proposals have been made. Most of them center around the view that males have been sexually

Précis 26.3.2.25 Playing with mechanical or construction objects (Original Table 17.5.3.7)

Age categories, number of studies, and sex	*Consist. score*	*Overviews*	*Countries*	*Time range*	*Non-human*	*Social role theory*			*Evolutionary theory*		*Theological*
						Origin	*Found*	*Biosoc*	*Select*	*ENA*	
Child 19M All Ages 31M	100.0 100.0	-0-	3	1910–2004	-0-	Silent	Silent	Silent	Expl-M	Expl-M	Silent

selected for being spatially oriented (GM Alexander & Hines 2002; GM Alexander 2003; Blakemore & Centers 2005; Levine, Foley et al. 2016).

Regarding ENA theory, various studies have suggested that brain exposure to testosterone promotes spatial reasoning, thereby influencing the tendency to be attracted to mechanical and construction objects. In this regard, prenatal exposure to testosterone has been found associated with interests in mechanical and construction toys (Knickmeyer, Wheelwright et al. 2005; GM Alexander & Saenz 2012; M Hines 2020). Also, a study of toddlers indicated that, among males, those with the highest urinary testosterone levels during infancy exhibited the greatest tendency to play with toy trains (Lamminmaki, Hines et al. 2012).

Assessment. Overall, we confidently declare the tendency to play with mechanical or construction objects (toys) to be a USD-M. The best theoretical explanations seem to be evolutionary in nature.

26.3.2.26 Feminine Toy Choices and Preferences

Evidence. Some toys have been found to be chosen more often by one sex rather than by the other. Examples include boys choosing to play with toy trucks and other mobile objects, while female are more likely to choose to play with dolls and pretend household spaces. Précis 26.3.2.26 shows that all 95 studies, conducted in ten different countries (plus one multi-national study) and published over a 79-year time frame, concluded that females prefer feminine-typical toys more than was the case for males. Two meta-analyses also reached this conclusion.

Explanation. Several of the studies of feminine toy preferences and choices involve infants (which is often prior to the learning of a language with which to distinguish males and females). Explaining these preferences with any of the three social role theories is all but impossible.

Evolutionary explanations have been offered for sex differences in preferences and choices for feminine-type toys. This include the view that females appear to be drawn more than males to faces, especially of infants and children. Theoretically, such face-oriented biases have been favored in order for females to eventually focus on caring for offspring by the time they become mothers (GM Alexander 2003; Jadva, Hines & Golombok 2010). Similarly, it may be argued that the tendency for females to gravitate toward pretend play with household space and appliances could also be the result of evolved predispositions toward performing household activities as a compliment to their child-rearing behavior (M Diamond 1976; Ellis 2011a:557; Bjorklund & Jordan 2013).

Regarding the possible involvement of brain exposure to androgens as contributing to sex differences in preferences and choices of dolls as objects of play, there has been some evidentiary support (Kolata 1979).

Précis 26.3.2.26 Feminine toy choices and preferences (Original Table 17.5.4.1)

Age categories, number of studies, and sex	*Consist. score*	*Overviews*	*Countries*	*Time range*	*Non-human*	*Social role theory*			*Evolutionary theory*		*Theological*
						Origin	*Found*	*Biosoc*	*Select*	*ENA*	
I/T 36F Child 59F All Ages 95F	100.0 100.0 100.0	F: 2Met	10 (1)	1932–2011	-0-	Silent	Silent	Silent	Expl-F	Expl-F	Silent

Probably the strongest evidence has come from a study of female children with a genetic condition known as *congenital adrenal hyperplasia* (*CAH*). This condition is associated with the adrenal glands (rather than the gonads) producing unusually high levels of testosterone. The study indicated that girls with CAH tend to play with dolls much less than other girls their age (Pasterski, Geffner et al. 2005).

Assessment. Tendencies to prefer or to choose feminine toys – dolls in particular – is all but certainly a USD-F. The most compelling explanations seem to be of an evolutionary nature.

26.3.2.27 Masculine Toy Choices and Preferences

Evidence. Preferences for masculine toys (particularly toys such as trucks, trains, and cars, as well as building material) have been widely studied. As shown in Précis 26.3.2.27, all 112 of these studies have concluded that males choose these toys more often than do females. The research has come from 13 different countries (plus one multi-cultural sample), and have been published over a time span of 125 years. Furthermore, two meta-analyses also concluded that males preferred (or spend more time playing with) masculine toys than to females.

Explanation. Even during infancy, boys tend to devote more time looking at or manipulating mechanical-type toys than do females (Lutchmaya & Baron-Cohen 2002). This is very difficult to explain with any of the three social role theories.

According to sexual selection theory, the inclination to play with most masculine toys (e.g., those with moving parts) are likely to reflect interest in things-oriented activities (Lippa 2016). Eventually, these things-oriented play patterns are likely to manifest themselves in terms of working in the so-called STEM fields of science, technology, engineering, and mathematics (Kung 2022), most of which are relatively high-salaried occupations. Because females tend to prefer mates with good earnings potential much more than males do (see Précis 26.2.7.19), sexual selection appears to favor males more than females who gravitate toward masculine things-oriented occupations (see Précis 26.3.5.19),

ENA theory predicts that these sex differences are substantially influenced by exposing the fetal brain to relatively high (male-typical) levels of testosterone or other androgens. Supportive evidence has been provided based on measuring amniotic fluid measures of testosterone (van de Beek, van Goozen et al. 2009).

Assessment. The tendency to choose or prefer masculine toys is all but certainly a USD-M. The strongest explanations appear to be those of an evolutionary nature.

Précis 26.3.2.27 Masculine toy choices and preferences (Original Table 17.5.4.2)

Age categories, number of studies, and sex	*Consist. score*	*Overviews*	*Countries*	*Time range*	*Non-human*	*Social role theory*			*Evolutionary theory*		*Theological*
						Origin	*Found*	*Biosoc*	*Select*	*ENA*	
I/T 38M Child 74M All Ages 112M	100.0 100.0 100.0	M: 2Met	13 (1)	1890–2015	-0-	Silent	Silent	Silent	Expl-M	Expl-M	Silent

26.3.2.28 *Choosing or Preferring Dolls*

Evidence. One can see in Précis 26.3.2.28 that all 42 studies of sex differences in choosing or preferring to look at or play with dolls have concluded that females do so more than do males. These studies were primarily based on studies of children, although research involving infants, toddlers, and adolescents have all reached the same conclusion. The relevant studies have come from samples drawn from eight different countries in studies published over a span of 105 years. It is worth noting that two studies of infant monkeys documented the same sex difference.

Explanation. Research has shown that, even before their first birthday, girls exhibit preferences for looking at pictures of dolls than do boys (e.g., Alexander, Wilcox & Woods 2009: Figure 1; Alexander & Saenz 2012:502). As noted above, dolls tend to epitomize what are considered feminine (or female-typical) toys.

Since it is all but impossible to argue that infants would have acquired sex stereotypes before the age of one in order to know that girls should look at dolls more than boys do, we will list the social role theories as being *silent* regarding this particular sex difference.

Evolutionary explanations for this sex difference center around the assertion that females have been sexually selected for nurturing and providing care to offspring, both in humans (Finch & Groves 2022) and in many other species (Tallamy 2000; Schuett, Tregenza & Dall 2010).

At least two studies have investigated the possibility of prenatal exposure to testosterone as playing a role in the tendency to play with dolls. Based on amniotic fluid levels of testosterone, two studies concluded that there was a negative correlation between prenatal levels of this hormone and their tendency to play with dolls as children (Knickmeyer, Wheelright et al. 2005; Lamminmäki, Hines et al. 2012). However, one study found no significant relationship between prenatal testosterone exposure (as assessed with a 2D:4D finger length measure) and time spent playing with dolls (Alexander & Saenz 2012).

Assessment. The tendency to orient toward, play with, or collect dolls can be considered a USD-F. The most persuasive theoretical explanations appear to be of an evolutionary nature.

26.3.2.29 *Choosing or Preferring Toys with Motion or That Can Be Used for Building*

Evidence. Given the results of Précis 26.3.2.26 on playing with mechanical or construction objects, it is not surprising to find that males often chose or prefer to play with toys that can be used for building more than was the case for females. Précis 26.3.2.29 shows that all 19 studies taking place in

Précis 26.3.2.28 Choosing or preferring dolls (Original Table 17.5.4.3)

Age categories, number of studies, and sex	*Consist. score*	*Overviews*	*Countries*	*Time range*	*Non-human*	*Social role theory*			*Evolutionary theory*		*Theological*
						Origin	*Found*	*Biosoc*	*Select*	*ENA*	
Infant/Toddler 12F	100.0	-0-	8	1910–2015	F: two primates	Silent	Silent	Silent	Expl-F	Expl-F	Silent
Child 28F	100.0										
All Ages 42F	100.0										

five countries. Furthermore, two studies of child-aged primates also found that males preferred these kinds of toys more than did females.

Explanation. These particular types of toy preferences overlap substantially with Précis 26.3.2.25, pertaining to sex differences in playing with mechanical or construction objects. Therefore, we will simply reiterate arguments and evidence presented there to also apply here.

Assessment. Choosing or preferring toys with motion or that can be used for building is a USD-M.

26.3.2.30 Playing Sports and Athletics in General

Evidence. Précis 26.3.2.30 summarizes findings regarding sex differences in participating in sports and athletics. One can see that all 48 studies reached the same basic conclusion: Male participation rates were higher than female participation rates. The studies were based on samples drawn from eight different countries along with four multi-national studies, all of which were published over a 30-year time frame. (Incidentally, in the following two précis, similar evidence is summarized, having to do with team sports and competitive sports respectively.)

Explanation. Because this précis, as well as the next two, all have to do with playing sports, and reach the same conclusion regarding sex differences, the explanation provided here will also applied to the next two précis.

Research has indicated that, compared to females, males are stereotyped as being more athletic (see Table 21.2.1.1) and more competitive (Table 21.2.7.16). From these stereotypes, social role theorists can argue that males are taught within a cultural context to be more interested in and more likely to play sports.

Proponents of the original version of social role theory would be hard pressed to account for the universality of sex differences in playing sports. Since it is likely that the earliest human society did not yet have any formal sports, proponents of the founder effect version of social role theory would also face a substantial challenge to account for universal sex differences. Regarding the biosocial version of social role theory, it could be argued that males play more sports due to their being larger and physically stronger than females. In fact, there is evidence that physical size and strength are positively correlated with performance in various sports (Bilsborough, Greenway et al. 2015:829; Ulbricht, Fernandez-Fernandez et al. 2016). However, these associations can also be explained using ENA theory (see below).

Turning to sexual selection theory, several researchers have proposed that playing physical sports has been sexually selected as a way of demonstrating to members of the opposite sex (G Miller 2000; De Block & Dewitte 2009; Lombardo 2012; Apostolou 2015; Deaner, Balish & Lombardo 2016). Performing well in sports arenas provides evidence of

Précis 26.3.2.29 Choosing or preferring toys with motion or that can be used for building (Original Table 17.5.4.4)

Age categories, number of studies, and sex	*Consist. score*	*Overviews*	*Countries*	*Time range*	*Non-human*	*Social role theory*			*Evolutionary theory*		*Theological*
						Origin	*Found*	*Biosoc*	*Select*	*ENA*	
I/T 10M All Ages 19M	100.0 100.0	-0-	5	1975–2017	M: two primates	Silent	Silent	Silent	Expl-F	Expl-F	Silent

Précis 26.3.2.30 Playing sports and athletics in general (Original Table 17.5.5.1)

Age categories, number of studies, and sex	*Consist. score*	*Overviews*	*Countries*	*Time range*	*Non-human*	*Social role theory*			*Evolutionary theory*		*Theological*
						Origin	*Found*	*Biosoc*	*Select*	*ENA*	
Adolescent 16M Adult 15M All Ages 48M	100.0 100.0 100.0	-0-	8 (4)	1985–2015	-0-	Contra	Contra	Poss-M	Expl-M	Expl-M	Silent

physical agility and a drive to be competitive, both traits associated with being indicative of competent resource provisioning capabilities later in life. Along these lines, one study of both male and female college students found that those who were actively involved in at least one sport reported having had more sex partners than did their non-athletic counterparts. Furthermore, among the athletes, top performers were especially likely to have had more sex partners (Faurie, Pontier & Raymond 2004).

Regarding the assertion by ENA theory that neuroandrogenic factors contribute to all sex differences, substantial supportive evidence has been reported in the case of sports participation and interest. In the case of prenatal testosterone, as inferred from 2D:4D ratios, two studies among male athletes have found evidence of higher testosterone exposure before birth when compared to non-athletes (Manning & Taylor 2001; Reed & Meggs 2017). However, another study of males reported no significant correlation (Voracek, Pum & Dressler 2009).

Turning to post-pubertal testosterone levels, one study of female rugby players found that their testosterone levels were elevated in anticipation of up-coming matches, although levels were unrelated to whether the players won or lost the particular match (Bateup, Booth et al. 2002). A study of male basketball players also found no significant differences in circulating testosterone associated with winning or losing a game (Gonzalez-Bono, Salvador et al. 1999). In a study of soccer players of both sexes, once again, winning and losing did not appear to significantly affect circulating testosterone levels. According to another study, soccer-playing athletes with the greatest skills (as rated by their teammates) had higher levels of testosterone than did the average or least-talented players (DA Edwards, Wetzel & Wyner 2006). Among females, one study reported higher levels of testosterone among athletes than non-athletes (Casto, Arthur et al. 2022).

Regarding brain exposure to prenatal androgens, most studies have found playing sports in general to be positively correlated with prenatal testosterone levels, as measured using 2D:4D data (*meta-analysis*: Honekopp & Schuster 2010). Overall, as predicted by ENA theory, testosterone appears to be associated with involvement in sports. At least some of this association could be due to testosterone promoting traits such as size, strength, and eye-hand coordination, all traits that would be conducive to better athletic performance (see Manning & Taylor 2001; Gettler, Agustin & Kuzawa 2010). However, as will be discussed elsewhere in this chapter (Précis 26.2.7.33), testosterone also appears to promote *interest* in sports.

Assessment. We confidently declare playing sports and athletics in general to be a USD-M. The strongest explanations seem to be along evolutionary lines.

26.3.2.31 Playing Team Sports

Evidence. When looking specifically at team sports, male participation rates were found to be higher than female participation rates. All 37 of the studies presented in Précis 26.3.2.31 found this pattern. The studies, including studies in five countries and one multi-national research project, published over a 23-year time span.

Explanation. Team sports are those in which two or more individuals cooperate to defeat two or more members of an opposing team. Theoretically explaining sex differences in playing team sports are usually subsumed under playing sports in general (see above). Three additional evolutionary proposals have been made: First, a major motivation for playing team sports (and possibly for watching them being played by others) involves satisfying desires to be competitive (Deaner, Balish & Lombardo 2016:77). Given that males enjoy competing more than females (see Précis 26.2.7.32) and actually *do* compete more (see Précis 26.3.1.5), the competitive elements of most team sports would help explain why males engage in them more. Second, team sports may have evolved as a byproduct of sexual selection for engaging in tribal combat as well as modern warfare (Scalise Sugiyama, Mendoza et al. 2018). Third, research has indicated that male athletes have more sex partners than do males in general (Deaner, Balish & Lombardo 2016:80).

Regarding possible brain exposure to androgens contributing to sex differences in playing team sports, no specific evidence was located. However, as noted in the narrative directly above, most studies have indicated that playing sports in general appears to be positively correlated with prenatal testosterone, as measured using 2D:4D data (*meta-analysis*: Honekopp & Schuster 2010).

Assessment. The evidence is strong that playing team sports is a USD-M. As with playing sports in general, the two evolutionary theories seem to have the strongest empirical support.

26.3.2.32 Playing Competitive Sports

Evidence. Nearly all sports have a competitive element, but some are very specific in this regard, while others are not. Most studies of sex differences in playing sports do not specifically distinguish between overtly competitive sports (e.g., basketball, baseball, track, hockey) and sports that focus on precision and grace (e.g., figure skating, gymnastics) (see Précis 26.3.2.30).

Findings of sex differences in overtly competitive sports are summarized in Précis 26.3.2.32. There, one can see that 37 studies were located, all of which concluded that males were more involved in playing competitive sports than were females. Samples for these studies came from

Précis 26.3.2.31 Playing team sports (Original Table 17.5.5.2)

Age categories, number of studies, and sex	*Consist. score*	*Overviews*	*Countries*	*Time range*	*Non-human*	*Social role theory*			*Evolutionary theory*		*Theological*
						Origin	*Found*	*Biosoc*	*Select*	*ENA*	
All Ages 37M	100.0	-0-	5 (1)	1989–2012	-0-	Contra	Contra	Silent	Poss-M	Poss-M	Silent

six different countries along with one multi-national study. These studies were published over an 83-year time frame.

Explanation. Readers can refer back to Précis 26.3.2.30 for a general treatment of theoretical work surrounding sex differences in the playing of sports. While most of these theoretical arguments did not specifically distinguish involvement in sports generally and competitive sports, at least one did (i.e., Apostolou 2015:10), arguing that females should be especially drawn toward males who play competitive sports. Support for this argument came from a study that found females were especially attracted to males playing competitive sports in especially aggressive ways (Brewer & Howarth 2012).

In the case of ENA theory, evidence supports the view that brain exposure to testosterone promotes tendencies to engage in competitive sports, at least regarding prenatal exposure (Manning & Taylor 2001). It is worth adding that, at least following puberty, testosterone levels appear to increase during competitive matches (DA Edwards & Casto 2013).

Assessment. The view that playing competitive sports is a USD-M is strongly supported. Evolutionary explanations appear to be much stronger than social role theories for explaining the sex differences in this regard.

26.3.2.33 Watching Pornography

Evidence. Findings from 62 studies have reported on sex difference in the watching pornography. Précis 26.3.2.33 shows that, without exception, greater proportions of males spend time watching pornography than did females. The findings were based on studies conducted over a 61-year time frame and involved sampling 18 different countries (plus two multi-national studies). Also, one literature review and two meta-analyses reached the same conclusion.

Explanation. To our knowledge, none of the three social role theories offers an explanation for why watching pornography would be more common among males. However, evolutionary explanations have been formulated (Salmon 2012:154; Wright & Vangeel 2019). In essence, these explanations involve noting that males have a stronger sex drive than do females, and this drive is not only expressed by engaging in sexual relationships, but also by masturbating while watching pornography.

Regarding the possibility that brain exposure to androgens could be contributing to sex differences in watching pornography more by males, most of the evidence is indirect. This indirect evidence includes studies showing that testosterone is positively correlated with the strength of people's sex drive (Brock, Heiselman et al. 2016; Wu, Zitzman et al. 2016; Arnocky, Carre et al. 2018). And, because males have much higher testosterone levels than do females, especially before birth and following

Précis 26.3.2.32 Playing competitive sports (Original Table 17.5.5.6)

Age categories, number of studies, and sex	*Consist. score*	*Overviews*	*Countries*	*Time range*	*Non-human*	*Social role theory*			*Evolutionary theory*		*Theological*
						Origin	*Found*	*Biosoc*	*Select*	*ENA*	
Child 19M Adolescent 12M All Ages 37M	100.0 100.0 100.0	M: 1Rev	6 (1)	1930–2013	-0-	Contra	Contra	Poss-M	Expl-M	Expl-M	Silent

Précis 26.3.2.33 Watching or enjoying pornography (Original Table 17.5.8.4)

Age categories, number of studies, and sex	*Consist. score*	*Overviews*	*Countries*	*Time range*	*Non-human*	*Social role theory*			*Evolutionary theory*		*Theological*
						Origin	*Found*	*Biosoc*	*Select*	*ENA*	
Adolescence 25M Adult 30M All Ages 62M	100,0 100,0 100,0	M: 1Rev M: 2Met	19 (2)	1960–2021	-0-	Silent	Silent	Silent	Expl-M	Expl-M	Silent

puberty (see Figure 25.1), males should exhibit greater average interest in sexual activity. With the assumption that accessing pornography basically reflects people's varying degrees of sexual interest, it is reasonable to infer that evolved neuroandrogenic factors contribute to the watching of pornography.

Assessment. Overall, we confidently conclude that the watching of pornography is a USD-M. In fact, given the large number of studies, and the wide range of countries in which they have been conducted, this variable may be one of the strongest USDs that our book has documented.

26.3.2.34 Watching or Enjoying Romantic Programs

Evidence. In contrast to the studies of pornography summarized directly above, Précis 26.3.2.34 shows that all 23 of the studies on sex differences in watching or enjoying romantic programs (e.g., so-called *soap operas*) found that females do so more than do males. While these studies were only drawn from two countries (plus two multi-national studies), they cover research published over an 82-year time frame.

Explanation. Evolutionary theory has been invoked to help explain sex differences in time spent watching mass media programs (including books) dealing with romantic relationships (Schmitt & Buss 2000; A Campbell 2013). The gist of the proposals has been that females have evolved a greater interest in the relationship aspects of mating, while the interest of most males is in the sexual aspects. The main reason for this sex difference is that females can rear more offspring if they have a committed long-term mate providing resources, while males can produce more offspring by having numerous sex partners.

Regarding the ENA version of evolutionary theory, high exposure to testosterone should diminish interests in the romantic aspects of mating (while increasing the overtly sexual aspects). We found no evidence either supporting or refuting this hypothesis (although, as noted directly above, there is substantial empirical support for the hypothesis that brain exposure to testosterone promotes the sexual aspects of mating).

Assessment. We confidently deem the watching and enjoyment of romantic programs (such as "soap operas" and romantic movies) to be a USD-F. This confidence comes partly from the strength of the theoretical evidence, the bulk of which supports sexual selection theory.

26.3.2.35 Watching or Enjoying Sports Programs

Evidence. Précis 26.3.2.35 shows that males have been found to watch (or enjoy watching) sports programs more than do females. These studies have been conducted in four countries (plus one multi-national study) published over a 79-year period.

Précis 26.3.2.34 Watching or enjoying romantic programs (Original Table 17.5.8.6)

Age categories, number of studies, and sex	*Consist. score*	*Overviews*	*Countries*	*Time range*	*Non-human*	*Social role theory*			*Evolutionary theory*		*Theological*
						Origin	*Found*	*Biosoc*	*Select*	*ENA*	
Adult 15F All Ages 23F	100.0 100.0	-0-	2 (2)	1929–2011	-0-	Silent	Silent	Silent	Expl-M	Poss-M	Silent

Précis 26.3.2.35 Watching or enjoying sports programs (Original Table 17.8.9)

Age categories, number of studies, and sex	*Consist. score*	*Overviews*	*Countries*	*Time range*	*Non-human*	*Social role theory*			*Evolutionary theory*		*Theological*
						Origin	*Found*	*Biosoc*	*Select*	*ENA*	
All Ages 14M	100.0	-0-	4 (1)	1932–2011	-0-	Silent	Silent	Silent	Expl-M	Expl-M	Silent

Explanation. Earlier in this chapter, evidence was summarized regarding sex differences in *playing* competitive sports. It showed that males do so more than females throughout the world (see Précis 26.3.2.32). The findings summarized here show that males also like to watching sports being played by others to a greater degree than do females. No explanation was found to be offered by any of the three social role theories for this sex difference.

In the case of the two evolutionary theories, they provide guidance for those seeking to explain sex differences in tendencies to watch and enjoy watching sports activities. This comes from noting that males express an interest in all forms of competition to a greater degree than do females (see Précis 26.2.7.32). Males are also more likely to engage in competitive activities (see Précis 26.3.1.5). From the perspective of sexual selection theory, this greater male focus on competition has been favored because it is at the heart of obtaining resources (Kodric-Brown & Brown 1984; Puts 2010). In other words, since the ability to acquire resources is important to females as a criterion for mate selection (see Précis 26.3.15.19), all types of competitiveness tendencies will have been favored among males more than among females.

Regarding the possibility that ENA theory also helps to explain why males watch and enjoy sports programs more than do females, several studies have indicated that watching one's favorite team when is associated with a rise in circulating testosterone (*review*: Raney 2009; *meta-analysis*: Geniole, Bird et al. 2017). However, does testosterone *promote* interests in and enjoyment of sporting events? Various lines of evidence point toward an affirmative answer (Deaner, Balish & Lombardo 2016), especially regarding prenatal exposure to this hormone (Berenbaum 1999; Frisen, Nordenstrom et al. 2009).

Assessment. Overall, watching and enjoying sports programs is a USD-M. The best supported explanations appear to be along evolutionary lines.

26.3.2.36 Playing Electronic or Video Games

Evidence. Considerable research has sought to determine if sex differences exist in the playing of electronic or video games. As shown in Précis 26.3.2.36, all but one of the 61 relevant studies found that males played such games more than did females. Four different countries were sampled in these studies plus one multi-country study, with studies having been published over a 34-year time range.

Explanation. None of the three social role theories have offered an explanation for why males spend more time playing electronic or video games more than females. This is especially true regarding why the sex difference appear to be universal.

Précis 26.3.2.36 Playing electronic or video games (Original Table 17.5.9.4)

Age categories, number of studies, and sex	*Consist. score*	*Overviews*	*Countries*	*Time range*	*Non-human*	*Social role theory*			*Evolutionary theory*		*Theological*
						Origin	*Found*	*Biosoc*	*Select*	*ENA*	
Child 11M Adolescent 18M Wide Age Range 15M All Ages 60M 1x	100.0 100.0 100.0 98.4	-0-	4 (1)	1983–2017	-0-	Silent	Silent	Silent	Expl-M	Poss-M	Silent

The evolutionary theories, however, seem to offer reasonable explanations. Regarding sexual selection theory, it is noteworthy that video games are comprised of both competitive and spatial reasoning elements. As discussed in the narratives for both competitiveness (see Précis 26.3.1.5) and spatial reasoning (see Précis 26.2.11.4), they are both higher among males and they both appear to have been favored by sexual selection more for males than for females.

In the case of ENA theory, considerable evidence points toward prenatal testosterone contributing to variations in people's playing electronic games. One was a study of males who spent more than two hours per day playing video games. When they were compared to males in general, their 2D:4D finger length ratios indicated that the "addicted" players had higher exposure than did other males (Kornhuber, Zenses et al. 2013). Another study reported the same pattern when they compared chronic online gamers to those who used the internet primarily for other purposes (Müller, Brand et al. 2017).

Assessment. A few decades ago, a research team noted that, although the research is very consistent in indicating that males play video games more than do females, "the origin of gender differences in game-playing habits has not yet been established" (Funk & Buchman 1996:27). At the present point in time, considerable evidence seems to point toward evolutionary theory as being especially helpful in accounting for this sex difference. In any case, extensive playing of electronic (or video) games is a well-established USD-M.

26.3.2.37 Direct Involvement in Warfare (Combat)

Evidence. Précis 26.3.2.37 shows that all 17 studies of sex differences in being involved in combat during wars, males are more involved than are females. These results were based on samples from six different countries in addition to six multi-national studies published over a 38-year time period. Two literature reviews reached the same conclusion. Furthermore, two studies of chimpanzees concluded that males were more likely to patrol their troop's territories and attack (and sometimes even kill) intruders from neighboring troops.

Explanation. As was noted in Table 25.17.16, findings of sex differences in most measures of physical aggression all agree that males are more involved than are females. As seriousness (usually in the form of harm to victims) increases, the proportion of male involvement increases. Nevertheless, in terms of overall physical aggression, the table shows that the consistency score was substantially below the 95.0 cutoff, thus denying overall physical aggression USD status. Regarding seriousness of physical aggression, additional evidence can be obtained by noting that criminal

Précis 26.3.2.37 Direct involvement in warfare (combat) (Original Table 17.6.1.10)

Age categories, number of studies, and sex	*Consist. score*	*Overviews*	*Countries*	*Time range*	*Non-human*	*Social role theory*			*Evolutionary theory*		*Theological*
						Origin	*Found*	*Biosoc*	*Select*	*ENA*	
Adult 14M All Ages 17M	100.0 100.0	M: 2Rev	6 (6)	1978–2016	M: two chimps	Contra	NoTest	Poss-M	Expl-M	Expl-M	Silent

assault (see Précis 26.3.4.2) and homicide (see Précis 26.3.4.2) are both well-established USD-Ms.

Turning to theoretically explaining sex differences in involved in combat during war, all five theories can be considered as having offered proposals in this regard. In the case of the three social role theories, one can cite evidence that males are stereotyped as being more physically aggressive (see Précis 16.3.21.6). This stereotype would incline societal leaders who decide to wage war to trait males to do most of the fighting. Of course, the original version of social role theory would not predict that that *all* societies would hold such a stereotype. The founder effect version of social role theory is essentially untestable, since there is no known empirical evidence about warfare in the first human societies.

As far as the biosocial version of the social role theory, it is reasonable to argue that the greater size and strength of males would prompt societal leaders to choose males more often than females to be involved in combat. This biosocial role theory would also lead one to think that societal leaders would want to maintain their populations by not risking death to fertile females in their societies.

Both of the evolutionary theories offer explanations for sex differences in being war combatants. In the case of sexual selection, two studies of pre-agrarian studies both indicated that males who engage in war-related combat have more offspring than males who do not (Chagnon 1988; Glowacki & Wrangham 2015). there should be some reproductive advantage to being willing to risk one's life on behalf of one's country. An interesting set of study of World War II veterans, researchers found that men who exhibited unusual heroism during the war went on to sire more offspring than did veterans in general (Rusch, Leunissen & Van Vugt 2015). In a second part to this latter study, a sample of women were asked to rate the degree to which they would be sexually attracted to three groups of men: those who avoided war, those who went to war, and those who went to war and were awarded a metal for heroism. The latter group received the highest ratings (Rusch, Leunissen & Van Vugt 2015:Figure 1). Overall, there is substantial evidence for sexual selection theory in terms of helping to explain sex differences in being a war combatant.

Regarding the possibility that brain exposure to testosterone contributes to combative war activities, the evidence has been mixed. On the supportive side is the fact that most studies have found that, even within each sex, circulating levels of testosterone are positively correlated with overall aggressive tendencies (*review*: Ellis, Farrington & Hoskin 2019:357).

Also, experiments in which testosterone has been administered intravenously have indicated that elevated testosterone modestly promotes various forms of aggression (O'Connor, Archer et al. 2002; Pope, Kouri & Hudson 2000). Another experimental study with male college students exposed half of them to a gun for 15 minutes, while the other half spent the same time examining a child's toy. Before the exposure, these two groups of research participants did not significantly differ in saliva testosterone, but afterwards, those who had spent time in the vicinity of the gun had elevated levels of testosterone compared to those exposed to the child's toy (Klinesmith, Kasser & McAndrew 2006). Given that military training usually involves extensive exposure to lethal weapons (such as guns), one can assume that testosterone will be elevated among solders compared to non-solders.

Non-supportive evidence has come from one study of men and women who took part in a simulated wargame. The experiment found no significant correlation between confidence in winning and circulating testosterone levels (Johnson, McDermott et al. 2006).

Assessment. Overall, one can confidently declare direct involvement in warfare to be a USD-M. The best theoretical explanation appears to involve sexual selection theory.

26.3.2.38 Masturbation

Evidence. Research published over 83 years has examined possible sex differences in the frequency of masturbation. Précis 26.3.2.38 shows that this research has sampled six different countries. Although one study found no significant sex difference, the remaining 49 studies, along with two meta-analyses, all concluded that males engaged in masturbation more frequently than do females. The same sex difference was found in five studies of non-human mammals.

Explanation. No proponents of social role theory were found to have offered an explanation for sex differences in rates of masturbation. Regarding evolutionary arguments, it is obviously difficult to argue that masturbation has any direct effects on reproduction. However, given that males produce thousands of sperm cells every hour and that sperm cells only live a few days (Reinhardt 2007; Regnerus 2017:107), it may actually be beneficial for these cells to be purged periodically so that relatively young healthy sperm cells can be used for insemination (Reinhardt & Turnell 2019). In any case, higher masturbation rates by males probably reflect their stronger sex drive (see Précis 26.2.7.6), which does appear to have been favored by sexual selection (RS Singh & Kulathinal 2005; Waterink 2014).

According to ENA theory, neuroandrogenic factors are largely responsible for greater sexual activity, including masturbation, among males. Most pertinent studies have found support for this hypothesis, at least among women (*review*: van Anders 2012). Furthermore, as abstinence from sex is prolonged, testosterone levels tend to rise, at least among males (Exton, Krueger et al. 2001). This elevation in testosterone is likely to promote sexual motivation.

Précis 26.3.2.38 Masturbation (Original Table 17.7.1.1)

Age categories, number of studies, and sex	*Consist. score*	*Overviews*	*Countries*	*Time range*	*Non-human*	*Social role theory*			*Evolutionary theory*		*Theological*
						Origin	*Found*	*Biosoc*	*Select*	*ENA*	
Adolescent 10M	100.0	M: 2Met	6	1937–2020	M: five mam-mals	Silent	Silent	Silent	Poss-M	Expl-M	Silent
Adult 32M 1x	97.0										
All Ages 49M 1x	98.0										

Assessment. We confidently conclude that masturbation is a USD-M. The strongest theoretical arguments seem to be of an evolutionary nature.

26.3.2.39 Number of Sex Partners

Evidence. Seventy-four studies were located based on samples of 16 countries (plus two multi-national samples) regarding sex differences in terms of the number of sex partners individuals have had. In addition, eight studies of various species of non-human mammals were also investigated regarding this variable. As shown in Précis 26.3.2.39, the majority of these 74 studies indicated that males have more sex partners than do females. However, only among the 15 studies of Wide Age Range research participants did the consistency score surpass the 95.0 cutoff.

Explanation. All studies of sex differences in the number of sex partners (among humans) have been based on self-reports. In this regard, it is worth noting that, at least in the case of heterosexual sex partners, one would expect there to be no sex difference if both sexes have been equally sampled and are being honest to the same degree. However, sex equality in terms of number of sex partners is obviously almost never achieved.

Researchers have used various methods to help verify these reports, looking in particular for evidence of exaggeration by males or underreporting by females. The evidence has indicated that honesty factors contribute at least slightly to the sex differences (Wiederman 1997; Alexander & Fisher 2003). However, additional factors responsible for sex differences in self-reported number of sex partners involve the inclusion of homosexual experiences especially by males along with heterosexual encounters with prostitutes by males.

As noted earlier in this chapter, the evidence clearly indicates that males *desire* more sex partners on average than do females, a sex difference that is likely to have an evolutionary basis (see Précis 26.2.7.7). However, if all heterosexual encounters were to be honestly reported, there is no mathematical reason to expect sex equality in terms of the number of sex partners individuals have had.

Assessment. Overall, the number of sex partners that individuals report having had will not be decreed a USD, nor will we offer an explanation for the observed differences with any of the five theories of sex differences.

26.3.2.40 Pushy Sexual Overtures

Evidence. Is one sex pushier than the other sex when it comes to sexual interactions? *Pushiness* in this context primarily refers to persistent verbal pestering and/or to trying to escalate petting behavior beyond mutual consent.

Précis 26.3.2.40 shows that all 11 studies of pushy sexual overtures found males engaged in such actions more than in the case of females.

Précis 26.3.2.39 Number of sex partners (Original Table 17.7.1.24)

Age categories, number of studies, and sex	*Consist. score*	*Overviews*	*Countries*	*Time range*	*Non-human*	*Social role theory*			*Evolutionary theory*		*Theological*
						Origin	*Found*	*Biosoc*	*Select*	*ENA*	
Adult 38M 10x 4F Wide Age Range 15M All Ages 60M 10x 4F	(67.9) 100.0 (76.9)	-0-	16 (2)	1982–2020	2M rodents; 6F non-mammals	Silent	Silent	Silent	Silent	Silent	Silent

However, the studies were limited to just two countries and were published over a time span of just 12 years.

Explanation. No explanation for pushy sexual overtures based on any of the three social role theories were found. Both of the evolutionary theories would hypothesis that males should seek to mate with multiple partners as a way of maximizing the number of offspring they produce. Females, on the other hand, should be more cautious regarding intimate sexual relationships because their reproductive potential depends more heavily on mates providing long-term resources during pregnancy and beyond (Ellis 1991a; Thornhill & Palmer 2001).

In the case of ENA theory, it asserts that testosterone should influence brain functioning in ways that promote pushy sexual behavior. Evidence bearing on this hypothesis varies. Basically, two studies found that, within samples of males, those who engaged in sexual violence had higher circulating testosterone than those who did not, but males who tended to be merely sexually pushy (e.g., child molesters) did not differ from males in general (Rada, Laws & Kellner 1976; Aromäki, Lindman & Eriksson 2002).

Assessment. While more countries need to be sampled in future studies, we consider pushy sexual overtures to be a USD-M.

26.3.2.41 Age at Marriage or During Courtship

Evidence. As shown in Précis 26.3.2.41, 126 out of 127 studies of sex differences in the relative age at marriage (or, in a few cases, age during courtship) concluded that males were older than females. One can see that the evidence came from 27 different countries (plus ten more studies of multiple countries). The studies were published over a 58-year time frame.

Explanation. The best explanation for males being older than their female mates at the time of marriage or during courtship centers around the fact that, throughout the world, males prefer younger mates (see Précis 26.2.7.17) and females prefer older mates (see Précis 26.2.7.18). For both of these sex differences in preferences, one finds that evolutionary explanations are most viable.

Assessment. Being older than one's marriage or dating partner is a USD-M. The most reasonable explanation appears to be evolutionary in nature.

26.3.2.42 Remarriage after Divorce

Evidence. Eleven studies examined which set was more likely to remarry after a divorce. As shown in Précis 26.3.2.41, all of these studies found that males remarried more often than females remarried. However, aside from one multinational study, all of the research was done in a single country.

Explanation. The greater tendency for males to remarry after being divorce may be explainable in evolutionary terms by noting that males

Précis 26.3.2.40 Pushy sexual overtures (Original Table 17.7.1.13)

Age categories, number of studies, and sex	*Consist. score*	*Overviews*	*Countries*	*Time range*	*Non-human*	*Social role theory*			*Evolutionary theory*		*Theo-logical*
						Origin	*Found*	*Biosoc*	*Select*	*ENA*	
Adult 11M	100.0M	-0-	2	1984–1996	-0-	Silent	Silent	Silent	Expl-M	Expl-M	Silent

Précis 26.3.2.41 Age at marriage or during courtship (Original Table 17.7.4.1)

Age categories, number of studies, and sex	*Consist. score*	*Overviews*	*Countries*	*Time range*	*Non-human*	*Social role theory*			*Evolutionary theory*		*Theological*
						Origin	*Found*	*Biosoc*	*Select*	*ENA*	
All Ages 126M 1x	97.7	-0-	27 (10)	1955–2013	-0-	Silent	Silent	Silent	Expl-M	Expl-M	Silent

have a higher potential reproductive ceiling than do females. In other words, for males, their so-called *reproductive success* depends more on the number of sex partners they have than is the case for females (Clutton-Brock 2017). Consequently, remarriage following a divorce usually provides a male with opportunities to produce additional offspring, especially if the new wife is still of reproductive age.

Regarding the possibility of neuroandrogenic factors contributing to the greater tendency by males to remarry, there is some empirical support. At least among men, the tendency to remarry after getting divorced has been found significantly positively correlated with circulating levels of testosterone (Pollet, van der Meij et al. 2011).

Assessment. Even though the number of countries sampled needs to be enhanced, we are confident that remarriage after divorce is a USD-M. This confidence comes from the fact that the theoretical explanations for the difference is well founded and the time frame during which studies have been conducted are substantial.

26.3.3 Acquiring, Selling, and Consuming Behavior

Chapter 18 pertains to sex differences in acquiring, selling activities along with most forms of consuming behavior. Many of these studies have been conducted in order to better understand consumer psychology and to refine marketing strategies.

26.3.3.1 Shopping in General (Except Online)

Evidence. Everyone shops, but does one sex do so more than the other? Précis 26.3.3.1 shows that the vast majority of studies have found that females engaged in more of this behavior than did males. The only exception was a study among adolescents that did not find a significant sex difference. Although no overviews were found on this topic, the studies were conducted in 12 countries. (Shopping online was considered separately and did not indicate consistent sex differences – see Table 18.1.1.2.)

Explanation. According to one U.S. study, females are stereotyped as devoting more time to shopping (South & Spitze 1994). With such a stereotype, social role theorizing would lead one to hypothesize that females would in fact shop more, at least in the United States, but not in all countries examined. One can doubt the relevance of both the founder effect version and the biosocial version because early human societies had no shopping in any modern sense, and shopping has no particular relevance to body size, strength, or child bearing.

In the case of evolutionary reasoning, the so-called *savanna hypothesis* (also sometimes termed the *hunting-gathering hypothesis*) has been offered

Précis 26.3.2.42 Remarriage after divorce (Original Table 17.7.4.14)

Age categories, number of studies, and sex	*Consist. score*	*Overviews*	*Countries*	*Time range*	*Non-human*	*Social role theory*			*Evolutionary theory*		*Theological*
						Origin	*Found*	*Biosoc*	*Select*	*ENA*	
Adult 11M All Ages 11M	100.0M 100.0M	-0-	1 (1)	1959–2021	-0-	Silent	Silent	Silent	Poss-M	Poss-M	Silent

Précis 26.3.3.1 Shopping in general (except online) (Original Table 18.1.1.1)

Age categories, number of studies, and sex	*Consist. score*	*Overviews*	*Countries*	*Time range*	*Non-human*	*Social role theory*			*Evolutionary theory*		*Theological*
						Origin	*Found*	*Biosoc*	*Select*	*ENA*	
Adult 26F All Ages 34F 1x	100.0 97.1	-0-	12	1982–2016	-0-	Contra	Silent	Silent	Poss-M	Poss-M	Silent

to help explain sex differences in shopping tendencies (Dennis & McCall 2005; Saad 2007; Kruger & Byker 2009; Cashdan, Marlowe et al. 2012; Dennis, Brakus et al. 2018). In other words, modern-day shopping can be considered an evolved vestige of predominantly gathering (as opposed to hunting) types of food obtaining methods.

According to this hypothesis, in pre-agrarian societies, females focused their food acquiring efforts primarily on gathering seasonal fruits, vegetables, and sometimes small easy-to-catch animals. Males, on the other hand focused primarily on hunting relatively large difficult-to-catch animals. When applied to studying modern-day shopping behavior, which females appear to do more of than males, researchers have argued that most forms of shopping more closely resemble ancient forms of foraging behavior rather than hunting behavior. Regarding any involvement of neuroandrogenic factors, as would be hypothesized by ENA theory, no research findings were located.

Assessment. Overall, we deem that frequency of shopping is a USD-F. The most reasonable explanation seems to be of an evolutionary nature.

26.3.3.2 Eating Meat

Evidence. Substantial amounts of research have compared males and females regarding the degree to which they consume meat (i.e., some sort of vertebrates). Précis 26.3.3.2 shows that all of the 18 studies on this topic found that males ate more meat than did females. The studies were conducted in six countries, along with two multi-national studies. The findings were also consistent with two reviews of the literature as well as with nine studies of various species of non-human primates.

Explanation. We are aware of no publication in which meat eating is stereotyped as being more common in males. Nor could we find any parts of the three social role theories that offered an explanation for sex differences in the consumption of meat.

Evolutionary explanations having to do with meat consumption have emphasized that this practice is a faster and more reliable way of consuming high protein and fat relative to adopting an all-vegetarian diet (Leroy & Praet 2015). Part of the reason is that animal products contain creatine, an amino acid that is located in muscle and brain tissue, although it is now possible to produce creatine synthetically. Most studies indicate that, when creatine is consumed, it promotes muscular coordination (SP Bird 2003; Kreider 2003), bone density (Mithal, Bonjour et al. 2013), and may even improve memory and promote intelligence (Rae, Digney et al. 2003).

From a reproductive standpoint, the effects of creatine could be especially beneficial to males performing high energy-consuming activities. (Note that sexual selection theory asserts that females are generally attracted to males with keen athletic skills and the ability to procure resources.) Following this

Précis 26.3.3.2 Eating meats (Original Table 18.1.3.4)

Age categories, number of studies, and sex	*Consist. score*	*Overviews*	*Countries*	*Time range*	*Non-human*	*Social role theory*			*Evolutionary theory*		*Theological*
						Origin	*Found*	*Biosoc*	*Select*	*ENA*	
Adult 13M All Ages 18M	100.0 100.0	M: 2Rev	6 (2)	1984–2015	M: nine primates	Silent	Silent	Silent	Expl-M	Expl-M	Silent

line of reasoning, one can deduce that males have been favored for gravitating toward activities that enhance coordination and cognitive ability. Meat consumption could be one of the dietary activities making this possible.

Regarding the possibility of neuroandrogenic factors being involved, as hypothesized by ENA theory, various lines of evidence are supportive. Specifically, both organizational and activational levels of testosterone has been linked to the consumption of meat (Wingfield, Jacobs & Hillgarth 1997; Asarian & Geary 2013).

Assessment. With substantial confidence, we declare meat eating to be a USD-M. This conclusion is consistent with two literature reviews, can be theoretically explained, and is even consistent with studies of other omnivorous primates such as capuchin monkeys and chimpanzees.

26.3.3.3 Consuming Therapeutic Medications in General

Evidence. Précis 26.3.3.3 shows the results of studies on the extent to which males and females consumed therapeutic medications. It shows that 27 out of 28 studies found that females consumed more of this type of medication than was true for males. The studies were carried out in three countries and in two multi-national studies.

Explanation. It is reasonable to assume that the consumption of therapeutic medications is to obtain treatment for some sort of health issue. If so, the evidence that females consume therapeutic drugs more than do males suggests that they have more health problems, or at least perceive themselves as having more. In this regard, most studies have indicated that females are more sensitive to pain (see Table 9.2.1.1) and are more likely to catastrophize in response to pain (see Table 4.2.2.1). With these sex differences in mind, it is reasonable to assume that female tendencies to consume more therapeutic medications at least partly reflects their greater sensitivity to pain.

From a theoretical standpoint, it is worth noting that most studies have concluded that females are more capable of tolerating pain than are males (see Table 21.2.6.27). This stereotype is actually the opposite of most of the empirical evidence (Table 9.2.1.1). Consequently, the ability of the three social role theories to explain sex differences in the consumption of therapeutic medications lacks merit.

In the case of the two evolutionary theories, both lead to the conclusion that females should be more sensitive to pain and likely to detect illness symptoms more readily than males. This is because males have been selected for their abilities to be competitive, aggressive, and willing to take risks more than is the case for females (Ross & Richerson 2014; Puts, Bailey & Reno 2015). Diminished sensitivity to pain is likely to be one of the main contributors to these types of traits.

Précis 26.3.3.3 Consuming therapeutic medications in general (Original Table 18.1.4.2)

Age categories, number of studies, and sex	*Consist. score*	*Overviews*	*Countries*	*Time range*	*Non-human*	*Social role theory*			*Evolutionary theory*		*Theological*
						Origin	*Found*	*Biosoc*	*Select*	*ENA*	
Adult 18F All Ages 27F 1x	100.0 96.4	-0-	3 (2)	1973–2004	-0-	Contra	Contra	Contra	Poss-F	Poss-F	Silent

Regarding ENA theory, substantial research points toward neuroandrogenic factors contributing to sex differences in pain sensitivity. Specifically, circulating testosterone levels have been found to inhibit pain sensitivity. This is true for males (Apkhazava, Kvachadze et al. 2018), for females (Bartley, Palit et al. 2015), and for both sexes combined (Fischer, Clemente & Tambeli 2007). Even studies of non-humans have concluded that elevated testosterone levels are inversely associated with pain sensitivity (Bai, Zhang et al. 2015; Kumar, Liu et al. 2015; Lee, Zhang et al. 2016).

Assessment. We conclude that the tendency to consume more therapeutic medications is a USD-F. This sex difference is likely to be largely reflective of a greater tendency for males to tolerate pain.

26.3.3.4 *Consuming Prescription Psychotropic Medications*

Evidence. There is also a sex difference when prescribed medications having a significant effect on mood and behavior are examined separately from other medications. The results of studies of these psychotropic medications are presented in Précis 26.3.3.4. It shows that studies in eight countries and three multi-national research projects all found that females consumed more prescribed psychotropic medications than did males.

Explanation. Given that prescription psychotropic medications are included in the category of therapeutic medications in general, the arguments presented in the narrative directly above can be applied here as well. It may also be relevant to mention that a sizable proportion of prescription psychotropic medications are for the treatment of mental illness, including depression. As noted earlier in this chapter, females appear to have more internalized mental illness symptoms (see Précis 26.2.6.2 and Précis 26.2.6.3), especially in the form of depression (see Précis 26.2.6.4).

Assessment. For the reasons stated in this section and in the previous one, as well as the consistency of the evidence, we conclude that consuming prescription psychotropic medications is a USD-F.

26.3.3.5 *Amount of Alcohol Consumed*

Evidence. There have been many studies concerning possible sex differences in the consumption of alcohol and problems related to that consumption. Regarding the overall tendency to consume alcohol, Précis 26.3.3.5 shows that the vast majority (76 out of 77) of studies have found that males consumed more alcohol than did females. The one exception found that females consumed more alcohol than did males. Overall, the studies spanned nearly 60 years and were conducted in 12 different countries. In addition, there were three multi-national research findings. Interestingly, among non-human mammals (mainly rodents), most findings have pointed toward greater consumption among females.

Précis 26.3.3.4 Consuming prescription psychotropic medications (original Table 18.1.4.6)

Age categories, number of studies, and sex	*Consist. score*	*Overviews*	*Countries*	*Time range*	*Non-human*	*Social role theory*			*Evolutionary theory*		*Theological*
						Origin	*Found*	*Biosoc*	*Select*	*ENA*	
Wide Age Range 18F	100.0	-0-	8 (3)	1960–2007	-0-	Silent	Silent	Silent	Poss-F	Poss-F	Silent

Précis 26.3.3.5 Amount of alcohol consumed (original Table 18.1.5.3)

Age categories, number of studies, and sex	*Consist. score*	*Overviews*	*Countries*	*Time range*	*Non-human*	*Social role theory*			*Evolutionary theory*		*Theological*
						Origin	*Found*	*Biosoc*	*Select*	*ENA*	
Adult 66M 1F All Ages 76M 1F	97.0 97.4	-0-	12 (3)	1953–2010	F: 5; x: one mammals	Poss-M	Silent	Silent	Poss-M	Expl-M	Silent

Explanation. Explaining sex differences in alcohol consumption is a theoretical challenge. From the perspective of social role theory, it could be argued that females are stereotyped as consuming less alcohol than males, or taught to consume less by their parents and peers, but we found no evidence directly pertaining to this speculation. However, there is evidence of greater social disapproval of excessive alcohol consumption by females than by males, at least in the United States (Huselid & Cooper 1992; Keefe 1994). Whether such disapproval is universal, thus explaining the apparent universality of more alcohol consumption by males, remains to be determined.

Turning attention to evolutionary theory, it was noted earlier in this chapter that males are more inclined to take risks than are females, a sex difference that appears to have been sexually selected (see Précis 26.3.1.2). If one assumes that alcohol consumption is a risky form of behavior (especially when consumed in substantial amounts), one can infer that males will consume alcohol more than females.

However, the explanation for greater alcohol consumption by males also seems to entail neurohormonal variables, as hypothesized by ENA theory. In this regard, studies have shown that males can consume more alcohol (even after adjusting for body weight) before exhibiting symptoms of intoxication (Lemle & Mishkind 1989; Nolen-Hoeksema & Hilt 2006:359). This sex differences appears to be at least partly due to males having higher exposure to testosterone. Based on an extensive review of research literature derived from samples of both humans as well as other animals, a team of researchers concluded that alcohol consumption is positively correlated with circulating testosterone levels, especially among males (*review*: Erol, Ho et al. 2019). This provides support for the neurohormonal component of ENA theory by suggesting that brain exposure to testosterone increases alcohol tolerance.

Assessment. The sizable number of studies, their consistency, and the substantial number of countries in which these studies were conducted provides considerable support for concluding that the tendency to gravitate toward consuming alcohol is a USD-M. Nevertheless, the fact that most studies of non-human mammals have concluded that females consume alcohol more, plus the minimal theoretical understanding of sex differences in this regard, calls for more research.

26.3.3.6 Abstinence from Alcohol Consumption

Evidence. Précis 26.3.3.6 shows that 39 out of 40 studies found that females abstained from alcoholism more than was the case for males. The single exception simply failed to find a significant sex difference. Eleven countries plus one multi-national study were found that had examined this sex-abstinence relationship over a 73-year time frame.

Précis 26.3.3.6 Abstinence from alcohol consumption (original Table 18.1.5.4)

Age categories, number of studies, and sex	*Consist. score*	*Overviews*	*Countries*	*Time range*	*Non-human*	*Social role theory*			*Evolutionary theory*		*Theological*
						Origin	*Found*	*Biosoc*	*Select*	*ENA*	
Adult 22F Wide Age Range 15F All Ages 39F 1x	100.0 100.0 97.5	-0-	11 (1)	1947–2020	-0-	Silent	Silent	Silent	Silent	Silent	Silent

Explanation. As noted directly above, most research has indicated that males consume greater amounts of alcohol than do females. It is rather predictable, therefore, that greater proportions of females would be abstainers. Explaining this sex difference with either social role theory or evolutionary theory is difficult (at least in terms of current knowledge).

Assessment. We confidently declare abstinence from alcohol consumption to be a USD-F. Nevertheless, theory development regarding the cause of this sex difference is needed.

26.3.3.7 Alcohol-Related Problems

Evidence. Consistent with the greater consumption of alcohol by males than by females is evidence that males have more alcohol-related problems than do females, Précis 26.3.3.7 shows that the vast majority of studies reported that males have more experiencing alcohol-related problems. Most of these problems have to do with their interpersonal relationships. The only two studies that did not reach this conclusion was one study that reported no significant sex difference and one study that reported females experiencing more dysphoria after consuming alcohol. The 51 studies pertaining to this matter came from seven countries plus one multi-national study.

Explanation. Explaining why males have more alcohol-related problems than do females with any of the five scientific theories is difficult. The only exception would involve the neurohormonal component of the ENA theory. When this component is combined with evidence cited above (see Précis 26.3.3.5) that brain exposure to testosterone seems to increase the chances of males consuming alcohol, their elevated risk of alcohol-related problems becomes predictable. Additional evidence pertinent to this matter can be found directly below. Specifically, males appear to abuse alcohol more (see 26.3.3.8) and are more prone to binge drink (see Précis 26.3.3.9).

Assessment. Even though the theoretical explanation for why males would experience more alcohol-related problems than do females, the evidence strongly supports this conclusion. Therefore, we deem this variable to be a USD-M.

26.3.3.8 Alcohol Abuse

Evidence. Précis 26.3.3.8 focuses specifically on alcohol-related problems involving frequent drunkenness and labeled alcohol abuse. It shows that 84 out of 88 studies found males abused alcohol more than females. The four other studies failed to find a sex difference. The study of alcohol abuse took place in 16 countries and there were two multi-national studies.

Explanation. As noted in the discussion surrounding Précis 26.3.3.5 having to do with sex differences in alcohol consumption in general,

Précis 26.3.3.7 Alcohol-related problems (original Table 18.1.5.7)

Age categories, number of studies, and sex	*Consist. score*	*Overviews*	*Countries*	*Time range*	*Non-human*	*Social role theory*			*Evolutionary theory*		*Theological*
						Origin	*Found*	*Biosoc*	*Select*	*ENA*	
Adult 26M 1x Wide Age Range 22M All Ages 49M 1x 1F	96.3 100.0 98.0	-0-	7 (1)	1947–2005	-0-	Silent	Silent	Silent	Silent	Poss-M	Silent

Précis 26.3.3.8 Alcohol abuse (original Table 18.1.5.9)

Age categories, number of studies, and sex	*Consist. score*	*Overviews*	*Countries*	*Time range*	*Non-human*	*Social role theory*			*Evolutionary theory*		*Theological*
						Origin	*Found*	*Biosoc*	*Select*	*ENA*	
Adult 65M 1x Wide Age Range 11M All Ages 84M 4x	98.5 100.0 95.5	-0-	16 (2)	1968–2010	-0-	Silent	Silent	Silent	Silent	Poss-M	Silent

evidence pointing toward brain exposure to testosterone contributing to greater consumption of alcohol has been published. It is possible that this greater consumption by males could be responsible for at least some of them to consume excessive amounts of alcohol.

Assessment. Alcohol abuse is a USD-M. More work on theoretically explaining this sex difference is needed.

26.3.3.9 Binge Drinking

Evidence. Binge drinking is an alcohol-related problem often defined as consuming four to six drinks within a few hours. Précis 26.3.3.9 shows that all of the 30 studies on binge drinking, conducted in a total of nine countries, found that it occurred more in males than in females.

Explanation. See the discussion surrounding Précis 26.3.3.5. As noted directly above, regarding alcohol abuse, the greater exposure of male brains to testosterone could be contributing to several forms of alcohol abuse, including binge drinking.

Assessment. Binge drinking is a USD-M. Theoretically explaining this sex difference is still needed.

26.3.3.10 Driving While under the Influence of Alcohol

Evidence. Driving while under the influence of alcohol is an alcohol related-problem that can be fatal to oneself and to others. Précis 26.3.3.10 shows that all of the studies found males drove while under the influence of alcohol more than did females. The studies were conducted in three different countries and published over a 40-year time frame.

Explanation. As noted above, alcohol consumption appears to be positively correlated with circulating testosterone levels (*review*: Erol, Ho et al. 2019). Both prenatal and circulating testosterone levels have also been found associated with risk-taking (see the narrative for Précis 26.3.1.2). Together, these two types of behavior – alcohol consumption and risk-taking – are likely contributors to the probability of driving while under the influence of alcohol.

Assessment. We deem driving while under the influence of alcohol to be a USD-M. Given that males have been shown to be more likely to consume alcohol, and to do so excessively, and to be more prone to take risks, all traits that appear to be promoted by brain exposure to testosterone, ENA theory is a good candidate for theoretically explaining this sex differences.

26.3.3.11 Owning or Possessing Weapons

Evidence. Précis 26.3.3.11 covers studies of weapon ownership. All of the 12 studies found that males were more likely to own weapons than were

Précis 26.3.3.9 Binge drinking (original Table 18.1.5.10)

Age categories, number of studies, and sex	*Consist. score*	*Overviews*	*Countries*	*Time range*	*Non-human*	*Social role theory*			*Evolutionary theory*		*Theological*
						Origin	*Found*	*Biosoc*	*Select*	*ENA*	
Adult 25M All Ages 30M	100.0 100.0	-0-	9	1971–2007	-0-	Silent	Silent	Silent	Silent	Poss-M	Silent

Précis 26.3.3.10 Driving while under the influence of alcohol (original Table 18.2.1.7)

Age categories, number of studies, and sex	*Consist. score*	*Overviews*	*Countries*	*Time range*	*Non-human*	*Social role theory*			*Evolutionary theory*		*Theological*
						Origin	*Found*	*Biosoc*	*Select*	*ENA*	
All Ages 14M	100.0	-0-	3	1968–2008	-0-	Silent	Silent	Silent	Poss-M	Expl-M	Silent

females. This finding was consistent with one literature review. However, it should be noted that all of the studies were conducted in just one country, the United States.

Explanation. No theoretical proposals for *sex differences* in weapons ownership and possessions were located. This may be partly due to the fact that such ownership can be for both aggressive or defensive purposes. We suspect that such behavior is associated with physical aggression and risk-taking, but, again, have no solid empirical basis for this suspicion.

Assessment. Because just one country was sampled in the 12 located studies of sex differences in weapons ownership and possession and no theoretical explanation was offered for the difference, we will make no declaration regarding the USD status of this variable.

26.3.4 Criminality, Near-Criminality, and Victimization

Except in the case of so-called *victimless offenses*, criminality causes harm to other (primarily in the form of physical injury, psychological trauma, and property confiscation). Similarly, traits that will be called *near-criminality* are those that cause harm to others, albeit usually harm that is more strictly emotional in nature. The main examples are known as *bullying* and *sexual harassment*). Attention will be given to the commission of criminal and near-criminal acts as well as to being victimized by these acts in the following section.

26.3.4.1 Officially Identified Criminal Offending in General

Evidence. According to the vast majority of studies, offenses recognized as crimes in formal legal codes are more likely to be committed by males than by females. Précis 26.3.4.1 shows that only two out of 191 studies failed to reach this conclusion. The studies have been conducted in 30 countries, along with ten multi-national studies. The studies have been published over a 174-year time frame. Two literature reviews and one meta-analysis also concluded that males committed more officially recognized crimes than did females.

Explanation. We found no scientific research on sex stereotypes in criminality, although, as will be discussed directly below, males are widely thought to be more aggressive than females. Therefore, will elaborate more on the possible effects of stereotyping on sex differences in criminality in the following section. Regarding criminality in general, we will simply note that no specific stereotype regarding any sex difference in *overall* criminality was found.

In the case of the evolutionary perspective, much has been published in recent years regarding the idea that males have been sexually selected for

Précis 26.3.3.11 Owning or possessing weapons (original Table 18.2.2.2)

Age categories, number of studies, and sex	*Consist. score*	*Overviews*	*Countries*	*Time range*	*Non-human*	*Social role theory*			*Evolutionary theory*		*Theological*
						Origin	*Found*	*Biosoc*	*Select*	*ENA*	
All Ages 12M	100.0	M: 1Rev	1	1991–2015	-0-	Silent	Silent	Silent	Silent	Silent	Silent

Précis 26.3.4.1 Officially identified criminal offending in general (original Table 19.1.1.1)

Age categories, number of studies, and sex	*Consist. score*	*Overviews*	*Countries*	*Time range*	*Non-human*	*Social role theory*			*Evolutionary theory*		*Theological*
						Origin	*Found*	*Biosoc*	*Select*	*ENA*	
Adolescent 100M 2x Adult 36M Wide Age Range 53M All Ages 189M 2x	98.0 100.0 100.0 99.0	M: 2Rev M: 1Met	30 (10)	1842–2016	-0-	Silent	Silent	Silent	Expl-M	Expl-M	Silent

criminality to a greater degree than females (Rowe 1996; Ellis & Walsh 1997; Ellis & Hoskins 2015; Paquette 2015; Wood 2017; Brown & Robb 2022). Some of this research points toward variation in an evolutionary concept known as *mating effort vs. parenting effort* (Perini, Ditzen, Fischvacher & Ehlert 2012). Regarding this concept, one finds species that reproduce by having numerous offspring while spending little time caring for each offspring. In other species (such as humans), the reproductive strategy usually focuses on having just a few offspring in a lifetime, while lavishing care on each one of them for decades.

Not only do *species* vary (on average) along this continuum, but, within species, variation also exists. Some of this variation is associated with sex. Specifically, males appear to exhibit more mating effort (and less parenting effort) than do female (Cabeza De Baca, Figueredo & Ellis 2012; Städele, Roberts et al. 2019). Some researchers have proposed that males who exhibit symptoms of psychopathy and/or persistent criminality throughout most of their lives are on the extreme mating-effort side of the continuum (Rowe 1996; Rowe, Vazsonyi & Figueredo 1997). Supporting this view, a Swedish study indicated that males with the most extensive criminal records had more children than did males in the general population (Yao, Langstrom et al. 2014).

ENA theory extends sexual selection theory to assert that neuroandrogenic factors contribute to criminality. In this regard, most, but not all research has indicated that brain exposure to both prenatal and post-pubertal (circulating) testosterone contribute to criminal tendencies, especially regarding crimes of a violent nature (*review*: Ellis, Farrington, & Hoskin 2019: 353–356). A particularly interesting Dutch study of more than a thousand boys found that their levels of testosterone at age 16 were predictive of their having a criminal record by age 21 (van Bokhoven, Van Goozen et al. 2006).

Assessment. The evidence that males are more criminal than females throughout the world is exceedingly strong, especially when focused on serious offenses that cause physical harm others, making officially designated criminal offending a USD-M. Evolutionary theory seems to provide the best explanation for this universal sex differences. It is also worth noting that research findings having to do with sex differences in psychopathy and antisocial behavior, often closely related to engaging in frequent criminal activity, are also well-documented USD-Ms (see Précis 26.2.6.9).

26.3.4.2 Officially Identified Violent Offending (Usually Excluding Homicide)

Evidence. Précis 26.3.4.2 examines the category of violent crimes, except homicide specifically. It shows that all 97 of these studies found violent crimes (primarily various forms of assault) are more often committed by males than by females. Twelve countries were samples, plus 11 multi-national studies that were published over a 54-year time frame. Furthermore, all six reviews of the literature on this topic came to the same conclusion.

Explanation. As noted in the section pertaining to criminality in general (directly above), no evidence was found that males are *stereotyped* as being more criminal than females. However, males are stereotyped as being more aggressive (Table 21.2.7.5). From this stereotype, one might use social role theory to infer that males would be more criminally violent. Such a belief concurs with a criminological theory known as *labeling theory*. In essence, this theory attributes criminal behavior to people stereotyping others as being criminal based on traits (such as their sex) or on past behavior. According to labeling theory, these stereotypes (or labels) can actually *cause* people to see themselves as "criminal" and/or cause them to be *treated* by others as criminals. In both cases, according to this theory, the chances of committing crimes in the future are increased (Wellford 1975; Paternoster & Iovanni 1989).

Social role theory resembles labelling theory except that the former is usually specific to sex differences. In the case of sex differences in criminality, both theories lead one to expect that males would be more criminal than females because of stereotypes (or labels). If countries could be found in which females engage in more crime than males, especially serious violent forms, one would expect stereotypes (or labels) to be the opposite of what all of the evidence in this précis suggests in the case. The fact that no such exceptions were found casts doubt on social role theory, especially its original version.

Regarding the founder effect version of the theory, it's relevance to sex differences in violent forms of criminality is questionable. This is because the earliest human society would not have yet developed criminal laws or a criminal justice system to enforce these laws.

According to the biosocial version of social role theory, sex differences in the commission of violent criminality should be the result of (a) males being larger and stronger and/or (b) females being the only sex capable of bearing offspring (Wood & Eagly 2012: 56). The fact that physical strength could enhance an individual's chances of causing bodily harm would be consistent with this line of reasoning.

Turning to the two evolutionary theories, we noted in the preceding narrative that many scientists have argued in recent decades that several forms of criminal behavior, especially those of a violent nature, may be comprehensible in evolutionary terms. This is especially true for sex differences in violent criminal behavior. A fundamental argument in these proposals has been that violent crime can be used not only to acquire resources, but also to gain sexual access to reluctant victims and fend off

Précis 26.3.4.2 Officially identified violent offending (usually excluding homicide) (original Table 19.1.1.2)

Age categories, number of studies, and sex	*Consist. score*	*Overviews*	*Countries*	*Time range*	*Non-human*	*Social role theory*			*Evolutionary theory*		*Theological*
						Origin	*Found*	*Biosoc*	*Select*	*ENA*	
Adolescent 15M Adult 14M Wide Age Range 68M All Ages 97M	100.0 100.0 100.0 100.0	M: 6Rev	12 (11)	1963–2017	-0-	Contra	NoTest	Poss-M	Expl-M	Expl-M	Silent

rivals for potential mates (Ellis & Walsh 1997; Thornhill & Palmer 2001; Quinsey 2002; Duntley & Shackelford 2008; Durrant & Ward 2015). As noted elsewhere, females choose mates to a substantial degree based on evidence of their resource acquiring abilities, which in turn puts evolutionary pressure on males to either comply with those female preferences or find other ways to pass their genes onto future generations (A Campbell, Muncer & Bibel 2001; Ellis & Hoskin 2015).

The ENA version of evolutionary theory asserts that brain exposure to testosterone promotes violent behavior. Based on samples drawn primarily from saliva or blood samples, the vast majority of studies have found circulating testosterone levels positively correlated with assaultive behavior, at least among males (Tarter, Kirisci et al. 2009; Armstrong, Boisvert et al. 2022; *review*: Ellis, Farrington & Hoskin 2019: 356).

It is worth adding that some studies have compared violent male criminals, not to males in general, but to other male criminals (e.g., mainly those involved in property and drug offenses). These studies have indicated that circulating testosterone is higher among those who have been involved in violent offenses than those involved in non-violent offenses (Dabbs, Jurkovic & Frady 1991; Dabbs, Carr et al. 1995; Brooks & Reddon 1996). A study of testosterone among prison inmates compared the average ratings they received from fellow inmates in terms of "toughness." This study concluded that the toughest inmates had higher saliva testosterone than those considered least tough (Dabbs, Frady et al. 1987). All in all, research points toward brain exposure to testosterone promoting violent forms of criminal behavior that result in arrest and conviction. (Findings having to do with testosterone and self-reported criminality are reported in a few précises below.)

One more aspect of testosterone that deserves attention in regard to criminality, especially of a violent nature, involves androgen receptors (ARs). These receptors are entirely under genetic control (Tirabassi, Cignarelli et al. 2015). In essence, in order for testosterone to influence bodily functioning, including that of the brain, there must be ARs located within the body of cells to lock onto androgen molecules. The more receptors there are, the more influence testosterone can have on brain functioning, and, thereby, on behavior. Some studies have indicated that genes regulating these receptors (known as AR CAG repeats) are associated with increased criminal behavior, especially for violent crimes (Aromaki, Lindman & Eriksson 1999; Cheng, Hong et al. 2006; Rajender, Pandu et al. 2008). While the average number of AR CAG repeats appears to be the same for males and females, these repeats can be much more consequential for males because males have much higher levels of testosterone and other androgens to lock into the available androgen receptors (see Précis 26.1.3.2).

Assessment. Based on data compiled by law enforcement, violent forms of crime (not usually including homicides) are a well-established USD-M. Both evolution-based theories seem to offer the strongest explanations for this sex difference. The next two entries will show that this same sex difference also holds true for officially registered homicide and homicide followed by suicide.

26.3.4.3 Officially Identified Homicide Perpetration

Evidence. Homicide is such a definitive type of violent crime that statistics are maintained regarding its occurrence in nearly all countries (Ellis, Farrington & Hoskin 2019:3). Findings having to do with sex differences in the commission of homicide are summarized in Précis 26.3.4.3. One can see that all 101 studies agree that males committed more homicides than do females. Twenty-five specific countries were sampled, along with 22 multi-national studies. The publication of these studies spanned a 70-year time frame.

Explanation. Because homicide is considered a form of violent crime, one can refer back to Précis 26.3.4.2 for theoretical explanations for sex differences in homicide. The narrative around this précis indicates that the two versions of evolutionary theory provide the most persuasive explanations.

Assessment. Homicide perpetration can be considered a USD-M with a high degree of confidence. As with violent crime more generally, the strongest explanations for this sex difference appears to be evolutionary in nature.

26.3.4.4 Officially Identified Homicide Followed by Suicide

Evidence. Some criminological research has conducted separate analyses of homicide perpetration followed in short order by suicide. These offenses are particularly common when family members are murdered.

Regarding sex differences in the commission of these homicides, Précis 26.3.4.4 shows that all but one of 29 studies concluded that males were significantly more likely to be the perpetrator. Findings from these studies were based on samples drawn from 12 countries plus one multi-national study. These studies were published over a 46-year time period.

Explanation. To theoretically explain sex differences in homicide followed by suicide, we will assume that the arguments provided for Précis 26.3.4.2 are relevant. Nevertheless, the fact that suicide is a stipulated part of this category of homicides presents an unusual challenge for understanding this type of violent crime from an evolutionary perspective. Regarding the ENA version of evolutionary theory, brain exposure to testosterone appears to contribute to both homicide (*review*: Ellis, Farrington & Hoskin 2019:353–357) and violent impulsive forms of suicide (Kavoussi,

Précis 26.3.4.3 Officially identified homicide perpetration (original Table 19.1.1.3)

Age categories, number of studies, and sex	*Consist. score*	*Overviews*	*Countries*	*Time range*	*Non-human*	*Social role theory*			*Evolutionary theory*		*Theological*
						Origin	*Found*	*Biosoc*	*Select*	*ENA*	
All Ages 101M	100.0	-0-	25 (22)	1950–2020	-0-	Contra	NoTest	Poss-M	Expl-M	Expl-M	Silent

Précis 26.3.4.4 Officially identified homicide followed by suicide (original Table 19.1.1.4)

Age categories, number of studies, and sex	*Consist. score*	*Overviews*	*Countries*	*Time range*	*Non-human*	*Social role theory*			*Evolutionary theory*		*Theological*
						Origin	*Found*	*Biosoc*	*Select*	*ENA*	
All Ages 28M 1x	96.6	M: 1Rev	12 (1)	1965–2011	-0-	Silent	Silent	Silent	Poss-M	Poss-M	Silent

Armstead & Coccaro 1997; G Lombardo 2021; Sher 2021). For this reason, the influence of this hormone may be pertinent to this particular form of criminality.

Assessment. While homicides that are followed by suicide are relatively rare, they have been studied in many countries, and can be considered a USD-M. Theoretical understanding of why males commit these acts more than do females is in need of further development.

26.3.4.5 Officially Identified Sex Offenses (Except Prostitution)

Evidence. Précis 26.3.4.5 summarizes findings regarding sex differences in the commission of sex offenses, including sexual assault (usually assessed on the basis of court convictions), albeit excluding prostitution. All of the 18 studies concluded that perpetrators were more likely to be males rather than females. Nine countries were sampled in these studies, along with one multi-national study. The studies were published over a 26-year time frame. Our findings are also consistent with one review of the literature and one meta-analysis.

Explanation. We found no evidence of sex stereotypes specifically pertaining to being involved in sex offenses. Therefore, the three social role theories are all listed as being silent to offering an explanation for males-female differences in the commission of (or conviction for) sex offenses.

Evolutionary theories have been frequently invoked to account for sex differences in these types of offenses. As noted elsewhere in this chapter, when compared to females, males on average in all societies appear to have stronger sex drives (see Précis 26.2.7.6), would like to have a greater number of sex partners (see Précis 26.2.7.7), and have less of a desire to be in love before having sex (see Précis 26.2.7.20). According to sexual selection theory, this is because males have a higher reproductive ceiling than females, given the time needed to gestate each offspring that can only be performed by females (West-Eberhard 1983; Wilson, Daly et al. 1996; Ecuyer-Dab & Robert 2004). Given this fundamental sex differences in reproductive potential, it is predictable that sizable numbers of males will resort to devious tactics (including the use of force) to access sex partners (Ellis 1991; Thornhill & Palmer 2001).

To the above evolutionary argument, ENA theory stipulates that brain exposure to androgens should be associated with an increased probability of sex offending. Some support for this assertion has been found, both regarding fetal exposure (Kruger, Sinke et al. 2019) and circulating levels (Aromaki, Lindman & Eriksson 2002; Giotakos, Markianos et al. 2004; Studer, Aylwin & Reddon 2005). Basically, when compared to males in general, sex offenders appear to have somewhat higher levels of testosterone, although, when they are compared to those who have committed

Précis 26.3.4.5 Officially identified sex offenses (except prostitution) (original Table 19.1.1.10)

Age categories, number of studies, and sex	*Consist. score*	*Overviews*	*Countries*	*Time range*	*Non-human*	*Social role theory*			*Evolutionary theory*		*Theological*
						Origin	*Found*	*Biosoc*	*Select*	*ENA*	
All Ages 18M	100.0	M: 1Rev M: 1Met	9 (1)	1994–2020	-0-	Silent	Silent	Silent	Expl-M	Expl-M	Silent

other serious offenses, particularly those of a violent nature, there are no significant differences (Wong & Gravel 2018). In addition, several studies have indicated that the risk of recidivating by sex offenders appears to be diminished by administering testosterone-suppressing drugs to parolees following prison release (*review*: Jordan, Fromberger et al. 2011).

Assessment. The overall conclusion is that the commission of officially defined sex offending (excluding prostitution) is a USD-M. Of course, sex offenses that are not officially detected should also be more common among males (as will be discussed in précises below). The best theoretical explanations seem to be evolutionary in nature.

26.3.4.6 *Officially Identified Drug Use/Possession/Sale*

Evidence. The use, possession, or sale of certain drugs are often a formally designated crime. As shown in Précis 26.3.4.6, this crime was found to be committed more by females than males in every study. The research was conducted in ten different countries and published over a 45-year time frame.

Explanation. No stereotype studies could be located pertaining to sex differences in illegal drug use, possession, or sale. Also, no specific attempt to explain such differences using social role theory were found.

Unless the use, possessing and sale of illegal drugs are thought of forms of risk-taking, it is also difficult to conceive of sexual selection theory as having a bearing on such behavior. In the case of ENA theory, it would predict that those who use, possess, and sell illegal drugs, should have had relatively high brain exposure to testosterone. Several studies have found support for this prediction by indicating that illegal drug use and/or possession are positively correlated with circulating testosterone levels, at least among males (Udry 1990; Reynolds, Tarter et al. 2007; Kirillova, Vanyukov et al. 2008; Tarter, Kirisci et al. 2007; Tarter, Kirisci et al. 2009).

Assessment. We conclude that the use, possession, and sale of illegal drugs (collectively known as drug offenses) are a USD-M. The most promising theory for explaining the universality of these offenses is ENA theory.

26.3.4.7 *Self-Reported Offending in General*

Evidence. Asking research participants to self-report any offenses they may have committed (nearly always anonymously) is a widely used alternative to gaining information on criminal and delinquent behavior from official sources. Précis 26.3.4.7 provides a summary of sex differences in all forms of self-reported crimes. It shows that all but six of 233 studies, conducted in 47 different countries along with seven multi-national studies, conduced that males self-report more criminal and delinquent behavior than do females. The studies were published over a 74-year time frame.

Précis 26.3.4.6 Officially identified drug use/possession/sale (original Table 19.1.1.11)

Age categories, number of studies, and sex	*Consist. score*	*Overviews*	*Countries*	*Time range*	*Non-human*	*Social role theory*			*Evolutionary theory*		*Theological*
						Origin	*Found*	*Biosoc*	*Select*	*ENA*	
All Ages 23M	100.0	-0-	10	1973–2018	-0-	Silent	Silent	Silent	Silent	Expl-M	Silent

Précis 26.3.4.7 Self-reported offending in general (original Table 19.1.2.1)

Age categories, number of studies, and sex	*Consist. score*	*Overviews*	*Countries*	*Time range*	*Non-human*	*Social role theory*			*Evolutionary theory*		*Theological*
						Origin	*Found*	*Biosoc*	*Select*	*ENA*	
Adolescent 196M 5x Wide Age Range 15M All Ages 227M 6x	97.5 100.0 97.4	-0-	46 (7)	1947–2021	-0-	Silent	Silent	Silent	Expl-M	Expl-M	Silent

Explanation. No specific stereotypes regarding sex differences in criminal and delinquent behavior were found. Nor did we locate any proposals from advocates of social role theory to use this perspective as a way to explain sex differences in such behavior. The only possible exception involves a criminological theory known as *labelling theory*. If used to explain sex differences in criminal behavior, it would argue that the primary way of preventing males from dominating in the commission of crimes is to treat them as though they are *not* more likely than females to commit crime (Chiricos, Barrick et al. 2007).

In the case of the evolutionary perspective, sexual selection theory asserts that males should reproductively benefit more than females by engaging in crime (including delinquency). As noted earlier regarding sex differences in official crime (see Précis 26.3.4.1), a substantial number of social scientists have argued along these lines in recent decades (Daly & Wilson 1997; Ellis & Walsh 1997; Quinsey 2002; Duntley & Shackelford 2008; Durrant & Ward 2015). Most have based their argument on evidence that males appear to have been sexually selective for being competitive risk-takers to a greater degree than females (Daly & Wilson 1997), followed by noting that male gain more mating opportunities than females do by competing over resources and warding off rivals, whether this competition fits within legal boundaries or not (Ellis 2005; A Campbell, Muncer & Bibel 2001).

Regarding ENA theory, several studies have found positive correlations between self-reported offending and exposure to testosterone, both in terms of prenatal exposure (Hoskin & Ellis 2015; Hoskin & Ellis 2021), and post-pubertal exposure (Van Bokhoven, Van Goozen et al. 2006). Most of these studies have indicated that the correlations are stronger when focusing on violent types of offenses than on property or drug offenses (Van Bokhoven, Van Goozen et al. 2006; Hoskin & Ellis 2021).

It is also worth mentioning that some of the research points toward testosterone making a contribution to aggressive criminality (or to physical aggression in general) when it interacts with cortisol (a stress hormone). Specifically, some people exhibit what is known as a *blunted cortisol reactivity response to stress* (Liu, Ein et al. 2017), a lifelong condition that appears to be influenced by testosterone, albeit in complex ways (Josephs, Cobb et al. 2017). Studies have indicated that, when this blunted reactivity is combined with high post-pubertal testosterone, males may be especially prone toward criminal/antisocial behavior (Portnoy, Raine et al. 2015; Fairchild, Baker & Eaton 2018; Bernard, Ackermann et al. 2022).

Assessment. The evidence is very strong that, as with official crime and delinquency evidence, self-reported offending is a USD-M. The theoretical explanations of an evolutionary nature appear to be the strongest.

26.3.4.8 Self-Reported Violent Offending

Evidence. Findings from studies of self-reported violent offending, sometimes excluding sexual assaults are summarized in Précis 26.3.4.8. It shows that 63 of the 64 studies concluded that these offenses were more often committed by males than by females. The remaining one study did not find a sex differences. Nine different countries were sampled in these studies (along with four multi-national studies), with publications occurring over a 40-year time frame.

Explanation. As noted in the narrative directly above, there is substantial evidence supporting both versions of evolutionary theory in terms of explaining sex differences in self-reported offending. This appears to be especially strong regarding violent offenses (Van Bokhoven, Van Goozen et al. 2006; Hoskin & Ellis 2021; Armstrong, Boisvert et al. 2022). Therefore, the studies summarized here can also be considered substantially explained in evolutionary terms.

Assessment. Self-reported violent offending, as with officially identified violent offending, can be considered a USD-M without any serious doubt. Evolutionary explanations appear to be helpful for understanding this sex difference.

26.3.4.9 Self-Reported Property Offending

Evidence. As shown in Précis 26.3.4.9, 68 studies of self-reported property offending were found, all but four of which indicated that males self-reported committing such acts more than was the case for females. Nevertheless, only amount adults (usually college students) were the consistency scores beyond the 95.0 cutoff. The research was conducted in 11 different countries plus seven multi-national studies. The publication of these studies occurred over a 93-year period of time. One literature review also concluded that more property offending was self-reported by males than by females.

Explanation. The two preceding précises on self-reported offending (one on all offending and the other specific to violent offending) both indicated that males are more likely to be involved. Therefore, it is rather predictable that males would be more often involved in committing property crimes, as shown here.

Furthermore, the explanations provided for self-reported offending in general (see Précis 26.3.4.7) can be applied here. Most, if not all, of the explanations appear to be evolutionary in nature. They involve noting that, because males are more likely than females to be chosen as mates if they have the ability to acquire resources – regardless of the methods used – they will have been sexually selected for using almost any means to obtain resources (Ellis 2005; Kanazawa 2008).

Précis 26.3.4.8 Self-reported violent offending (original Table 19.1.2.2)

Age categories, number of studies, and sex	*Consist. score*	*Overviews*	*Countries*	*Time range*	*Non-human*	*Social role theory*			*Evolutionary theory*		*Theological*
						Origin	*Found*	*Biosoc*	*Select*	*ENA*	
Adolescent 46M 1x Wide Age Range 10M All Ages 63M 1x	97.9 100.0 98.4	-0-	9 (4)	1979–2019	-0-	Silent	Silent	Silent	Expl-M	Expl-M	Silent

Précis 26.3.4.9 Self-reported property offending (original Table 19.1.2.3)

Age categories, number of studies, and sex	*Consist. score*	*Overviews*	*Countries*	*Time range*	*Non-human*	*Social role theory*			*Evolutionary theory*		*Theological*
						Origin	*Found*	*Biosoc*	*Select*	*ENA*	
Adolescent 35M 2x Adult 21M Wide Age Range 8M 2x All Ages 64M 4x	(94.6) 100.0 (80.0) (94.1)	M: 1Rev	11 (7)	1926–2019	-0-	Silent	Silent	Silent	Expl-M	Expl-M	Silent

Regarding ENA theory, brain exposure to testosterone should promote property offending, given that theft is one way to acquire resources. In other words, while testosterone should promote violent criminality the most, it should also promote property offending. Evidence bearing on this hypothesis has been supported, at least regarding prenatal testosterone exposure (Hoskin & Ellis 2015). In the case of circulating testosterone, a direct positive relationship with self-reported violent criminal behavior was found but the relationship between "income generating crime" and testosterone was only significant after controlling for levels of the stress hormone, cortisol (Armstrong, Boisvert et al. 2022).

Assessment. At least in the case of adults, self-reported property offenses were significantly more prevalent among males than females. The strongest theoretical explanations appear to be evolutionary in nature.

26.3.4.10 Self-Reported (or Victim-Identified) Sexual Assault

Evidence. Sex differences in the commission of sexual assault has been studied by self-report and/or victim-report (as well as by official data, as reviewed earlier). When these two types of studies are combined, all of the studies indicate that males committed sexual assault more often than females. These studies were conducted in seven different countries and were published over a 37-year time frame. One literature review also concluded that males were more likely to be the offender in these assaults.

Explanation. As was noted earlier regarding official data on the commission of sexual assault and rape (see Précis 26.3.4.5), the best explanations for why males are universally more likely to engage in sexual assault is evolutionary in nature. To briefly re-characterize this line of reasoning, males have evolved all forms of behavioral tactics for accessing mating opportunities. For a substantial minority of males, these tactics involve the use of force as well as being very pushy and pestering (Ellis 1991a).

At a neurohormonal level, testosterone appears to contribute to the use of these tactics by enhancing the sex drive (see Précis 26.2.7.6). The tendency for testosterone to promote risk-taking (see Précis 26.3.1.2) could also promote sexual assault. Since male brains are exposed to more testosterone than are females, their sex drive and risk-taking tendencies should be stronger (again see Précis 26.2.7.6 and Précis 26.3.1.2).

Assessment. The evidence is strong that self-reported (or victim-identified) sexual assault is a USD-M. This evidence is bolstered by data compiled by the criminal justice system all over the world.

26.3.4.11 Age of Onset of Illegal Drug Use

Evidence. Précis 26.3.4.11 summarizes findings having to do with sex difference in the onset of illegal drug use. It shows that all 12 of the located

Précis 26.3.4.10 Self-reported (or victim-identified) sexual assault (original Table 19.1.2.4)

Age categories, number of studies, and sex	*Consist. score*	*Overviews*	*Countries*	*Time range*	*Non-human*	*Social role theory*			*Evolutionary theory*		*Theological*
						Origin	*Found*	*Biosoc*	*Select*	*ENA*	
Adult 14M All Ages 18M	100.0 100.0	M: 1Rev	7	1982–2021	-0-	Silent	Silent	Silent	Expl-M	Expl-M	Silent

Précis 26.3.4.11 Age of onset illegal drug use (Original Table 19.1.3.3)

Age categories, number of studies, and sex	*Consist. score*	*Overviews*	*Countries*	*Time range*	*Non-human*	*Social role theory*			*Evolutionary theory*		*Theological*
						Origin	*Found*	*Biosoc*	*Select*	*ENA*	
All Ages 12M	100.0	-0-	1	1985–2001	-0-	Silent	Silent	Silent	Silent	Silent	Silent

studies found that males began using illegal drugs at a younger age than did females. However, it should be noted that all of these studies were obtained in just one country, i.e., the United States, and were published over a fairly short 16-year time period.

Explanation. Why would males be prone to begin using illegal drugs sooner than females? None of the theories of sex differences seem to directly address this issue unless one assumes that using illegal drugs is a form of risk-taking. With this assumption, one could predict that males would be more likely to use illegal drugs than females. We suspect that sex differences in illegal drug use is more complex than simply reflecting risk-taking, and, therefore, will not try to develop it here, but interested readers can refer to information on sex differences in risk-taking in Précis 26.3.1.2.

Assessment. Because just one country was sampled over a limited time period, and there seems to be no way to explain this sex difference, we will not declare the age of onset for illegal drug use to be a USD.

26.3.4.12 Perpetration of Violent Crime Based on Victim Reports

Evidence. Beginning in the 1970s, national surveys began to be conducted to help assess crime victimization in a way that bypassed (and thereby helped to verify) official crime statistics. These surveys have come to be known as *victimization surveys*. In them, a representative sample of several thousand people are asked if they were victimized by a crime (usually in a prescribed time frame such as the past six months). For those who answer yes, the interviewer usually asks if they were able to identify the sex of the perpetrator (particularly in the case of assaults and robberies).

As shown in Précis 26.3.4.12, ten findings point toward males being perpetrators of crime to a greater degree than females. This conclusion was also reached by a meta-analysis on the topic. The studies were conducted in four countries and a 21-year time period.

Explanation. The narrative surrounding Précis 26.3.4.2 has already presented a review of evidence having to do with why males would be universally more involved in violent crime than are females (also see Précis 26.3.4.8). For this reason, readers may refer to this précis to see why the strongest evidence seems to support evolutionary arguments.

Assessment. Persons involved in committing violent crime (based on victim reports) appears to be a USD-M. Evolutionary theory appears to provide the best explanation for this universal sex difference.

26.3.4.13 Cruelty and Sadistic Behavior

Evidence. Cruel and sadistic behavior comes in many forms, i.e., both physical and psychological, and can be imposed not only on humans but on other animals (such as pets). Précis 26.3.4.13 summarizes findings from

Précis 26.3.4.12 Perpetration of violent crime based on victim reports (original Table 19.1.4.1)

Age categories, number of studies, and sex	*Consist. score*	*Overviews*	*Countries*	*Time range*	*Non-human*	*Social role theory*			*Evolutionary theory*		*Theological*
						Origin	*Found*	*Biosoc*	*Select*	*ENA*	
All Ages 10M	100.0	M: 1Met	4	1990–2011	-0-	Contra	NoTest	Poss-M	Expl-M	Expl-M	Silent

13 studies of sex differences in cruel and sadistic behavior. One can see that these studies unanimously agree that such behavior is more common among males than it is among females. Samples for these studies came from four countries, published over a 49-year time frame. One literature review also concluded that males engaged in more cruel and sadistic behavior.

Explanation. No stereotype study was located regarding sex differences in traits such as cruelty and being sadistic. However, there is evidence that females are stereotyped as being more concerned about others, more loving, and more soft-hearted (see Tables 21.2.4.6, 21.2.6.18 & 21.2.6.23 in Volume III). All three of these stereotypes can be considered as being near opposites of cruelty and sadism. Consequently, at least the biosocial version of social role theory could explain why males would be more prone toward cruelty and sadism for the same reason that females are more concerned about others, more loving, and soft-hearted. Specifically, because women are expected and trained to care for offspring more than males are, and males are more muscular, people may stereotype both boys and men to be less concerned about the welfare of others.

From an evolutionary perspective, the apparent universal sex difference in cruelty and sadism could be considered an extreme expression of traits such as competitiveness and criminal aggression, both of which have been found to be more common among males than females (see Précis 26.3.1.5 and Précis 26.3.4.2). Males also appear to be more antisocial and psychopathic (see Précis 26.2.6.9) as well as being more accepting of the use of military power to settle international disputes (see Précis 26.3.15.53). When all of these sex differences are considered together, they point toward males being less concerned with doing harm to others.

Sexual selection theory offers a way to explain such sex differences. Specifically, males who are in the low (female-typical) range of having concern for harming others will have fewer offspring (perhaps because they are less often chosen as mating partners) than males in the high (male-typical) range. This prediction obviously needs to be tested. However, if it is confirmed, the ranges for both sexes are likely to be normally distributed. This would mean that many more males than female will be at the extremely high male-typical end of the continuum. In other words, a greater minority of males will be very unconcerned about harming others, all but assuring that they will be more prone toward cruel and sadistic behavior.

In the case of ENA theory, the sexual selection argument provided above would be merged with the assertion that cruel and sadistic behaviors are promoted by exposing the brain to high testosterone levels. Examination of the narratives for the précises cited two paragraphs above for competitiveness, violent criminality, and antisocial behavior all indicate that these traits are promoted by high testosterone exposure.

Précis 26.3.4.13 Cruelty and sadistic behavior (original Table 19.2.1.3)

Age categories, number of studies, and sex	*Consist. score*	*Overviews*	*Countries*	*Time range*	*Non-human*	*Social role theory*			*Evolutionary theory*		*Theological*
						Origin	*Found*	*Biosoc*	*Select*	*ENA*	
All Ages 13M	100.0	M: 1Rev	4	1971–2020	-0-	Contra	NoTest	Poss-M	Poss-M	Expl-M	Silent

Assessment. Substantial evidence suggests that cruel and sadistic behavior are USD-Ms. Evolutionary theory appears to offer the most promising explanations.

26.3.4.14 *Commission of Domestic Violence (Spousal Violence) with Serious Injury*

Evidence. Findings regarding sex differences in the commission of minor forms of domestic (usually spousal) violence have been mixed (Table 19.2.1.6). However, as shown in Précis 26.3.4.14, 13 studies of violence resulting in serious injury (i.e., usually requiring medical treatment) have all agreed that offenders are mostly males. Pertinent studies come from two countries that were published over a 27-year time frame. A meta-analysis reached the same conclusion.

Explanation. The apparent universal tendency for males to commit more domestic violence resulting in serious injury (not those resulting in little to no injury) could be explained by the biosocial version of social role theory. This is in light of the theory stipulating that at least some behavioral sex differences are universal because of universal sex differences in physical traits such as physical strength (Wood & Eagly 2012).

Various proposals along evolutionary lines have been made regarding sex differences in the commission of domestic violence. Most of these have to do with males having greater risks than females of rearing non-biological offspring (Tracy & Crawford 2019).

Regarding ENA theory, findings having to do with any relationships between circulating androgen levels and domestic violence have been quite mixed, but seem to provide little support for any simple connection (Lindman, von der Pahlen et al. 1992; von der, Sarkola et al. 2002). One possible exception is when high androgen levels are combined with high alcohol consumption (McKenry, Julian & Gavazzi 1995; Soler, Vinayak & Quadagno 2000; Daly 2017).

Assessment. While data from more than two countries would be desirable before making a firm decision, the commission of domestic violence resulting in serious injury appears to be a USD-M. Theoretically explanations of this sex difference are not yet well developed and tested.

26.3.4.15 *Recipient of Teacher Discipline or School Discipline*

Evidence. Twenty-three studies of sex differences in tendencies to be a discipline problem in school were located. Précis 26.3.4.15 shows that all of these studies concluded that males were more likely to be disciplined by teachers or other school officials than were males. The studies were conducted in two two countries over a 50-year time frame.

Précis 26.3.4.14 Commission of domestic violence (spousal violence) with serious injury (original Table 19.2.1.7)

Age categories, number of studies, and sex	*Consist. score*	*Overviews*	*Countries*	*Time range*	*Non-human*	*Social role theory*			*Evolutionary theory*		*Theological*
						Origin	*Found*	*Biosoc*	*Select*	*ENA*	
All Ages 13M	100.0	M: 1Met	2	1980–2007	-0-	Silent	Silent	Poss-M	Poss-M	Poss-M	Silent

Précis 26.3.4.15 Recipient of teacher discipline or school discipline (original Table 19.2.2.2)

Age categories, number of studies, and sex	*Consist. score*	*Overviews*	*Countries*	*Time range*	*Non-human*	*Social role theory*			*Evolutionary theory*		*Theological*
						Origin	*Found*	*Biosoc*	*Select*	*ENA*	
Child 11M	100.0	-0-	2	1952–2002	-0-	Silent	Silent	Silent	Expl-M	Expl-M	Silent
Adolescent 10M	100.0										
All Ages 23M	100.0										

Explanation. Elsewhere in this chapter, it has been shown that males are more prone to be physically active (see Précis 26.2.7.6). The tendency for testosterone to promote risk-taking (see Précis 26.3.1.6) and to exhibit ADHD symptoms (see Précis 26.2.6.22). They are also more likely to curse and swear (see Précis 26.3.2.14), and to be delinquent (see Précis 26.3.4.7). All of these forms of behavior, when exhibited in school are likely to result in some form of discipline by teachers or other school officials.

As documented by their respective précises, the two evolutionary theories offer better explanations for males exhibiting more physically active, ADHD, cursing and swearing, and delinquent. Therefore, these theories also appear to be helpful for explaining why males are more likely to be disciplined in school than is the case for females.

Assessment. Even though it would be desirable to have data from more than two countries, the 100% consistency of the findings, along with the strength of the theoretical explanations, leads to the conclusion that this variable to be a USD-M.

26.3.4.16 Suspension/Expulsion from School

Evidence. When it comes to being suspended or expelled from school, males appear to be the recipients of such extreme punishment more often than females. As shown in Précis 26.3.4.16, the relevant studies have been conducted in two countries published over a 28-year time frame.

Explanation. It is worth noting that findings on school suspensions and expulsions complement findings noted directly above regarding sex differences in being the recipient of school discipline. Therefore, we regard theoretical explanations to be essentially the same.

Assessment. Because the number of pertinent studies was just ten and the number of countries sampled were only two, despite a fairly solid theoretical explanation for this sex difference, we will not declare it to be a USD.

26.3.4.17 Unsafe (Reckless) Driving in General

Evidence. Unsafe or reckless driving injures and kills many people. Many studies have been undertaken to determine if such behavior is more common among one sex than the other. Précis 26.3.4.17 reveals that 22 studies have sought to address this issue, all of which have concluded that males engaged in dangerous driving more than females. This research was conducted in 11 countries plus one multi-national study. A meta-analysis also found the same sex difference.

Explanation. Stereotype research has indicated that males are thought to take more risks than are females (Table 21.2.4.7). Based on this evidence, one could argue from the perspective of social role theory that males would be more likely to exhibit unsafe driving. The original version

Précis 26.3.4.16 Suspension/expulsion from school (original Table 19.2.2.3)

Age categories, number of studies, and sex	*Consist. score*	*Overviews*	*Countries*	*Time range*	*Non-human*	*Social role theory*			*Evolutionary theory*		*Theological*
						Origin	*Found*	*Biosoc*	*Select*	*ENA*	
Adolescent 10M	100.0	-0-	2	1986–2014	-0-	Silent	Silent	Poss-M	Expl-M	Expl-M	Silent

Précis 26.3.4.17 Unsafe (reckless) driving in general (original Table 19.2.2.5)

Age categories, number of studies, and sex	*Consist. score*	*Overviews*	*Countries*	*Time range*	*Non-human*	*Social role theory*			*Evolutionary theory*		*Theological*
						Origin	*Found*	*Biosoc*	*Select*	*ENA*	
Adolescent 16M	100.0	M: 1Met	11 (1)	1973–2019	-0-	Contra	Contra	NoTest	Expl-M	Expl-M	Silent
All Ages 22M	100.0										

of social role theory would predict that sex differences in unsafe driving is due to males learning to reflect these stereotypes. However, the original version of this theory would not lead one to expect females to be victimized more in all societies (as evidence summarized in Précis 26.3.4.17 suggests is the case). Regarding the founder effect version of social role theory, there is no way to determine if the very first human society had sex differences in risk-taking behavior that was "faithfully passed down thousands of times over" ever since (Fausto-Sterling 1992: 199). However, such an explanation for unsafe driving per se would made no sense given that vehicles were not invented yet.

In the case of the biosocial role theory, attributes universal cognitive and behavioral sex differences to the fact that males are larger and stronger and/or only females can bear offspring. The relevance of any of these variables to sex differences in unsafe driving is hard to grasp.

The two evolutionary theories both offer ways of explaining sex differences in unsafe driving by assuming that such behavior primarily affect risk-taking behavior. As noted in Précis 26.3.1.2, males appear to have been sexually selected for their willingness to take greater risks than females take. This sex difference is partly due to the fact that risk-taking can benefit males more because doing so increases their chances of mating with multiple members of the opposite sex, thereby passing their genes on at relatively high rates (Greitemeyer, Kastenmüller & Fischer 2013; Engqvist, Cordes & Reinhold 2015; Puts 2016).

Regarding the ENA version of sextual selection theory, while there is no evidence linking unsafe driving per se with neurohormonal factors, studies have shown that risk-taking in general is associated with high circulating levels of testosterone (*review*: Apicella, Carré & Dreber 2015). Furthermore, based on 2D:4D finger length, studies have indicated that risk-taking is elevated later in life by high prenatal testosterone exposure, particularly among males (Coates, Gurnel & Rustichini 2009; Stenstrom, Saad et al. 2011; Aycinena, Baltaduonis & Rentschler 2014; Brañas-Garza, Galizzi & Nieboer 2018).

Assessment. Given the sizable number of studies along with the number of countries sampled for these studies, and the fact that it can be theoretically explained, we unsafe driving (along with risk-taking generally) as being a USD-M.

26.3.4.18 Victim of Rape or Sexual Assault

Evidence. Ninety-three studies have examined whether one sex is more likely to be the victim of sexual assault and/or rape. Précis 26.3.4.18 reveals that all of those studies found that more females than males were the victims of such acts. The research samples were obtained from 11 countries (plus one multi-country study) and the dates of publication spanned 55 years.

Précis 26.3.4.18 Victim of rape or sexual assault (original Table 19.3.1.10)

Age categories, number of studies, and sex	*Consist. score*	*Overviews*	*Countries*	*Time range*	*Non-human*	*Social role theory*			*Evolutionary theory*		*Theological*
						Origin	*Found*	*Biosoc*	*Select*	*ENA*	
All Ages 91F	100.0	-0-	11 (1)	1964–2019	-0-	Silent	Silent	Silent	Expl-F	Poss-F	Silent

Explanation. Invoking social role theory to explain the apparent universal sex differences in rape or sexual assault victimization is rather difficult. Perhaps, one could use the biosocial version to argue that, due to the greater size and strength of males, they would be able to impose their desire for sex more than is the case for females.

Evolutionary theorizing seems to provide a more solid foundation upon which to explain why females are more often the victims of rape or sexual assault than males. Such theorizing begins by noting that males can pass more genes onto future generations by having sex with multiple sex partners, whereas the number of sex partners for females usually does little to promote their reproduction (Archer 1996; Pedersen, Miller et al. 2002).

Among the tactics males can use to acquire sex partners is varying degrees of force, and, obviously, the victim must be a female. So, especially before the widespread availability of contraception and abortion, pregnancy would be a fairly common outcome of rape victimization, similar to voluntary sexual intercourse (Holmes, Resnick et al. 1996; Basile, Smith et al. 2018). In other words, males who used assaultive tactics as one of their methods in having sexual intercourse are likely to have more offspring than males who only use non-coercive tactics (Thornhill & Thornhill 1983; Ellis 1991a; Thornhill & Palmer 2001). While rape victimization before puberty or after menopause occurs, the vast majority of rape victims are females between 15 and 35, when pregnancy is most likely to occur (Felson & Cundiff 2014; Conroy & Cotter 2017).

The ENA theory would add to the above reasoning that rape victimization should be associated with neuroandrogenic factors. Evidence in this regard is mixed. Specifically, when rapists are compared to other violent offenders, there appear to be no significant differences, but, when compared to males in general, rapists and other violent offenders appear to be somewhat higher circulating testosterone levels (see Précis 26.3.4.5). Regarding the possibility of androgens affecting females sexual assault *victims*, no evidence was found.

Assessment. The overall conclusion that sexual assault victimization is a USD-F is inescapable. Evolutionary theory seems to offer the best explanation for this sex difference.

26.3.4.19 Victim of Sexual Abuse in General (Excluding Violent Forms)

Evidence. Being the victim of sexual abuse here refers to all non-violent forms of sexual victimization. Among the most common are acts of molestation and incest. Précis 26.3.4.19 summarizes the results of studies concerning the sex of victims of non-violent forms of sexual abuse. It shows that all of the studies, and two literature reviews, concluded that females were more likely to be the victims of non-violent forms of sexual

Précis 26.3.4.19 Victim of sexual abuse in general (excluding violent forms) (original Table 19.3.1.11)

Age categories, number of studies, and sex	*Consist. score*	*Overviews*	*Countries*	*Time range*	*Non-human*	*Social role theory*			*Evolutionary theory*		*Theological*
						Origin	*Found*	*Biosoc*	*Select*	*ENA*	
Chile 10F 1x Adolescent 17F Wide Age Range 12F All Ages 39F 3x	90.9 100.0 100.0 92.9	F: 2Rev	10	1979–2020	-0-	Silent	Silent	Silent	Poss-F	Poss-F	Silent

abuse than were males. The studies were based on samples drawn from five different countries published over a 41-year time frame.

Explanation. As with sexual assault victimization (see directly above), the most persuasive explanation for why females would be victimized more in sexual terms involves the concept of *sexual selection*. Specifically, given that males can produce offspring so much more rapidly than can females, they are said to have a much higher reproductive ceiling (M Daly & Wilson 1980; Ellis, Widmayer & Palmer 2009). But, to realize any of this greater potential (i.e., leaving more offspring in future generations), males need to have sex with multiple females.

One argument that might be made against an evolutionary theory is that quite a few victims of sexual abuse are not of reproductive age (i.e., ~13–45). However, it should be noted that, while sexual abuse does often begin before victims reach puberty, pre-pubertal victims have been found to have more than a fivefold increased probability of being victimized again later in life, either by the initial assailant or by someone else (Papalia, Mann & Ogloff 2021). This suggests that there are often long-term reproductive consequences to being sexually abused that could have a bearing on the reproductive potential of sexual abusers.

In addition to asserting that natural selection has favored more male perpetration as well as more female victimization, ENA theory leads to the hypothesis that brain exposure to testosterone should increase the chances of committing sexual assault, and could even affect the chances of sexual abuse victimization (Ellis 2005). Supporting such reasoning, findings have been mixed, at least regarding the commission of sexual abuse (Studer, Aylwin & Reddon 2005; *review*: JS Wong & Gravel 2018). Regarding the possibility of testosterone elevating the risk of sexual abuse victimization, no research findings were located, although one study did report that single females with high female-typical testosterone tended to have greater interest in sociosexual interactions with males (Edelstein, Chopik & Kean 2011).

Assessment. While more evidence is needed to better understand the phenomenon of sexual abuse victimization from a theoretical standpoint, we found no reason to doubt that it is a USD-F.

26.3.4.20 Victim of Physical Aggression in General (Except Criminal)

Evidence. Précis 26.3.4.20 shows that, among children, boys are more likely to be the victims of non-criminal physical aggression than are girls. However, findings for all other age categories did not have consistency scores in the 95.0 to 100.0 range. The studies were conducted in five countries and were published over a 53-year time frame. Studies of sex differences in non-human physical aggression have produced mixed results, as was the case for post-pubertal humans.

Précis 26.3.4.20 Victim of physical aggression in general (except criminal) (original Table 19.3.2.6)

Age categories, number of studies, and sex	*Consist. score*	*Overviews*	*Countries*	*Time range*	*Non-human*	*Social role theory*			*Evolutionary theory*		*Theological*
						Origin	*Found*	*Biosoc*	*Select*	*ENA*	
Child 11M Adolescent 8M 3x Adult 11M 1x 1F All Ages 30M 4x 1F	100.0 (72.7) (78.6) (83.3)	-0-	5	1966–2019	F: 2 M: 3	Silent	Silent	Silent	Poss-M	Poss-M	Silent

Explanation. Research has indicated that females are *stereotyped* as being more likely to be the victims of aggression than are males (Table 21.2.9.18). As shown in Précis 26.3.4.20, this stereotype is actually the opposite of what most studies of physical aggression have found. For this reason, we will not attempt to apply any of the three social role theories to explaining this variable.

Regarding evolutionary reasoning, the explanation would likely have to do with evidence that male victimization from physical aggression is often the result of males engaging in physical aggression more than females (Table 25.17.16). In other words, a high proportion of physical aggression victims are actually one of the combatants before the injury occurred. If this assumption is true, then one can explain sex differences in physical aggression in essentially the same way as one explains *engaging* in physical aggression. Considerable evidence has suggested that males have been naturally selected for being physically aggressive as a part of their being more competitive with regard to acquiring and controlling resources (Charlesworth & La Freniere 1983; Johnson, Burk & Kirkpatrick 2007).

ENA theory leads to the prediction that, not only does brain exposure to testosterone promote physical aggression, but it also promotes victimization from physical aggression. This predict comes from noting that most victims of physical aggression (excluding that which is sexually motivated) were physically fighting with the assailant before being injured (Cheng, Johnson et al. 2006; RA Powers 2015). Supportive evidence, at least regarding prenatal testosterone, was located. Specifically, two studies concluded that high prenatal exposure (based on 2D:4D measures) was positively correlated with having sustained aggression-induced injuries (Joyce, O'Regan et al. 2017 among males; O'Briain, Dawson et al. 2017 among both sexes).

Assessment. Overall, the evidence of males being the victims of physical aggression is associated with a 95.0+ consistency score only in the case of children. Since, none of the five theories seems to offer any obvious explanation for this specific sex difference, we will not declare non-criminal physical aggression victimization to be a USD.

26.3.4.21 Being Pressured to Have Sex

Evidence. As shown in Précis 26.3.4.21, 17 out of 18 studies have concluded that females are pressured to have sex more often than are males. These studies do not include the actual use of physical force (the latter being covered in Précis 26.3.4.10). However, the relevant studies all came from one country, the United States, that were published over a 32-year time frame.

Explanation. Given the similarity between being pressured to have sex and being sexually assaulted, the arguments surrounding sexual assault

Précis 26.3.4.21 Being pressured to have sex (original Table 19.3.2.11)

Age categories, number of studies, and sex	*Consist. score*	*Overviews*	*Countries*	*Time range*	*Non-human*	*Social role theory*			*Evolutionary theory*		*Theological*
						Origin	*Found*	*Biosoc*	*Select*	*ENA*	
Adolescent 2F 1x Adult 15F All Ages 17F 1x	(66.7) 100.0 94.4	-0-	1	1987–2019	-0-	Silent	Silent	Silent	Expl-F	Expl-F	Silent

victimization can also be applied to being pressured to have sex, albeit to a less extreme degree. In this regard, both sexes should be willing to have sex given that doing so is all but essential for sexual reproduction. However, females appear to have been favored for being more cautious in this regard in order to restrict their sexual activity to a few males who seem willing to help provide resources to any offspring that are produced, while males have been favored for having numerous sex partners (see Précis 26.2.7.7). Among the regrettable consequences of this sex difference is that males tend to be pushier than females when it comes to sexual intercourse.

Another evolutionary argument regarding males using all manners of pressure and force to have sex is that males should primarily victimize females of reproductive age (Ellis 1989; Thornhill & Palmer 2000; Malamuth, Huppin & Bryant 2005). Evidence strongly supports this argument (R Bachman 1998; Rennison 2001; Shi, Lu et al. 2021).

According to the ENA version of evolutionary theory, sex differences in tendencies to pressure others to have sex should be substantially influenced by brain exposure to testosterone and other androgens (Hoskin & Ellis 2015). This hypothesis is consistent with evidence males tend to have stronger sex drives (see Précis 26.2.7.6) and appear to be more physically aggressive (see Table 25.17.16) than females. Evidence supporting this line of reasoning is strong in the case of violent sexual assaults (Longpré, Guay & Knight 2020; also see Précis 26.3.4.10), but no evidence was found specifically regarding pushy (but non-assaultive) forms of pressure to have sex.

Assessment. Being pressured to have sex appears to be a USD-F. However, more research is needed, particularly regarding the number of countries sampled.

26.3.5 Education, Work, Social Status, and Territorial Behavior

Education, work, social status, and territorial behavior have in common the fact that they pertain to how individuals make a living and often stratify themselves in the process of doing so. This section reviewed all of the variables having to do with sex differences in these phenomena with consistency scores of 95.0 or higher.

Among the topics covered are those having to do with the types of occupations males and females tend to occupy. Obviously, sex differences in this regard are likely to vary a great deal, depending on the proportion of each sex who happened to hold jobs. For example, if only 5% of all the employed persons in a particular country (and point in time) are females, one would expect far fewer females in be in the vast majority of occupations than would be the case in countries where females comprise nearly half of the paid workforce. Understandably, the proportion of each sex in the paid workforce is rarely included in studies comparing sex differences in specific occupations. As a result, statements such as "With the entry of women into previously male-dominated occupations it has become obvious that many traditional gender expectations are the product of stereotypes rather than evolution or biology" (Hamel 2020:241) are very difficult to assess.

Similar reasoning should be kept in mind as sex differences in people's college majors are considered. In other words, if only 5% of all college graduates are female, one would expect the proportion of females in nearly all fields of study to be below those of males than in countries where over half of all college students are female. Pertinent to this point, in quite a few countries in recent years, more females than males have been graduating from college (see Table 20.1.2.9). Of course, this means that the proportion of males majoring in male-dominated fields will be substantially reduced as the proportion of college graduates increase, even to the point of surpassing male college graduates.

26.3.5.1 Time Spent Studying and Doing Homework

Evidence. Précis 26.3.5.1 shows that all 20 studies found females spent more time studying and doing homework than males. The studies were conducted in eight countries over 80 years.

Explanation. Why would females spend more time studying than do males? In some ways, it seems to defy evolutionary logic, since the theory asserts that males have been sexually selected for their abilities to provide resources to their families (or future families). What better way to provide resources than to become well educated? Numerous studies have in fact indicated that academic performance is positively correlated with long term financial success (*review*: Ellis, Hoskin, & Ratnasingam 2018:7). Furthermore, studying and doing homework have been shown to be positively correlated with academic performance (Cooper, Lindsay et al. 1998; Kitsantas & Zimmerman 2009).

Despite the evidence that studying and doing homework will help to promote academic performance, and thereby promote the ability to make a good living, other aspects of human evolution may have bearing on sex differences in studying and homework completion. One of these aspects involves physical activity, which nearly all studies have found to be more pronounced among males (see Table 16.4.3.2). Physical activity could lead to difficulty studying. Sex differences in activity levels are especially noticeable when one considers extreme manifestations known as *hyperactivity* or *hyperkinesis* (Tables 14.4.8.1 & 14.4.8.2). In fact, both ADHD and ADD are well established USD-Ms (see Précis 26.2.6.22 and Précis 26.2.6.23).

Another noteworthy sex difference that may serve to lower male tendencies to spend time studying involves language development. While males appear to acquire language skills on par with females by late adolescence,

Précis 26.3.5.1 Time spent studying and doing homework (original Table 20.1.1.4)

Age categories, number of studies, and sex	*Consist. score*	*Overviews*	*Countries*	*Time range*	*Non-human*	*Social role theory*			*Evolutionary theory*		*Theological*
						Origin	*Found*	*Biosoc*	*Select*	*ENA*	
All Ages 21F	100.0	-0-	8	1929–2013	-0-	Silent	Silent	Silent	Poss-F	Poss-F	Silent

nearly all studies of children have found males being slower than females to vocalize and acquire language (Précis 26.2.1.11). Males are also more often diagnosed as having speech and language disorders (Précis 26.2.6.17). Because males are more physically active and are slower to attain comparable language proficiency (both tendencies that seem to have evolutionary neurohormonal roots), it is predictable that females would surpass males in time spent studying and diligently doing homework.

Regarding the possibility that brain exposure to testosterone enhances physical activity levels, the evidence has been mixed, particularly based on studies using the rather crude 2D:4D indicator of prenatal exposure (Lemiere, Boets & Danckaerts 2010; Roberts & Martel 2013; Wang, Chou et al. 2017; Wernicke, Zabel et al. 2020). In the case of delaying language learning, the evidence for neuroandrogenic influences is substantially supportive (Lutchmaya, Baron-Cohen & Raggatt 2001; Hollier, Mattes et al. 2013; Kung, Browne et al. 2016).

Assessment. Overall, time spent studying and doing homework is almost certainly a USD-F. The best theoretical explanations seem to be of an evolutionary nature, although more detailed development is needed, especially regarding activity levels being affected.

26.3.5.2 Majoring (or Taking Advanced Courses) in Computer Science

Evidence. Many studies have assessed sex differences in taking courses, especially advanced courses, in computer science. All 38 studies found that males majored in, or took more courses in, computer science than was true for females. This research included three multi-national studies along with samples drawn from five individual countries.

Explanation. The tendency for males to major in computer science is likely to involve their greater interest in things-oriented occupations rather than people-oriented occupations (see Précis 26.2.7.44) along with their greater visual-spatial reasoning abilities (see Précis 26.2.3.4) and their tendencies to engage in systemizing thought more than females (see Précis 26.2.3.8). All three of these précises lead one to expect males to be drawn to take courses in computer science and to do better in them than females.

In the case of ENA theory, it predicts that brain exposure to testosterone will enhance both interest and ability in things-oriented fields. Supporting this deduction, one study correlated grades in a college course in Java programming with 2D:4D finger length. It indicated that students with the lowest ratios (indicating the highest prenatal testosterone) received significantly higher grades (Brosnan, Gallop et al. 2011).

Assessment. We declare the tendency to major in computer science to be a USD-M. However, more work is needed to further refine theoretical explanations of this sex difference.

Précis 26.3.5.2 Majoring (or taking advanced courses) in computer science (original Table 20.1.4.9)

Age categories, number of studies, and sex	*Consist. score*	*Overviews*	*Countries*	*Time range*	*Non-human*	*Social role theory*			*Evolutionary theory*		*Theological*
						Origin	*Found*	*Biosoc*	*Select*	*ENA*	
Adult 35M All Ages 38M	100.0 100.0	-0-	5 (3)	1983–2017	-0-	Silent	Silent	Silent	Expl-M	Expl-M	Silent

26.3.5.3 Majoring (or Taking Advanced Courses) in Economics

Evidence. Based on findings from 11 studies, Précis 26.3.5.3 indicates that males studied economics more than did females. Research on this matter was conducted in four countries over a 21-year time frame.

Explanation. No evidence was found that males are stereotyped as taking or liking to take courses in economics. The closest any stereotype that was located came involved males being stereotyped as being better than females at math (see Table 21.2.5.11). Incidentally, this stereotype was empirically supported by a majority of studies, albeit not to the point of being a USD (see Table 12.2.3.1).

Given that economics tends to be more heavily laden with mathematics than most other college majors, one would expect a tendency for more males to major in economics. The original version of social role theory would not predict that this pattern would be universal, as is indicated by the evidence. In the case of the founder effect social role theory, its ability to predict more males majoring in economics is seriously undercut by the fact that primordial human societies had no college, let alone college majors in economics. Because economics and mathematics have no discernable connection with body size, strength, or being able to have offspring, the biosocial role theory also does not seem to predict that males would dominate in economics.

Both of the evolutionary theories offer explanations for why males appear to universally be more likely than females to major in economics. Specifically, because economics relies heavily on statistics and other forms of mathematics, it tends to be more things-oriented than people-oriented, and to do so in systemizing ways. As noted in Précis 26.2.7.44, males appear to have been sexually selected for being relatively things-oriented, and, in Précis 26.2.3.8, evidence is summarized showing that males are more prone to systemizing thinking than are females.

Regarding the ENA version of evolutionary theory, no evidence directly linking brain exposure to testosterone with interest in or majoring in economics was found. However, there is some evidence suggesting an association between interests in mathematics and prenatal testosterone exposure (Brosnan 2006, also see Valla & Ceci 2011).

Assessment. While more work needs to be conducted assess connection between interest in economics and brain exposure to testosterone (both prenatally and post-pubertally), majoring (or taking advanced courses) in economics is a USD-M.

26.3.5.4 Majoring (or Taking Advanced Courses) in Education

Evidence. Majoring in, or taking advanced courses in, the field of education is predominately in preparation for a career in teaching at the elementary or secondary level. As seen in Précis 26.3.5.4, females were

Précis 26.3.5.3 Majoring (or taking advanced courses) in economics (original Table 20.1.4.13)

Age categories, number of studies, and sex	*Consist. score*	*Overviews*	*Countries*	*Time range*	*Non-human*	*Social role theory*			*Evolutionary theory*		*Theological*
						Origin	*Found*	*Biosoc*	*Select*	*ENA*	
All Ages 11M	100.0	-0-	4	1986–2007	-0-	Contra	Contra	Contra	Poss-M	Poss-M	Silent

Précis 26.3.5.4 Majoring (or taking advanced courses) in education (original Table 20.1.4.14)

Age categories, number of studies, and sex	*Consist. score*	*Overviews*	*Countries*	*Time range*	*Non-human*	*Social role theory*			*Evolutionary theory*		*Theological*
						Origin	*Found*	*Biosoc*	*Select*	*ENA*	
Adult 11F All Ages 12F	100.0 100.0	-0-	2 (1)	1971–2017	-0-	Silent	Silent	Silent	Expl-M	Expl-M	Silent

found to major in education more than did males. Two countries were sampled along with one multi-national study. The studies were published over a 46-year time frame.

Explanation. People who become elementary and secondary teachers are usually motivated by desires to be helpful to upcoming generations of children and adolescents (Karadağ & Mutafçılar 2009). Such altruistic non-competitive occupations rarely result in high salaries (Gicheva 2022). While altruism is not a USD, a substantial majority of studies have concluded that this trait is more characteristic of females than of males (Table 17.2.1.1 in Volume II). Given the fact that most studies have found females to be more altruistic as well as empathetic (Table 10.2.8.3 in Volume I), it is rather predictable that females would be more interested in becoming elementary and secondary teachers.

Evolutionary theory leads one to hypothesize that people's motivations to take up teaching, especially at the primary levels, reflect a desire to care for children (Nicolson 1991). As noted earlier in this chapter, interest in caring for children is a USD-F that is likely to have been sexually selected (see Précis 26.2.7.28).

A number of hormones appear to operate on the brain to incline humans (as well as other primates) to care for children (*review*: Saltzman & Maestripieri 2011). ENA theory suggests that testosterone should be negatively correlated with such tendencies. Among the supportive evidence are indications that circulating testosterone levels for new mothers and fathers tend to decline (Barrett, Tran et al. 2013; Gettler, McDade et al. 2013).

Assessment. Overall, majoring in education in college appears to be a USD-F. It can be explained primarily in evolutionary terms by noting that education majors primarily become primary and secondary teachers, which are types of profession that closely resemble parenting behavior.

26.3.5.5 Majoring (or Taking Advanced Courses) in Engineering

Evidence. All of the studies summarized in Précis 26.3.5.5 found that males studied engineering more than females. This topic was studied in six countries and six multi-national projects.

Explanation. Research has shown that, at least in Western countries, males are stereotyped as having greater interests in science and engineering than are females (Table 21.2.4.3). Given this stereotype, proponents of the original social role theory could explain why males in Western countries are more likely than females to major in engineering. However, this sex difference appears to be worldwide, which presents a serious challenge to the original version of social role theory. Regarding the founder effect version, it too seems to be unable to explain the universal tendency for males to major in engineering since the field of engineering would not have

Précis 26.3.5.5 Majoring (or taking advanced courses) in engineering (original Table 20.1.4.15)

Age categories, number of studies, and sex	*Consist. score*	*Overviews*	*Countries*	*Time range*	*Non-human*	*Social role theory*			*Evolutionary theory*		*Theological*
						Origin	*Found*	*Biosoc*	*Select*	*ENA*	
Adult 43M All Ages 44M	100.0 100.0	-0-	6 (6)	1981–2017	-0-	Contra	Contra	Contra	Expl-M	Expl-M	Silent

existed in the "small progenital stock" of early humans. Likewise, the biosocial version of social role theory would have difficulty explaining why sex differences in physical size, strength, or the ability to bear offspring would impact the tendency to major in engineering.

Turning to the two evolutionary theories, they both offer credible insights into the universal tendency for males to major in engineering. Regarding sexual selection theory, a strong case has been made for males having been selected for their abilities to track and hunt wild game, thereby producing males with greater spatial reasoning abilities (Joseph 2000: 56–57; Jones, Braithwaite & Healy 2003). In contemporary time, these spatial reasoning abilities are likely to have provided males with greater interests and abilities in engineering (Browne 2006; Halpern, Benbow et al. 2007).

In the case of the ENA version of sexual selection theory, there is evidence that male-typical brain exposure to testosterone promotes spatial reasoning (Hier & Crowley 1982; Hampson, Rovet & Altmann 1998). However, one study indicated that, while testosterone does appear to be positively correlated with spatial reasoning among females at all levels of exposure, among males, this is true only up to intermediate male-typical testosterone levels, beyond which there may actually be diminished associations (Gouchie & Kimura 1991).

Assessment. Evidence strongly supports concluding that majoring in engineering is a USD-M. More research is called for in terms of fully assessing the evolutionary neurohormonal basis for this sex difference.

26.3.5.6 Majoring (or Taking Advanced Courses) in Fine Arts

Evidence. The fine arts primarily have to do with theatrical and musical performances. As shown in Précis 26.3.5.6, based on samples drawn from three different countries, all ten studies concluded that females are more likely than males to major in (or take advanced courses in) the field of fine arts.

Explanation. At least in Western countries, females are stereotyped as being more interested in artistic endeavors than are males (Table 21.2.4.2). Given this sex stereotype, the three social role theories would lead one to expect females to be more likely to pursue advanced training in the fine arts, at least in Western countries. It so happens that all ten studies of sex differences in majoring in the fine arts (e.g., art, acting, music, and other fields of creativity) were in fact conducted in Western countries. Therefore, the original social role theory could explain why females are more likely to take courses in fine arts (i.e., in Western countries, females are stereotyped this way). Both of the alternative social role theories could make similar predictions, although they would imply that the sex differences would extend into all human cultures. At the present time, we will simply consider the

Précis 26.3.5.6 Majoring (or taking advanced courses) in fine arts (original Table 20.1.4.16)

Age categories, number of studies, and sex	*Consist. score*	*Overviews*	*Countries*	*Time range*	*Non-human*	*Social role theory*			*Evolutionary theory*		*Theological*
						Origin	*Found*	*Biosoc*	*Select*	*ENA*	
All Ages 10F	100.0	-0-	3	1980–2007	-0-	Poss-F	NoTest	NoTest	Poss-F	Poss-F	Silent

latter two social role theories untested with regard to explaining sex differences in being fine art majors.

The evolution of artistic tendencies among humans has been challenge in terms of providing a scientific explanation (Dutton 2003). However, the preservation of cave art, has indicated that artistic tendencies go back tens of thousands of years (Morriss-Kay 2010). Regarding sex differences in tendency to be artistic, a great deal is still unknown. A major element in artistic tendencies involves creativity, and, in this regard, research findings suggest that there are no overall sex differences (see Table 11.2.7.3).

One study indicated that there may be average sex differences in certain regions of the brain when males and females are asked to make judgments having to do with aesthetic appreciation (Cela-Conde, Ayala et al. 2009). Regarding the possible involvement of testosterone, a couple of studies reported that musicians, particularly those who were the most creative, usually had intermediate levels, i.e., in the low range for males but in the high range for females (Hassler & Nieschlag 1989; Hassler 1992). While fragmentary, such evidence suggests that there may be both evolutionary and neurohormonal underpinnings to sex differences in people's aesthetic interests and abilities, and that these differences may be responsible for females being more likely than males to major in fine arts.

Assessment. We deem majoring in fine arts to be a USD-F. Nevertheless, more work is needed to confirm that this sex difference is found outside of Western cultures, and to provide a scientific understanding of why this sex difference would exist.

26.3.5.7 Majoring (or Taking Advanced Courses) in the Humanities

Evidence. The humanities have to do with courses in the arts, language, and history. Précis 26.3.5.7 shows that females were more likely to major in, or take advanced courses in, the humanities than was the case for males. This was found in 18 different studies based on samples from seven countries (plus two multi-national studies) ranging over a 36-year time frame.

Explanation. The humanities tend to be heavily reliant on language for their transmission, and that they are minimally reliant on spatial reasoning. This would lead one to expect that more females than males would be drawn to the humanities as a field of study. Furthermore, as the name itself implies, the *humanities* reflect a focus on human activities, both present and past. Given the well-documented tendency for females to be more interested in people than things (see Précis 26.2.7.30), one would expect that females would be more likely to major in the humanities than is the case for males.

Assessment. Tendencies to major in the humanities relative to other disciplines of study can be considered a USD-F. Given that this appears to fit with the things-versus-people distinction, the two evolutionary theories

Précis 26.3.5.7 Majoring (or taking advanced courses) in the humanities (Original Table 4.1.4.19)

Age categories, number of studies, and sex	*Consist. score*	*Overviews*	*Countries*	*Time range*	*Non-human*	*Social role theory*			*Evolutionary theory*		*Theological*
						Origin	*Found*	*Biosoc*	*Select*	*ENA*	
All Ages 18F	100.0	-0-	7 (2)	1980–2016	-0-	Silent	Silent	Silent	Expl-F	Expl-F	Silent

appear to be in the best position to explain why such a sex difference would exist.

26.3.5.8 Majoring (or Taking Advanced Courses) in Language

Evidence. According to Précis 26.3.5.8, females took more courses in language than did males. The studies of this topic were conducted in five countries.

Explanation. Most of the pertinent evidence has indicated that females are more interested than males in language. This fits with stereotypes that females are more language focused (see Table 21.2.5.8), so the original social role theory would predict that females would be more likely to major in the field of languages, at least in cultures where the stereotype is prevalent. However, all three of the social role theories would have difficulty explaining why there appears to be a *universal* tendency for females to major in language. Wood and Eagly's (2012) biosocial role theory, for example, would predict that such a major has something to do with males being larger or stronger and/or females being the only sex who bears offspring. We found no such theoretical arguments made by social role theory proponents. The approach of the two evolutionary theories would focus on understanding why males appear to be less drawn to language studies in the first place. Obviously, human language has been an extremely advantageous trait to have evolved for both sexes. However, there may be certain tradeoffs with other cognitive skills that are required in order for language skills to reach their maximum proficiency. In this regard, the development of keen visual-spatial reasoning skills may compete with the development of language abilities. Such a line of reasoning would lead one to expect that, because males have evolved keener visual-spatial reasoning, it could be at the expense of slower development in language skills. At the extreme, sex differences in autism may be illustrative here. Research has shown that a common symptom of this predominantly-male disorder (see Précis 26.2.6.24) is a failure to develop language skills (or at least a developmental delay in doing so), while affected individuals often exhibit keen interests in manipulating objects in three-dimensional space (Tiegerman & Primavera 1982; Baranek, Barnett et al. 2005).

If the above reasoning is correct, ENA theory would lead one to extend sexual selection theory by hypothesizing that exposing the brain to high (male-typical) levels of testosterone is likely to promote spatial reasoning, and slowing down the development of language skills. Support for this line of reasoning is substantial (Lust, Geuze et al. 2010; Whitehouse, Mattes et al. 2012; Schaadt, Hesse & Friederici 2015), including findings having to do with autism (Knickmeyer & Baron-Cohen 2006; Auyeung, Taylor et al. 2010).

Précis 26.3.5.8 Majoring (or taking advanced courses) in language (original Table 20.1.4.20)

Age categories, number of studies, and sex	*Consist. score*	*Overviews*	*Countries*	*Time range*	*Non-human*	*Social role theory*			*Evolutionary theory*		*Theological*
						Origin	*Found*	*Biosoc*	*Select*	*ENA*	
Adult 13F	100.0	-0-	5	1975–2007	-0-	Silent	Silent	Silent	Expl-F	Expl-F	Silent

Assessment. Overall, we conclude that females are more likely to major in language-intensive subjects, making this a USD-F. The best explanations appear to be along evolutionary neurohormonal lines.

26.3.5.9 Majoring (or Taking Advanced Courses) in Mathematics

Evidence. Précis 26.3.5.9 presents a summary of the evidence surrounding sex differences in tendencies to major in (or take advanced courses in) mathematics. For adolescents, there were exceptions in terms of taking advanced courses in math, but for adults, all 42 of the studies on this topic concluded that males are more likely to major in math. This comports with one published literature review, and was found to be the case in five multi-national studies and nine studies of individual countries published over a 45-year time frame.

Explanation. To begin theoretically explaining sex differences in tendencies to major in math, it is helpful to put the evidence in a broader context. Studies have indicated that males are stereotyped as being better at mathematics than are females (see Table 21.2.5.11). Is this stereotype accurate? In fact, this stereotype is highly questionable if one uses grades that students receive in math courses, in that more studies report females receiving higher grades than males receive (see Table 11.1.2.1 in Volume II). But, when scores on college admission tests are considered, the vast majority of studies indicate that males score higher (see Table 11.1.3.6 in Volume II). It is also relevant to note that many tests of basic math knowledge have reached very inconsistent conclusions regarding sex differences (see Table 12.2.3.1). The only exception has to do with differences in obtaining exceptionally high scores, where males do surpass females very consistently (see Table 12.2.3.4 in Volume II). However, this too needs to be qualified by noting that males also appear to score extremely low on tests of math knowledge (see Table 12.2.3.3 in Volume II). In other words, males seem to exhibit greater overall *variability* in math reasoning ability. The bottom line seems to be that sex differences in math ability only slightly favor males. The same is true for *interests* in math: Only a slight majority of studies point toward males surpassing females (see Table 15.9.2.3 in Volume II), albeit two meta-analyses found the differences to be significant (Hyde, Fennema, Ryan et al. 1990; Su, Rounds & Armstrong 2009).

Social role theorists have argued that the stereotype that males are better at math is responsible for why more males major in math than do females (Steele 2003; Beasley & Fischer 2012). One can see from our précis review that this sex difference appears to be universal. If so, the original version of social role theory would not account for sex differences in math majoring, since it assumes that mere cultural stereotypes are responsible for the sex difference. Regarding the founder effect version, it

Précis 26.3.5.9 Majoring (or taking advanced courses) in mathematics (original Table 20.1.4.23)

Age categories, number of studies, and sex	*Consist. score*	*Overviews*	*Countries*	*Time range*	*Non-human*	*Social role theory*			*Evolutionary theory*		*Theological*
						Origin	*Found*	*Biosoc*	*Select*	*ENA*	
Adolescent 33M 5x Adult 42M All Ages 75.5	(86.8) 100.0 (93.8)	M: 1Rev	9 (5)	1972–2017	-0-	Contra	NoTest	Contra	Poss-M	Poss-M	Silent

would have difficulty explaining the worldwide tendency for males to major in math more, since the original (progenital) human society this theory hypothesizes would not have had such educational training in math or any other subject area (Fausto-Sterling 1992). In the case of the biosocial version of social role theory, the theory would be hard pressed to explain why the slight male superiority in math (at least at the most advanced levels) could be the result of males being taller or stronger or of females being the only sex capable of bearing offspring (Wood & Eagly 2012).

Can an evolutionary perspective shed light on why, throughout the world, males are more likely to major in math than are females? One can begin to explore this possibility by noting that substantial evidence indicates that many animals possess rudimentary mathematical skills (Agrillo, Piffer et al. 2012). Nonetheless, humans are superior, at least beyond late childhood, partly due to their abilities to represent numerical concepts with symbols (Beran 2008).

As to why minor sex differences might exist in math interests and skills, it may have to do with sexual selection for the ability to calculate probabilities (such as when competing with other males and when tracking prey) as well as when judging distances (such as when throwing projectiles). Both of these abilities would have favored males more than in females (see Geary 1996).

In the case of the ENA version of evolutionary theory, there is considerable evidence that brain exposure to testosterone can affect math ability, but not in a straight-forward linear fashion. Instead, it appears that such exposure promotes math ability when levels of exposure rise to a low-to-moderate male-typical range; beyond that range, however, testosterone may actually have detrimental effects (Kimura 2003:35). Regarding interests in math, one study found indirect evidence that brain exposure to prenatal androgens was positively correlated among females, but not correlated to a significant degree among males (Ellis & Das 2009: Table 4). Another line of evidence that suggests a positive association between math ability and brain exposure to testosterone comes from studying autism. In this regard, one study indicated that college students with at least modest degrees of autism (i.e., autism spectrum disorder) are more likely to major in math than are college students in general (Baron-Cohen, Wheelwright et al. 2007).

Assessment. We conclude that the evidence strongly supports concluding that majoring in mathematics is a USD-M, and that evolutionary theorizing offers the best explanations for this sex difference.

26.3.5.10 Majoring (or Taking Advanced Courses) in Physics, Chemistry, or Other Physical Sciences and Technology

Evidence. Précis 26.3.5.10 shows that the vast amount of research has found males taking more advanced courses in, or were more likely to

Précis 26.3.5.10 Majoring (or taking advanced courses) in physics, chemistry, or other physical sciences and technology (original Table 20.1.4.26)

Age categories, number of studies, and sex	*Consist. score*	*Overviews*	*Countries*	*Time range*	*Non-human*	*Social role theory*			*Evolutionary theory*		*Theological*
						Origin	*Found*	*Biosoc*	*Select*	*ENA*	
Adolescence 30M 1x 1F	(85.7)	-0-	11 (5)	1963–2013	-0-	Silent	Silent	Silent	Expl-M	Expl-M	Silent
Adult 68M 2x	97.1										
All Ages 98M 3x 1F	95.1										

major in the fields of physics, chemistry, or other physical sciences. There were 102 studies on this topic conducted in 11 countries, along with five multi-national studies.

Explanation. Much of what was stated directly above regarding sex differences in the tendency to major in (or take advanced courses in) mathematics can also be stated with reference to majoring in the physical sciences. This is because, like mathematics, the physical sciences tend to focus on things rather than on people or other living things. As a result, the same conclusion can be reached regarding the best theoretical explanations.

Assessment. Majoring (or taking advanced courses) in the physical sciences can be considered a USD-M. The most persuasive explanation appears to come from reasoning along evolutionary lines.

26.3.5.11 Participation in Paid Workforce in General

Evidence. Nearly all of 138 studies were found having to do with sex differences in participating in the paid workforce. Précis 26.3.5.11 shows that the relevant studies came from 40 different countries plus 20 multinational studies. Only in the case of adults was the consistency score above 95.0 (mainly because findings were mixed in the case of studies involving children and adolescents). The four studies of adults that concluded there were no significant sex differences had to do with young adults.

Explanation. The only stereotype we could locate bearing on sex differences in participating in the paid workforce was one indicating that males were stereotyped as earning more money (Table 21.2.11.5). This vaguely implies that the sex ratio of persons in the paid workforce would be equal if there were no sex difference in average pay. Biosocial role theory might explain why females are less involved in the paid workforce due to pregnancy interrupting their careers.

Both evolutionary theories predict that males would be more involved in the paid workforce than are females due to sexual selection. Specifically, females, especially those who produce the most offspring, are assumed to have evolved tendencies to prefer mates who are relatively capable of provisioning resources so that the female can concentrate on caring for offspring (Hamilton 1984; Gangestad & Thornhill 1997; Miller 1998; Ellis 2011b).

Regarding ENA theory, it adds to asserting that sexual selection has produced sex differences in traits the belief that neuroandrogenic factors are involved. In the present case, testosterone should promote participation in the paid workforce, even within each sex. Just one study was found that provided somewhat indirect evidence bearing on this hypothesis. It indicated that prenatal testosterone was positively correlated with earnings among males but had no apparent effects in the case of females (Gielen, Holmes & Myers 2016).

Précis 26.3.5.11 Participation in paid workforce in general (original Table 20.2.1.1)

Age categories, number of studies, and sex	*Consist. score*	*Overviews*	*Countries*	*Time range*	*Non-human*	*Social role theory*			*Evolutionary theory*		*Theological*
						Origin	*Found*	*Biosoc*	*Select*	*ENA*	
Adult 112M 4x All Ages 127M 6x 5F	96.6 (88.8)	-0-	40 (20)	1962–2020	-0-	Silent	NoTest	Poss-M	Poss-M	Poss-M	Poss-M

Finally, it is worth noting that theological explanations for sex differences in the paid workforce may be offered. For example, the Koran (IV:4) states that the husband has primary responsibility for "providing economic support for his wife(s) and family." Similarly, according to Mormon scriptures, "Women have claim on their husbands for their maintenance" (Doctrine and Covenants 83:2). These religious edicts could help to explain why more males are employed than females, although the explanations may only apply to affiliates of specific religions.

Assessment. Sex differences in the tendency to participate in the paid workforce, especially full-time participation (see below), has been extensively investigated all over the world. Findings leave no serious doubt that this variable is a USD-M, at least among adults. While one can offer a scriptural explanation for this sex difference, such an explanation would be unable to account for why the differences has been found in all countries. The biosocial role theory and the two evolutionary theories appear to offer the most likely explanations.

26.3.5.12 Full-Time Paid Employment

Evidence. Not everyone who is employed works full-time (typically defined as 40 hours per week). As shown in Précis 26.3.5.12, 80 out of 82 studies of persons involved in full-time paid employment have found that males are more often employed than females. This sex difference was found in 25 different countries along with 12 multi-national studies. Two exceptions were reported. One Japanese study of young adult (entry-level) workers concluded that there were no sex differences, and one U.S. study indicated that, among blacks, more females than males were employed full time.

Explanation. No sex stereotypes bearing directly on full-time paid employment were found. The strongest theoretical proposals seem to be evolutionary in nature. Specifically, as noted in the preceding précis, females appear to have been sexually selected for choosing mates based on their mates' abilities to provision resources, while males are favored for using such criteria in mate selection much less. Consequently, the males who are most likely to pass genes on to future generations are those with full-time employment (or with some other means of making a living).

In addition to recognizing the role of sexual selection in promoting full-time paid employment, ENA theory would hypothesize that testosterone promotes such behavior. No evidence either supporting or refuting this hypothesis was located.

Regarding the possibility of a theological explanation for sex differences in being a full-time worker, some supportive evidence was found. As noted in the preceding narrative, passages from both the Koran and Mormon

Précis 26.3.5.12 Full-time paid employment (original Table 20.2.2.1)

Age categories, number of studies, and sex	*Consist. score*	*Overviews*	*Countries*	*Time range*	*Non-human*	*Social role theory*			*Evolutionary theory*		*Theological*
						Origin	*Found*	*Biosoc*	*Select*	*ENA*	
Adult 79M 1x 1F All Ages 80M 1x 1F	96.3 96.4	-0-	25 (12)	1941–2020	-0-	Silent	Silent	Silent	Poss-M	Poss-M	Poss-M

scriptures both indicate that males must be responsible for providing financial support to their wives and offspring.

Assessment. With the possible exception of a few select subpopulations within countries, full-time paid employment is a USD-M.

26.3.5.13 Employed Part Time

Evidence. When research findings shift from full-time employment to part-time employment, the sex difference reverses from the shown in the previous précis. All of the studies presented in Précis 26.3.5.13 found that females participated in part-time employment more than did males. These studies included two multi-national research projects and studies in three countries.

Explanation. The only theoretical explanation for why females would be more likely to work part time than is the case for males is along evolutionary lines. Specifically, as noted above, males are likely to have been sexually selected for working full time as a way of making a stable income. Females, on the other hand, would only be likely to work full time before having children, or after completing child-rearing responsibilities. Theoretically, women who remain childless should be most likely to resemble males in holding full-time employment. No evidence was located regarding this specific hypothesis.

Assessment. We decree part-time employment to be a USD-F, even though theoretical explanations are sketchy at best.

26.3.5.14 Employment in High-Paying Businesses or Occupations

Evidence. Is there a sex difference in regard to holding high paying jobs? Précis 26.3.5.15 shows that all 13 of the studies that were located bearing on this question concluded that greater proportions of males held particularly high-paying jobs than was true for females. The studies covered four different countries plus two multi-national studies.

Explanation. Males are stereotyped as earning more than females (Table 21.2.11.5). Nevertheless, aside from simply saying that this stereotype appears to be true, we found no effort by any of the three social role theories to provide an explanation.

Both evolutionary theories offer explanations for sex differences in the tendency to strive toward high paying business and occupational opportunities. In essence, males should spend more time and energy seeking such high-status positions because a primary characteristic a female look for in a mate is evidence of a stable comfortable income so that she can focus on producing and caring for offspring (Flinn & Low 1986; Mulder 2017). This should be especially strong in areas of the world where environmental conditions are harsh.

Précis 26.3.5.13 Employed part time (original Table 20.2.2.2)

Age categories, number of studies, and sex	*Consist. score*	*Overviews*	*Countries*	*Time range*	*Non-human*	*Social role theory*			*Evolutionary theory*		*Theological*
						Origin	*Found*	*Biosoc*	*Select*	*ENA*	
All Ages 13F	100.0	-0-	3 (2)	1987–2018	-0-	Silent	Silent	Silent	NoTest	NoTest	Silent

According to ENA theory, brain exposure to testosterone should promote traits such as competitiveness and risk-taking in males to make employment in high-paying occupations more probable. Considerable evidence supports the conclusion that status-striving is positively correlated with circulating testosterone is positively correlated with status-striving (Mazur & Booth 1998; Nave, Nadler et al. 2018).

Finally, in the way of a possible explanation, it might be noted that a biblical passage states that males should be paid more than females for performing the same job (Leviticus 27:1–4). This edict might underly the finding that males end up in higher-paying jobs than females.

Assessment. Substantial support for concluding that employment in high-paying businesses or occupations is a USD-M.

26.3.5.15 *Centrality of One's Job or Career (Outside the Home) to One's Life*

Evidence. Is one's employment more important to the average male or the average female? According to Précis 26.3.5.15, all of the studies found that a job or career outside of the home to be more important to males than it was to females. The relevant research was based on sampled drawn from three different countries; a meta-analysis reached the same conclusion.

Explanation. No research evidence of stereotypes having to do with sex differences in the importance of one's job as a part of life. Also, no effort to explain sex differences in this regard were found either.

Because females are much more likely to choose mates with the ability to make a living than are males (see Précis 26.2.7.19), from a sexual selection perspective, males should place a greater emphasis on their jobs than is true for females. In addition, ENA theory leads to the prediction that that the perceived importance of one's job should be promoted by exposing the brain to androgens. No evidence either supporting or refuting this hypothesis was found.

Assessment. Evidence that centrality of one's job to one's life is a USD-M is substantial. So far, the best theoretical explanation involves sexual selection.

26.3.5.16 *Duration of Lifetime Spent in the Paid Labor Market*

Evidence. Précis 26.3.5.16 shows that, according to all 20 pertinent studies, males devote more of their times in paid employment than do females. This sex difference was found in five countries, plus two multinational studies published over a 44-year time span.

Explanation. As with the explanation for the centrality of one's job (discussed directly above), sexual selection theory seems to offer the most obvious explanation for this sex difference. In other words, males who

Précis 26.3.5.14 Employed in high-paying businesses or occupations (original Table 20.2.2.3)

Age categories, number of studies, and sex	*Consist. score*	*Overviews*	*Countries*	*Time range*	*Non-human*	*Social role theory*			*Evolutionary theory*		*Theological*
						Origin	*Found*	*Biosoc*	*Select*	*ENA*	
All Ages 13M	100.0	-0-	4 (2)	1980–2004	-0-	Silent	Silent	Silent	Expl-M	Expl-M	Poss-M

Précis 26.3.5.15 Centrality of one's job or career (outside the house) to one's life (original Table 20.2.3.3)

Age categories, number of studies, and sex	*Consist. score*	*Overviews*	*Countries*	*Time range*	*Non-human*	*Social role theory*			*Evolutionary*		*Theological*
						Origin	*Found*	*Biosoc*	*Select*	*ENA*	
All Ages 18M	100.0	M: 1Met	3	1966–1997	-0-	Silent	Silent	Silent	Expl-M	Poss-M	Silent

Précis 26.3.5.16 Duration of lifetime spent in the paid labor market (original Table 20.2.4.1)

Age categories, number of studies, and sex	*Consist. score*	*Overviews*	*Countries*	*Time range*	*Non-human*	*Social role theory*			*Evolutionary theory*		*Theological*
						Origin	*Found*	*Biosoc*	*Select*	*ENA*	
All Ages 20M	100.0	-0-	5 (2)	1970–2014	-0-	Silent	Silent	Silent	Expl-M	Poss-M	Poss-M

spend more time in paid employment have generally out-reproduced males in general. Comparable relationships are not found for females.

Assessment. Greater time spent in the paid labor market is a USD-M. Sexual selection theory offers the best explanation.

26.3.5.17 Average Hours Worked per Week

Evidence. Given that males are more likely to be employed in full-time jobs (Précis 26.3.5.12) and females with jobs are more likely to work part-time (Précis 26.3.5.13), it is not surprising to find that males work more hours per week than do females. Précis 26.3.5.17 shows that this was true except in two studies that failed to find a sex difference. It is worth noting that several of these studies were based on samples drawn from various specific professions, such as physicians, lawyers, and college professors.

Explanation. The evolutionary perspective seems to provide the most obvious explanation for why males work more hours per week outside the home than do females (thus, without counting housework or childcare). Specifically, when choosing mates, females appear to discriminate in favor of males who appear to be ambitious and capable of making a living to a greater degree than do males (see Table 15.6.5.14 in Volume II). This type of sexual selection is likely to have produced males with tendencies to focus on paid employment opportunities to a greater degree than do females.

Assessment. Average hours worked per week is a USD-M. Sexual selection theory seems to offer the most reasonable explanation.

26.3.5.18 Manual/Blue-Collar Occupations

Evidence. Fifteen studies from six different countries all found that males were more often involved in manual labor or so-called *blue-collar jobs* than were females. As shown in Précis 26.3.5.18, these studies were published over a 47-year time frame.

Explanation. Why would the sexes differ regarding involvement in blue-collar jobs? Part of the explanation is likely to involve the fact that males tend to be physically stronger (see Précis 26.1.2.8), and sex differences in strength is a part of the biosocial social role theory (Wood & Eagly 2012:123).

However, because sex differences in physical strength can also be accounted for in evolutionary terms, sexual selection theory and ENA theory could provide a more complete explanation. In this regard, when blue-collar jobs are compared to white-collar jobs, there are at least three differences besides physical strength that could make blue-collar jobs more appealing to males. First, blue-collar jobs are more things-oriented (rather than people-oriented). Second, these jobs usually involve greater physical activity. Third, they often entail relatively high risk of physical injury.

Précis 26.3.5.17 Average hours worked per week (original Table 20.2.4.5)

Age categories, number of studies, and sex	*Consist. score*	*Overviews*	*Countries*	*Time range*	*Non-human*	*Social role theory*			*Evolutionary theory*		*Theological*
						Origin	*Found*	*Biosoc*	*Select*	*ENA*	
All Ages 44M 2x	95.7	-0-	4 (3)	1980–2014	-0-	Silent	Silent	Silent	Expl-M	Poss-M	Silent

Précis 26.3.5.18 Manual/blue-collar occupations (original Table 20.2.5.3)

Age categories, number of studies, and sex	*Consist. score*	*Overviews*	*Countries*	*Time range*	*Non-human*	*Social role theory*			*Evolutionary theory*		*Theological*
						Origin	*Found*	*Biosoc*	*Select*	*ENA*	
All Ages 15M	100.0	-0-	6	1971–2006	-0-	Silent	Silent	Poss-M	Expl-M	Expl-M	Silent

Regarding all three of the above features, males surpass females. Specifically, males are more things-oriented (see Précis 26.2.7.44), more physically active (see Précis 26.3.1.6), and more prone to take risks (see Précis 26.3.1.2). In other words, evolutionary theory can explain why males are more involved in manual occupations than females not only in terms of sex differences in physical strength, but also in terms of males being more things-oriented, physically active, and willing to take risks.

Assessment. We conclude that employment in manual/blue-collar jobs is a USD-M. More work on theoretically explaining this sex difference is needed.

26.3.5.19 Masculine Occupations/Male-Typical Occupations

Evidence. Some occupations are considered more masculine or male-typical than others. As discussed elsewhere, including directly above, masculine jobs are more likely to involve strenuous physical labor and spatial reasoning. Précis 26.3.5.19 shows that 95 studies conducted in 11 different countries, plus ten multi-national studies, have all concluded that males worked in these types of jobs more than did females.

Explanation. Social role theory has obvious bearing on why males would gravitate toward masculine occupations. However, in the case of the original version, one would not expect universality in what are considered masculine (or male-typical) occupations. In the case of the founder effect version, it is reasonable to assume that there were no occupations per se in the earliest human societies. Regarding the biosocial version of social role theory, it would account for sex differences in traits that appear to be universal by asserting that the differences are due to males being larger or stronger or to the fact that only females can give birth. While these sex differences would account for some occupational sex differences (such as blue-collar jobs involving heavy labor (see directly above), it would not address sex differences in the majority of professional white-collar jobs.

The evolutionary explanations for why there would be universally recognized masculine (male-typical) occupations and why males would fill them more than females would involve examining specific characteristics of masculine occupations. As noted above, regarding manual labor jobs, they tend to be things-oriented (rather than people-oriented), which males have been shown to prefer more than do females (see Précis 26.2.7.44). Also, males have been shown to gravitate toward risk-taking activities more than is the case for females, including financial risk-taking (see Précis 26.3.1.2). Both of these sex differences could help to explain why a number of so-called *masculine occupations* are perceived as such throughout the world and attract more males throughout the world as well.

In the case of ENA theory, it asserts that brain exposure to testosterone contributes to universal sex differences. Research has indicated that at

Précis 26.3.5.19 Masculine occupations/male-typical occupations (Original Table 20.2.6.1)

Age categories, number of studies, and sex	*Consist. score*	*Overviews*	*Countries*	*Time range*	*Non-human*	*Social role theory*			*Evolutionary theory*		*Theological*
						Origin	*Found*	*Biosoc*	*Select*	*ENA*	
All Ages 95M	100.0	-0-	11 (10)	1973–2012	-0-	Contra	NoTest	Poss-M	Expl-M	Expl-M	Silent

least prenatal testosterone exposure is predictive of things-oriented occupational preferences (Hell & Pabler 2011) as well as interests in male-typical occupations (Manning, Reimers et al. 2010; Manning, Trivers & Fink 2017). Furthermore, considerable evidence suggests that brain exposure to testosterone contributes to risk-taking activities. This is true for prenatal testosterone, based on the 2D:4D finger length indicator (Vermeersch, T'sjoen et al. 2008; Coates, Gurnel & Rustichini 2009; Ronay & von Hippel 20010b; Stenstrom, Saad et al. 2011; Aycinena, Baltaduonis & Rentschler 2014; Brañas-Garza, Galizzi & Nieboer 2018).

An additional line of evidence regarding prenatal testosterone exposure comes from a study of women with a condition known as *congenital adrenal hyperplasia* (*CAH*). This condition is associated with abnormally high production of androgens, including testosterone. The study of CAH women concluded that they expressed significantly more interested in male-typical (things-oriented) occupations than did women in general (Beltz, Swanson & Berenbaum 2011).

Another line of evidence comes from studies of post-pubertal circulating testosterone. Overall, these studies have indicated that, within each sex, employment in male-typical occupations is positively associated with working in male-typical jobs (*review*: Apicella, Carré & Dreber 2015; *meta-analysis*: Kurath & Mata 2018).

Assessment. Overall, the evidence that very similar male-typical (masculine) occupations are universally recognized and that males are more likely to work in these occupations is exceedingly strong, making these occupations collectively a USD-M. The best supported explanations appear to be evolutionary in nature.

26.3.5.20 Feminine Occupations/Female-Typical Occupations

Evidence. When compared to male-typical occupations (see directly above), only about half as many studies were located pertaining to sex differences in working in female-typical occupations. Nonetheless, Précis 26.3.5.20 shows that the findings were equally consistent, i.e., all 45 studies unanimously indicate that females fill these types of jobs more than was the case for males. Nine different occupations were sampled along with seven multi-national samples. Findings were published over a 33-year time frame.

Explanation. The basic theoretical explanations offered for feminine occupations would be essentially the reverse of the explanations for masculine occupations (as provided directly above). Probably the single most notable feature of feminine occupations is that they are focused on people (rather than things) and that they are focused on providing help to others (rather than competing with others). In this regard, studies have indicated

Précis 26.3.5.20 Feminine occupations/female-typical occupations (Original Table 20.2.6.2)

Age categories, number of studies, and sex	*Consist. score*	*Overviews*	*Countries*	*Time range*	*Non-human*	*Social role theory*			*Evolutionary theory*		*Theological*
						Origin	*Found*	*Biosoc*	*Select*	*ENA*	
All Ages 45F	100.0	-0-	9 (7)	1970–2003	-0-	Contra	NoTest	Poss-F	Expl-F	Expl-F	Silent

that females are more people-oriented (see Précis 26.2.7.30) and more interested in jobs that emphasize helping others (see Précis 36.2.15.45).

According to evolutionary thinking, females have been sexually selected for being people-oriented (Lippa 2010a; Stoet & Geary 2022) and for being helpful to others (Tay, Ting & Tan 2019). To this basic sexual selection argument, ENA theory adds the hypothesis that brain exposure to testosterone will inhibit most feminine traits, including those that are associated with occupational preferences. On most cases, both prenatal and post-pubertal (circulating) testosterone should have similar depressing effects on female-typical traits, including those pertaining to occupational preferences (Ellis 2011a). In this regard, studies have indicated that prenatal testosterone exposure lowers tendencies to exhibit feminine preferences later in life (SN Davis & Risman 2015; Manning, Trivers & Fink 2017). Also, a study of circulating testosterone among females found the levels of this hormone inversely correlated among women with their self-ratings of having conventional feminine personality traits (Baucom, Besch & Callahan 1985).

Assessment. Most people in all countries appear to recognize feminine (as opposed to masculine) occupations, and largely agree on what constitutes these samples. Common examples are nursing (see Précis 26.3.5.30) and social worker (Précis 26.3.5.35), as well as teaching young children (Perra & Ruspini 2013).

Overall, the evidence is strong that feminine occupations collectively constitute a USD-F. The explanations that appear to have the most empirical support are evolutionary in nature.

26.3.5.21 Diversity (Variability) in Occupations

Evidence. Do the members of one sex work in a wider diversity of occupations than members of the other sex do? Précis 26.3.5.21 shows that all ten studies have answered affirmatively, with males doing so more than females. Two publications also offered the generalization that this was the case. The pertinent studies came from two countries plus one multi-national study.

Explanation. Only the evolutionary theories seem to provide an explanation for why males would work in a more diversified number of occupations than females. Regarding sexual selection, the explanation centers around proposals that males have been sexually selected for providing resources to spouses (see Précis 26.2.7.19), so that females can focus attention on bearing and rearing children (see Précis 26.2.7.28). This line of reasoning implies that males would be selected for their willingness to work at almost any occupation, even those that are unpleasant and dangerous, provided the compensation is sufficient, a conclusion discussed in connection with Précis 26.2.7.47.

Précis 26.3.5.21 Diversity (variability) in occupations (original Table 20.2.7.1)

Age categories, number of studies, and sex	*Consist. score*	*Overviews*	*Countries*	*Time range*	*Non-human*	*Social role theory*			*Evolutionary theory*		*Theological*
						Origin	*Found*	*Biosoc*	*Select*	*ENA*	
All Ages 10M	100.0	M: 2Gen	2 (1)	1976–2001	-0-	Silent	Silent	Silent	Expl-M	Expl-M	Silent

Regarding the involvement of brain exposure to androgens, while the impact appears to be complex, findings suggest that testosterone helps to promote occupational diversity (Dabbs, de La Rue & Williams 1990; Dabbs 1992; Geary 1996:245; Sapienza, Zingales & Maestripieri 2009). One of the ways this may occur is by various regimens of testosterone altering the hemispheres in ways that affect language skills (Brosnan 2006).

Assessment. Overall, while samples drawn from a greater number of countries would be desirable, diversity in occupations can be considered a USD-M. An evolutionary perspective appears to be stronger than a social role perspective in accounting for this diversity.

26.3.5.22 *Clerical/Service Occupations*

Evidence. Précis 26.3.5.22 summarizes research undertaken to determine if there are sex differences in regard to employment in clerical and/or service occupations. All 18 of the pertinent studies have concluded that females are more likely to be in these occupations than are males. Eight different countries were sampled and the results were published over a 30-year time frame.

Explanation. Among the reasons females are more likely than males to work in clerical/service occupations involves the fact that these types of jobs are people-oriented, a job characteristic that females usually prefer (see Précis 26.2.7.43). Another reason is that females express greater interests in being helpful and serving the needs of others (see Précis 26.3.2.7, also see Précis 26.2.7.45). As noted in the narrative pertaining to these sex differences in interests, they all seem to be inhibited by exposing the brain to testosterone, which is more characteristic of females than of males.

Assessment. Working in clerical/service occupations is a USD-F. The best explanation appears to be in evolutionary terms.

26.3.5.23 *Computer-Related Occupations*

Evidence. Précis 26.3.5.23 shows that all 11 studies found males held computer-related jobs more than did females. Unfortunately, these studies were all based on samples drawn from just one country, with their publication spanning a narrow 17 years.

Explanation. Some of the tendency for males to work in computer-related jobs more than females involves the fact that they are more likely than females to major in computer science (see Précis 26.3.5.2). Additional sex difference traits that are likely to contribute are the facts that males gravitate toward things-oriented occupations than do females (see Table 15.10.2.6 of Volume II). Additional contributors are likely to involve males being better at spatial reasoning (see Précis 26.2.3.4) and at systemizing thinking

Précis 26.3.5.22 Clerical/service occupations (Original Table 20.2.7.13)

Age categories, number of studies, and sex	*Consist. score*	*Overviews*	*Countries*	*Time range*	*Non-human*	*Social role theory*			*Evolutionary theory*		*Theological*
						Origin	*Found*	*Biosoc*	*Select*	*ENA*	
All Ages 18F	100.0	-0-	8	1976–2006	-0-	Silent	Silent	Silent	Expl-F	Expl-F	Silent

Précis 26.3.5.23 Computer-related occupations (Original Table 20.2.7.14)

Age categories, number of studies, and sex	*Consist. score*	*Overviews*	*Countries*	*Time range*	*Non-human*	*Social role theory*			*Evolutionary theory*		*Theological*
						Origin	*Found*	*Biosoc*	*Select*	*ENA*	
All Ages 11M	100.0	-0-	1	1990–2007	-0-	Silent	Silent	Silent	Expl-M	Expl-M	Silent

(see Précis 26.2.3.8). As documented in the narratives surrounding these précises, they all appear to be promoted by exposing the brain to high (male-typical) levels of testosterone.

Assessment. There is little reason to doubt that working in computer-related occupations is a USD-M, although the fact that just one country was sampled in the pertinent study that were located calls for research in a wider range of countries. The sex difference can be explained best in evolutionary terms.

26.3.5.24 Corporate Executive Officers (CEOs/Managers)

Evidence. Précis 26.3.5.24 shows that 18 studies have sought to determine if males are more likely than females to be corporate executive officers or managers of companies. Based on samples drawn from three countries as well as two multi-country studies, all findings agreed that males held these types of positions more than did females. Publication of these studies occurred over a 21-year time frame.

Explanation. Many studies were reviewed in Chapter 21 (in Volume III) of sex stereotypes that could have bearing on sex differences in tendencies to oversee the functioning of large companies or groups of workers. These stereotypes included the belief that males are more ambitious (see Table 21.2.7.8), autocratic (see Table 21.2.7.11), competitive (see Table 21.2.7.16), enterprising (see Table 21.2.7.29), forceful and insistent (see Table 21.2.7.30) and dominant (Table 21.2.9.7) than females. From this evidence, social role theorists can argue that these stereotypes are responsible for males being more likely than females to ascend to high supervisory positions within companies. However, the original version of the social role theory would not predict universality in this sex difference, and, because CEO positions would not have existed in early human societies, the founder effect version would not be able to account for sex differences in the holding of these positions either. In the case of the biosocial version of social role theory, it asserts that universal sex differences in traits have to do with males being larger and stronger or to females being the only sex capable of bearing offspring (Wood & Eagly 2012:56). It is reasonable to believe that the greater size and strength of men and the time devoted to bearing children by women could give males an edge in securing powerful corporate positions.

Evolutionary explanation for sex differences in holding CEO/managerial positions would most likely focus on attitudinal and personality traits that differ between the sexes. For example, males have been found to value power more (see Précis 26.2.7.1) and to be more competitive (see Précis 26.3.1.5). Both of these universal sex differences could be sexually selected as part of male efforts to attract mates, since access to

Précis 26.3.5.24 Corporate executive officers (CEOs/managers) (Original Table 20.2.7.17)

Age categories, number of studies, and sex	*Consist. score*	*Overviews*	*Countries*	*Time range*	*Non-human*	*Social role theory*			*Evolutionary theory*		*Theological*
						Origin	*Found*	*Biosoc*	*Select*	*ENA*	
All Ages 18M	100.0	-0-	3 (2)	1977–2001	-0-	Silent	Silent	Poss-M	Expl-M	Poss-M	Silent

resources is a major priority for females when they select long-term sex partners (see Précis 26.2.7.19).

To the basic argument offered by sexual selection theory, ENA theory adds the premise that brain exposure to testosterone drives sex differences. Is there support for this asserting regarding people being CEOs and managers? The evidence is fragmentary and mixed. On the affirmative side, a study of women who were in managerial jobs found that their average circulating testosterone levels were significantly higher than the levels for women who were clerical workers and housewives (Purifoy & Koopmans 1979). Another study of women reported that those who exhibited tendencies to be dominant had higher testosterone in their blood than did women who were least dominant (VJ Grant & France 2001).

Research studies among males have revealed mixed findings, however. One study indicated that circulating testosterone levels were positively correlated with a measure of entrepreneurship (White, Thornhill & Hampson 2006). However, another study concluded that males of high social status exhibited *lower* circulating testosterone levels than males of relatively low social status (Dabbs 1992).

Assessment. Overall, being a corporate executive or manager is a USD-M. The best explanation appears to involve sexual selection.

26.3.5.25 Engineering

Evidence. Engineers design, build, and maintain machines as well as other structures designed to perform a wide array of mechanical tasks. Précis 26.3.5.25 sums up the studies seeking to determine if there is a sex difference in this type of occupation. The answer appears to be yes because all of the 40 studies, and two literature reviews, concluded that more males than females were engineers. The studies took place in six different countries.

Explanation. The most persuasive theoretical explanations for sex differences in the field of engineering is likely to entail the fact that these occupations nearly always entail considerable spatial reasoning. As noted in Précis 26.2.3.4, males tend to surpass females in spatial reasoning. Also, engineering jobs tend to be more things-oriented rather than people-oriented, a fact that may help to attract high proportions of males (see Table 15.10.2.6 of Volume II). Evolutionarily speaking, both spatial reasoning and things-orientation appear to have been sexually selected (Byrd-Craven, Massey et al. 2015) and promoted by exposing the brain to high levels of testosterone (Gouchie & Kimura 1991; Janowsky, Oviatt & Orwol 1994; Berenbaum, Korman & Leveroni 1995; Grimshaw, Sitarenios & Finegan 1995; Aleman, Bronk et al. 2004; Hooven, Chabris et al. 2004; Zitzmann 2006; Valla & Cech 2011).

Précis 26.3.5.25 Engineering (Original Table 20.2.7.27)

Age categories, number of studies, and sex	*Consist. score*	*Overviews*	*Countries*	*Time range*	*Non-human*	*Social role theory*			*Evolutionary theory*		*Theological*
						Origin	*Found*	*Biosoc*	*Select*	*ENA*	
All Ages 40M	100.0	M: 2Rev	6	1969–2012	-0-	Silent	Silent	Silent	Expl-M	Expl-M	Silent

Assessment. Employment in the field of engineering is a USD-M. The best theoretical explanation appears to be in evolutionary terms.

26.3.5.26 Geographer or Mapping Scientist

Evidence. Creating and interpreting maps are skills crucial to the field of geography. Précis 26.3.5.26 shows that all 17 studies of sex differences in being geographers concluded that greater proportions of males were in this profession than were females. The studies were in four countries over a 32-year time frame.

Explanation. Geography involves substantial spatial reasoning, which males appear to be more capable of than females (see Précis 26.2.3.4). For this reason, both evolutionary theories lead one to expect males to gravitate toward this line of work.

Assessment. With substantial confidence, being a geographer (or mapping scientist) is a USD-M. The best explanation seems to be evolutionary in nature.

26.3.5.27 Law Enforcement/Police Officers

Evidence. Précis 26.3.5.27 shows that males were employed in law enforcement more often than were females. All 29 studies of this topic, conducted in four countries and published over a 53-year time frame.

Explanation. Given that law enforcement and police work is people-oriented occupation and often involves trying to help others, one might think that females would be more likely to engage in this line of work than males. However, this line of work is often confrontational in nature. This could make police work more appealing to people with competitive and risk-taking tendencies, both of which males exhibit more than females (see Précis 26.3.1.5 and Précis 26.3.1.2). Furthermore, males have been shown to exhibit stronger preferences for aggressive responses to conflict than females (see Précis 26.2.7.2). As noted in their respective précis narratives, all three of these tendencies – competitiveness, risk-taking, and aggressive responding to conflict – appear to have been sexually selected and promoted by exposing the brain to testosterone.

Assessment. Being a police officer can be considered a USD-M. Evolutionary theory appears to provide the most compelling explanation.

26.3.5.28 Lawyer

Evidence. Précis 26.3.5.28 shows that 28 studies, conducted in five different countries all concluded that more lawyers were males than females. The studies were published over a 39-year time frame.

Précis 26.3.5.26 Geographer or mapping scientist (Original Table 20.2.7.30)

Age categories, number of studies, and sex	*Consist. score*	*Overviews*	*Countries*	*Time range*	*Non-human*	*Social role theory*			*Evolutionary theory*		*Theological*
						Origin	*Found*	*Biosoc*	*Select*	*ENA*	
All Ages 17M	100.0	-0-	4	1973–2005	-0-	Silent	Silent	Silent	Expl-M	Expl-M	Silent

Précis 26.3.5.27 Law enforcement/police officers (original Table 20.2.7.35)

Age categories, number of studies, and sex	*Consist. score*	*Overviews*	*Countries*	*Time range*	*Non-human*	*Social role theory*			*Evolutionary theory*		*Theological*
						Origin	*Found*	*Biosoc*	*Select*	*ENA*	
All Ages 29M	100.0	-0-	4	1953–2006	-0-	Silent	Silent	Silent	Poss-M	Poss-M	Silent

Précis 26.3.5.28 Lawyer (original Table 20.2.7.36)

Age categories, number of studies, and sex	*Consist. score*	*Overviews*	*Countries*	*Time range*	*Non-human*	*Social role theory*			*Evolutionary theory*		*Theological*
						Origin	*Found*	*Biosoc*	*Select*	*ENA*	
All Ages 28M	100.0	-0-	5	1965–2004	-0-	Silent	Silent	Silent	Silent	Silent	Silent

Explanation. No theoretical explanation for why more males than females would be lawyers were identified. Therefore, all of the explanatory cells are listed as "silent."

Assessment. Even though no theoretical explanation was located, being a lawyer must be considered a USD-M.

26.3.5.29 *Mechanic or Machine Operator*

Evidence. Is working with machines as a mechanic or machine operator an occupation held by one sex more often than the other sex? Précis 26.3.5.29 indicates that males held such positions more often than did females. All ten studies, conducted in three countries and published from 1984–2006, reached that conclusion.

Explanation. Most research has found that males are better at spatial reasoning (see Table 11.2.4.1 in Volume II) and more interested in things than in people (see Table 15.10.2.6 in Volume II). Such evidence leads one to predict that more males than females will become mechanics or machine operators. By examining the explanations offered for sex differences in spatial reasoning and being more interest in things, one will see that the best explanations appear to be evolutionary in nature.

Assessment. One can say with substantial confidence that being mechanics and machine operators constitute a USD-M. Evolutionary theory appears to provide the best explanations.

26.3.5.30 *Nursing*

Evidence. A large number of people in the field of health care are employed as nurses. Précis 26.3.5.30 shows that all 19 studies of sex differences in being nurses have concluded that more females than males work in this profession. The studies were conducted in seven different countries and published over a 39-year time frame.

Explanation. As shown elsewhere in this chapter, research has indicated that females are more likely than males to prefer people-oriented jobs (see Précis 26.2.7.43). Also, females express greater interests in being helpful and serving the needs of others (see Précis 26.3.2.7, also see Précis 26.2.7.45). Narratives for these précises indicate that evolutionary reasoning seems to offer more persuasive explanations than does reasoning along the lines of social role theory for these pro-socially oriented sex differences. In essence, the arguments suggest that pro-sociality is favored among females more because they have been sexually selected for providing care to offspring more than males. In the case of ENA theory, it adds to the sexual selection premise that exposure to low levels of testosterone helps to promote pro-social caring behavior.

Précis 26.3.5.29 Mechanic or machine operator (Original Table 20.2.7.41)

Age categories, number of studies, and sex	*Consist. score*	*Overviews*	*Countries*	*Time range*	*Non-human*	*Social role theory*			*Evolutionary theory*		*Theological*
						Origin	*Found*	*Biosoc*	*Select*	*ENA*	
All Ages 10M	100.0	-0-	3	1984–2006	-0-	Silent	Silent	Silent	Expl-M	Expl-M	Silent

Précis 26.3.5.30 Nursing (original Table 20.2.7.44)

Age categories, number of studies, and sex	*Consist. score*	*Overviews*	*Countries*	*Time range*	*Non-human*	*Social role theory*			*Evolutionary theory*		*Theological*
						Origin	*Found*	*Biosoc*	*Select*	*ENA*	
All Ages 19F	100.0	-0-	7	1969–2008	-0-	Silent	Silent	Silent	Expl-F	Expl-F	Silent

Assessment. Being involved in the nursing profession can be considered a USD-F. Various lines of evidence point toward evolutionary theory as providing the most persuasive explanations for this sex difference.

26.3.5.31 Physical Scientist

Evidence. Individuals such as physicists, astronomers, and geologists are collectively classified as *physical scientists.* Précis 26.3.5.31 shows that all of the 19 studies on employment as physical scientists concluded that males hold positions in these fields more than do females. The research on this topic included three countries plus one multi-national study.

Explanation. Social role theorists can point to studies suggesting that males are stereotyped as being more interested in physical science occupations (see Table 21.2.4.3 in Volume III) and more likely working in science-related occupations (see Table 21.2.12.18 in Volume III). From such evidence, social role theory predicts that, through social pressure and expectations, males learn to prefer such occupations to greater degrees than females. However, if social learning is responsible, one would expect to find some cultures in which the opposite sex difference would be found. None were found. Regarding the founder effect version of this theory, it's relevance to explaining sex differences in being physical scientists is questionable, since the first human societies would have had no such occupations. The biosocial version of social role theory does not seem to account for the apparent universal tendency for males being involved in this line of work since sex differences in size, physical strength, or the ability to bear children has no obvious connection to being physical scientists.

In the case of both evolutionary theories, the best explanation for more males being in the physical sciences than females seem to involve evidence that males are better at spatial reasoning than females. Studies have indeed indicated that spatial reasoning underlies tendencies to be both interested in, and proficient at, the physical sciences (Humphreys, Lubinski & Yao 1993; Wai, Lubinski & Benbow 2009; Stieff, Dixon et al. 2014). As explained earlier (see Précis 26.2.3.4), the most widely espoused evolutionary explanation for sex differences in spatial reasoning is that males were sexually selected over millennia for focusing their food provisioning on hunting animals of various forms (Jones, Braithwaite & Healy 2003). In more recent centuries, these spatial reasoning traits have been maintained by males using them in many technology-oriented occupations.

Regarding the possibility of neuroandrogenic factors being involved, as hypothesized by ENA theory, findings are mixed. While some studies point toward brain exposure to prenatal testosterone contributing to spatial reasoning (Hines 2006; Knickmeyer & Baron-Cohen 2006), other studies have failed to confirm this association (Herlitz & Lovén 2009;

Précis 26.3.5.31 Physical scientist (original Table 20.2.7.46)

Age categories, number of studies, and sex	*Consist. score*	*Overviews*	*Countries*	*Time range*	*Non-human*	*Social role theory*			*Evolutionary theory*		*Theological*
						Origin	*Found*	*Biosoc*	*Select*	*ENA*	
All Ages 19M	100.0	-0-	3 (1)	1969–2004	-0-	Contra	NoTest	Contra	Expl-M	Expl-M	Silent

Toivainen, Pannini et al. 2018). Part of the reason for the inconsistent evidence could involve the possibility that the relationship is not entirely linear. Specifically, some evidence indicates that prenatal exposure to testosterone that is in the intermediate male-typical range promotes spatial reasoning the most (Berenbaum, Korman & Leveroni 1995).

Circulating testosterone among adult males have also been investigated as possibly being associated with spatial reasoning. Studies in this case are more consistently supporting of a positive relationship (Hooven, Chabris et al. 2004).

Assessment. Our overall assessment is that sex differences in occupations having to do with physical science is a USD-M. The empirical evidence is consistent and reasonable theoretical explanations have been formulated.

26.3.5.32 Politician: Appointed Office Holder

Evidence. When an individual is elected into office, he/she typically appoints others to lead various agencies in the new government. They are often referred to as *cabinet secretaries*. Précis 26.3.5.32 shows that ten studies of sex differences in persons appointed to these government positions were located, all of which concluded that males held them in higher proportions than did females. The studies were based on samples from four countries plus four multi-national studies, and were published over a 48-year time period.

Explanation. No evidence was found of sex stereotypes being applied to these positions. Therefore, no attempt will be made to explain why males would be more prominent based on social role theory. Also, regarding sexual selection theory as well as ENA theory, no evidence was found directly bearing on sex differences in appointed political office holders. Therefore, all five versions of these two theories are listed as silent on this matter.

Assessment. Being appointed to political posts of new governmental administrations appears to be a USD-M. Theoretical explanations need to be developed.

26.3.5.33 Politician: Candidate for Elected Office

Evidence. As shown in Précis 26.3.5.33, research from ten studies carried out in three different countries all concluded that males are more likely to run for political elected office than are females. The pertinent studies were published over a 28-year time frame.

Explanation. Many factors are likely to contribute to sex differences in running for political office, including societal factors (Eagly & Carli 2004). Nevertheless, the possibility of universal sex difference in this regard seems

Précis 26.3.5.32 Politician: appointed office holder (Original Table 20.2.7.49)

Age categories, number of studies, and sex	*Consist. score*	*Overviews*	*Countries*	*Time range*	*Non-human*	*Social role theory*			*Evolutionary theory*		*Theological*
						Origin	*Found*	*Biosoc*	*Select*	*ENA*	
All Ages 10M	100.0	-0-	4 (4)	1955–2003	-0-	Silent	Silent	Silent	Silent	Silent	Silent

Précis 26.3.5.33 Politician: Candidate for elected office (Original Table 20.2.7.50)

Age categories, number of studies, and sex	*Consist. score*	*Overviews*	*Countries*	*Time range*	*Non-human*	*Social role theory*			*Evolutionary theory*		*Theological*
						Origin	*Found*	*Biosoc*	*Select*	*ENA*	
All Ages 10M	100.0	-0-	3	1976–2004	-0-	Silent	Silent	Silent	Poss-M	Poss-M	Silent

difficult to explain other than with an evolutionary perspective. Male tendencies to be more overtly competitive (see Précis 26.3.1.5) and their tendency to value power more than females (see Précis 26.2.7.1) could be among the contributing factors.

Assessment. Running for an elective political office appears to be a USD-M, although more evidence from a wider range of countries is needed to verify this tentative finding. If confirmed, more exploration of this variable at a theoretical level is needed.

26.3.5.34 Politician: Holder of Elected Office

Evidence. Précis 26.3.5.34 summarizes evidence of sex differences in tendencies to hold an elected office. One can see that 127 of the 129 studies concluded that more males held elected office than was true for females. The two exceptions failed to find a sex difference. The studies examined 22 different countries in addition to 22 multi-national studies. The research findings were published over a 90-year time period.

Explanation. Why males would be more likely to hold elected political offices than females could partly have to do with males generally prefer performing leadership roles more than do females (see Table 15.11.2.29 in Volume II). Most studies have also concluded that dominance striving (Table 20.4.1.1 in Volume III) and seeking to fill leadership roles (see Table 20.4.3.1 in Volume III) are both more characteristic of males than of females.

From an evolutionary standpoint, evidence suggests that many females are more sexually attracted to mates with dominant tendencies than are males (Weisfeld & Dillon 2012:25). This suggests that dominance is at least a rough indication of sex differences in the ability to acquire resources, and females throughout the world appear to prefer mates with higher earnings than do males (see Précis 26.2.7.19). Males also appear to value power as a personal attribute more than do females (see Précis 26.2.7.1), a sex differences that may also have been sexual selected (Buss & Kenrick 1998:983).

Regarding the possible involvement of brain exposure to testosterone, a number of studies are relevant. As hypothesized by ENA theory, there appears to be a positive correlation between circulating testosterone and male tendencies to pursue power and to achieve social status (Tremblay 1998; Dabbs & Dabbs 2000; Määttänen, Jokela et al. 2013; Arnocky, Albert et al. 2018). At least one study of females reached the same conclusion (Cashdan 1995). In a complementary fashion, high exposure to *prenatal* testosterone also appears to be positively associated with achieving high social status later in life (Neave, Laing et al. 2003; Moffit & Swanik 2011).

Précis 26.3.5.34 Politician: Holder of elected office (original Table 20.2.7.51)

Age categories, number of studies, and sex	*Consist. score*	*Overviews*	*Countries*	*Time range*	*Non-human*	*Social role theory*			*Evolutionary theory*		*Theological*
						Origin	*Found*	*Biosoc*	*Select*	*ENA*	
All Ages 127M 2x	98.5	-0-	22 (22)	1930–2019	-0-	Silent	Silent	Silent	Expl-M	Expl-M	Silent

Assessment. A vast amount of evidence supports concluding that holding an elected political office is a USD-M. Evolutionary explanations appear to be the best supported by empirical evidence.

26.3.5.35 Social/Welfare Worker

Evidence. Précis 26.3.5.35 summarizes findings of sex differences in employment as a social worker or welfare worker. It indicates that all 11 available studies, conducted in four countries, lead to the conclusion that females were more likely than males to be employed in this line of work. The evidence was published over a 34-year time period.

Explanation. Accounting for why females are more likely than males to be social workers is likely to include the fact that females are more people-oriented (see Précis 26.2.7.43), more empathetic (see Précis 26.2.2.9) and are more inclined to provide help to others (see Précis 26.3.2.7). By referring back to the narrative pertaining to of these documented sex differences, one can see that they appear to have evolutionary and neurohormonal underpinnings.

Assessment. One can tentatively conclude that being a social worker is a USD-F. The best explanations appear to be of an evolutionary nature.

26.3.5.36 Teacher: College Level

Evidence. The findings from 29 studies of sex differences in being college teachers were located. As one can see, all of these studies concluded that males were more involved than females. The studies were conducted in eight countries and an additional four multi-national projects, and were published over a 43-year time period.

Explanation. The consistency of findings regarding males being more involved in teaching at the college level than females is somewhat surprising, given that the vast majority of studies of primary and secondary college teachers have concluded that they are females (see Table 20.2.7.66 in Volume III). While exceptions may one day be found, especially for teachers in certain areas of study, such as the humanities, it seems reasonable to assume that the majority of studies of sex differences in college teaching have to do with subject areas such as the sciences, engineering, and mathematics. These are areas in which males are themselves most likely to major (see Précises 26.3.5.2 through 26.3.5.10). As noted in the narratives surrounding these précises, most of the evidence points toward evolutionary explanations for sex differences in both interest in and tendencies to take advanced courses in science (especially physical science), engineering, and mathematics.

Assessment. Being a college teacher (professor) has been found to be a USD-M. Evolutionary theory may provide a way to explain these sex differences.

Précis 26.3.5.35 Social/welfare worker (original Table 20.2.7.60)

Age categories, number of studies, and sex	*Consist. score*	*Overviews*	*Countries*	*Time range*	*Non-human*	*Social role theory*			*Evolutionary theory*		*Theological*
						Origin	*Found*	*Biosoc*	*Select*	*ENA*	
All Ages 11F	100.0	-0-	4	1969–2003	-0-	Silent	Silent	Silent	Expl-F	Expl-F	Silent

Précis 26.3.5.36 Teacher: College level (original Table 20.2.7.64)

Age categories, number of studies, and sex	*Consist. score*	*Overviews*	*Countries*	*Time range*	*Non-human*	*Social role theory*			*Evolutionary theory*		*Theological*
						Origin	*Found*	*Biosoc*	*Select*	*ENA*	
All Ages 29M	100.0	-0-	8 (4)	1969–2012	-0-	Silent	Silent	Silent	Poss-M	Poss-M	Silent

26.3.5.37 *Performing Indoor Household Cleaning and Chores in General*

Evidence. Well over 300 studies were located having to do with performing household cleaning and maintenance chores. Précis 26.3.5.37 shows that all but three of these studies concluded that females were significantly more likely than males to routinely do housework and indoor household chores. These three exceptions involved samples of children, and were not very specific about the types of household tasks involved (e.g., clean your room, take out the trash). Thirty-one different countries were sampled along with 41 multi-national studies. The publication of the studies spanned 51 years. It is also worth noting that 11 literature reviews reached the same conclusion, i.e., that females perform household chores more than do males.

Explanation. Neither the original social role theory nor the founder effect version of the theory seems to predict a universal sex difference in performing household cleaning and maintenance chores. One might be able to argued that the biosocial version of the social role theory could offer an explanation due to the fact that only females can give birth. In other words, becoming pregnant and giving birth nearly always obliges females to spend more time at home, which in turn would at least partially account for their spending more time performing household chores. However, some of the research on sex differences in performing housework chores has found sex differences even among childless couples, dual-earning couples, and men and women living along (Parkman 2004; Shirley & Wallace 2004; Kanazawa 2005:283).

Evolutionary theory offers a rather persuasive set of arguments for explaining sex differences in time spent engaging in routine housework. The arguments can be derived partly from the fact that all multi-cellular animals need to avoid prolonged exposure to harmful bacterial and viral parasites, contaminants, and toxins in their environment (Curtis, Aunger & Rabie 2004; Kavaliers & Choleris 2018). Given that females spend more time at home, while males are more physically active (Précis 26.3.1.6) and roam about their environment more widely (see Table 20.5.1.3 in Volume III), females should have been favored for taking greater care to maintain household cleanliness, both for themselves and for any offspring. Female tendencies toward cleanliness could also be linked to their having more refined aesthetic tastes (Varella, Valentova & Fernández 2017:14) and greater tendencies toward avoiding risks (see Précis 26.3.1.2 and Table 16.2.3.3 in Volume III) and desires to maintain health (see Table 16.2.3.10 in Volume III). Furthermore, while not all studies agree, most indicate that females have superior olfaction than do males (see Table 9.1.3.2 and Table 9.1.3.3 in Volume II), thus making them more

Précis 26.3.5.37 Performing indoor household chores (original Table 20.3.3.1)

Age categories, number of studies, and sex	*Consist. score*	*Overviews*	*Countries*	*Time range*	*Non-human*	*Social role theory*			*Evolutionary theory*		*Theological*
						Origin	*Found*	*Biosoc*	*Select*	*ENA*	
Child 31F 3x Adolescent 26F Adult 22F Wide Age Range 255F All Ages 334F 3x	91.2 100.0 100.0 100.0 99.1	F: 11Rev	31 (41)	1967–2018	-0-	Silent	Silent	Poss-F	Expl-F	Poss-F	Poss-F

likely to detect foul odors. Overall, evolutionary reasoning leads one to offer an account for sex differences in performing household chores by noting that females have higher standards of cleanliness than do males.

ENA theory would lead one to expect that brain exposure to testosterone and other androgens would diminish tendencies toward personal standards of cleanliness. Unfortunately, no specific evidence was located to either confirm or refute this hypothesis.

There may also be a way to account for sex differences in performing household chores in religious terms. One author has argued that various biblical passages assert that females should be responsible for housework more than males (Voicu 2009:158; Voicu 2019). Similar arguments have been made from the perspective of the Koran (Islam & Karim 2012). Therefore, one could make the case that theological factors are responsible for sex differences in performing household chores, at least among adherents to religions of Middle East origins.

Assessment. The overall conclusion that performing more household maintenance chores (such as cleaning and organizing one's home) is that this is a USD-F. Given the pan-cultural nature of this sex difference and the explanatory arguments outlined above, evolutionary reasoning appears strong. Regarding the possibility of neuroandrogenic factors contributing to this sex differences, research is needed.

26.3.5.38 Performing Outdoor or Construction Household Chores

Evidence. In contrast to routine *indoor* household chores, where females have been shown to dominate (as discussed above), performing *outdoor* chores (such as maintain the yard and keeping motor vehicles in working order as well as doing household construction and maintenance), all ten pertinent studies concluded that males are more involved. Précis 26.3.5.38 shows that these studies were based on samples drawn from four different countries that were published over a 25-year time range.

Explanation. The biosocial version of social role theory could offer a reasonable explanation for why males do more outdoor chores than is the case for females. In particular, this theory asserts that universal sex differences are due to traits such as males being larger and physically stronger, while females are the only sex capable of bearing offspring (Wood & Eagly 2012:56). Given the overwhelming evidence that males are physically stronger (see Précis 26.1.2.8), one can deduce that, because outdoor chores usually require more physical exertion and strength than indoor chores, males would be more likely to take outdoor tasks.

From an evolutionary perspective, there are several ways to account for sex differences in tendencies to perform outdoor or construction-type household tasks. They included the one mentioned above in defense of the

Précis 26.3.5.38 Performing outdoor or construction household chores (Original Table 20.3.3.3)

Age categories, number of studies, and sex	*Consist. score*	*Overviews*	*Countries*	*Time range*	*Non-human*	*Social role theory*			*Evolutionary theory*		*Theological*
						Origin	*Found*	*Biosoc*	*Select*	*ENA*	
All Ages 10M	100.0	-0-	4	1981–2006	-0-	Silent	Silent	Poss-M	Expl-M	Expl-M	Silent

biosocial version of social role theory, i.e., that outdoor tasks usually require more physical exertion and strength than is true for indoor tasks. At least five additional explanations are as follows: First, studies have indicated that, even as children, males are more interested in mechanical and construction-like objects (see Précis 26.2.7.38) and in vehicles (see Précis 26.2.7.39) than are females. Second, as children, when compared to females, males are also more interested in stories about adventures (see Précis 26.2.7.40) and exhibit more exploratory behavior (see Précis 26.3.1.7) and physical activity (see Précis 26.3.1.6) than do females. Third, especially following puberty, males appear to be more likely to navigate their surroundings with little more than a basic sense of direction and distance (see Précis 26.2.3.7). Fourth, throughout life, males have been shown to be more willing to take risks (see Précis 26.3.1.2). Fifth, because *outdoor* home chores are usually less sanitary than those performed *indoors*, females may have stronger preferences for spending time in relatively clean and safe environments (Manhardt 1972; AM Konrad 2003). Taken together, these rather basic sex differences could be largely responsible for more complex sex differences, such as those involving chores and tasks outside the home (rather than chores and tasks inside the home).

Regarding the involvement of brain exposure to testosterone, as hypothesized by ENA theory, readers can refer back to all of the précises cited in the above paragraph. In each of these précises, evidence is cited pertaining to these more basic traits being influenced by either prenatal or post-pubertal (or both). If so, it is reasonable to infer that male tendencies to do more household chores of an outdoor or construction nature is androgen-influenced.

Assessment. Performing outdoor or construction-type household chores can be considered a USD-M. The best explanations appear to be along evolutionary lines.

26.3.5.39 Earnings/Salaries for Workers

Evidence. The sex difference variable for which the single largest number of studies were located had to do with earnings or salaries for workers. So numerous were the studies that they were presented in two tables, one for workers outside of North America (Table 20.3.1.1a) and the other for Canada and the United States (Table 20.3.1.1b). Here, findings from these two tables are combined.

One can see in Précis 26.3.5.39 that all but eight of the 585 studies of sex differences in earnings reached the conclusion that, on average, males earn more than do females. If one looks back to the original tables (located in Chapter 20), one sees that the exceptions all involved some rather atypical research designs and/or samples. In the case of studies reporting

Précis 26.3.5.39 Earnings/salaries for workers (Original Table 20.3.1.1)

Age categories, number of studies, and sex	*Consist. score*	*Overviews*	*Countries*	*Time range*	*Non-human*	*Social role theory*			*Evolutionary theory*		*Theological*
						Origin	*Found*	*Biosoc*	*Select*	*ENA*	
All Ages 579M 6x 2F	98.0	M: 3Met	47 (58)	1961–2021	-0-	Contra	Contra	Poss-M	Expl-M	Expl-M	Poss-M

no significant sex differences, all were conducted in the United States. Three of them had to do with starting salaries for college graduates, two others were hypothetical in the sense of presenting resumes to research participants in order to determine if they would discriminate between the sexes simply on the basis of their having male-typical and female-typical names. The other study involved salaries for mentally retarded workers.

Regarding the three studies reporting that females actually received higher salaries than males, two studies were conducted in Britain and had to do with black workers. Again, because all of these eight exceptional findings involve hypothetical circumstances or subsamples of national populations, we do not consider them genuine contradicts to the generalization that, in all countries, males earn more than females.

It is also informative to examine Table 20.3.1.2 for further insight into the universality of sex differences in average wages between countries. This table pertains to sex differences in wages within specific occupational areas. As shown in Table 24.20.11, row 4 (of the present volume), there were a total of 170 studies, 161 of which indicated that males received higher wages than female, eight indicated that there were no significant differences and one study actually reported higher wages for females (the latter having been conducted among meat processing workers). The consistency score for this set of studies was 94.2, slightly shy of the 95.0 cutoff. It is worth noting that seven of the eight studies reporting no significant sex differences within occupational categories were conducted in the United States, all of which involved fairly high-status occupational areas such as college professors, engineers, and corporate executive officers. Otherwise, the most distinctive feature of these seven studies is that they had to do with entry level hiring. In other words, when studies focus within occupational categories on entry level wages, at least in the United States, there are few if any sex differences in earnings. But, after a few years of experience, this table indicates that males invariably pull ahead.

Explanation. All five theories offer explanations for sex differences in earnings. It is even possible to explain this difference in theological terms. Therefore, we will examine each of these explanations one by one.

1 In the case of the original social role theory, it asserts that sex differences in cognitive and behavioral traits are the result of culturally varied stereotypes and discrimination. If so, there should be cultures in which women earn more than men or, at least, where there are no significant sex differences. Having found no such studies casts serious doubt on this particular version of social role theory.
2 The founder effect version of social role theory contends that some universal sex differences exist because some sex roles began in the very earliest human society that was then "faithfully passed down thousands

of times over" (Fausto-Sterling 1992:199). A major problem with this explanation has to do with the term *primordial human cultures*. These would be foraging societies dating back tens of thousands of years ago, and the earliest known use of money (with which to pay workers) only dates back about 3,000 years (G Davies 2010; Peneder 2022).

3 The biosocial version of social role theory asserts that whatever *universal* sex differences exist are the result of males being larger and stronger or females being the only sex capable of child birth (Wood & Eagly 2012). The strength factor could explain why males receive higher pay for jobs involving strenuous manual labor, but not for most white-collar occupations. The fact that pregnancy can sometimes interrupt career stability could also play a role in diminishing the average salaries of women according to the biosocial role theory.

   It is worth noting that nearly all of the studies that have failed to document significant sex differences in wages involve entry-level employment, where six studies reported no significant sex differences in earnings (see Table 20.3.1.2 in Volume III). Biosocial role theorists could argue that this has to do with women often leaving their jobs, at least temporarily or periodically, in order to care for newborn offspring.

4 According to the evolutionary sexual selection theory, males earn more than females in all societies because, in all societies, a female can reproduce more reliably if they mate with a male who provides financial support to her while she bears and cares for offspring (Emlen & Oring 1977; Ellis 2001). Without this sexual division of labor, females will often limit their reproductive potential. If one assumes that females have evolved a tendency to prefer males with the ability to be a good provider, males should evolve a wide range of resource procuring abilities (as evidence suggests, see Précis 26.3.5.21).

5 Building atop the evolutionary sexual selection theory, ENA theory hypothesizes that brain exposure to androgens describe the primary biochemical/physiological mechanisms that produce average sex differences in traits. In the case of earnings, some evidence points toward a positive correlation between prenatal exposure to testosterone and earnings in adulthood among males (but not among females) (Gielen, Holmes & Myers 2016). According to another study, circulating testosterone among males was positively correlated with income (Luoto, Krama et al. 2021).

6 Finally, average male-female differences in earnings may be a sex difference for which there may be a discernable religious explanation. In a Biblical passage, it is written that Israelites should pay males and females (at least between the ages of 20 and 50) differently. The guidance stated in Leviticus (27:1–4) is that males should receive 50 silver shekels for work while females are given 30 shekels. The policy of

paying females 60% of what males are paid, presumably for an equivalent amount of work, could be the result of wanting to follow this Biblical guidance. However, one would think that those who would be most likely to do so would be in Judeo-Christian countries. Thus, it does not seem to provide an explanation for the world-wide sex differences in wages. Instead, this passage may simply *reflect* a much broader recognition that males end up earning more than females.

Assessment. Given the exceedingly large number of studies (i.e., 587) along with the number of countries in which these studies were conducted (i.e., 47, plus 58 multi-country studies), there is little doubt that the average earning by workers is a USD-M.

26.3.5.40 *Effects of Worker Sex Ratio on an Occupation's Average Salary*

Evidence. A large number of studies have investigated the relationship between the proportion of workers who are women in an occupation and the average salaries paid to those working in the occupation. As shown in Précis 26.3.5.40, of the 51 studies, 50 concluded that as the ratio of females in an occupation increase, the average salary decreases; the one remaining study simply found no significant relationship. Five different countries were sampled (plus one multi-country study) published over a 25-year time frame.

Explanation. For this particular variable, none of the six explanations appear to be particularly relevant. Instead, we would propose that primary factor has to do with the fact that as the number of candidates for a job of any type increases, the wages that need to be offered to people tend to be less.

Assessment. The tendency for an occupation's worker sex ratio to affect salaries is a USD-F.

26.3.5.41 *Financial Well-Being*

Evidence. Ten studies of sex differences in people's feelings of financial well-being were located that were conducted in seven different countries. As shown in Précis 26.3.5.41, all of the studies agree that males consider themselves better off in financial terms than do females.

Explanation. None of the five theories seems to offer any specific explanation for sex differences in feelings of financial well-being. If financial well-being is assessed on the basis of some objective criteria, such as earnings, one can refer back to Précis 26.3.5.39.

Assessment. Even though only the minimum of ten studies were located, and no theory of sex differences seems to pertain to sex differences in

Précis 26.3.5.40 Effects of a worker sex ratio on an occupation's average salary (original Table 20.3.1.16)

Age categories, number of studies, and sex	*Consist. score*	*Overviews*	*Countries*	*Time range*	*Non-human*	*Social role theory*			*Evolutionary theory*		*Theological*
						Origin	*Found*	*Biosoc*	*Select*	ENA	
All Ages 50F decrease 1x	98.0	-0-	5 (1)	1979–2004	-0-	Silent	Silent	Silent	Silent	Silent	Silent

Précis 26.3.5.41 Financial well-being (original Table 20.3.1.18)

Age categories, number of studies, and sex	*Consist. score*	*Overviews*	*Countries*	*Time range*	*Non-human*	*Social role theory*			*Evolutionary theory*		*Theological*
						Origin	*Found*	*Biosoc*	*Select*	ENA	
All Ages 10M	100.0	M: 1Gen	7	1992–2003	-0-	Silent	Silent	Silent	Silent	Silent	Silent

financial well-being, feelings of financial well-being can be tentatively considered a USD-M.

26.3.5.42 *Achieving Prominence in One's Occupational Field*

Evidence. Précis 26.3.5.42 has to do with studies of sex differences in the achievement of high occupational status. Based on 138 studies drawn from 22 individual countries (plus 20 multi-country studies), one can see that the evidence strongly supports concluding that males tend to achieve higher job status, authority, and eminence than is the case for females. The pertinent studies were conducted in four different countries (plus one multi-country study), published over a 76-year time frame.

Explanation. The original version of social role would not predict that a sex differences in status, authority, and eminence would be universal, but both of the other two versions could do so. The founder effect version could assert that the first human societies exhibited such sex differences, and that all human societies ever since have simply maintained this sociocultural tradition through stereotyping and learning. In the case of the biosocial version of social role theory, stereotyping and learning, of course, are also central to explaining the sex difference. However, proponents of this theory also argue that body size, strength, and ability to give birth are also involved. Thus, they could argue that that males are more likely to attain high job status, authority, and eminence because they are larger and physically stronger, thereby more intimidating to other high-status seekers. Furthermore, female pregnancy and subsequent child care responsibilities could periodically interrupt the time they devote to striving for such status.

Turning to evolutionary explanations, the sexual selection version would assert that females have evolved a tendency to prefer mates of high status as a means of obtaining resources with which to reproduce (Ellis 1991b). According to this theoretical line of reasoning, males have evolved traits such a greater body size and strength, along with greater tendencies to be aggressive and competitive, and greater willingness to take risks, in order to achieve status (Van Kleef, Heerdink et al. 2021). Of course, the primary evolutionary "payoff" to all of these risky strategies by males is a greater chance of obtaining mating opportunities and any resulting offspring (Gangestad & Simpson 2000; Kruger, Wang & Wilke 2007). Additional arguments along these lines were made earlier regarding sex differences in income (see Précis 26.3.5.41).

Regarding the ENA version of evolutionary theory, the sexual selection arguments are combined with the view that brain exposure to androgens is also be involved. Specifically, enhanced status striving should be at least partly due to brain exposure to testosterone or other androgens. Evidence supporting this premise include studies indicating that, at least among

Précis 26.3.5.42 Achieving prominence in one's occupational field (original Table 20.4.3.4)

Age categories, number of studies, and sex	*Consist. score*	*Overviews*	*Countries*	*Time range*	*Non-human*	*Social role theory*			*Evolutionary theory*		*Theological*
						Origin	*Found*	*Biosoc*	*Select*	*ENA*	
Adult 146M 2x	98.6	M: 1Met	22 (20)	1930–2006	-0-	Silent	Silent	Silent	Expl-M	Expl-M	Silent

males, circulating testosterone is positively correlated with status-striving (Mazur & Booth 1998; Nave, Nadler et al. 2018).

Other studies suggest that testosterone interacts with stress hormones (especially cortisol), to impact human status-striving and achievement. Basically, status-striving is positively correlated with circulating testosterone, but inversely correlated with cortisol (Mehta & Josephs 2010; Casto & Edwards 2016; Grebe, Del Gildice et al. 2019).

It is also worth noting that *dominance* among non-humans and *social status* in humans are essentially parallel concepts (Cheng, Tracy & Henrich 2010; Chen Zeng, Cheng & Henrich 2022). Accordingly, studies have found positive correlations between testosterone and dominance-striving among various non-human species (MN Muller 2017; Coyne 2018).

Assessment. The evidence strongly supports concluding that achieving high job status, authority, and eminence (i.e., becoming prominent in one's occupational field) is a USD-M. The strongest theoretical arguments seem to involve evolutionary argument that males who attain relatively high-status pass on their genes at higher rates than do males who make minimal effort in this regard.

26.3.5.43 Upward Social Status Mobility

Evidence. Social status can be measured in many ways, but it usually has to do with accessing resources in one form or another (Ellis, Hoskin & Ratnasingam 2019: 1–5). Social status mobility has to do with whether or not an individual's social status early in life (usually based on that of his or her family) is higher or lower than the individual's social status later in life.

One can see in Précis 26.3.5.43 that all but one of 99 pertinent studies of upward social status mobility (often simply called *upward social mobility*) have concluded that males tend to advance in social status to greater degree than do females. Most assessments of such mobility compare an individual's status of origin (i.e., the social status of one or both parents) to his or her own status at a comparable age based on years of education or income. The studies were conducted in 12 different countries along with four multi-national studies. The research was published over a 52-year time frame.

Explanation. Males are stereotyped as being more aggressive (Table 21.2.7.6) and competitive (Table 21.2.7.16). From this, all three social role theories would predict that males would at least *strive* for higher social status than would females. However, the original version of the theory would not predict cultural universality in this regard. The founder effect version would explain the universality of this sex difference by stating that it happened to be the case in the first human society and has continued to be "passed down thousands of times over" (Fausto-Sterling

Précis 26.3.5.43 Upward social status mobility (original Table 20.4.3.5)

Age categories, number of studies, and sex	*Consist. score*	*Overviews*	*Countries*	*Time range*	*Non-human*	*Social role theory*			*Evolutionary theory*		*Theological*
						Origin	*Found*	*Biosoc*	*Select*	*ENA*	
Adult 98M 1x	99.0	-0-	12 (4)	1964–2016	-0-	Contra	Contra	Poss-M	Expl-M	Expl-M	Silent

1992:199). However, it is unlikely that the earliest human society would have had a social hierarchy in the sense found in today's societies. In the case of the biosocial version of social role theory, the basic argument would be that either due to males being larger or stronger or females being uniquely capable of bearing children (Wood & Eagly 2012), males in all societies are able to be more upwardly mobile than females.

Both evolutionary theories focus on the fact that males have more to gain in reproductive terms by striving for, and especially actually achieving, high social status than do females. Specifically, because females prefer high status mates more than do males (see Précis 26.2.7.19), it is likely that the most status-striving males will attract more mates than females when the two sexes strive for status (or actually achieve status) to the same degree.

In the case of ENA theory, it adds to sexual selection theory the assertion that neuroandrogenic factors are crucial for status striving. In this regard, there is considerable evidence that high testosterone is linked to status-striving (Mazur & Booth 1998; VJ Grant & France 2001; Archer 2006:332; Terburg & van Honk 2013; van Honk, Bos & Terburg 2014; Vermeer, Krol et al. 2020; also see Simmons & Roney 2011).

Assessment. Overall, upward social status mobility can be considered a USD-M. Evolutionary explanation appears to be stronger than social role explanations.

26.3.6 Sex Stereotypes

Substantial research has been conducted since the 1930s on sex stereotypes. These refer to people's *beliefs* about the existence of sex differences. Only ten sex stereotypes were for which at least ten studies were located.

The purpose of listing sex stereotypes here is not to offer theoretical explanations for them, but simply to determine if these stereotypes appear to be true. In other words, are any of the well-documented stereotypes consistent with research findings summarized elsewhere in this book? Since no attempt will be made to theoretically explain sex stereotypes, the "Explanation" portion of the narratives are omitted and the shaded portion of each précis is reduced down to a single column.

26.3.6.1 Sex Stereotyped Math Ability

Evidence. As shown in Précis 26.3.6.1, 21 studies assessed people's stereotypes regarding sex differences in math ability, all of which concluded that males were more often stereotyped as being better at math than females. The studies were conducted in eight countries over a 36-year time frame. This conclusion was also reached by one meta-analysis.

Précis 26.3.6.1 Sex stereotyped math ability (original Table 21.1.2.3)

Age categories, number of studies, and sex	*Consist. score*	*Overviews*	*Countries*	*Time range*	*Empirical assessment*
All Ages 21M	100.0	M: 1Met	8	1976–2012	Questionable

Assessment. The *stereotype* that males are better at math has been found in eight different countries. However, from an *empirical* standpoint, the accuracy of this stereotype is subject to question (see Table 25.12.4 in this volume).

26.3.6.2 Sex Stereotyped Emotionality

Evidence. As shown in Précis 26.3.6.2, all 27 studies having to do with sex stereotypes in emotionality concluded that females are more emotional than males. These studies were based on samples drawn from two specific countries in addition to two multi-national research projects published over a 34-year time frame.

Précis 26.3.6.2 Sex stereotyped emotionality (original Table 21.2.5.1)

Age categories, number of studies, and sex	*Consist. score*	*Overviews*	*Countries*	*Time range*	*Empirical assessment*
Adult 21F All Ages 27F	100.0 100.0	-0-	2 (2)	1968–2002	Somewhat questionable

Assessment. While females are very consistently stereotyped as being more emotional than males, the empirical evidence is somewhat mixed. Specifically, by referring back to Table 25.10.1, one will see that the consistency score for emotionality in general is only 80.0, substantially below the 95.0 score being used here to warrant being considered a universal sex difference.

26.3.6.3 Sex Stereotyped Emotional Expressiveness

Evidence. When it comes to stereotypes about sex differences in emotional expressiveness, Précis 26.3.6.3 indicates that females are stereotyped as being so more expressive than males. The pertinent studies were conducted in two countries, with publication dates spanning 31 years.

Précis 26.3.6.3 Sex stereotyped emotional expressiveness (original Table 21.2.6.2)

Age categories, number of studies, and sex	*Consist. score*	*Overviews*	*Countries*	*Time range*	*Empirical assessment*
Adult 12F All Ages 12F	100.0 100.0	-0-	2	1972–2003	Somewhat questionable

Assessment. The universality of *actual* sex differences in emotional expressiveness are summarized in Table 25.10.9. This table indicates that a substantial majority of studies have indicated that females are more emotionally expressive than males. However, the consistency score was just 87.5, which was below the 95.0 cutoff score for being considered a cultural universal sex difference.

26.3.6.4 *Sex Stereotyped Fearfulness*

Evidence. Findings have to do with stereotypes about sex differences in fearfulness are summarized in Précis 26.3.6.5. One can see that results from ten studies were located, all of which agreed that females were thought to be more fearful than males.

Assessment. If one refers back to Table 25.10.3, one will see that all but one of the 97 studies of sex differences in fearfulness in general concluded that females surpassed males. This table also shows that nearly all studies have concluded that females are more fearful of animals, of crime victimization, and of being negatively evaluated by others.

Précis 26.3.6.4 Sex stereotyped fearfulness (Original Table 21.2.6.14)

Age categories, number of studies, and sex	*Consist. score*	*Overviews*	*Countries*	*Time range*	*Empirical assessment*
All Ages 10F	100.0	-0-	1 (1)	1957–2003	Substantially confirmed

26.3.6.5 *Sex Stereotyped Adventurousness*

Evidence. Ten studies examined stereotypes having to do with sex differences in adventurousness. Précis 26.3.6.5 shows that all of these studies found males being stereotyped as being more adventurous than females. Four countries were sampled, plus two multi-national studies. The publication of these studies spanned 44 years.

Précis 26.3.6.5 Sex stereotyped adventurousness (Original Table 21.2.7.3)

Age categories, number of studies, and sex	*Consist. score*	*Overviews*	*Countries*	*Time range*	*Empirical assessment*
All Ages 10M	100.0	-0-	4 (2)	1957–2001	Largely confirmed

Assessment. According to most studies, males appear to actually be more adventurous than females (see Table 16.2.4.2 in Volume III). Therefore, this stereotype can be considered largely confirmed.

26.3.6.6 *Sex Stereotyped Physical Aggressiveness*

Evidence. Précis 26.3.6.6 shows that 26 studies of stereotyped sex differences in physical aggression were located, and that they all agreed that males were more physically aggressive than females. Four countries were sampled in these studies along with one one multi-country study. The studies were published over a 58-year time frame.

Précis 26.3.6.6 Sex stereotyped physical aggressiveness (Original Table 21.2.7.5)

Age categories, number of studies, and sex	*Consist. score*	*Overviews*	*Countries*	*Time range*	*Empirical assessment*
Adult 18M All Ages 26M	100.0 100.0	-0-	4 (1)	1957–2015	Largely confirmed

Assessment. So, is this stereotype true? The answer is not entirely clear-cut. If one goes back to Table 25.17.16 in this volume, one will see that, while a substantial majority of studies have concluded that males are more physically aggressive than females, only in the case of warfare was the consistency score beyond the 95.0 cutoff point. Nevertheless, it is worth adding that researchers in some of these studies used rather minor or vague definitions of *physical aggression*.

Another way of assessing sex differences in physical aggression comes from the field of criminology. Nearly all forms of violent criminality have to do with serious forms of physical aggression. In this regard, Table 25.19.1 (in this volume) includes summaries of sex differences in violent crime, including homicide. One can see that there is unanimous agreement that males are more often convicted of these offenses in every country sampled. Furthermore, Table 25.19.2 (in this volume) shows that males are more likely self-report committing violent offenses than are females.

Overall, it seems far to conclude that the stereotype of males being more physically aggressive appears to be true, especially regarding serious (even deadly) forms of aggression.

26.3.6.7 *Sex Stereotyped Being Independent*

Evidence. As shown in Précis 26.3.6.7, males are stereotyped as being more independent than females according to a total of 17 studies conducted in three countries (plus two multi-country studies). The pertinent studies were published over a 51-year time frame.

Précis 26.3.6.7 Sex stereotyped being independent (Original Table 21.2.7.39)

Age categories, number of studies, and sex	*Consist. score*	*Overviews*	*Countries*	*Time range*	*Empirical assessment*
Adult 13M All Ages 17M	100.0 100.0	-0-	3 (2)	1957–2008	Substantially confirmed

Assessment. This stereotype is likelihood to be true even though the consistency score for sex differences in traits such as self-sufficiency, resourcefulness, and independence was below the 95.0 cutoff. Specifically, for all age groups combined, the score was 89.2 based, while for adults, it was 93.7 (see Table 25.16.4 in this volume).

26.3.6.8 *Sex Stereotyped Being Nurturing*

Evidence. Females are stereotyped as being more nurturing than are males. Précis 26.3.6.8 shows that all ten of the studies on this topic came to this conclusion. The studies were based on samples from two countries spanning 57-years in terms of their publication.

Précis 26.3.6.8 Sex stereotyped being nurturing (Original Table 21.2.9.10)

Age categories, number of studies, and sex	*Consist. score*	*Overviews*	*Countries*	*Time range*	*Empirical assessment*
All Ages 10F	100.0	-0-	2	1957–2014	Very well documented

Assessment. Actual studies of sex differences in nurturing have concluded that, when they are compared to males, females provide more care to offspring (Précis 26.3.2.6), more care to family members other than offspring (Précis 26.3.2.6), and more care to others in general (Précis 26317.7). Therefore, the stereotype of females being more nurturing is an extremely well-documented stereotype.

26.3.6.9 Sex Stereotype About Dominance and Leadership

Evidence. Thirty-one studies of sex stereotypes having to do with traits such as dominance and leadership were found. Précis 26.3.6.9 shows that all but one of these studies concluded that males surpassed females. The only exception failed to find a sex difference. Three countries were sampled in these studies, that were published over a 51-year time frame. One meta-analysis also concluded males were stereotyped as posing more dominance or leadership tendencies than females.

Précis 26.3.6.9 Sex stereotype about dominance and leadership (Original Table 21.2.11.2)

Age categories, number of studies, and sex	*Consist. score*	*Overviews*	*Countries*	*Time range*	*Empirical assessment*
Adult 29M 1x All Ages 30M 1x	96.7 96.8	M: 1Met	3	1957–2008	Generally supportive

Assessment. Considerable research on actual sex differences in dominance and leadership traits has been published over the years. As shown in Table 25.20.14 as well as Table 25.20.14, one can see that most of the findings point toward males exhibiting these traits more, although most of the consistency scores were below the 95.0 cutoff that was stipulated for considering these traits universal sex differences. The only two exceptions involved occupational achievement and social mobility, both of which were higher in males.

26.3.6.10 Sex Stereotyped Higher in Earnings or Wealth

Evidence. Précis 26.3.6.10 reveals that all of the ten studies on this topic found that males were more likely to be stereotyped as earning more or as being wealthier than females. The studies were conducted in three countries.

Précis 26.3.6.10 Sex stereotyped higher in earnings or wealth (Original Table 21.2.11.15)

Age categories, number of studies, and sex	*Consist. score*	*Overviews*	*Countries*	*Time range*	*Empirical assessment*
All Ages 10M	100.0	-0-	3	1982–2014	Supportive

Assessment. The assertion that males earn more or are wealthier than females has been overwhelmingly supported (see Précis 26.3.5.39 and Précis 26.3.5.41). Therefore, this stereotype is supported by essentially all empirical evidence.

26.3.7 *Attitudes and Actions toward Others, and Portrayals in the Mass Media*

When people interact, they often do so in different ways, depending on the sex of the other person (or persons). Also, people are often portrayed differently in the mass media depending on their sex. Findings in this regard for which consistency scores of 95.0 or higher are summarized below. Readers will see that we were unable to identify any theoretical explanation for several of these traits.

26.3.7.1 *Correctly Judging Sex Based on Writing Style*

Evidence. Setting aside the issue of penmanship, can people correctly guess who wrote a short essay simply by assessing the style of writing? Précis 26.3.7.1 shows that 15 studies conducted in four different countries all found that most people can to a level that is beyond mere chance. The studies were published over a 95-year time frame.

Explanation. As with sex stereotypes, none of the five theories of sex differences offers any explanation for why people seem to be able to correctly guess the sex of authors of written information.

Assessment. Correctly judging the sex of a writer based on writing style can be considered a USD-M.

26.3.7.2 *Choice of Social Interactants, Own Sex*

Evidence. Is there a sex difference regarding wanting to socially interact with members of one's own sex? Précis 26.3.7.24 reveals that 24 studies suggests that such a sex difference does exist. The studies were conducted in five countries published over a 72-year time frame. All 24 studies concluded that males preferred social interactions with their own sex more than do females. One study of primates found the same result.

Explanation. No theoretical explanation was located.

Assessment. Given the substantial number of studies extending over five countries and a substantial publication time frame, we declare the tendency to choose one's own sex as a social interactant more to be a USD-M.

26.3.7.3 *Risk Assessment*

Evidence. Précis 26.3.7.3 shows that 14 studies all concluded that females rated the risk involved in various activities as being higher than the

Précis 26.3.7.1 Correctly judging sex based on writing style (original Table 22.1.1.1)

Age categories, number of studies, and sex	*Consist. score*	*Overviews*	*Countries*	*Time range*	*Non-human*	*Social role theory*			*Evolutionary theory*		*Theological*
						Origin	*Found*	*Biosoc*	*Select*	*ENA*	
All Ages above chance, 15 studies	100.0	-0-	4	1910–2005	-0-	Silent	Silent	Silent	Silent	Silent	Silent

Précis 26.3.7.2 Choice of social interactants, own sex (original Table 22.1.5.2)

Age categories, number of studies, and sex	*Consist. score*	*Overviews*	*Countries*	*Time range*	*Non-human*	*Social role theory*			*Evolutionary theory*		*Theological*
						Origin	*Found*	*Biosoc*	*Select*	*ENA*	
Child 21M All Ages 24M	100.0 100.0	-0-	5	1927–1999	1M primates	Silent	Silent	Silent	Silent	Silent	Silent

Précis 26.3.7.3 Risk assessment (original Table 22.1.6.8)

Age categories, number of studies, and sex	*Consist. score*	*Overviews*	*Countries*	*Time range*	*Non-human*	*Social role theory*			*Evolutionary theory*		*Theological*
						Origin	*Found*	*Biosoc*	*Select*	*ENA*	
All Ages 14F	100.0	-0-	6	1984–2014	-0-	Silent	Silent	Silent	Poss-F	Poss-F	Silent

assessments provided by males. The studies were conducted in six different countries, published over a 30-year time frame.

Explanation. This variable may be explainable, at least in part, based on both evolutionary theories. As noted elsewhere, males are willing to take risks to a greater degree than are females (see Précis 26.3.1.2). This sex difference seems to be at least partly due to the fact that males of nearly all species have more to gain in reproductive terms by taking at least moderate risks (Kruger & Nesse 2004). Also, substantial evidence supports the hypothesis that brain exposure to testosterone promotes risk-taking (*review*: Apicela, Carre & Dreber 2015; *meta-analysis*: Kurath & Mata 2018). Both of these lines of reasoning suggest that females tend to be more cautions and inclined to perceive risks as being greater than do males.

Assessment. We declare risk assessment to be a USD-F based on the strength of the evidence and the fact that this sex difference can be theoretically explained.

26.3.7.4 Physical Punishment of Offspring by Parents

Evidence. Sex differences in children being physically punished by their parents have been the subject of considerable research. Précis 26.3.7.4 shows that all but one of 37 studies found that males received more physical punishment than do females. The only exception found no significant sex difference. Studies of this variable were conducted in five countries extending over a 45-year publication time frame. One meta-analysis also found male offspring received more parental physical punishment than female offspring.

Explanation. Only the evolutionary theories seem to offer explanations for this sex difference. These theories do so by noting that males have been selected for being more physically active (see Précis 26.3.1.6), more externalizing (see Précis 26.2.6.1) and antisocial (see Précis 26.2.6.9). Presumably, when these traits are exhibited during childhood, particularly in the form of conduct disorders (see Précis 26.2.6.8), they often illicit disciplinary behavior by parents.

It may also be relevant to note that most research indicates that males are able to tolerate pain to a greater degree than females (see Table 4.2.1.3). If so, males would be likely to withstand harsher punishment before seeing the connection between their own troublesome behavior and any adverse consequences. Along related lines, females are more fearful than males (see Précis 26.2.2.2), which would make them more likely to behave in ways that limit physical punishment by parents.

As to the possible involvement of brain exposure to testosterone, as hypothesized by ENA theory, considerable supportive evidence indicates that testosterone promotes behavior that promotes physical activity,

Précis 26.3.7.4 Physical punishment of offspring by parents (original Table 22.2.1.12)

Age categories, number of studies, and sex	*Consist. score*	*Overviews*	*Countries*	*Time range*	*Non-human*	*Social role theory*			*Evolutionary theory*		*Theological*
						Origin	*Found*	*Biosoc*	*Select*	*ENA*	
Child 28M All Ages 36M 1x	100.0 97.3	M: 1Met	5	1961–2006	3M 3F primates	Silent	Silent	Silent	Expl-M	Expl-M	Silent

externalizing, and antisocial behavior (see the narratives pertaining to the précises cited above). In addition, testosterone exposure appears to promote pain tolerance (Kerem, Akbayrak et al. 2002; Apkhazava, Kvachadze et al. 2018) and reduce fearfulness (Boissy & Bouissou 1994; Bos, van Honk et al. 2013).

Assessment. Given that the pattern of findings regarding sex differences in offspring being physically punished by parents and the reasonableness of the theoretical explanations, we declare this to be a USD-M.

26.3.7.5 *Parental Supervision of Offspring*

Evidence. Does the sex of an offspring influence the amount of parental supervision that the offspring receives? Précis 26.3.7.5 shows that all 18 studies found female offspring received more parental supervision than male offspring. The studies were conducted in four countries and there was one multi-national study.

Explanation. Accounting for sex differences regarding being supervised by one's parents does not seem to be specifically addressed by any of the five theories of sex differences. However, some of the explanation may involve evidence that females are more fearful than males (see Précis 26.2.2.2). Parents may detect this fearfulness and provide more supervision and protection to females to help reassure them.

Assessment. Despite the lack of any specific theoretical explanation, we declare parental supervision of offspring a USD-F in light of the number of studies reporting differences and the number of countries sampled.

26.3.7.6 *Age of People Featured in Mass Media*

Evidence. Précis 26.3.7.6 shows that males portrayed in the media tend to be older than females portrayed in the media. That conclusion was reached by all 19 studies located on this topic. The studies were conducted in five countries along with two multi-national studies.

Explanation. To explain age disparities in the portrayals of males and females in the mass media from an evolutionary perspective, one can first recognize that many mass media portrayals have a romantic theme. In this regard, nearly all studies have concluded that males are older than the females they date and marry (Précis 26.3.2.41). Given that most romantic depictions usually resemble typical patterns in terms of the ages of the couples being portrayed, one would expect that males would be older on average than the females they date and marry in the mass media. Then, to account for why such a sex difference exists in real-world male-female relationships, one can notice that it appears to reflect universal preferences. Specifically, males prefer younger mates (see Précis 26.2.7.17) and females

Précis 26.3.7.5 Parental supervision of offspring (Original Table 22.2.1.23)

Age categories, number of studies, and sex	*Consist. score*	*Overviews*	*Countries*	*Time range*	*Non-human*	*Social role theory*			*Evolutionary theory*		*Theological*
						Origin	*Found*	*Biosoc*	*Select*	*ENA*	
All Ages 18F	100.0	-0-	4 (1)	1968–2013	-0-	Silent	Silent	Silent	Silent	Silent	Silent

Précis 26.3.7.6 Age of people featured in mass media (original Table 22.5.2.1)

Age categories, number of studies, and sex	*Consist. score*	*Overviews*	*Countries*	*Time range*	*Non-human*	*Social role theory*			*Evolutionary theory*		*Theological*
						Origin	*Found*	*Biosoc*	*Select*	*ENA*	
Adult 11M All Ages 19M	100.0 100.0	-0-	5 (2)	1972–2005	-0-	Silent	Silent	Silent	Expl-M	Poss-M	Silent

prefer older mates (see Précis 26.2.7.18). If one returns to either of these latter two précises, one will see that substantially well-documented proposals along evolutionary lines have been offered.

Assessment. Many studies from a range of different countries have found that older age persons featured in mass media is a USD-M. Evolutionary explanations appear to be the most persuasive.

26.3.7.7 Portrayals of Aggression and Violence

Evidence. Males are more likely to be portrayed as violent and aggressive than are females. Précis 26.3.7.7 shows that this was true for all 20 studies located. The studies were conducted in two specific countries plus one multi-national study, published over a 30-year time span.

Explanation. If one assumes that sex differences in mass media portrayals of aggression and violence are roughly reflective of sex differences in real life, it is to be expected that males would be portrayed as being the aggressor more. Of course, this begs the question of why there are sex differences in aggression and violence in real life. At least regarding physical aggression, the vast majority of studies have concluded that males surpass females (see Table 25.17.16). This is especially true when aggression of a criminal nature is considered (see Précis 26.3.4.2 through Précis 26.3.4.4 and Précis 26.3.4.98). Even among non-human species, nearly all studies have concluded that males are more physically aggressive than females (see Table 17.6.1.1e in Volume III).

Males are stereotyped as being more aggressive and violent than females (see Précis 26.3.6.6). Therefore, social role theorists can argue that this sex difference is the result of sex role training and expectations. However, the original version of this theory would not lead one to expect the sex difference to be found in all societies (as evidence collected so far indicates that it is).

The founder effect version could assert that the sex difference was established in the first human societies and then has been maintained ever sense. Unfortunately, there is no basis for empirically testing this proposal.

Regarding the biosocial version of the social role theory, it asserts that universal sex differences in behavior are the result of males being larger and stronger and/or females being the only sex capable of giving birth. Why any of these sex differences would affect the probability of physical aggression and violence is not obvious. In fact, one could even argue that greater body size and strength might actually *lessen* the need the need to be aggressive and violent, not enhance it.

Overall, evolutionary theory offers the best explanations for why males are more physically aggressive and violent than female in nearly all species, including our own. Sexual selection theory asserts that aggression is part

Précis 26.3.7.7 Portrayals of aggression and violence (original Table 22.5.3.5)

Age categories, number of studies, and sex	*Consist. score*	*Overviews*	*Countries*	*Time range*	*Non-human*	*Social role theory*			*Evolutionary theory*		*Theological*
						Origin	*Found*	*Biosoc*	*Select*	*ENA*	
Wide Age Range 16M	100.0	-0-	2 (1)	1974–2004	-0-	Contra	NoTest	Silent	Poss-M	Poss-M	Silent
All Ages 20M	100.0										

of the behavior patterns that help males compete for and maintain control over resources, thereby attracting females (Smuts 1995; Geary & Flinn 2001; Watts 2010). Evolutionary theory's ENA version adds to this base argument that brain exposure to testosterone promotes aggressive and violent behavior (Ellis 2001; Ellis & Hoskin 2015).

Ultimately, one can say that males are portrayed as more physically aggressive and violent in the mass media because they *are* more physically aggressive and violent. And, males are that way mainly because such behavior tends to serve their reproductive interests more than the same behavior by females serves their reproductive interests.

Assessment. We declare mass media portrayals of greater aggression and violence to be a USD-M. The most persuasive explanation for this sex difference is along evolutionary lines.

26.3.7.8 Portrayals of Dominance and Social Status

Evidence. Précis 26.3.7.8 shows that males are more likely to be portrayed as being dominant and having high status. The research on this sex difference were based on samples from four different countries in studies published over a 27-year time frame.

Explanation. Nearly all studies of dominance and social status have concluded that males surpass females (see Précis 26.3.5.39 through Précis 29.3.20.45). In evolutionary terms, this sex difference is due to females exercising a preference for mates who have the ability to provide resources that can be used to bear and rear offspring (Ellis 2001). If one assumes that mass media portrayals *usually* reflect reality, it can be expected that males would be portrayed as being higher in dominance or social status than is the case for females. While there are good theoretical explanations for why social status indicators are higher among males than females, there is no theoretical explanation for why males would be portrayed in the mass media as higher other than the fact that to assert that most of these portrayals tend to reflect reality.

Assessment. We declare the portrayals of dominance and social status to be a USD-M even though there is not actual theoretical explanation for this portrayal.

26.3.7.9 Portrayals of Working Inside the Home

Evidence. When people are portrayed working inside the home, are they more likely to be males or females? Précis 26.3.7.9 shows that all but one of 50 studies concluded that the answer is females. The results were based on studies of 15 different countries plus one multi-national study covering a 43-year time frame.

Précis 26.3.7.8 Portrayals of dominance and social status (original Table 22.5.3.7)

Age categories, number of studies, and sex	*Consist. score*	*Overviews*	*Countries*	*Time range*	*Non-human*	*Social role theory*			*Evolutionary theory*		*Theological*
						Origin	*Found*	*Biosoc*	*Select*	*ENA*	
Wide Age Range 12M	100.0	-0-	4	1979–2006	-0-	Silent	Silent	Silent	Silent	Silent	Silent

Précis 26.3.7.9 Portrayals of working inside the home (original Table 22.5.6.1)

Age categories, number of studies, and sex	*Consist. score*	*Overviews*	*Countries*	*Time range*	*Non-human*	*Social role theory*			*Evolutionary theory*		*Theological*
						Origin	*Found*	*Biosoc*	*Select*	*ENA*	
Adult 36F 1x Wide Age Range 11F All Ages 49F 1x	97.3 100.0 98.0	-0-	15 (1)	1971–2014	-0-	Contra	NoTest	Poss-F	Silent	Silent	Silent

Explanation. Two of the three social role theories can be interpreted as addressing this issue if one accepts the premise that work inside the home is learned in sociocultural settings. The original (basic) version leads one to expect to find cultural variations in this sex difference, a conclusion that is contrary to the evidence. Regarding the founder effect version of the theory, there seems to be no way of determining whether this sex difference was found in the earliest human societies. In the case of the biosocial version of social role theory, it might be argued that females learn to stay at home and do housework more than males because this is how all societies portray females.

Assessment. Overall, while we can confidently declare the portrayal of work inside the home to be a USD-F, more work is needed to theoretically understanding, unless one simply declares this sex difference to be a roughly accurate representation of reality.

26.3.7.10 Portrayals of Working Outside the Home

Evidence. When people are portrayed working outside the home, are they more likely to be portrayed as male or female? Précis 26.3.7.11 shows that 22 of the 23 studies of this variable found this sex difference, with the remaining study simply reporting no significant difference. These studies were conducted in six countries in studies (plus one multi-national study) published over a 20-year time frame.

Explanation. Unlike finding of sex differences in those being portrayed as working *in*side the home, males are more often portrayed as working *out*side the home. Two of the three social role theories can be interpreted as making predictions in this regard. The original version predicts that this should vary between cultures, which does not seem to be the case. The biosocial version of social role theory predicts cultural consistency in this sex difference, and might draw on the fact that pregnancy may inhibit females from working outside the home as much as males.

Assessment. The portrayal of the sex who works outside the home appears to be a USD-M. The biosocial version of social role theory can be used to explain why males are depicted as working outside the home more. However, it may be more reasonable to simply attribute this sex difference to a reflection of reality.

26.3.7.11 Ecologically Based Sex Differences

In recent years, considerable research attention has been given to ecologically based sex differences. The majority of these studies compare nations throughout the world regarding average sex differences in various traits, e.g., sex differences in personality traits or in aggression. Then the average sex difference found in each country for one of these sex difference traits is

Précis 26.3.7.10 Portrayals of working outside the home (original Table 22.5.6.3)

Age categories, number of studies, and sex	*Consist. score*	*Overviews*	*Countries*	*Time range*	*Non-human*	*Social role theory*			*Evolutionary theory*		*Theological*
						Origin	*Found*	*Biosoc*	*Select*	*ENA*	
All Ages 22M 1x	95.7	-0-	6 (1)	1971–2003	-0-	Contra	NoTest	Poss-M	Silent	Silent	Silent

compared to some other characteristic of each country, such as each country's degree of economic development or the extent to which each country has laws and customs favoring gender equality.

The purpose of these studies has involved assessing which average sex differences traits in countries appear to be correlated with national customs and traditions of the countries themselves. As indicated throughout Chapter 25, findings from these studies have been quite interesting, partly because many have produced a number of counter-intuitive findings. None of the tables in this chapter are summarized here, however, because none of them have yet reached the minimum threshold of ten deemed necessary for calculating a consistency score.

27 Tabular Listings of the Universal Sex Differences in Order of the Evidentiary Strength

This four-volume book was undertaken with two overarching goals in mind. The first objective was to identify sex differences that have been consistently documented by scientific research in all countries studied, i.e., discover universal sex differences (USDs). Once identified, it should be possible to provide a tentative list of USDs (i.e., sex differences that characterize the human species as a whole). The second goal was to examine each USD with respect to scientific theories that have been offered to account for universal sex differences.

To achieve the first of these two goals, Volumes I through III contain thousands of tables in which close to 50,000 findings of sex differences are cited. These citations are in according to the age groups and the countries sampled. The tables were organized into 23 different chapters subsumed under the three volumes, with the first volume focused on basic biology, the second volume on cognition, and the third volume on behavior. This fourth volume provides a summary of what was found in the first three volumes in terms of the variables that appear to be the most promising universal sex differences (USDs) and seeks to identify the most reasonable theoretical interpretations.

Except for a brief epilogue, this Chapter 27 will be Volume IV's final chapter. In it, we will provide highly condensed summaries of conclusions reached in Chapter 26. As a particularly lengthy chapter, Chapter 26 considered each likely USD from two perspectives: (a) the strength of the empirical evidence and (b) what theories offered the most reasonable explanations. Occasionally, even theological (scriptural) explanations were identified as being worth of consideration.

To provide this final condensed summary, a number of tables will be presented. All but one of these tables pertains to variables that will be identified as *likely USDs*. In the last remaining table, 18 nearly certain USDs are presented.

Regarding the first set of tables, i.e., those pertaining to likely USDs, we will deal with variables in three sections. Section 27.1 pertains to basic biological variables, Section 27.2 is concerned with cognitive variables, and Section 27.3 has to do primarily with behavioral variables.

DOI: 10.4324/9781003405290-4

Each table in these three sections is comprised of six columns, with one column divided into six sub-columns. The information listed in the columns are (a) the variable name or description, (b) the number of supportive findings divided by the total number of findings, (c) the number of countries sampled, (d) the six possible explanations, and (e) the précis number appearing in Chapter 26 for each variable. The subdivided fifth column identifies each of the five potential theories plus one column for scriptural explanations.

Variables within each of these tables are arranged according to the total number of findings located. In the case of ties, the number of countries sampled was used as a secondary criterion for ordering the variables.

Regarding the explanation sub-columns, an "x" is used to indicate that this particular theory or religious explanation seems to offer no bearing on a particular USD. If a theory or religious passage predicts a given USD, but with little independent empirical support, a single checkmark (✓) appears. In the case of theories that not only predicted a USD, but also have some independent empirical support, a double checkmark (✓✓) is used. Occasionally, a triple checkmark (✓✓✓) appears to indicate that the amount of corroborating evidence for this particular theoretical explanation is quite strong. In all cases, readers can find more details by returning to each USD's corresponding précis in Chapter 26.

Two comments are worth stating before examining these tables. First, our interpretations of how well each of the five theories meshes with a particular USD may not always be the same as others. Therefore, readers may sometimes be aware of other reasonable interpretations or of supportive evidence that we failed to identify. Overall, should future editions of this book be published, some of the information here is almost certain to change.

Second, in a few cases, the number of countries sampled may actually surpass the number of findings reported. This is because the number of findings only pertain to age groups with consistency scores of 95.0 or higher, but the number of countries sampled refers to the entirety of all the findings report, *regardless* of age.

27.1 Likely Universal Sex Differences: Basic Biology

As noted earlier, all sex differences are ultimately biological. This is because sex determination is largely under genetic control and all thinking and behaving is predominantly controlled by the brain. Even the most complex culturally learned thoughts and behaviors are neurologically regulated. Nevertheless, in recognition of the separation that can be made between basic biology and biology responsible for thought and behavior, this book's first volume was devoted to documenting the most basic aspects of biology.

Tables 27.1-M. Twenty-four USDs of a basic biological nature (including those dealing with physical health) were located in which males

Table 27.1-M Male-higher USDs for basic biological variables

Variable name or description	*Supportive findings*	*Age groupings*	*Countries sampled*	*Supported theory or theories (or scriptures)*						*Précis #*
				Origin. social role	*Founder effect social role*	*Biosocial social role*	*Sexual selection*	*Evol. neuroan-drogenic*	*Theological*	
Height (following mid to late adolescence)	221/221	A-A	58 (12)	x	x	x	✓	✓✓✓	x	26.1.1.4
Accidental injuries and fatalities in general	130/134	All	17 (11)	x	x	x	✓✓	✓✓	x	26.1.5.4
Brain volume (brain size)	120/122	Adult	23 (1)	x	x	x	✓	✓✓	x	26.1.4.1
Testosterone	105/105	P-A-A	16 (1)	x	x	x	x	✓✓✓	x	26.1.3.2
Body weight	72/72	Adult	26 (4)	x	x	x	x	✓✓	x	26.1.1.5
Muscularity	68/68	All	9 (1)	x	x	x	✓✓	✓✓✓	x	26.1.2.7
Physical strength	56/58	All	22 (3)	x	x	x	✓✓	✓✓✓	x	26.1.2.8
Skin color (skin darkness)	54/56	All	22 (8)	x	x	x	x	✓✓	x	26.1.2.9
Bone density (bone mass)	51/52	A-A	10 (2)	x	x	x	✓✓	✓✓	x	26.1.2.1
Accidental drownings	46/47	All	9 (7)	x	x	x	✓✓	✓✓	x	26.1.5.7
Physical maturation rate	41/41	All	8	x	x	x	x	✓✓	x	26.1.1.3
Neocortex, size of	36/37	All	8 (1)	x	x	x	x	✓✓	x	26.1.4.3

Stillbirth and early infant death	31/31	NA	10 (3)	x	x	x	x	✓✓	x	26.1.1.1
Accidental lightning strike injuries or fatalities	25/25	All	14 (3)	x	x	x	✓	✓	x	26.1.5.8
Larynx (vocal cords) size	25/25	A-A	3	x	x	x	✓✓	✓✓	x	26.1.2.10
Anogenital distance	24/24	All	12 (1)	x	x	x	x	✓✓✓	x	26.1.2.6
Leukemia and non-Hodgkin's lymphoma	18/18	All	6 (1)	x	x	x	x	x	x	26.1.5.10
Mortality following puberty	18/18	A-A	5 (1)	x	x	x	✓✓	✓✓	x	26.1.5.3
Bicycle injuries	16/16	WAR	7 (1)	x	x	x	✓	✓	x	26.1.5.6
Finger length or hand size	14/14	All	8 (1)	x	x	x	✓	✓✓	x	26.1.2.2
Sudden infant death syndrome	13/13	Infant	3 (1)	x	x	x	x	✓✓	x	26.1.5.2
Brain ventricles, size of	12/12	All	6	x	x	x	x	x	x	26.1.4.2
DHEA-S	12/12	Adult	5	x	x	x	x	✓✓	x	26.1.3.1
Motor vehicle injuries	10/10	Adol.	3	x	x	x	✓	✓	x	26.1.5.5

Note: A-A is an abbreviation for adolescent and adult. P-A-A is an abbreviation for prenatal, adolescent, and adult. WAR is an abbreviation for wide age range.

obtained higher scores. Topping the list in terms of the number of pertinent findings was the variable of height. In this regard, one can see that, following middle adolescence (i.e., around age 15), all 221 studies agreed that males are taller, on average, than females.

The next three variables in this table were based on more than 100 pertinent findings. They indicate that males are more likely to sustain accidental injuries, including those having to do with motor vehicles, bicycles, and drownings. Males also have higher levels of testosterone, at least prior to birth, during adolescence, and throughout adulthood. Studies also indicated that males have larger brains and neocortices (even after adjusting for sex differences in body size). For the latter variable, the sex difference was only a USD regarding three age groupings: before birth, during adolescence, and throughout adulthood.

The best explanations for this sex difference appears to be evolutionary in nature (i.e., sexual selection plus testosterone).

Regarding theoretically explaining sex differences in most of these basic biological variables, both versions of evolutionary theory predominate. Nevertheless, there are several instances in which ENA theory (which attributes sex differences to androgen exposure) appears to be likely without having any sexual selection explanation for the sex difference. This can often be attributed to the fact that sexual selection sometimes operates in ways that are not easily explained or tested.

Table 27.1-F. The basic biological variables that appear to be USD-Fs (i.e., females being higher than males) are shown in Table 27.1-F. Nine such variables were identified. The two best documented sex differences were percent body fat and seeking or utilizing healthcare services more. Regarding the latter of these two, females surpassed males most consistently only when ob/gyn services were included. Other well-documented USD-Fs indicated that females reach puberty at significantly younger ages and had higher leptin levels. Females are more likely to suffer from breast cancer, exhibit greater joint laxity, have longer QT intervals, higher 2D:4D finger length ratios, and lose bone density with age more rapidly than males.

27.2 Likely Universal Sex Differences: Cognition

Variables listed in Chapters 9 through 15 (i.e., all of Volume II) primarily pertain to sex differences of a cognitive nature. In this section of the present chapter, a summary of the USDs covered in Volume II are specified in the following four categories: (a) perceptual abilities, motor skills, and emotions, (b) intellectual abilities and self-reflections, (c) mental health, and (d) attitudes and preferences. Two tables are presented for each of these four categories, one for variables in which males are higher (or more prevalent) than females, and the other for the opposite pattern.

Table 27.1-F Female-higher USDs for basic biological variables

Variable name or description	*Supportive findings*	*Age groupings*	*Countries sampled*	*Supported theory or theories (or scriptures)*						*Précis #*
				Origin. social role	*Founder effect social role*	*Biosocial social role*	*Sexual selection*	*Evol. neuroandro-genic*	*Theological*	
Percent body fat	77/78	A-A	14 (3)	x	x	x	x	✓✓	x	26.1.1.6
Seeking or utilizing healthcare services (including ob/gyn)	72/75	All	9 (1)	x	x	x	✓✓	✓✓	x	26.1.5.1
Puberty, age of onset	39/39	All	11	x	x	x	x	✓	x	26.1.1.2
Leptin	34/34	Adult	8	x	x	x	x	✓✓	x	26.1.3.3
Breast cancer	27/27	Adult	11 (2)	x	x	x	x	✓	x	26.1.5.9
Joint laxity	20/20	All	11 (1)	x	x	x	✓	✓✓	x	26.1.2.5
QT interval	20/20	Adult	5	x	x	x	x	✓	x	26.1.2.11
2D:4D finger length ratio	15/15	WAR	7 (4)	x	x	x	x	✓✓	x	26.1.2.3
Loss of bone density with age	10/10	Adult	3	x	x	x	✓	✓✓	x	26.1.2.4

Note: A-A is an abbreviation for adolescent and adult. WAR is an abbreviation for wide age range.

27.2a USDs for Perceptual Abilities, Motor Skills, and Emotions

The USDs having to do with perceptual or motor skills nature or with emotions are summarized Table 27.2a. Six of these variables indicate that males are higher or more numerous, while females were higher or more numerous for 16 of the variables.

Table 27.2a-M. One can see in Table 27.2a-M that males surpass females in five aspects of athletic ability. In addition to a general ability measure, males are faster runners, throw at higher velocities, and jump higher and over longer distances. One can see that the number of countries sampled to reach these conclusions are considerable, and that evolutionary theories provide likely explanations for the sex differences. Parenthetically, it is worth noting that these five indicators of athletic ability can be seen as indicating more about sex differences of a basic biological nature than of a cognitive or behavioral nature. These are among several examples of hard-to-classify variables.

The final variable in this table is of an emotional nature. It indicates that males are more prone to seek revenge when others have caused them harm or betrayed them. No theoretical explanation was identified.

Table 27.2a-F. One can see in Table 27.2a-F that most of the 16 USDs of a perceptual, motor, and emotional nature pertain to emotions. With the exception of empathy, all of these emotions are of a negative nature (e.g., some type of fear, stress, or sadness). The five variables of a non-emotional nature in which females were highly involved: superior color discrimination, exhibiting more left/right confusion, having keener auditory acuity, being more likely to vocalize during infancy, and having more voice nasality. In most cases, the best explanations for these female-higher variables appear to be of an evolutionary nature.

27.2b USDs for Intellectual and Self-Reflective Variables

Findings of USDs of an intellectual or self-reflective nature are summarized in the following two tables. The, the first table pertains to variables for which exhibited higher values, while the second were for variables for which high values were found among females.

Tables 27.2b-M. Twenty variables are listed in Table 27.2-M in descending order of the number of studies located. The top two variables indicates that males have more body-satisfaction and self-confidence. They also seem to have more knowledge, especially in subject areas that emphasize spatial reasoning and mechanical problem solving.

However, males are also more likely to be mentally retarded, to have learning disabilities, and to repeat a grade in school. Regarding dreams, males are more likely to recall ones involving physical objects and sexual themes.

Table 27.2a-M Male-higher USDs for perceptual, motor, and emotional variables

Variable name or description	*Supportive findings*	*Age groupings*	*Countries sampled*	*Supported theory or theories (or scriptures)*						*Précis #*
				Origin. social role	*Founder effect social role*	*Biosocial Social Role*	*Sexual selection*	*Evol. neuroandrogenic*	*Theological*	
Athletic ability/ performance in general	30/30	A-A	15 (4)	x	x	x	✓✓	✓✓✓	x	26.2.9.5
Running speed	29/29	All	8 (11)	x	x	x	✓✓	✓✓	x	26.2.9.10
Throwing velocity or distance	29/29	All	8	x	x	x	✓✓	✓✓	x	26.2.9.6
Jumping height	24/24	All	12	x	x	x	✓✓	✓✓	x	26.2.9.8
Jumping distance	16/16	All	7 (1)	x	x	x	✓✓	✓✓	x	26.2.9.7
Being vengeful or spiteful	11/11	All	4	x	x	x	x	x	x	26.2.10.1

Note: *A-A* is an abbreviation for adolescent and adult.

Table 27.2a-F Female-higher USDs for perceptual, motor, and emotional variables

Variable name or description	*Supportive findings*	*Age groupings*	*Countries sampled*	*Supported theory or theories (or scriptures)*						*Précis #*
				Origin. social role	*Founder effect social role*	*Biosocial social role*	*Sexual selection*	*Evol. neuroandrogenic*	*Theological*	
Fearfulness in general	96/97	All	16 (2)	x	x	x	✓	✓	x	26.2.10.2
Crying	70/71	Adult	36 (3)	x	x	✓	✓	✓	x	26.2.10.10
Fear of crime victimization	57/59	All	8 (1)	x	x	x	✓	✓	x	26.2.10.4
Fundamental frequency (F_0) when vocalizing	41/41	Adult	11	x	x	x	✓✓	✓✓	x	26.2.9.12
Color discrimination/ recognition (naming)	35/35	All	7 (1)	x	x	x	x	x	x	26.2.9.2
Estimating/ perceiving environmental hazards	34/35	All	6	x	x	x	x	x	x	26.2.9.3
Stress/anxiety associated with technology	27/28	Adult	5	x	x	x	✓	✓	x	26.2.10.8
Left/right confusion	23/24	All	8	x	x	x	✓	✓	x	26.2.9.4
Empathy, feelings of	19/20	Adolescent	12 (3)	x	x	x	✓✓	✓✓	x	26.2.10.9

Stress/anxiety associated with providing care to others	19/19	All	2	x	x	x	✓	✓	x	26.2.10.7
Sadness (dysphoria) in general	14/14	All	2 (2)	x	x	x	✓✓	✓✓	x	26.2.10.6
Auditory acuity	13/13	Adult	2	x	x	x	x	✓✓	x	26.2.9.1
Fear of animals	12/12	All	3 (1)	x	x	x	✓	✓	x	26.2.10.3
Vocalizing behavior	11/11	Infant/Tod	3	x	x	x	✓	✓	x	26.2.9.11
Nasality of voice during speech	11/11	All	2	x	x	x	✓✓	x	x	26.2.9.13
Fearing being negatively evaluated by others	10/10	All	1	x	x	x	✓	✓	x	26.2.10.5

Table 27.2b-M Male-higher USDs regarding intellectual and self-reflective variables

Variable name or description	*Supportive findings*	*Age groupings*	*Countries sampled*	*Supported theory or theories (or scriptures)*						*Précis #*
				Origin. social role	*Founder effect social role*	*Biosocial social role*	*Sexual selection*	*Evol. neuroandrogenic*	*Theological*	
Body satisfaction	78/80	All	11	x	x	x	✓	✓	x	26.2.13.2
Fantasizing about sexual behavior	39/41	All	9	x	x	x	✓✓	✓✓	x	26.2.12.10
Mechanical problem solving	39/40	All	7	x	x	x	✓✓	✓✓	x	26.2.11.6
Exceptionally high math scores	37/37	All	7 (4)	x	x	x	x	x	x	26.2.12.3
Physical science knowledge	36/37	All	6 (4)	x	x	✓	✓	✓	x	26.2.12.5
Learning disabilities or difficulties in general	28/28	All	4	x	x	x	x	✓✓	x	26.2.11.2
Overall self-confidence	22/23	All	8	x	x	✓✓	✓✓	✓✓	x	26.2.13.1
Geography knowledge	21/22	All	4	x	x	✓	✓✓	✓✓	x	26.2.12.6
Systemizing thinking	20/20	All	8 (1)	x	x	x	✓✓	✓✓	x	26.2.11.8
Self-assessed computer competency	19/19	All	5	x	x	x	✓✓	✓✓	x	26.2.13.5

Knowledge in general	18/18	Adult	12 (2)	x	x	x	x	✓✓	x	26.2.12.1
Spatial reasoning ability, self-rated	17/17	All	5 (3)	x	x	✓	✓✓	✓✓	x	26.2.11.4
Estimating distances and projectile landings	17/17	All	8	x	x	x	✓✓	✓✓	x	26.2.11.5
Math ability, intra-sex variability in	14/14	All	4 (3)	x	x	x	✓	✓	x	26.2.12.2
Dream content: Sexual themes	13/13	All	4 (1)	x	x	x	✓	✓	x	26.2.12.9
Institutionalized for low intelligence or mental disabilities	13/13	All	7	x	x	x	x	✓✓	x	26.2.11.3
Assessment of one's ability in general	13/13	Adult	1	x	x	✓	✓	✓	x	26.2.13.4
Science knowledge in general	12/12	Adult	58 (20)	x	x	x	✓	✓	x	26.2.12.4
Dream content: Physical objects and artifacts	11/11	All	3 (1)	x	x	x	✓	✓	x	26.2.12.8
Repeating (failing to pass) a grade in school	11/11	All	2	x	x	x	x	✓✓	x	26.2.11.1

The two versions of evolutionary theory appear to offer the most persuasive explanations for most of these variables. Nevertheless, the biosocial version of social role theory also appears to have some explanatory potential for at least three of these variables.

Table 27.2b-F. Turning to the USDs of an intellectual or self-reflective nature for which females appear to be higher than males, Table 27.2b-F identifies four variables. The one with the greatest evidentiary support involves self-assessed body weight. Nearly all studies have indicated that females consider their weight to be greater than desired more than do males.

Females also appear to have greater language ability, at least during childhood. Regarding the third most consistently documented variable, females recall dreams more often than do males. Finally, while intelligence scores and academic performance are positively correlated with each other among both sexes, the strength of this relationship has been found to be stronger for females than for males.

One or both of the evolutionary theories appear to be helpful for understanding three of these sex differences, and the biosocial version of social role theory shows promise for one of them. In the case of the variable having to do with the association between IQ scores and academic ability, none of the five theories sees relevant.

27.2c USDs for Mental Disorders and Diseases

A substantial number of variables pertaining to mental disorders and diseases appear to be universally more common among either males or females. The relevant findings are summarized in the following two tables.

Table 27.2c-M. Fifteen mental disorder-type variables that are more prevalent or pronounced among males are listed in Table 27.2c-M. The two most thoroughly documented disorders in this regard are those of conduct disorders and attention deficit hyperactivity disorders (ADHD). Following these two disorders are three variables having to do with alcohol: Males are more likely to be alcoholic, have drinking problems, and to have an early age of alcoholism onset. Other well-documented male-higher mental disorders are autism spectrum disorder, behavioral disorders, externalizing behavior, antisocial personality disorder, and schizophrenia. Stuttering and other language disorders along with gender identity disorder also appear to be more common among males.

Two of these types of disorders have no readily available explanation in terms of any of the five theories. When theories were identified, all of them point toward one of the evolutionary theories and none toward the social role theories.

Table 27.2c-F. Eleven mental disorders were found to be universally more common among females than males. Table 27.2c-F shows that the two

Table 27.2b-F Female-higher USDs regarding intellectual and self-reflection variables

Variable name or description	*Supportive findings*	*Age groupings*	*Countries sampled*	*Supported theory or theories (or scriptures)*						*Précis #*
				Origin. social role	*Founder effect social role*	*Biosocial social role*	*Sexual selection*	*Evol. neuroandrogenic*	*Theological*	
Self-assessment of being overweight (or wanting to lose weight)	94/95	All	12	x	x	x	✓	✓	x	26.2.13.3
Language ability	37/38	Child	16	x	x	✓	x	✓	x	26.2.11.10
Dreaming, frequency of recall	18/18	Adult	5	x	x	x	✓	✓	x	26.2.12.7
Relationship between IQ scores and academic performance	10/10	Adult	5	x	x	x	x	x	x	26.2.11.9

Table 27.2c-M Male-higher USDs for mental disorders and diseases

Variable name or description	*Supportive findings*	*Age groupings*	*Countries sampled*	*Supported theory or theories (or scriptures)*						*Précis #*
				Origin. social role	*Founder effect social role*	*Biosocial social role*	*Sexual selection*	*Evol. neuroandrogenic*	*Theological*	
Conduct disorder	126/131	Child	19 (1)	x	x	x	✓✓	✓✓	x	26.2.14.8
Attention deficit hyperactivity disorder (ADHD)	124/126	All	18	x	x	x	✓✓	✓✓	x	26.2.14.22
Alcoholism (alcohol dependence)	78/78	All	13 (2)	x	x	x	x	✓✓	x	26.2.14.11
Autism spectrum disorder (ASD)	68/69	All	16	x	x	x	✓✓	✓✓	x	26.2.14.24
Behavioral disorders and externalizing behavior	64/65	All	11	x	x	x	✓✓	✓✓	x	26.2.14.1
Completed suicide	34/34	Adol	34 (44)	x	x	x	x	✓	x	26.2.14.26
Antisocial personality disorder, including psychopathy	34/34	WAR	14 (1)	x	x	x	✓✓	✓✓	x	26.2.14.9
Schizophrenia	28/29	Adult	15 (3)	x	x	x	✓	✓	x	26.2.14.5
Problem drinking/ alcohol abuse	25/26	All	5	x	x	x	x	✓	x	26.2.14.10

Stuttering	21/21	All	8	x	x	x	x	✓	x	26.2.14.18
Alcoholism, age of onset	19/19	All	1	x	x	x	x	✓	x	26.2.14.12
Speech and language disorders in general	11/11	All	2	x	x	x	✓	✓	x	26.2.14.17
Attention deficit disorder (ADD)	10/10	All	4	x	x	x	x	x	x	26.2.14.23
Compulsive gambling, age of onset	10/10	All	3	x	x	x	x	x	x	26.2.14.13
Gender identity disorder (GID)	10/10	All	3	x	x	x	x	✓✓	x	26.2.14.25

Note: WAR is an abbreviation for wide age range.

Table 27.2c-F Female-higher USDs for mental disorders and diseases

Variable name or description	*Supportive findings*	*Age groupings*	*Countries sampled*	*Supported theory or theories (or scriptures)*						*Précis #*
				Origin. social role	*Founder effect social role*	*Biosocial social role*	*Sexual selection*	*Evol. neuroandrogenic*	*Theological*	
General depression, self-diagnosed	71/73	WAR	48 (14)	x	x	✓	✓	✓✓	x	26.2.14.3
Major (clinical) depression	71/73	Adol	46 (5)	x	x	✓	✓✓	✓✓	x	26.2.14.4
Anorexia nervosa	53/54	All	9	x	x	x	x	x	x	26.2.14.15
Bulimia nervosa	47/47	All	12	x	x	x	x	✓	x	26.2.14.16
Eating disorders in general	32/32	All	10	x	x	x	✓	✓	x	26.2.14.14
Internalizing behavior	29/30	All	9 (3)	x	x	x	x	✓	x	26.2.14.2
Panic disorder	27/27	All	5 (1)	x	x	x	x	✓	x	26.2.14.20
Phobias in general	20/21	WAR	10 (1)	x	x	x	✓✓	✓✓	x	26.2.14.21
Psychological problems in general	15/15	All	5	x	x	✓	✓	✓✓	x	26.2.14.7
Post-traumatic stress disorder (PTSD)	13/13	Adol	13 (1)	x	x	x	x	x	x	26.2.14.19
Social competence of schizophrenics	12/12	All	4	x	x	x	x	x	x	26.2.14.6

Note: WAR is an abbreviation for wide age range.

most heavily documented are those of self-diagnosed general depression and clinically diagnosed major depression, the latter being more common particularly among adolescent females. Also, well-documented mental disorders that are universally more prevalent among females are anorexia, bulimia, and eating disorders in general. Other female-higher mental disorders are those of internalizing behavior, panic disorders, and phobias, as well as psychological problems generally. Post-traumatic stress disorder (PTSD) also appears to be more common among females, at least among adolescents.

The last entry in this table had to do with schizophrenia, a disorder that, as noted above, is actually more common among males. However, according to all 12 studies of persons with schizophrenia, females afflicted with this disease seem to be more socially competent than males with the disease.

In terms of theoretical explanations, no explanations for three of the USD-F mental disorders. For the remaining eight variables, the biosocial version of social role theory offered a possible explanation for three of them, while the evolutionary theories appeared promising for all of them. ENA theory appeared to be particularly strong for four of these variables.

27.2d USDs for Attitudinal and Preference Variables

The number of attitudinal and preference variables qualifying as USDs was surprisingly numerous. For the USD-Ms, 33 were located, and for USD-Fs, 26 were found. Although most social scientists consider attitudes and preferences to be culturally learned, our search of the literature on sex differences located 59 instances in with the same basic differences are found in all societies yet investigated. This suggests that some underlying factors present in all human societies are driving males and females toward consistently different attitudes and preferences regarding a wide range of topics.

Table 27.2d-M. Leading off the variables listed in Table 27.2d-M are the first four, which each have more than 50 independent findings. At the top of the list are findings that males are more accepting or supporting of sexual intercourse outside of marriage than are females. The second listed variable indicates that males preferred mates who were younger than themselves; out of 17 countries samples (plus seven multi-country studies), no exception to this generalization was found.

According to the third-best documented sex difference in which males had higher scores in this table, males have a stronger sex drive. The fourth variable in this list indicated that males express stronger desires for promiscuous sexual relationships than do females.

Without giving individual treatment to each of the remaining 29 USD-Ms, one can see that males have greater interest in competition and in

Table 27.2d-M Male-higher USDs for attitudinal and preference variables

Variable name or description	*Supportive findings*	*Age groupings*	*Countries sampled*	*Supported theory or theories (or scriptures)*						*Précis #*
				Origin. social role	*Founder effect social role*	*Biosocial social role*	*Sexual selection*	*Evol. neuroandrogenic*	*Theological*	
Attitude towards premarital/casual sexual intercourse	65/66	All	12 (1)	x	x	x	✓✓	✓✓	x	26.2.15.21
Preference for a mate who is younger than oneself	57/57	All	17 (7)	x	x	x	✓✓	✓✓	x	26.2.15.17
Sex drive, strength of (desire for sex)	54/55	All	11 (3)	x	x	x	✓✓	✓✓	x	26.2.15.6
Desire for promiscuous sexual relationship	53/54	All	8 (4)	x	x	x	✓✓	✓✓	x	26.2.15.7
Interest in or enjoyment of competitive activities	41/42	All	10 (4)	x	x	x	✓	✓	x	26.2.15.32
Interest in sports participation	37/37	All	8 (1)	x	x	x	✓	✓	x	26.2.15.35
Preference for male-typical occupations	37/37	All	7 (2)	x	x	x	x	x	x	26.2.15.42
Attribute more of the responsibility for rape to the victim	36/36	All	3	x	x	x	x	x	x	26.2.15.56

Preference for things-oriented occupations	31/31	All	8	x	x	x	✓✓	✓✓	x	26.2.15.44
Preference for mechanical and building objects of play	28/28	All	7	x	x	x	✓✓	✓✓	x	26.2.15.38
Preference for, or acceptance of, aggressive responses to conflicts	28/28	All	4	x	x	✓	✓	✓✓	x	26.2.15.2
Attribution of sexual meaning/motivation to male-female interactions	27/27	All	2	x	x	x	✓	✓	x	26.2.15.11
Interest in spectator sporting activities	24/24	All	5	x	x	x	✓✓	✓	x	26.2.15.34
Preference for vehicles as objects of play	23/23	All	5	x	x	x	✓✓	✓✓	x	26.2.15.39
Preference regarding monetary compensation of jobs	22/22	Adol	4 (1)	x	x	x	✓	✓	x	26.2.15.48
Interest in sports in general	17/17	All	3 (2)	x	x	x	✓	✓	x	26.2.15.33
Preference for a mate who is relatively short	16/16	All	8	x	x	x	✓	✓	x	26.2.15.16

(Continued)

Table 27.2d-M (Continued)

Variable name or description	*Supportive findings*	*Age groupings*	*Countries sampled*	*Supported theory or theories (or scriptures)*						*Précis #*
				Origin. social role	*Founder effect social role*	*Biosocial social role*	*Sexual selection*	*Evol. neuroandrogenic*	*Theological*	
Preference for a physically attractive mate	13/13	WAR	23 (14)	x	x	x	✓✓	✓	x	26.2.15.14
Interest in being a computer programmer or analyst	12/12	All	5	x	x	x	✓✓	✓✓	x	26.2.15.50
Excusing or tolerating rape	12/12	Adult	3	x	x	x	✓✓	✓✓	x	26.2.15.55
Attitude toward aggression and violence	12/12	All	1 (1)	x	x	✓	✓	✓	x	26.2.15.59
Breadth of occupational interest	12/12	All	1	x	x	x	✓✓	✓	x	26.2.15.47
Tolerance of sexual harassment	12/12	All	1	x	x	x	✓	✓	x	26.2.15.57
Attitude toward war and the use of military power	11/11	WAR	8 (3)	x	x	✓	x	x	x	26.2.15.53
Valuing power (*Scriptural explanation possible*)	11/11	All	4 (2)	x	x	✓	✓	✓✓	x	26.2.15.1

Preference for a shorter time between sexual attraction and sexual intimacy	11/11	All	1	x	x	x	✓✓	✓✓	x	26.2.15.12
Interest in or enjoyment of science in general	10/10	Child	34 (7)	x	x	x	✓	✓	x	26.2.15.25
Preference for the color blue	10/10	All	5	x	x	x	x	x	x	26.2.15.5
Attitude toward social dominance	10/10	All	3 (1)	x	x	x	✓	✓	x	26.2.15.52
Attitude toward nuclear power	10/10	All	2 (1)	x	x	x	✓✓	✓✓	x	26.2.15.54
Preference for adventure stories	10/10	All	3	x	x	x	✓	✓	x	26.2.15.40
Attitude towards nudity and sexually explicit material	10/10	All	2	x	x	x	✓✓	✓✓	x	26.2.15.22
Relationship between self-reported sexual arousal and physiological measures	10/10	All	2	x	x	x	x	x	x	26.2.15.13

Note: WAR is an abbreviation for wide age range.

sports throughout the world. Males are also more things-oriented when choosing toys (as children) and when choosing occupations (as adolescents and adults). Research also shows that males appear to be more inclined to attribute sexual motivation to interactions between the sexes. They also have more favorable attitudes toward nudity. Regarding the use of violence when disputes arise, males are more supportive, and they are more likely to believe that rape victims bear some responsibility for being raped. Regarding mate selection, males prioritize a prospective mate's physically attractive and want their mate to be shorter than themselves more than do females.

In the case of theoretically explaining these universal sex differences, the two versions of evolutionary theory seem have offered the greatest explanatory power. Nevertheless, the biosocial version of sex role theory offered reasonable explanations for five of the 33 USD-Ms.

Table 27.2d-F. The 26 attitudinal and preference variables wherein females scored higher appear in Table 27.2d-F. Three of these variables were addressed by more than 50 separate findings, making them the best supported USD-Fs in this set of variables. They indicate that, when compared to males, females prefer mates with resources more, are more accepting of homosexuality, and prefer people-oriented occupations more. The number of countries sampled in all three of these traits were sizable, but especially so in the case of preferences for mates with resource (or resource potential). One can see that 27 different countries were sampled, along with 12 additional multi-country studies.

Other USD-Fs documented in this table are the following: Females enjoy school more, and, as children they play with dolls more. As adolescents and adults, females desire to interact with infants and toddlers more and to socialize more generally. In terms of mate selection, females prefer mates who are taller and older than themselves. They are also more likely to be bisexual or have ambivalent sexual orientations, and, when asked to identify what constitutes sexual harassment, they are more inclusive than are males. Females also express stronger desires to be in love before having sex. Regarding attitudes, females consider humane treatment of animals to be more important, hold more pro-family attitudes, and are more health-conscious than males.

Turning to theoretically explaining USD-Fs having to do with attitudes and preferences, either one or both of the evolutionary theories seem to better account for the differences. Nevertheless, there are three instances in which the biosocial version of social role theory appeared to have potential. There was even one variable where religious scriptures seemed to have the potential for explaining an attitudinal sex difference.

Table 27.2d-F Female-higher USDs for attitudinal and preference variables

Variable name or description	*Supportive findings*	*Age groupings*	*Countries sampled*	*Supported theory or theories (or scriptures)*						*Précis #*
				Origin. social role	*Founder effect social role*	*Biosocial social role*	*Sexual selection*	*Evol. neuroandrogenic*	*Theological*	
Preference for a mate with resources (wealth, high income)	104/107	All	27 (12)	x	x	x	✓✓	✓✓	x	26.2.15.19
Favorable or accepting attitudes toward homosexuals or homosexual behavior	86/89	All	14 (1)	x	x	x	x	x	✓	26.2.15.10
Preference for people-oriented occupations	50/51	All	9	x	x	x	✓✓	✓✓	x	26.2.15.43
Enjoyment of school in general	39/39	All	10 (2)	x	x	x	✓✓	✓✓	x	26.2.15.24
Preference for playing with dolls	28/28	All	4	x	x	✓	✓✓	✓✓	x	26.2.15.37
Preference for a mate who is relatively tall	26/27	All	10 (1)	x	x	x	✓✓	✓✓	x	26.2.15.15
Preference for a mate who is older than oneself	26/26	All	10 (4)	x	x	x	✓✓	✓✓	x	26.2.15.18

(*Continued*)

Table 27.2d-F (Continued)

Variable name or description	*Supportive findings*	*Age groupings*	*Countries sampled*	*Supported theory or theories (or scriptures)*						*Précis #*
				Origin. social role	*Founder effect social role*	*Biosocial social role*	*Sexual selection*	*Evol. neuroandrogenic*	*Theological*	
Importance of humane treatment of animals	26/26	All	4	x	x	x	✓	✓✓	x	26.2.15.3
Inclusiveness in assessing what constitutes sexual harassment	26/26	All	2	x	x	x	✓✓	✓✓	x	26.2.15.58
Interest in spending time with infants and toddlers	23/24	Adult	3	x	x	✓	✓✓	✓✓	x	26.2.15.28
Preference for socializing in general	20/20	All	4	x	x	x	✓✓	✓✓	x	26.2.15.31
Bisexual preference (sexual attraction to both sexes)	19/20	All	4 (2)	x	x	x	x	✓✓	x	26.2.15.8
Interest in artistic and creative/ expressive occupations	19/20	All	5	x	x	x	✓	x	x	26.2.15.49
Ambivalent sexual orientation	17/17	All	4	x	x	x	x	✓	x	26.2.15.9
Range in play and reading interest	17/17	All	3	x	x	x	x	x	x	26.2.15.36

Being health conscious	16/16	All	6 (2)	x	x	x	✓✓	✓✓	x	26.2.15.23
Interest in forming and maintaining close interpersonal relationships	16/16	All	4 (1)	x	x	x	✓✓	✓✓	x	26.2.15.29
Attitudes toward female participation in the paid workforce	16/16	Adult	1	x	x	x	x	x	x	26.2.15.41
Pro-family attitudes and values	14/14	All	4 (1)	x	x	✓	✓✓	✓✓	x	26.2.15.27
Interest in or enjoyment of the biological and health sciences	12/12	All	5	x	x	x	✓	✓	x	26.2.15.26
Interest in helping-oriented occupations	11/11	Adult	4	x	x	x	✓✓	✓✓	x	26.2.15.45
Interest in people more than things	10/10	All	4 (1)	x	x	x	✓✓	✓✓	x	26.2.15.30
Belief in astrology or palm reading	10/10	Adult	5	x	x	x	x	x	x	26.2.15.4
Interest in being teachers	10/10	All	3	x	x	x	✓	✓	x	26.2.15.51
Wanting to be in love before having sex	10/10	All	3	x	x	x	✓	✓	x	26.2.15.20
Preference for jobs with flexible hours	10/10	Adult	1 (1)	x	x	x	✓✓	x	x	26.2.15.46

27.3 Likely Universal Sex Differences: Behavior

Volume III is comprised of Chapters 16 through 23 and primarily has to do with sex differences in behavior. Below, the USDs pertaining to these behavioral variables are listed in terms of six categories. These categories are: (a) behavioral and personality traits in general; (b) recreation, play, and sexual behavior; (c) acquiring and consuming resources; (d) education-related variables; (e) work-related variables; and (f) sex stereotypes, mass-media portrayals, and treatment of others according to their sex. For each category, two tables are presented, one pertaining male-higher USDs and the other to female-higher USDs.

27.3a USDs in Various Behavioral and Personality Traits

The first group of behavioral USDs is wide ranging and includes both personality traits as well as fairly specific behavioral traits. In the two summary tables combined, a total of 30 variables are identified as USDs.

Male-Higher USDs. Eleven general behavioral and personality traits that appear to be more universally found in males than in females are listed in Table 27.3a-M. The four most extensively documented traits are that males exercise more, take more risks, are more physically active, and are more competitive. In terms of more specific types of behavior, males surpass females in terms of belonging to gangs, in frequency of swearing and cursing, in carry weapons, and in being homeless. Males are also more likely to communicate in mix-sex social settings and to issue commands to others.

Turning to theoretically explaining these 11 USD-Ms, the biosocial version of social role theory offers possible explanations for sex differences in three of the variables. The sexual selection provided possible explanations for five of the variables and likely explanations for three others. Seven of the variables in this list appeared to be explainable with ENA theory, and one other was possibly explainable with this theory. Finally, three of these variables do not appear to be explainable with any of the five theories.

Female-Higher USDs. As shown in Table 27.3a-F, there are 19 different behavioral and personality traits that appear to be more common or pronounced in females than in males. Leading this list in terms of the number of supportive findings is the variable of caring for one's offspring. This is followed by trying to lose weight, and then by providing care to family members generally. Females also appear to confide and share secrets with others, to socially interact in small groups, provide care and comfort to others, to have more intimate friendships, to be more gregarious. In terms of religiosity, females appear to be more involve in religious activities and to pray more often.

Table 27.3a-M Male-higher USDs regarding various behavioral and personality traits

Variable name or description	*Supportive findings*	*Age groupings*	*Countries sampled*	*Supported theory or theories (or scriptures)*						*Précis #*
				Origin. social role	*Founder effect social role*	*Biosocial social role*	*Sexual selection*	*Evol. neuroandrogenic*	*Theological*	
Physical exercise	142/149	All	18 (4)	x	x	x	✓	✓✓	x	26.3.16.4
Risk-taking (recklessness) in general	104/109	Adult	14 (2)	x	x	x	✓✓	✓✓✓	x	26.3.16.2
Activity levels	63/66	Adol	20 (7)	x	x	✓	✓	✓✓	x	26.3.16.6
Gang membership	50/50	All	6 (1)	x	x	x	✓✓	✓✓	x	26.3.17.3
Competitiveness	40/41	Adult	12 (2)	x	x	✓	✓	✓✓	x	26.3.16.5
Swearing and cursing	39/41	All	6	x	x	x	✓	✓	x	26.3.17.14
Exploratory behavior in general	40/40	All	9 (1)	x	x	x	✓	✓✓	x	26.3.16.7
Carrying weapons	30/30	All	5	x	x	✓	✓✓	✓✓	x	26.3.16.9
Mixed-sex (vs. same-sex) linguistic communication	23/24	All	4	x	x	x	x	x	x	26.3.17.16
Being homeless	10/10	All	5	x	x	x	x	x	x	26.3.17.9
Issuing commands and instructions or assertive speech in general	10/10	Adult	3	x	x	x	x	x	x	26.3.17.18

Table 27.3a-F Female-higher USDs regarding various behavioral and personality traits

Variable name or description	*Supportive findings*	*Age groupings*	*Countries sampled*	*Supported theory or theories (or scriptures)*						*Précis #*
				Origin. social role	*Founder effect social role*	*Biosocial social role*	*Sexual selection*	*Evol. neuroandrogenic*	*Theological*	
Providing care to one's own offspring	92/92	All	13 (10)	x	x	✓	✓✓	✓✓	x	26.3.17.5
Trying to lose weight (dieting)	70/71	All	12 (2)	x	x	x	✓	✓✓	x	26.3.16.3
Providing care to family members other than offspring	70/70	All	10 (2)	x	x	✓	✓✓	✓✓	x	26.3.17.6
Confiding and sharing secrets	41/41	A-A	4	x	x	x	x	x	x	26.3.17.17
Social interactions in small groups	27/27	All	4	x	x	x	x	x	x	26.3.17.8
Involvement in religious activities in general	26/26	All	5	x	x	x	✓	✓	x	26.3.16.10
Praying	22/23	All	5 (3)	x	x	x	✓	✓✓	x	26.3.16.11
Providing care and comfort to others in general	21/22	All	5 (2)	x	x	✓	✓✓	✓✓	x	26.3.17.7
Having intimate friendships	21/22	Adol	6 (1)	x	x	x	✓	✓✓	x	26.3.17.2

Seeking or receiving help when under stress	21/21	Adult	3 (1)	x	x	x	x	x	x	26.3.17.11
Gregariousness (friendliness)	19/20	Adult	4 (7)	x	x	✓	✓	✓✓	x	26.3.16.1
Having primary custody of children after divorce	19/19	All	3	x	x	✓	✓	✓✓	x	26.3.17.10
Using a cradling-like book-carrying style	14/14	All	2	x	x	x	✓	✓	x	26.3.16.8
Socializing tendency	13/13	All	2	x	x	x	✓	✓✓	x	26.3.17.1
Supportive and inclusive communication	13/13	All	2	x	x	✓	✓	✓	x	26.3.17.19
Help-seeking when interpersonal dispute arises	12/12	Adol	3 (1)	x	x	x	x	x	x	26.3.17.12
Smiling	11/11	Adol	3 (1)	x	x	✓	✓✓	✓✓	x	26.3.17.13
Parent-offspring conflict	10/10	All	3	x	x	x	x	x	x	26.3.17.4
Using intensifying adverbs	10/10	All	1	x	x	x	x	x	x	26.3.17.15

Note: A-A is an abbreviation for adolescent and adult.

Most of the remaining USD-Fs in this table have to do with socializing behavior. Some specific exceptions are that females are more likely than males to use a cradling-like book-carrying style, to smile more, to experience more parent-offspring conflicts (at least based on self-reports), and to use more intensifying adverbs when communicating.

Turning to the possibility of theoretically explaining these particular sex differences, none of the theories seemed to bear on 6 of the 19 variables. For the remaining 13 variables ENA theory appears to provide a way of providing a substantial explanation for ten and at least a potential for accounting for the remaining three. Next in strength was sexual selection theory, followed by the biosocial social role theory. Of course, for the details, one should refer back to the précis for each of these variables.

27.3b USDs in Recreational, Play, and Sexual Behavior

Quite a few variables having to do with sex differences in recreational, play, and sexual behavior were found to be universal. As shown in the two tables below, most of these variables appear to be more pronounced in males than in females.

Male-Higher USDs. In Table 27.3b-M, one can see that well over 100 studies have indicated that males are older on average than the females they date or marry. A very large number of studies also indicate that males choose to play with masculine toys more than do females; these toys primarily are ones with moving parts such as wheels or components for building things.

Numerous studies have also documented that males watch (or enjoy) pornography and masturbate more than do females. The playing of electronic or video games, playing sports and competitive games, and playing with mechanical and construction objects are all more common among males. Studies have shown that males are more directly involved in military combat than females. Additional universal male-higher sex differences have to do with being more likely to remarry after divorce and being pushier when it comes to sexual encounters.

One can see in the shaded portion of this table that the two evolutionary theories seem to offered the greatest number of probable explanations for male-higher USDs having to do with recreational, play, and sexual behavior. Nonetheless, the biosocial version of social role theory also offers possible explanations for three of these variables.

Female-Higher USDs. Just five USD-Fs were located pertaining to recreational, play, and sexual behavior. As one can see in Table 27.3b-F, four of the five variables have to do with play. In addition to feminine toys in general, females are more likely to play with dolls and to play house. Also, when females play, they do so cooperatively, rarely competitively. The only USD-F of a sexual nature involved watching (or enjoying) romantic programs, such as those on television or in movies.

Table 27.3b-M Male-higher USDs regarding recreational, play, and sexual behavior

Variable name or description	*Supportive findings*	*Age groupings*	*Countries sampled*	*Supported theory or theories (or scriptures)*						*Précis #*
				Origin. social role	*Founder effect social role*	*Biosocial social role*	*Sexual selection*	*Evol. neuroandrogenic*	*Theological*	
Older age at marriage or during courtship	126/127	All	27 (10)	x	x	x	✓✓	✓✓	x	26.3.17.41
Masculine toy choices and preferences	112/112	All	13 (1)	x	x	x	✓✓	✓✓	x	26.3.17.27
Watching or enjoying pornography	62/62	All	19 (2)	x	x	x	✓✓	✓✓	x	26.3.17.33
Playing electronic or video games	60/61	All	4 (1)	x	x	x	✓✓	✓	x	26.3.17.36
Masturbation	49/50	All	6	x	x	x	✓	✓✓	x	26.3.17.38
Playing sports and athletics in general	48/48	All	8 (4)	x	x	✓	✓✓	✓✓	x	26.3.17.30
Playing competitive sports	37/37	All	6 (1)	x	x	✓	✓✓	✓✓	x	26.3.17.32
Playing team sports	37/37	All	5 (1)	x	x	x	✓	✓	x	26.3.17.31
Playing with mechanical or construction objects	31/31	All	3	x	x	x	✓✓	✓✓	x	26.3.17.25
Competitive social play	21/22	All	4	x	x	✓	✓	✓	x	26.3.17.22

(*Continued*)

Table 27.3b-M (Continued)

Variable name or description	*Supportive findings*	*Age groupings*	*Countries sampled*	*Supported theory or theories (or scriptures)*						*Précis #*
				Origin. social role	*Founder effect social role*	*Biosocial social role*	*Sexual selection*	*Evol. neuroandrogenic*	*Theological*	
Choosing or preferring toys with motion or that can be used for building	19/19	All	5	x	x	x	x	x	x	26.3.17.29
Direct involvement in warfare (combat)	17/17	All	6 (6)	x	x	x	✓✓	✓✓	x	26.3.17.37
Number of sex partners	15/15	WAR	16 (2)	x	x	x	x	x	x	26.3.17.39
Play or recreating outside (instead of indoors)	15/15	All	3	x	x	x	✓✓	✓✓	x	26.3.17.21
Conversational focus on work and/ or leisure	15/15	All	2	x	x	x	✓	✓	x	26.3.17.20
Watching or enjoying sports programs	14/14	All	4 (1)	x	x	x	✓✓	✓✓	x	26.3.17.35
Remarriage after divorce	11/11	All	1 (1)	x	x	x	✓	✓	x	26.3.17.42
Pushy sexual overtures	11/11	Adult	2	x	x	x	✓✓	✓✓	x	26.3.17.40

Note: WAR is an abbreviation for wide age range.

Table 27.3b-F Female-higher USDs regarding recreational, play, and sexual behavior

Variable name or description	*Supportive findings*	*Age groupings*	*Countries sampled*	*Supported theory or theories (or scriptures)*						*Précis #*
				Origin. social role	*Founder effect social role*	*Biosocial social role*	*Sexual selection*	*Evol. neuroandrogenic*	*Theological*	
Feminine toy choices and preferences	95/95	All	10 (1)	x	x	x	✓✓	✓✓	x	26.3.17.26
Choosing or preferring dolls to play with or collect	42/42	All	8	x	x	x	✓✓	✓✓	x	26.3.17.28
Watching or enjoying romantic programs	23/23	All	2 (2)	x	x	x	✓✓	✓	x	26.3.17.34
Playing house	19/19	All	3	x	x	x	✓✓	✓✓	x	26.3.17.24
Cooperative social play	10/10	Child	1	x	x	x	✓	✓✓	x	26.3.17.23

Theoretically explaining these USD-Fs was best achieved with one or both of the evolutionary theories. In essence, it appears that female preferences for play objects and thinking along sexual lines have been favored by sexual selection, and are promoted by relatively low brain exposure to androgens, especially prenatally.

27.3c ***USDs in Criminality, Resource Acquisition, and Consuming Behavior***

Many apparent universal sex differences having to do with criminality, resource acquisition, and consuming behavior were identified. These variables are summarized in the two tables below along with information regarding which theories best explain *why* they are more common in one sex or the other.

Male-Higher USDs. Twenty-five variables that studies suggest are more prevalent in males than in females are listed in Table 27.3c-M. The first four of these variables document that males are more likely to engage in criminality (often including delinquency). This is based both on self-reports (usually on anonymous questionnaires) as well as official crime statistics, both being true especially for violent crime. As one scrolls further down in the table, one can see that several other types of criminal behavior, or near criminality (such as reckless driving, cruelty, and illegal drug use), are also more prevalent among males.

The table also summarily documents that males exhibit various types of alcohol-related variables, including binge drinking, and driving while drunk. Essentially, the only non-offensive types of behavior listed in this table had to do with meat eating and owning weapons, both of which are more prevalent among males than females.

Regarding theoretically explaining the male-higher USDs listed in this table, the shaded portion shows that the two evolutionary theories appear to have the greatest power. A large proportion of this power comes from noting that males appear to be favored for exhibiting greater competitiveness and risk-taking.

Female-Higher USDs. Table 27.3c-F lists seven variables having to do with recreational, play, and sexual behavior. The summarized information show that females are more often victimized sexually, both in physically assaultive terms and in non-violent terms. Females have been found to spend more time shopping, except in the case of online shopping. They also consume more medications, including ones for treating psychological symptoms, other than alcohol and other illegal drugs.

As shown in the shaded section of this table, only the two evolutionary theories appear to offer explanations for most of these sex differences. Of these, only being pressured to have sex appears to be supported by significant empirical evidence.

Table 27.3c-M Male-higher USDs regarding criminality, resource acquisition, and consuming behavior

Variable name or description	*Supportive findings*	*Age groupings*	*Countries sampled*	*Supported theory or theories (or scriptures)*						*Précis #*
				Origin. social role	*Founder effect social role*	*Biosocial social role*	*Sexual selection*	*Evol. neuroandrogenic*	*Theological*	
Self-reported offending in general	227/233	All	46 (7)	x	x	x	✓✓	✓✓	x	26.3.19.7
Officially identified criminal offending in general	189/191	All	30 (10)	x	x	x	✓✓	✓✓	x	26.3.19.1
Officially identified homicide perpetration	101/101	All	25 (22)	x	x	✓	✓✓	✓✓	x	26.3.19.3
Officially identified violent offending (usually excluding homicide)	97/97	All	12 (11)	x	x	✓	✓✓	✓✓	x	26.3.19.2
Alcohol abuse	84/88	All	16 (2)	x	x	x	x	✓✓	x	26.3.18.8
Amount of alcohol consumed	76/77	All	12 (3)	x	x	x	✓	✓✓	x	26.3.18.5
Self-reported violent offending	63/64	All	9 (4)	x	x	x	✓✓	✓✓	x	26.3.19.8
Alcohol-related problems	49/51	All	11/1	x	x	x	x	✓	x	26.3.18.7

(*Continued*)

Table 27.3c-M (Continued)

Variable name or description	*Supportive findings*	*Age groupings*	*Countries sampled*	*Supported theory or theories (or scriptures)*						*Précis #*
				Origin. social role	*Founder effect social role*	*Biosocial social role*	*Sexual selection*	*Evol. neuroandrogenic*	*Theological*	
Binge drinking	30/30	All	9	x	x	x	x	✓	x	26.3.18.9
Officially identified homicide followed by suicide	28/29	All	12 (1)	x	x	x	✓	✓	x	26.3.19.4
Officially identified drug use/possession/ sale	23/23	All	10	x	x	x	x	✓✓	x	26.3.19.6
Recipient of teacher discipline or school discipline	23/23	All	2	x	x	x	✓✓	✓✓	x	26.3.19.15
Unsafe (reckless) driving in general	22/22	All	11 (1)	x	x	x	✓✓	✓✓	x	26.3.19.17
Self-reported property offending	21/21	Adult	11 (7)	x	x	x	✓✓	✓✓	x	26.3.19.9
Officially identified sex offenses (except prostitution)	18/18	All	9 (1)	x	x	x	✓✓	✓	x	26.3.19.5
Eating meat	18/18	All	6 (2)	x	x	x	✓✓	✓✓	x	26.3.18.2

Self-reported (or victim-identified) sexual assault	18/18	All	7	x	x	x	✓✓	✓✓	x	26.3.19.10
Driving while under the influence of alcohol	14/14	All	3	x	x	x	✓	✓✓	x	26.3.18.10
Cruelty and sadistic behavior	13/13	All	4	x	x	✓	✓✓	✓✓	x	26.3.19.13
Commission of domestic violence (spousal violence) with serious injury	13/13	All	2	x	x	✓	✓	✓	x	26.3.19.14
Illegal drug use, age of onset	12/12	All	1	x	x	x	x	x	x	26.3.19.11
Owning or possessing weapons	12/12	All	1	x	x	x	x	x	x	26.3.18.11
Victim of physical aggression in general (except criminal)	11/11	Child	5	x	x	x	✓	✓	x	26.3.19.20
Perpetration of violent crime based on victim reports	10/10	All	4	x	x	✓	✓✓	✓✓	x	26.3.19.12
Recipient of teacher discipline or school discipline	10/10	All	2	x	x	x	✓✓	✓✓	x	26.3.19.16

Table 27.3c-F Female-higher USDs regarding recreational, play, and sexual behavior

Variable name or description	*Supportive findings*	*Age groupings*	*Countries sampled*	*Supported theory or theories (or scriptures)*						*Précis #*
				Origin. social role	*Founder effect social role*	*Biosocial social role*	*Sexual selection*	*Evol. neuroandrogenic*	*Theological*	
Victim of rape or sexual assault	91/91	All	11 (1)	x	x	x	✓	✓	x	26.3.19.18
Abstinence from alcohol consumption	39/40	All	11/1	x	x	x	x	x	x	26.3.18.6
Shopping in general (except online)	34/35	All	12	x	x	x	✓	✓	x	26.3.18.1
Victim of sexual abuse in general (excluding violent forms)	29/29	All	10	x	x	x	✓	✓	x	26.3.19.19
Consuming therapeutic medications in general	27/28	All	3 (2)	x	x	x	✓	✓	x	26.3.18.3
Consuming prescription psychotropic medications	18/18	WAR	8 (3)	x	x	x	✓	✓	x	26.3.18.4
Being pressured to have sex	15/15	Adult	1	x	x	x	✓✓	✓✓	x	26.3.19.21

Note: WAR is an abbreviation for wide age range.

27.3d USDs in Education-Related Behavior

Universal sex differences of an educational nature have been identified for both males and females. These are summarized in the following two tables.

Male-Higher USDs. As shown in Table 27.3d-M, all over the world, males have been found to be more likely to major (or take advanced courses) in the physical sciences, engineering, computer science, economics, and mathematics. With the possible exception of economics, which is usually considered a social science, these majors are subsumed under the acronym *STEM* (for science, technology, engineering, and mathematics).

The shaded portion of this table shows that the most powerful explanations for these sex differences are the two evolutionary theories. In this regard, males seem to have been sexually selected for a wide range of spatial reasoning interests, which in turn cause males to develop their spatial reasoning abilities. Considerable evidence reviewed under these précises indicate that exposing the brain to high (male-typical) levels of testosterone promotes such interests.

Female-Higher USDs. Table 27.3d-F summarizes findings having to do with education-related behavior and course specialization that females exhibit more than males. Leading the list are 21 studies indicating that females spend more time studying and doing homework. The remainder of the female-typical variables in this table are ones indicating that females are more likely to major (or take advanced courses) in four areas of study: the humanities, language, education, and fine arts.

Three of the five theoretical explanations bearing on why these sex differences would exist were identified. As explained in their respective précises, females appear to have evolved greater interests in language than is the case for males (with the latter being more interested in spatial reasoning). As to why females would be more inclined to spend time studying, it is noteworthy that nearly all studying is a sedentary activity. Therefore, male tendencies to be more physically active than females (see Précis 26.3.16.4) would cause males to be more resistant to prolonged studying.

We also noted that females in Western societies are stereotyped as being more interested in artistic endeavors than are males (Table 21.2.4.2) and that all ten of the studies of sex differences in persons majoring in the fine arts involved Western populations. Therefore, the original version of social role theory is currently consistent with the evidence.

Table 27.3d-M Male-higher USDs for education-related behavior

Variable name or description	*Supportive findings*	*Age groupings*	*Countries sampled*	*Supported theory or theories (or scriptures)*						*Précis #*
				Origin. social role	*Founder effect social role*	*Biosocial social role*	*Sexual selection*	*Evol. neuroandrogenic*	*Theological*	
Majoring (or taking advanced courses) in physics, chemistry, or other physical sciences and technology	98/102	All	11 (5)	x	x	x	✓✓	✓✓	x	26.3.20.10
Majoring (or taking advanced courses) in engineering	44/44	All	6 (6)	x	x	x	✓✓	✓✓	x	26.3.20.5
Majoring (or taking advanced courses) in mathematics	42/42	Adult	9 (5)	x	x	x	✓	✓	x	26.3.20.9
Majoring (or taking advanced courses) in computer science	38/38	All	5 (3)	x	x	x	✓✓	✓✓	x	26.3.20.2
Majoring (or taking advanced courses) in economics	11/11	All	4	x	x	x	✓	✓✓	x	26.3.20.3

Table 27.3d-F Female-higher USDs for education-related behavior

Variable name or description	*Supportive findings*	*Age groupings*	*Countries sampled*	*Supported theory or theories (or scriptures)*						*Précis #*
				Origin. social role	*Founder effect social role*	*Biosocial social role*	*Sexual selection*	*Evol. neuroandrogenic*	*Theological*	
Time spent studying and doing homework	21/21	All	8	x	x	x	✓	✓	x	26.3.20.1
Majoring (or taking advanced courses) in the humanities	18/18	All	7 (2)	x	x	x	✓✓	✓✓	x	26.3.20.7
Majoring (or taking advanced courses) in language	13/13	All	5	x	x	x	✓✓	✓✓	x	26.3.20.8
Majoring (or taking advanced courses) in education	12/12	All	2 (1)	x	x	x	✓✓	✓✓	x	26.3.20.4
Majoring (or taking advanced courses) in fine arts	10/10	All	3	✓	x	x	✓	✓	x	26.3.20.6

27.3e USDs in Work-Related Behavior

Many work-related behavior patterns have been studied in connection with sex differences. As shown in the following section, this research has made it possible to identify a substantial number of USDs, especially for males.

Male-Higher USDs. Table 27.3a-M shows that massive amounts of research throughout the world have been conducted on a wide range of sex difference variables having to do with work. One can see that the first listing has to do with sex differences in people's salaries, and that all but 8 out of 587 studies (drawn from 47 individual countries plus 58 multi-country studies) agreed that males were paid more than females. If Table 20.3.1.1 in Volume III is examined, one sees that all of the eight exceptions pertain to studies of sub-groups within more general populations, e.g., among newly hired MBAs. Sex differences in salaries is, in fact, the single most extensively documented USD in this entire book.

Other aspects of work in which males are more numerous have to do with achieving prominence in one's particular occupation and in terms of upward social mobility. Research has also indicated that males work longer hours per week and spend greater proportions of their lives in paid employment than do females.

Many types of occupations have been found to be filled by males more than by females. They include the following: engineers, lawyers, physical scientists, law enforcement, college-level teachers, politicians (both elected and appointed), corporate executive officers, and mechanics, along with other forms of manual labor. Finally, it is worth mentioning that males perform more outdoor work around the home (including construction and maintenance work) than do females.

Regarding theoretical explanations for USD-Ms, those with the greatest power are sexual selection theory and ENA theory. It is worth mentioning that theological (scriptural) explanations appear possible for at least four of these USD-Ms. In particular, scriptural passages indicating that indicate that males should assume more responsibility providing family support and should receive higher wages than females can possibly account for the universality of some of these sex differences (Table 27.3e-M).

Female-Higher USDs. The seven female-higher USDs are summarized in Table 27.3e-F. The single best documented of these USDs has to do with performing indoor household chores. Based on samples drawn from 31 countries and 41 multiple countries, all but three of the 337 studies concluded that females surpass males. These three exceptions all involved studies of pre-adolescent children. Therefore, it appears to be very certain that, particularly among adults, females do more indoor housework than do males (see Table 20.3.3.1 in Volume III for the complete documentation).

Table 27.3e-M Male-higher USDs for work-related behavior

Variable name or description	*Supportive findings*	*Age groupings*	*Countries sampled*	*Supported theory or theories (or scriptures)*						*Précis #*
				Origin. social role	*Founder effect social role*	*Biosocial social role*	*Sexual selection*	*Evol. neuroandrogenic*	*Theological*	
Earnings/salaries for workers	579/587	All	47 (58)	x	x	✓	✓✓	✓✓	✓	26.3.20.39
Achieving prominence in one's occupational field	146/148	All	22 (20)	x	x	✓	✓✓	✓✓	x	26.3.20.42
Politician: Holder of elected office	127/129	All	22 (22)	x	x	x	✓✓	✓✓	x	26.3.20.34
Participation in paid workforce in general	112/116	Adult	40 (20)	x	x	x	✓	✓	✓	26.3.20.11
Upward social status mobility	98/99	All	12 (4)	x	x	✓	✓✓	✓✓	x	26.3.20.43
Masculine occupations/male-typical occupations	95/95	All	11 (10)	x	x	✓	✓✓	✓✓	x	26.3.20.19
Full-time paid employment	80/82	All	25 (12)	x	x	x	✓	✓	✓	26.3.20.12
Average hours worked per week	44/46	All	4 (3)	x	x	x	✓✓	✓	x	26.3.20.17
Engineering	40/40	All	6	x	x	x	✓✓	✓✓	x	26.3.20.25

(*Continued*)

Table 27.3e-M (Continued)

Variable name or description	*Supportive findings*	*Age groupings*	*Countries sampled*	*Supported theory or theories (or scriptures)*						*Précis #*
				Origin. social role	*Founder effect social role*	*Biosocial social role*	*Sexual selection*	*Evol. neuroandrogenic*	*Theological*	
Teacher: College level	29/29	All	8 (4)	x	x	x	✓	✓	x	26.3.20.36
Law enforcement/ police officers	29/29	All	4	x	x	x	✓	✓	x	26.3.20.27
Lawyer	28/28	All	5	x	x	x	x	x	x	26.3.20.28
Duration of lifetime spent in the paid labor market	20/20	All	5 (2)	x	x	x	✓✓	✓	x	26.3.20.16
Physical scientist	19/19	All	3 (1)	x	x	x	✓✓	✓✓	x	26.3.20.31
Corporate executive officers	18/18	All	3 (2)	x	x	✓	✓	✓✓	x	26.3.20.24
Centrality of one's job or career (outside the house) to one's life	18/18	All	3	x	x	x	✓✓	✓	x	26.3.20.15
Geographer or mapping scientist	17/17	All	4	x	x	x	✓✓	✓✓	x	26.3.20.26
Manual/blue-collar occupations	15/15	All	6	x	x	✓	✓✓	✓✓	x	26.3.20.18

Employed in high-paying businesses or occupations	13/13	All	4 (2)	x	x	x	✓✓	✓✓	✓	26.3.20.14
Computer-related occupations	11/11	All	1	x	x	x	✓✓	✓✓	x	26.3.20.23
Politician: Appointed office holder	10/10	All	4 (4)	x	x	x	x	x	x	26.3.20.32
Financial well-being	10/10	All	7	x	x	x	x	x	x	26.3.20.41
Diversity (variability) in occupations	10/10	All	2 (1)	x	x	x	✓✓	✓✓	x	26.3.20.21
Performing outdoor or construction household chores	10/10	All	4	x	x	x	✓✓	✓✓	x	26.3.20.38
Mechanic or machine operator	10/10	All	3	x	x	x	✓✓	✓✓	x	26.3.20.29
Politician: Candidate for elected office	10/10	All	3	x	x	x	✓	✓	x	26.3.20.33

Table 27.3e-F Female-higher USDs for work-related behavior

Variable name or description	*Supportive findings*	*Age groupings*	*Countries sampled*	*Supported theory or theories (or scriptures)*						*Précis #*
				Origin. social role	*Founder effect social role*	*Biosocial social role*	*Sexual selection*	*Evol. neuroandrogenic*	*Theological*	
Performing indoor household chores	334/337	All	31 (41)	x	x	✓	✓	✓✓	x	26.3.20.37
Effects of the proportion of females in a job sector and the average salary for its workers	50/51	All	5 (1)	x	x	x	x	x	x	26.3.20.40
Feminine occupations/ female-typical occupations	45/45	All	9 (7)	x	x	✓	✓✓	✓✓	x	26.3.20.20
Nursing	19/19	All	7	x	x	x	✓✓	✓✓	x	26.3.20.30
Clerical/service occupations	18/18	All	8	x	x	x	✓✓	✓✓	x	26.3.20.22
Employed part time	13/13	All	3 (2)	x	x	x	✓	✓	x	26.3.20.13
Social/welfare worker	11/11	All	4	x	x	x	✓✓	✓✓	x	26.3.20.35

The second-best documented USD-F having to do with work involves studies of the proportion of females in a given job sector and the average salary paid for the workers in that sector. All but one of the 51 pertinent studies concluded that, as the proportion of females in a job sector increases, the average wages paid to its workers decreases. One can see that none of the five theories seemed to have any bearing on this particular sex difference. Instead, a rather simple law of supply and demand could be largely responsible for the phenomenon.

The table shows that employed females are more likely to be employed part time (rather than full time). This could help to account for the more limited number of occupations that are dominated by females. In this regard, while a number of stereotypically feminine occupations are universal, only three occupations were found to be specifically more often filled by females than by males. These are nursing, clerical/service occupations, and social work.

Both sexual selection theory and ENA theory appear to provide explanations for six of the seven USD-Fs. The biosocial social role theory also does so in two cases.

27.3f USDs in Sex Stereotypes and Portrayals and Sex Differences in Expectations and Treatment

The final category of sex differences is an odd mixture of traits. They include sex stereotypes, portrayals of males and females in the mass media, and sex differences in how people are treated by others. As a special note, in the case of sex stereotypes and most of the mass media portrayals, no attempt is made to theoretically explain them. Instead, we simply offer assessment as to whether or not the stereotypes and portrayals appear to conform with reality. One can refer back to the respective précises for more details.

Male-Higher USDs. Regarding traits that are more pronounced among males than females, Table 27.3f-M shows that 11 variables fit into this final category. The first variable indicates that males are punished more often or more severely than females are as children.

In terms of stereotypes, males are thought to be more dominant (or to have more leadership ability), to be more physically aggressive, to be better at math, to be more independent, to be more adventurous, and to be have higher earnings than is the case for females. One can see that most of these stereotypes appear to be at least generally confirmed by empirical observations. The exception has to do with math ability. As documented in Table 12.2.3.1 of Volume II, the findings on math performance tests have been quite mixed in terms of sex differences. The main exception has to do with exceptionally high scores on math tests (i.e., scores above two standard

Table 27.3f-M Male-higher USDs for sex stereotypes, mass media portrayals, and sex differences in expectations and treatment

Variable name or description	*Supportive findings*	*Age groupings*	*Countries sampled*	*Supported theory or theories (or scriptures)*						*Précis #*
				Origin. social role	*Founder effect social role*	*Biosocial social role*	*Sexual selection*	*Evol. neuroandrogenic*	*Theological*	
Physical punishment of offspring by parents	36/37	All	25	x	x	x	✓✓	✓✓	x	26.3.22.4
Sex stereotype about dominance and leadership	30/31	All	3	Empirically confirmed stereotype						26.3.21.9
Sex stereotyped physical aggressiveness	26/26	All	4 (1)	Empirically confirmed stereotype						26.3.21.6
Portrayals of working outside the home	22/23	All	6 (1)	Empirically verified portrayal						26.3.22.10
Sex stereotyped math ability	21/21	All	8	Empirically somewhat questionable stereotype						26.3.21.1
Portrayals of aggression and violence	20/20	All	2 (1)	Empirically verified portrayal						26.3.22.7
Age of persons featured in mass media	19/19	All	5 (2)	x	x	x	✓✓	✓	x	26.3.22.6

Sex stereotyped being independent	17/17	All	3 (2)	Empirically confirmed stereotype	26.3.21.7
Portrayals of dominance and social status	12/12	WAR	4	Empirically verified portrayal	26.3.22.8
Sex stereotyped adventurousness	10/10	All	4 (2)	Empirically confirmed stereotype	26.3.21.5
Sex stereotyped higher in earnings or wealth	10/10	All	3	Empirically confirmed stereotype	26.3.21.10

Note: WAR is an abbreviation for wide age range.

deviations from the mean). For these scores, there is little doubt that males score higher in all countries sampled (Table 12.2.3.4 of Volume II).

Regarding sex differences in how people are portrayed in the mass media, males are portrayed as working away from home, as being more aggressive, and as being more dominant or of higher social status. These portrayals seem to conform to reality, as documented in empirical studies. The last mass media portrayal involved the age of persons featured in the mass media; where all 19 studies concluded that males are older on average than females.

Turning to theoretical explanations, only two of the 11 variables listed in this table called for explanations based on any of the five theories. To account for why males are physically punished more by their parents than are females, both of the evolutionary theories seem to be relevant (given that physical punishment is often imposed for childhood aggression and disobedience). In the case of the age of persons featured in the mass media, several factors having to do with sexual selection (e.g., males being older than females during courtship, and males of high social status usually being featured in the mass media more than high status females).

Female-Higher USDs. As shown in Table 27.3f-F, the best documented sex difference pertaining to sex stereotypes, mass media portrayals, expectations and treatment involves portrayals of persons working inside the home. In all but one out of 50 studies, females were found being portrayed as doing so more than males. Most of these portrayals occurred in television commercials, with samples drawn from 15 different countries (plus one multi-country sample). These depictions, of course, comport with findings summarized in Table 27.3e-F, indicating that females *are* more involved in performing household chores than are males.

The second most well documented sex difference in this final table has to do with emotionality. All 27 studies of stereotypes concluded that females were considered more emotional than males. This stereotypic generalization is somewhat questionable in light of the empirical evidence presented in Table 10.1.1.1 (Volume II).

The third variable in this table's list indicates that girls are supervised more closely by their parents than are boys. We found no obvious explanation for this sex difference using any of the five theories.

In terms of perceiving risks in their environment, females have been found to perceive more than males. This difference could be at least partly attributable to male tendencies to be more *prone* to take risks, i.e., by downplaying the degree of risk involved (Précis 26.3.16.2). If so, evolutionary theory may provide an explanation.

The table's last three entries all pertain to sex stereotypes. They indicate that females are more emotional expressive, fearful, and nurturing. The first of these three is somewhat questionable in light of findings summarized in

Table 27.3f-F Female-higher USDs for sex stereotypes, mass media portrayals, and sex differences in expectations and treatment

Variable name or description	*Supportive findings*	*Age groupings*	*Countries sampled*	*Supported theory or theories (or scriptures)*						*Précis #*
				Origin. social role	*Founder effect social role*	*Biosocial social role*	*Sexual selection*	*Evol. neuroandrogenic*	*Theological*	
Portrayals of working inside the home	49/50	All	15 (1)	Empirically verified portrayal						26.3.22.9
Sex stereotyped emotionality	27/27	All	2 (2)	Empirically somewhat questionable stereotype						26.3.21.2
Parental supervision of offspring	18/18	All	4 (1)	x	x	x	x	x	x	26.3.22.5
Risk assessment	14/14	All	6	x	x	x	✓	✓	x	26.3.22.3
Sex stereotyped emotional expressiveness	12/12	All	2	Empirically somewhat questionable stereotype						26.3.21.3
Sex stereotyped fearfulness	10/10	All	1 (1)	Empirically generally confirmed stereotype						26.3.21.4
Sex stereotyped being nurturing	10/10	All	2	Empirically generally confirmed stereotype						26.3.21.8

Table 10.2.9.1 of Volume II. While no studies were located indicating that males were more emotional expressive than females, several studies have reported no significant sex differences. In the case of fearfulness and nurturing tendencies, the stereotypes of females being greater than males appears to be true.

27.4 Nearly Certain Universal Sex Differences

Most scientific research, including that which has been covered in this book, is based on samples. These samples are usually drawn with the idea of roughly reflecting a wider population, such as a country. The people comprising these samples typically range in numbers from a few dozen to several hundred, although samples in the thousands are becoming somewhat more common.

No matter how large samples are, few *precisely* reflect a population from which they were drawn, even within a prescribed age range. For this reason, and because no variables can be measured with perfect accuracy, conclusions reached by scientific research must be considered *estimates* of reality, not reality itself. Nevertheless, as research on sex differences continues to accumulate, the reality of universality in some of these differences will continue to become clearer. This principle of sampling is worth bearing in mind as we draw this book to a close.

In this, last part of the final chapter (except for a brief epilogue), we will go beyond identifying *likely USDs* in order to draw attention to *nearly-certain USDs*. These USDs consist of those already listed above for which the number of pertinent findings surpassed 100 (rather than the minimum of ten used to help identify likely USDs). Seventeen such USDs were identified.

Before considering these 18 virtually certain USDs, we will briefly describe the five scientific theories that can shed light on *why* any USDs might exist. Doing so is in recognition that if *any* USDs exist, these 18 nearly certain ones call for a theoretical accounting to the greatest degree.

27.4.1 Reiterating the Five Theories of USDs

As first documented in Chapter 24, there appear to be just five theories proposed so far that address the issue of possible universal sex differences in traits, particularly cognitive and behavioral traits. Three of these theories attribute sex differences in cognition and behavior to sociocultural stereotyping and learning, while the remaining two theories incorporate evolutionary concepts into their explanations. In brief terms, these five theories can be summarized as follows:

1 *Original social role theory* maintains that sex differences in cognitive and behavioral traits are learned through the socialization process.

Regarding the possibility of any universal sex differences, this theory leads one to expect few if any to exist given the wide diversity of human cultures.

2 *Founder effect social role theory* asserts that all sex differences in cognitive and behavioral traits are learned in sociocultural contexts. However, some sex differences may be universal due to sex differences arbitrarily appearing in the earliest human societies simply being "faithfully passed down a thousand times over" to the present day (Fausto-Sterling 1992: 199).

3 The *biosocial version of social role theory* agrees that all sex differences in cognition and behavior are culturally learned, but adds that this learning may result in some sex differences being universal due to certain biological factors. Those factors are that males are (a) males are taller and stronger, and (b) only females are able to bear offspring (Wood & Eagly 2012).

4 *Sexual selection theory* explains universal sex differences in evolutionary terms. This theory states that each sex has evolved certain combinations of traits to the extent that these traits have affected their probabilities of reproducing. For example, Table 28.1 shows that one of the nearly certain USDs (i.e., number 15) is the variable of preferring mates with resources, a trait for which 104 out of 107 studies concluded that females are higher (the three exceptions simply found differences to be non-significant). According to sexual selection theory, females with these types of preferences have usually left more copies of their genes in subsequent generations than females without such preferences. Furthermore, whatever tendencies males may have to comply with these female preferences may have helped them to leave descendents in future generations as well. Note that this theory is entirely compatible with the view that learning is involved in producing male-female differences, since sexual selection for *preferring* to learn some things more than other things may be at the heart of how the brain has been sexually differentiated.

5 ENA (*evolutionary neuroandrogenic*) *theory* is basically an extension of sexual selection theory. It asserts that bodily exposure to androgens (especially testosterone) plays a central role in producing sexually selected traits. Regarding cognitive and behavioral traits, one part of the body, i.e., the brain, is crucial in producing universal sex differences. In other words, varying the degree to which the brain is exposed to androgens throughout life is responsible for much of the male-typical and female-typical thought patterns and behavior patterns.

27.4.2 The 18 Nearly Certain USDs

The 18 nearly-certain USDs are listed in Table 27.4 in order according to the number of studies located. This table has six columns, the first one of which identifies the trait along with the sex exhibiting the trait more (in parentheses). In the second column, the number of findings confirming this sex difference divided by the total number of findings is indicated, followed by a "% support" column. After that, a column showing the number of countries sampled (plus the number of studies based on multi-national samples appears in parentheses) is provided. The last two columns identify (a) the table in which the citations are located, and (b) the précis in which the theoretical significance of each USD is discussed and documented.

Table 27.4 The 18 nearly certain USDs

Variables	*# Findings*	*% Support*	*# Countries*	*Table of citations*	*Précis summaries*
1. Earnings/salaries for workers (M)	579/587	98.0	47 (58)	20.3.1.1	26.3.20.39
2. Performing routine indoor household chores (F)	334/337	99.1	31 (41)	20.3.3.1	26.3.20.37
3. Self-reported offending in general (M)	227/233	97.4	46 (7)	19.1.2.1	26.3.19.7
4. Height (following mid to late adolescence) (M)	221/221	100.0	58 (12)	1.2.1.1	26.1.1.4
5. Officially identified criminal offending in general (M)	189/191	99.0	30 (10)	19.1.1.1	26.3.19.1
6. Physical exercise (M)	142/149	95.3	18 (4)	16.3.2.2	26.3.16.4
7. Achieving prominence in one's occupational field (M)	146/148	98.6	22 (20)	20.4.3.4	26.3.20.42
8. Accidental injuries and fatalities in general (M)	130/134	97.0	17 (11)	5.2.1.1	26.1.5.4
9. Politician: Holder of elected office (M)	127/129	98.4	22 (22)	20.2.7.51	26.3.20.34
10. Attention deficit hyperactivity disorder (ADHD) (M)	124/126	98.4	18	14.4.8.1	26.2.14.22
11. Conduct disorder (M)	126/131	96.2	19 (1)	14.4.3.1	26.2.14.8
12. Brain volume (brain size) (M)	120/122	98.4	23 (1)	4.1.1.1	26.1.4.1

(*Continued*)

Table 27.4 (Continued)

Variables	*# Findings*	*% Support*	*# Countries*	*Table of citations*	*Précis summaries*
13. Participation in paid workforce in general (M)	112/116	96.6	40 (20)	20.2.1.1	26.3.20.11
14. Risk-taking (recklessness) in general (M)	104/109	95.4	14 (2)	16.2.3.1	26.3.16.2
15. Preference for a mate with resources (wealth, high income) (F)	104/107	97.2	27 (12)	15.6.5.9	26.2.15.19
16. Testosterone (M)	105/105	100.0	16 (1)	3.1.1.1	26.1.3.2
17. Majoring (or taking advanced courses) in physics, chemistry, or other physical sciences and technology (M)	98/102	96.1	11 (5)	20.1.4.26	26.3.20.10
18. Officially identified homicide perpetration (M)	101/101	100.0	25 (22)	19.1.1.3	26.3.19.3

Variable 1. The first variable listed in Table 27.4 constitutes the most highly replicated USD that was found. It involved sex differences in earnings (among full-time workers). Specifically, 579 such studies were located, all but eight of which concluded that males surpassed females. If one returns to Table 20.3.1.1 (in Volume III), one will see that all eight of these exceptions were of questionable relevance to overall nationwide sex differences. Two involved studies of blacks living in Britain, two others were experimental studies in hypothetical hiring situations, one involved studies of people who were mentally retarded, and the remaining three had to do with first-hirings of college graduates. With 47 different countries represented in these studies along with 58 additional multi-national studies, there can be little basis for lingering doubt that males on average earn more than do females throughout the world. In other words, the variable of sex differences in earnings (among full-time workers) is all but certainly an established USD-M.

Regarding how best to explain this USD-M, if one refers back to Précis 26.2.15.19, one will see that the theories with the greatest support are the evolutionary theories. In particular, female tendencies to prefer mates with resources (or who appear to have the potential for stable resource acquisition) is another USD-F (see entry 15 in Table 27.4). It is reasonable to deduce that this universal female preference will have the effect of increasing male tendencies to focus considerable effort on "earning a living."

Variable 2. Second in the list of the nearly certain USDs involved performing indoor household chores. For this variable, 334 studies concluded that females surpass males. Three exceptional findings were reported. However, all of these exceptions involved studies of children. Among adolescents and adults, there were no exceptions, i.e., females always surpassed males (see Table 20.3.3.1 in Volume III). Thirty-one different countries were sampled in these studies, along with 41 other multi-national studies.

Possibilities for theoretically explaining this USD-F were explored in Précis 26.3.20.37. Based on various lines of reasoning, including the possibility that females may have evolved a keener sense of neatness and cleanliness, this précis indicated that sexual selection theory has the greatest promise for explaining why females perform more indoor household maintenance chores than do males.

Variable 3. Sex differences in self-reported criminality (including delinquency) were the third in the table's list. For this variable, out of 233 studies, only six studies failed to find males higher (and none found females higher). Five of the exceptions involved self-reported delinquency during adolescence. The remaining study was based on self-reported offending among adult prisoners (see Table 19.1.2.1 in Volume III). Forty-seven countries were sampled along with seven multi-national studies.

It is all but impossible to explain the universality of sex differences in self-reported criminality simply in terms of social learning. Accordingly, the 3 social role theories are essentially silent to this matter. As explained more in the narrative surrounding Précis 26.3.19.7, evolutionary reasoning does seem to offer an account. It does so by assuming that, when males are compared to females, the former have been sexually selected for being more overtly competitive for resources and mating opportunities. Nonetheless, most successful forms of socially acceptable competition require considerably more learning than most criminal forms of competition. Thus, if most forms of criminality are seen as relatively ineffective forms of competitiveness, both of the evolutionary theories predict that criminal and delinquent behavior will be more prevalent among males than among females.

Variable 4. The fourth most frequently documented USD in Table 27.4 has to do with height. Two hundred and 21 studies pertaining to this variable were found regarding sex differences for research participants older than age of 15. Based on samples drawn from 58 individual countries along with 12 additional studies of multiple countries, all of these studies agreed that males were taller than females. It is interesting to note that this sex difference is a *given* in Wood and Eagly's (2012) biosocial learning theory, i.e., the difference is *assumed* without explanation. Both of the evolutionary theories, on the other hand, offers specific explanations for the difference. In essence, sexual selection theory asserts that increased body size (height in the case of humans) often gives males an edge in competitive activities. From the standpoint of ENA theory, bodily exposure to testosterone promotes bone growth and height (see Précis 26.1.1.4 for details).

Variable 5. According to the fifth entry in Table 27.4, more males than females are involved in the commission of crime, based on what are known as *official statistics* (i.e., data compiled by the criminal justice system). All but two of the 191 pertinent studies agreed on this generalization. As shown in Table 19.1.1.1 (in Volume III), both of these exceptions involved studies of adolescents; among adults, there were no exceptions.

Attempts to theoretically explain this USD are summarized in Précis 26.3.19.1. One can see that only the two evolutionary theories seem to be promising in this regard. These explanations assert that males have been sexually selected for greater competitiveness and tendencies to take risks. Especially in the decade or so following puberty (when male brain exposure to testosterone is highest), males are especially likely to express these competitive and risk-taking tendencies in many particularly crude ways. Full adults, particularly those responsible for instituting criminal/delinquency statutes, find many of these crude expressions of competitive and risk-taking traits to be so objectionable as to have instituted the criminal justice system to help prevent such behavior.

Variable 6. Table 27.4 shows that 142 out of 149 studies indicated that males engage in physical exercise than do females. All seven of the exceptional finds simply found no significant sex differences, and were focused on moderate degrees of exercise, rather than on vigorous exercise. Eighteen different countries were sampled, along with four multiple-country studies, were included in this set of observations.

Theoretically speaking, Précis 26.3.16.4 argued that ENA theory offers the best explanation for this particular sex difference. It does so by asserting that testosterone has multiple effects on sex differences in behavior, including overall tendencies to be physically active.

Variable 7. Sex differences in achieving prominence in one's occupation has received a great deal of research attention over the years. As shown in Table 27.4, 146 out of 148 studies concluded that males do so more than females. The two exceptions simply failed to identify a sex difference that was statistically significant. Twenty-two individual countries were studies, along with 20 multiple country studies.

In Précis 26.3.20.42, one can see that both of the evolutionary theories offer explanations for this sex difference. It primarily involves noting that females prefer mates with control over resources, and that one way to obtain and maintain such control involves achieving high social status. Furthermore, brain exposure to testosterone appears to enhance status-striving.

Variable 8. Being the victim of accidental injuries (including ones that are fatal) has been shown to be more common in males. All but three of the 134 pertinent studies reached this conclusion. Studies covered 17 different countries along with 11 multi-national studies.

Regarding possible theoretical explanations for this sex difference, Précis 26.1.5.4 shows that all of the evidence points toward the two evolutionary theories. In particular, the apparent male tendencies to take more risks throughout their lives (see Précis 26.3.16.2) appear to have adverse health consequences. Furthermore, brain exposure to testosterone appears to promote such risk-taking.

Variable 9. Many studies have assessed sex differences in being political office holders (usually based on the results of an election). As shown in Table 20.2.7.51, 127 out of 129 studies indicated that males are more likely to hold these positions than females, the remaining two simply reporting no significant sex difference.

Précis 26.3.20.34 provides information regarding possible theoretical explanations for this USD. It indicates that social role theories seem to be silent to the issue. However, both of the two evolutionary theories offer strong explanations. Regarding sexual selection theory, given that male tendencies to strive for high social status often provides a pathway to stable access to resources, thereby attracting mates. In the case of ENA theory, brain exposure to testosterone seems to be a major driver for status striving behavior.

Variable 10. Many studies of sex differences in attention deficit hyperactivity disorder (ADHD) were located. Based on samples drawn from 18 countries, all but two of the 126 studies concluded that males surpass females in being diagnosed with this disorder. Both of the two exceptions simply found no significant sex difference.

To account for ADHD as a USD-M, no social role explanations were located. As shown in Précis 26.2.14.22, the two evolutionary theories do appear to be relevant. Sexual selection theory argues that males have been favored for being more physically active than females. Regarding ENA theory, evidence was located indicating that brain exposure to testosterone promotes tendencies to be physically active.

Variable 11. Even though *conduct disorder* is a rather vague term, it has been widely studied, with nearly all studies finding it to be substantially more prevalent among males than females, particularly in the case of children and the most aggressive forms of the disorder. As shown in Table 27.4, 126 out of 132 studies, based on samples from 19 different countries (plus one multi-country study) came to this conclusion. The exceptional findings all reported no significant difference.

Regarding explanations, only the two evolutionary theories seem to offer reasonable proposals. Much of this comes from noting that males have been sexually selected for being more competitive, aggressive, and prone to take risks. Also, the tendency for females to acquire (and therefore attend to) language more readily could also contribute to male conduct disorders. In the case of brain exposure to testosterone, research was reviewed in the narrative surrounding Précis 26.2.14.8 that supports the hypothesis that this exposure is higher among children with conduct disorders than with children generally.

Variable 12. Over the years, many studies have reported that males have larger brains even after adjusting for sex differences in body size. At least among adults, Table 4.1.1.1 (Volume I) cites 122 studies, only two of which failed to document a significant difference. Twenty-three different countries (plus one multi-cultural study) were sampled.

As to why such a sex difference would exist, only the two evolutionary theories seem to offer viable explanations (Précis 26.1.4.1). Regarding sexual selection theory, it may be that some cognitive tendencies such as spatial reasoning – at which males usually attain higher scores (see Table 11.2.4.1) – may be enhanced by somewhat larger brains. If so, these types of brain-size promoted cognitive tendencies may enhance the tendency for males to be chosen as mates. In the case of ENA theory, considerable evidence indicates that prenatal exposure to testosterone results in increased brain size.

Variable 13. Sex differences in the amount of time spent working in the paid workforce has been widely studies. As shown in 20.2.1.1, among adults, 116 studies were located, all but four of which concluded that males surpassed females (the four exceptions all reported no significant sex difference). Samples for these studies were derived from 40 different countries, along with 20 multi-national studies.

No specific arguments for or against the idea of sex differences in time spent working as paid employees were found having to do with social role theories. However, the biosocial version of this social learning theory asserts that only females bear offspring. From this, one could infer that, at least women who have children will spend less time in the in the paid workforce than their male spouses.

As shown in Précis 26.3.20.11, both evolutionary theories bear on time spent in the paid workforce. Regarding sexual selection theory, it argues that males who are stable providers of resources will be more likely to attract mates, especially mates who will bear and rear offspring than males who are not reliable in this regard. According to ENA theory, male tendencies to be stable provisioners of resources should be promoted by exposing the brain to high levels of testosterone. Empirical evidence bearing on both of these hypotheses were supportive, but fairly weak.

It is also worth mentioning that various religious passages seem to be pertinent to sex differences in the paid workforce. Specifically, but Muslim and Mormon scriptures state that males are primarily responsible for providing financial support to their spouses and children.

Variable 14. A great deal of research has assessed sex differences in risk-taking tendencies. As shown in Table 16.2.3.1 (of Volume II), the vast majority of studies have concluded that males take more risks than do females, especially among adults. Specifically, out of the 109 studies of adults, 104 concluded that males are more prone toward risk-taking than are females. All of the exceptions simply reported no significant differences. Studies came from 14 different countries along with two multi-national studies.

Regarding theoretical explanations, only the two evolutionary theories appear to offer reasonable accountings for average sex differences in risk-taking. Both of these theories argue that, because females prefer mates who are stable provisioners of resources, males have evolved tendencies to engage in riskier behavior patterns in order to effectively comply with female preferences. In addition, ENA theory stipulates that the greater tendency for males to take risks are driven to a substantial degree by exposing the brain to testosterone. As discussed in the narrative surrounding Précis 26.3.16.2, both of these evolutionary lines of reasoning have empirical support.

Variable 15. One hundred and seven studies of sex differences in preferences for mates with resources were located, all but three of which concluded that females have stronger preferences in this regard than do males. The three exceptional all concluded that the differences were not statistically significant. Studies were based on samples drawn from 27 countries as well as 12 multi-country studies.

No predictions by any of the three social role theories were found regarding this particular sex difference. However, in evolutionary terms, one of the most fundamental driving forces behind sex differences in cognitive and behavioral tendencies is the need by both sexes to pass their genes onto future generations. In this regard, to a much greater degree than male, females appear to have been sexually selected for preferring mates with stable access to resources (Bereczkei, Voros et al. 1997; Ellis 2001; Zhu, & Chang 2019). In other words, women who prefer mates who appear capable of a "good provider" will be more successful at bearing and rearing offspring than women without this type of preference (Brooks, Blake & Fromhage 2022).

Besides predicting that females would prefer mates with resources more than males would, ENA theory also predicts that neurohormonal factors should be involved. Results from at least three studies have indicated that testosterone levels are positively correlated with income among males (Gielen, Holmes & Myers 2016; Hughes & Kumari 2019; Harrison, Davies et al. 2021). However, no evidence was found that low brain exposure to testosterone was associated with increased preferences for mates with high income potential.

Variable 16. Studies of sex differences in testosterone levels have found consistent results for three groupings: before birth, during the first 3–4 months after birth, and after the onset of puberty (see Figure 24.1). The 105 studies of these age groups all agreed that males had higher levels. Findings came from 16 different countries along with one multi-national study.

The only theory that specifically asserts that males would have higher levels of testosterone than females is ENA theory. In fact, this assertion is at the heart of the theory's neurohormonal component. Several other researchers have argued that testosterone is an evolved biochemical for producing sex differences in cognitive and behavioral traits (Cohen-Bendahan, van de Beek & Berenbaum 2005; Archer 2006; Mehta & Beer 2010; Volman, Toni et al. 2011; Celec, Ostatníková & Hodosy 2015; Nave, Nadler et al. 2018).

Variable 17. One hundred and two studies of sex differences in persons majoring (or taking advanced courses) in what are known as the physical sciences or technology were cited in Table 20.1.4.26 (in Volume III). All but four of these studies concluded that males surpass females in this regard. The exceptions reported no significant differences, especially when chemistry was included as one of the physical sciences.

Both of the evolutionary theories seem to offer well supported arguments to explain this sex differences. The arguments build on evidence that females have been sexually selected more than males for being people-oriented in their interests, thereby inclining females to be more inclined to care for offspring. Given that majoring in the physical sciences or technology are more things-oriented than people-oriented, one would expect that males would gravitate more toward these types of course offerings. The narrative surrounding Précis 26.3.20.10 briefly reviewed evidence that, as predicted by ENA theory, that interests in subject areas such as the physical sciences and engineering are enhanced by exposing the brain to testosterone.

Variable 18. The last nearly certain USD with total agreement had to do with the commission of homicide. All 101 findings regarding this variable agreed that males were more likely to commit homicide than were females. Samples obtained in this regard came from 25 countries plus 22 other studies of multiple countries.

As noted in Précis 26.3.19.3, both of the evolutionary theories appear to provide the best explanations for why males would dominate in the commission of homicide throughout the world. In essence, such aggression and violence reflect sexual selection for extreme competitiveness, both for resources and for mating opportunities. Considerable evidence points toward brain exposure to testosterone as contributing to such aggression.

27.5 Summarizing Findings Regarding Universal Sex Differences

The two goals of this book were to (a) identify universal sex differences, should any exist, and (b) explore all reasonable ways of explaining those differences in theoretical terms. In this regard, we even widened the range of possible explanations to include theological scriptures.

To accomplish these goals, the book began by organizing findings of sex differences into several thousand tables, with each table devoting a separate variable. Because of the massive number of findings that were located (i.e., roughly 50,000 findings located in approximately 40,000 studies). The tables were presented in 3 volumes. The first volume pertained to basic biology, the second to cognition, and the third to behavior, although a sharp distinction between these three categories was not always possible.

27.5.1 Organization of Volume IV

In this, the book's fourth volume, Chapter 24 began with a description of scientific theories having to do with the possibility of sex differences existing, not just in some "sexist" societies, but in all societies. Five such theories were found that made predictions in this regard, especially concerning cognitive and behavioral traits. Three of these theories focused on sociocultural learning and sex stereotyping as causing sex differences. These were the *original social role theory*, *founder effect social role theory*, and *biosocial role theory*. The remaining two theories were evolutionary in nature. They were *sexual selection theory* and *evolutionary neuroandrogenic theory*. Attention was also given to the possibility of religious scriptures sometimes helping to account for universal sex differences.

Chapter 25 was devoted to identifying all of the variables located throughout the book's first three volumes for with at least ten relevant findings had been cited. These variables were organized into tables along with scores for each variable indicating the degree of agreement regarding which sex exhibited the trait to the greatest degree (called *consistency scores*).

In Chapter 26, only the variables with nearly perfect consistency scores (i.e., 95.0 or higher) were given further consideration. This particularly lengthy chapter considered each high consistency score variables one at a time regarding its qualifications for being a USD (universal sex difference). Among the main factors considered were the number of studies located, the number of countries sampled, and how well each sex difference could be theoretically explained.

The present chapter condenses information provided in Chapter 26 down to two final listings. One list consisted of 308 variables with high consistency scores that were deemed *likely USDs*. The second listing was much shorter. It identified 18 variables for which at least 100 pertinent studies were cited, also with high consistency scores. These variables were deemed to be *nearly certain USDs.*

27.5.2 Theories with the Greatest Explanatory Power

The second goal of this book – that of identifying the most promising theoretical explanations for USDs – will now be specifically addressed. Table 27.5 provides a summary of how well each of the five scientific theories appears to explain why USDs would exist for each of the three categories of variables (i.e., basic biology, cognition, and behavior). Possible theological explanations of these variables are also listed.

For interpretation, "x" means that no theoretical explanation appeared to exist. One checkmark (✓) means that a particular theory provides a plausible explanation for a variable being different for males and females. Two checkmarks (✓✓) indicate that a particular theory offers a reasonable explanation with at least a modest degree of independent empirical support, and 3 checkmarks (✓✓✓) means that strong independent empirical support for this theoretical or theological explanation was found.

Basic Biology. To orient oneself, first, consider how well the original version of sex role theory explains the USDs of a basic biological nature. As one would expect, it was unable to explain any of the 33 USDs having to do with basic biology. Similarly, neither the founder effect version nor the biosocial version of social role theory accounted for any of the 33 basic biology USDs.

Turning to the two versions of evolutionary theory, one can see that sexual selection theory seemed to offer no explanation for 17 of the basic biological USDs, but can account for 16. The ENA theory fairs even better. It explains all but two of these 33 USDs. Regarding theological explanations for the 33 basic biological USDs, none were found.

In summary, as nearly everyone would expect, the relative power of the theories for explaining universal sex differences of a basic biological nature tilts strongly toward the two evolutionary theories. Keeping in mind that ENA theory is simply an extension of sexual selection theory in which contributions by sex hormones to the evolutionary process of sexual differentiation occurs, it is also not particularly surprising to find that this theory is somewhat stronger than sexual selection theory in predicting basic biological USDs.

Cognition. Now consider the 131 cognitive USDs. Table 27.5 shows that the original sexual role theory and the founder effect social role theory were both unable to account for any of these universal sex differences. The biosocial version of social role theory, however, did offer an explanation for 17 of the 131 differences.

In the case of the two evolutionary theories, they both performed relatively well. Regarding the sexual selection theory, it was about to explain all

Table 27.5 A summary of how well the five theories (and various theological scriptures) explain likely USDs

Scientific theories and theological explanations	*Three categories of likely USD variables*												*All likely USD variables combined*			
	Basic biological USDs				*Cognitive USDs*				*Behavioral USDs*							
	x	✓	✓✓	✓✓✓	*x*	✓	✓✓	✓✓✓	*x*	✓	✓✓	✓✓✓	*x*	✓	✓✓	✓✓✓
Original Social Role Theory	33	0	0	0	131	0	0	0	143	1	0	0	307	1	0	0
Founder effect social role Theory	33	0	0	0	131	0	0	0	144	0	0	0	308	0	0	0
Biosocial Role Theory	33	0	0	0	114	17	1	0	118	26	0	0	265	43	0	0
Sexual selection Theory	17	8	8	0	35	41	56	0	37	41	66	0	89	90	131	0
Evolutionary Neuroandrogenic Theory	2	6	20	5	22	51	58	0	33	34	76	1	57	91	154	6
Theological Explanations	33	0	0	0	130	1	0	0	141	3	0	0	304	4	0	0

but 35 of the likely USDs of a cognitive nature. ENA theory did even better. It offered an explanation in the correct direction in all but 22 cases. Also worth noting is that a sizable proportion of the explanations offered by the evolutionary theories pointed toward substantial empirical evidence.

Behavior. Turning to the columns in Table 27.5 having to do with the 144 likely USDs, one can see that, once again, neither of the first two social role theories offer much to account for likely behavioral USDs. The biosocial role theory is more persuasive in this regard, having attained 26 single checkmarks.

As with both the basic biological and the cognitive variables, the evolutionary theories clearly have greatest explanatory power when it comes to understanding likely USDs of a behavioral nature. This overall pattern comports with conclusions reached by Archer (1996) and Lippa (2010a) that evolutionary approaches to understanding well-documented sex differences in cognition and behavior appear to be better explained with evolution-based theories than with social role theories.

The power of evolutionary arguments does not mean that learning has little to do with sex differences in cognition and behavior. Instead, it implies that males and females, on average, have been reproductively favored for learning different things. And, if ENA theory is correct, brain exposure to testosterone has helped to drive these average learning inclinations is sex divergent directions.

Of course, readers should reach their own conclusions regarding the relative strength of each theory. However, throughout this volume, we tried to make assessments using as objective of methods as possible. For the greatest detail in this regard, the arguments presented in Chapter 26 should be consulted. Nevertheless, misinterpretations and/or misrepresentations cannot be ruled out.

Special comments are in order regarding the original social role theory and the founder effect version. One can see that these two theories were found to be almost totally unable to account for why USDs exist.

The greatest obstacle for the original social role theory is that it essentially only explains sex differences that are *not* culturally universal. While such sex differences surely exist (as indicated by many of the tables throughout Volumes I, II, and III), the focus of this last volume was on variables for which universal sex differences do appear to exist.

Regarding the founder effect version of social role theory, it pivots on the concept of the "human progenitor stock" as being the source of universal sex differences (Fausto-Sterling 1992: 199). This term appears to refer to the earliest human culture that ever existed. Unfortunately, there is no empirical data regarding sex differences in this culture. Therefore, there is simply no way of ever knowing if today's USDs have been passed down "thousands of times over" from this hypothesized culture to the present day.

27.5.3 Surprises

Three noteworthy surprises based on the analyses performed in this final volume are worth sharing. The first surprise had to do with the sheer number of likely USDs (i.e., 308). Even if the basic biological variables were to be excluded, 275 likely USDs remain, 131 of which pertained to cognitive traits and the remaining 144 had to do with behavioral traits.

Another unexpected finding involved the number of likely USDs specifically pertaining to interests, attitudes, and preferences. By referring back to Table 27.2d, one can see that there were 59 such USDs. This means that, of all the likely USDs we located, almost 20% of them had to do with people's interests, attitudes, and preferences. In other words, throughout the world, there are many culturally universal differences in the average interests, attitudes, and preferences that people have.

If attention is limited to just the 131 *cognitive* USDs, almost half (i.e., 45%) of them involved interests, attitudes, and preferences. Most of the other cognitive sex different traits involved various types of *abilities*. We interpret this finding to indicate that many sex differences in people's abilities are at least partially the result of underlying sex differences in interests and attitudes. Such an interpretation seems to comport with a study of sex differences in STEM occupations. Evidence from this study suggested that interest seemed to be driving most people's occupational choices to a greater degree than actual abilities (Wang, Eccles & Kenny 2013).

The third surprise involved the accuracy of sex stereotypes. Most of the stereotypes for which USDs were located basically supported their corresponding stereotypes. Specifically, in essentially all countries sampled, males appear to be more physically aggressive, independent, adventurous, dominant, and likely to receive higher incomes (see Table 27.3f-M). Females, on the other hand, appear to be more fearful and prone to be nurturing as stereotypes have indicated (see Table 27.3f-F). However, these tables also indicate that some sex differences are still too tenuous to be considered supportive of the stereotypes people hold regarding them. For example, males are stereotyped as being better at math and females are stereotyped as being more emotional. The empirical evidence, however, indicates that sex differences in these two variables are too inconsistent to warrant their being classified as USDs.

28 Epilogue

The two overarching goals of this book were to locate universal sex differences (USDs) and then identify the most reasonable theoretical explanations for these differences. To accomplish the first of these goals, close to 50,000 citations are provided in the book's first 23 chapters (contained in Volumes I through III). In this, the book's final volume, we have consolidated the information contained in the first three volumes in various ways and then matched the evidence for the most promising USDs with the five theories offered to explain sex differences. We even considered the possibility that religious scriptures might help to explain some of the differences. The process undertaken in this final chapter involved using criteria recommended by Archer (2019: 1384) for identifying universal sex differences, keeping in mind that age can often affect findings.

Finally, we wish to emphasize that, despite the progress made by this book, the search for USDs is far from complete. As more studies are located and citations to them are compiled and organized, more USDs are almost certain to be located; and, perhaps, some of the USDs tentatively identified here will be called into question as new research findings are found. Hopefully, the search for USDs will continue indefinitely. Toward that end, we have hopes of being a part of at least one more edition of this book. Accordingly, interested readers are invited to provide input for future editions. Two types of input would be especially useful.

First, if errors in the citations made anywhere in this book are found, please let one of us know, so that they can be corrected in any future edition. Also, citations to specific omissions would be useful to know for inclusion in the next edition (specific to the appropriate tables, if possible). Our email addresses are as follows: Lee Ellis (lee.ellis@hotmail.com), Craig T. Palmer (palmerct@missouri.edu), Rosemary Hopcroft (rlhopcro@uncc.edu), and Anthony W. Hoskin (anthonyhoskin@isu.edu).

DOI: 10.4324/9781003405290-5

Second, readers with an interest in actively contributing to a future edition of this book by locating studies of sex differences for entry into appropriate tables are invited to contact us. Of course, such individuals should have a keen interest in reading and an ability to interpret empirical research findings on sex differences.

References

Abbott, A. D., Colman, R. J., Tiefenthaler, R., Dumesic, D. A., & Abbott, D. H. (2012). Early-to-mid gestation fetal testosterone increases right hand 2D: 4D finger length ratio in polycystic ovary syndrome-like monkeys. *PLoS One 7*(8), e42372.

Abramov, I., Gordon, J., Feldman, O., & Chavarga, A. (2012). Sex and vision I: Spatio-temporal resolution. *Biology of Sex Differences 3*, 1–20.

Achilli, C., Pundir, J., Ramanathan, P., Sabatini, L., Hamoda, H., & Panay, N. (2017). Efficacy and safety of transdermal testosterone in postmenopausal women with hypoactive sexual desire disorder: A systematic review and meta-analysis. *Fertility and Sterility 107*, 475–482.

Agrillo, C., Piffer, L., Bisazza, A., & Butterworth, B. (2012). Evidence for two numerical systems that are similar in humans and guppies. *PLoS One 7*(2), e31923.

Ahlborg, H. G., Johnell, O., Turner, C. H., Rannevik, G., & Karlsson, M. K. (2003). Bone loss and bone size after menopause. *New England Journal of Medicine 349*, 327–334.

Alacreu-Crespo, A., Costa, R., Abad-Tortosa, D., Hidalgo, V., Salvador, A., & Serrano, M. Á. (2019). Hormonal changes after competition predict sex-differentiated decision-making. *Journal of Behavioral Decision Making 32*, 550–563.

Aleman, A., Bronk, E., Kessels, R. P., Koppeschaar, H. P., & van Honk, J. (2004). A single administration of testosterone improves visuospatial ability in young women. *Psychoneuroendocrinology 29*, 612–617.

Alexander, G. M. (2003). An evolutionary perspective of sex-typed toy preferences: Pink, blue, and the brain. *Archives of Sexual Behavior 32*, 7–14.

Alexander, G. M., & Hines, M. (2002). Sex differences in response to children's toys in non-human primates (*Cercopithecus aethiops sabaeus*). *Evolution & Human Behavior 23*, 467–469.

Alexander, G. M., & Saenz, J. (2012). Early androgens, activity levels and toy choices of children in the second year of life. *Hormones and Behavior 62*, 500–504.

Alexander, G. M., Wilcox, T., & Woods, R. (2009). Sex differences in infants' visual interest in toys. *Archives of Sexual Behavior 38*(3), 427–433.

Alexander, M. G., & Fisher, T. D. (2003). Truth and consequences: Using the bogus pipeline to examine sex differences in self-reported sexuality. *Journal of Sex Research 40*(1), 27–35.

Ålgars, M., Santtila, P., Varjonen, M., Witting, K., Johansson, A., Jern, P., & Sandnabba, N. K. (2009). The adult body: How age, gender, and body mass index are related to body image. *Journal of Aging and Health 21*, 1112–1132.
Almanza-Sepulveda, M. L., Fleming, A. S., & Jonas, W. (2020). Mothering revisited: A role for cortisol? *Hormones and Behavior 121*, 104679.
Almeida, O. P. (2011). Evolution, depression and the interplay between chance and choices. *International Psychogeriatrics 23*, 1021–1025.
Almeida, O. P., Yeap, B. B., Hankey, G. J., Jamrozik, K., & Flicker, L. (2008). Low free testosterone concentration as a potentially treatable cause of depressive symptoms in older men. *Archives of General Psychiatry 65*, 283–289.
Alvergne, A., Faurie, C., & Raymond, M. (2009). Variation in testosterone levels and male reproductive effort: Insight from a polygynous human population. *Hormones and Behavior 56*, 491–497.
Anderson, K. G., Kaplan, H., & Lancaster, J. B. (2007). Confidence of paternity, divorce, and investment in children by Albuquerque men. *Evolution and Human Behavior 28*, 1–10.
Anderson, T. M., Lavista Ferres, J. M., Ren, S. Y., Moon, R. Y., Goldstein, R. D., Ramirez, J.-M., & Mitchell, E. A. (2019). Maternal smoking before and during pregnancy and the risk of sudden unexpected infant death. *Pediatrics 143*(4), e20183325.
Andreasen, N. C., Flaum, M., Victor Swayze, I., O'Leary, D. S., Alliger, R., & Cohen, G. (1993). Intelligence and brain structure in normal individuals. *American Journal of Psychiatry 150*, 130–134.
Apicella, C. L., Carré, J. M., & Dreber, A. (2015). Testosterone and economic risk taking: A review. *Adaptive Human Behavior and Physiology 1*, 358–385.
Apicella, C. L., Dreber, A., Gray, P. B., Hoffman, M., Little, A. C., & Campbell, B. C. (2011). Androgens and competitiveness in men. *Journal of Neuroscience, Psychology, and Economics 4*, 54–67.
Apkhazava, M., Kvachadze, I., Tsagareli, M., Mzhavanadze, D., & Chakhnashvili, M. (2018). The relationship between thermal pain sensation, free testosterone, Trpv1, Mor levels and various degrees of hostility in young healthy males. *Georgian Medical News 283*, 109–114.
Apostolou, M. (2015). The athlete and the spectator inside the man: A cross-cultural investigation of the evolutionary origins of athletic behavior. *Cross-Cultural Research 49*, 151–173.
Archer, J. (1996). Sex differences in social behavior: Are the social role and evolutionary explanations compatible? *American Psychologist 51*, 909–917.
Archer, J. (2006). Cross-cultural differences in physical aggression between partners: A social-role analysis. *Personality and Social Psychology Review 10*, 133–153.
Archer, J. (2006). Testosterone and human aggression: An evaluation of the challenge hypothesis. *Neuroscience & Biobehavioral Reviews 30*, 319–345.
Archer, J. (2009). Does sexual selection explain human sex differences in aggression? *Behavioral and Brain Sciences 32*, 249–266.
Archer, J. (2019). The reality and evolutionary significance of human psychological sex differences. *Biological Reviews 94*, 1381–1416.

Arden-Close, E., Eiser, C., & Pacey, A. (2011). Sexual functioning in male survivors of lymphoma: A systematic review. *Journal of Sexual Medicine 8*, 1833–1840.

Armstrong, T. A., Boisvert, D. L., Wells, J., Lewis, R. H., Cooke, E. M., Woeckener, M., ... Harper, J. M. (2022). Testosterone, cortisol, and criminal behavior in men and women. *Hormones and Behavior 146*, 105260.

Arnocky, S., Albert, G., Carré, J. M., & Ortiz, T. L. (2018). Intrasexual competition mediates the relationship between men's testosterone and mate retention behavior. *Physiology & Behavior 186*, 73–78.

Arnocky, S., Carré, J. M., Bird, B. M., Moreau, B. J., Vaillancourt, T., Ortiz, T., & Marley, N. (2018). The facial width-to-height ratio predicts sex drive, sociosexuality, and intended infidelity. *Archives of Sexual Behavior 47*, 1375–1385.

Aromäki, A. S., Lindman, R. E., & Eriksson, C. P. (1999). Testosterone, aggressiveness, and antisocial personality. *Aggressive Behavior 25*, 113–123.

Aromäki, A. S., Lindman, R. E., & Eriksson, C. P. (2002). Testosterone, sexuality and antisocial personality in rapists and child molesters: A pilot study. *Psychiatry Research 110*, 239–247.

Asarian, L., & Geary, N. (2013). Sex differences in the physiology of eating. *American Journal of Physiology-Regulatory, Integrative and Comparative Physiology 305*(11), R1215–R1267.

Askovic, B., & Kirchengast, S. (2012). Gender differences in nutritional behavior and weight status during early and late adolescence. *Anthropologischer Anzeiger 69*, 289–304.

Asperholm, M., Högman, N., Rafi, J., & Herlitz, A. (2019). What did you do yesterday? A meta-analysis of sex differences in episodic memory. *Psychological Bulletin 145*, 785–799.

Assari, S., Caldwell, C. H., & Zimmerman, M. A. (2014). Sex differences in the association between testosterone and violent behaviors. *Trauma Monthly 19*(3), e18040.

Aubin, H.-J., Berlin, I., & Kornreich, C. (2013). The evolutionary puzzle of suicide. *International Journal of Environmental Research and Public Health 10*, 6873–6886.

Auyeung, B., Baron-Cohen, S., Ashwin, E., Knickmeyer, R., Taylor, K., Hackett, G., & Hines, M. (2009a). Fetal testosterone predicts sexually differentiated childhood behavior in girls and in boys. *Psychological Science 20*, 144–148.

Auyeung, B., Baron-Cohen, S., Ashwin, E., Knickmeyer, R., Taylor, K., & Hackett, G. (2009b). Fetal testosterone and autistic traits. *British Journal of Psychology 100*, 1–22.

Auyeung, B., Baron-Cohen, S., Chapman, E., Knickmeyer, R., Taylor, K., & Hackett, G. (2006). Foetal testosterone and the child systemizing quotient. *European Journal of Endocrinology 155*(suppl 1), S123–S130.

Auyeung, B., Baron-Cohen, S., Wheelwright, S., & Allison, C. (2008). The autism spectrum quotient: Children's version (AQ-Child). *Journal of Autism and Developmental Disorders 38*, 1230–1240.

Auyeung, B., Taylor, K., Hackett, G., & Baron-Cohen, S. (2010). Foetal testosterone and autistic traits in 18 to 24-month-old children. *Molecular Autism 1*(1), 1–8.

Auyeung, T. W., Lee, J. S. W., Kwok, T., Leung, J., Ohlsson, C., Vandenput, L., ... Woo, J. (2011). Testosterone but not estradiol level is positively related to muscle strength and physical performance independent of muscle mass: A cross-sectional study in 1489 older men. *European Journal of Endocrinology 164*, 811–817.

Avidime, O. M., Avidime, S., Olorunshola, K. V., & Dikko, A. A. (2011). Anogenital distance and umbilical cord testosterone level in newborns in Zaria, Northern Nigeria. *Nigerian Journal of Physiological Sciences 26*(1), 23–28.

Aycinena, D., Baltaduonis, R., & Rentschler, L. (2014). Risk preferences and prenatal exposure to sex hormones for ladinos. *PLoS One 9*, e103332.

Bachevalier, J., & Hagger, C. (1991). Sex differences in the development of learning abilities in primates. *Psychoneuroendocrinology 16*, 177–188.

Bachman, R. (1998). The factors related to rape reporting behavior and arrest: New evidence from the National Crime Victimization Survey. *Criminal Justice and Behavior 25*, 8–29.

Bai, X., Zhang, X., Li, Y., Lu, L., Li, B., & He, X. (2015). Sex differences in peripheral mu-opioid receptor mediated analgesia in rat orofacial persistent pain model. *PLoS One 10*(3), e0122924.

Baker, J. H., Lichtenstein, P., & Kendler, K. S. (2009). Intrauterine testosterone exposure and risk for disordered eating. *British Journal of Psychiatry 194*, 375–376.

Baker, M. D., & Maner, J. K. (2008). Risk-taking as a situationally sensitive male mating strategy. *Evolution and Human Behavior 29*, 391–395.

Baloyi, E. (2008). The Biblical exegesis of headship: A challenge to Patriarchal understanding that impinges on women's rights in the church and society. *Verbum et Ecclesia 29*, 1–13.

Bancroft, J. (2002). Sexual effects of androgens in women: Some theoretical considerations. *Fertility and Sterility 77*, 55–59.

Bandelow, B., Sengos, G., Wedekind, D., Huether, G., Pilz, J., Broocks, A., ... Rüther, E. (1997). Urinary excretion of cortisol, norepinephrine, testosterone, and melatonin in panic disorder. *Pharmacopsychiatry 30*, 113–117.

Baranek, G. T., Barnett, C. R., Adams, E. M., Wolcott, N. A., Watson, L. R., & Crais, E. R. (2005). Object play in infants with autism: Methodological issues in retrospective video analysis. *American Journal of Occupational Therapy 59*, 20–30.

Barber, N. (1995). The evolutionary psychology of physical attractiveness: Sexual selection and human morphology. *Ethology and Sociobiology 16*, 395–424.

Barel, E., Shahrabani, S., & Tzischinsky, O. (2017). Sex hormone/cortisol ratios differentially modulate risk-taking in men and women. *Evolutionary Psychology 15*(1), 1474704917697333.

Baron-Cohen, S. (2006). The hyper-systemizing, assortative mating theory of autism. *Progress in Neuro-Psychopharmacology and Biological Psychiatry 30*, 865–872.

Baron-Cohen, S. (2008). Autism, hyper-systemizing, and truth. *Quarterly Journal of Experimental Psychology 61*, 64–75.

Baron-Cohen, S., Knickmeyer, R. C., & Belmonte, M. K. (2005). Sex differences in the brain: Implications for explaining autism. *Science 310*, 819–823.

Baron-Cohen, S., Wheelwright, S., Burtenshaw, A., & Hobson, E. (2007). Mathematical talent is linked to autism. *Human Nature 18*, 125–131.

Barrett, E. S., Tran, V., Thurston, S., Jasienska, G., Furberg, A.-S., Ellison, P. T., & Thune, I. (2013). Marriage and motherhood are associated with lower testosterone concentrations in women. *Hormones and Behavior 63*, 72–79.

Barrett-Connor, E., von Mühlen, D. G., & Kritz-Silverstein, D. (1999). Bioavailable testosterone and depressed mood in older men: The Rancho Bernardo Study. *Journal of Clinical Endocrinology & Metabolism 84*, 573–577.

Barth, C., Villringer, A., & Sacher, J. (2015). Sex hormones affect neurotransmitters and shape the adult female brain during hormonal transition periods. *Frontiers in Neuroscience 9*, 37–49.

Bartley, E. J., Palit, S., Kuhn, B. L., Kerr, K. L., Terry, E. L., DelVentura, J. L., & Rhudy, J. L. (2015). Natural variation in testosterone is associated with hypoalgesia in healthy women. *Clinical Journal of Pain 31*, 730–739.

Basile, K. C., Smith, S. G., Liu, Y., Kresnow, M.-j., Fasula, A. M., Gilbert, L., & Chen, J. (2018). Rape-related pregnancy and association with reproductive coercion in the US. *American Journal of Preventive Medicine 55*, 770–776.

Bateup, H. S., Booth, A., Shirtcliff, E. A., & Granger, D. A. (2002). Testosterone, cortisol, and women's competition. *Evolution and Human Behavior 23*, 181–192.

Batty, G. D., Shipley, M. J., Gunnell, D., Huxley, R., Kivimaki, M., Woodward, M., ... Smith, G. D. (2009). Height, wealth, and health: An overview with new data from three longitudinal studies. *Economics & Human Biology 7*, 137–152.

Baucom, D. H., Besch, P. K., & Callahan, S. (1985). Relation between testosterone concentration, sex role identity, and personality among females. *Journal of Personality and Social Psychology 48*, 1218–1223.

Baumeister, R. F., Catanese, K. R., & Vohs, K. D. (2001). Is there a gender difference in strength of sex drive? Theoretical views, conceptual distinctions, and a review of relevant evidence. *Personality and Social Psychology Review 5*, 242–273.

Baylis, J. R. (1981). The evolution of parental care in fishes, with reference to Darwin's rule of male sexual selection. *Environmental Biology of Fishes 6*, 223–251.

Beasley, M. A., & Fischer, M. J. (2012). Why they leave: The impact of stereotype threat on the attrition of women and minorities from science, math and engineering majors. *Social Psychology of Education 15*, 427–448.

Bedgood, D., Boggiano, M. M., & Turan, B. (2014). Testosterone and social evaluative stress: The moderating role of basal cortisol. *Psychoneuroendocrinology 47*, 107–115.

Behre, H. M., Simoni, M., & Nieschlag, E. (1997). Strong association between serum levels of leptin and testosterone in men. *Clinical Endocrinology 47*, 237–240.

Beijers, R., Breugelmans, S., Brett, B., Willemsen, Y., Bos, P., & de Weerth, C. (2022). Cortisol and testosterone concentrations during the prenatal and postpartum period forecast later caregiving quality in mothers and fathers. *Hormones and Behavior 142*, 105177.

Belsky, J. (2012). The development of human reproductive strategies: Progress and prospects. *Current Directions in Psychological Science 21*, 310–316.

Beltz, A. M., Swanson, J. L., & Berenbaum, S. A. (2011). Gendered occupational interests: Prenatal androgen effects on psychological orientation to Things versus People. *Hormones and Behavior 60*, 313–317.

Benbow, C. P., & Lubinski, D. (2007). *Psychological profiles of the mathematically talented: Some sex differences and evidence supporting their biological basis*. Paper presented at the Ciba Foundation Symposium 178-The Origins and Development of High Ability: The Origins and Development of High Ability: Ciba Foundation Symposium 178.

Beran, M. J. (2008). The evolutionary and developmental foundations of mathematics. *PLoS biology 6*(2), e19.

Bereczkei, T., Voros, S., Gal, A., & Bernath, L. (1997). Resources, attractiveness, family commitment; reproductive decisions in human mate choice. *Ethology 103*, 681–699.

Berenbaum, S. A. (1999). Effects of early androgens on sex-typed activities and interests in adolescents with congenital adrenal hyperplasia. *Hormones and Behavior 35*, 102–110.

Berenbaum, S. A., Korman, K., & Leveroni, C. (1995). Early hormones and sex differences in cognitive abilities. *Learning and Individual Differences 7*, 303–321.

Berger, J. (1989). Female reproductive potential and its apparent evaluation by male mammals. *Journal of Mammalogy 70*, 347–358.

Bernhard, A., Ackermann, K., Martinelli, A., Chiocchetti, A. G., Vllasaliu, L., González-Madruga, K., … Jansen, L. M. (2022). Neuroendocrine stress response in female and male youths with conduct disorder and associations with early adversity. *Journal of the American Academy of Child & Adolescent Psychiatry 61*, 698–710.

Bhasin, S., Parker, R. A., Sattler, F., Haubrich, R., Alston, B., Umbleja, T., & Shikuma, C. M. (2007). Effects of testosterone supplementation on whole body and regional fat mass and distribution in human immunodeficiency virus-infected men with abdominal obesity. *Journal of Clinical Endocrinology & Metabolism 92*, 1049–1057.

Bhasin, S., Storer, T. W., Berman, N., Callegari, C., Clevenger, B., Phillips, J., … Casaburi, R. (1996). The effects of supraphysiologic doses of testosterone on muscle size and strength in normal men. *New England Journal of Medicine 335*, 1–7.

Bhasin, S., Storer, T. W., Berman, N., Yarasheski, K. E., Clevenger, B., Phillips, J., … Casaburi, R. (1997). Testosterone replacement increases fat-free mass and muscle size in hypogonadal men. *Journal of Clinical Endocrinology & Metabolism 82*, 407–413.

Bigham, A., Bauchet, M., Pinto, D., Mao, X., Akey, J. M., Mei, R., … López Herráez, D. (2010). Identifying signatures of natural selection in Tibetan and Andean populations using dense genome scan data. *PLoS Genetics 6*(9), e1001116.

Bilsborough, J. C., Greenway, K. G., Opar, D. A., Livingstone, S. G., Cordy, J. T., Bird, S. R., & Coutts, A. J. (2015). Comparison of anthropometry, upper-body strength, and lower-body power characteristics in different levels of Australian football players. *Journal of Strength & Conditioning Research 29*, 826–834.

Bird, S. P. (2003). Creatine supplementation and exercise performance: A brief review. *Journal of Sports Science & Medicine 2*(4), 123–131.

Bjorklund, D. F., & Brown, R. D. (1998). Physical play and cognitive development: Integrating activity, cognition, and education. *Child Development 69*, 604–606.

Bjorklund, D. F., & Jordan, A. C. (2013). Human parenting from an evolutionary perspective. In K. K. Kline & W. B. Wilcox (Eds.), *Gender and parenthood: Biological and social scientific perspectives* (pp. 61–90). New York: Columbia University Press.

Bjorklund, D. F., & Yunger, J. (2001). Evolutionary developmental psychology: A useful framework for evaluating the evolution of parenting. *Parenting 1*, 63–66.

Blakemore, J. E. O., & Centers, R. E. (2005). Characteristics of boys' and girls' toys. *Sex Roles 53*, 619–633.

Blekhman, R., Man, O., Herrmann, L., Boyko, A. R., Indap, A., Kosiol, C., ... Przeworski, M. (2008). Natural selection on genes that underlie human disease susceptibility. *Current Biology, 18*, 883–889.

Bleske-Rechek, A., & Gunseor, M. M. (2021). Gendered perspectives on sharing the load: Men's and women's attitudes toward family roles and household and childcare tasks. *Evolutionary Behavioral Sciences 16*, 201–219.

Blum, W. F., Englaro, P., Hanitsch, S., Juul, A., Hertel, N. T., Müller, J., ... Attanasio, A. M. (1997). Plasma leptin levels in healthy children and adolescents: Dependence on body mass index, body fat mass, gender, pubertal stage, and testosterone. *Journal of Clinical Endocrinology & Metabolism 82*, 2904–2910.

Bogaert, A. F., & Fisher, W. A. (1995). Predictors of university men's number of sexual partners. *Journal of Sex Research 32*, 119–130.

Bogaert, A. F., & Skorska, M. N. (2020). A short review of biological research on the development of sexual orientation. *Hormones and Behavior 119*, 104659.

Boissy, A., & Bouissou, M. F. (1994). Effects of androgen treatment on behavioral and physiological responses of heifers to fear-eliciting situations. *Hormones and Behavior 28*, 66–83.

Bönte, W., Procher, V. D., Urbig, D., & Voracek, M. (2017). Digit ratio (2D: 4D) predicts self-reported measures of general competitiveness, but not behavior in economic experiments. *Frontiers in Behavioral Neuroscience 11*, 238–249.

Booth, A., Johnson, D. R., & Granger, D. A. (1999). Testosterone and men's depression: The role of social behavior. *Journal of Health and Social Behavior 40*, 130–140.

Booth, A., Johnson, D. R., Granger, D. A., Crouter, A. C., & McHale, S. (2003). Testosterone and child and adolescent adjustment: The moderating role of parent-child relationships. *Developmental Psychology 39*, 85–98.

Bos, P. A., Hechler, C., Beijers, R., Shinohara, K., Esposito, G., & de Weerth, C. (2018). Prenatal and postnatal cortisol and testosterone are related to parental caregiving quality in fathers, but not in mothers. *Psychoneuroendocrinology 97*, 94–103.

Bos, P. A., Panksepp, J., Bluthé, R.-M., & Van Honk, J. (2012). Acute effects of steroid hormones and neuropeptides on human social–emotional behavior: A review of single administration studies. *Frontiers in Neuroendocrinology 33*, 17–35.

Bos, P. A., van Honk, J., Ramsey, N. F., Stein, D. J., & Hermans, E. J. (2013). Testosterone administration in women increases amygdala responses to fearful and happy faces. *Psychoneuroendocrinology 38*, 808–817.

Box, H. O. (1984). Behavioural responses to change—Natural events II. In *Primate Behaviour and Social Ecology* (pp. 98–146). Springer.

Brañas-Garza, P., Galizzi, M. M., & Nieboer, J. (2018). Experimental and self-reported measures of risk taking and digit ratio (2D: 4D): Evidence from a large, systematic study. *International Economic Review 59*, 1131–1157.

Brazil, K. J., & Volk, A. A. (2022). Cads in dads' clothing? Psychopathic traits and men's preferences for mating, parental, and somatic investment. *Evolutionary Psychological Science 8*, 299–315.

Brewer, G., & Howarth, S. (2012). Sport, attractiveness and aggression. *Personality and Individual Differences 53*, 640–643.

Brock, G., Heiselman, D., Maggi, M., Kim, S. W., Rodríguez Vallejo, J. M., Behre, H. M., ... Knorr, J. (2016). Effect of testosterone solution 2% on testosterone concentration, sex drive and energy in hypogonadal men: Results of a placebo controlled study. *Journal of Urology 195*(3), 699–705.

Brooks, J. H., & Reddon, J. R. (1996). Serum testosterone in violent and non-violent young offenders. *Journal of Clinical Psychology 52*, 475–483.

Brooks, R., Blake, K., & Fromhage, L. (2022). Effects of gender inequality and income inequality on within-sex mating competition under hypergyny. *Evolution and Human Behavior 43*, 501–509, 10.1016/j.evolhumbehav.2022.1008.1006.

Brosnan, M. J. (2006). Digit ratio and faculty membership: Implications for the relationship between prenatal testosterone and academia. *British Journal of Psychology 97*, 455–466.

Brosnan, M., Gallop, V., Iftikhar, N., & Keogh, E. (2011). Digit ratio (2D: 4D), academic performance in computer science and computer-related anxiety. *Personality and Individual Differences 51*, 371–375.

Brown, G. A., Vukovich, M. D., Sharp, R. L., Reifenrath, T. A., Parsons, K. A., & King, D. S. (1999). Effect of oral DHEA on serum testosterone and adaptations to resistance training in young men. *Journal of Applied Physiology 87*, 2274–2283.

Brown, S. L. (2009). *Play: How it shapes the brain, opens the imagination, and invigorates the soul.* New York: Penguin.

Brown, S. L., & Robb, C. (2022). Understanding female crime and antisocial behavior through a biosocial and evolutionary lens. In S. L. Brown & L. Gelsthorpe (Eds.), *The Wiley handbook on what works with girls and women in conflict with the law: A critical review of theory, practice, and policy* (pp. 46–61). New York: Wiley.

Browne, K. R. (2006). Evolved sex differences and occupational segregation. *Journal of Organizational Behavior 27*, 143–162.

Browne, K. R. (2013). Biological sex differences in the workplace: Reports of the end of men are greatly exaggerated (as are claims of women's continued inequality). *Boston University Law Review 93*, 769.

Burkitt, J., Widman, D., & Saucier, D. M. (2007). Evidence for the influence of testosterone in the performance of spatial navigation in a virtual water maze in women but not in men. *Hormones and Behavior 51*, 649–654.

Buss, D. M., & Barnes, M. (1986). Preferences in human mate selection. *Journal of Personality and Social Psychology 50*(3), 559–568.

Buss, D. M., & Kenrick, D. T. (1998). Evolutionary social psychology. In D. Gilbert, S. T. Fiske, & G. Lindzey (Eds.), *Evolutionary social psychology* (Vol. 2, pp. 982–1026). New York: McGraw-Hill.

Buss, D. M., & Schmitt, D. P. (1993). Sexual strategies theory: An evolutionary perspective on human mating. *Psychological Review 100*, 204–232.

Buss, D. M., & Shackelford, T. K. (2008). Attractive women want it all: Good genes, economic investment, parenting proclivities, and emotional commitment. *Evolutionary Psychology 6*(1), 147470490800600116.

Byrd-Craven, J., & Geary, D. (2013). An evolutionary understanding of sex differences. In M. K. Ryan & N. R. Branscombe (Eds.), *The Sage handbook of gender and psychology* (pp. 100–114). Sage.

Byrd-Craven, J., Massey, A. R., Calvi, J. L., & Geary, D. C. (2015). Is systemizing a feature of the extreme male brain from an evolutionary perspective? *Personality and Individual Differences 82*, 237–241.

Cabeza De Baca, T., Figueredo, A. J., & Ellis, B. J. (2012). An evolutionary analysis of variation in parental effort: Determinants and assessment. *Parenting 12*, 94–104.

Campbell, A. (2013). *A mind of her own: The evolutionary psychology of women*. Oxford, England: Oxford.

Campbell, A., Muncer, S., & Bibel, D. (2001). Women and crime: An evolutionary approach. *Aggression and Violent Behavior 6*, 481–497.

Campbell, B. C., Dreber, A., Apicella, C. L., Eisenberg, D. T., Gray, P. B., Little, A. C., ... Lum, J. K. (2010). Testosterone exposure, dopaminergic reward, and sensation-seeking in young men. *Physiology & Behavior 99*, 451–456.

Campbell, M. C., & Tishkoff, S. A. (2010). The evolution of human genetic and phenotypic variation in Africa. *Current Biology 20*, R166–R173.

Carani, C., Qin, K., Simoni, M., Faustini-Fustini, M., Serpente, S., Boyd, J., ... Simpson, E. R. (1997). Effect of testosterone and estradiol in a man with aromatase deficiency. *New England Journal of Medicine 337*, 91–95.

Carlo, G., Raffaelli, M., Laible, D. J., & Meyer, K. A. (1999). Why are girls less physically aggressive than boys? Personality and parenting mediators of physical aggression. *Sex Roles 40*, 711–729.

Carré, J. M., & Mehta, P. H. (2011). Importance of considering testosterone–cortisol interactions in predicting human aggression and dominance. *Aggressive Behavior 37*, 489–491.

Carruth, L. L., Reisert, I., & Arnold, A. P. (2002). Sex chromosome genes directly affect brain sexual differentiation. *Nature Neuroscience 5*, 933–934.

Carter, S. K. (2017). Body-led mothering: Constructions of the breast in attachment parenting literature. *Women's Studies International Forum 62*, 17–24.

Casey, B. M., & Ganley, C. M. (2021). An examination of gender differences in spatial skills and math attitudes in relation to mathematics success: A bio-psycho-social model. *Developmental Review 60*, 100963.

Cashdan, E. (1995). Hormones, sex, and status in women. *Hormones and Behavior 29*, 354–366.

Cashdan, E., Marlowe, F. W., Crittenden, A., Porter, C., & Wood, B. M. (2012). Sex differences in spatial cognition among Hadza foragers. *Evolution and Human Behavior 33*, 274–284.

Casimir, A., Chukwuelobe, M. C., & Ugwu, C. (2014). The church and gender equality in Africa: Questioning culture and the theological paradigm on women oppression. *Open Journal of Philosophy 4*, Article ID:46171.

Casto, K. V., & Edwards, D. A. (2016). Testosterone, cortisol, and human competition. *Hormones and Behavior 82*, 21–37.

Casto, K. V., Arthur, L. C., Hamilton, D. K., & Edwards, D. A. (2022). Testosterone, athletic context, oral contraceptive use, and competitive persistence in women. *Adaptive Human Behavior and Physiology 8*, 52–78.

Cauley, J. A., Lucas, F. L., Kuller, L. H., Stone, K., Browner, W., Cummings, S. R., & Group, S. o. O. F. R. (1999). Elevated serum estradiol and testosterone concentrations are associated with a high risk for breast cancer. *Annals of Internal Medicine 130*(4_Part_1), 270–277.

Cela-Conde, C., Ayala, F., Munar, E., Maestú, F., Nadal, M., Capó, M., ... Marty, G. (2009). Sex-related similarities and differences in the neural correlates of beauty. *Proceedings of the National Academy of Sciences 106*(10), 3847–3852.

Celec, P., Ostatníková, D., & Hodosy, J. (2015). On the effects of testosterone on brain behavioral functions. *Frontiers in Neuroscience 9*, 10.3389/fnins.2015.00012.

Chagnon, N. A. (1988). Life histories, blood revenge, and warfare in a tribal population. *Science 239*, 985–992.

Chan, Y.-C. (2016). Neural correlates of sex/gender differences in humor processing for different joke types. *Frontiers in Psychology 7*, 536–548.

Chapman, E., Baron-Cohen, S., Auyeung, B., Knickmeyer, R., Taylor, K., & Hackett, G. (2006). Fetal testosterone and empathy: Evidence from the empathy quotient (EQ) and the "reading the mind in the eyes" test. *Social Neuroscience 1*, 135–148.

Charlesworth, W. R., & La Freniere, P. (1983). Dominance, friendship, and resource utilization in preschool children's groups. *Ethology and Sociobiology 4*, 175–186.

Charlton, B. G., & Rosenkranz, P. (2016). Evolution of empathizing and systemizing: Empathizing as an aspect of social intelligence, systemizing as an evolutionarily later consequence of economic specialization. *The Winnower, April 29*, 1–15.

Charness, G., & Gneezy, U. (2012). Strong evidence for gender differences in risk taking. *Journal of Economic Behavior & Organization 83*, 50–58.

Chen Zeng, T., Cheng, J. T., & Henrich, J. (2022). Dominance in humans. *Philosophical Transactions of the Royal Society B 377*(1845), 20200451.

Chen, J.-F., Lin, P.-W., Tsai, Y.-R., Yang, Y.-C., & Kang, H.-Y. (2019). Androgens and androgen receptor actions on bone health and disease: From androgen deficiency to androgen therapy. *Cells 8*, 1318–1345.

Cheng, D., Hong, C.-J., Liao, D.-L., & Tsai, S.-J. (2006). Association study of androgen receptor CAG repeat polymorphism and male violent criminal activity. *Psychoneuroendocrinology 31*, 548–552.

Cheng, J. T., Tracy, J. L., & Henrich, J. (2010). Pride, personality, and the evolutionary foundations of human social status. *Evolution and Human Behavior 31*, 334–347.

Cheng, T. L., Johnson, S., Wright, J. L., Pearson-Fields, A. S., Brenner, R., Schwarz, D., ... Scheidt, P. C. (2006). Assault-injured adolescents presenting to the emergency department: Causes and circumstances. *Academic Emergency Medicine 13*, 610–616.

Chichinadze, K., & Chichinadze, N. (2008). Stress-induced increase of testosterone: Contributions of social status and sympathetic reactivity. *Physiology & Behavior 94*, 595–603.

Chin, K.-Y., Soelaiman, I.-N., Naina Mohamed, I., Shahar, S., Teng, N. I. M. F., Suhana Mohd Ramli, E., ... Zurinah Wan Ngah, W. (2012). Testosterone is associated with age-related changes in bone health status, muscle strength and body composition in men. *Aging Male 15*, 240–245.

Chiricos, T., Barrick, K., Bales, W., & Bontrager, S. (2007). The labeling of convicted felons and its consequences for recidivism. *Criminology 45*, 547–581.

Christov-Moore, L., Simpson, E. A., Coudé, G., Grigaityte, K., Iacoboni, M., & Ferrari, P. F. (2014). Empathy: Gender effects in brain and behavior. *Neuroscience & Biobehavioral Reviews 46*, 604–627.

Chu, E. M.-Y., O'Neill, M., Purkayastha, D. D., & Knight, C. (2019). Huntington's disease: A forensic risk factor in women. *Journal of Clinical Movement Disorders 6*, 1–6.

Clarke, B. L., & Khosla, S. (2009). Androgens and bone. *Steroids 74*, 296–305.

Clutton-Brock, T. (2017). Reproductive competition and sexual selection. *Philosophical Transactions of the Royal Society B: Biological Sciences 372*(1729), 20160310.

Clutton-Brock, T. H. (1985). Size, sexual dimorphism, and polygyny in primates. In W. L. Jungers (Ed.), *Size and scaling in primate biology* (pp. 51–60). New York: Springer.

Coates, J. M., & Herbert, J. (2008). Endogenous steroids and financial risk taking on a London trading floor. *Proceedings of the National Academy of Science 105*, 6167–6172.

Coates, J. M., Gurnell, M., & Rustichini, A. (2009). Second-to-fourth digit ratio predicts success among high-frequency financial traders. *Proceedings of the National Academy of Sciences 106*, 623–628.

Cohen-Bendahan, C. C. C., Buitelaar, J. K., van Goozen, S. H. M., & Cohen-Kettenis, P. T. (2004). Prenatal exposure to testosterone and functional cerebral lateralization: A study in same-sex and opposite-sex twin girls. *Psychoneuroendocrinology 29*, 911–916.

Cohen-Bendahan, C. C., van de Beek, C., & Berenbaum, S. A. (2005). Prenatal sex hormone effects on child and adult sex-typed behavior: Methods and findings. *Neuroscience & Biobehavioral Reviews 29*, 353–384.

Comings, D. E., Chen, C., Wu, S., & Muhleman, D. (1999). Association of the androgen receptor gene (AR) with ADHD and conduct disorder. *Neuroreport 10*, 1589–1592.

Connellan, J., Baron-Cohen, S., Wheelwright, S., Batki, A., & Ahluwalia, J. (2000). Sex differences in human neonatal social perception. *Infant Behavior and Development 23*, 113–118.

Conroy, S., & Cotter, A. (2017). *Self-Reported Sexual Assault in Canada, 2014*. Ottawa, Canada: Ministry of Industry, Statistics Canada.

Cooper, H., Lindsay, J. J., Nye, B., & Greathouse, S. (1998). Relationships among attitudes about homework, amount of homework assigned and completed, and student achievement. *Journal of Educational Psychology 90*(1), 70–82.

Corona, G., Isidori, A. M., Buvat, J., Aversa, A., Rastrelli, G., Hackett, G., … Mannucci, E. (2014). Testosterone supplementation and sexual function: A meta-analysis study. *Journal of Sexual Medicine 11*, 1577–1592.

Cotton, S., Small, J., & Pomiankowski, A. (2006). Sexual selection and condition-dependent mate preferences. *Current Biology*, *16*, R755–R765.

Cox, R. M., Stenquist, D. S., & Calsbeek, R. (2009). Testosterone, growth and the evolution of sexual size dimorphism. *Journal of Evolutionary Biology 22*, 1586–1598.

Coyne, S. P. (2018). The endocrinology of dominance relations in non-human primates. In *Routledge international handbook of social neuroendocrinology* (pp. 99–112). Routledge.

Crocchiola, D. (2014). Art as an indicator of male fitness: Does prenatal testosterone influence artistic ability? *Evolutionary Psychology 12*(3), 147470491401200303.

Cross, S., & Madson, L. (1997). Models of the self: Self-construals and gender. *Psychological Bulletin 122*, 5–37.

Culbert, K. M., Breedlove, S. M., Burt, S. A., & Klump, K. L. (2008). Prenatal hormone exposure and risk for eating disorders: A comparison of opposite-sex and same-sex twins. *Archives of General Psychiatry 65*, 329–336.

Culbert, K. M., Breedlove, S. M., Sisk, C. L., Burt, S. A., & Klump, K. L. (2013). The emergence of sex differences in risk for disordered eating attitudes during puberty: A role for prenatal testosterone exposure. *Journal of Abnormal Psychology 122*, 420–432.

Culbert, K. M., Burt, S. A., & Klump, K. L. (2017). Expanding the developmental boundaries of etiologic effects: The role of adrenarche in genetic influences on disordered eating in males. *Journal of Abnormal Psychology 126*(5), 593–604.

Culbert, K. M., Burt, S. A., McGue, M., Iacono, W. G., & Klump, K. L. (2009). Puberty and the genetic diathesis of disordered eating attitudes and behaviors. *Journal of Abnormal Psychology 118*, 788–799.

Cummins, D. (2015). Dominance, status, and social hierarchies. In D. Buss (Ed.), *The handbook of evolutionary psychology* (pp. 676–697) New York: Wiley.

Curtis, V., Aunger, R., & Rabie, T. (2004). Evidence that disgust evolved to protect from risk of disease. *Proceedings of the Royal Society of London. Series B: Biological Sciences 271*(suppl 4), S131–S133.

Czajka-Oraniec, I., & Simpson, E. R. (2010). Aromatase research and its clinical significance. *Endokrynologia Polska 61*, 126–134.

da Silva, W. R., Marôco, J., & Campos, J. A. D. B. (2021). Examination of the factorial model of a scale developed to assess body satisfaction in the Brazilian context: A study with people 18 to 40 years old. *Eating and Weight Disorders-Studies on Anorexia, Bulimia and Obesity* 26, 2701–2712.

Dabbs Jr, J. M., Carr, T. S., Frady, R. L., & Riad, J. K. (1995). Testosterone, crime, and misbehavior among 692 male prison inmates. *Personality and Individual Differences 18*, 627–633.

Dabbs Jr, J. M., Strong, R., & Milun, R. (1997). Exploring the mind of testosterone: A beeper study. *Journal of Research in Personality*, *31*, 577–587.

Dabbs, J. M. (1992). Testosterone and occupational achievement. *Social Forces 70*, 813–824.

Dabbs, J. M. (1997). Testosterone, smiling, and facial appearance. *Journal of Nonverbal Behavior 21*, 45–55.

Dabbs, J. M., & Dabbs, M. G. (2000). *Heroes, rogues and lovers: Testosterone and behavior*. New York: McGraw-Hill.

Dabbs, J. M., & Mallinger, A. (1999). High testosterone levels predict low voice pitch among men. *Personality and Individual Differences* 27, 801–804.

Dabbs, J. M., de La Rue, D., & Williams, P. M. (1990). Testosterone and occupational choice: Actors, ministers, and other men. *Journal of Personality and Social Psychology 59*, 1261–1265.

Dabbs, J. M., Frady, R. L., Carr, T. S., & Besch, N. F. (1987). Saliva testosterone and criminal violence in young adult prison inmates. *Psychosomatic Medicine 49*, 174–182.

Dabbs, J. M., Jurkovic, G. J., & Frady, R. L. (1991). Salivary testosterone and cortisol among late adolescent male offenders. *Journal of Abnormal Child Psychology 19*, 469–478.

Dalton, P. S., & Ghosal, S. (2018). Self-confidence, overconfidence and prenatal testosterone exposure: Evidence from the lab. *Frontiers in Behavioral Neuroscience 12*, 5.

Daly, M. (2017). *Killing the competition: Economic inequality and homicide*. London: Routledge.

Daly, M., & Wilson, M. (1978). *Sex, evolution, and behavior*. North Scituate, MA: Duxbury Press.

Daly, M., & Wilson, M. (1980). Discriminative parental solicitude: A biological perspective. *Journal of Marriage and the Family* 42, 277–288.

Daly, W., Seegers, C. A., Rubin, D. A., Dobridge, J. D., & Hackney, A. C. (2005). Relationship between stress hormones and testosterone with prolonged endurance exercise. *European Journal of Applied Physiology 93*, 375–380.

Dariotis, J. K., Schen, F. R., & Granger, D. A. (2016). Latent trait testosterone among 18–24 year-olds: Methodological considerations and risk associations. *Psychoneuroendocrinology* 67, 1–9.

Darwin, C. (1871). *The descent of man, and selection in relation to sex*. London: John Murray.

Darwin, C. (1874). *The descent of man: Selection in relation to sex*. New York: A. Burt.

Darwin, C. (1876). Sexual selection in relation to monkeys. *Nature 15*, 18–19.

Davies, G. (2010). *History of money*. University of Wales Press.

Davis, A. C., & Arnocky, S. (2022). An evolutionary perspective on appearance enhancement behavior. *Archives of Sexual Behavior*, *51*, 13–37.

Davis, S. N., & Risman, B. J. (2015). Feminists wrestle with testosterone: Hormones, socialization and cultural interactionism as predictors of women's gendered selves. *Social Science Research 49*, 110–125.

Davison, S. L., & Davis, S. R. (2003). Androgens in women. *Journal of Steroid Biochemistry and Molecular Biology 85*, 363–366.

De Block, A., & Dewitte, S. (2009). Darwinism and the cultural evolution of sports. *Perspectives in Biology and Medicine 52*, 1–16.

de Bruijn, A. T., van Bakel, H. J., & van Baar, A. L. (2009). Sex differences in the relation between prenatal maternal emotional complaints and child outcome. *Early Human Development 85*, 319–324.

De Bruin, E. I., De Nijs, P. F., Verheij, F., Verhagen, D. H., & Ferdinand, R. F. (2009). Autistic features in girls from a psychiatric sample are strongly associated with a low 2D: 4D ratio. *Autism 13*, 511–521.

de Bruin, E. I., Verheij, F., & Ferdinand, R. F. (2006). WISC-R subtest but no overall VIQ–PIQ difference in Dutch children with PDD-NOS. *Journal of Abnormal Child Psychology 34*, 254–262.

de Vries, E. E., van der Pol, L. D., Vermeer, H. J., Groeneveld, M. G., Fiers, T., & Mesman, J. (2019). Testosterone and fathers' parenting unraveled: Links with the quantity and quality of father-child interactions. *Adaptive Human Behavior and Physiology 5*, 297–316.

De Waal, F. B. (2012). The antiquity of empathy. *Science 336*(6083), 874–876.

Deaner, R. O., Balish, S. M., & Lombardo, M. P. (2016). Sex differences in sports interest and motivation: An evolutionary perspective. *Evolutionary Behavioral Sciences 10*, 73–97.

Deaner, R. O., Dunlap, L. C., & Bleske-Rechek, A. (2022). Sex differences in competitiveness in massively multiplayer online role-playing games (MMORPGs). *Evolutionary Psychology 20*, 14747049221109388.

Deaton, A., & Arora, R. (2009). Life at the top: The benefits of height. *Economics & Human Biology* 7, 133–136.

Deaux, K. (1985). Sex and gender. *Annual Review of Psychology 36*, 49–81.

Dekkers, T. J., van Rentergem, J. A. A., Meijer, B., Popma, A., Wagemaker, E., & Huizenga, H. M. (2019). A meta-analytical evaluation of the dual-hormone hypothesis: Does cortisol moderate the relationship between testosterone and status, dominance, risk taking, aggression, and psychopathy? *Neuroscience & Biobehavioral Reviews 96*, 250–271.

Del Giudice, M. (2010). Reduced fertility in patients' families is consistent with the sexual selection model of schizophrenia and schizotypy. *PLoS One 5*(12), e16040.

Del Giudice, M., & Belsky, J. (2010). Sex differences in attachment emerge in middle childhood: An evolutionary hypothesis. *Child Development Perspectives 4*, 97–105.

Del Giudice, M., Angeleri, R., & Manera, V. (2009). The juvenile transition: A developmental switch point in human life history. *Developmental Review 29*, 1–31.

Del Giudice, M., Angeleri, R., Brizio, A., & Elena, M. R. (2010). The evolution of autistic-like and schizotypal traits: A sexual selection hypothesis. *Frontiers in Psychology 1*, 41–52.

Delgado, P. F., Maya-Rosero, E., Franco, M., Montoya-Oviedo, N., Guatibonza, R., & Mockus, I. (2020). Testosterona y homicidio: Aspectos neuroendocrinos de la agresión. *Revista de la Facultad de Medicina 68*, 283–294.

Demers, L. M. (2010). Androgen deficiency in women; role of accurate testosterone measurements. *Maturitas 67*, 39–45.

Denegar, C. R., Hertel, J., & Fonseca, J. (2002). The effect of lateral ankle sprain on dorsiflexion range of motion, posterior talar glide, and joint laxity. *Journal of Orthopaedic & Sports Physical Therapy 32*, 166–173.

Dennis, C., & McCall, A. (2005). The Savannah hypothesis of shopping. *Business Strategy Review 16*, 12–16.

Dennis, C., Brakus, J. J., Ferrer, G. G., McIntyre, C., Alamanos, E., & King, T. (2018). A cross-national study of evolutionary origins of gender shopping styles: She gatherer, he hunter? *Journal of International Marketing 26*, 38–53.

Derntl, B., Windischberger, C., Robinson, S., Kryspin-Exner, I., Gur, R. C., Moser, E., & Habel, U. (2009). Amygdala activity to fear and anger in healthy young males is associated with testosterone. *Psychoneuroendocrinology 34*, 687–693.

Diamond, M. (1976). Human sexual development: Biological foundations for social development. In F. A. Beach (Ed.), *Human sexuality in four perspectives* (pp. 22–61).

DiPietro, J. A., & Voegtline, K. M. (2017). The gestational foundation of sex differences in development and vulnerability. *Neuroscience 342*, 4–20.

Doll, L. M., Cárdenas, R. A., Burriss, R. P., & Puts, D. A. (2016). Sexual selection and life history: Earlier recalled puberty predicts men's phenotypic masculinization. *Adaptive Human Behavior and Physiology 2*, 134–149.

Dönmez, Y. E., Özcan, Ö., Bilgiç, A., & Miniksar, D. Y. (2019). The relationship between prenatal testosterone and developmental stuttering in boys. *Turkish Journal of Pediatrics 61*, 193–199.

Dreher, J.-C., Dunne, S., Pazderska, A., Frodl, T., Nolan, J. J., & O'Doherty, J. P. (2016). Testosterone causes both prosocial and antisocial status-enhancing behaviors in human males. *Proceedings of the National Academy of Sciences 113*(41), 11633–11638.

Drummond, P. D., & Bailey, T. (2013). Eye contact evokes blushing independently of negative affect. *Journal of Nonverbal Behavior 37*, 207–216.

Duckworth, A. L., & Seligman, M. E. P. (2005). Self-discipline outdoes IQ in predicting academic performance of adolescents. *Psychological Science* 16, 939–944.

Dugatkin, L. A. (2013). The evolution of risk-taking. *Cerebrum 1* (Jan-Feb), 23516663.

Duncan, L. E., & Peterson, B. E. (2010). Gender and motivation for achievement, affiliation–intimacy, and power. In J. C. Chrisler & D. R. McCreary (Eds.), *Handbook of gender research in psychology* (pp. 41–62). New York: Springer.

Dunsworth, H. M. (2020). Expanding the evolutionary explanations for sex differences in the human skeleton. *Evolutionary Anthropology: Issues, News, and Reviews 29*, 108–116.

Duntley, J. D., & Shackelford, T. K. (2008). Darwinian foundations of crime and law. *Aggression and Violent Behavior 13*, 373–382.
Durdiaková, J., Celec, P., Laznibatová, J., Minárik, G., Lakatošová, S., Kubranská, A., & Ostatníková, D. (2015). Differences in salivary testosterone, digit ratio and empathy between intellectually gifted and control boys. *Intelligence 48*, 76–84.
Durrant, R., & Ward, T. (2015). *Evolutionary criminology: Towards a comprehensive explanation of crime*. New York: Academic Press.
Dutton, D. (2003). Aesthetics and evolutionary psychology. In J. Levinson (Ed.), *The oxford handbook for aesthetics*. New York: Oxford University Press.
Eagly, A. H. (1987). *Sex differences in social behavior: A social-role interpretation*. Hillsdale, NJ: Erlbaum.
Eagly, A. H. (2013). *Sex differences in social behavior: A social-role interpretation*. Washington, DC Psychology Press.
Eagly, A. H., & Steffen, V. J. (1984). Gender stereotypes stem from the distribution of women and men into social roles. *Journal of Personality and Social Psychology 46*, 735–754.
Eagly, A. H., & Steffen, V. J. (1986). Gender and aggressive behavior: A meta-analytic review of the social psychological literature. *Psychological Bulletin 100*, 309–330.
Eagly, A. H., & Wood, W. (1991). Explaining sex differences in social behavior: A meta-analytic perspective. *Personality and Social Psychology Bulletin 17*, 306–315.
Eagly, A. H., & Wood, W. (2005). Universal sex differences across patriarchal cultures≠ evolved psychological dispositions. *Behavioral and Brain Sciences 28*, 281–283.
Eagly, A. H., & Wood, W. (2012). Social role theory. In A. K. P. van Lange, & E. T. Higgins (Eds.), *Handbook of theories of social psychology* (Vol. 2, pp. 458–476). Thousand Oaks, CA: Sage.
Eagly, A. H., & Wood, W. (2013). The nature–nurture debates: 25 years of challenges in understanding the psychology of gender. *Perspectives on Psychological Science 8*, 340–357.
Eagly, A. H., Wood, W., & Diekman, A. B. (2000). Social role theory of sex differences and similarities: A current appraisal. In T. Eckes & H. M. Trautner (Eds.), *The developmental social psychology of gender* (pp. 123–174). New York: Taylor & Francis.
Eastwick, P. W., & Finkel, E. J. (2008). Sex differences in mate preferences revisited: Do people know what they initially desire in a romantic partner? *Journal of Personality and Social Psychology 94*, 245–264.
Eastwick, P. W., Eagly, A. H., Finkel, E. J., & Johnson, S. E. (2011). Implicit and explicit preferences for physical attractiveness in a romantic partner: A double dissociation in predictive validity. *Journal of Personality and Social Psychology 101*(5), 993–1011.
Ebeling, P. R., Atley, L. M., Guthrie, J. R., Burger, H. G., Dennerstein, L., Hopper, J. L., & Wark, J. D. (1996). Bone turnover markers and bone density across the menopausal transition. *Journal of Clinical Endocrinology & Metabolism 81*, 3366–3371.

Ebinger, M., Sievers, C., Ivan, D., Schneider, H. J., & Stalla, G. K. (2009). Is there a neuroendocrinological rationale for testosterone as a therapeutic option in depression? *Journal of Psychopharmacology 23*, 841–853.

Ecuyer-Dab, I., & Robert, M. (2004). Have sex differences in spatial ability evolved from male competition for mating and female concern for survival? *Cognition 91*, 221–257.

Edelstein, R. S., & Chin, K. (2018). Hormones and close relationship processes: Neuroendocrine bases of partnering and parenting. In O. C. Schultheiss & P. H. Mehta (Eds.), *Routledge international handbook of social neuroendocrinology* (pp. 281–297). London: Routledge.

Edelstein, R. S., Chin, K., Saini, E. K., Kuo, P. X., Schultheiss, O. C., & Volling, B. L. (2019). Adult attachment and testosterone reactivity: Fathers' avoidance predicts changes in testosterone during the strange situation procedure. *Hormones and Behavior 112*, 10–19.

Edelstein, R. S., van Anders, S. M., Chopik, W. J., Goldey, K. L., & Wardecker, B. M. (2014). Dyadic associations between testosterone and relationship quality in couples. *Hormones and Behavior 65*, 401–407.

Edwards, D. A., & Casto, K. V. (2013). Women's intercollegiate athletic competition: Cortisol, testosterone, and the dual-hormone hypothesis as it relates to status among teammates. *Hormones and Behavior 64*, 153–160.

Edwards, D. A., Wetzel, K., & Wyner, D. R. (2006). Intercollegiate soccer: Saliva cortisol and testosterone are elevated during competition, and testosterone is related to status and social connectedness with teammates. *Physiology & Behavior 87*, 135–143.

Edwards, E. A., Hamilton, J. B., Quimby Duntley, S., & Hubert, G. (1941). Cutaneous vascular and pigmentary changes in castrate and eunuchoid men. *Endocrinology 28*, 119–128.

Ehrenkranz, J., Bliss, E., & Sheard, M. H. (1974). Plasma testosterone: Correlation with aggressive behavior and social dominance in man. *Biological Psychiatry 55*, 546–552.

Eibl-Eibesfeldt, I. (1989). *Human ethology*. New York: Aldine de Gruyter.

Eikenaar, C., Whitham, M., Komdeur, J., Van der Velde, M., & Moore, I. T. (2011). Endogenous testosterone is not associated with the trade-off between paternal and mating effort. *Behavioral Ecology 22*, 601–608.

Eisenegger, C., Kumsta, R., Naef, M., Gromoll, J., & Heinrichs, M. (2017). Testosterone and androgen receptor gene polymorphism are associated with confidence and competitiveness in men. *Hormones and Behavior 92*, 93–102.

Elgar, F. J., Roberts, C., Tudor-Smith, C., & Moore, L. (2005). Validity of self-reported height and weight and predictors of bias in adolescents. *Journal of Adolescent Health 37*, 371–375.

Elias, A., & Kumar, A. (2007). Testosterone for schizophrenia. *Cochrane Database of Systematic Reviews, Issue 3*, Art. No. CD006197.

Elliot, D. L., Cheong, J., Moe, E. L., & Goldberg, L. (2007). Cross-sectional study of female students reporting anabolic steroid use. *Archives of Pediatrics & Adolescent Medicine 161*, 572–577.

Ellis, B. J. (1992). The evolution of sexual attraction: Evaluative mechanisms in women. In J. H. Barkow, L. Cosmides, & J. Tooby (Eds.), *The adapted mind: Evolutionary psychology and the generation of culture* (pp. 267–288). New York: Oxford University Press.

Ellis, B. J., & Symons, D. (1990). Sex differences in sexual fantasy: An evolutionary psychological approach. *Journal of Sex Research 27*, 527–555.

Ellis, B. J., Del Giudice, M., Dishion, T. J., Figueredo, A. J., Gray, P., Griskevicius, V., ... Volk, A. A. (2012). The evolutionary basis of risky adolescent behavior: Implications for science, policy, and practice. *Developmental Psychology 48*, 598–613.

Ellis, L. (1989). *Theories of rape: Inquiries into the causes of sexual aggression.* New York: Hemisphere.

Ellis, L. (1991a). A synthesized (biosocial) theory of rape. *Journal of Consulting and Clinical Psychology 59*, 631–642.

Ellis, L. (1991b). A biosocial theory of social stratification derived from the concepts of pro/antisociality and r/K selection. *Politics and the Life Sciences 10*, 5–22.

Ellis, L. (1993). A biosocial theory of social stratification: An alternative to functional theory and conflict theory. In L. Ellis (Ed.), *Social stratification and socioeconomic inequality, Volume 1: A comparative biosocial analysis* (pp. 159–174). New York: Praeger.

Ellis, L. (1994a). Height, health, and social status (plus birth weight, mental health, intelligence, brain size, and fertility): A broad theoretical integration. In L. Ellis (Ed.), *Social stratification and socioeconomic inequality, volume II: Reproductive and interpersonal aspects of dominance and status* (pp. 145–163). Westport, CT: Praeger.

Ellis, L. (1994b). The high and the mighty among man and beast: How universal is the relationship between height (or body size) and social status? In L. Ellis (Ed.), *Social stratification and socioeconomic inequality, volume II: Reproductive and interpersonal aspects of dominance and status* (pp. 93–111). Westport, CT: Praeger.

Ellis, L. (1995). Dominance and reproductive success among nonhuman animals: A cross-species comparison. *Ethology and Sociobiology 16*, 257–333.

Ellis, L. (2001). The biosocial female choice theory of social stratification. *Biodemography and Social Biology 48*, 298–320.

Ellis, L. (2005). A theory explaining biological correlates of criminality. *European Journal of Criminology 2*, 287–315.

Ellis, L. (2006). Gender differences in smiling: An evolutionary neuroandrogenic theory. *Physiology & Behavior 88*, 303–308.

Ellis, L. (2011a). Identifying and explaining apparent universal sex differences in cognition and behavior. *Personality and Individual Differences 51*(5), 552–561.

Ellis, L. (2011b). Evolutionary neuroandrogenic theory and universal gender differences in cognition and behavior. *Sex Roles 64*, 707–722.

Ellis, L., & Ames, M. A. (1987). Neurohormonal functioning and sexual orientation: A theory of homosexuality–heterosexuality. *Psychological Bulletin 101*, 233–258.

Ellis, L., & Das, S. (2009). Androgen-promoted traits and intellectual abilities and interests. *Personality and Individual Differences 47*, 647–651.

Ellis, L., & Hoskin, A. W. (2015). The evolutionary neuroandrogenic theory of criminal behavior expanded. *Aggression and Violent Behavior 24*, 61–74.

Ellis, L., & Hoskin, A. W. (2022). National gender equality and AR CAG repeats among resident males. *Evolutionary Psychological Science 9*, 61–70.

Ellis, L., & Ratnasingam, M. (2015). Naturally selected mate preferences appear to be androgen-influenced: evidence from two cultures. *Evolutionary Psychological Science 1*, 103–122.

Ellis, L., & Walsh, A. (1997). Gene-based evolutionary theories in criminology. *Criminology 35*, 229–276.

Ellis, L., Farrington, D. P., & Hoskin, A. W. (2019). *Handbook of crime correlates, 2nd edition.* San Diego, CA: Academic Press.

Ellis, L., Hoskin, A. W., & Buker, H. (2021). National variations in CAG repeats of men's androgen receptor gene: A tabulated review. *Mankind Quarterly 61*, 430–461.

Ellis, L., Hoskin, A. W., & Ratnasingam, M. (2016). Testosterone, risk taking, and religiosity: Evidence from two cultures. *Journal for the Scientific Study of Religion 55*, 153–173.

Ellis, L., Hoskin, A. W., & Ratnasingam, M. (2018). *Handbook of social status correlates.* Amsterdam: Elsevier.

Ellis, L., Lykins, A., Hoskin, A., & Ratnasingam, M. (2015). Putative androgen exposure and sexual orientation: Cross-cultural evidence suggesting a modified neurohormonal theory. *Journal of Sexual Medicine 12*(12), 2364–2377.

Ellis, L., Widmayer, A., & Palmer, C. T. (2009). Perpetrators of sexual assault continuing to have sex with their victims following the initial assault: Evidence for evolved reproductive strategies. *International Journal of Offender Therapy and Comparative Criminology 53*, 454–463.

Ellison, P. T., Lipson, S. F., & Meredith, M. D. (1989). Salivary testosterone levels in males from the Ituri forest of Zaire. *American Journal of Human Biology 1*, 21–24.

Else-Quest, N. M., Higgins, A., Allison, C., & Morton, L. C. (2012). Gender differences in self-conscious emotional experience: A meta-analysis. *Psychological Bulletin 138*, 947–981.

Else-Quest, N. M., Hyde, J. S., & Linn, M. C. (2010). Cross-national patterns of gender differences in mathematics: A meta-analysis. *Psychological Bulletin 136*, 103–127.

Eme, R. (2016). Evolutionary roots of the sex difference in the prevalence of severe anti-social behavior: A literature review. *Psychology and Cognitive Sciences 2*(2), 49–53.

Emery, M. J., Krous, H. F., Nadeau-Manning, J. M., Marck, B. T., & Matsumoto, A. M. (2005). Serum testosterone and estradiol in sudden infant death. *Journal of Pediatrics 147*, 586–591.

Emlen, S. T., & Oring, L. W. (1977). Ecology, sexual selection, and the evolution of mating systems. *Science 197*(4300), 215–223.

Engqvist, L., Cordes, N., & Reinhold, K. (2015). Evolution of risk-taking during conspicuous mating displays. *Evolution 69*, 395–406.

Enter, D., Spinhoven, P., & Roelofs, K. (2016). Dare to approach: Single dose testosterone administration promotes threat approach in patients with social anxiety disorder. *Clinical Psychological Science 4*, 1073–1079.

Erol, A., Ho, A. M. C., Winham, S. J., & Karpyak, V. M. (2019). Sex hormones in alcohol consumption: A systematic review of evidence. *Addiction Biology 24*, 157–169.

Eshed, V., Gopher, A., Galili, E., & Hershkovitz, I. (2004). Musculoskeletal stress markers in Natufian hunter-gatherers and Neolithic farmers in the Levant: The upper limb. *American Journal of Physical Anthropology: The Official Publication of the American Association of Physical Anthropologists 123*, 303–315.

Ethington, C. A., & Wolfle, L. M. (1986). Sex differences in quantitative and analytical GRE performance: An exploratory study. *Research in Higher Education 25*, 55–67.

Evans, K. L., & Hampson, E. (2014). Does risk-taking mediate the relationship between testosterone and decision-making on the Iowa Gambling Task? *Personality and Individual Differences 61*, 57–62.

Evans, S., Neave, N., & Wakelin, D. (2006). Relationships between vocal characteristics and body size and shape in human males: An evolutionary explanation for a deep male voice. *Biological Psychology 72*, 160–163.

Evans, S., Neave, N., Wakelin, D., & Hamilton, C. (2008). The relationship between testosterone and vocal frequencies in human males. *Physiology & Behavior 93*, 783–788.

Evardone, M., & Alexander, G. M. (2009). Anxiety, sex-linked behaviors, and digit ratios (2D: 4D). *Archives of Sexual Behavior 38*, 442–455.

Exton, M. S., Krueger, T. H., Bursch, N., Haake, P., Knapp, W., Schedlowski, M., & Hartmann, U. (2001). Endocrine response to masturbation-induced orgasm in healthy men following a 3-week sexual abstinence. *World Journal of Urology 19*, 377–382.

Fairchild, G., Baker, E., & Eaton, S. (2018). Hypothalamic-pituitary-adrenal Axis function in children and adults with severe antisocial behavior and the impact of early adversity. *Current Psychiatry Reports 20*(10), 1–9.

Fares, N. E., Ramirez, J. M., Cabrera, J. M., Lozano, F., & Salas, F. (2011). Justification of physical and verbal aggression in Uruguayan children, and adolescents. *Open Psychology Journal 4*, 45–54.

Faurie, C., Pontier, D., & Raymond, M. (2004). Student athletes claim to have more sexual partners than other students. *Evolution and Human Behavior 25*(1), 1–8.

Fausto-Sterling, A. (1992). *Myths of gender: Biological theories about women and men, 2nd edition*. New York: Basic Books.

Federman, D. D. (2006). The biology of human sex differences. *New England Journal of Medicine 354*(14), 1507–1514.

Feinberg, D. R., Jones, B. C., Little, A. C., Burt, D. M., & Perrett, D. I. (2005). Manipulations of fundamental and formant frequencies influence the attractiveness of human male voices. *Animal Behaviour 69*, 561–568.

Feingold, A. (1992). Sex differences in variability in intellectual abilities: A new look at an old controversy. *Review of Educational Research 62*, 61–84.

Felson, R. B., & Cundiff, P. R. (2014). Sexual assault as a crime against young people. *Archives of Sexual Behavior 43*, 273–284.

Fergusson, D. M., Horwood, L. J., Ridder, E. M., & Beautrais, A. L. (2005). Sexual orientation and mental health in a birth cohort of young adults. *Psychological Medicine 35*, 971–981.

Fetchenhauer, D., & Buunk, B. P. (2005). How to explain gender differences in fear of crime: Towards an evolutionary approach. *Sexualities, Evolution & Gender 7*, 95–113.

Finch, J., & Groves, D. (2022). *A labour of love: Women, work and caring*. London: Taylor & Francis.

Finegan, J.-A. K., Niccols, G. A., & Sitarenios, G. (1992). Relations between prenatal testosterone levels and cognitive abilities at 4 years. *Developmental Psychology 28*, 1075–1086.

Fink, B. R. (1963). Larynx and speech as determinants in the evolution of man. *Perspectives in Biology and Medicine 7*, 85–93.

Finkelstein, S. R. (2018). Swearing and the brain. In K. Allan (Ed.), *Oxford handbook of taboo words and language*. Oxford, England: Oxford University Press.

Fischer, L., Clemente, J. T., & Tambeli, C. H. (2007). The protective role of testosterone in the development of temporomandibular joint pain. *Journal of Pain 8*, 437–442.

Flannelly, K. J., Flannelly, S., & Santos, d. (2017). *Religious beliefs, evolutionary psychiatry, and mental health in America*. New York: Springer.

Flinn, M. V., & Low, B. S. (1986). Resource distribution, social competition and mating patterns in human societies. In D. I. Rubenstein & R. W. Wrangham (Eds.), *Ecology and social evolution*. Princeton, NJ: Princeton University Press.

Focquaert, F., Steven, M. S., Wolford, G. L., Colden, A., & Gazzaniga, M. S. (2007). Empathizing and systemizing cognitive traits in the sciences and humanities. *Personality and Individual Differences 43*, 619–625.

Fogel, M. L., Tuross, N., Johnson, B. J., & Miller, G. H. (1997). Biogeochemical record of ancient humans. *Organic Geochemistry 27*, 275–287.

Folland, J. P., Mc Cauley, T. M., Phypers, C., Hanson, B., & Mastana, S. S. (2012). The relationship of testosterone and AR CAG repeat genotype with knee extensor muscle function of young and older men. *Experimental Gerontology 47*, 437–443.

Fonberg, E. (1988). Dominance and aggression. *International Journal of Neuroscience 41*, 201–213.

Foong, T. L., Arshat, Z., & Juhari, R. (2020). Sex differences in intellectual ability among preschool children in Putrajaya. *International Journal of Education, Psychology and Counselling 5*, 40–59.

Forquer, M. R., Hashimoto, J. G., Roberts, M. L., & Wiren, K. M. (2011). Elevated testosterone in females reveals a robust sex difference in altered androgen levels during chronic alcohol withdrawal. *Alcohol 45*, 161–171.

Foster, C. M., Olton, P. R., Racine, M. S., Phillips, D. J. , & Padmanabhan, V. (2004). Sex differences in FSH-regulatory peptides in pubertal age boys and girls and effects of sex steroid treatment. *Human Reproduction* 19, 1668–1676.

Fowler, C. D., Freeman, M. E., & Wang, Z. (2003). Newly proliferated cells in the adult male amygdala are affected by gonadal steroid hormones. *Journal of Neurobiology 57*, 257–269.

Frankenbach, J., Weber, M., Loschelder, D. D., Kilger, H., & Friese, M. (2022). Sex drive: Theoretical conceptualization and meta-analytic review of gender differences. *Psychological Bulletin 148*, 621–661.

Frederick, D. A., & Haselton, M. G. (2007). Why is muscularity sexy? Tests of the fitness indicator hypothesis. *Personality and Social Psychology Bulletin 33*, 1167–1183.

Frederick, D. A., & Jenkins, B. N. (2015). Height and body mass on the mating market: Associations with number of sex partners and extra-pair sex among heterosexual men and women aged 18–65. *Evolutionary Psychology 13*(3), 1474704915604563.

Frisen, L., Nordenstrom, A., Falhammar, H., Filipsson, H., Holmdahl, G., Janson, P. O., ... Nordenskjold, A. (2009). Gender role behavior, sexuality, and psychosocial adaptation in women with congenital adrenal hyperplasia due to CYP21A2 deficiency. *Journal of Clinical Endocrinology & Metabolism 94*, 3432–3439.

Fukumori, N., Yamamoto, Y., Takegami, M., Yamazaki, S., Onishi, Y., Sekiguchi, M., ... Fukuhara, S. (2015). Association between hand-grip strength and depressive symptoms: Locomotive Syndrome and Health Outcomes in Aizu Cohort Study (LOHAS). *Age and Ageing 44*, 592–598.

Funk, J. B., & Buchman, D. D. (1996). Playing violent video and computer games and adolescent self-concept. *Journal of Communication 46*, 19–32.

Gangestad, S. W., & Simpson, J. A. (2000). The evolution of human mating: Trade-offs and strategic pluralism. *Behavioral and Brain Sciences 23*, 573–587.

Gangestad, S. W., & Thornhill, R. (1997). Human sexual selection and developmental stability. *Evolutionary Social Psychology 10*, 169–196.

Gaulin, S. J. (1993). How and why sex differences evolve, with spatial ability as a paradigm example. In M. Haug, R. E. Whalen, C. Aron, & K. L. Olsen (Eds.), *The development of sex differences and similarities in behavior* (pp. 111–130). New York: Springer.

Gayef, A., Oner, C., & Telatar, B. (2014). Is asking same question in different ways has any impact on student achievement? *Procedia-Social and Behavioral Sciences 152*, 339–342.

Geary, D. C. (1995). Sexual selection and sex differences in spatial cognition. *Learning and Individual Differences 7*, 289–301.

Geary, D. C. (1996). Sexual selection and sex differences in mathematical abilities. *Behavioral and Brain Sciences 19*, 229–247.

Geary, D. C. (2006b). Sex differences in social behavior and cognition: Utility of sexual selection for hypothesis generation. *Hormones and Behavior 49*, 273–275.

Geary, D. C. (2010). *Male, female: The evolution of human sex differences*. Washington, DC: American Psychological Association.

Geary, D. C. (2016). Evolution of sex differences in trait-and age-specific vulnerabilities. *Perspectives on Psychological Science 11*, 855–876.

Geary, D. C., & Flinn, M. V. (2001). Evolution of human parental behavior and the human family. *Parenting 1*, 5–61.
Geary, D. C., Byrd-Craven, J., Hoard, M. K., Vigil, J., & Numtee, C. (2003). Evolution and development of boys' social behavior. *Developmental Review 23*, 444–470.
Gegenhuber, B., Weinland, C., Kornhuber, J., Mühle, C., & Lenz, B. (2018). OPRM1 A118G and serum β-endorphin interact with sex and digit ratio (2D: 4D) to influence risk and course of alcohol dependence. *European Neuropsychopharmacology 28*, 1418–1428.
Geher, G., & Kaufman, S. B. (2013). *Mating intelligence unleashed: The role of the mind in sex, dating, and love.* Oxford, England: Oxford University Press.
Geniole, S. N., & Carré, J. M. (2018). Human social neuroendocrinology: Review of the rapid effects of testosterone. *Hormones and Behavior 104*, 192–205.
Geniole, S. N., Bird, B. M., McVittie, J. S., Purcell, R. B., Archer, J., & Carré, J. M. (2020). Is testosterone linked to human aggression? A meta-analytic examination of the relationship between baseline, dynamic, and manipulated testosterone on human aggression. *Hormones and Behavior 123*, 104644.
Geniole, S. N., Bird, B. M., Ruddick, E. L., & Carré, J. M. (2017). Effects of competition outcome on testosterone concentrations in humans: An updated meta-analysis. *Hormones and Behavior* 92, 37–50.
Genovese, J. E. (2008). Physique correlates with reproductive success in an archival sample of delinquent youth. *Evolutionary Psychology* 6(3), 147470490800600301
Georgiev, A. V., Klimczuk, A. C., Traficonte, D. M., & Maestripieri, D. (2013). When violence pays: A cost-benefit analysis of aggressive behavior in animals and humans. *Evolutionary Psychology 11*, 147470491301100313.
Gettler, L. T. (2020). Exploring evolutionary perspectives on human fatherhood and paternal biology: Testosterone as an exemplar. In H. E. Fitzgerald, K. Klitzing, N. J. Cabrera, J. S. Mendonça, & T. Skjøthaug (Eds.), *Handbook of fathers and child development* (pp. 137–152). New York: Springer.
Gettler, L. T., Agustin, S. S., & Kuzawa, C. W. (2010). Testosterone, physical activity, and somatic outcomes among Filipino males. *American Journal of Physical Anthropology 142*, 590–599.
Gettler, L. T., McDade, T. W., Agustin, S. S., & Kuzawa, C. W. (2011). Short-term changes in fathers' hormones during father–child play: Impacts of paternal attitudes and experience. *Hormones and Behavior 60*, 599–606.
Gettler, L. T., McDade, T. W., Agustin, S. S., Feranil, A. B., & Kuzawa, C. W. (2013). Do testosterone declines during the transition to marriage and fatherhood relate to men's sexual behavior? Evidence from the Philippines. *Hormones and Behavior 64*, 755–763.
Gicheva, D. (2022). Altruism and burnout: Long hours in the teaching profession. *ILR Review 75*, 427–457.
Gielen, A. C., Holmes, J., & Myers, C. (2016). Prenatal testosterone and the earnings of men and women. *Journal of Human Resources 51*, 30–61.
Giery, S. T., & Layman, C. A. (2019). Ecological consequences of sexually selected traits: An eco-evolutionary perspective. *Quarterly Review of Biology 94*, 29–74.

Gillespie-Lynch, K., Greenfield, P. M., Lyn, H., & Savage-Rumbaugh, S. (2014). Gestural and symbolic development among apes and humans: Support for a multimodal theory of language evolution. *Frontiers in Psychology 5*, 1228.

Giltay, E. J., Enter, D., Zitman, F. G., Penninx, B. W., van Pelt, J., Spinhoven, P., & Roelofs, K. (2012). Salivary testosterone: Associations with depression, anxiety disorders, and antidepressant use in a large cohort study. *Journal of Psychosomatic Research*, 72, 205–213.

Giordano, S. H. (2018). Breast cancer in men. *New England Journal of Medicine 378*(24), 2311–2320.

Giotakos, O., Markianos, M., Vaidakis, N., & Christodoulou, G. N. (2004#). Sex hormones and biogenic amine turnover of sex offenders in relation to their temperament and character dimensions. *Psychiatry Research 127*, 185–193.

Giri, D., Patil, P., Blair, J., Dharmaraj, P., Ramakrishnan, R., Das, U., ... Senniappan, S. (2017). Testosterone therapy improves the first year height velocity in adolescent boys with constitutional delay of growth and puberty. *International Journal of Endocrinology and Metabolism 15*, e42311.

Gladden, P. R., Figueredo, A. J., & Jacobs, W. J. (2009). Life history strategy, psychopathic attitudes, personality, and general intelligence. *Personality and Individual Differences 46*, 270–275.

Glaser, R., & Dimitrakakis, C. (2015). Testosterone and breast cancer prevention. *Maturitas 82*, 291–295.

Glaudas, X., Rice, S. E., Clark, R. W., & Alexander, G. J. (2020). The intensity of sexual selection, body size and reproductive success in a mating system with male–male combat: Is bigger better? *Oikos 129*, 998–1011.

Glowacki, L., & Wrangham, R. (2015). Warfare and reproductive success in a tribal population. *Proceedings of the National Academy of Sciences 112*, 348–353.

Glutton-Brock, T. H., & Vincent, A. C. (1991). Sexual selection and the potential reproductive rates of males and females. *Nature*, *351*(6321), 58–60.

Goetz, S. M., Weisfeld, G. Z., & Zilioli, S. (2019). Reproductive behavior in the human male. In L. L. M. Welling & T. K. Shackelford (Eds.), *The Oxford handbook of evolutionary psychology and behavioral endocrinology* (pp. 125–141). New York: Oxford University Press.

Golan, O., & Baron-Cohen, S. (2006). Systemizing empathy: Teaching adults with Asperger syndrome or high-functioning autism to recognize complex emotions using interactive multimedia. *Development and Psychopathology 18*, 591–617.

Golby, J., & Meggs, J. (2011). Exploring the organizational effect of prenatal testosterone upon the sporting brain. *Journal of Sports Science & Medicine 10*, 445–449.

Gonzalez-Bono, E., Salvador, A., Serrano, M. A., & Ricarte, J. (1999). Testosterone, cortisol, and mood in a sports team competition. *Hormones and Behavior 35*, 55–62.

Gordon, I., Pratt, M., Bergunde, K., Zagoory-Sharon, O., & Feldman, R. (2017). Testosterone, oxytocin, and the development of human parental care. *Hormones and Behavior 93*, 184–192.

Gorostiaga, E. M., Izquierdo, M., Iturralde, P., Ruesta, M., & Ibáñez, J. (1999). Effects of heavy resistance training on maximal and explosive force production, endurance and serum hormones in adolescent handball players. *European Journal of Applied Physiology and Occupational Physiology 80*, 485–493.

Gouchie, C., & Kimura, D. (1991). The relationship between testosterone levels and cognitive ability patterns. *Psychoneuroendocrinology 16*, 323–334.

Goudriaan, A. E., Lapauw, B., Ruige, J., Feyen, E., Kaufman, J.-M., Brand, M., & Vingerhoets, G. (2010). The influence of high-normal testosterone levels on risk-taking in healthy males in a 1-week letrozole administration study. *Psychoneuroendocrinology 35*, 1416–1421.

Goy, R. W., & McEwen, B. S. (1980). *Sexual differentiation of the brain.* Cambridge, MA: The MIT Press.

Goyal, R. O., Sagar, R., Ammini, A. C., Khurana, M. L., & Alias, A. G. (2004). Negative correlation between negative symptoms of schizophrenia and testosterone levels. *Annals of the New York Academy of the Sciences 1032*, 291–294.

Grammer, K., Fink, B., Møller, A. P., & Thornhill, R. (2003). Darwinian aesthetics: Sexual selection and the biology of beauty. *Biological Reviews 78*, 385–407.

Grant, V. J., & France, J. T. (2001). Dominance and testosterone in women. *Biological Psychology 58*, 41–47.

Gray, P. B., Kahlenberg, S. M., Barrett, E. S., Lipson, S. F., & Ellison, P. T. (2002). Marriage and fatherhood are associated with lower testosterone in males. *Evolution and Human Behavior 23*, 193–201.

Gray, P. B., Straftis, A. A., Bird, B. M., McHale, T. S., & Zilioli, S. (2020). Human reproductive behavior, life history, and the challenge hypothesis: A 30-year review, retrospective and future directions. *Hormones and Behavior 123*, 104530.

Grebe, N. M., Del Giudice, M., Thompson, M. E., Nickels, N., Ponzi, D., Zilioli, S., … Gangestad, S. W. (2019). Testosterone, cortisol, and status-striving personality features: A review and empirical evaluation of the Dual Hormone hypothesis. *Hormones and Behavior 109*, 25–37.

Greitemeyer, T., Kastenmüller, A., & Fischer, P. (2013). Romantic motives and risk-taking: An evolutionary approach. *Journal of Risk Research 16*, 19–38.

Grimshaw, G. M., Bryden, M. P., & Finegan, J.-A. K. (1995). Relations between prenatal testosterone and cerebral lateralization in children. *Neuropsychology 9*, 68–81.

Grimshaw, G. M., Sitarenios, G., & Finegan, J.-A. K. (1995). Mental rotation at 7 years-relations with prenatal testosterone levels and spatial play experiences. *Brain and cognition 29*, 85–100.

Gross, M. R. (1996). Alternative reproductive strategies and tactics: Diversity within sexes. *Trends in Ecology & Evolution 11*, 92–98.

Gugatschka, M., Kiesler, K., Obermayer-Pietsch, B., Schoekler, B., Schmid, C., Groselj-Strele, A., & Friedrich, G. (2010). Sex hormones and the elderly male voice. *Journal of Voice 24*, 369–373.

Gur, R. C., Turetsky, B. I., Matsui, M., Yan, M., Bilker, W., Hughett, P., & Gur, R. E. (1999). Sex differences in brain gray and white matter in healthy young adults: Correlations with cognitive performance. *Journal of Neuroscience 19*, 4065–4072.

Gutierrez, G., Wamboldt, R., & Baranchuk, A. (2021). The impact of testosterone on the QT interval: A systematic review. *Current Problems in Cardiology 47*, 100882.

Güvendir, E. (2015). Why are males inclined to use strong swear words more than females? An evolutionary explanation based on male intergroup aggressiveness. *Language Sciences 50*, 133–139.

Hadziselimovic, F., Verkauskas, G., Vicel, B., & Stadler, M. (2019). Cryptorchid boys with abrogated mini-puberty display differentially expressed genes involved in sudden infant death syndrome. *ESPE Abstracts 28*, 1–18, 10.1186/s12610-019-0097-3.

Hagen, E. H. (2011). Evolutionary theories of depression: A critical review. *Canadian Journal of Psychiatry 56*, 716–726.

Hagen, E. H., & Rosenström, T. (2016). Explaining the sex difference in depression with a unified bargaining model of anger and depression. *Evolution, Medicine, and Public Health 2016*, 117–132.

Hahn, A. C., Fisher, C. I., Cobey, K. D., DeBruine, L. M., & Jones, B. C. (2016). A longitudinal analysis of women's salivary testosterone and intrasexual competitiveness. *Psychoneuroendocrinology 64*, 117–122.

Hahnel-Peeters, R. K., & Goetz, A. T. (2022). Development and validation of the Rape Excusing Attitudes and Language Scale. *Personality and Individual Differences 186*, 111359.

Hald, G. M. (2006). Gender differences in pornography consumption among young heterosexual Danish adults. *Archives of Sexual Behavior 35*, 577–585.

Halpern, C. T., Udry, R., & Suchindran, C. (1998). Monthly measures of salivary testosterone predict sexual activity in adolescent males. *Archives of Sexual Behavior 27*, 445–465.

Halpern, D. F., Benbow, C. P., Geary, D. C., Gur, R. C., Hyde, J. S., & Gernsbacher, M. A. (2007). Sex, math and scientific achievement: Why do men dominate the fields of science, engineering and mathematics? *Scientific American Mind 18*(6), 44–55.

Hamel, J. (2020). Explaining symmetry across sex in intimate partner violence: Evolution, gender roles, and the will to harm. *Partner Abuse 11*, 228–267.

Hames, R. B., & Babchuk, W. A. (1985). Sex differences in the recognition of infant facial expressions of emotion: The primary caretaker hypothesis. *Ethology and Sociobiology*, 6, 89–101.

Hamilton, M. E. (1984). Revising evolutionary narratives: A consideration of alternative assumptions about sexual selection and competition for mates. *American Anthropologist 86*, 651–662.

Hampson, E., Ellis, C. L., & Tenk, C. M. (2008). On the relation between 2D:4D and sex-dimorphic personality traits. *Archives of Sexual Behavior 37*(1), 133–144.

Hampson, E., Rovet, J. F., & Altmann, D. (1998). Spatial reasoning in children with congenital adrenal hyperplasia due to 21-hydroxylase deficiency. *Developmental Neuropsychology 14*, 299–320.

Hampson, E., van Anders, S. M., & Mullin, L. I. (2006). A female advantage in the recognition of emotional facial expressions: Test of an evolutionary hypothesis. *Evolution and Human Behavior 27*, 401–416.

Han, C., Bae, H., Lee, Y.-S., & Won, S.-D. (2016). The ratio of 2nd to 4th digit length in Korean alcohol-dependent patients. *Clinical Psychopharmacology and Neuroscience 14*, 148–159.

Hanafy, H. M. (2007). Testosterone therapy and obstructive sleep apnea: Is there a real connection? *Journal of Sexual Medicine 4*, 1241–1246.

Handelsman, D. J., Hirschberg, A. L., & Bermon, S. (2018). Circulating testosterone as the hormonal basis of sex differences in athletic performance. *Endocrine Reviews 39*(5), 803–829.

Hankinson, S. E., & Eliassen, A. H. (2007). Endogenous estrogen, testosterone and progesterone levels in relation to breast cancer risk. *Journal of Steroid Biochemistry and Molecular Biology 106*, 24–30.

Hansen, L., Bangsbo, J., Twisk, J., & Klausen, K. (1999). Development of muscle strength in relation to training level and testosterone in young male soccer players. *Journal of Applied Physiology 87*, 1141–1147.

Hansen, T. F., & Price, D. K. (1995). Good genes and old age: Do old mates provide superior genes? *Journal of Evolutionary Biology*, *8*, 759–778.

Haring, R., Völzke, H., Steveling, A., Krebs, A., Felix, S. B., Schöfl, C., … Wallaschofski, H. (2010). Low serum testosterone levels are associated with increased risk of mortality in a population-based cohort of men aged 20–79. *European Heart Journal 31*, 1494–1501.

Harris, J. A., Rushton, J. P., Hampson, E., & Jackson, D. N. (1996). Salivary testosterone and self-report aggressive and pro-social personality characteristics in men and women. *Aggressive Behavior 22*, 321–331.

Harrison, D. W., Gorelczenko, P. M., & Cook, J. (1990). Sex differences in the functional asymmetry for facial affect perception. *International Journal of Neuroscience 52*, 11–16.

Harrison, S., Davies, N. M., Howe, L. D., & Hughes, A. (2021). Testosterone and socioeconomic position: Mendelian randomization in 306,248 men and women in UK Biobank. *Science Advances 7*(31), 8257.

Hartley, R. (2009). Variability in the behavior of women in violent inter-group conflict. *Mankind Quarterly 49*, 320–331.

Harvey, P. H., & Krebs, J. R. (1990). Comparing brains. *Science 249*(4965), 140–146.

Hassler, M. (1992). Creative musical behavior and sex hormones: Musical talent and spatial ability in the two sexes. *Psychoneuroendocrinology 17*, 55–70.

Hassler, M., & Nieschlag, E. (1989). Masculinity, femininity, and musical composition: Psychological and psychoendocrinological aspects of musical and spatial faculties. *Archives of Psychology 141*, 71–84.

Hassler, M., Gupta, D., & Wollmann, H. (1992). Testosterone, estradiol, ACTH and musical, spatial and verbal performance. *International Journal of Neuroscience 65*, 45–60.

Hastrup, J. L., Kraemer, D. T., Bornstein, R. F., & Trezza, G. R. (2001). *Adult crying: A biopsychosocial approach*. London: Routledge.
Hau, M. (2007). Regulation of male traits by testosterone: Implications for the evolution of vertebrate life histories. *BioEssays 29*, 133–144.
Heany, S. J., van Honk, J., Stein, D. J., & Brooks, S. J. (2016). A quantitative and qualitative review of the effects of testosterone on the function and structure of the human social-emotional brain. *Metabolic Brain Disease 31*, 157–167.
Hedges, L. V., & Friedman, L. (1993). Gender differences in variability in intellectual abilities: A reanalysis of Feingold's results. *Review of Educational Research 63*, 94–105.
Heggen, C. H. (2019). Religious beliefs and abuse. In C. C. Kroeger & J. R. Beck (Eds.), *Women, abuse, and the Bible: How scripture can be used to hurt or heal* (pp. 15–27). Eugene, OR: Wipf and Stock Publishers.
Hell, B., & Päßler, K. (2011). Are occupational interests hormonally influenced? The 2D: 4D-interest nexus. *Personality and Individual Differences 51*, 376–380.
Henss, R. (1995). Waist-to-hip ratio and attractiveness. Replication and extension. *Personality and Individual Differences 19*, 479–488.
Herlitz, A., & Lovén, J. (2009). Sex differences in cognitive functions. *Acta Psychologica Sinica 41*, 1081–1090.
Hermans, E. J., Putman, P., & Van Honk, J. (2006). Testosterone administration reduces empathetic behavior: A facial mimicry study. *Psychoneuroendocrinology 31*, 859–866.
Hermans, E. J., Putman, P., Baas, J. M., Gecks, N. M., Kenemans, J. L., & van Honk, J. (2007). Exogenous testosterone attenuates the integrated central stress response in healthy young women. *Psychoneuroendocrinology 32*, 1052–1061.
Hermans, E. J., Putman, P., Baas, J. M., Koppeschaar, H. P., & van Honk, J. (2006). A single administration of testosterone reduces fear-potentiated startle in humans. *Biological Psychiatry 59*, 872–874.
Hewlett, B. S. (1992). Husband-wife reciprocity and the father-infant relationship among Aka pygmies. In B. S. Hewlett (Ed.), *Father-child relations: Cultural and biosocial contexts* (pp. 153–176). London: Routledge.
Hier, D. B., & Crowley, W. F. (1982). Spatial ability in androgen-deficient men. *New England Journal of Medicine 306*, 1202–1205.
Hill, A. K., Bailey, D. H., & Puts, D. A. (2017). Gorillas in our midst? Human sexual dimorphism and contest competition in men. In M. Tibayrenc & J. Ayala (Eds.), *On human nature* (pp. 235–249). New York: Elsevier.
Hill, A. K., Hunt, J., Welling, L. L., Cardenas, R. A., Rotella, M. A., Wheatley, J. R., ... Puts, D. A. (2013). Quantifying the strength and form of sexual selection on men's traits. *Evolution and Human Behavior 34*(5), 334–341.
Hiller-Sturmhöfel, S., & Bartke, A. (1998). The endocrine system: An overview. *Alcohol: Health and Research World 22*, 153–165.
Hines, M. (2006). Prenatal testosterone and gender-related behaviour. *European Journal of Endocrinology 155*(suppl_1), S115–S121.
Hines, M. (2010). Sex-related variation in human behavior and the brain. *Trends in Cognitive Sciences 14*, 448–456.

Hines, M. (2020). Human gender development. *Neuroscience & Biobehavioral Reviews 118*, 89–96.

Hines, M., Golombok, S., Rust, J., Johnston, K. J., Golding, J., Parents, & Children Study Team, A. L. S. O. (2002). Testosterone during pregnancy and gender role behavior of preschool children: A longitudinal, population study. *Child Development 73*, 1678–1687.

Hintikka, J., Niskanen, L., Koivumaa-Honkanen, H., Tolmunen, T., Honkalampi, K., Lehto, S. M., & Viinamäki, H. (2009). Hypogonadism, decreased sexual desire, and long-term depression in middle-aged men. *Journal of Sexual Medicine 6*, 2049–2057.

Hirschenhauser, K. (2012). Testosterone and partner compatibility: Evidence and emerging questions. *Ethology 118*, 799–811.

Hittelman, J. H., & Dickes, R. (1979). Sex differences in neonatal eye contact time. *Merrill-Palmer Quarterly of Behavior and Development 25*, 171–184.

Hjelmervik, H., Westerhausen, R., Hirnstein, M., Specht, K., & Hausmann, M. (2015). The neural correlates of sex differences in left–right confusion. *NeuroImage* 113, 196–206.

Hodges-Simeon, C. R., Gaulin, S. J., & Puts, D. A. (2011). Voice correlates of mating success in men: Examining "contests" versus "mate choice" modes of sexual selection. *Archives of Sexual Behavior 40*, 551–557.

Hodges-Simeon, C. R., Gurven, M., & Gaulin, S. J. (2015). The low male voice is a costly signal of phenotypic quality among Bolivian adolescents. *Evolution and Human Behavior 36*, 294–302.

Hodosy, J., Páleš, J., Ostatníková, D., & Celec, P. (2010). The effects of exogenous testosterone on spatial memory in rats. *Open Life Sciences 5*, 466–471.

Hollier, L. P., Mattes, E., Maybery, M. T., Keelan, J. A., Hickey, M., & Whitehouse, A. J. (2013). The association between perinatal testosterone concentration and early vocabulary development: A prospective cohort study. *Biological Psychology 92*, 212–215.

Holloway, V., & Wylie, K. (2015). Sex drive and sexual desire. *Current Opinion in Psychiatry 28*, 424–429.

Holmes, M. M., Resnick, H. S., Kilpatrick, D. G., & Best, C. L. (1996). Rape-related pregnancy: Estimates and descriptive characteristics from a national sample of women. *American Journal of Obstetrics and Gynecology 175*, 320–325.

Hönekopp, J., & Watson, S. (2010). Meta-analysis of digit ratio 2D: 4D shows greater sex difference in the right hand. *American Journal of Human Biology 22*(5), 619–630.

Hönekopp, J., Rudolph, U., Beier, L., Liebert, A., & Müller, C. (2007). Physical attractiveness of face and body as indicators of physical fitness in men. *Evolution and Human Behavior 28*, 106–111.

Hooven, C. K., Chabris, C. F., Ellison, P. T., & Kosslyn, S. M. (2004). The relationship of male testosterone to components of mental rotation. *Neuropsychologia 42*, 782–790.

Hopcroft, R. L. (2002). The evolution of sex discrimination. *Psychology, Evolution & Gender 4*, 43–67.

Hoskin, A. W., & Ellis, L. (2015). Fetal testosterone and criminality: Test of evolutionary neuroandrogenic theory. *Criminology 53*, 54–73.
Hoskin, A. W., & Ellis, L. (2021). Androgens and offending behavior: Evidence based on multiple self-reported measures of prenatal and general testosterone exposure. *Personality and Individual Differences 168*, 110282.
Hotchkiss, A. K., Lambright, C. S., Ostby, J. S., Parks-Saldutti, L., Vandenbergh, J. G., & Gray Jr, L. E. (2007). Prenatal testosterone exposure permanently masculinizes anogenital distance, nipple development, and reproductive tract morphology in female Sprague-Dawley rats. *Toxicological Sciences 96*, 335–345.
Hoyenga, K. B., & Wallace, B. (1979). Sex differences in the perception of autokinetic movement of an afterimage. *Journal of General Psychology 100*, 93–101.
Hughes, A., & Kumari, M. (2019). Testosterone, risk, and socioeconomic position in British men: Exploring causal directionality. *Social Science & Medicine 220*, 129–140.
Humphreys, L. G., Lubinski, D., & Yao, G. (1993). Utility of predicting group membership and the role of spatial visualization in becoming an engineer, physical scientist, or artist. *Journal of applied psychology 78*(2), 250–262.
Huppin, M., Malamuth, N. M., & Linz, D. (2019). An evolutionary perspective on sexual assault and implications for interventions. In W. T. O'Donohue & P. Schewe (Eds.), *Handbook of sexual assault and sexual assault prevention* (pp. 17–44). New York: Springer.
Husak, J. F., Fuxjager, M. J., Johnson, M. A., Vitousek, M. N., Donald, J. W., Francis, C. D., ... Knapp, R. (2021). Life history and environment predict variation in testosterone across vertebrates. *Evolution 75*, 1003–1010.
Huselid, R. F., & Cooper, M. L. (1992). Gender roles as mediators of sex differences in adolescent alcohol use and abuse. *Journal of Health and Social Behavior 33*, 348–362.
Huxley, J. S. (1938). Darwin's theory of sexual selection and the data subsumed by it, in the light of recent research. *American Naturalist* 72(742), 416–433.
Hyde, J. S., & Mertz, J. E. (2009). Gender, culture, and mathematics performance. *Proceedings of the National Academy of Sciences 106*, 8801–8807.
Hyde, J. S., Fennema, E., & Lamon, S. J. (1990). Gender differences in mathematics performance: A meta-analysis. *Psychological Bulletin 107*, 139–155.
Hyland, W. E., Hoff, K. A., & Rounds, J. (2022). Interest–ability profiles: An integrative approach to knowledge acquisition. *Journal of Intelligence 10*, 43–57.
Iannuzzi-Sucich, M., Prestwood, K. M., & Kenny, A. M. (2002). Prevalence of sarcopenia and predictors of skeletal muscle mass in healthy, older men and women. *Journals of Gerontology Series A: Biological Sciences and Medical Sciences 57*, M772–M777.
Imre, P. D. (1963). A correlation study of verbal IQ and grade achievement. *Journal of Clinical Psychology 19*, 218–219.
Islam, M. M., & Karim, K. R. (2012). Men's views on gender and sexuality in a Bangladesh village. *International Quarterly of Community Health Education 32*, 339–354.

Jackson, L. A., & McGill, O. D. (1996). Body type preferences and body characteristics associated with attractive and unattractive bodies by African Americans and Anglo Americans. *Sex Roles 35*, 295–307.

Jacobs, L. F. (1996). Sexual selection and the brain. *Trends in Ecology & Evolution 11*, 82–86.

Jadva, V., Hines, M., & Golombok, S. (2010). Infants' preferences for toys, colors, and shapes: Sex differences and similarities. *Archives of Sexual Behavior 39*, 1261–1273.

Jain, V. G., Goyal, V., Chowdhary, V., Swarup, N., Singh, R. J., Singal, A., & Shekhawat, P. (2018). Anogenital distance is determined during early gestation in humans. *Human Reproduction 33*, 1619–1627.

James, W. H. (2008). Further evidence that some male-based neurodevelopmental disorders are associated with high intrauterine testosterone concentrations. *Developmental Medicine & Child Neurology 50*, 15–18.

Jäncke, L. (2018). Sex/gender differences in cognition, neurophysiology, and neuroanatomy. *F1000 Research 7*, 1–10, 10.12688/f1000research.13917.1

Jäncke, L., Mérillat, S., Liem, F., & Hänggi, J. (2015). Brain size, sex, and the aging brain. *Human brain mapping 36*, 150–169.

Janowsky, J. S., Oviatt, S. K., & Orwoll, E. (1994). Testosterone influences spatial cognition in older men. *Behavioral Neuroscience 108*, 325–332.

Jardí, F., Laurent, M. R., Kim, N., Khalil, R., De Bundel, D., Van Eeckhaut, A., … Schollaert, D. (2018). Testosterone boosts physical activity in male mice via dopaminergic pathways. *Scientific Reports 8*, 1–14.

Jaros, D., & White, E. S. (1983). Sex, endocrines, and political behavior. In M. Watts (Ed.), *Biopolitics and gender* (pp. 129–145). New York: Haworth Press.

Jeffery, A. J., Shackelford, T. K., Zeigler-Hill, V., Vonk, J., & McDonald, M. (2019). The evolution of human female sexual orientation. *Evolutionary Psychological Science 5*, 71–86.

Jockenhövel, F., Minnemann, T., Schubert, M., Freude, S., Hübler, D., Schumann, C., … Ernst, M. (2009). Timetable of effects of testosterone administration to hypogonadal men on variables of sex and mood. *Aging Male 12*, 113–118.

Johnson, D. D., McDermott, R., Barrett, E. S., Cowden, J., Wrangham, R., McIntyre, M. H., & Peter Rosen, S. (2006). Overconfidence in wargames: Experimental evidence on expectations, aggression, gender and testosterone. *Proceedings of the Royal Society B: Biological Sciences 273*(1600), 2513–2520.

Johnson, M. W., Anch, A. M., & Remmers, J. E. (1984). Induction of the obstructive sleep apnea syndrome in a woman by exogenous androgen administration. *American Review of Respiratory Disease 129*(6), 1023–1025.

Johnson, R. T., Burk, J. A., & Kirkpatrick, L. A. (2007). Dominance and prestige as differential predictors of aggression and testosterone levels in men. *Evolution and Human Behavior 28*(5), 345–351.

Jones, C. M., Braithwaite, V. A., & Healy, S. D. (2003). The evolution of sex differences in spatial ability. *Behavioral Neuroscience 117*, 403–412.

Jordan, K., Fromberger, P., Stolpmann, G., & Müller, J. L. (2011). The role of testosterone in sexuality and paraphilia: A neurobiological approach. Part II: Testosterone and paraphilia. *Journal of Sexual Medicine 8*, 3008–3029.

Joseph, R. (2000). The evolution of sex differences in language, sexuality, and visual–spatial skills. *Archives of Sexual Behavior 29*, 35–66.

Josephs, L., & Shimberg, J. (2010). The dynamics of sexual fidelity: Personality style as a reproductive strategy. *Psychoanalytic Psychology 27*(3), 273–284.

Josephs, R. A., Cobb, A. R., Lancaster, C. L., Lee, H.-J., & Telch, M. J. (2017). Dual-hormone stress reactivity predicts downstream war-zone stress-evoked PTSD. *Psychoneuroendocrinology 78*, 76–84.

Joyce, C. W., O'Regan, A., Kelly, J. L., & O'Shaughnessy, M. (2017). Fight bite injuries: Aggressive tendencies associated with smaller second to fourth digit ratio. *Journal of Hand Surgery (Asian-Pacific Volume) 22*, 452–456.

Judge, T. A., & Cable, D. M. (2004). The effect of physical height on workplace success and income: Preliminary test of a theoretical model. *Journal of Applied Psychology 89*(3), 428–439.

Kanazawa, S. (2005). Is "discrimination" necessary to explain the sex gap in earnings? *Journal of Economic Psychology 26*, 269–287.

Kanazawa, S. (2008). An evolutionary psychological perspective on theft. In J. Duntley & T. K. Shackelford (Eds.), *Evolutionary Forensic Psychology*. New York: Oxford University Press.

Kanazawa, S., & Vandermassen, G. (2005). Engineers have more sons, nurses have more daughters: An evolutionary psychological extension of Baron–Cohen's extreme male brain theory of autism. *Journal of Theoretical Biology 233*, 589–599.

Kane, J. M., & Mertz, J. E. (2012). Debunking myths about gender and mathematics performance. *Notices of the AMS 59*(1), 10–21.

Kaplan, H., Hill, K., Lancaster, J., & Hurtado, A. M. (2000). A theory of human life history evolution: Diet, intelligence, and longevity. *Evolutionary Anthropology 9*, 156–185.

Karadağ, E., & Mutafçılar, I. (2009). A research on altruism levels of the primary and secondary school teachers. *Ondokuz Mayıs Üniversitesi Eğitim Fakültesi Dergisi 28*, 75–92.

Kardos, M., Luikart, G., Bunch, R., Dewey, S., Edwards, W., McWilliam, S., ... Kijas, J. (2015). Whole-genome resequencing uncovers molecular signatures of natural and sexual selection in wild bighorn sheep. *Molecular Ecology 24*, 5616–5632.

Karlović, D., Serretti, A., Marčinko, D., Martinac, M., Silić, A., & Katinić, K. (2012). Serum testosterone concentration in combat-related chronic posttraumatic stress disorder. *Neuropsychobiology 65*, 90–95.

Katznelson, L., Finkelstein, J. S., Schoenfeld, D. A., Rosenthal, D. I., Anderson, E. J., & Klibanski, A. (1996). Increase in bone density and lean body mass during testosterone administration in men with acquired hypogonadism. *Journal of Clinical Endocrinology & Metabolism 81*, 4358–4365.

Kaufman, J. M., & T'sjoen, G. (2002). The effects of testosterone deficiency on male sexual function. *Aging Male 5*, 242–247.

Kavaliers, M., & Choleris, E. (2018). The role of social cognition in parasite and pathogen avoidance. *Philosophical Transactions of the Royal Society B: Biological Sciences 373*(1751), 20170206.

Kavoussi, R. J., Armstead, P., & Coccaro, E. (1997). The neurobiology of impulsive aggression. *Psychiatric Clinics of North America 20*, 395–403.

Keefe, K. (1994). Perceptions of normative social pressure and attitudes toward alcohol use: Changes during adolescence. *Journal of Studies on Alcohol 55*, 46–54.

Kelava, I., Chiaradia, I., Pellegrini, L., Kalinka, A. T., & Lancaster, M. A. (2022). Androgens increase excitatory neurogenic potential in human brain organoids. *Nature 602*(7895), 112–116.

Kempenaers, B., & Sheldon, B. C. (1996). Why do male birds not discriminate between their own and extra-pair offspring? *Animal Behaviour 51*, 1165–1173.

Kenny, A. M., Prestwood, K. M., Marcello, K. M., & Raisz, L. G. (2000). Determinants of bone density in healthy older men with low testosterone levels. *Journals of Gerontology Series A: Biological Sciences and Medical Sciences 55*, M492–M497.

Kenrick, D. T., & Keefe, R. C. (1992). Age preferences in mates reflect sex differences in human reproductive strategies. *Behavioral and Brain Sciences*, *15*, 75–91.

Kenrick, D. T., & Luce, C. L. (2012). An evolutionary life-history model of gender differences and similarities. In T. Eckes & H. M. Trautner (Eds.), *The developmental social psychology of gender* (pp. 49–78). San Diego: Psychology Press.

Kerem, M., Akbayrak, T., Bumin, G., Yigiter, K., Armutlu, K., & Kerimoglu, D. (2002). A correlation between sex hormone levels and pressure pain threshold and tolerance in healthy women. *Pain Clinic 14*, 43–47.

Kerry, N., & Murray, D. R. (2021). Physical strength partly explains sex differences in trait anxiety in young Americans. *Psychological Science 32*, 809–815.

Keski-Rahkonen, A., Neale, B. M., Bulik, C. M., Pietilainen, K. H., Rose, R. J., Kapiro, J., & Rissanen, A. M. (2005). Intentional weight loss in young adults: Sex-specific genetic and environmental effects. *Obesity Research 13*, 745–753.

Khonicheva, N. M., Livanova, L. M., Tsykunov, S. G., Osipova, T., Loriya, M., Élbakidze, A., ... Airapetyants, M. (2008). Blood testosterone in rats: Correlation of the level of individual anxiety and its impairment after "death threat". *Neuroscience and Behavioral Physiology 38*, 985–989.

Kim, J. M., Shin, S. C., Park, G. C., Lee, J. C., Jeon, Y. K., Ahn, S. J., ... Lee, B. J. (2020). Effect of sex hormones on extracellular matrix of lamina propria in rat vocal fold. *The Laryngoscope 130*, 732–740.

Kimura, D. (2003). Sex differences in the brain. *Scientific American 287*, 32–37.

Kious, B. M., Kondo, D. G., & Renshaw, P. F. (2019). Creatine for the treatment of depression. *Biomolecules 9*, 406–417.

Kirbas, G., Abakay, A., Topcu, F., Kaplan, A., Ünlu, M., & Peker, Y. (2007). Obstructive sleep apnoea, cigarette smoking and serum testosterone levels in a male sleep clinic cohort. *Journal of International Medical Research 35*, 38–45.

Kirchengast, S., & Marosi, A. (2008). Gender differences in body composition, physical activity, eating behavior and body image among normal weight adolescents–an evolutionary approach. *Collegium Antropologicum 32*, 1079–1086.

Kirillova, G. P., Vanyukov, M. M., Kirisci, L., & Reynolds, M. (2008). Physical maturation, peer environment, and the ontogenesis of substance use disorders. *Psychiatry Research 158*, 43–53.

Kirkpatrick, S. W., Campbell, P. S., Wharry, R. E., & Robinson, S. L. (1993). Salivary testosterone in children with and without learning disabilities. *Physiology & Behavior 53*, 583–586.

Kitsantas, A., & Zimmerman, B. J. (2009). College students' homework and academic achievement: The mediating role of self-regulatory beliefs. *Metacognition and Learning 4*, 97–110.

Klimas, C., Ehlert, U., Lacker, T. J., Waldvogel, P., & Walther, A. (2019). Higher testosterone levels are associated with unfaithful behavior in men. *Biological Psychology, 146*, 107730.

Klimek, M., Galbarczyk, A., Nenko, I., Alvarado, L. C., & Jasienska, G. (2014). Digit ratio (2D: 4D) as an indicator of body size, testosterone concentration and number of children in human males. *Annals of Human Biology 41*, 518–523.

Klinesmith, J., Kasser, T., & McAndrew, F. T. (2006). Guns, testosterone, and aggression: An experimental test of a mediational hypothesis. *Psychological Science 17*, 568–571.

Klomberg, K. F., Garland Jr, T., Swallow, J. G., & Carter, P. A. (2002). Dominance, plasma testosterone levels, and testis size in house mice artificially selected for high activity levels. *Physiology & Behavior 77*, 27–38.

Klump, K. L., Gobrogge, K. L., Perkins, P. S., Thorne, D., Sisk, C. L., & Breedlove, S. M. (2006). Preliminary evidence that gonadal hormones organize and activate disordered eating. *Psychological medicine 36*, 539–546.

Klump, K. L., Perkins, P. S., Burt, S. A., McGue, M., & Iacono, W. G. (2007). Puberty moderates genetic influences on disordered eating. *Psychological medicine 37*, 627–634.

Knafo, A., & Spinath, F. M. (2011). Genetic and environmental influences on girls' and boys‘ gender-typed and gender-neutral values. *Developmental Psychology 47*, 726–731.

Knickmeyer, R. C., & Baron-Cohen, S. (2006). Fetal testosterone and sex differences. *Early Human Development 82*, 755–760.

Knickmeyer, R. C., Wheelwright, S., Taylor, K., Raggatt, P., Hackett, G., & Baron-Cohen, S. (2005). Gender-typed play and amniotic testosterone. *Developmental Psychology 41*(3), 517–528.

Knickmeyer, R. C., Woolson, S., Hamer, R. M., Konneker, T., & Gilmore, J. H. (2011). 2D: 4D ratios in the first 2 years of life: Stability and relation to testosterone exposure and sensitivity. *Hormones and Behavior 60*, 256–263.

Knickmeyer, R., Baron-Cohen, S., Raggatt, P., & Taylor, K. (2005). Foetal testosterone, social relationships, and restricted interests in children. *Journal of Child Psychology and Psychiatry 46*, 198–210.

Knight, E. L., Morales, P. J., Christian, C. B., Prasad, S., Harbaugh, W. T., Mehta, P. H., & Mayr, U. (2022). The causal effect of testosterone on men's competitive behavior is moderated by basal cortisol and cues to an opponent's status: Evidence for a context-dependent dual-hormone hypothesis. *Journal of Personality and Social Psychology 123*(4), 693–716, 10.31234/osf.io/y31234hfu.

Kodric-Brown, A., & Brown, J. H. (1984). Truth in advertising: The kinds of traits favored by sexual selection. *American Naturalist 124*, 309–323.

Kokko, H., & Jennions, M. D. (2008). Parental investment, sexual selection and sex ratios. *Journal of Evolutionary Biology 21*, 919–948.

Kokko, H., & Lindström, J. (1996). Kin selection and the evolution of leks: Whose success do young males maximize? *Proceedings of the Royal Society of London. Series B: Biological Sciences 263*(1372), 919–923.

Kolakowski, D., & Malina, R. M. (1974). Spatial ability, throwing accuracy and man's hunting heritage. *Nature 251*, 410–412.

Kolata, G. B. (1979). Sex hormones and brain development: What goes on early in development when sex hormones act on the brain? *Science 205*(4410), 985–987.

Komlos, J. (1990^). Height and social status in eighteenth-century Germany. *Journal of Interdisciplinary History 20*, 607–621.

Konrad, A. M. (2003). Family demands and job attribute preferences: A 4-year longitudinal study of women and men. *Sex Roles*, *49*, 35–46.

Kordsmeyer, T. L., & Penke, L. (2019). Effects of male testosterone and its interaction with cortisol on self-and observer-rated personality states in a competitive mating context. *Journal of Research in Personality 78*, 76–92.

Kornhuber, J., Erhard, G., Lenz, B., Kraus, T., Sperling, W., Bayerlein, K., ... Stoessel, C. (2011). Low digit ratio 2D : 4D in alcohol dependent patients. *PLoS One 6*(4), e19332.

Kornhuber, J., Zenses, E.-M., Lenz, B., Stoessel, C., Bouna-Pyrrou, P., Rehbein, F., ... Mößle, T. (2013). Low 2D: 4D values are associated with video game addiction. *PLoS One 8*(11), e79539.

Koyama, N. (1967). On dominance rank and kinship of a wild Japanese monkey troop in Arashiyama. *Primates 8*, 189–216.

Kramer, S. S., & Russell, R. (2022). A novel human sex difference: Male sclera are redder and yellower than female sclera. *Archives of Sexual Behavior 51*, 2733–2740.

Kreider, R. B. (2003). Species-specific responses to creatine supplementation. *American Journal of Physiology-Regulatory, Integrative and Comparative Physiology 285*, R725–R726.

Kruger, D. J., & Nesse, R. M. (2004). Sexual selection and the male:female mortality ratio. *Evolutionary Psychology 2*, 66–85.

Kruger, D. J., & Nesse, R. M. (2006). An evolutionary life-history framework for understanding sex differences in human mortality rates. *Human Nature 17*(1), 74–97.

Kruger, D. J., Wang, X.-T., & Wilke, A. (2007). Towards the development of an evolutionarily valid domain-specific risk-taking scale. *Evolutionary Psychology 5*(3), 147470490700500306.

Kruger, D., & Byker, D. (2009). Evolved foraging psychology underlies sex differences in shopping experiences and behaviors. *Journal of Social, Evolutionary, and Cultural Psychology 3*, 328–333.

Kruger, T. H., Sinke, C., Kneer, J., Tenbergen, G., Khan, A. Q., Burkert, A., ... von Wurmb-Schwark, N. (2019). Child sexual offenders show prenatal and epigenetic alterations of the androgen system. *Translational Psychiatry 9*(1), 1–11.

Kumar, A., Liu, N.-J., Madia, P. A., & Gintzler, A. R. (2015). Contribution of endogenous spinal endomorphin 2 to intrathecal opioid antinociception in rats is agonist dependent and sexually dimorphic. *Journal of Pain 16*, 1200–1210.

Kung, K. T., Browne, W. V., Constantinescu, M., Noorderhaven, R. M., & Hines, M. (2016). Early postnatal testosterone predicts sex-related differences in early expressive vocabulary. *Psychoneuroendocrinology 68*, 111–116.

Kupperman, H. S. (1944). Hormone control of a dimorphic pigmentation area in the golden hamster (Criceus auratus). *Anatomical Record 88*, 442–444.

Kurath, J., & Mata, R. (2018). Individual differences in risk taking and endogeneous levels of testosterone, estradiol, and cortisol: A systematic literature search and three independent meta-analyses. *Neuroscience & Biobehavioral Reviews 90*, 428–446.

Kuzawa, C. W., Gettler, L. T., Huang, Y.-y., & McDade, T. W. (2010). Mothers have lower testosterone than non-mothers: Evidence from the Philippines. *Hormones and Behavior 57*, 441–447.

Kuzawa, C. W., Gettler, L. T., Muller, M. N., McDade, T. W., & Feranil, A. B. (2009). Fatherhood, pair-bonding and testosterone in the Philippines. *Hormones and Behavior 56*, 429–435.

Lamminmäki, A., Hines, M., Kuiri-Hänninen, T., Kilpeläinen, L., Dunkel, L., & Sankilampi, U. (2012). Testosterone measured in infancy predicts subsequent sex-typed behavior in boys and in girls. *Hormones and Behavior 61*, 611–616.

Lane, C. J. (2006). *Evolution of gender differences in adult crying.* (PhD thesis). Arlington, TX: Department of Anthropology, University of Texas at Arlington.

Lange, B. P., Hennighausen, C., Brill, M., & Schwab, F. (2016). Only cheap talk after all? New experimental psychological findings on the role of verbal proficiency in mate choice. *Psychology of Language and Communication 20*(1), 1–12.

Larson, E. J. (2003). *Trial and error: The American controversy over creation and evolution.* New York: Oxford University Press.

Lawson, A. E., & Worsnop, W. A. (1992). Learning about evolution and rejecting a belief in special creation: Effects of reflective reasoning skill, prior knowledge, prior belief and religious commitment. *Journal of Research in Science Teaching 29*, 143–166.

Leblanc, A., Taylor, B. A., Thompson, P. D., Capizzi, J. A., Clarkson, P. M., Michael White, C., & Pescatello, L. S. (2015). Relationships between physical activity and muscular strength among healthy adults across the lifespan. *Springerplus 4*(1), 1–10.

Lee, K. S., Zhang, Y., Asgar, J., Auh, Q.-S., Chung, M.-K., & Ro, J. Y. (2016). Androgen receptor transcriptionally regulates μ-opioid receptor expression in rat trigeminal ganglia. *Neuroscience 331*, 52–61.

Leeb, R. T., & Rejskind, F. G. (2004). Here's looking at you, kid! A longitudinal study of perceived gender differences in mutual gaze behavior in young infants. *Sex Roles 50*, 1–14.

Lemiere, J., Boets, B., & Danckaerts, M. (2010). No association between the 2D: 4D fetal testosterone marker and multidimensional attentional abilities in children with ADHD. *Developmental Medicine & Child Neurology 52*(9), e202–e208.

Lemle, R., & Mishkind, M. E. (1989). Alcohol and masculinity. *Journal of Substance Abuse Treatment* 6, 213–222.

Lenz, B., & Kornhuber, J. (2018). Cross-national gender variations of digit ratio (2D: 4D) correlate with life expectancy, suicide rate, and other causes of death. *Journal of Neural Transmission 125*, 239–246.

Lenz, B., Bouna-Pyrrou, P., Mühle, C., & Kornhuber, J. (2018). Low digit ratio (2D: 4D) and late pubertal onset indicate prenatal hyperandrogenziation in alcohol binge drinking. *Progress in Neuro-Psychopharmacology and Biological Psychiatry 86*, 370–378.

Lenz, B., Mühle, C., Braun, B., Weinland, C., Bouna-Pyrrou, P., Behrens, J., ... Saigali, S. (2017). Prenatal and adult androgen activities in alcohol dependence. *Acta Psychiatrica Scandinavica 136*, 96–107.

Lenz, B., Thiem, D., Bouna-Pyrrou, P., Mühle, C., Stoessel, C., Betz, P., & Kornhuber, J. (2016). Low digit ratio (2D: 4D) in male suicide victims. *Journal of Neural Transmission 123*(12), 1499–1503.

Leroy, F., & Praet, I. (2015). Meat traditions. The co-evolution of humans and meat. *Appetite 90*, 200–211.

Leuze, K., & Strauß, S. (2016). Why do occupations dominated by women pay less? How 'female-typical' work tasks and working-time arrangements affect the gender wage gap among higher education graduates. *Work, Employment and Society 30*, 802–820.

Levine, S. C., Foley, A., Lourenco, S., Ehrlich, S., & Ratliff, K. (2016). Sex differences in spatial cognition: Advancing the conversation. *Wiley Interdisciplinary Reviews: Cognitive Science* 7, 127–155.

Lidborg, L. H., Cross, C. P., & Boothroyd, L. G. (2020). Masculinity matters (but mostly if you're muscular): A meta-analysis of the relationships between sexually dimorphic traits in men and mating/reproductive success. *BioRxiv*, 10.1101/2020.03.06.980896.

Ligon, J. D., Thornhill, R., Zuk, M., & Johnson, K. (1990). Male-male competition, ornamentation and the role of testosterone in sexual selection in red jungle fowl. *Animal Behaviour 40*, 367–373.

Lindeman, M., & Svedholm-Häkkinen, A. M. (2016). Does poor understanding of physical world predict religious and paranormal beliefs? *Applied Cognitive Psychology 30*, 736–742.

Lindenfors, P., & Tullberg, B. S. (2011). Evolutionary aspects of aggression: The importance of sexual selection. *Advances in Genetics 75*, 7–22.

Lindman, R., Von der Pahlen, B., Öst, B., & Eriksson, C. P. (1992). Serum testosterone, cortisol, glucose, and ethanol in males arrested for spouse abuse. *Aggressive Behavior 18*, 393–400.

Lippa, R. A. (2002). *Gender, nature, and nurture*. Mahwah, NJ: Lawrence Erlbaum Associates.

Lippa, R. A. (2003a). Are 2D:4D finger-length ratios related to sexual orientation? Yes for men, no for women. *Journal of Personality and Social Psychology 85*, 179–191.

Lippa, R. A. (2009). Sex differences in sex drive, sociosexuality, and height across 53 nations: Testing evolutionary and social structural theories. *Archives of Sexual Behavior 38*(5), 631–651.

Lippa, R. A. (2010a). Gender differences in personality and interests: When, where, and why? *Social and Personality Psychology Compass 4*, 1098–1110.

Lippa, R. A. (2010b). Sex differences in personality traits and gender-related occupational preferences across 53 nations: Testing evolutionary and social-environmental theories. *Archives of Sexual Behavior 39*, 619–636.

Lippa, R. A. (2016). Biological influences on masculinity. In Y. J. Wong (Ed.), *APA handbook of men and masculinities* (pp. 187–209). Washington, DC: American Psychological Association.

Lippa, R. A., Preston, K., & Penner, J. (2014). Women's representation in 60 occupations from 1972 to 2010: More women in high-status jobs, few women in things-oriented jobs. *PLoS One 9*(5), e95960.

Lips, H., & Lawson, K. (2009). Work values, gender, and expectations about work commitment and pay: Laying the groundwork for the "motherhood penalty"? *Sex Roles 61*, 667–676.

Little, A. C., Jones, B. C., & DeBruine, L. M. (2011). Facial attractiveness: Evolutionary based research. *Philosophical Transactions of the Royal Society B: Biological Sciences 366*(1571), 1638–1659.

Liu, J. J., Ein, N., Peck, K., Huang, V., Pruessner, J. C., & Vickers, K. (2017). Sex differences in salivary cortisol reactivity to the Trier Social Stress Test (TSST): A meta-analysis. *Psychoneuroendocrinology 82*, 26–37.

Liu, T.-C., Lin, C.-H., Huang, C.-Y., Ivy, J. L., & Kuo, C.-H. (2013). Effect of acute DHEA administration on free testosterone in middle-aged and young men following high-intensity interval training. *European Journal of Applied Physiology 113*, 1783–1792.

Llaurens, V., Raymond, M., & Faurie, C. (2009). Why are some people left-handed? An evolutionary perspective. *Philosophical Transactions of the Royal Society B: Biological Sciences 364*, 881–894.

Loewenthal, K. M., MacLeod, A. K., & Cinnirella, M. (2002). Are women more religious than men? Gender differences in religious activity among different religious groups in the UK. *Personality and Individual Differences 32*, 133–139.

Lofeu, L., Brandt, R., & Kohlsdorf, T. (2017). Phenotypic integration mediated by hormones: Associations among digit ratios, body size and testosterone during tadpole development. *BMC Evolutionary Biology 17*, 175–182.

Lombardo, M. P. (2012). On the evolution of sport. *Evolutionary Psychology 10*(1), 1–28, 147470491201000101.

Lombardo, M. V., Ashwin, E., Auyeung, B., Chakrabarti, B., Taylor, K., Hackett, G., … Baron-Cohen, S. (2012). Fetal testosterone influences sexually dimorphic gray matter in the human brain. *Journal of Neuroscience 32*, 674–680.

Long, T. A., Agrawal, A. F., & Rowe, L. (2012). The effect of sexual selection on offspring fitness depends on the nature of genetic variation. *Current Biology 22*, 204–208.

Longpré, N., Guay, J. P., & Knight, R. A. (2020). Sadistic sexual aggressors. In J. Proulx, F. Cortoni, L. A. Craig, & E. J. Letourneau (Eds.), *The Wiley handbook of what works with sexual offenders: Contemporary perspectives in theory, assessment, treatment, and prevention* (pp. 387–409). New York: Wiley.

Louissaint Jr, A., Rao, S., Leventhal, C., & Goldman, S. A. (2002). Coordinated interaction of neurogenesis and angiogenesis in the adult songbird brain. *Neuron 34*, 945–960.

Lummaa, V., Merila, J., & Kause, A. (1998). Adaptive sex ratio variation in preindustrial human (Homo sapiens) populations? *Proceeding of the Royal Society of London, Biological Science 265*, 563–568.

Lummaa, V., Vuorisalo, T., Barr, R. G., & Lehtonen, L. (1998). Why Cry? Adaptive Significance of Intensive Crying in Human Infants. *Evolution and Human Behavior* 19(3), 193–202.

Lunardini, A. (1989). Social organization in a confined group of Japanese macaques (Macaca fuscata): An application of correspondence analysis. *Primates 30*, 175–185.

Luoto, S., & Varella, M. A. C. (2021). Pandemic leadership: Sex differences and their evolutionary–developmental origins. *Frontiers in Psychology 12* (618), 1–23.

Luoto, S., Krama, T., Rubika, A., Borráz-León, J. I., Trakimas, G., Elferts, D., ... Birbele, E. (2021). Socioeconomic position, immune function, and its physiological markers. *Psychoneuroendocrinology 127*, 105202.

Lust, J. M., Geuze, R. H., Van de Beek, C., Cohen-Kettenis, P. T., Groothuis, A. G. G., & Bouma, A. (2010). Sex specific effect of prenatal testosterone on language lateralization in children. *Neuropsychologia 48*, 536–540.

Lutchmaya, S., & Baron-Cohen, S. (2002). Human sex differences in social and non-social looking preferences, at 12 months of age. *Infant Behavior & Development 25*, 319–325.

Lutchmaya, S., Baron-Cohen, S., & Raggatt, P. (2001). Foetal testosterone and vocabulary size in 18-and 24-month-old infants. *Infant Behavior and Development 24*, 418–424.

Lutchmaya, S., Baron-Cohen, S., & Raggatt, P. (2002a). Fetal testosterone and eye contact in 12-month-old human infants. *Infant Behavior and Development 25*, 327–335.

Lutchmaya, S., Baron-Cohen, S., & Raggatt, P. (2002b). Fetal testosterone and vocabulary size in 18- and 24-month-old infants. *Infant Behavior & Development 24*, 418–424.

Luukkaa, V., Pesonen, U., Huhtaniemi, I., Lehtonen, A., Tilvis, R., Tuomilehto, J., ... Huupponen, R. (1998). Inverse correlation between serum testosterone and leptin in men. *Journal of Clinical Endocrinology & Metabolism 83*, 3243–3246.

Lydecker, J. A., Pisetsky, E. M., Mitchell, K. S., Thornton, L. M., Kendler, K. S., Reichborn-Kjennerud, T., ... Mazzeo, S. E. (2012). Association between co-twin sex and eating disorders in opposite sex twin pairs: Evaluations in North American, Norwegian, and Swedish samples. *Journal of Psychosomatic Research 72*, 73–77.

Määttänen, I., Jokela, M., Hintsa, T., Firtser, S., Kähönen, M., Jula, A., ... Keltikangas-Järvinen, L. (2013). Testosterone and temperament traits in men: Longitudinal analysis. *Psychoneuroendocrinology 38*, 2243–2248.

Madsen, K. L., Adams, W. C., & Van Loan, M. D. (1998). Effects of physical activity, body weight and composition, and muscular strength on bone density in young women. *Medicine and Science in Sports and Exercise 30*, 114–120.

Maione, L., Pala, G., Bouvattier, C., Trabado, S., Papadakis, G., Chanson, P., … Maghnie, M. (2020). Congenital hypogonadotropic hypogonadism/Kallmann syndrome is associated with statural gain in both men and women: A monocentric study. *European Journal of Endocrinology 182*, 185–195.

Malamuth, N. M. (1996). Sexually explicit media, gender differences and evolutionary theory. *Journal of Communication 46*(3), 8–31.

Malamuth, N., Huppin, M., & Bryant, P. (2005). Sexual coercion. In D. Buss (Ed.), *The evolutionary psychology handbook* (pp. 394–418). New York: Wiley.

Manhardt, P. (1972). Job orientation of male and female college graduates in business. *Personal Psychology 25*, 361–369.

Manning, J. T., & Fink, B. (2011). Digit ratio (2D: 4D) and aggregate personality scores across nations: Data from the BBC internet study. *Personality and Individual Differences 51*, 387–391.

Manning, J. T., & Taylor, R. P. (2001). Second to fourth digit ratio and male ability in sport: Implications for sexual selection in humans. *Evolution and Human Behavior 22*, 61–69.

Manning, J. T., Baron-Cohen, S., Wheelwright, S., & Sanders, G. (2001). The 2nd to 4th digit ratio and autism. *Developmental Medicine and Child Neurology 43*, 160–164.

Manning, J. T., Bundred, P. E., Newton, D. J., & Flanagan, B. F. (2003). The second to fourth digit ratio and variation in the androgen receptor gene. *Evolution and Human Behavior 24*, 399–405.

Manning, J. T., Morris, L., & Caswell, N. (2007). Endurance running and digit ratio (2D:4D): Implications for fetal testosterone effects on running speed and vascular health. *American Journal of Human Biology 19*, 416–421.

Manning, J. T., Reimers, S., Baron-Cohen, S., Wheelwright, S., & Fink, B. (2010). Sexually dimorphic traits (digit ratio, body height, systemizing–empathizing scores) and gender segregation between occupations: Evidence from the BBC internet study. *Personality and Individual Differences 49*, 511–515.

Manning, J. T., Scutt, D., Wilson, J., & Lewis-Jones, D. I. (1998). The ratio of 2nd to 4th digit length: A predictor of sperm numbers and concentrations of testosterone, luteinizing hormone and oestrogen. *Human Reproduction 13*, 3000–3004.

Manning, J. T., Trivers, R., & Fink, B. (2017). Is digit ratio (2D: 4D) related to masculinity and femininity? Evidence from the BBC internet study. *Evolutionary Psychological Science 3*, 316–324.

Manno, F. A. M. (2008). Measurement of the digit lengths and the anogenital distance in mice. *Physiology & Behavior 93*, 364–368.

Manson, J. H., Chua, K. J., Rodriguez, N. N., Barlev, M., Durkee, P. K., & Lukaszewski, A. W. (2022). Sex differences in fearful personality traits are mediated by physical strength. *Social Psychological and Personality Science*, 19485506221094086.

Mari, G., & Luijkx, R. (2020). Gender, parenthood, and hiring intentions in sex-typical jobs: Insights from a survey experiment. *Research in Social Stratification and Mobility 65*, 100464.

Martel, M. M. (2013). Sexual selection and sex differences in the prevalence of childhood externalizing and adolescent internalizing disorders. *Psychological Bulletin 139*, 1221–1259.

Martel, M. M., Gobrogge, K. L., Breedlove, S. M., & Nigg, J. T. (2008). Masculinized finger-length ratios of boys, but not girls, are associated with attention-deficit/hyperactivity disorder. *Behavioral Neuroscience 122*(2), 273–284.

Martin, L. A., & Ter-Petrosyan, M. (2020). Positive affect moderates the relationship between salivary testosterone and a health behavior composite in university females. *International Journal of Behavioral Medicine 27*, 305–315.

Masdrakis, V. G., Papageorgiou, C., & Markianos, M. (2019). Associations of plasma testosterone with clinical manifestations in acute panic disorder. *Psychoneuroendocrinology 101*, 216–222.

Mason, J. W., Giller, E. L., Kosten, T. R., & Wahby, V. S. (1990). Serum testosterone levels in post-traumatic stress disorder inpatients. *Journal of Traumatic Stress 3*, 449–457.

Mazlan, N., Najmuddin, N. H. A., Ismail, S. Z. I., Rashid, N. A. A., Maliki, N., & Shariff, N. S. M. (2020). Human hormonal effects on criminal behaviour. *International Journal of Medical Toxicology & Legal Medicine 23*, 160–164.

Mazur, A. (1995). Biosocial models of deviant behavior among male army veterans. *Biological Psychology 41*, 271–293.

Mazur, A., & Booth, A. (1998). Testosterone and dominance in men. *Behavioral and Brain Sciences 21*, 353–363.

McCord, J. (1999). Understanding childhood and subsequent crime. *Aggressive Behavior 25*, 241–253.

McElduff, A., Wilkinson, M., Ward, P., & Posen, S. (1988). Forearm mineral content in normal men: Relationship to weight, height and plasma testosterone concentrations. *Bone 9*, 281–283.

McFadden, D. (1998). Sex differences in the auditory system. *Developmental Neuropsychology 14*, 261–298.

McGlothlin, J. W., Jawor, J. M., & Ketterson, E. D. (2007). Natural variation in a testosterone-mediated trade-off between mating effort and parental effort. *American Naturalist 170*, 864–875.

McGuire, M. T., Troisi, A., & Raleigh, M. M. (1997). Depression in evolutionary context. In S. Baron-Cohen (Ed.), *The maladapted mind: Classic readings in evolutionary psychopathology* (pp. 255–282). London: Taylor & Francis.

McHenry, H. M., & Coffing, K. (2000). Australopithecus to Homo: Transformations in body and mind. *Annual Review of Anthropology 29*, 125–146.

McIntyre, M. H., Ellison, P. T., Lieberman, D. E., Demerath, E., & Towne, B. (2005). The development of sex differences in digital formula from infancy in the Fels Longitudinal Study. *Proceedings of the Royal Society B: Biological Sciences 272*(1571), 1473–1479.

McKenry, P. C., Julian, T. W., & Gavazzi, S. M. (1995). Toward a biopsychosocial model of domestic violence. *Journal of Marriage and the Family 57*, 307–320.

McNamara, P. (1996). REM sleep: A social bonding mechanism. *New Ideas in Psychology 14*, 35–46.

McPhedran, S. (2009). A review of the evidence for associations between empathy, violence, and animal cruelty. *Aggression and Violent Behavior 14*, 1–4.
Međedović, J., Wertag, A., & Sokić, K. (2018). Can psychopathic traits be adaptive? Sex differences in relations between psychopathy and emotional distress. *Psychological Topics 27*, 481–497.
Meggs, J., Chen, M., & Mounfield, D. (2019). The organizational effect of prenatal testosterone upon gender role identity and mental toughness in female athletes. *Women in Sport and Physical Activity Journal 27*, 37–44.
Mehta, P. H., & Beer, J. (2010). Neural mechanisms of the testosterone–aggression relation: The role of orbitofrontal cortex. *Journal of Cognitive Neuroscience 22*, 2357–2368.
Mehta, P. H., & Josephs, R. A. (2010). Testosterone and cortisol jointly regulate dominance: Evidence for a dual-hormone hypothesis. *Hormones and Behavior 58*, 898–906.
Mehta, P. H., Jones, A. C., & Josephs, R. A. (2008). The social endocrinology of dominance: Basal testosterone predicts cortisol changes and behavior following victory and defeat. *Journal of Personality and Social Psychology 94*(6), 1078–1092.
Mehta, P. H., Welker, K. M., Zilioli, S., & Carré, J. M. (2015). Testosterone and cortisol jointly modulate risk-taking. *Psychoneuroendocrinology 56*, 88–99.
Mehu, M., & Dunbar, R. I. (2008). Naturalistic observations of smiling and laughter in human group interactions. *Behaviour 145*, 1747–1780.
Mehu, M., Little, A. C., & Dunbar, R. I. (2008). Sex differences in the effect of smiling on social judgments: An evolutionary approach. *Journal of Social, Evolutionary, and Cultural Psychology 2*, 103–114.
Meijer, W. M., van IJzendoorn, M. H., & Bakermans-Kranenburg, M. J. (2019). Challenging the challenge hypothesis on testosterone in fathers: Limited meta-analytic support. *Psychoneuroendocrinology 110*, 104435.
Meyer, H. E., & Selmer, R. (1999). Income, educational level and body height. *Annals of Human Biology 26*, 219–227.
Miller, C. W., & Svensson, E. I. (2014). Sexual selection in complex environments. *Annual Review in Entomology 59*, 427–445.
Miller, E. M. (1994). Paternal provisioning versus mate seeking in human populations. *Personality and Individual Differences 17*, 227–255.
Miller, E. M. (2000). Homosexuality, birth order, and evolution: Toward an equilibrium reproductive economics of homosexuality. *Archives of Sexual Behavior 29*, 1–34.
Miller, G. (2000). *The mating mind*. New York: Anchor Books.
Miller, G. F. (1998). How mate choice shaped human nature: A review of sexual selection and human evolution. In C. Crawford & D. L. Krebs (Eds.), *Handbook of evolutionary psychology: Ideas, issues, and applications* (pp. 87–129). LEA.
Miller, G. F. (2001). Aesthetic fitness: How sexual selection shaped artistic virtuosity as a fitness indicator and aesthetic preferences as mate choice criteria. *Bulletin of Psychology and the Arts 2*, 20–25.
Minnis, A. M., Moore, J., Doherty, I., Rodas, C., Auerswald, C., Shiboski, S., & Padian, N. S. (2008). Gang exposure and pregnancy incidence among female

adolescents in San Francisco: Evidence for the need to integrate reproductive health with violence prevention efforts. *American Journal of Epidemiology 167*, 1102–1109.
Mischel, W. (2015). A social learning view of sex differences in behavior. In V. Burr (Ed.), *Gender and psychology* (pp. 108–129). New York: Taylor & Francis.
Mishra, S., Barclay, P., & Sparks, A. (2017). The relative state model: Integrating need-based and ability-based pathways to risk-taking. *Personality and Social Psychology Review 21*, 176–198.
Misiak, B., Frydecka, D., Loska, O., Moustafa, A. A., Samochowiec, J., Kasznia, J., & Stańczykiewicz, B. (2018). Testosterone, DHEA and DHEA-S in patients with schizophrenia: A systematic review and meta-analysis. *Psychoneuroendocrinology 89*, 92–102.
Mithal, A., Bonjour, J.-P., Boonen, S., Burckhardt, P., Degens, H., El Hajj Fuleihan, G., ... Rizzoli, R. (2013). Impact of nutrition on muscle mass, strength, and performance in older adults. *Osteoporosis International 24*, 1555–1566.
Moffat, S. D., & Hampson, E. (1996). A curvilinear relationship between testosterone and spatial cognition in humans: Possible influence of hand preference. *Psychoneuroendocrinology 21*(3), 323–337.
Moffit, D. M., & Swanik, C. B. (2011). The association between athleticism, prenatal testosterone, and finger length. *Journal of Strength & Conditioning Research 25*, 1085–1088.
Moisey, R., Swinburne, J., & Orme, S. (2008). Serum testosterone and bioavailable testosterone correlate with age and body size in hypogonadal men treated with testosterone undecanoate (1000 mg IM–Nebido®). *Clinical Endocrinology 69*, 642–647.
Mondragón-Ceballos, R., García Granados, M. D., Cerda-Molina, A. L., Chavira-Ramírez, R., & Hernández-López, L. E. (2015). Waist-to-hip ratio, but not body mass index, is associated with testosterone and estradiol concentrations in young women. *International Journal of Endocrinology 2015*, 654046. doi:10.1155/2015/654046
Montag, C., Bleek, B., Breuer, S., Prüss, H., Richardt, K., Cook, S., ... Reuter, M. (2015). Prenatal testosterone and stuttering. *Early Human Development 91*, 43–46.
Monteleone, P., Luisi, M., Colurcio, B., Casarosa, E., Monteleone, P., Ioime, R., ... Maj, M. (2001). Plasma levels of neuroactive steroids are increased in untreated women with anorexia nervosa or bulimia nervosa. *Psychosomatic Medicine 63*, 62–68.
Moore, A. J., & Moore, P. J. (1988). Female strategy during mate choice: Threshold assessment. *Evolution 42*, 387–391.
Moore, L., Kyaw, M., Vercammen, A., Lenroot, R., Kulkarni, J., Curtis, J., ... Weickert, T. W. (2013). Serum testosterone levels are related to cognitive function in men with schizophrenia. *Psychoneuroendocrinology 38*, 1717–1728.
Moraes, Y. L., Valentova, J. V., & Varella, M. A. C. (2022). The evolution of playfulness, play and play-like phenomena in relation to sexual selection. *Frontiers in Psychology, 13*, doi: 10.3389/fpsyg.2022.925842.

Morriss-Kay, G. M. (2010). The evolution of human artistic creativity. *Journal of Anatomy 216*, 158–176.

Mossey, J. M., & Shapiro, E. (1982). Self-rated health: A predictor of mortality among the elderly. *American Journal of Public Health* 72, 800–808.

Mouridsen, S. E., & Hauschild, K.-M. (2010). The sex ratio of siblings of individuals with a history of developmental language disorder. *Logopedics Phoniatrics Vocology 35*, 144–148.

Mueller, A. E., & Thalmann, U. R. S. (2000). Origin and evolution of primate social organization: A reconstruction. *Biological Reviews 75*, 405–435.

Mulder, M. B. (2017). Reproductive decisions. In E. A. Smith (Ed.), *Evolutionary ecology and human behavior* (pp. 339–374). London: Routledge.

Muller, M. N. (2017). Testosterone and reproductive effort in male primates. *Hormones and Behavior 91*, 36–51.

Müller, M., Brand, M., Mies, J., Lachmann, B., Sariyska, R. Y., & Montag, C. (2017). The 2D: 4D marker and different forms of Internet use disorder. *Frontiers in Psychiatry 8*, 213–225.

Muraleedharan, V., & Hugh Jones, T. (2014). Testosterone and mortality. *Clinical Endocrinology 81*, 477–487.

Murphy, N. (2018). *Theology in the age of scientific reasoning*. Ithica, NY: Cornell University Press.

Myer, G. D., Ford, K. R., Paterno, M. V., Nick, T. G., & Hewett, T. E. (2008). The effects of generalized joint laxity on risk of anterior cruciate ligament injury in young female athletes. *American Journal of Sports Medicine 36*, 1073–1080.

Mysterud, A., Langvatn, R., & Stenseth, N. C. (2004). Patterns of reproductive effort in male ungulates. *Journal of Zoology 264*, 209–215.

Nadler, A., Camerer, C. F., Zava, D. T., Ortiz, T. L., Watson, N. V., Carré, J. M., & Nave, G. (2019). Does testosterone impair men's cognitive empathy? Evidence from two large-scale randomized controlled trials. *Proceedings of the Royal Society B 286* (1910), 20191062.

Nakamura, K., Hoshino, Y., Kodama, K., & Yamamoto, M. (1999). Reliability of self-reported body height and weight of adult Japanese women. *Journal of Biosocial Science 31*, 555–558.

Nave, G., Nadler, A., Dubois, D., Zava, D., Camerer, C., & Plassmann, H. (2018). Single-dose testosterone administration increases men's preference for status goods. *Nature Communications* 9(1), 1–8.

Neave, N., Laing, S., Fink, B., & Manning, J. T. (2003). Second to fourth digit ratio, testosterone and perceived male dominance. *Proceedings of the Royal Society of London B: Biological Sciences 270*(1529), 2167–2172.

Neff, K. D., & Terry-Schmitt, L. N. (2002). Youths' attributions for power-related gender differences: Nature, nurture, or God? *Cognitive Development 17*, 1185–1202.

Nelsen, H. M., & Potvin, R. H. (1981). Gender and regional differences in the religiosity of protestant adolescents. *Review of Religious Research* 22, 268–285.

Nepomuceno, M. V., de Aguiar Pastore, C. M., & Stenstrom, E. (2021). Prenatal hormones (2D:4D), intrasexual competition, and materialism in women. *Psychology & Marketing 38*, 239–248.

Nesse, R. (1999). Proximate and evolutionary studies of anxiety, stress and depression: Synergy at the interface. *Neuroscience & Biobehavioral Reviews 23*, 895–903.

Nettle, D. (2001). *Strong imagination: Madness, creativity and human nature.* Oxford, England: Oxford University Press.

Nettle, D. (2004). Evolutionary origins of depression: A review and reformulation. *Journal of Affective Disorders 81*, 91–102.

Nettle, D. (2007). Empathizing and systemizing: What are they, and what do they contribute to our understanding of psychological sex differences? *British Journal of Psychology 98*, 237–255.

Nettle, D., & Clegg, H. (2006). Schizotypy, creativity and mating success in humans. *Proceedings of the Royal Society B: Biological Sciences 273*(1586), 611–615.

Neyse, L., Bosworth, S., Ring, P., & Schmidt, U. (2016). Overconfidence, incentives and digit ratio. *Scientific Reports 6*(1), 1–8.

Nguyen, T. V. (2018). Developmental effects of androgens in the human brain. *Journal of Neuroendocrinology 30*(2), e12486.

Nicolson, N. A. (1991). Maternal behavior in human and nonhuman primates. In J. D. Loy & C. B. Peters (Eds.), *Understanding behavior: What primate studies tell us about human behavior* (pp. 17–50). Oxford, England: Oxford University Press.

Nielsen, T. (2012). Variations in dream recall frequency and dream theme diversity by age and sex. *Frontiers in Neurology 3*, 106.

Niemann, C., Wegner, M., Voelcker-Rehage, C., Holzweg, M., Arafat, A. M., & Budde, H. (2013). Influence of acute and chronic physical activity on cognitive performance and saliva testosterone in preadolescent school children. *Mental Health and Physical Activity 6*, 197–204.

Nieschlag, E., Behre, H. M., Bouchard, P., Corrales, J. J., Jones, T. H., Stalla, G. K., … Wu, F. C. (2004). Testosterone replacement therapy: Current trends and future directions. *Human Reproduction Update 10*, 409–419.

Nitschke, J. P., & Bartz, J. A. (2020). Lower digit ratio and higher endogenous testosterone are associated with lower empathic accuracy. *Hormones and Behavior 119*, 104648.

Nolen-Hoeksema, S., & Hilt, L. (2006). Possible contributors to the gender differences in alcohol use and problems. *Journal of General Psychology 133*, 357–374.

Nowak, M. A., & Krakauer, D. C. (1999). The evolution of language. *Proceedings of the National Academy of Sciences 96*, 8028–8033.

O'Briain, D. E., Dawson, P. H., Kelly, J. C., & Connolly, P. (2017). Assessment of the 2D: 4D ratio in aggression-related injuries in children attending a paediatric emergency department. *Irish Journal of Medical Science 186*(2), 441–445.

O'Connor, D. B., Archer, J., Hair, W. M., & Wu, F. C. (2002). Exogenous testosterone, aggression, and mood in eugonadal and hypogonadal men. *Physiology & Behavior 75*, 557–566.

O'Connor, S. M., Culbert, K. M., Mayhall, L. A., Burt, S. A., & Klump, K. L. (2020). Differences in genetic and environmental influences on body weight and shape concerns across pubertal development in females. *Journal of Psychiatric Research 121*, 39–46.

Oinonen, K. A., & Bird, J. L. (2012). Age at menarche and digit ratio (2D: 4D): Relationships with body dissatisfaction, drive for thinness, and bulimia symptoms in women. *Body Image 9*, 302–306.

Okami, P., & Shackelford, T. K. (2001). Human sex differences in sexual psychology and behavior. *Annual Review of Sex Research 12*, 186–241.

O'Keefe, J. H., Vogel, R., Lavie, C. J., & Cordain, L. (2011). Exercise like a hunter-gatherer: A prescription for organic physical fitness. *Progress in Cardiovascular Diseases 53*, 471–479.

Olsson, H. (1984). *Epidemiological Studies in Malignant Lymphoma with Special Reference to Occupational Exposure to Organic Solvents and to Reproductive Factors*. (PhD). Lund, Sweden: Lund University.

Olvera-Hernández, S., Tapia-Rodríguez, M., Swaab, D. F., & Fernández-Guasti, A. (2017). Prenatal administration of letrozole reduces SDN and SCN volume and cell number independent of partner preference in the male rat. *Physiology & Behavior 171*, 61–68.

O'Neal, D. M., Reichard, D. G., Pavilis, K., & Ketterson, E. D. (2008). Experimentally-elevated testosterone, female parental care, and reproductive success in a songbird, the Dark-eyed Junco (Junco hyemalis). *Hormones and Behavior 54*, 571–578.

Orians, G. (1980). Habitat selection: General theory and applications to human behavior. In J. S. Lockard (Ed.), *The evolution of human social behavior* (pp. 49–66). Chicago, IL: Elsevier.

Osofsky, J. D., & O'Connell, E. J. (1977). Patterning of newborn behavior in an urban population. *Child Development 48*, 532–536.

Ostatníková, D., Celec, P., Putz, Z., Hodosy, J., Schmidt, F., Laznibatová, J., & Kúdela, M. (2007). Intelligence and salivary testosterone levels in prepubertal children. *Neuropsychologia 45*, 1378–1385.

Ostatníková, D., Laznibatová, J., & Dohnányiová, M. (1996). Testosterone influence on spatial ability in prepubertal children. *Studia Psychologica 38*, 237–244.

Ou, J., Wu, Y., Hu, Y., Gao, X., Li, H., & Tobler, P. N. (2021). Testosterone reduces generosity through cortical and subcortical mechanisms. *Proceedings of the National Academy of Sciences 118*, e2021745118.

Owens, I. P., & Short, R. V. (1995). Hormonal basis of sexual dimorphism in birds: Implications for new theories of sexual selection. *Trends in Ecology & Evolution 10*, 44–47.

Owen-Smith, N. (1977). On territoriality in ungulates and an evolutionary model. *Quarterly Review of Biology 52*, 1–38.

Palanza, P. (2001). Animal models of anxiety and depression: How are females different? *Neuroscience & Biobehavioral Reviews 25*, 219–233.

Palermo, G. B. (2010). Biological and environmental correlates of aggressive behavior. *Journal of Forensic Psychology Practice 10*, 300–324.

Palmen, D. G. C., Kolthoff, E. W., & Derksen, J. J. L. (2021). The need for domination in psychopathic leadership: A clarification for the estimated high prevalence of psychopathic leaders. *Aggression and Volent Behavior 61*, 101650.

Palmer, C. T., & Tilley, C. F. (1995). Sexual access to females as a motivation for joining gangs: An evolutionary approach. *Journal of Sex Research 32*, 213–217.

Palmer-Hague, J. L., Wong, S. T. S., Wassersug, R. J., Kingstone, A., & Wibowo, E. (2021). Hormones and visual attention to sexual stimuli in older men: An exploratory investigation. *Aging Male 24*, 106–118.

Papalia, N., Mann, E., & Ogloff, J. R. (2021). Child sexual abuse and risk of revictimization: Impact of child demographics, sexual abuse characteristics, and psychiatric disorders. *Child maltreatment 26*, 74–86.

Paquette, D. (2015). An evolutionary perspective on antisocial behavior: Evolution as a foundation for criminological theories. In J. Morizot & L. Kazemian (Eds.), *The development of criminal and antisocial behavior* (pp. 315–330). New York: Springer.

Parkman, A. M. (2004). Bargaining over housework: The frustrating situation of secondary wage earners. *American Journal of Economics and Sociology 63*, 765–794.

Parolini, M., Romano, A., Possenti, C. D., Caprioli, M., Rubolini, D., & Saino, N. (2017). Contrasting effects of increased yolk testosterone content on development and oxidative status in gull embryos. *Journal of Experimental Biology 220*, 625–633.

Pascual-Sagastizabal, E., Azurmendi, A., Sánchez-Martín, J. R., Braza, F., Carreras, M. R., Muñoz, J. M., & Braza, P. (2013). Empathy, estradiol and androgen levels in 9-year-old children. *Personality and Individual Differences 54*, 936–940.

Pascual-Sagastizabal, E., Del Puerto, N., Cardas, J., Sanchez-Martin, J. R., Vergara, A. I., & Azurmendi, A. (2019). Testosterone and cortisol modulate the effects of empathy on aggression in children. *Psychoneuroendocrinology 103*, 118–124.

Passe, T. J., Rajagopalan, P., Tupler, L. A., Byrum, C. E., Macfall, J. R., & Krishnan, K. R. R. (1997). Age and sex effects on brain morphology. *Progress in Neuro-Psychopharmacology and Biological Psychiatry 21*, 1231–1237.

Pasterski, V., Geffner, M. E., Brain, C., Hindmarsh, P., Brook, C., & Hines, M. (2011). Prenatal hormones and childhood sex segregation: Playmate and play style preferences in girls with congenital adrenal hyperplasia. *Hormones and Behavior 59*, 549–555.

Paternoster, R., & Iovanni, L. (1989). The labeling perspective and delinquency: An elaboration of the theory and an assessment of the evidence. *Justice Quarterly 6*, 359–394.

Paus, T., Nawaz-Khan, I., Leonard, G., Perron, M., Pike, G. B., Pitiot, A., … Pausova, Z. (2010). Sexual dimorphism in the adolescent brain: Role of testosterone and androgen receptor in global and local volumes of grey and white matter. *Hormones and Behavior 57*, 63–75.

Pawlowski, B., Atwal, R., & Dunbar, R. I. (2008). Sex differences in everyday risk-taking behavior in humans. *Evolutionary Psychology 6*(1), 147470490800600104.

Pedersen, C. A. (2004). Biological aspects of social bonding and the roots of human violence. *Annals of the New York Academy of Sciences 1036*, 106–127.

Pedersen, W. C., Miller, L. C., Putcha-Bhagavatula, A. D., & Yang, Y. (2002). Evolved sex differences in the number of partners desired? The long and the short of it. *Psychological Science 13*, 157–161.

Pellegrini, A. D., & Bartini, M. (2001). Dominance in early adolescent boys: Affiliative and aggressive dimensions and possible functions. *Merrill-Palmer Quarterly 47*, 142–163.

Peneder, M. (2022). Digitization and the evolution of money as a social technology of account. *Journal of Evolutionary Economics 32*, 175–203.

Penner, A. M., & Cadwallader Olsker, T. (2012). Gender differences in mathematics and science achievement across the distribution: What international variation can tell us about the role of biology and society. In H. Forgasz & F. Rivera (Eds.), *Towards equity in mathematics education* (pp. 441–468). New York: Springer.

Pennington, B. F., Filipek, P. A., Lefly, D., Chhabildas, N., Kennedy, D. N., Simon, J. H., ... DeFries, J. C. (2000). A twin MRI study of size variations in the human brain. *Journal of Cognitive Neuroscience 12*, 223–232.

Peper, J. S., Brouwer, R. M., van Baal, G. C. M., Schnack, H. G., van Leeuwen, M., Boomsma, D. I., ... Pol, H. E. H. (2009). Does having a twin brother make for a bigger brain? *European Journal of Endocrinology 160*, 739–746.

Peper, J. S., Koolschijn, P. C. M., & Crone, E. A. (2013). Development of risk taking: Contributions from adolescent testosterone and the orbito-frontal cortex. *Journal of Cognitive Neuroscience 25*, 2141–2150.

Perez-Rodriguez, M. M., Lopez-Castroman, J., Martinez-Vigo, M., Diaz-Sastre, C., Ceverino, A., Nunez-Beltran, A., ... Baca-Garcia, E. (2011). Lack of association between testosterone and suicide attempts. *Neuropsychobiology 63*, 125–130.

Perini, T., Ditzen, B., Fischbacher, S., & Ehlert, U. (2012). Testosterone and relationship quality across the transition to fatherhood. *Biological Psychology 90*, 186–191.

Perra, M. S., & Ruspini, E. (2013). Men who work in 'non-traditional' occupations. *International Review of Sociology 23*, 265–270.

Perrin, J. S., Hervé, P.-Y., Leonard, G., Perron, M., Pike, G. B., Pitiot, A., ... Paus, T. (2008). Growth of white matter in the adolescent brain: Role of testosterone and androgen receptor. *Journal of Neuroscience 28*, 9519–9524.

Perry, D. G., & Bussey, K. (1979). The social learning theory of sex differences: Imitation is alive and well. *Journal of Personality and Social Psychology 37*(10), 1699–1712.

Petersen, E. (2016). Working with religious leaders and faith communities to advance culturally informed strategies to address violence against women. *Agenda 30*, 50–59.

Petersen, J. L., & Hyde, J. S. (2011). Gender differences in sexual attitudes and behaviors: A review of meta-analytic results and large datasets. *Journal of Sex Research 48*(2-3), 149–165.

Petrie, T. A., Greenleaf, C., & Martin, S. (2010). Biopsychosocial and physical correlates of middle school boys' and girls' body satisfaction. *Sex Roles 63*, 631–644.

Pickering, G. J., Anger, N., Baird, J., Dale, G., & Tattersall, G. J. (2022). Use of crowdsourced images for determining 2D:4D and relationship to pro-environmental variables. *Acta Ethologica 25*, 165–178.

Pietraszkiewicz, A., Kaufmann, M., & Formanowicz, M. M. (2017). Masculinity ideology and subjective well-being in a sample of Polish men and women. *Polish Psychological Bulletin 48*, 79–86.

Pillay, M. N. (2013). The Anglican Church and Feminism: Challenging the patriarchy of our faith. *Journal of Gender and Religion in Africa 19*, 53–71.

Pinker, S. (2002). *The blank slate: The modern denial of human nature.* New York: Penguin (Viking).

Pinker, S., & Bloom, P. (1990). Natural language and natural selection. *Behavioral and Brain Sciences 13*, 707–727.

Pintzka, C. W., Evensmoen, H. R., Lehn, H., & Håberg, A. K. (2016). Changes in spatial cognition and brain activity after a single dose of testosterone in healthy women. *Behavioural Brain Research 298*, 78–90.

Pittam, J. (1990). The relationship between perceived persuasiveness of nasality and source characteristics for Australian and American listeners. *Journal of Social Psychology 130*, 81–87.

Platek, S. M., & Singh, D. (2010). Optimal waist-to-hip ratios in women activate neural reward centers in men. *PLoS One 5*(2), e9042.

Plavcan, J. M. (2001). Sexual dimorphism in primate evolution. *American Journal of Physical Anthropology 116*, 25–53.

Pollet, T. V., Cobey, K. D., & van der Meij, L. (2013). Testosterone levels are negatively associated with childlessness in males, but positively related to offspring count in fathers. *PLoS One 8*(4), e60018.

Pope, H. G., Kouri, E. M., & Hudson, J. I. (2000). Effects of supraphysiologic doses of testosterone on mood and aggression in normal men: A randomized controlled trial. *Archives of General Psychiatry 57*, 133–140.

Portnoy, J., Raine, A., Glenn, A. L., Chen, F. R., Choy, O., & Granger, D. A. (2015). Digit ratio (2D: 4D) moderates the relationship between cortisol reactivity and self-reported externalizing behavior in young adolescent males. *Biological Psychology 112*, 94–106.

Poston Jr, D. L., & Baumle, A. K. (2010). Patterns of asexuality in the United States. *Demographic Research 23*, 509–530.

Powers, R. A. (2015). Consequences of using self-protective behaviors in nonsexual assaults: The differential risk of completion and injury by victim sex. *Violence and Victims 30*, 846–869.

Preston, S. D., & de Waal, F. B. M. (2002). Empathy: Its ultimate and proximate bases. *Behavioral and Brain Sciences 25*, 1–72.

Pruetz, J. D., Ontl, K. B., Cleaveland, E., Lindshield, S., Marshack, J., & Wessling, E. G. (2017). Intragroup lethal aggression in West African chimpanzees (Pan troglodytes verus): Inferred killing of a former alpha male at Fongoli, Senegal. *International Journal of Primatology 38*, 31–57.

Purifoy, F. E., & Koopmans, L. H. (1979). Androstenedione, testosterone, and free testosterone concentration in women of various occupations. *Social Biology 26*, 179–188.

Purifoy, F. E., Koopmans, L. H., & Tatum, R. W. (1980). Steroid hormones and aging: Free testosterone, testosterone and androstenedione in normal females aged 20-87 years. *Human Biology 52*, 181–191.

Puts, D. A. (2010). Beauty and the beast: Mechanisms of sexual selection in humans. *Evolution and Human Behavior 31*, 157–175.

Puts, D. A. (2016). Human sexual selection. *Current opinion in psychology 7*, 28–32.

Puts, D. A., Bailey, D. H., & Reno, P. L. (2015). Contest competition in men. In *The handbook of evolutionary psychology* (pp. 1–18). New York: Wiley.

Puts, D. A., Hill, A. K., Bailey, D. H., Walker, R. S., Rendall, D., Wheatley, J. R., & Ramos-Fernandez, G. (2016). Sexual selection on male vocal fundamental frequency in humans and other anthropoids. *Proceedings of the Biological Sciences, 283*, doi: 10.1098/rspb.2015.2830.

Puts, D. A., Jones, B. C., & DeBruine, L. M. (2012). Sexual selection on human faces and voices. *Journal of Sex Research 49*, 227–243.

Quinsey, V. L. (2002). Evolutionary theory and criminal behaviour. *Legal and Criminological Psychology 7*, 1–13.

Quinton, S. J., Smith, A. R., & Joiner, T. (2011). The 2nd to 4th digit ratio (2D: 4D) and eating disorder diagnosis in women. *Personality and Individual Differences 51*, 402–405.

Rada, R. T., Laws, D. R., & Kellner, R. (1976). Plasma testosterone levels in the rapist. *Psychosomatic Medicine 38*, 257–268.

Rae, C., Digney, A. L., McEwan, S. R., & Bates, T. C. (2003). Oral creatine monohydrate supplementation improves brain performance: A double–blind, placebo–controlled, cross–over trial. *Proceedings of the Royal Society of London. Series B: Biological Sciences 270*(1529), 2147–2150.

Raisanen, J. C., Chadwick, S. B., Michalak, N., & van Anders, S. M. (2018). Average associations between sexual desire, testosterone, and stress in women and men over time. *Archives of Sexual Behavior 47*, 1613–1631.

Raj, A., Gupta, B., Chowdhury, A., & Chadha, S. (2010). A study of voice changes in various phases of menstrual cycle and in postmenopausal women. *Journal of Voice 24*, 363–368.

Rajender, S., Pandu, G., Sharma, J., Gandhi, K., Singh, L., & Thangaraj, K. (2008). Reduced CAG repeats length in androgen receptor gene is associated with violent criminal behavior. *International Journal of Legal Medicine 122*, 367–372.

Randall, T. M., & Desrosiers, M. (1980). Measurement of supernatural belief: Sex differences and locus of control. *Journal of Personality Assessment 44*, 493–498.

Raney, A. A. (2009). The effects of viewing televised sports. In R. L. Nabi & M. B. Oliver (Eds.), *The Sage handbook of media processes and effects* (pp. 439–453). Thousand Oaks, CA: Sage.

Ranson, R., Stratton, G., & Taylor, S. R. (2015). Digit ratio (2D: 4D) and physical fitness (Eurofit test battery) in school children. *Early Human Development 91*, 327–331.

Rashad, I. (2008). Height, health, and income in the US, 1984–2005. *Economics & Human Biology 6*, 108–126.

Rasika, S., Nottebohm, F., & Alvarez-Buylla, A. (1994). Testosterone increases the recruitment and/or survival of new high vocal center neurons in adult female canaries. *Proceedings of the National Academy of Sciences 91*, 7854–7858.

Re, D. E., O'Connor, J. J., Bennett, P. J., & Feinberg, D. R. (2012). Preferences for very low and very high voice pitch in humans. *PLoS One* 7(3), e32719.

Reed, S., & Meggs, J. (2017). Examining the effect of prenatal testosterone and aggression on sporting choice and sporting longevity. *Personality and Individual Differences 116*, 11–15.

Regnerus, M. (2017). *Cheap sex: The transformation of men, marriage, and monogamy*. London: Oxford University Press.

Reichborn-Kjennerud, T., Bulik, C. M., Tambs, K., & Harris, J. R. (2004). Genetic and environmental influences on binge eating in the absence of compensatory behaviors: A population-based twin study. *International Journal of Eating Disorders 36*, 307–314.

Reijnen, A., Rademaker, A. R., Vermetten, E., & Geuze, E. (2015). Prevalence of mental health symptoms in Dutch military personnel returning from deployment to Afghanistan: A 2-year longitudinal analysis. *European Psychiatry*, *30*, 341–346.

Reilly, R. R. (2014). *The closing of the Muslim mind: How intellectual suicide created the modern Islamist crisis*. Open Road Media.

Reinhardt, K. (2007). Evolutionary consequences of sperm cell aging. *Quarterly Review of Biology 82*, 375–393.

Reinhardt, K., & Turnell, B. (2019). Sperm ageing: A complex business. *Functional Ecology 33*, 1188–1189.

Rennison, C. M. (2001). *Intimate partner violence and age of victim, 1993-99 (BJS Publication No. 187635)*. Washington, DC: US Department of Justice, Office of Justice Programs, Bureau of Justice Statistics.

Reno, P. L., Meindl, R. S., McCollum, M. A., & Lovejoy, C. O. (2003). Sexual dimorphism in Australopithecus afarensis was similar to that of modern humans. *Proceedings of the National Academy of Sciences 100*, 9404–9409.

Reynolds, A. J., Temple, J. A., Ou, S. R., Robertson, D., Mersky, J. P., Topitzes, J. W., & Niles, M. D. (2007#). Effects of a school-based, early childhood intervention on adult health and well-being: A 19-year follow-up of low-income families. *Archives of Pediatric and Adolescent Medicine 161*, 730–739.

Rhode, D. L. (1988). Occupational inequality. *Duke Law Journal*, *1988:1207*, 1207–1241.

Rhoden, E. L., & Morgentaler, A. (2004). Risks of testosterone-replacement therapy and recommendations for monitoring. *New England Journal of Medicine 350*, 482–492.

Rice, T. R., & Sher, L. (2017). Adolescent suicide and testosterone. *International Journal of Adolescent Medicine and Health*, *29*(4), 10.1515/ijamh-2015-0058.

Richards, G., & Browne, W. V. (2022). Prenatal testosterone and sexually differentiated childhood play preferences: A meta-analysis of amniotic fluid studies. *Current Psychology*, 1–14.

Richards, G., Davies, W., Stewart-Williams, S., Bellin, W., & Reed, P. (2018). 2D: 4D digit ratio and religiosity in university student and general population samples. *Transpersonal Psychology Review 20*, 23–36.

Rieger, G., & Savin-Williams, R. C. (2012). The eyes have it: Sex and sexual orientation differences in pupil dilation patterns. *PLoS One* 7(8), doi:10.1371/journal.pone.0040256.

Roberts, B. A., & Martel, M. M. (2013). Prenatal testosterone and preschool disruptive behavior disorders. *Personality and Individual Differences 55*, 962–966.

Roellke, E., Raiss, M., King, S., Lytel-Sternberg, J., & Zeifman, D. M. (2019). Infant crying levels elicit divergent testosterone response in men. *Parenting 19*, 39–55.

Rogers, D. (1969). *Child psychology*. Belmont, CA: Brooks/Cole.

Rogol, A. D., Tkachenko, N., & Bryson, N. (2016). Natesto, a novel testosterone nasal gel, normalizes androgen levels in hypogonadal men. *Andrology 4*, 46–54.

Rohr, U. D. (2002). The impact of testosterone imbalance on depression and women's health. *Maturitas 41*, 25–46.

Roitman, S., Green, T., Osher, Y., Karni, N., & Levine, J. (2007). Creatine monohydrate in resistant depression: A preliminary study. *Bipolar Disorders 9*, 754–758.

Ronay, R., & von Hippel, W. (2010a). The presence of an attractive woman elevates testosterone and physical risk taking in young men. *Social Psychological and Personality Science 1*, 57–64.

Ronay, R., & Von Hippel, W. (2010b). Power, testosterone, and risk-taking. *Journal of Behavioral Decision Making 23*, 473–482.

Roney, J. R., & Gettler, L. T. (2015). The role of testosterone in human romantic relationships. *Current Opinion in Psychology 1*, 81–86.

Roney, J. R., Simmons, Z. L., & Gray, P. B. (2011). Changes in estradiol predict within-women shifts in attraction to facial cues of men's testosterone. *Psychoneuroendocrinology 36*, 742–749.

Roof, R. L., & Havens, M. D. (1992). Testosterone improves maze performance and induces development of a male hippocampus in females. *Brain Research 572*(1-2), 310–313.

Roof, R. L., Zhang, Q., Glaiser, M. M., & Stein, D. G. (1993). Gender-specific impairment on Morris water maze tasks after entorhinal cortex lesion. *Behavioural Brain Research 57*, 47–51.

Rosenfield, K. A., Sorokowska, A., Sorokowski, P., & Puts, D. A. (2020). Sexual selection for low male voice pitch among Amazonian forager-horticulturists. *Evolution and Human Behavior 41*, 3–11.

Ross, C. T., & Richerson, P. J. (2014). New frontiers in the study of human cultural and genetic evolution. *Current Opinion in Genetics & Development 29*, 103–109.

Roughgarden, J. (2017). Homosexuality and evolution: A critical appraisal. In M. Tibayrenc & F. J. Ayala (Eds.), *On human nature* (pp. 495–516). Amsterdam: Elsevier.

Rowe, D. C. (1996). An adaptive strategy theory of crime and delinquency. In J. D. Hawkins (Ed.), *Delinquency and crime: Current theories* (pp. 268–314). Cambridge: Cambridge University Press.

Rowe, D. C., Vazsonyi, A. T., & Figueredo, A. J. (1997). Mating-effort in adolescence: A conditional or alternative strategy. *Personality and Individual Differences 23*, 105–115.

Rowe, L., & Rundle, H. D. (2021). The alignment of natural and sexual selection. *Annual Review of Ecology, Evolution, and Systematics 52*, 499–517.

Rowe, R., Maughan, B., Worthman, C. M., Costello, E. J., & Angold, A. (2004). Testosterone, antisocial behavior, and social dominance in boys: Pubertal development and biosocial interaction. *Biological Psychiatry 55*(5), 546–552.

Rupp, H. A., & Wallen, K. (2007). Relationship between testosterone and interest in sexual stimuli: The effect of experience. *Hormones and Behavior 52*(5), 581–589.

Rusch, H., Leunissen, J. M., & Van Vugt, M. (2015). Historical and experimental evidence of sexual selection for war heroism. *Evolution and Human Behavior 36*, 367–373.

Russock, H. (2011). An evolutionary interpretation of the effect of gender and sexual orientation on human mate selection preferences, as indicated by an analysis of personal advertisements. *Behaviour 148*, 307–323.

Ryan, M. J. (2021). Darwin, sexual selection, and the brain. *Proceedings of the National Academy of Sciences 118*(8), e2008194118.

Ryzhavskii, B. Y. (2002). Effect of injection of testosterone derivatives to pregnant rats on the brain of their one-day offspring. *Bulletin of Experimental Biology and Medicine 134*(5), 509–511.

Saad, G. (2007). *The evolutionary bases of consumption*. Mahwah, NJ: Lawrence Erlbaum.

Saad, G., & Gill, T. (2000). Applications of evolutionary psychology in marketing. *Psychology & Marketing 17*(12), 1005–1034.

Sadr, M., Khorashad, B. S., Talaei, A., Fazeli, N., & Hönekopp, J. (2020). 2D: 4D suggests a role of prenatal testosterone in gender dysphoria. *Archives of Sexual Behavior 49*, 421–432.

Saenz, J., & Alexander, G. M. (2013). Postnatal testosterone levels and disorder relevant behavior in the second year of life. *Biological Psychology 94*, 152–159.

Sagi, A., & Hoffman, M. L. (1976). Empathic distress in the newborn. *Developmental Psychology 12*, 175–182.

Salmon, C. (2012). The pop culture of sex: An evolutionary window on the worlds of pornography and romance. *Review of General Psychology 16*, 152–160.

Saltzman, W., & Maestripieri, D. (2011). The neuroendocrinology of primate maternal behavior. *Progress in Neuro-Psychopharmacology and Biological Psychiatry 35*, 1192–1204.

Sánchez-Martın, J., Fano, E., Ahedo, L., Cardas, J., Brain, P. F., & Azpíroz, A. (2000). Relating testosterone levels and free play social behavior in male and female preschool children. *Psychoneuroendocrinology 25*, 773–783.

Sanders, G. (2013). Sex differences in motor and cognitive abilities predicted from human evolutionary history with some implications for models of the visual system. *Journal of Sex Research 50*(3-4), 353–366.

Sapienza, P., Zingales, L., & Maestripieri, D. (2009). Gender differences in financial risk aversion and career choices are affected by testosterone. *Proceedings of the National Academy of Sciences 106*(36), 15268–15273.

Savic, I., Frisen, L., Manzouri, A., Nordenstrom, A., & Lindén Hirschberg, A. (2017). Role of testosterone and Y chromosome genes for the masculinization of the human brain. *Human Brain Mapping 38*, 1801–1814.

Scalise Sugiyama, M., Mendoza, M., White, F., & Sugiyama, L. (2018). Coalitional play fighting and the evolution of coalitional intergroup aggression. *Human Nature 29*, 219–244.

Schaadt, G., Hesse, V., & Friederici, A. D. (2015). Sex hormones in early infancy seem to predict aspects of later language development. *Brain and Language 141*, 70–76.

Scheib, J. E. (2001). Context-specific mate choice criteria: Women's trade-offs in the contexts of long-term and extra-pair mateships. *Personal Relationships 8*, 371–389.

Schmitt, D. P. (2003). Universal sex differences in the desire for sexual variety: Tests from 52 nations, 6 continents, and 13 islands. *Journal of Personality and Social Psychology 85*, 85–104.

Schmitt, D. P. (2015). The evolution of culturally-variable sex differences: Men and women are not always different, but when they are ... it appears not to result from patriarchy or sex role socialization. In T. K. Shackelford & R. D. Hanson (Eds.), *The evolution of sexuality* (pp. 221–256). New York: Springer.

Schmitt, D. P., & Buss, D. M. (2000). Sexual dimensions of person description: Beyond or subsumed by the big five? *Journal of Research in Personality 34*, 141–177.

Schorer, J., Rienhoff, R., Westphal, H., & Baker, J. (2013). Digit ratio effects between expertise levels in American football players. *Talent Development and Excellence 5*, 113–116.

Schuett, W., Tregenza, T., & Dall, S. R. (2010). Sexual selection and animal personality. *Biological Reviews 85*, 217–246.

Schwartz, S. H., & Bilsky, W. (1987). Toward a universal psychological structure of human values. *Journal of Personality and Social Psychology 53*(3), 550–561.

Sedlak, T., Shufelt, C., Iribarren, C., & Merz, C. N. B. (2012). Sex hormones and the QT interval: A review. *Journal of Women's Health 21*, 933–941.

Seidman, S. N. (2003). Testosterone deficiency and mood in aging men: Pathogenic and therapeutic interactions. *World Journal of Biological Psychiatry 4*, 14–20.

Selcuk, E. B., Erbay, L. G., Özcan, Ö. Ö., Kartalci, Ş., & Batcioğlu, K. (2015). Testosterone levels of children with a diagnosis of developmental stuttering. *Therapeutics and Clinical Risk Management 11*, 793–802.

Self, C. M., & Golledge, R. G. (2018). Sex, gender, and cognitive mapping. In *Cognitive mapping* (pp. 197–220). Routledge.

Sell, A., Hone, L. S., & Pound, N. (2012). The importance of physical strength to human males. *Human Nature 23*, 30–44.

Sell, A., Lukazsweski, A. W., & Townsley, M. (2017). Cues of upper body strength account for most of the variance in men's bodily attractiveness. *Proceedings of the Royal Society B: Biological Sciences 284*(1869), 20171819.

Senberg, A., Münchau, A., Münte, T., Beste, C., & Roessner, V. (2021). Swearing and coprophenomena–A multidimensional approach. *Neuroscience & Biobehavioral Reviews 126*, 12–22.

Seney, M. L., & Sibille, E. (2014). Sex differences in mood disorders: Perspectives from humans and rodent models. *Biology of Sex Differences 5*(1), 1–10.

Shelley-Tremblay, J. F., & Rosen, L. A. (1996). Attention deficit hyperactivity disorder: An evolutionary perspective. *Journal of Genetic Psychology 157*, 443–453.

Sher, L. (2013). Low testosterone levels may be associated with suicidal behavior in older men while high testosterone levels may be related to suicidal behavior in adolescents and young adults: A hypothesis. *International Journal of Adolescent Medicine and Health 25*, 263–268.

Sher, L. (2021). *Ambient temperature, testosterone, and suicide*. San Paulo: SciELO Brasil.

Sherry, D. F., & Hampson, E. (1997). Evolution and the hormonal control of sexually-dimorphic spatial abilities in humans. *Trends in Cognitive Sciences 1*, 50–56.

Shi, L., Lu, T., Li, Y., & Deng, Z.-h. (2021). Sexual assault against women: A retrospective study of 292 cases in Fujian Province, China. *Journal of Forensic Nursing 17*, 210–218.

Shilling, A. A., & Brown, C. M. (2016). Goal-driven resource redistribution: An adaptive response to social exclusion. *Evolutionary Behavioral Sciences 10*(3), 149–161.

Shirley, C., & Wallace, M. (2004). Domestic work, family characteristics, and earnings: Reexamining gender and class differences. *Sociological Quarterly 45*, 663–690.

Shirtcliff, E. A., Dahl, R. E., & Pollak, S. D. (2009). Pubertal development: Correspondence between hormonal and physical development. *Child Development 80*, 327–337.

Shores, M. M., Moceri, V. M., Sloan, K. L., Matsumoto, A. M., & Kivlahan, D. R. (2005). Low testosterone levels predict incident depressive illness in older men: Effects of age and medical morbidity. *Journal of Clinical Psychiatry 66*, 7–14.

Sih, A., & Bell, A. M. (2008). Insights for behavioral ecology from behavioral syndromes. *Advances in the Study of Behavior 38*, 227–281.

Silverman, I., Choi, J., Mackewn, A., Fisher, M., Moro, J., & Olshansky, E. (2000). Evolved mechanisms underlying wayfinding: Further studies on the hunter-gatherer theory of spatial sex differences. *Evolution and Human Behavior 21*, 201–213.

Simmons, Z. L., & Roney, J. R. (2011). Variation in CAG repeat length of the androgen receptor gene predicts variables associated with intrasexual competitiveness in human males. *Hormones and Behavior 60*, 306–312.

Simpson, E. A., Nicolini, Y., Shetler, M., Suomi, S. J., Ferrari, P. F., & Paukner, A. (2016). Experience-independent sex differences in newborn macaques: Females are more social than males. *Scientific Reports 6*(1), 1–7.

Sinervo, B., & Zamudio, K. R. (2001). The evolution of alternative reproductive strategies: Fitness differential, heritability, and genetic correlation between the sexes. *Journal of Heredity 92*, 198–205.

Singh, D. (1994). Ideal female body shape: Role of body weight and waist-to-hip ratio. *International Journal of Eating Disorders 16*, 283–288.

Singh, D. (2006). Universal allure of the hourglass figure: An evolutionary theory of female physical attractiveness. *Clinics in Plastic Surgery 33*, 359–370.

Singh, D., & Luis, S. (1995). Ethnic and gender consensus for the effect of waist-to-hip ratio on judgment of women's attractiveness. *Human Nature 6*, 51–65.

Singh, R. S., & Kulathinal, R. J. (2005). Male sex drive and the masculinization of the genome. *BioEssays 27*, 518–525.

Sinnesael, M., Boonen, S., Claessens, F., Gielen, E., & Vanderschueren, D. (2011). Testosterone and the male skeleton: A dual mode of action. *Journal of Osteoporosis* 2011, Article ID 240328, 10.4061/2011/240328.

Smedley, K. D., McKain, K. J., & McKain, D. N. (2014). 2D: 4D digit ratio predicts depression severity for females but not for males. *Personality and Individual Differences 70*, 136–139.

Smith, A. R., Hawkeswood, S. E., & Joiner, T. E. (2010). The measure of a man: Associations between digit ratio and disordered eating in males. *International Journal of Eating Disorders 43*, 543–548.

Smith, L. M., Cloak, C. C., Poland, R. E., Torday, J., & Ross, M. G. (2003). Prenatal nicotine increases testosterone levels in the fetus and female offspring. *Nicotine & Tobacco Research 5*, 369–374.

Smith, M. D., Jones, L. S., & Wilson, M. A. (2002). Sex differences in hippocampal slice excitability: Role of testosterone. *Neuroscience 109*, 517–530.

Smuts, B. (1995). The evolutionary origins of patriarchy. *Human Nature* 6, 1–32.

Snyder, P. J., Peachey, H., Hannoush, P., Berlin, J. A., Loh, L., Lenrow, D. A., ... Rosen, C. J. (1999). Effect of testosterone treatment on body composition and muscle strength in men over 65 years of age. *Journal of Clinical Endocrinology & Metabolism 84*(8), 2647–2653.

Sol, D., Duncan, R. P., Blackburn, T. M., Cassey, P., & Lefebvre, L. (2005). Big brains, enhanced cognition, and response of birds to novel environments. *Proceedings of the National Academy of Sciences 102*(15), 5460–5465.

Soler, H., Vinayak, P., & Quadagno, D. (2000). Biosocial aspects of domestic violence. *Psychoneuroendocrinology 25*, 721–739.

Somerville, L. H., Jones, R. M., & Casey, B. J. (2010). A time of change: Behavioral and neural correlates of adolescent sensitivity to appetitive and aversive environmental cues. *Brain and Cognition* 72, 124–133.

Sorge, R. E., & Totsch, S. K. (2017). Sex differences in pain. *Journal of Neuroscience Research 95*(6), 1271–1281.

South, S. J., & Spitze, G. (1994). Housework in marital and nonmarital households. *American Sociological Review 59*, 327–347.

Souza, A. L., Conroy-Beam, D., & Buss, D. M. (2016). Mate preferences in Brazil: Evolved desires and cultural evolution over three decades. *Personality and Individual Differences 95*, 45–49.

Sowers, M., Beebe, J., McConnell, D., Randolph, J., & Jannausch, M. (2001). Testosterone concentrations in women aged 25–50 years: Associations with lifestyle, body composition, and ovarian status. *American Journal of Epidemiology 153*, 256–264.

Spikins, P., Wright, B., & Scott, C. (2018). Autism spectrum conditions affect preferences in valued personal possessions. *Evolutionary Behavioral Sciences 12*(2), 99–108.

Spritzer, M. D., Daviau, E. D., Coneeny, M. K., Engelman, S. M., Prince, W. T., & Rodriguez-Wisdom, K. N. (2011). Effects of testosterone on spatial learning and memory in adult male rats. *Hormones and Behavior 59*, 484–496.

Spritzer, M. D., Fox, E. C., Larsen, G. D., Batson, C. G., Wagner, B. A., & Maher, J. (2013). Testosterone influences spatial strategy preferences among adult male rats. *Hormones and Behavior 63*, 800–812.

Städele, V., Roberts, E. R., Barrett, B. J., Strum, S. C., Vigilant, L., & Silk, J. B. (2019). Male–female relationships in olive baboons (Papio anubis): Parenting or mating effort? *Journal of Human Evolution 127*, 81–92.

Stanton, S. J., Liening, S. H., & Schultheiss, O. C. (2011). Testosterone is positively associated with risk taking in the Iowa Gambling Task. *Hormones and Behavior 59*, 252–256.

Stanton, S. J., Welker, K. M., Bonin, P. L., Goldfarb, B., & Carré, J. M. (2021). The effect of testosterone on economic risk-taking: A multi-study, multi-method investigation. *Hormones and Behavior 134*, 105014.

Stark, R. (2002). Physiology and faith: Addressing the "universal" gender difference in religious commitment. *Journal for the Scientific Study of Religion 41*, 495–507.

Steele, J. (2003). Children's gender stereotypes about math: The role of stereotype stratification 1. *Journal of Applied Social Psychology 33*, 2587–2606.

Stenstrom, E., Saad, G., Nepomuceno, M. V., & Mendenhall, Z. (2011). Testosterone and domain-specific risk: Digit ratios (2D: 4D and rel2) as predictors of recreational, financial, and social risk-taking behaviors. *Personality and Individual Differences 51*, 412–416.

Stern, C., & Madison, G. (2022). Sex differences and occupational choice Theorizing for policy informed by behavioral science. *Journal of Economic Behavior & Organization 202*, 694–702.

Stewart-Williams, S., & Halsey, L. G. (2021). Men, women and STEM: Why the differences and what should be done? *European Journal of Personality 35*, 3–39.

Stieff, M., Dixon, B. L., Ryu, M., Kumi, B. C., & Hegarty, M. (2014). Strategy training eliminates sex differences in spatial problem solving in a stem domain. *Journal of Educational Psychology 106*(2), 390.

Stijak, L., Kadija, M., Djulejić, V., Aksić, M., Petronijević, N., Marković, B., ... Filipović, B. (2015). The influence of sex hormones on anterior cruciate ligament rupture: Female study. *Knee Surgery, Sports Traumatology, Arthroscopy 23*, 2742–2749.

Stoet, G., & Geary, D. C. (2022). Sex differences in adolescents' occupational aspirations: Variations across time and place. *Plus One 17*(1), e0261438.

Storer, T. W., Woodhouse, L., Magliano, L., Singh, A. B., Dzekov, C., Dzekov, J., & Bhasin, S. (2008). Changes in muscle mass, muscle strength, and power but not physical function are related to testosterone dose in healthy older men. *Journal of the American Geriatrics Society 56*, 1991–1999.

Studer, L. H., Aylwin, A. S., & Reddon, J. R. (2005). Testosterone, sexual offense recidivism, and treatment effect among adult male sex offenders. *Sexual Abuse: A Journal of Research and Treatment 17*, 171–181.

Stulp, G., & Barrett, L. (2016). Evolutionary perspectives on human height variation. *Biological Reviews 91*, 206–234.

Styne, D. M., & Grumbach, M. M. (2002). Puberty in boys and girls. In D. Pfaff, A. Arnold, A. M. Etgen, S. Fahrbach, & R. T. Rubin (Eds.), *Hormones, brain and behavior* (pp. 661–716). San Diego, CA: Elsevier.

Su, R., Rounds, J., & Armstrong, P. I. (2009). Men and things, women and people: A meta-analysis of sex differences in interests. *Psychological Bulletin 135*, 859–872.

Sudhakar, H. H., Majumdar, P., Umesh, V., & Panda, K. (2014). Second to fourth digit ratio is a predictor of sporting ability in elite Indian male kabaddi players. *Asian Journal of Sports Medicine 5*(3), e23073.

Suija, K., Timonen, M., Suviola, M., Jokelainen, J., Järvelin, M.-R., & Tammelin, T. (2013). The association between physical fitness and depressive symptoms among young adults: Results of the Northern Finland 1966 birth cohort study. *BMC Public Health 13*, 1–7.

Sundblad, C., Bergman, L., & Eriksson, E. (1994). High levels of free testosterone in women with bulimia nervosa. *Acta Psychiatrica Scandinavica 90*, 397–398.

Sundblad, C., Landén, M., Eriksson, T., Bergman, L., & Eriksson, E. (2005). Effects of the androgen antagonist flutamide and the serotonin reuptake inhibitor citalopram in bulimia nervosa: A placebo-controlled pilot study. *Journal of Clinical Psychopharmacology 25*, 85–88.

Svedberg, P., Lichtenstein, P., & Pedersen, N. L. (2001). Age and sex differences in genetic and environmental factors for self-rated health. *Journals of Gerontology Series B: Psychological Sciences and Social Sciences 56*, S171–S178.

Svedholm-Häkkinen, A. M., & Lindeman, M. (2016). Testing the Empathizing-Systemizing theory in the general population: Occupations, vocational interests, grades, hobbies, friendship quality, social intelligence, and sex role identity. *Personality and Individual Differences 90*, 365–370.

Tallamy, D. W. (2000). Sexual selection and the evolution of exclusive paternal care in arthropods. *Animal Behaviour 60*, 559–567.

Tan, Ü. (1994). The grasp reflex from the right and left hand in human neonates indicates that the development of both cerebral hemispheres in males, but only the right hemisphere in females, is favoured by testosterone. *International Journal of Psychophysiology 16*, 39–47.

Tandon, R., Keshavan, M. S., & Nasrallah, H. A. (2008). Schizophrenia, "just the facts" what we know in 2008. 2 Epidemiology and etiology. *Schizophrenia Research 102*, 1–18.

Tarter, B. C., Kirisci, L., Tarter, R. E., Weatherbee, S., Jamnik, V., McGuire, E., & Gledhill, N. (2009). Use of aggregate fitness indicators to predict transition into the National Hockey League. *Journal of Strength & Conditioning Research 23*, 1828–1832.

Tarter, R. E., Kirisci, L., Kirillova, G. P., Gavaler, J., Giancola, P., & Vanyukov, M. M. (2007). Social dominance mediates the association of testosterone and neurobehavioral disinhibition with risk for substance use disorder. *Psychology of Addictive Behaviors 21*, 462–476.

Tay, P. K. C., Ting, Y. Y., & Tan, K. Y. (2019). Sex and care: The evolutionary psychological explanations for sex differences in formal care occupations. *Frontiers in Psychology 10*, 867–881.

Tecumseh Fitch, W., & Reby, D. (2001). The descended larynx is not uniquely human. *Proceedings of the Royal Society of London. Series B: Biological Sciences 268*(1477), 1669–1675.

Terburg, D., & van Honk, J. (2013). Approach–avoidance versus dominance–submissiveness: A multilevel neural framework on how testosterone promotes social status. *Emotion Review 5*, 296–302.

Thompson, R. G., Rodriguez, A., Kowarski, A., Migeon, C., & Blizzard, R. M. (1972). Integrated concentrations of growth hormone correlated with plasma testosterone and bone age in preadolescent and adolescent males. *Journal of Clinical Endocrinology & Metabolism 35*, 334–337.

Thong, F. S., McLean, C., & Graham, T. E. (2000). Plasma leptin in female athletes: Relationship with body fat, reproductive, nutritional, and endocrine factors. *Journal of Applied Physiology 88*, 2037–2044.

Thornhill, R., & Palmer, C. T. (2001). *A natural history of rape: Biological bases of sexual coercion*. Cambridge, MA: MIT Press.

Thornhill, R., & Thornhill, N. (1983). Human rape: An evolutionary analysis. *Ethology and Sociobiology 4*, 137–173.

Tiegerman, E., & Primavera, L. (1982). Object manipulation: An interactional strategy with autistic children. *Journal of Autism and Developmental Disorders 11*, 427–438.

Tierney, J. E., deMenocal, P. B., & Zander, P. D. (2017). A climatic context for the out-of-Africa migration. *Geology 45*, 1023–1026.

Tirabassi, G., Cignarelli, A., Perrini, S., Furlani, G., Gallo, M., Pallotti, F., … Gandini, L. (2015). Influence of CAG repeat polymorphism on the targets of testosterone action. *International Journal of Endocrinology 2015*, 298107, 10.1155/2015/298107.

Toivainen, T., Pannini, G., Papageorgiou, K. A., Malanchini, M., Rimfeld, K., Shakeshaft, N., & Kovas, Y. (2018). Prenatal testosterone does not explain sex differences in spatial ability. *Scientific Reports 8*(1), 1–8.

Tomonaga, M. (2006). Development of chimpanzee social cognition in the first 2 years of life. In T. Matsuzawa, M. Tomonaga, & M. Tanaka (Eds.), *Cognitive development in chimpanzees* (pp. 182–197). New York: Springer.

Torrance, J. S., Hahn, A. C., Kandrik, M., DeBruine, L. M., & Jones, B. C. (2018). No evidence for associations between men's salivary testosterone and responses on the intrasexual competitiveness scale. *Adaptive Human Behavior and Physiology 4*, 321–327.

Tracy, J. L., Robins, R. W., Schriber, R. A., & Solomon, M. (2011). Is emotion recognition impaired in individuals with autism spectrum disorders? *Journal of Autism and Developmental Disorders 41*, 102–109.

Tracy, K. K., & Crawford, C. B. (2019). Wife abuse: Does it have an evolutionary origin? In D. A. Counts, J. K. Brown, & J. C. Campbell (Eds.), *Sanctions and sanctuary* (pp. 19–32). London: Routledge.

Tramo, M. J., Loftus, W. C., Stukel, T. A., Green, R. L., Weaver, J. B., & Gazzaniga, M. S. (1998). Brain size, head size, and intelligence quotient in monozygotic twins. *Neurology 50*, 1246–1252.

Tremblay, R. E. (1998). Testosterone, physical aggression, dominance, and physical development in early adolescence. *International Journal of Behavioral Development 22*, 753–777.

Tremblay, R. E., Schaal, B., Boulerice, B., Arseneault, L., Soussignan, R. G., Paquette, D., & Laurent, D. (1998). Testosterone, physical aggression, dominance, and physical development in early adolescence. *International Journal of Behavioral Development 22*, 753–777.

Trivers, R. L. (2017). Parental investment and sexual selection. In *Sexual selection and the descent of man* (pp. 136–179). London: Routledge.

Trivers, R., Hopp, R., & Manning, J. (2013). A longitudinal study of digit ratio (2D: 4D) and its relationships with adult running speed in Jamaicans. *Human Biology 85*, 623–626.

Trivers, R., Manning, J., & Jacobson, A. (2006). A longitudinal study of digit ratio (2D: 4D) and other finger ratios in Jamaican children. *Hormones and Behavior 49*, 150–156.

Troisi, A. (2001). Gender differences in vulnerability to social stress: A Darwinian perspective. *Physiology & Behavior 73*, 443–449.

Turan, B., Guo, J., Boggiano, M. M., & Bedgood, D. (2014). Dominant, cold, avoidant, and lonely: Basal testosterone as a biological marker for an interpersonal style. *Journal of Research in Personality 50*, 84–89.

Udry, J. R. (1990). Biosocial models of adolescent problem behaviors. *Social Biology 37*, 1–10.

Udry, J. R. (2000). Biological limits of gender construction. *American Sociological Review 65*, 443–457.

Udry, R. J., Morris, N. M., & Kovenock, J. (1995). Androgen effects on women's gendered behaviour. *Journal of Biosocial Science 27*, 359–368.

Ulbricht, A., Fernandez-Fernandez, J., Mendez-Villanueva, A., & Ferrauti, A. (2016). Impact of fitness characteristics on tennis performance in elite junior tennis players. *Journal of Strength & Conditioning Research 30*, 989–998.

Urban, R. J., Bodenburg, Y. H., Gilkison, C., Foxworth, J., Coggan, A. R., Wolfe, R. R., & Ferrando, A. (1995). Testosterone administration to elderly men increases skeletal muscle strength and protein synthesis. *American Journal of Physiology-Endocrinology and Metabolism 269*, E820–E826.

Valla, J. M., & Ceci, S. J. (2011). Can sex differences in science be tied to the long reach of prenatal hormones? Brain organization theory, digit ratio (2D/4D), and sex differences in preferences and cognition. *Perspectives on Psychological Science 6*, 134–146.

van Anders, S. M. (2012). Testosterone and sexual desire in healthy women and men. *Archives of Sexual Behavior 41*, 1471–1484.

Van Anders, S. M., & Hampson, E. (2005). Waist-to-hip ratio is positively associated with bioavailable testosterone but negatively associated with sexual desire in healthy premenopausal women. *Psychosomatic Medicine 67*, 246–250.

van Anders, S. M., & Watson, N. V. (2006). Relationship status and testosterone in North American heterosexual and non-heterosexual men and women: Cross-sectional and longitudinal data. *Psychoneuroendocrinology 31*, 715–723.

Van Anders, S. M., Goldey, K. L., & Kuo, P. X. (2011). The steroid/peptide theory of social bonds: Integrating testosterone and peptide responses for classifying social behavioral contexts. *Psychoneuroendocrinology 36*, 1265–1275.

Van Anders, S. M., Tolman, R. M., & Volling, B. L. (2012). Baby cries and nurturance affect testosterone in men. *Hormones and Behavior 61*, 31–36.

Van Bokhoven, I., Van Goozen, S. H., Van Engeland, H., Schaal, B., Arseneault, L., Séguin, J. R., ... Tremblay, R. E. (2006). Salivary testosterone and aggression, delinquency, and social dominance in a population-based longitudinal study of adolescent males. *Hormones and Behavior 50*, 118–125.

van de Beek, C., van Goozen, S. H. M., Buitelaar, J. K., & Cohen-Kettenis, P. T. (2009). Prenatal sex hormones (maternal and amniotic fluid) and gender-related play behavior in 13-month-old infants. *Archives of Sexual Behavior 38*, 6–15.

Van Goozen, S. H., Cohen-Kettenis, P. T., Gooren, L. J., Frijda, N. H., & Van De Poll, N. E. (1995). Gender differences in behaviour: Activating effects of cross-sex hormones. *Psychoneuroendocrinology 20*, 343–363.

van Honk, J., Bos, P. A., & Terburg, D. (2014). Testosterone and dominance in humans: Behavioral and brain mechanisms. In J. Decety & I. Christen (Eds.), *New frontiers in social neuroscience* (pp. 201–214). Springer.

Van Honk, J., Peper, J. S., & Schutter, D. J. (2005). Testosterone reduces unconscious fear but not consciously experienced anxiety: Implications for the disorders of fear and anxiety. *Biological Psychiatry 58*, 218–225.

Van Kleef, G. A., Heerdink, M. W., Cheshin, A., Stamkou, E., Wanders, F., Koning, L. F., ... Georgeac, O. A. M. (2021). No guts, no glory? How risk-taking shapes dominance, prestige, and leadership endorsement. *Journal of Applied Psychology 106*, 1673–1694.

van Milligen, B. A., Lamers, F., Guus, T., Smit, J. H., & Penninx, B. W. (2011). Objective physical functioning in patients with depressive and/or anxiety disorders. *Journal of Affective Disorders 131*, 193–199.

van Milligen, B. A., Vogelzangs, N., Smit, J. H., & Penninx, B. W. (2012). Physical function as predictor for the persistence of depressive and anxiety disorders. *Journal of Affective Disorders 136*, 828–832.

Van Slyke, J. A. (2017). Can sexual selection theory explain the evolution of individual and group-level religious beliefs and behaviors? *Religion, Brain & Behavior 7*, 335–338.

Van Tilburg, M. A., Unterberg, M. L., & Vingerhoets, A. J. (2002). Crying during adolescence: The role of gender, menarche, and empathy. *British Journal of Developmental Psychology 20*, 77–87.

Van Vugt, M. (2009). Sex differences in intergroup competition, aggression, and warfare: The male warrior hypothesis. *Annals of the New York Academy of Sciences 1167*, 124–134.

Vanderschueren, D., & Bouillon, R. (1995). Androgens and bone. *Calcified Tissue International 56*, 341–346.

Vanderschueren, D., Gaytant, J., Boonen, S., & Venken, K. (2008). Androgens and bone. *Current Opinion in Endocrinology, Diabetes and Obesity 15*, 250–254

Varella, M. A. C., Valentova, J. V., & Fernández, A. M. (2017). Evolution of artistic and aesthetic propensities through female competitive ornamentation. In M. L. Fisher (Ed.), *The Oxford handbook of women and competition* (pp. 757–783). Oxford, England: Oxford University Press.

Varga, S. (2012). Evolutionary psychiatry and depression: Testing two hypotheses. *Medicine, Health Care and Philosophy 15*, 41–52.

Ventura, T., Gomes, M. C., Pita, A., Neto, M. T., & Taylor, A. (2013). Digit ratio (2D: 4D) in newborns: Influences of prenatal testosterone and maternal environment. *Early Human Development 89*, 107–112.

Vermeer, A. L., Krol, I., Gausterer, C., Wagner, B., Eisenegger, C., & Lamm, C. (2020). Exogenous testosterone increases status-seeking motivation in men with unstable low social status. *Psychoneuroendocrinology 113*, 104552.

Vermeersch, H., T'sjoen, G., Kaufman, J.-M., & Vincke, J. (2008). The role of testosterone in aggressive and non-aggressive risk-taking in adolescent boys. *Hormones and Behavior 53*(3), 463–471.

Versola-Russo, J. M. (2005). Cultural and demographic factors of Schizophrenia. *International Journal of Psychosocial Rehabilitation 10*(2), 89–103.

Verster, J. C., Mackus, M., van de Loo, A. J., Garssen, J., & Roth, T. (2017). Insomnia, total sleep time and the 2D: 4D digit ratio. *Current Psychopharmacology 6*, 158–161.

Vettor, R., De Pergola, G., Pagano, C., Englaro, P., Laudadio, E., Giorgino, F., … Federspil, G. (1997). Gender differences in serum leptin in obese people: Relationships with testosterone, body fat distribution and insulin sensitivity. *European Journal of Cinical Investigation 27*, 1016–1024.

Vigil, J. M., & Strenth, C. (2014). No pain, no social gains: A social-signaling perspective of human pain behaviors. *World Journal of Anesthesiology 3*, 18–30.

Vignozzi, L., Corona, G., Petrone, L., Filippi, S., Morelli, A., Forti, G., & Maggi, M. (2005). Testosterone and sexual activity. *Journal of Endocrinological Investigation 28*(3 Suppl), 39–44.

Vingerhoets, A. J., & Bylsma, L. M. (2016). The riddle of human emotional crying: A challenge for emotion researchers. *Emotion Review 8*, 207–217.

Vingerhoets, A. J., & Scheirs, J. (2000). Sex differences in crying: Empirical findings and possible explanations. In A. H. Fischer (Ed.), *Gender and emotion: Social psychological perspectives* (pp. 143–165). Cambridge, England: University of Cambridge Press.

Vlazneva, S., & Androsova, O. (2021). *Multiple-choice questions and essays in assessing economics*. Paper presented at the SHS Web of Conferences.

Vogel, D. L., Wester, S. R., Heesacker, M., & Madon, S. (2003). Confirming gender stereotypes: A social role perspective. *Sex Roles 48*, 519–528.

Voicu, M. (2009). Religion and gender across Europe. *Social Compass 56*, 144–162.

Voicu, M. (2019). Religion and housework division: The interplay between religious and gender identity. *AnALize: Revista de Studii Feministe 13*(27), 23–41.

Voisin, D. R., Salazar, L. F., Crosby, R., DiClemente, R. J., Yarber, W. L., & Staples-Horne, M. (2004). The association between gang involvement and sexual behaviours among detained adolescent males. *Sexually Transmitted Infections 80*, 440–442.

Volman, I., Toni, I., Verhagen, L., & Roelofs, K. (2011). Endogenous testosterone modulates prefrontal–amygdala connectivity during social emotional behavior. *Cerebral Cortex 21*, 2282–2290.

von der Pahlen, B., Sarkola, T., Seppa, K., & Eriksson, C. P. (2002). Testosterone, 5 alpha-dihydrotestosterone and cortisol in men with and without alcohol-related aggression. *Journal of Studies on Alcohol 63*, 518–526.

Vongas, J. G., & Al Hajj, R. (2015). The evolution of empathy and women's precarious leadership appointments. *Frontiers in Psychology 6*, 1751.

Voracek, M., Pum, U., & Dressler, S. G. (2010). Investigating digit ratio (2D: 4D) in a highly male-dominated occupation: The case of firefighters. *Scandinavian Journal of Psychology 51*, 146–156.

Votinov, M., Knyazeva, I., Habel, U., Konrad, K., & Puiu, A. A. (2022). A Bayesian modeling approach to examine the role of testosterone administration on the endowment effect and risk-taking. *Frontiers in Neuroscience 16*, 858168.

Wai, J., Lubinski, D., & Benbow, C. P. (2009). Spatial ability for STEM domains: Aligning over 50 years of cumulative psychological knowledge solidifies its importance. *Journal of Educational Psychology 101*, 817–826.

Wakabayashi, A., & Nakazawa, Y. (2010). On relationships between digit ratio (2D: 4D) and two fundamental cognitive drives, empathizing and systemizing, in Japanese sample. *Personality and Individual Differences 49*, 928–931.

Wallen, K., & Parsons, W. A. (1997). Sexual behavior in same-sexed nonhuman primates: Is it relevant to understanding human homosexuality? *Annual Review of Sex Research 8*, 195–223.

Wang, C., Alexander, G., Berman, N., Salehian, B., Davidson, T., McDonald, V., ... Swerdloff, R. S. (1996). Testosterone replacement therapy improves mood in hypogonadal men—A clinical research center study. *Journal of Clinical Endocrinology & Metabolism 81*, 3578–3583.

Wang, L.-J., Chou, M.-C., Chou, W.-J., Lee, M.-J., Lee, S.-Y., Lin, P.-Y., ... Yen, C.-F. (2017). Potential role of pre-and postnatal testosterone levels in attention-deficit/hyperactivity disorder: Is there a sex difference? *Neuropsychiatric Disease and Treatment 13*, 1331–1343.

Wang, Y., Wang, H. L., Li, Y. H., Zhu, F. L., Li, S. J., & Ni, H. (2016). Using 2D: 4D digit ratios to determine motor skills in children. *European Review of Medical Pharmacological Science 20*, 806–809.

Warrington, N. M., Shevroja, E., Hemani, G., Hysi, P. G., Jiang, Y., Auton, A., ... Kemp, J. P. (2018). Genome-wide association study identifies nine novel loci for 2D: 4D finger ratio, a putative retrospective biomarker of testosterone exposure in utero. *Human Molecular Genetics 27*, 2025–2038.

Waterink, W. (2014). In steady heterosexual relationships men masturbate more than women because of gender differences in sex drive. *New Voices in Psychology 10*, 96–108.

Watts, D. P. (2010). Dominance, power, and politics in nonhuman and human primates. In P. M. Kappeler & J. B. Silk (Eds.), *Mind the gap* (pp. 109–138). New York: Springer.

Weisfeld, G. E., & Dillon, L. M. (2012). Applying the dominance hierarchy model to pride and shame, and related behaviors. *Journal of Evolutionary Psychology 10*, 15–41.

Weisfeld, G. E., Weisfeld, C. C., & Goetz, S. M. (2017). Towards a model of marriage. In G. E. Weisfeld, C. C. Weisfeld, & L. M. Dillon (Eds.), *The psychology of marriage: An evolutionary and cross-cultural view* (pp. 317–334). Lanham, MD: Lexington.

Weiss, E., Siedentopf, C. M., Hofer, A., Deisenhammer, E. A., Hoptman, M. J., Kremser, C., ... Delazer, M. (2003). Sex differences in brain activation pattern during a visuospatial cognitive task: A functional magnetic resonance imaging study in healthy volunteers. *Neuroscience Letters 344*, 169–172.

Weitz, S. (1977). *Sex roles.* New York: Oxford University Press.

Welker, K. M., Gruber, J., & Mehta, P. H. (2015). A positive affective neuroendocrinology approach to reward and behavioral dysregulation. *Frontiers in Psychiatry* 6, 93–105.

Wellford, C. (1975). Labelling theory and criminology: An assessment. *Social Problems 22*, 332–345.

Welling, H. (2003). An evolutionary function of the depressive reaction: The cognitive map hypothesis. *New Ideas in Psychology 21*, 147–156.

Wernicke, J., Zabel, J. T., Zhang, Y., Becker, B., & Montag, C. (2020). Association between tendencies for attention-deficit/hyperactivity disorder (ADHD) and the 2D: 4D digit ratio: a cross-cultural replication in Germany and China. *Early Human Development 143*, 104943.

West, M. M., & Konner, M. J. (1976). The role of the father: An anthropological perspective. In M. E. Lamb (Ed.), *The role of the father in child development.* New York: Plenum Press.

West-Eberhard, M. J. (1983). Sexual selection, social competition, and speciation. *Quarterly Review of Biology 58*, 155–183.

Westneat, D. F., & Sherman, P. W. (1993). Parentage and the evolution of parental behavior. *Behavioral Ecology 4*, 66–77.

White, R. E., Thornhill, S., & Hampson, E. (2006). Entrepreneurs and evolutionary biology: The relationship between testosterone and new venture creation. *Organizational Behavior and Human Decision Processes 100*, 21–34.

Whitehouse, A. J., Mattes, E., Maybery, M. T., Sawyer, M. G., Jacoby, P., Keelan, J. A., & Hickey, M. (2012). Sex-specific associations between umbilical cord blood testosterone levels and language delay in early childhood. *Journal of Child Psychology and Psychiatry 53*, 726–734.

Whitehouse, A. J., Maybery, M. T., Hart, R., Sloboda, D. M., Stanley, F. J., Newnham, J. P., & Hickey, M. (2010). Free testosterone levels in umbilical-cord blood predict infant head circumference in females. *Developmental Medicine & Child Neurology 52*, e73–e77.

Wiederman, M. W. (1997). Extramarital sex: Prevalence and correlates in a national survey. *Journal of Sex Research 34*, 167–174.

Wilhelm, K., Parker, G., & Hadzi-Pavlovic, D. (1997). Fifteen years on: Evolving ideas in researching sex differences in depression. *Psychological Medicine 27*, 875–883.

Williams, J. H. G., Greenhalgh, K. D., & Manning, J. T. (2003). Second to fourth finger ratio and possible precursors of developmental psychopathology in preschool children. *Early Human Development 72*, 57–65.

Williamson, T. T., Zhu, X., Pineros, J., Ding, B., & Frisina, R. D. (2020). Understanding hormone and hormone therapies' impact on the auditory system. *Journal of Neuroscience Research 98*(9), 1721–1730.

Wilson, G. D. (1997). Gender differences in sexual fantasy: An evolutionary analysis. *Personality and Individual Differences 22*(1), 27–31.

Wilson, M. J., & Spaziani, E. (1976). The melanogenic response to testosterone in scrotal epidermis: Effects on tyrosinase activity and protein synthesis. *European Journal of Endocrinology 81*, 435–448.

Wilson, M. L., Miller, C. M., & Crouse, K. N. (2017). Humans as a model species for sexual selection research. *Proceedings of the Royal Society B: Biological Sciences 284*(1866), 20171320.

Wilson, M., Daly, M., & Pound, N. (2002). An evolutionary psychological perspective on the modulation of competitive confrontation and risk-taking. *Hormones, Brain and Behavior 5*, 381–408.

Wilson, M., Daly, M., Gordon, S., & Pratt, A. (1996). Sex differences in valuations of the environment? *Population and Environment 18*, 143–159.

Winegard, B. M., & Winegard, B. M. (2018). The emerging science of evolutionary criminology. *Journal of Criminal Justice 59*, 122–126.

Wingfield, J. C., Hegner, R. E., Dufty Jr, A. M., & Ball, G. F. (1990). The "challenge hypothesis": Theoretical implications for patterns of testosterone secretion, mating systems, and breeding strategies. *American Naturalist 136*(6), 829–846.

Wingfield, J. C., Jacobs, J., & Hillgarth, N. (1997). Ecological constraints and the evolution of hormone-behavior interrelationships. *Annals of the New York Academy of the Sciences 807*, 22–41.

Wiren, K. M. (2005). Androgens and bone growth: It's location, location, location. *Current Opinion in Pharmacology 5*, 626–632.

Witelson, S. F., Beresh, H., & Kigar, D. L. (2006). Intelligence and brain size in 100 postmortem brains: Sex, lateralization and age factors. *Brain 129*, 386–398.

Witt, E. D. (2007). Puberty, hormones, and sex differences in alcohol abuse and dependence. *Neurotoxicology and Teratology 29*, 81–95.

Wittert, G. (2014). The relationship between sleep disorders and testosterone. *Current Opinion in Endocrinology, Diabetes and Obesity 21*, 239–243.

Wondergem, T. R., & Friedlmeier, M. (2012). Gender and ethnic differences in smiling: A yearbook photographs analysis from kindergarten through 12th grade. *Sex Roles 67*, 403–411.

Wong, J. S., & Gravel, J. (2018). Do sex offenders have higher levels of testosterone? Results from a meta-analysis. *Sexual Abuse 30*, 147–168.

Wong, W. I., & Hines, M. (2015a). Preferences for pink and blue: The development of color preferences as a distinct gender-typed behavior in toddlers. *Archives of Sexual Behavior 44*, 1243–1254.

Wong, W. I., & Hines, M. (2015b). Effects of gender color-coding on toddlers' gender-typical toy play. *Archives of Sexual Behavior 44*, 1233–1242.

Wood, J. C. (2017). Future agendas for research on violent crime: The challenge to history from evolutionary psychology. *Crime, Histoire & Sociétés/Crime, History & Societies, 21*, 351–359.

Wood, W., & Eagly, A. H. (2002). A cross-cultural analysis of the behavior of women and men: Implications for the origins of sex differences. *Psychological Bulletin 128*, 699–727.

Wood, W., & Eagly, A. H. (2012). Biosocial construction of sex differences and similarities in behavior. In M. P. Zanna (Ed.), *Advances in experimental social psychology* (Vol. 46, pp. 55–123). Elsevier.

Wren, B., Launer, J., Reiss, M. J., Swanepoel, A., & Music, G. (2019). Can evolutionary thinking shed light on gender diversity? *British Journal of Psychology Advances 25*(6), 351–362.

Wright, D. B., Eaton, A. A., & Skagerberg, E. (2015). Occupational segregation and psychological gender differences: How empathizing and systemizing help explain the distribution of men and women into (some) occupations. *Journal of Research in Personality 54*, 30–39.

Wright, P. J., & Vangeel, L. (2019). Pornography, permissiveness, and sex differences: An evaluation of social learning and evolutionary explanations. *Personality and Individual Differences 143*, 128–138.

Wu, F., Zitzmann, M., Heiselman, D., Donatucci, C., Knorr, J., Patel, A. B., & Kinchen, K. (2016). Demographic and clinical correlates of patient-reported improvement in sex drive, erectile function, and energy with testosterone solution 2%. *Journal of Sexual Medicine 13*, 1212–1219.

Wynn, T. G., Tierson, F. D., & Palmer, C. T. (1996). Evolution of sex differences in spatial cognition. *American Journal of Physical Anthropology 101*(S23), 11–42.

Xu, Y., & Zheng, Y. (2015). The digit ratio (2D: 4D) in China: A meta-analysis. *American Journal of Human Biology 27*, 304–309.

Yao, S., Långström, N., Temrin, H., & Walum, H. (2014). Criminal offending as part of an alternative reproductive strategy: Investigating evolutionary hypotheses using Swedish total population data. *Evolution and Human Behavior 35*, 481–488.

Yildirim, B. O., & Derksen, J. J. (2012a). A review on the relationship between testosterone and the interpersonal/affective facet of psychopathy. *Psychiatry Research 197*, 181–198.

Yildirim, B. O., & Derksen, J. J. L. (2012b). A review on the relationship between testosterone and life-course persistent antisocial behavior. *Psychiatry Research 200*, 984–1010.

Yuksel, T., Sizer, E., & Durak, H. (2019). 2D: 4D ratios as an indicator of intrauterine androgen exposure in children who stutter. *Early Human Development 135*, 27–31.

Zaidi, Z. F. (2010). Gender differences in human brain: A review. *Open Anatomy Journal 2*, 37–55.

Zarrouf, F. A., Artz, S., Griffith, J., Sirbu, C., & Kommor, M. (2009). Testosterone and depression: Systematic review and meta-analysis. *Journal of Psychiatric Practice 15*, 289–305.

Zhang, J., Jia, C.-X., & Wang, L.-L. (2015). Testosterone differs between suicide attempters and community controls in men and women of China. *Physiology & Behavior 141*, 40–45.

Zhang, X., Tworoger, S. S., Eliassen, A. H., & Hankinson, S. E. (2013). Postmenopausal plasma sex hormone levels and breast cancer risk over 20 years of follow-up. *Breast Cancer Research and Treatment 137*, 883–892.

Zhu, N., & Chang, L. (2019). Evolved but not fixed: A life history account of gender roles and gender inequality. *Frontiers in Psychology 10*(1709), 1–16.

Zilioli, S., & Bird, B. M. (2017). Functional significance of men's testosterone reactivity to social stimuli. *Frontiers in Neuroendocrinology 47*, 1–18.

Zilioli, S., & Watson, N. V. (2014). Testosterone across successive competitions: Evidence for a 'winner effect' in humans? *Psychoneuroendocrinology 47*, 1–9.

Zitzmann, M. (2006). Testosterone and the brain. *Aging Male 9*, 195–199.

Zitzmann, M. (2020). Testosterone, mood, behaviour and quality of life. *Andrology 8*, 1598–1605.

Zöttl, M., Vullioud, P., Goddard, K., Torrents-Ticó, M., Gaynor, D., Bennett, N. C., & Clutton-Brock, T. (2018). Allo-parental care in Damaraland mole-rats is female biased and age dependent, though independent of testosterone levels. *Physiology & Behavior 193*, 149–153.

Index